DATE DUE

APR 1 6 1994	

UPI 261-2

D1214571

Plant
Biochemistry
THIRD EDITION

Contributors

T. Akazawa

Peter Albersheim

Leonard Beevers

A. A. Benson

James Bonner

R. W. Breidenbach

J. K. Bryan

R. H. Burris

J. E. Gander

M. D. Hatch

Peter K. Helper

E. J. Hewitt

David Tuan-Hua Ho

D. P. Hucklesby

Alice Tang Jokela

Arthur L. Karr

Joel L. Key

Bessel Kok

Tsune Kosuge

Abraham Marcus

Ph. Matile

B. A. Notton

Roderic B. Park

Jack Preiss

Peter H. Quail

D. W. Rains

Ziva Reuveny

P. K. Stumpf

Erhard Stutz

J. E. Varner

Lloyd G. Wilson

Plant Biochemistry

Third Edition

Edited by

James Bonner
California Institute of Technology

Joseph E. Varner
Washington University

ACADEMIC PRESS   New York • San Francisco • London

A Subsidiary of Harcourt Brace Jovanovich, Publishers

ACADEMIC PRESS, INC.
111 Fifth Avenue, New York, New York 10003

United Kingdom Edition published by
ACADEMIC PRESS, INC. (LONDON) LTD.
24/28 Oval Road, London NW1

Library of Congress Cataloging in Publication Data

Bonner, James Frederick,   Date   ed.
     Plant biochemistry.

     Includes bibliographies.
     1.   Botanical chemistry.   I.   Varner, J. E.
II.   Title.
QK861.B6   1976   581.1'92   76-21693
ISBN 0–12–114860–2

Contents

v

25 Photosynthesis: The Path of Energy
BESSEL KOK

26 Nitrogen Fixation
R. H. BURRIS

Index

List of Contributors

Numbers in parentheses indicate the pages on which the authors' contributions begin.

T. AKAZAWA   (381)   Research Institute for Biochemical Regulation, School of Agriculture, Nagoya University, Chikusa, Nagoya, Japan

PETER ALBERSHEIM   (225)   Department of Chemistry, University of Colorado, Boulder, Colorado

LEONARD BEEVERS   (771)   Department of Botany and Microbiology, University of Oklahoma, Norman, Oklahoma

A. A. BENSON   (65)   Scripps Institution of Oceanography, University of California at San Diego, La Jolla, California

JAMES BONNER   (3, 37)   Division of Biology, California Institute of Technology, Pasadena, California

R. W. BREIDENBACH   (91)   Plant Growth Laboratory, Department of Agronomy and Range Science, University of California, Davis, California

J. K. BRYAN   (525)   Department of Biology, Syracuse University, Syracuse, New York

R. H. BURRIS   (887)   Department of Biochemistry, University of Wisconsin, Madison, Wisconsin

J. E. GANDER   (337)   Department of Biochemistry, College of Biological Sciences, University of Minnesota, St. Paul, Minnesota

M. D. HATCH   (797)   Division of Plant Industry, Commonwealth Scientific and Industrial Research Organization, Canberra City, Australia

PETER K. HELPER   (147)   Department of Biological Sciences, Stanford University, Stanford, California

E. J. HEWITT   (633)   Plant Physiology and Research Station, University of Bristol, Long Ashton, England

DAVID TUAN-HUA HO   (713)   Department of Biology, Washington University, St. Louis, Missouri

D. P. HUCKLESBY (633) Plant Physiology and Research Station, University of Bristol, Long Ashton, England

ALICE TANG JOKELA (65) Department of Microbiology, San Diego State University, San Diego, California

ARTHUR L. KARR (405) Department of Plant Pathology, University of Missouri-Columbia, Columbia, Missouri

JOEL L. KEY (463) Botany Department, University of Georgia, Athens, Georgia

BESSEL KOK (845) Martin Marietta Laboratories, Baltimore, Maryland

TSUNE KOSUGE (277) Department of Plant Pathology, University of California, Davis, California

ABRAHAM MARCUS (507) The Institute for Cancer Research, Philadelphia, Pennsylvania

Ph. MATILE (189) Department of General Botany, Swiss Federal Institute of Technology, Zurich, Switzerland

B. A. NOTTON (633) Plant Physiology and Research Station, University of Bristol, Long Ashton, England

RODERIC B. PARK (115) Department of Botany, University of California, Berkeley, California

JACK PREISS (277) Department of Biochemistry and Biophysics, University of California, Davis, California

PETER H. QUAIL (683) Research School of Biological Sciences, Australian National University, Canberra, Australia

D. W. RAINS (561) Department of Agronomy and Range Science, University of California, Davis, California

ZIVA REUVENY* (599) MSU/ERDA Plant Research Laboratory, Michigan State University, East Lansing, Michigan

P. K. STUMPF (427) Department of Biochemistry and Biophysics, University of California, Davis, California

ERHARD STUTZ† (15) Department of Biological Sciences, Northwestern University, Evanston, Illinois

J. E. VARNER (714) Department of Biology, Washington University, St. Louis, Missouri

LLOYD G. WILSON (599) MSU/ERDA Plant Research Laboratory, Michigan State University, East Lansing, Michigan

* Present address: Biology Division, Oak Ridge National Laboratory, Oak Ridge, Tennessee.

† Present address: Laboratoire de Physiologie Végétale et Biochemie, Université de Neuchâtel, Neuchâtel, Switzerland.

Preface

This treatise is intended for the advanced student or professional worker in the plant sciences. It is directed to the biochemist who desires information in areas of biochemistry that are unique to plants, for example, cell wall matters, photosynthesis, or nitrogen fixation, or who is interested in the degree to which plants share biochemical pathways found in other organisms. This work will also be valuable to plant biologists in general. Biochemistry can and does contribute to the understanding and solution of the problems involved in many of the more specialized aspects of plant biology—taxonomy, morphology, ecology, horticulture, agronomy, phytopathology, to name a few. We believe this book can help students and research workers in these diverse fields by providing them with a ready source of biochemical information directly applicable to plants. Finally, we feel that it can be used successfully as a text in plant biochemistry courses. The student in such a course would need some background in organic chemistry, but previous study of biochemistry would not necessarily be required.

We have tried to present each topic comprehensively in the sense that we have started with general principles and ended with the current state of the subject. We hope that the reader, after having studied a topic in this book, will find himself qualified to go into his laboratory and start investigations possessing the latest knowledge available in that field. To assist the research worker we have included references pertinent to the original literature; to assist the student we have also included suggestions for more general reading on each topic. The student without previous knowledge of biochemistry will find such reading desirable—perhaps necessary.

In the eleven years that have elapsed since the publication of the first edition of "Plant Biochemistry" there have been, of course, advances in all areas of this subject. Some are due to the increasing emphasis of plant physiology on plant biochemistry as a means of better understanding the individual physiological processes; others to the understanding of subjects not comprehended eleven years ago. Examples of the latter can be found throughout the volume.

We trust this edition will prove useful to its readers. We thank our colleagues for their contributions and intellectual support of this volume. We are also indebted to the staff of Academic Press for their continuous and skillful help in the preparation of this work.

James Bonner
Joseph Varner

Plant Biochemistry

THIRD EDITION

I

Plant Cell: Substructures and Subfunctions

1

Cell and Subcell

JAMES BONNER

I. Introduction

One of the most powerful generalizations of biochemistry is that cells of all kinds and of all creatures possess the same, rather small, number of kinds of subcellular components. These subcellular entities are similar between the different kinds of cells, not only in their morphology and submicroscopic structure but also in chemical composition and most importantly in chemical function, each kind contributing its own mite to the overall functioning of the cell. The untangling of the biochemical pathways of metabolism and the development of our understanding of the strategy of life has been due very largely to the technology developed since approximately 1950 which has made it possible to separate several subcellular components from one another and to identify the enzyme systems associated with each. We turn our attention, therefore, first to the subcellular components of the plant cell.

II. The Subcellular Components of the Plant Cell

The principal subcellular components of the plant cell, and those with which biochemistry is principally concerned, are the nucleus, the chloroplast, the mitochondria, the lysosomes and other vacuoles, the ribosomes, messenger RNA, and the individual soluble enzymes. Table I summarizes the number of each of these kinds of entities found in a typical or average cell. The vast majority of plant cells contains, of

3

course, one nucleus, although many instances of multinucleate cells are known—even in higher plants, for example, latex vessels or the sieve tubes, cells that are multinucleate by virtue of dissolution of transverse cell walls. Chloroplasts in the photosynthetic portion of the plant number in general a few tens—50 is the rough average per cell. To this number should be added the proplastids from which the mature chloroplasts arise, but since today we still have no good estimate of the number of proplastids to be found in a typical cell, we will not consider them further. The proplastids are found in the nonphotosynthetic as well as in the photosynthetic portions of the plant—for example, in roots.

Mitochondria characteristically occur in the plant cell in the order of the 100's—500 to 1000 being a typical number. Lysosomes and other vacuoles are present in about equal numbers. Ribosomes, the next smaller category of particles, occur in vastly greater numbers than chloroplasts or mitochondria. A growing functional plant cell might perhaps contain a few hundred thousand ribosomes, although this number varies greatly with age, state of activity, and so on. The bulk of the cytoplasmic protein, the portion to which we refer in general as the nonparticulate cytoplasm, is of course composed of enzyme molecules, in fact, a great number of kinds of enzyme molecules. The total number of enzyme molecules in a typical cell would be of the order of 1,000,000,000. These consist of several thousand, perhaps 1000 to 10,000, different species of enzyme molecules, each qualified to catalyze one specific kind of chemical reaction. A typical plant cell might then contain 1,000,000,000 enzyme molecules of 10,000 different kinds, 100,000 being representative of each of the 10,000 different species present. The proportion that any individual kind of enzyme molecule constitutes of the total soluble cytoplasmic protein often departs widely, however, from the average 0.001%, which would be expected on the basis of the above calculation. Thus, we know that particular kinds of enzyme molecules in particular kinds of cells may constitute from a few tenths to as much as several percent of total soluble protein. As a general rule, however, we must expect, because there are so many kinds of enzyme molecules, that each one will constitute but a small proportion of the total, and it is not surprising, therefore, that in the purification of enzymes, enrichments of 10,000-fold or more are not uncommonly needed to achieve pure material.

We have referred above to messenger RNA as a typical component of the plant cell. Messenger RNA may indeed be isolated and characterized by methods to be considered in a later chapter. For the time being, it may be noted that messenger RNA may most easily and characteristically be detected in the plant cell by virtue of its interaction with the ribosome. Ribosomes interact with and attach to messenger RNA, and since a single messenger RNA strand may simultaneously bind many ribosomes, the great bulk of the ribosomes of the plant cell are often detected as large aggregates of so-called polysomes. Transfer RNA is characterized by small molecular size like the soluble enzymes and is a component of the nonparticulate cytoplasmic material. For this reason, it is often known in the literature of the cell as soluble RNA, although the preferred name is tRNA.

The entire assemblage of subcellular structure as outlined above is, of course, contained within the membrane system, and we might properly include the membrane system as one of the most characteristic of cellular components. This membrane system comprises not only the protoplasmic membrane itself but also the membranes surrounding the nucleus, chloroplasts, and vacuole, as well as the membrane elements of the mitochondria, the endoplastic reticulum, and other structures outlined below. The plant cell is characterized also by the cell wall external to the protoplasmic membrane, and in a sense the wall might, too, be considered as a subcellular component characteristic of the plant cell.

The plant cell contains still further subcellular systems. These entities are, of course, important subcellular components, but they are either less universal or less understood than those enumerated in Table I. They include, for example, lysosomes that contain hydrolytic enzymes and conduct the autolysis of injured or aging cells. They include also the Golgi apparatus that concerns itself with concentration, chemical modification, and secretion of enzymes and substrates that are to be secreted into the outside of the cell (as for example, cell wall-forming materials). The glyoxysomes are particularly well known in germinating fatty seeds, where they are responsible for the transformation of fatty acids into sugar precursors. Peroxisomes are responsible for photorespiration. Into this category also we must place the microtubules and related structures responsible for photoplasmic streaming, and also we may place the

TABLE I

Numbers and Sizes of Subcellular Particles of Various Classes Present in a Typical Plant Cell

Subcellular particle	Diameter	Number per cell
Nucleus	5–20 μm	1
Chloroplasts	5–20 μm	50–200
Mitochondria	1–5 μm	500–2000
Ribosomes	250 Å	5–50×10^5
Enzyme molecules	20–100 Å	5–50×10^8

spindle fibers that are responsible for the movements of chromosomes in mitosis and meiosis as well as the spindle fiber generating organelles, which, although they are clear in animals (the centrioles), are not clear in plants and may be generated by genetic material included in the nuclear genome. Further subcellular components of more evident function are the starch grains, fat droplets, calcium oxalate crystals characteristic of many cells, and the aleurone grains, which are dense protein bodies in which the reserve protein of seeds are characteristically deposited.

III. The Logic of Cell Life

Before we consider in detail the operation of individual subcellular systems, it will be helpful to consider the overall logic or strategy that the cell uses to conduct its affairs. The enzyme molecules conduct, of course, the transformation of available substrates into the kinds of molecules, the building blocks, from which further cell components are to be made. It is a basic law of biology that for each kind of chemical reaction conducted in the cell there is a species of enzyme molecule that catalyzes that reaction. It is by these means that the living organism selects, from all thermodynamically possible chemical reactions, those reactions that it will use in its cellular metabolism. The enzyme molecules of various kinds are thus the basic elements of cellular transformations, and their presence is a basic requirement for life and growth. All the other subsystems of the cell are associated directly or indirectly with the production of enzyme molecules. Since enzyme molecules are not alive and cannot reproduce themselves, they must be synthesized, that is assembled, from their constituent amino acids. The function of enzyme synthesis is shared by ribosomes, messenger RNA, plus miscellaneous adjunct species of enzyme molecules as outlined in Chapter 16. The long chain molecules of messenger RNA may be likened to punch tapes containing information about the sequence in which amino acids are to be assembled to make that particular kind of enzyme molecule. Ribosomes decode this information and with the assistance of transfer RNA and appropriate specialized kinds of enzyme molecules assemble further enzyme molecules. The production of ribosomes then is an important task and it is one function of the nucleolus to produce the ribosomes. The generation of messenger RNA is the function of the chromosomes of the nucleus. We know for plants as for all cells that the formation of each kind of enzyme is controlled by a gene or genes of the genetic material. The genetic material made of DNA possesses the ability to print out copies of itself. These copies each contain the information of one or a few genes of the messenger

RNA molecules, which are then available for the decoding by the ribosomes and transfer RNA.

Finally, it is the function of the nucleus not only to produce ribosomes and messenger RNA but in addition to replicate the genetic material. Such replication of the DNA is required for cell division, and by means of such replication it is assured that each daughter cell gets a complete copy of the genetic information—the information about how to make all the kinds of enzyme molecules required in a cellular economy (Fig. 1).

Where in the logic of life of the cell do chloroplasts and mitochondria fit in? Chloroplasts and mitochondria are both relatively large bodies, surrounded by membranes, as is the cell itself. Chloroplasts, as we shall see in a later chapter, appear today to be themselves essentially small cells complete with their own genetic material, their own ribosomes, their own ability to produce enzymes characteristic of chloroplasts, enzymes

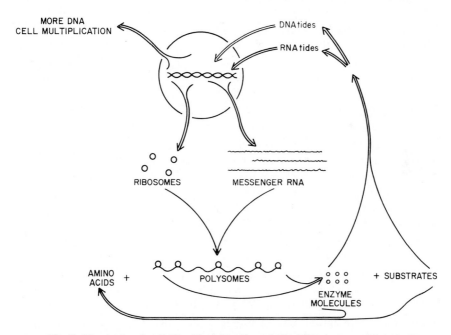

Fig. 1. The logic of cell life. The objective of the DNA is to multiply itself. To do so requires the deoxyriboside triphosphates. These are made from available substrates by enzymatic reactions. To make enzymes the DNA makes ribosomes and messenger RNA. These acting in concert make enzyme molecules from amino acids. The enymes then make not only deoxyribonucleotides (DNAtides) but also the amino acids and ribonucleotides (RNAtides) required to keep the system in operation.

suitable for the conduct of the photosynthetic process. Mitochondria, too, not only contain their own ribosomes, but also their own DNA, DNA which codes for the production of mitochondrial ribosomes and apparently for mitochondrial membrane structural protein, a species of protein that acts as host and receptor for the mitochondrial enzymes that are coded for in the nuclear DNA, produced by nuclearly transcribed messenger RNA, and which assemble themselves on the mitochondrial membrane. There is considerable evidence, as we shall see below, that mitochondria, like chloroplasts, are semiautonomous living entities within the cell, and that they possess the power of self-replication. The function of the complete mitochondrion is, of course, to oxidize available substrate and provide the energy in the form of ATP required for the energy consuming reactions of the cell.

IV. Methods of Cell Fractionation

The first task in separation of subcellular components is always the rupturing of the cell wall and the protoplasmic membranes. This is conventionally done by use of a blender. The plant tissue, immersed in an equal weight of grinding medium as specified below, is ground for 30 to 120 seconds at full speed in a blender. This results, at least with non-fibrous tissue, in the disruption of the majority of cells and liberation of their cell content. Such grinding must, of course, be done in the cold to minimize enzymatic changes in the homogenate which result from bringing together substrates and enzymes not accessible to one another in the intact cell (as, for example, rupture of lysosomes and liberation of the lysosomal hydrolytic enzymes). Cold room or ice bucket temperatures of 2°–4°C are in general suitable for and used for this purpose. Although the blender is a fast and convenient instrument, it does exert a great deal of shear, not only upon the cell walls of the tissue but also upon the subcellular components. Such grinding commonly results, for example, in destruction of the nuclear membranes and, therefore, destruction of the nucleus, in a considerable amount of lysis of lysosomes, and in almost complete disruption of glyoxysomes and peroxysomes. If one wishes to isolate intact nuclei, intact lysosomes, or intact glyoxysomes or peroxysomes, a gentler kind of grinding procedure is required for cell rupture. One such method is grinding in a glass homogenizer with loosely fitting pestle. The homogenization is carried out for a period of one to a few minutes with the pestle rotating at a speed of 100 to a few hundred revolutions per minute. Still gentler is hand grinding by mortar and pestle. We have earlier recommended the rupture of cells under conditions of near zero shear by the enuclear reactor described by Rho and Chipchase (1962). In this device, the tissues pass between counterrotating rollers, and each cell is subjected to increased hydrostatic pressure of the cell contents as result of compression of one end. The enuclear reactor has proved to be useful in the isolation of intact nuclei from plant tissue, although it does so with very low yield.

In all methods of extraction, it is necessary to use an appropriate grinding medium. The considerations that apply to the selection of such a grinding medium are as follows: (1) It must be isotonic with the cell contents so as to minimize changes in structure of cell components that are surrounded by semipermeable membranes, such as chloroplasts, mitochondria, lysosomes. This involves in general the use of 0.25 to 0.45 M sucrose or mannitol as the osmotically reactive agent of the grinding medium. (2) It must possess a buffering capacity to minimize changes in the pH of the cytoplasm that result from a release of organic acids from the vacuole of the plant cell. For this purpose Tris(hydroxymethyl)-aminomethane (Tris) buffer is often used at a pH of 7–8, that is the pH of cytoplasm itself. Such buffer in the concentration of 0.05 M is sufficient for the adjustment of the pH of all but the most acid of plant tissues. (3) If ribosomes are to be isolated intact, it is necessary that the grinding medium contain magnesium ions in a concentration of at least 0.001 M, since in lower concentrations of magnesium, ribosomes dissociate into their subcomponents. (4) If nuclei are to be isolated intact not aggregated, it is essential that calcium ions be present in the grinding medium, and the optimum concentration for this ion is also about 0.001 M. (5) Many enzymes possess as an essential portion of their active site an amino acid containing a sulfhydryl group. Preservation of the activity of such enzymes requires therefore that the grinding medium has in it a sulfhydryl-containing compound whose presence ensures that the sulfhydryl group of the enzyme remains in the reduced state. β-Mercaptoethanol 0.01 M is often used in this function as are also cysteine and glutathione. There may well be other special requirements for grinding media in the isolation of particular and specialized enzymes. The above are, however, the general considerations that govern the selection of grinding media.

Once the tissue has been ground, it is possible to proceed with separation of the subcellular components. It is convenient first to remove cell wall fragments. A simple and generally useful way is filtration of the homogenate through silicone-treated paper (Miracloth). Cell wall fragments are retained; nuclei and chloroplasts pass through unimpeded, as do the smaller subcelluar components. Centrifugation in a basket centrifuge lined with sharkskin filter paper has also been used. The cell wall fragments thus obtained may be treated as described later for the further study of cell wall matters.

The cell wall-free homogenate is in general then subjected to differential centrifugation for the separation first of the heaviest and largest component—the nucleus. Nuclei, if intact, may be removed from the homogenate by centrifugation of 100 g to a few hundred g (gravity). The nuclear pellet may be purified by the methods outlined in a later chapter. It may be remarked parenthetically that since homogenization of plant tissue in a blender ruptures most of the nuclei, the isolation of intact nuclei from plant tissue is often a difficult task. It is, however, a very simple task to prepare the interphase chromosomes of plant tissue, and since these are the most interesting components of the nucleus, it is of interest to isolate them. Plant tissue ground in a blender and filtered through Miracloth is centrifuged at 4000 g for 15 minutes. Under these conditions, starch and interphase chromatin is pelleted, while mitochondria and lysosomes, e.g., remain in the supernate (to be sure, chloroplasts will pellet, and the procedure described applies only to chloroplast-free organs). At the end of such a 4000 g centrifugation, the loosely packed chromatin is scraped from the underlying starch layer and washed several times by centrifugation in 0.01 M Tris buffer pH 8. It can then be purified by sucrose density gradient centrifugation as described in a later chapter.

We now return to the main line of differential centrifugation for the separation of plant organelles. After intact chloroplasts have been removed by centrifugation at 100 to 400 g, the chloroplasts may now be pelleted by centrifugation at 500 to 2000 g. Although differential centrifugation as described above suffices for the separation of chloroplasts, it ordinarily yields nuclear fractions contaminated with chloroplasts and chloroplast fractions contaminated with nuclei. More clean-cut separation is required. To this end, sucrose density gradient centrifugation is useful. For example, a centrifugation tube is filled with a solution whose composition is continuously changed during filling. The bottom of the tube might, for example, contain sucrose of a concentration of 1.7 M, while the top contains sucrose in the concentration 0.25 M. The homogenate is then layered above the sucrose gradient and centrifuged at 150,000 to 250,000 g in a swinging bucket rotor of an ultracentrifuge. Nuclei or even fragmented nuclear materials, such as chromatin, sediment to the bottom of the gradient, that is, they are sufficiently large and dense that they pellet through 1.7 M sucrose. Chloroplasts, on the other hand, do not travel completely to the bottom of the gradient, but remain suspended in the region of about 1.3 M sucrose. Although the densities and rates of centrifugation of chloroplasts of individual species vary somewhat, still such sucrose density gradient centrifugation does suffice for clear-cut separation of nuclei and chloroplasts in a large number of instances (Fig. 2).

The homogenate, freed of nuclei and chloroplasts, may now be used for the separation of mitochondria and lysosomes. Mitochondria are pelleted completely by centrifugation at 10,000 g periods of about 10–20 minutes. The mitochondrial preparation that is obtained contains, of course, proplastids as a contaminating fraction, as well as lysosomes. These entities may be separated from one another by sucrose density gradient centrifugation with a shallow gradient of sucrose, about 1.35 to 1.25 M sucrose. Lysosomes band at a lower density than mitochon-

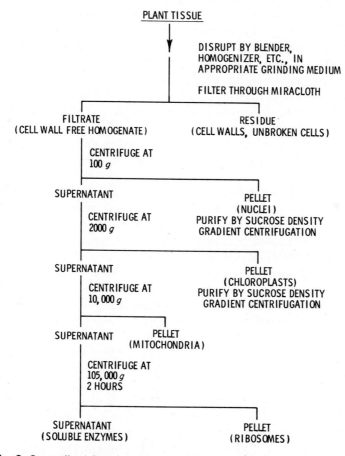

Fig. 2. Generalized flowchart for separation of plant subcellular constituents. If the nuclei have been ruptured by the grinding process, the nuclear fragments will pellet with the chloroplasts but are separated from the latter by sucrose density gradient centrifugation.

dria. They may be identified by lysis of the fraction in 0.01 M Tris pH 8 and studying the increase in activity of specific hydrolytic enzymes released by such rupture, such as acid phosphatase.

Ribosomes are next sedimented from the mitochondrial-free homogenate. It is characteristic of free ribosomes that they are sedimented by 2 hours of centrifugation at 105,000 g. The thus pelleted ribosomes may be purified by the methods outlined in a later chapter. The isolation of polyribosomes requires centrifugation through a sucrose gradient.

The cellular homogenate, now freed of the larger or particulate subcellular enzymes, contains only soluble enzymes, transfer RNA, and metabolites of small molecular weight. Among the soluble enzymes contained in such a homogenate is a large or major proportion of the enzymes contained in glyoxysomes or peroxysomes. Separation of the individual enzymes from one another is a general problem of enzymology. Isolation of intact glyoxysomes and peroxysomes is a special case. In tissues that contain glyoxysomes, they are more numerous than mitochondria. In tissues that contain peroxysomes, as leaves, particularly leaves of those plants that use the Calvin–Benson cycle for photosynthesis, peroxysomes are about one-third as numerous as mitochondria. Both peroxysomes and glyoxysomes are particularly subject to rupture during grinding. Their isolation, therefore, requires careful grinding, such as grinding in a blender for a few seconds in isoosmotic solution or hand grinding in a mortar and pestle, all in isoosmotic solution. They can then be separated, although in a very low yield because of the low rupture of cells caused by the mild grinding, on a 1.5–2.2 M sucrose gradient. These entities band at 1.24 to 1.26 gm/cm^3 (density). They band just below mitochondria, whereas lysosomes band just above mitochondria.

The isolation of other specialized subcellular components, such as microtubules and spindle fibers, will be left for later specialized chapters. We will note here that there are two principal problems that attend the separation of subcellular components. The first is a possibility of the generation of artifacts. During the separation process new entities, not present in the original cell, may be formed. These include, for example, the cleavage of individual subcellular components into smaller units, the loss of biological activity of subcellular components by enzymatic or other degradations during the isolation process, and nonspecific complexes of proteins with nucleic acid. Although there is no uniform way in which the possibility of artifact formation may be avoided, still we may state in general that the generation of artifacts is minimized by rapid execution of the fractionation and by conduct of the entire procedure at low temperature, as well as by the utilization of procedures that are as gentle as possible. This includes, for example, grinding with a shear no greater than that required for the purpose.

The second general problem of cell fractionation is the contamination of each fraction by others. Such contamination has been extensively studied, and many ingenious techniques have been devised for measuring it. For example, cytochrome c is present in mitochondria but not in chloroplasts. The presence of cytochrome c in the chloroplast separation indicates therefore that it is contaminated by mitochondria. Nuclei do not contain chlorophyll; chloroplasts do. The presence of chlorophyll in nuclear preparation is a guarantee of its contamination. Microscopic examination of particular fractions also may be used to reveal contamination of one component by another. Still another general method of procedure used by many students of cell fractionation is the deliberate addition to a homogenate of a radioactively labeled preparation of the contaminant under study. For example, to a homogenate from which nuclei are to be isolated, labeled ribosomes are added. The nuclei are then isolated and the presence or absence of labeled ribosomal contamination is detected. The impurity of supposedly pure subcellular fractions is a constant threat to the plant biochemist.

V. Separation of the Golgi Apparatus

The Golgi apparatus, or Golgi bodies, are particularly fragile and particularly difficult to separate by differential centrifugation from other membrane-containing subcellular particulates. One procedure of value is to first treat the gently homogenized homogenate with glutaraldehyde. This bifunctional reagent links protein molecules to one another and makes protein-containing membranous components more stable. The Golgi bodies may then be separated from other membranous components by virtue of the fact that the Golgi bodies are larger than other membrane component fractions that survive blending or grinding.

VI. Cell as a Community

The plant cell, as indeed all cells, is a highly organized entity containing in it large populations of numerous kinds of subcellular entities, membranes, nucleus, ribosomes, enzymes, Golgi bodies, lysosomes, glyoxysomes, and peroxysomes. These form one organized system of the cell. But the plant cell is an even more complex community. We see it now as a community of several different kinds of separate organized systems living symbiotically with one another. The chloroplasts with their own DNA and ribosomes and their own enzyme generating capacity constitute an organized subsystem within the plant cell. The same may be said of mitochondria with their own DNA, their own ribosomes, and their own capability to reproduce by fission. All other plant cell subcellular entities appear to be formed directly under the control of the nuclear information. In any case, the implications of our view of the plant cell as a community of separate semiautonomous but interdependent subsystems for our understanding of the evolution of plants is considerable. When during the course of evolution did the plant cell become infected by those cells that we now know as chloroplast? At what stage during evolution did plant cells become infected by those creatures that we now know as mitochondria? By what steps have chloroplasts and mitochondria evolved into particles totally dependent on their host cell for their supplies of amino acids, nucleotides, and even some species of enzyme molecules, These are fascinating questions to be resolved by the evolutionary studies of plants.

GENERAL REFERENCES

Beevers, H. (1969). *Ann. N.Y. Acad. Sci.* **168**, 313.

de Duve, C. (1969). *Proc. Roy. Soc., Ser. B* **173**, 71.

Frederic, S., and Newcomb, E. (1969). *Science* **163**, 1353.

Granik, S. (1964). *In* "The Cell" (J. Brachet and A. E. Mirsky, eds.), Vol. 6, p. 245. Academic Press, New York.

Millerd, A. (1956). *In* "Handbuch der Pflanzenphysiologie" (W. Ruhland, ed.), Vol. 2, p. 573. Springer-Verlag, Berlin and New York.

Mollenhauer, H. H., and Morre, D. J. (1966). *Annu. Rev. Plant Physiol.* **17**, 27.

Muhlenthaler, K. (1961). *In* "The Cell" (J. Brachet and A. E. Mirsky, eds.), Vol. 2, p. 85. Academic Press, New York.

Rho, J. H., and Chipcase, M. (1962). *J. Cell Biol.* **14**, 183.

Stumpf, P. J. (1969). *Annu. Rev. Biochem.* **38**, 159.

Tewari, K. K. (1971). *Annu. Rev. Plant Physiol.* **22**, 141.

Tolbert, N. E. (1963). *Nat. Acad. Sci.—Nat. Res. Counc. Publ.* **1145**, 648.

Tolbert, N. E. (1971). *Annu. Rev. Plant Physiol.* **22**, 45.

Voeller, B. R. (1964). *In* "The Cell" (J. Brachet and A. E. Mirsky, eds.), Vol. 6, p. 245. Academic Press, New York.

Zelitch, I. (1964). *Annu. Rev. Plant Physiol.* **15**, 121.

2

Ribosomes

ERHARD STUTZ

I. Introduction

The history of the ribosome in general and the plant ribosome in particular goes back to the early 1950's when a novel type of particle was discovered in the cytoplasm of cells that had diameters in the range of 200–300 Å (Robinson and Brown, 1953). The function of these granules was still obscure in those days, but soon it was recognized that these particles were involved in protein synthesis (Littlefield *et al.*, 1955; Webster, 1955). Once it was established that these particles, named ribosomes, were the universal "machine" whereupon the correct assembly of amino acids into proteins occurs, the race for isolation and structural and functional analysis of this enigmatic particle was on (for historical account of protein synthesis, see Zamecnik, 1969). The central position of the particle in protein synthesis required not only the study of ribosomes per se but also the study of its interaction with the other components of the protein synthesis process (translation), in particular the transfer RNA (tRNA), messenger RNA (mRNA), and a host of enzymes and proteins which in one or more steps are catalytically or stoichiometrically involved in this most complex reaction. Although considerable differences exist between the translational systems of bacteria, animals, and plants, the basic features are the same, and during evolution, nature has been rather conservative and parsimonious in

15

introducing entirely new concepts. Therefore, although those aspects
typical for the plant kingdom shall be stressed, references will be made
to representatives of other phyla, especially since many of the more
advanced aspects of ribosome structure and function were obtained with
nonplant ribosomes.

Before focusing on the ribosome particles as such, a survey of its
role in the translational reaction seems necessary. In this context,
translation means the assembly of amino acids into a defined protein
molecule according to a message written in a three letter code on the
mRNA. The mRNA itself is a replica of certain region (cistron) of one
of the DNA strands. The code is deciphered and translated into proteins
by a sequential interaction of aminoacyl-tRNA with mRNA, the
ribosomal subunits, and several specific proteins (factors). One may
consider the ribosome as the center particle toward which information
flows along with the assembly parts and from which the final product
(protein) is dispatched. The finer details of translation and especially
its regulation are not yet fully understood. The following is a simplified
version of the readout process. It is generally accepted that translation
can be broken down into three consecutive steps: (1) initiation, (2)
elongation, and (3) termination (release).

1. *Initiation.* Initiation of protein synthesis in prokaryotes seems to
start with the binding of mRNA to the 30 S subunit. Several ribosomal
proteins, in particular S1 (see Table V), seem to be required for this
binding step. The 30 S particle is able to recognize a particular base
sequence of the mRNA (initiation region). The first activated amino
acid to attach to the 30 S mRNA complex is the *N*-formylmethionyl-
tRNA (fMet-tRNA). It recognizes the initiator codon AUG or GUG
of the mRNA. This binding step is very complex and requires three
complementary initiation factors (proteins IF-1, IF-2, IF-3) and GTP.
Finally, a 50 S particle joins the 30 S mRNA–(fMet-tRNA)–IF complex yielding the functional 70 S ribosome, while the IF proteins are
released (catalytic function).

Chloroplast and mitochondrial ribosomes also use fMet-tRNA as
initiator amino acid, and the sequence of events seems to be identical to
that in bacterial cells.

The protein synthesizing system of eukaryotic cells (cell sap,
cytoplasm) uses a similar initiation mechanism. However, the first
amino acid is unformylated methionine esterified to a special tRNA
that also recognizes the AUG or GUG codon. Up to five different initiation factors (IF-E_1, . . . , IF-E_5) are required. The precise function,
however, of each of the many factors is still disputed.

2. *Elongation.* In this repetitive reaction, the translating ribosome

carries the growing peptide chain attached via tRNA first in the P site (in this position peptidyl–tRNA reacts with puromycin) accepting in the A site (peptidyl–tRNA not reactive with puromycin) the new aminoacyl–tRNA. Peptide bond formation occurs between the carboxyl group of the N-terminal amino acid esterified to tRNA (growing peptide chain) and the free amino group of the incoming aminoacyl-tRNA. Ester hydrolysis yields sufficient energy to allow peptide bond formation. This reaction is catalyzed by peptidyl transferase, an activity localized on the large subunit. After peptide bond formation, the nascent protein chain, now in the A site, is moved relative to the large subunit into the P site; in other words, the 70 S ribosome moves one codon in the 3′ direction of the mRNA (translocation) releasing the deacylated tRNA. The A site is now free to accept another aminoacyl-tRNA. This complex reaction requires several specific elongation factors (EF-Tu, EF-Ts, and EF-G for prokaryotes; EF-1 and EF-2 for eukaryotes) and GTP. Detailed operational schemes have been worked out.

3. *Termination* (*Release*). Termination as well as initiation is a single event, occurring once for every protein molecule synthesized. Peptide synthesis comes to an end when the termination signal, UAG, UAA, or UGA, of the mRNA reaches the translating ribosome (A site). Sometimes, more than one terminator signal is inserted in the mRNA at the end of a cistron. All indications are that prokaryotes and eukaryotes use the same terminator signals and a similar release mechanism. There seems to be no special tRNA involved in this last step; rather, protein factors RF-1 and RF-2 (release factor) are bound somewhere on the ribosome and convert the peptidyl transferase into a hydrolytic enzyme, which is able to split the finished protein chain from the last tRNA (P site). The run-off ribosomes (having read the entire message) can dissociate, and the small subunit can attach again to another mRNA initiator region. This dissociation step is controlled by IF proteins (Lengyel, 1969; Bretscher, 1971; Lucas-Lenard and Lipman, 1971; Caskey et al., 1972; Haselkorn and Rothman-Denes, 1973).

II. Structure of Ribosomes

A. General Properties of Ribosomes

Ribosomes from all sources so far studied are made up of two nucleoprotein particles (subunits), unequal in size, shape, and chemical composition. The subunits contain between 30 to 50% by weight of protein, the remainder being RNA and a minor amount of inorganic molecules. It became convenient to classify ribosomes and their subunits according to S values, and, as a rule, we may say that ribosomes from prokaryotes are of the 70 S type with 30 S and 50 S subunits, while ribosomes from eukaryotes are of the 80 S type with 40 S and 60 S subunits. Plant cells may contain up to three different kinds of ribosomes, namely, 80 S in the cytoplasm (cell sap) and 70 S within the chloroplasts and mitochondria—the two sets of 70 S ribosomes not being identical. Such a gross classification certainly does not take into account the many smaller ribosome size differences reported; however, several functional characteristics, e.g., antibiotic susceptibility, translation mechanism, and ion requirements, parallel and validate this classification (Spirin and Gavrilova, 1969).

The large subunits (50 S to 60 S) from almost all sources contain one large RNA molecule (23 S to 28 S) and one 5 S RNA molecule—ribosomes from mitochondria are exceptional in not having a 5 S RNA. The small subunits contain only one RNA molecule (16 S to 19 S). Large subunits of 80 S type ribosomes (eukaryotes) contain in addition one smaller RNA molecule (5.5 S, 5.8 S, 7 S) that is noncovalently linked to the large RNA molecule.

The rRNA molecules are highly folded *in situ*, and double helical regions alternate with single-stranded nucleotide segments. A certain number of different ribosomal proteins are in specific and intimate contact with the rRNA yielding nucleoproteins of characteristic shape.

The spatial dimensions (shape) of the ribosome and its subunits have been estimated by electron microscopy, X-ray analysis, viscosity measurements, neutron scattering, and fluorescence spectroscopy to name a few. The results from many such analyses lead to the conclusion that the 70 S ribosomes and the 80 S ribosomes are very similar in the overall architecture. Plant ribosomes were among the first to be physicochemically characterized. In Table I, earlier and some more recent results from electron microscopic studies are compiled. Comparing the dimensions of the three 80 S type ribosomes, considerable differences can be seen. However, this seems to be more due to different preparative conditions, and less to actual size differences. A main problem in shape determination is the sensitivity of the strongly hydrated globules to the nonphysiological preparative conditions. From detailed studies of *E. coli* ribosomes and rat liver ribosomes, it was concluded that the small and large subunits are oblate or prolate spheroids. The small subunit fits over the ends of the large subunit in such a way as to form a "tunnel" between the particles. This tunnel could accommodate the mRNA, tRNA, and protein synthesis factors. A model of a 80 S type ribosome is given in Fig. 1. The small subunit covers the large subunit. No claim is made

TABLE I

Shape and Size of Some Plant Ribosomes

Source	S value approx.	Length (Å)	Width (Å)	Mass ($\times 10^{-6}$)	Reference
Pea cell sap	80	350	160	4.1	Ts'o et al. (1958)
	60	290–330	250–300	— ⎫	Amelunxen and Spiess
	40	250–300	90–130	— ⎭	(1971)
Yeast cell sap	80	250	206	3.6 ⎫	Mazelis and Petermann
	60	202	211	2.5 ⎬	(1973)
	40	52	185	1.0 ⎭	
Tobacco cell sap	80	286 ± 28	222 ± 25	— ⎫	Miller et al. (1966)
Tobacco chloroplast	70	268 ± 28	214 ± 20	— ⎭	

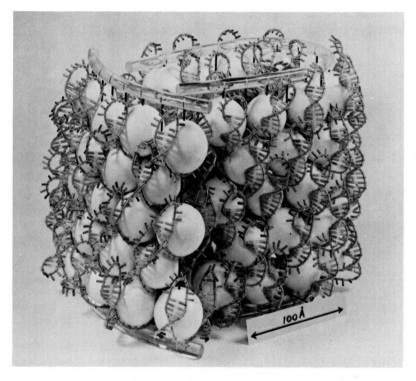

Fig. 1. Model of an 80 S ribosome. (After Cox and Bonanou, 1969.)

by the authors (Cox and Bonanou, 1969) to match the *in situ* ribosome structure.

The shape of the ribosomes (subunits) strongly depends on the ionic conditions of the milieu. Of particular importance is the concentration in Mg^{2+} ions and the ratio of monovalent to divalent ions. Removal of Mg^{2+} by strong chelating agents (e.g., EDTA) leads to ribosome dissociation and unfolding of the subunits (changes in S values and viscosity).

Under normal growth conditions, the majority of the ribosomes is engaged in protein synthesis. Since several ribosomes may translate the same messenger molecule simultaneously, a certain number of translating ribosomes are connected with each other through the mRNA strand. Such an aggregate of ribosomes is called a polysome (polyribosome). Clark *et al.* (1964) first reported the presence of polysomes in higher plants, and they inferred from sedimentation analyses (analytical centrifugation) that the leaves of chinese cabbage contain two populations of polysomes, one originating from cell sap, the other from the chloroplasts. This was later confirmed by Stutz and Noll (1967), who isolated polysomes from the cytoplasm and the chloroplasts of bean leaves and compared their sedimentation characteristics in preparative sucrose density gradients. The respective recordings are shown in Fig. 2 for the cytoplasmic polysomes (A), the chloroplast polysomes before (B) and after (C) ribonuclease treatment, and a mixture of cytoplasmic and chloroplast ribosomes (D). The last profile clearly shows the difference in sedimentation rate between the cytoplasmic (80 S) and the chloroplast (70 S) ribosomes.

Ribosomes, polysomes, and ribosomal subunits are usually recovered from the so-called S-30 supernatant (centrifugation at 30,000 g for 20 to 30 minutes) of a cell homogenate by centrifugation at 100,000 g for 2–4 hours. The ribosomes are often pelleted through 1 M buffered sucrose in order to remove contaminating proteins and other cellular lighter components. A good ribosome preparation should have a UV absorbance ratio of 1.6 to 1.7 at $\lambda = 260/280$ nm. Usually the preparation will consist of variable amounts of polysomes, monomers, and subunits, the ratios depending on the physiological stage of the cell (tissue) and the buffer and salt conditions used during extraction and analysis. Plant tissues contain several potent ribonucleases that rapidly hydrolyze the mRNA strand, this being a main reason for often poor recovery of polysomes.*

---

*The newest aspect of ribosome structure, function, and assembly are competently treated by several authors in M. Nomura, A. Tissieres, and P. Lengyel, eds. (1974). "Ribosomes." Cold Spring Harbor Lab. Publ., Cold Spring Harbor, New York.

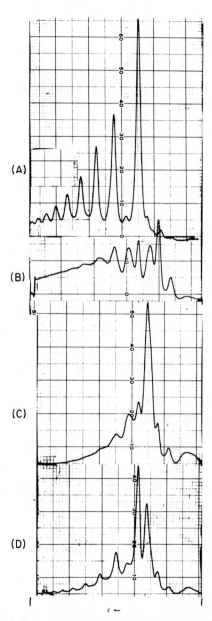

Fig. 2. Sedimentation patterns of polysomes from bean cytoplasm and chloroplasts. Polysomes from (A) cytoplasm; (B) crude chloroplast fraction; (C) crude chloroplast fraction treated with 0.1 mg/ml crystalline bovine ribonuclease, 5 minutes at 0°C; (D) mixture of equal parts of preparation (A) and (C). After Stutz and Noll (1967).

B. Ribosomal RNA

With respect to size, plant cytoplasmic rRNA are between the bacterial and the higher animal rRNA. In Table II, a few molecular weights as calculated from gel electrophoretic measurements are compiled, and for comparative reasons the rRNA from bacteria and mice are included. Loening (1968) showed that the logarithm of the molecular weight of the rRNA correlates linearly with the distance traveled in a gel under standardized conditions. One has to keep in mind, however, that any such correlation is only true under the assumption that the secondary structure of the various classes of rRNA is approximately the same.

In Table III, the overall base composition of the RNA from the large and small subunits are listed for a few representatives of the plant kingdom. The values for *E. coli* are included for comparative reasons. The differences in base composition are relatively small, especially when considering the vast evolutionary distance between prokaryotes, fungi, algae, and higher plants. Similar analyses were done with the 5 S RNA from lower and higher plants (Payne and Dyer, 1971).

The overall base composition, of course, allows little conclusion concerning the secondary structure of the molecule. Physical measurements, such as X-ray diffraction, optical rotation, and thermal melting

TABLE II

Molecular Weights of Ribosomal RNA of Different Organisms[a]

Origin	RNA from large subunit $(MW \times 10^{-6})$	RNA from small subunit $(MW \times 10^{-6})$
Bacteria	1.08	0.56
Actinomycetes	1.12	0.56
Blue-green algae	1.07	0.56
Chloroplasts	1.07 to 1.11	0.56
Mitochondria (yeast)	1.27	0.72
Higher plants	1.27 to 1.31	0.70
Algae, ferns	1.28 to 1.34	0.70
Fungi	1.28 to 1.3	0.70
Euglena	1.3	0.85
Mice	1.75	0.70

[a] Excerpt from Loening (1968).

TABLE III

Ribonucleotide Composition (in moles%) of Cytoplasmic rRNA

Origin[a]		CMP	AMP	GMP	UMP	$\dfrac{A+U}{C+G}$
Potato tuber[b]	(l)	22.0	25.1	31.7	21.2	0.86
	(s)	22.2	25.4	27.2	25.2	1.02
Pea seedlings[b]	(l)	22.6	23.6	32.1	21.6	0.82
	(s)	20.1	23.7	31.1	25.1	0.96
Neurospora[c]	(l)	21.9	24.8	29.4	23.9	0.95
	(s)	21.6	25.3	27.7	25.4	1.03
Aspergillus[d]	(l)	25.0	21.5	29.5	24.0	0.83
	(s)	23.0	23.5	29.0	24.5	0.92
Euglena gracilis[e]	(l)	23.2	21.5	32.8	22.5	0.86
	(s)	25.7	21.1	31.3	21.9	0.76
E. coli[f]	(l)	21.5	25.4	33.5	19.6	0.82
	(s)	22.7	24.8	31.0	21.5	0.86

[a] l, large; s, small.
[b] Click and Hackett (1966).
[c] Rifkin et al. (1967).
[d] Verma et al. (1971).
[e] Rawson and Stutz (1969).
[f] Midgley (1962).

studies (optical absorbance), indicate that considerable stretches of the macromolecule (approximately 77%) occur in short double helical regions that alternate with single-stranded regions. In order to find out which part of the strand could fold on itself (bases paired according to the rule G = C and A = U), it is necessary to establish the base sequence. For such large molecules (1700 to 5000 nucleotides), the sequencing is a formidable task. So far, only the 16 S RNA and 23 S from *E. coli* are sequenced to any extent (Fellner *et al.*, 1972). The sequences of the analyzed fragments show potential for extensive pairing and hairpin loop formation. The regions so far sequenced do not show any extensive repetitions (however, there are short repetitions), meaning that rRNA lacks important symmetrical elements that could be exploited as possible handles to study RNA protein assembly. The first rRNA totally sequenced was the 5 S RNA from *E. coli* (120 nucleotides) (Brownlee and Sanger, 1967). Other examples of rRNA totally sequenced are the 5 S and 5.8 S RNA from yeasts (Hindley and Page, 1972; Rubin, 1973).

The rRNA isolated from a variety of 80 S plant ribosomes or 70 S chloroplast ribosomes show sedimentation and gel electrophoretic characteristics similar to those of bacterial or animal rRNA. An interesting exception, however, was reported by Edelman *et al.* (1970). They compared the mitochondrial rRNA of *Aspergillus nidulans* with its cytoplasmic rRNA and *E. coli* rRNA in a series of experiments including sucrose gradient analysis, gel electrophoresis, and thermal melting studies. They conclude from the results that the mitochondrial rRNA is different, having a unique base sequence (Verma *et al.*, 1971) and an unusually low melting midpoint. They estimate the percent of $G + C$ in ordered regions to be in the range of 30% compared to 50% for the cytoplasmic rRNA. The unusual base composition (see Section IV) seems to be characteristic for mitochondrial rRNA.

Cell sap rRNA is the transcriptional product of a particular genome region morphologically correlated to the nucleolus. The number of copies for rRNA varies in eukaryotes from several hundred to several thousand, allowing the cell to synthesize ribosomes rapidly when needed. The proportion of the genome that codes for rRNA greatly varies; it is, e.g., 2% in yeast (Schweizer *et al.*, 1969) and 0.1% in tobacco (Tewari and Wildman, 1968). Independent of the number of copies or the relative proportion of the ribosomal DNA to total DNA, the sequence coding for rRNA is polycistronic, yielding a first transcription product (precursor) that is considerably larger than any of the final rRNA; this precursor RNA undergoes posttranscriptional tailoring in several steps. For some plants, the formation of these RNA molecules has been studied by labeling experiments and subsequent analysis by gel electrophoresis. The precursors for plant cytoplasmic rRNA have an approximate molecular weight of 2.3×10^6 to 2.6×10^6. This molecule is cleaved yielding a precursor for the large component and the small component, leaving excess RNA of 0.3×10^6 to 0.6×10^6 (Loening, 1970). *Euglena gracilis* is exceptional by having a larger precursor (3.5×10^6) with a 0.85×10^6 piece cleaved off (small rRNA), leaving a hypothetical 2.7×10^6 piece that is further broken down to a rather stable 2.2×10^6 RNA. This molecule is finally cut to a size of 1.35×10^6 (large rRNA) (Brown and Haselkorn, 1971). The 5.5 S (5.8 S, 7 S) RNA is derived from the large RNA precursor and linked to the 24 S to 28 S RNA in a noncovalent form (Udem and Warner, 1972). The 5 S RNA is a transcript from an entirely different cistron.

Precursors for chloroplast rRNA seem to be only in slight excess of the final size. For *Euglena* chloroplast rRNA, the precursors are equivalent to 1.2×10^6 and 0.66×10^6 daltons (Heizmann, 1974). They are converted to 1.1×10^6 and 0.55×10^6 dalton molecules, respectively, during maturation. However, a common and larger precursor RNA, similar to that found in *Neurospora* mitochondria (Kuriyama and Luck, 1973), may also occur in chloroplasts.

All rRNA carry a small number of methyl groups. Methylation occurs at the precursor level. It is mainly the nonmethylated sequences that are discarded during maturation. Ribosomal RNA transcribed from nuclear DNA is preferentially methylated at the ribosyl moeity, while prokaryotic rRNA is preferentially methylated at the bases.

C. Ribosomal Proteins

From the ribosome model shown in Fig. 1, it is evident that many protein molecules are attached to the folded RNA molecule. It was thought earlier that multiple copies of no more than a few different proteins make up the protein "coat" of the ribosome. However, later work clearly showed that a ribosome contains many different proteins. The proteins can be removed from the RNA backbone, e.g., by urea treatment, and have been separated by gel electrophoresis. The various kinds of proteins show up as distinct bands, the distance traveled being a function of the molecular weight and the net charge of the molecule. In Fig. 3, the gel patterns of various plant ribosomes have been compared (Gualerzi and Cammarano, 1970), by the so-called split gel technique. Coincidence of bands indicates similar or identical types of proteins, and in Table IV the evaluation of such a study is compiled. Obviously, the chloroplast 70 S ribosomal proteins poorly match the 80 S ribosomal proteins (Fig. 3A), which is further evidence for the existence of separate ribosome classes in chloroplasts and cell sap, respectively. It also seems that the dissimilarity percentage increases the more distantly related the specimens are (Fig. 3B and Table IV).

Once it was recognized that each subunit contains a set of different proteins, a multitude of questions had to be answered: (a) the exact number of proteins per ribosome, their molecular weights, the amino acid composition and sequence; (b) the site of each of these proteins relative to each other and relative to the RNA molecule; (c) the function of each of these proteins in ribosome assembly and translation.

The best studied ribosome so far is the *E. coli* 70 S ribosome. For the 30 S *E. coli* subunit, the number of proteins is 21, for the large subunit, 34 proteins. In terms of amino acid sequence (partially hydrolyzed by specific proteolytic enzymes) and molecular weight, they are all different. A similar diversity seems to hold also for lower and higher plant 70 S and 80 S ribosomes (Gualerzi *et al.*, 1974).

A real breakthrough in elucidating the topology of the ribosome and the function of the various proteins occurred with the discovery that it is possible to remove proteins stepwise from the subunits and that a functional particle can be reconstituted under controlled con-

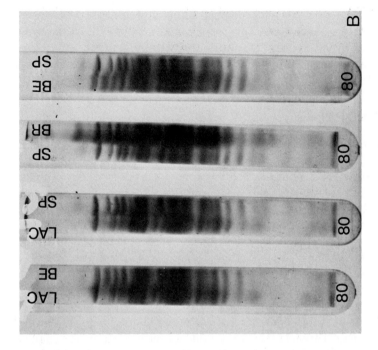

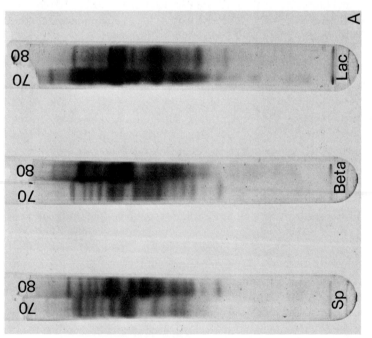

Fig. 3. Panel (A) Split gel comparison of ribosomal proteins from chloroplast (70 S) and cytoplasm (80 S) ribosomes of different plants. (B) Split gel comparison of ribosomal proteins from 80 S ribosomes. Sp, *Spinacia oleracea*; Beta, *Beta vulgaris*; Br, *Brassica oleracea*; Lac, *Lactuca dioica*. After Gualerzi and Cammarano (1970).

TABLE IV

Degree of Electrophoretic Dissimilarity among Ribosomal Proteins Isolated from Chloroplastic (70 S) and Cytoplasmic (80 S) Particles of Differently Related Plants[a]

Ribosome classes compared with	Plants	Taxonomic difference	Degree of electrophoretic dissimilarity[b] (%)
80 S–80 S	B. oleracea–B. rapa	Species	None
	Beta–Spinacia	Genus	27
	Spinacia–Brassica	Order	35
	Spinacia–Lactuca	Subclass	57
	Beta–Lactuca	Subclass	47
70 S–70 S	Beta–Spinacia	Genus	48
	Spinacia–Lactuca	Subclass	72
	Beta–Lactuca	Subclass	73
70 S–80 S	Spinacia		65
	Beta		64
	Lactuca		60

[a] From Gualerzi and Cammarano (1970).

[b] Estimated from densitometer recordings of gel patterns. The degree of electrophoretic dissimilarity between two samples was calculated as the percentage ratio between total noncoinciding bands and total number of electrophoretically resolved bands.

ditions. A first step in this direction was the finding that the 30 S ribosomal subunit loses some proteins (split proteins) in a CsCl density gradient leaving a "core particle," containing the RNA and residual proteins (Staehelin and Meselson, 1966). Neither of the separated components was functional in protein synthesis, but when recombined, function was restored. Split proteins from the 50 S subunit could not replace the split proteins from the 30 S subunit and *vice versa*. These observations have two fundamental implications: (a) The proteins could spontaneously assemble to a functional ribosome under proper conditions. (b) The various proteins had specific functions in translation. In other words, the ribosomal proteins would recognize their place within the subunits. Nomura's group in further experiments succeeded in totally reconstituting both the 30 S subunit and the 50 S subunit (Nomura and Erdman, 1970). Finally, it was recognized that 23 S and 16 S rRNA could be mixed with the protein components and under proper ionic and temperature conditions, the functional subunits would be reformed. The *in vitro* assembly reaction is first order with respect to the formation of the active subunits. The rate is strongly dependent on the incubation temperature (30°C is optimal for the small subunit, 50°C is optimal for the large subunit). The 5 S RNA is necessary to obtain a functional 50 S particle. So far, no plant ribosomes have been reconstituted.

In vivo, it is the precursor RNA that associates with the various proteins to yield preribosomal particles larger in size than the mature subunits. During maturation, RNA sequences and proteins are removed. The subunits are released from the nuclei (eukaryotes) into the cytoplasm where they join the translation machinery (polysomes). The older version that rRNA has a messenger function, coding for ribosomal proteins, no longer holds.

Some results about the *E. coli* ribosomal proteins are compiled as an example in Table V. All 21 proteins have been isolated and purified, and the molecular weights were determined. Protein S1 (Berlin nomenclature) has by far the highest molecular weight of the 30 S ribosomal

TABLE V

Some Characteristics of Proteins from the 30 S Subunit (*E. coli*)[a]

Protein	Molecular weight	Required for assembly	Stoichiometric classification[b]
S1	65,000		F[c]
S2	30,000		F
S3	31,000		
S4	26,000	+	U[c]
S5	24,000		
S6	18,000		
S7	21,500	+	U
S8	17,000	+	U
S9	21,600	+	U
S10	16,000		
S11	18,300		U
S12	19,000		F
S13			
S14	15,600		F
S15	13,200		F
S16	13,500	+	U
S17	10,700	+	U
S18	14,600		
S19	15,000		F
S20	14,000		F
S21	13,000		F

[a] After Kurland (1971, 1972).
[b] Blanks are left where data are uncertain.
[c] U, unit protein; F, fractional protein.

proteins. It occurs in only 10–30% of the small particles and is a good example for a F-type protein. It is not necessary for the *in vitro* reconstitution of the 30 S particle; however, it seems to be necessary for mRNA binding. This means that the 30 S subunit population is not homogeneous *in vivo*. On the other hand, there exist U-type proteins, occurring once per particle. From Table V, it is evident that all those proteins necessary for the assembly (reconstitution experiments) are U-type proteins. Present research, especially with *E. coli* ribosomes (prokaryotes) and mammalian cell ribosomes (eukaryotes) is concentrated on elucidating the relative position (e.g., inside, outside, attachment to rRNA) and function (A-site or P-site) of each of the proteins. Unfortunately, research of this type on plant ribosomes considerably lags behind.

III. The Ribosome Cycle

The functional unit in translation is the ribosome (monomer), and it was generally accepted that the ribosome after synthesis of a protein would move to another mRNA and resume translation. More recent studies however show that the so-called run-off ribosomes (having completed one round of translation) dissociate into the two subunits rejoining the pool of subunits before being attached to another mRNA. Strong evidence for such a cycle was given by Kaempfer (1968), who first showed in the bacterial system and then in the yeast cell sap by heavy isotope transfer experiments that subunit exchange occurs. Kaempfer (1969) grew yeast cells (*Candida crusei*) for nine generations (generation time 3 hours) in heavy isotope (^{15}N) medium containing [^{3}H]uridine. When the mixture reached 2×10^7 cells/ml, unlabeled uridine was added. Twenty minutes later, the cells were washed with light medium (^{14}N) and resuspended in light medium for further growth. Ribosomes were isolated 0.1, 4, and 8 generations after transfer into the light medium, and the sedimentation distribution of the monomers was measured in isokinetic sucrose gradients. Since the heavy ribosomes (^{15}N) sedimented approximately 20% faster than the light ribosomes, the exchange process, as a function of generation time, could be monitored by a corresponding shift in sedimentation rate. In this particular experiment, the monomers sedimented almost exclusively as hybrids after 4 generations of growth. This means that heavy ribosomes had dissociated, entered a subunit pool and recombined with newly synthesized subunits. Similar experiments were done *in vitro*, and under these better defined conditions, it was clearly shown that ribosomes (70 S) dissociate after each round of protein synthesis (Kaempfer and Meselson, 1969). It is, of course, possible that a certain subunit exchange occurs that is not mediated by initiation. More recent work gives evidence that the IF proteins are crucial in this dissociation step and in selecting the next mRNA. We should keep in mind, however, that for eukaryotes (80 S) the subunit exchange may be much less frequent and other mechanisms may be in operation, especially when considering that different classes of ribosomes can exist, e.g., strongly membrane-bound versus free ribosomes which seem to have different functions (see, e.g., Rosbash and Penman, 1971).

IV. Ribosome Hybrids

The reversible dissociation of ribosomal subunits under proper ionic conditions allows one to investigate the exchange ability of subunits from different origin. Such hybrid ribosomes were first made with bacterial subunits, and they were shown to be functional in poly(U) mediated *in vitro* translation (Takeda and Lipman, 1966). The dissociation of 80 S ribosomes into functional subunits turned out to be more difficult, but it was achieved by relatively high KCl concentration (0.8 M) (Martin and Wool, 1969). In Table VI, a series of hybrids,

TABLE VI

Percent Formation of 80 S Monomers from Ribosomal Subunits[a,b]

Small subunit origin	Large subunit from					
	Tetra-hymena	Saccharo-myces	Chlamy-domonas	Wheat	Pea	Mouse
Saccharomyces	5	90–100	NT	90–100	90–100	90–100
Chlamydomonas	NT	NT	80–90	NT	80–90	70–90
Wheat	5	40–60	NT	70–90	80–90	70–80
Pea	5	30–50	80–90	80–90	70–90	80–90
Mouse	5	90–100	90–100	90–100	90–100	90–100
Tetrahymena	90–100	90–100	NT	90–100	80–100	90–100

[a] Excerpt from T. E. Martin et al. (1970). Biochem. Genet. **4**, 603. Reprinted by permission of Plenum Publishing Corp.

[b] 80 S ribosomes were dissociated by high salt into their subunits and the appropriate mixture was dialyzed against 50 mM Tris HCl, pH 7.8; 12.5 mM MgCl$_2$; 80 mM KCl. The content in 80 S ribosomes was calculated from the sucrose gradients. NT, not tested.

mainly of plant 80 S ribosomes, are listed along with the percentage of recombination (Martin *et al.*, 1970). The degree of recombination is reciprocal in most cases, a definite exception being the combination of pea ribosome subunits with *Saccharomyces* subunits. The pea small subunit recombines poorly with the *Saccharomyces* large subunit, but the reciprocal mix does very well. On the other hand, *Tetrahymena* ribosomal subunits show poor recombination with all others. This may be taken as an indication that such recombinations measure specific interactions, and possibly some evolutionary differences.

V. Chloroplast and Mitochondrial Ribosomes

Lyttleton (1962) showed for the first time that chloroplasts from spinach leaves contain a special class of ribosomes sedimenting with a lower rate than did the cytoplasmic ribosomes. This observation was confirmed in many subsequent studies, and all chloroplast ribosomes so far studied fall into the 70 S class (Smillie and Scott, 1969). The two main RNA components have S values of 23 S/16 S corresponding to 1.1×10^6 and 0.56×10^6 molecular weights, and the large subunit has a 5 S RNA. The ribosome needs a relatively high Mg^{2+} concentration, e.g., up to 20 mM for functioning, and it easily dissociates into its subunits if the Mg^{2+} concentration drops below 4 mM. The 23 S RNA component is very unstable and easily broken down to smaller fragments (Leaver and Ingle, 1971). The ribosomal proteins (see Fig. 3) are different from those of the corresponding 80 S ribosomes. Wittmann *et al.* (1969) studied the mutual relationship of chloroplast and cytoplasmic ribosomes by serological techniques, and according to this, the chloroplast ribosomes from pea, bean, wheat, spinach, and tobacco are less related to each other than the corresponding 80 S cytoplasmic ribosomes; they could not detect any serological relationship between these chloroplast ribosomes and ribosomes from bacteria or a blue green algae (*Anacystis*). Nevertheless, the chloroplast ribosomes still have sufficient similarity to bacterial ribosomes in order to use, e.g., *E. coli* supernatant (protein factors) to perform *in vitro* protein synthesis (Eisenstadt and Brawerman, 1966). It is even possible to make functional hybrid ribosomes (70 S) consisting of a 30 S *Euglena* chloroplast ribosomal subunits and 50 S *E. coli* subunits using the *E. coli* polymerizing enzymes (Lee and Evans, 1971). The reciprocal mixture does not seem to work.

Plant mitochondrial ribosomes are difficult to isolate, and the knowledge about higher plant mitochondrial ribosomes is scanty, although their existence is beyond doubt (Vasconcelos and Bogorad, 1971). Comprehensive studies were done with mitochondrial ribosomes from *Neurospora*, yeast, and *Aspergillus*. It is accepted now that these ribosomes are of the 70 S type (Borst and Grivell, 1971). They differ, however, in subunit structure and in the RNA composition from a "usual" 70 S ribosome. In particular, the rRNA is low in cytosin and very high in uracil. In Table VII, the base composition from three different mitochondrial RNA are compiled and compared with *Euglena* chloroplast RNA.

Fungal mitochondrial ribosomes, although different in many respects from bacterial ribosomes, were shown to be capable of using the initiation and elongation factors along with the N-formylmethionine from bacteria. For example, it was shown that *Neurospora* mitochondrial ribosomes can recognize, bind, and translate *E. coli* N-formylmethionyl-tRNA in response to the initiator triplet AUG, and they can translate poly(U) with the bacterial supernatant fraction (Sala and Küntzel, 1970). Polymerizing enzymes from the cytoplasmic (80 S) system seem not or only partly to support organellar mitochondrial or chloroplast (70 S) protein synthesis (Ciferri and Parisi, 1970). Such generalizing statements are tentative at best.

The biosynthesis and assembly of ribosomes of organelles is a problem in itself. A green plant cell has three different protein synthesizing systems separated by organellar membranes. The synthesis of the three types of ribosomes and their assembly could theoretically happen in various loci. Certainly, it is reasonable to assume that the 80 S ribosomes are manufactured with the help of the cytoplasmic translational apparatus, the genes for the rRNA and the various proteins being transcribed from the nuclear DNA. However, for the organellar 70 S ribosomes, the situation seems to be more complex: The most simple assumption of total independence does not hold. Rather, the

TABLE VII

Base Composition (in moles%) of Organellar rRNA

Source[a]	CMP	AMP	GMP	UMP	Reference
Neurospora crassa, M	15	27	23	30	Küntzel and Noll (1967)
Aspergillus nidulans,[b] M	13	35	18	34	Verma *et al.* (1971)
Euglena gracilis, M	12	39	15	33	Krawiec and Eisenstadt (1970)
Euglena gracilis, C	18	27	28	26	Crouse *et al.* (1974)

[a] M, mitochondria; C, chloroplast.
[b] Averaged from the heavy and light components that were determined separately.

mitochondrial 70 S ribosomal proteins are synthesized outside of the organelle, at least to such an extent that the formation of new 70 S ribosomes is immediately stopped in the presence of cycloheximide, which inhibits protein synthesis on 80 S ribosomes (Küntzel, 1969). There are most likely various degrees of independence, the mitochondria being less autonomous in this respect than the chloroplast. On the other hand, there is good evidence from DNA/RNA hybridization studies that the 23 S/16 S organellar rRNA originates from the organellar genome, as shown, e.g., for yeast mitochondria (Morimoto et al., 1971) and for Euglena chloroplasts (Stutz and Rawson, 1970). Genetic studies, especially with Chlamydomonas, indicate that some genes affecting the 70 S ribosomal proteins are inherited uniparentally, which suggests that the respective cistrons are located on the chloroplast genome (Sager, 1972). A tentative conclusion is that some components of the 70 S ribosome are coded for and synthesized inside (rRNA, some mRNA) and that the majority of ribosomal proteins are synthesized outside using the respective mRNA transcribed either from the nuclear or organellar genome. Another question concerns the locus for the assembly of ribosomal particle: Perhaps the organellar outer membrane is the assembly place.

REFERENCES

Amelunxen, F., and Spiess, E. (1971). *Cytobiologie* **4**, 293.
Borst, P., and Grivell, L. A. (1971). *FEBS (Fed. Eur. Biochem. Soc.) Lett.* **13**, 73.
Bretscher, M. S. (1971). *In* "Protein Synthesis" (E. H. McConkey, ed.), Vol. I, pp. 89–120. Dekker, Inc. New York.
Brown, R. D., and Haselkorn, R. (1971). *J. Mol. Biol.* **59**, 491.
Brownlee, G. G., and Sanger, F. (1967). *J. Mol. Biol.* **23**, 337.
Caskey, T., Leder, P., Moldave, K., and Schlessinger, D. (1972). *Science* **176**, 195.
Ciferri, O., and Parisi, B. (1970). *Progr. Nucl. Acid Res. Mol. Biol.* **10**, 121–144.
Clark, M. F., Matthews, R. E. F., and Ralph, R. K. (1964). *Biochim. Biophys. Acta* **91**, 289.
Click, R. E., and Hackett, D. P. (1966). *J. Mol. Biol.* **16**, 279.
Cox, R. A., and Bonanou, S. A. (1969). *Biochem. J.* **114**, 769.
Crouse, E., Vandry, J. P., and Stutz, E. (1974). *Proc. 3rd Int. Congress Photosynthesis*, pp. 1775–1786.
Edelman, M., Verma, I. M., and Littauer, U. Z. (1970). *J. Mol. Biol.* **49**, 67.
Eisenstadt, J. M., and Brawerman, G. (1966). *Biochemistry* **5**, 2777.
Fellner, P., Ehresman, C., Stiegler, P., and Ebel, J. P. (1972). *Nature (London), New Biol.* **239**, 1.
Gillham, N. W., Boynton, J. E., and Burkholder, B. (1970). *Proc. Nat. Acad. Sci. U.S.* **67**, 1026.
Gualerzi, C., and Cammarano, T. (1970). *Biochim. Biophys. Acta* **199**, 203.
Gualerzi, C., Janda, H. G., Passow, H., and Stöffler, G. (1974). *J. Biol. Chem.* **249**, 3347.

Haselkorn, R., and Rothman-Denes, L. B. (1973). *Annu. Rev. Biochem.* **42**, 397.

Heizmann, P. (1974). *Biochem. Biophys. Res. Commun.* **56**, 112.

Hindley, J., and Page, S. M. (1972). *FEBS (Fed. Eur. Biochem. Soc.) Lett.* **26**, 157.

Kaempfer, R. (1968). *Proc. Nat. Acad. Sci. U.S.* **61**, 106.

Kaempfer, R. (1969). *Nature (London)* **222**, 590.

Kaempfer, R., and Meselson, M. (1969). *Cold Spring Harbor Symp. Quant. Biol.* **34**, 209.

Krawiec, S., and Eisenstadt, J. M. (1970). *Biochim. Biophys. Acta* **217**, 132.

Küntzel, H. (1969). *Nature (London)* **222**, 142.

Küntzel, H., and Noll, H. (1967). *Nature (London)* **215**, 1340.

Kuriyama, Y., and Luck, D. J. L. (1973). *J. Mol. Biol.* **73**, 425.

Kurland, G. C. (1971). *In* "Protein Synthesis" (E. H. McConkey, ed.), Vol. I, pp. 179–228. Dekker, New York.

Kurland, G. C. (1972). *Annu. Rev. Biochem.* **41**, 337.

Leaver, C. J., and Ingle, J. (1971). *Biochem. J.* **123**, 235.

Lee, S. G., and Evans, W. R. (1971). *Science* **173**, 241.

Lengyel, P. (1969). *Cold Spring Harbor Symp. Quant. Biol.* **34**, 828.

Littlefield, J. W., Keller, E. B., Gross, J., and Zamecnik, P. C. (1955). *J. Biol. Chem.* **217**, 111.

Loening, U. E. (1968). *J. Mol. Biol.* **38**, 355.

Loening, U. E. (1970). *Symp. Soc. Gen. Microbiol.* **20**, 77–106.

Lucas-Lenard, J., and Lipman, F. (1971). *Annu. Rev. Biochem.* **40**, 409.

Lyttleton, J. W. (1962). *Exp. Cell Res.* **26**, 312.

Martin, T. E., and Wool, I. L. (1969). *J. Mol. Biol.* **43**, 151.

Martin, T. E., Bicknell, J. N., and Kumar, A. (1970). *Biochem. Genet.* **4**, 603.

Mazelis, A. G., and Petermann, M. L. (1973). *Biochim. Biophys. Acta* **312**, 111.

Midgley, J. E. M. (1962). *Biochim. Biophys. Acta* **61**, 513.

Miller, A., Karlsson, U., and Boardman, N. K. (1966). *J. Mol. Biol.* **17**, 487.

Morimoto, H., Scragg, A. H., Nekhorocheff, J., Villa, V., and Halvorson, H. O. (1971). *In* "Autonomy and Biogenesis of Mitochondria and Chloroplasts" (N. K. Boardman, A. W. Linnane, and R. M. Smillie, eds.), pp. 282–292. North-Holland Publ., Amsterdam.

Nomura, M., and Erdman, V. A. (1970). *Nature (London)* **228**, 744.

Nomura, M., Tissières, A., and Lengyel, P. (eds.) (1974). "Ribosomes." Cold Spring Harbor Laboratory Publ., Cold Spring Harbor, New York.

Rawson, J. R., and Stutz, E. (1969). *Biochim. Biophys. Acta* **190**, 368.

Rifkin, M. R., Wood, D. D., and Luck, D. J. C. (1967). *Proc. Nat. Acad. Sci. U.S.* **58**, 1025.

Robinson, E., and Brown, R. (1953). *Nature (London)* **171**, 313.

Rosbash, M., and Penman, S. (1971). *J. Mol. Biol.* **59**, 243.

Rubin, G. M. (1973). *J. Biol. Chem.* **248**, 3860.

Sager, R. (1972). "Cytoplasmic Genes and Organelles." Academic Press, New York.

Sala, F., and Küntzel, H. (1970). *Eur. J. Biochem.* **15**, 280.

Schweizer, E., Mackechnie, C., and Halvorson, H. O. (1969). *J. Mol. Biol.* **40**, 261.

Smillie, R. M., and Scott, N. S. (1969). *Prog. Mol. Subcell. Biol.* **1**, 136.

Spirin, A. S., and Gavrilova, L. P. (1969). "The Ribosome." Springer Publ., New York.

Staehelin, T., and Meselson, M. (1966). *J. Mol. Biol.* **16**, 245.

Stutz, E., and Noll, H. (1967). *Proc. Nat. Acad. Sci. U.S.* **57**, 744.

Stutz, E., and Rawson, J. R. (1970). *Biochim. Biophys. Acta* **209**, 16.

Takeda, M., and Lipman, J. (1966). *Proc. Nat. Acad. Sci. U.S.* **56**, 1875.
Tewari, K. K., and Wildman, S. G. (1968). *Proc. Nat. Acad. Sci. U.S.* **59**, 569.
Ts'o, P.O.P., Bonner, J., and Vinograd, J. (1958). *Biochim. Biophys. Acta* **30**, 570.
Udem, S. A., and Warner, J. R. (1972). *J. Mol. Biol.* **65**, 227.
Vasconcelos, A. C. L., and Bogorad, L. (1971). *Biochim. Biophys. Acta* **228**, 492.
Verma, I. M., Edelman, M., and Littauer, U. Z. (1971). *Eur. J. Biochem.* **19**, 124.
Webster, G. C. (1955). *Plant Physiol.* **30**, Suppl., 28.
Wittmann, H. G., Stöffler, G., Kaltschmidt, E., Rudloff, V., Janda, H. G., Dzionara, M., Donner, D., Nierhaus, K., Cech, M., Hindennach, I., and Wittman, B. (1970). *Proc. FEBS Symp.* **21**, 33.
Zamecnik, P. C. (1969). *Cold Spring Harbor Symp. Quant. Biol.* **34**, 1.

3

The Nucleus

JAMES BONNER

I. Introduction

Within the nucleus are the chromosomes containing, in their genetic DNA, the directions for making all of the other subcellular components. In the nucleus, too, is the nucleolus, a chromosomal product with specialized functions, in particular the function of manufacture of ribosomes.

Nuclei as seen by electron microscopy are surrounded by two unit double membranes of the type described in Chapter 4. This structure in turn possesses pores or holes of the order of 500 to 1000 Å in diameter. The membrane serves then to retain the larger subnuclear components, e.g. chromosomes, nucleolus, within the nucleus. However, the pores permit traffic between cytoplasm and nucleus. This traffic includes, on the one hand, the passage into the nucleus of small molecules, such as building blocks for the making of more chromosomes and more RNA, and, on the other hand, the passage from the nucleus to the cytoplasm of nuclear products, such as messenger RNA and ribosomal subunits.

II. Preparative Procedures

A. Isolation of Plant Nuclei

In the isolation of plant nuclei, high recovery must be sacrificed in the interest of nuclear integrity. This is because the shear forces required to break open the cell wall of most plant cells is more than enough to shear the nuclear membrane as well. Cells with weak walls, as in many tissue-cultured cells, may be disrupted with a glass homogenizer without extensive damage to their nuclei. For other tissues, however, hand grinding of such materials, such as leaf and seedling with mortar and pestle, is recommended by Ts'o and Sato (1959) and used by many others (Jaworski and Key, 1974).

A low shear cell disrupter has been described by Rho and Chipchase (1962) and by Birnstiel *et al.* (1963) and used successfully by others. In this device the tissue is loaded onto a moving nylon gauze belt. It passes first through a continuously chopping guillotine, which cuts the tissue into fragments about 1 mm in length. The fragments then pass through two counterrotating spring-loaded rollers. The shear is adjusted to pop open the cells and release their contents, which then flow through the nylon gauze into a collector. The yield of nuclei by this procedure is about 5%.

In all of the above described methods a suitable grinding medium must be used. This contains, in general, 0.25 to 0.5 M sucrose to make the medium isotonic to the cell contents and 2 to 10 mM $MgCl_2$ or $CaCl_2$ to stabilize nuclei and to stiffen the cell walls so they are more readily sheared. In addition, buffer of about 50 mM Tris, pH 8 should be used.

An alternative and attractive method for disruption of plant tissue without undue nuclear rupture is the use of very high speed blenders for very short periods of time. Thus, grinding of tissue in the VirTis 45 blender for 5 to 10 seconds at 45,000 rpm accomplishes the same end result as hand grinding with less trouble. Scott and Ingle (1973) include in the grinding medium high molecular weight dextran polymers (2.5% Ficoll, 5% dextran) to increase the viscosity of the medium and hence the shear forces exerted on the cell wall.

The plant homogenate, however prepared, is next filtered through cheesecloth and/or Miracloth (siliconized paper) to remove cell wall fragments, unground tissue, etc. Next the nuclei are pelleted at 500–1000 g.

The principal contaminants at this stage are starch grains, ruptured nuclei, and chloroplasts, if present. Starch can be removed by gently scraping the nuclei from the underlying starch and repelleting the former.

The nuclei may be further purified by layering them on 60% sucrose and forcing them through this dense medium (1 hour at 24,000 rpm in the Spinco SW 25.1 rotor). The integrity of the nuclei should be checked at all points by inspection with the phase contrast microscope.

B. Separation of Subnuclear Components

The physical fractionation of the subnuclear systems requires that the nuclear membrane is ruptured. This is best done by making the nuclear preparation 4% in Triton-X (a detergent). This ruptures the nuclear membrane and permits the subnuclear structures to be separated from one another. Although there have been many reports of methods for separation of nucleoli, there is no one method that works well with all tissues. In general, nucleoli however, separated from the intact nucleus are contaminated by much non-nucleolar DNA. Since the principal function of the nucleolus is to produce ribosomal RNA and ribosomal subunits (see below) much more fruitful paths of attack on the isolation of nucleolar DNA have been discovered, using methods other than the isolation of nucleoli. This is in particular true of the use of the anucleolar mutant of *Xenopus laevis*. The genome of this mutant contains no ribosomal DNA (Birnstiel *et al.*, 1966).

C. Direct Isolation of Chromatin

As an alternative to the isolation of chromatin (interphase chromosomes) from previously purified nuclei, it is also possible to isolate chromatin directly from plant tissue (Bonner *et al.*, 1968a). Such methods have the appeal that in contrast to the 5% or so of nuclear DNA that can be recovered from the direct isolation of nuclei, some 80–95% or more of nuclear DNA can be isolated from plant cells by the direct isolation of chromatin. The reason for this paradox is that although nuclei are easily ruptured during the homogenization of plant cells, the chromatin of plant cells nonetheless remains large and dense, and are among the most easily pelleted of plant subcellular components.

Plant tissue is ground in a Waring blender at full speed in an appropriate isotonic grinding medium for 1 minute. The grinding medium could include, for example, 0.25 M sucrose, 0.05 M Tris buffer pH 8, and 1 mM $MgCl_2$, (Huang *et al.*, 1960; Bonner *et al.*, 1968a). The blending not only disrupts the cell walls of the individual cells but also the nuclear membrane, releasing the subnuclear components. Nucleoli are also largely disintegrated by the blending process. The filtered homogenate is next centrifuged in a centrifugal field that is too slight to bring down mitochondria (4000 g for 30 minutes). The resulting pellet contains starch grains upon which are layered chromatin and nuclear debris. Chloroplasts and chloroplast fragments, if present in the initail tissue, will also be present in the pellet. The gelatinous layer is removed from the underlying starch, resuspended in the above grinding medium, and recentrifuged through several cycles to remove the starch. The chromosomal layer is scraped each time from the underlying starch pellet. The crude chromatin can then be purified by sucrose density gradient centrifugation. The crude chromatin, either in 0.01 M Tris or in the original grinding medium, is layered on 1.8 M sucrose, buffered with 0.01 M Tris pH 8. The interface is stirred slightly; the preparation is then centrifuged at 22,000 rpm (Spinco 25.1 rotor), and the chromatin in greatly swollen form is found at the bottom of the gradient (Huang and Bonner, 1962; Bonner and Huang, 1963). Nonchromosomal material is largely removed by this last step, which may be repeated. The direct isolation of chromatin is rapid, recovers the great bulk of the DNA present in the original tissues, and yields material that is as pure as material isolated from previously isolated nuclei.

D. Isolation and Determination of DNA and RNA

There are many methods for the preparation of pure DNA from either plant nuclei or plant chromatin separated as described above. One such method of deproteinization of chromatin to produce pure DNA is that of Marmur (1961). In this procedure, the chromatin is dispersed in 1 M NaCl and in the presence of a detergent, for example, 1% sodium dodecyl sulfate. The protein liberated from the chromatin, by the high ionic strength, complexes with the detergent, which in turn is held at the interface between the aqueous phase and a nonpolar (chloroform:isoamyl alcohol 24:1) phase. The two phases are separated by brief centrifugation at 10,000 g. The procedure is repeated until no protein is recovered at the interface between the aqueous phase and the chloroform:isoamyl alcohol phase. The DNA is then precipitated from the aqueous phase with 70% ethanol and recovered by winding it out as fibers.

A second procedure, which has much to recommend it, it is that of Bonner *et al.* (1968a). Chromatin as prepared above is dissolved in 4 M CsCl and centrifuged for 18 hours at 40,000 rpm in the Ti-50 rotor (Spinco). Under these circumstances, the DNA pellets, while the protein removed from the DNA by the high salt concentration floats to the surface of this high density medium. The DNA in the pellet, which contains 2–4% protein, can then be further purified by shaking with redistilled phenol saturated with 10 mM Tris pH 8). Again, after separation of the two phases, the aqueous phase is precipitated with 2 volumes of alcohol, and the DNA wound out from it.

DNA prepared as described above is still contaminated by trace amounts of protein and RNA. The principal modes of attack to remove these contaminants are (1) to treat the DNA in solution with DNase-free RNase. This is followed by (2) treatment of the DNA with pronase, a highly nonselective protease that destroys the RNase and in turn destroys itself. The DNA is then again purified by phenol extraction to remove all protein.

Often it is desired to isolate nuclear RNA as contrasted with whole cell RNA. For the isolation of nuclear RNA there is no short-cut to the isolation of nuclei themselves. There are no guidelines for the preparation of whole nuclear RNA from plant cells. However, the methods described for animal cells (Holmes and Bonner, 1973, 1974a,b) presumably will serve this end. In the isolation of nuclear RNA every effort must be made to minimize RNA degradation. Among the precautions advocated are the use of gloves throughout (hands secrete RNase), low temperatures, and rapid procedure.

Often it is desired merely to separate DNA from RNA for the determination of each. For this purpose the selective hydrolysis of Schmidt and Tannhauser (1945; Ts'o and Sato, 1959) is useful. The extract containing RNA and DNA is incubated for 18 hours at 37°C in 0.3 N KOH. This quantitatively hydrolyzes RNA to ribonucleotides. The remaining DNA is precipitated by acidification of the solution with perchloric acid. This may in turn be hydrolyzed by heating for 10 minutes at 90°C in 1 N perchloric acid, and non-nucleic acid contaminants are again precipitated by neutralization with KOH. The ribonucleotides and deoxyribonucleotides of the two hydrolysates may be determined by their ultraviolet absorption or by chemical determination. For RNA, the orcinol method (Dische and Schwarz, 1937) is convenient. For DNA, the standard assay is the diphenylamine procedure of Burton (see Bonner, 1968a). Extensive studies of procedures for the determination of plant nucleic acids have been made by many people and are summarized in "Methods in Enzymology," for example, Vol. 12, Part B (Bonner et al., 1968a).

III. Findings

A. Amount of DNA per Nucleus

The amount of DNA contained in the plant cell nucleus is in principle (apart from meiosis and polyploidy) constant for each individual

TABLE I
Amount of DNA per Genome of Varied Plant Species[a]

Species	DNA per genome (gm $\times 10^{12}$)	Base pairs (10^9)
Escherichia coli	0.004	0.004
Saccharomyces sp.	0.02	0.02
Neurosopora sp.	0.04	0.04
Cucurbita pepo	0.85	0.85
Cucumis melo	0.85	0.85
Pisum sativum	5.0	5.0
Nicotiana sp.	5.0	5
Zea maize	5.0	5
Helianthus annus	5.0	5
Vicia faba	9.0	9
Allium cepa	20.0	20
Triticum cerealae (hexaploid)	33.0	33
Lilium longiflorum	55.0	55
Trillium erectum	60.0	60

[a] Values taken from varied sources. It now appears that all are somewhat suspect of error due to replication, polyploidy, multi-nucleation, and other problems. The value for E. coli is correct. See discussion of Pisum in text.

species. Table I gives the value of DNA per cell which has been found for various plants. Pea plants (*Pisum sativum*) contain on the order of 5×10^9 base pairs of DNA per genome [about 5×10^{-12} pg (picograms)] per nucleus. Lower plants, such as yeast, *Neurospora*, and *Aspergillus*, contain less, and other species, notably some monocotyledonous ones, contain a great deal more—10 times the above amount or even more. It is possible that many higher plants are polyploid in origin and that those with very high DNA content do not necessarily contain more genetic information than those with lower DNA content per haploid genome. It is also possible that large-sized genomes contain large proportions of repetitive sequences. We shall see that the chemically measured sizes of plant genomes are often at fault owing to polypoloidy of somatic cells.

B. Properties of DNA

DNA is, then, in many respects the most interesting and most important component of the nucleus, and indeed of the cell, constituting as it does the basic recipe for life. We will therefore briefly review the principal facts concerning the chemistry and physicochemistry of the DNA molecule. DNA is a long chain polymer made of four kinds of repeating units, the four nucleotides: deoxyadenylic acid, deoxguanylic acid, deoxycytidylic acid, and deoxythymidylic acid (Fig. 1). Higher plants may contain an additional nucleotide that replaces a portion of the deoxycytidylic acid, namely, 5-methyldeoxycytidylic acid. These four kinds of nucleotide building blocks are linked to one another through phosphodiester linkages between the 5'-hydroxyl group of one molecule and the 3'-hydroxyl group of the next. A DNA strand, therefore, possesses polar-

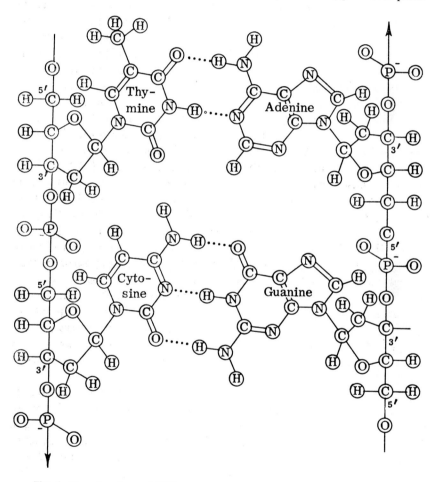

Fig. 1. The structure of DNA. The two strands are held together by hydrogen bonds between adenine and thymine and between guanine and cytosine pairs. Each strand in turn is held together along this length by phosphodiester linkages between deoxyriboside monophosphate residues. Note that the polarities of the two strands are opposite.

ity (Fig. 1). We may travel along it in the 5′ to 3′ direction, or in the 3′ to 5′ direction. The DNA molecule as it exists in nature is, with few exceptions, double-stranded, and although the two strands are of opposite polarity, the sequence of bases in the first strand determines the sequence of the bases in the second. The two strands are said to be complementary to one another. The complementarity rule that governs the biology of the nucleic acids and which is apparently the most basic rule of all biology is simply this: Wherever there is a deoxyadenylic acid (dA) in strand number one of DNA, there must be at the same level a deoxy-thymidylic acid (dT) in strand number two. Wherever there is a deoxy-cytidylic acid (dC) in the strand number one, there must be a deoxy-guanilic acid (dG) in strand number two. dA pairs with dT; dG pairs with dC. When this rule is observed, dA forms hydrogen bonds with dT, and dG forms hydrogen bonds with dC. The whole double-stranded struc-ture can assume the well-known and highly characteristic double helix, first proposed as the structure of DNA by Watson and Crick in 1953 (Fig. 2). The complementarity rule has as a consequence that in native double-stranded DNA the content of dA always equals the content of dT, and the content of dG always equals the content of dC. DNA is said to be base complementary in composition. There are, however, no restrictions in the composition of DNA on the ratio of dA + dT to dG + dC, and indeed DNA's of very different dA + dT to dG + dC ratios are known in nature. Thus in crabs of several species of the genus *Cancer*, the ratio of dA + dT to dG + dC is extraordinarily high (Sueoka and Cheng, 1962). This is due to the fact that such crabs possess in their genome long stretches of poly(dAT) with the base sequence dAdTdAdT · · · ·. Such DNA is not known in plants. Particular micro-organisms, however, contain DNA rich in either dG + dC or dA + dT. The ribosomal cistrons of plant DNA are, in general, richer in dG + dC than are other portions of the plant genome as we will see below. The compositions of whole DNA from typical higher plants is given in Table II.

DNA as isolated from sheared plant cells, as described above, is double-stranded and possesses a molecular weight on the order of 5×10^6 to 10^7. It is therefore about 8×10^3 to 1.5×10^4 base pairs in length. This is no indication of the length of the DNA in the cell itself, since the DNA strands are sheared during isolation. In the bacterium *E. coli* the whole genomal DNA of the single chromosome characteristic of this organism is present as one gigantic molecule, approximately 4×10^6 base pairs in length (Cairns, 1963). There is much evidence that the DNA of each chromosome of higher organisms is also present as one giant mole-cule. This has been shown with a considerable degree of certainty in

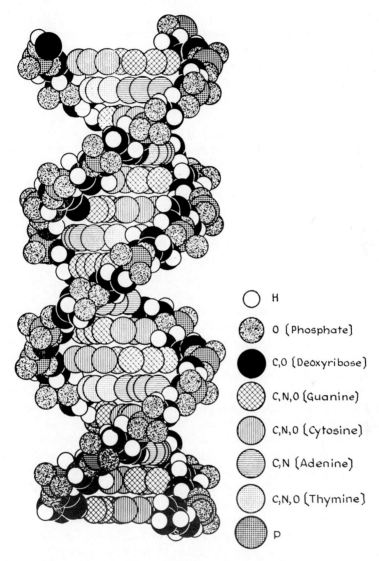

H

O (Phosphate)

C,O (Deoxyribose)

C,N,O (Guanine)

C,N,O (Cytosine)

C,N (Adenine)

C,N,O (Thymine)

P

Fig. 2. The helical structure of DNA. In this structure, which is the one assumed in solution, the molecule is held together not only by hydrogen bonds between base pairs but also by stacking interactions between base pairs.

Drosophila, the fruit fly (Kavenoff and Zimm, 1973). It has not yet been shown to be true for higher plants, but we suspect that it will be shown to be true of all living cells. We know already that the more gently the DNA is prepared, the longer the chain length of such DNA.

TABLE II
Base Compositions of DNA's of Various Higher Plants[a]

Species	Composition (moles/100 moles)					Total dC + dC-Me
	dA	dT	dG	dC	dC-Me	
Pisum sativum	30.8	30.5	19.2	13.5	5.0	18.5
Pinus siberica	29.2	30.5	20.8	14.6	4.9	19.5
Papaver somniferum	29.6	29.8	20.6	14.8	5.3	20.1
Cucurbita pepo	30.2	29.0	21.0	16.1	3.7	19.8
Phaeseolus vulgaris	29.7	29.6	20.6	14.9	5.2	20.1
Arachis hypogaea	29.3	29.8	20.3	14.4	6.1	20.5
Allium cepa	31.8	31.3	18.4	12.8	5.4	18.2
Triticum vulgare	25.6	26.0	23.8	18.2	6.4	24.6
Gossypium hirsatum	32.8	33.0	17.0	12.7	4.6	17.3
Nicotiana tobacum	29.6	30.7	19.8	14.0	5.6	19.6

[a] Data from various sources.

There are many physical properties of DNA which can be usefully studied by the physical biochemist. One is hyperchromicity. In a double-stranded DNA molecule, the bases of each nucleotide absorb less ultra-violet light than do the free individual nucleotides. This is due to the stacking of the base pairs, one above another, in a double helical structure. This results in a $\pi-\pi$ interaction between the stacked bases, causing the extinction coefficient of each nucleotide to be diminished. Nucleotides assembled in double helical DNA, therefore, exhibit what is known as hypochromicity. When a DNA solution is gradually heated, a temperature is ultimately reached at which thermal energy becomes sufficient to disrupt base pairing and base stacking forces so that the DNA molecule melts. The structure collapses, yielding random coil single strands. Thus, melting of the DNA double helical structure is accompanied by hyperchromicity and an increase in optical density of the solution. The increase in optical density of DNA on melting is 35–40% (Fig. 3). The melting temperature or temperature at which half of all hyperchromicity has occurred depends upon both the characteristics of the DNA (that is its base composition and fragment length) and also upon the composition of the solution in which it is melted. Thus, melting temperature of DNA depends upon ionic strength of the medium in which the DNA is melted, the melting temperature increasing with ionic strength. The melting temperature is particularly sensitive to magnesium ion concentrations, since magnesium stabilizes DNA against melting better than monovalent cations. DNA containing only dA and dT base pairs melts at a lower temperature

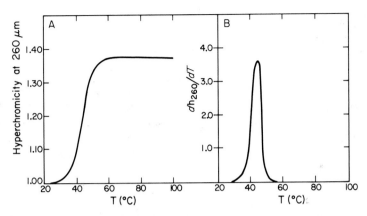

Fig. 3. Melting of pea plant DNA. In (A) the optical density of a solution (1 O.D.) is followed as a function of temperature. In the ionic strength used (2.5 × 10⁻⁴ *M* EDTA), half of the resulting increase in optical density has occurred by a temperature of 42°C (T_m). (B) is the derivative of (A). d(hyperchromicity$_{260}$)/dT (where T is temperature) is plotted as a function of temperature. Pea DNA melts in a single melting peak.

than does DNA containing all four bases. Association of DNA with any kind of polycation also stabilizes DNA against melting. This matter is discussed below in connection with the stabilization of DNA by the histone component of chromatin.

C. Composition of Chromatin

The composition of the chromatin of a typical plant is shown in Table III. It is composed principally of DNA and histone. Lesser components of nonhistone chromosomal proteins and small amounts of RNA are also contained in chromatin. The histones are basic proteins that occur universally in association with DNA in the chromosomes of higher creatures. The mass ratio of histone to DNA in the chromosomes of the organs of different kinds of plant tissues and plant species varies from about 0.8 to about 1.3, with an average of about 1.1. Similar histone to DNA ratios are found in the different chromatins of the different organs and tissues of animals and protozoa. In general, true histones, interestingly enough, appear to be absent from the chromatin of fungi (Leighton *et al.*, 1971).

It has been known for years that histones are components of chromatin (Huang and Bonner, 1962). The interest in histones lies in the fact that these proteins, by complexing with DNA, prevent DNA from being transcribed by RNA polymerase and thus prevent primary tran-

TABLE III
Chemical Composition of Typical Plant Chromatin[a]

	Content relative to DNA (%)				Template activity of chromatin[b]
Source of chromatin	DNA	Histone	Nonhistone protein	RNA	
Pea vegetative bud	1.0	1.30	0.10	0.11	6
Pea embryonic axis	1.0	1.03	0.29	0.26	12
Pea growing cotyledon	1.0	0.76	0.36	0.13	32

[a] After Bonner et al. (1968b) and Bekhor et al. (1969).

[b] Template activity is rate of RNA synthesis supported by a given amount of DNA as chromatin, relative to the rate supported by the same amount of pure DNA. Template activity is measured in the presence of an excess of added purified RNA polymerase and all of the substrates required for RNA synthesis.

scripts. This has been shown by a wide variety of methods. Thus, the template activity (that is, the ability of chromatin to support RNA synthesis by added RNA polymerase) is an inverse function of the histone to DNA ratio of the chromatin. In addition, histone–DNA complexes, made either by recomplexing purified histone to purified DNA or by selective removal of histones from DNA, have template activities for RNA synthesis which are in inverse linear function to the histone to DNA ratio of the chromatin (Fig. 4) (Bonner et al., 1973). Ability to restrict transcription of DNA by RNA polymerase is not a trivial matter. Only one other class of protein that can perform this function is known—the class known as protamines. These peptides replace histone in chromatin during the maturation of the sperm of certain fish.

Histones are, therefore, interesting components of chromatin. They are characterized by high contents of basic amino acids, about 1 residue in 4 being either arginine or lysine. The cationic groups of the histone bind to the anionic phosphate resolutes of the DNA, forming what is known as the nucleohistone complex of chromatin. Intensive work on the chemistry of histones of both plants and animals (Fambrough and Bonner, 1966, 1968, 1969; Fambrough et al., 1968; Bonner and Garrard, 1974) have shown that there are 5 principal species of plant (as well as animal) histones. These are known, respectively, as histone I (in which a majority of the basic groups are lysine), histones IIb1 and IIb2 (in which lysine and arginine are present in more nearly equal amounts), and histones III and IV (in which arginine predominates as the basic amino acid). Histones IIb1, IIb2, III, and IV have been subjected to determination

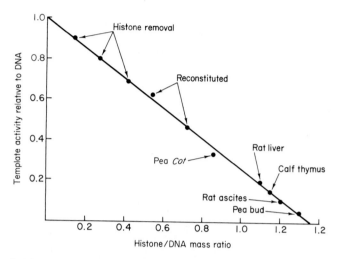

Fig. 4. The histone to DNA ratio of a histone–DNA complex is linearly and inversely related to the ability of that complex to support transcription into RNA by RNA polymerase. This relation suggests that only that DNA not complexed with histones is available for transcription. After Bonner *et al.*, 1973.

of their primary structure (Johnson *et al.*, 1974). The first histone for which such amino acid sequencing was accomplished was histone IV, the smallest and most easily isolated histone (DeLange *et al.*, 1968). It was found that the primary structures of pea and bovine histone IV are almost identical (Fig. 5). In both cases the histone is 102 amino acids long, and only 2 conservative amino acid substitutions appear between the 2 species of histone molecules. In pea histone IV, at one position arginine replaces the lysine in the bovine histone IV (basic for basic amino acid), and, at a second position in the peptide chain, isoleucine replaces valine (hydrophobic for hydrophobic amino acid). The pea and bovine histones III are nearly as related (Fig. 5). Not surprisingly, the histones IV of other mammals that have thus far been sequenced are identical to those of the bovine histone. Hopefully, many more comparative amino acid sequences of histone will be made in the years to come. This tool may provide an excellent one for the study of genetic relationships among organisms, since primary structure of histones would appear to be one of the most conserved characteristics of higher organisms.

Histones can be modified by phosphorylation at several positions, in each case the phosphoryl group appearing in a serine or threonine hydroxyl group. Histones can also be acetylated, the N-terminal amino group as well as others being subject to alteration in this case. The bio-

<pre>
                                       10   16   20
Ac-Ser-Gly-Arg-Gly-Lys-Gly-Gly-Lys-Gly-Leu-Gly-Lys-Gly-Gly-Ala-Lys(Ac)-Arg-His-Arg-Lys(Me)-
                                                                                    Lys   1, 2
                                       30   40
Val-Leu-Arg-Asp-Asn-Ile-Gln-Gly-Ile-Thr-Lys-Pro-Ala-Ile-Arg-Arg-Leu-Ala-Arg-Arg-Gly-Gly-Val-

                                  50   60
Lys-Arg-Ile-Ser-Gly-Leu-Ile-Tyr-Glu-Glu-Thr-Arg-Gly-Val-Leu-Lys-Val-Phe-Leu-Glu-Asn-Val-Ile-

              70   77   80
Arg-Asp-Ala-Val-Thr-Tyr-Thr-Glu-His-Ala-Lys-Arg-Lys-Thr-Val-Thr-Ala-Met-Asp-Val-
                                               Arg
              90   100
Val-Tyr-Ala-Leu-Lys-Arg-Gln-Gly-Arg-Thr-Leu-Tyr-Gly-Phe-Gly-Gly-COOH
</pre>

(A)

<pre>
                    10   20
Nrl-Ala-Arg-Thr-Ala-Arg-Lys(Me)-Ser-Thr-Gly-Gly-Lys-Ala-Pro-Arg-Lys-Glr-Leu-Ala-Thr-Lys-Ala-
  2   0-2
      27   30   40
Ala-Arg-Lys(Me)-Ser-Ala-Pro-Ala-Thr-Gly-Gly-Val-Lys-Lys-Pro-His-Arg-Phe-Arg-Pro-Gly-Thr-
      0-2   Tyr-cow
           50   60
Val-Ala-Leu-Arg-Glu-Ile-Arg-Lys-Tyr-Glu-Lys-Ser-Thr-Glu-Leu-Leu-Ile-Arg-Lys-Leu-Pro-Phe-
                  Arg-cow
      70   80
Glu-Arg-Leu-Val-Arg-Glu-Ile-Ala-Alr-Asp-Phe-Lys-Thr-Asp-Len-Arg-Phe-Gln-Ser-Ser-Ala-Val-

  90   [Ser]   100   110
Ser-Ala-Leu-Gln-Glu-Ala-[Ala]-Glu-Ala-Tyr-Leu-Val-Gly-Leu-Phe-Glu-Asp-Thr-Asn-Leu-Cys-Ala-Ile-
Met-cow   Cys-cow
                   120   130   135
His-Ala-Lys-Arg-Val-Thr-Ile-Met-Pro-Lys-Asp-Ile-Gln-Leu-Ala-Arg-Arg-Ile-Arg-Gly-Glu-Arg-Ala-COOH
</pre>

(B)

Fig. 5. (A) Comparison of the amino acid sequences of calf thymus and pea seedling histone IV. The continuous sequence is that of the calf histone, with the residues in pea histone, which differ from the calf histone, shown below the continuous sequence. From DeLange, Fambrough, Smith, and Bonner, 1969. (B) Structures of pea and bovine histone III. The continuous sequence is that for pea. There are two kinds of histone III in pea. They differ at residue 96; 60% contain Ala; 40% Ser. Bovine and pea histone III differ in reference only at 4 positions which are noted above; residues 41, 53, 90, and 96. All are conservative replacements. After Patthy et al., 1973.

logical significance of such chemical modification of histones is not yet apparent.

A further interesting property of histones which is apparent in the structure shown in Fig. 5 is that the basic residues are not randomly distributed along the peptide chain. This is true of all species of histone molecules. In histones II, III, and IV the N-terminal half of the molecule contains a great concentration of basic groups, while the remainder of the molecule contains the major portion of the hydrophobic groups. In histone I, on the other hand, the C-terminal half of the molecule contains a majority of the basic groups, while the N-terminal end contains the majority of the hydrophobic groups. Histone molecules are quite evidently designed to interact at one end with the phosphate molecules of DNA and at the other end, through hydrophobic stacking interactions, with other protein molecules. These properties of histone molecules become apparent in the properties of the DNA–histone or nucleohistone complex.

Association of histone with DNA confers upon the DNA properties that are different from those of DNA itself; that is, not only is the DNA complexed with histone not transcribable by RNA polymerase (although it is replicatable by DNA polymerase) but the DNA of the complex is also stabilized against melting, as shown in Fig. 6. The DNA is clearly highly stabilized against melting by virtue of the proximity of the charged groups of the histone. More subtle effects of histone on the melting profile of DNA are apparent if we plot the data as a derivative melting profile as shown in Fig. 6. Here the slope of the melting curve at each point along the latter is plotted as a function of temperature. It is clear that there are four major melting components: the first is a very

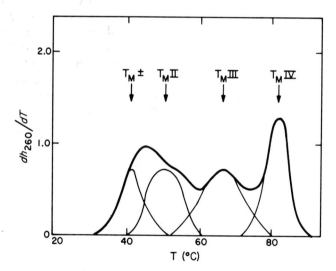

Fig. 6. Melting profile in derivative form of pea and nucleohistone. The nucleohistone was dissolved and melted in 2.5×10^{-4} M EDTA. The several melting peaks of the original data are also resolved into the 4 individual peaks, which together constitute the observed profile. Peak T_MI is due to DNA which melts as five DNA. Peak T_MII is thought to be due to DNA complexed with nonhistone proteins, while peaks T_MIII and T_MIV are due to DNA complexed with histone. h_{260} is fractional increase per °C in optical density at 260 nm. dh_{260}/dT is rate of change of h with T. After Li and Bonner, 1971.

small amount that melts at the temperature of free DNA. This is often absent in usual chromatin preparations. A further, slightly more stabilized component melts at a temperature intermediate between that of DNA and nucleohistone itself. This may well be DNA covered with nonhistone proteins as we will see below. Two final melting peaks, separated by melting temperatures of about 13°C in the solvent used, are both due to the complexing of histone to DNA, since removal of histone, bit by bit, results in successive lowering of these two peaks and the DNA thus liberated is shifted to the peak that melts as pure DNA (Li and Bonner, 1971). Separation of histone molecules into two halves, namely, the more basic and the less basic, shows that that portion of the DNA complexed with a less basic half of the histone molecule represents the low melting peak (III) of Fig. 6, while DNA complexed with the highly basic end of a histone molecule is the source of the higher melting peak (IV) of Fig. 6.

D. Replication of Chromosomes

The DNA of chromatin must replicate before cell division so that each daughter cell can receive a full complement of the genetic information. One enzyme that may be involved with such replication is DNA polymerase I, first isolated from *E. coli* by Kornberg (1957) and Lehman *et al.* (1958) and described in plants by Mory *et al.* (1974) and others. DNA polymerase catalyzes the polymerization of deoxyriboside triphosphates to new DNA chains provided only that (1) DNA be present to be used as a template and (2) that a primer molecule bound to single-stranded DNA is present. DNA polymerase I can elongate preexisting primers but cannot itself initiate new DNA chains. We imagine that in DNA replication the two strands of the double-stranded DNA molecules become separated from one another, perhaps by the presence of DNA melting proteins, which have been found in bacteria and in animal cells. We then imagine that an RNA polymerase molecule initiates transcription over a short distance and that the DNA polymerase then not only elongates from the RNA primer site but also, because of the fact that the DNA polymerase I contains an endonuclease activity, removes the RNA and replaces it by DNA. An additional complication is that even though DNA polymerase I can replicate in one direction only, we know that in chromosomal DNA of bacteria and mammals both strands are replicated, although they run in opposite directions. A resolution of this matter with plant cells has not yet been achieved. In *E. coli* and in animal cells, it is quite clear that both strands are replicated in the manner outlined above with by RNA polymerase initiation. The DNA polymerase

I elongation of the RNA chains thus initiated by RNA polymerase and the joining of the short segments of the DNA thus formed by DNA ligase (which has the property of forming covalent unions between adjoining DNA chains) appears to be a general feature of DNA replication. In any case, at the end of the replication, two new DNA double helical molecules have been produced, each with the same nucleotide sequence composition of the original template molecule.

DNA complexed with histones (pea nucleohistone) is as active in the function of supporting the replication of DNA by DNA polymerase as is pure DNA itself (Schwimmer and Bonner, 1965).

That each strand of the DNA double helix remains intact throughout the replication was first shown by Meselson and Stahl (1958) in an experiment with *E. coli*. This experiment makes use of the fact that DNA molecules of different densities can be separated by gradient density centrifugation in CsCl gradients A similar experiment was done by Filner (1965) with the cultured cells of tobacco, which increase in solution culture with a doubling time of 2 days. The cells were first grown through several cell generations in medium containing $^{15}NO_3^-$ as a nitrogen source. The DNA of such cells possesses a density of 1.711 (in approximately 8 M CsCl). In contrast, the density of the DNA of cells grown in $^{14}NO_3^-$ is 1.696. Cells containing heavy DNA were then transferred to medium containing $^{14}NO_3^-$ and allowed to grow until they had doubled their DNA. Their DNA was found to be all half heavy, that is, to possess a density of 1.703 (Fig. 7). After a further doubling in $^{14}NO_3^-$ medium, one-half of the DNA was found to be light, one-half to be half heavy. These findings are exactly like those made with *E. coli* and support the hypothesis that DNA replication in higher plants, too, is semiconservative, and that the indivdiual DNA strands are immortal.

E. Synthesis of Histones

Although histones are nuclear proteins, they are not synthesized in the nucleus but rather in the cytoplasm and are thereafter transported to the nucleus (see, e.g., Robbins and Brown, 1967; Kedes and Gross, 1969). Synthesis of histones is orthodox; that is, mRNA's for histone synthesis are transcribed from nuclear DNA, and they are translated by ribosomes in the usual manner for protein synthesis. The only unusual thing about the mRNA for histones is that these mRNA's contain no poly(A) sequences at their 3' end, in contrast to other animal mRNA's (see below). In some well-studied systems, the genes that code for histone synthesis are highly reiterated, as in *Xenopus laevis* (Kedes and Birnstiel,

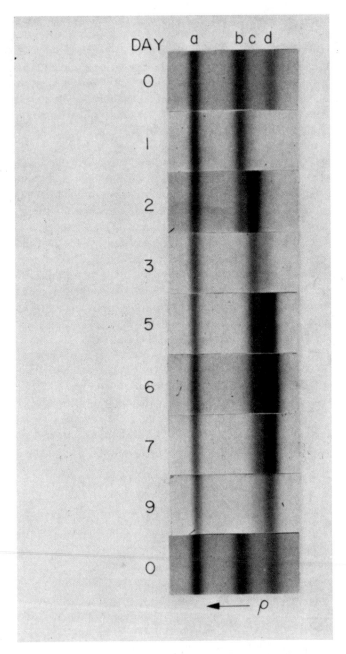

Fig. 7. Transfer experiments showing the semiconservative replication of DNA in exponentially growing cells of higher plants. The cells (callus cells of *Nicotiana tobacum* growing in solution culture) are grown either in [15]N (heavy) nitrogen or in [14]N (light) nitrogen. In each case the DNA sample is centrifuged to

1971), in which several hundred copies of the gene for each species of histone are present, clustered as are the genes for rRNA. Whether this is also true in plants is not yet known.

F. Transcription of Chromatin

The DNA of chromatin is transcribed into RNA by RNA polymerase, an enzyme discovered simultaneously in 1960 by several groups, including one group that worked with the pea plant (Huang *et al.*, 1960). Plant RNA polymerases have been purified to homogeneity or near homogeneity by several groups, including those of Steiner *et al.* (1970), Mondol *et al.* (1970, 1972), and Horgen and Key (1973). RNA polymerase is (1) large (molecular weight about 8×10^5 daltons) and (2) a complex of four or more subunits. In any case, RNA polymerase binds to DNA at specific binding sites of unknown nature and appears locally to melt the DNA. It progresses down the molecule, transcribing only one of the two strands (the so-called "sense" strand). Such transcription continues until the polymerase comes to a stop site, also of unknown nature. At this juncture the growing RNA chain is released. We know that in animal cells, a large portion of the initial nuclear transcripts are large, 15,000 to 30,000 bases in length (Holmes and Bonner, 1973), and that they contain both repeated and single copy sequences (see below). Whether these transcripts include segments that become mRNA or whether the latter are separately transcribed is a matter of current debate. In any case, the situation in plant nuclei is completely unknown.

In animal cells, the mRNA transcript is next polyadenylated at the 3'-hydroxyl end. A segment of 100–200 polyadenylic acid is added to this end. The same is true for at least some of the mRNA's of plant cells (Verma *et al.*, 1974; Tobin and Klein, 1974). In addition, an enzyme

---

buoyant equilibrium on a CsCl density gradient by centrifugation. The positions of the heavy and light nitrogen-labeled DNA is determined by a photograph taken in ultraviolet light in the analytical ultracentrifuged cell and in the CsCl gradient. Higher densities are at the left, lesser densities at the right. The ultraviolet absorbing band at the extreme right is a standard heavy marker DNA of known density. Day 0: This gradient contains DNA from ^{15}N-labeled cells and from ^{14}N-labeled cells in approximately equal amounts. The separation of the heavy (left) from the light (right) DNA is clearly visible. Day 1: On this day the cells are transferred to ^{14}N. All of the DNA at the time of the transfer is heavy nitrogen which bands under the band labeled b. Day 2: One replication of DNA has occurred. Almost all of the DNA is half heavy, a small remaining amount is not replicated and is still heavy. Day 3: All of the DNA is replicated, all is half heavy. Day 5: After further replication half of the DNA is half heavy and half is light. Days 7 and 9: As replication continues for a further round the majority of the DNA becomes light. After Filner, 1965.

capable of poly(A) synthesis has been found in plant tissues (Mans, 1974).

G. The Nucleolus and Ribosomal RNA

Ribosomal RNA hybridizes to a considerable fraction of denatured higher plant DNA. This was first shown by Chipchase and Birnstiel (1963) for the pea plant. In this species the two ribosomal RNA species, 28 S and 18 S, hybridize to 0.3% of whole genomal DNA. This is interesting because it shows that each ribosomal cistron (28 S + 18 S) is repeated several times per genome (about 300 times, see below).

That the ribosomal cistrons are in general present in many copies per genome in higher plants has been shown by Matsuda and Siegel (1967). Their data and those of others are shown in Table IV. It appears that in higher plants the ribosomal cistrons are represented by a few hundred to many thousands of copies per genome.

An interesting specific case is that of pumpkin (*Cucurbita pepo*). Thornburg and Siegel (1973) have found that the ribosomal RNA's (28 S + 18 S) hybridize to about 2.5% of the pumpkin genome, which represents about 3400 copies of each component per genome.

It may be noted in passing that the rRNA cistrons in their multiple copies are clustered and, therefore, in many species constitute a fraction of the DNA with a density difference (generally greater than) compared with other nuclear DNA's when banded in CsCl. On this basis, the DNA that codes for rRNA may be physically separated from other nuclear

TABLE IV

Fraction of Genome Hybridized by Cytoplasmic Ribosomal RNA in Various Organisms Together with Number of Ribosomal DNA Cistrons per Genome

Source	Genome size base pairs (haploid)	% of genome hybridized to 28 S + 18 S rRNA	No. of ribosomal cistrons per genome
Pisum sativum[a]	5.0×10^9	0.3	2600
Pisum sativum[b]	0.7×10^9	0.3	300
Nicotiana tobacum	5.0×10^9	0.007	230
Cucumis melc[c]	0.85×10^9	2.4	3300
Cucurbita pepo[d]	0.85×10^9	2.5	3400

[a] From Chipchase and Birnstiel, 1963.
[b] From Fig. 9.
[c] From Matsuda and Siegel, 1967.
[d] From Thornburg and Siegel, 1973.

DNA's in many plant species (Scott and Ingle, 1973; Ingle *et al.*, 1973; Thornburg and Siegel, 1973; Benditch and Anderson, 1974), just as was first done by Birnstiel *et al.* (1968) for *Xenopus*.

Why do plants have this need to produce and conserve so many rRNA cistrons per genome when bacteria (which grow more rapidly by a factor of 25 or more than do plants) have one or a few rRNA cistrons per genome. Even higher animals possess, in somatic cells, only 0.1 to 0.01 of the ribosomal RNA producing capacity of higher plants. There are many possible answers. None is rigorously shown to be correct. Perhaps plants produce many long-lasting mRNA's, which require a large number of ribosomes for their translation. Perhaps plants produce only a few copies of each species of each mRNA, and each must be saturated with ribosomes to produce the required enzyme products. The developing oocytes of many animals engage in selective replication of ribosomal cistrons so that these cistrons are amplified by about 1000-fold over their proportion in somatic cells (Brown and Weber, 1968). Perhaps the plant strategy is a compromise between the ribosomal needs of the egg and developing embryo and those of the more mature individual. In any case, no one has yet shown that all rRNA cistrons are continuously exprssed in the higher plant somatic cell.

The nucleotide sequences of ribosomal cistrons are remarkably similar in different higher plants. This is expressed by the fact that the ribosomal RNA's of two different species cross hybridize to the DNA's of the same two species just as do the homologous rRNA's. These aspects of the matter are shown for two forms in Table V.

The genes for transfer RNA (tRNA) are also highly reiterated in plants. Thus, in *Cucurbita pepo*, Thornburg and Siegel (1973) found about 8600 tRNA cistrons per genome. This should, therefore, include about 135 copies of each of the 64 sequences that code for tRNA. This

TABLE V

Cross Hybridizations of rRNA's of Two Species with Denatured DNA's of the Same Two Species[a]

Source of DNA	Source of rRNA	% hybridization of rRNA to DNA
Tobacco	Tobacco	0.07
Tobacco	Chinese cabbage	0.10
Chinese cabbage	Chinese cabbage	0.93
Chinese cabbage	Tobacco	0.82

[a] From Matsuda and Siegel, 1967.

is in contrast to prokaryote and animal genomes, which contain fewer tRNA cistrons by one to two orders of magnitude.

The nucleolus is not only the site of production of rRNA but also of the assemblage of the two ribosomal subunits; that is, the 60 S and 40 S subunits of the 80 S ribosome. In the nucleolus, the rRNA subunits appear to become complexed with the ribosomal proteins (Birnstiel *et al.*, 1961, 1962, 1963). Since intact ribosomes are in general not found in nuclei, it is probable that the ribosomal proteins are made in the cytoplasm as are other proteins and migrate to the nucleolus for assembly. Although much is known about the assembly of the ribosomal RNA's and proteins of *E. coli* (Nomura, 1973), little is known about this subject in higher plants.

H. Structure of the Plant Genome

The genomes of most higher organisms contain sequences that are repeated only once per genome. These are the unique or single copy sequences. They also include others that are repeated hundreds, thousands, or, even in some instances, one million times. These are the repetitive or reiterated sequences. The way in which these facts have been discovered and quantified is by use of so-called reassociation kinetics (Britten and Kohne, 1968; Wetmur and Davidson, 1968). Whole genomal DNA is sheared generally to a standard length of 350 base pairs (hydraulic or French press). It is then converted to the single-stranded form by heating or by brief exposure to alkali. The denatured DNA is then neutralized and incubated, for example, in 0.12 M phosphate buffer, at a temperature such that only perfectly paired or near perfectly paired duplexes are stable. Such a temperature is in general about 20°C below the T_m of the native DNA in the same solvent. Thus, if the T_m of the DNA in 0.12 M phosphate buffer is 86°C, reannealing should be done at a temperature of 66°C. The rate of reassociation is then followed. This may be done in either of two ways. The first is optical. When DNA is melted to the single-stranded form, its optical density increases as outlined above. As it reassociates, its optical density drops, and on complete reassociation the original optical density of native DNA will be restored. Thus, we may follow reassociation by spectrophotometry. The second method is based on the fact that although DNA is bound to hydroxyapatite in 0.05 M phosphate buffer, single-stranded material is eluted by washing with 0.12 M phosphate buffer, while double-stranded DNA is eluted only by 0.48 M phosphate buffer (Britten and Kohne, 1968). The latter method is preferable particularly for higher plants and animals having large genomes that require long times for total reannealing of their DNA's.

Rate of reassociation of DNA fragments is a second-order reaction. Two fragment molecules must interact to consumate reassociation. Rate of reassociation follows the general formulation $-d(S)/dt = kS^2$, where S is the concentration of unpaired nucleotides and k is a second-order rate constant; k is dependent on fragment length, base composition, and salt concentration, as well as on sequence complexity.

It is clear at once that the rate at which DNA of any given cell will reanneal or reassociate is a function of how many different sequences are contained in the sample. Thus, if a given amount of DNA contains 10 sequences, the probability of any given bimolecular collision being a fruitful one (yielding a perfect duplex) is 10 times as great as it would be if the same quantity of DNA were to contain 100 different sequences. The rate of reannealing is, therefore, inversely proportion to sequence complexity.

In the case of viral and bacterial DNA's, the rate of reassociation is inversely proportional to genome size (Wetmur and Davidson, 1968). This is because in such simple organisms, all of the genome is single copy. An example of the reassociation profile for the DNA of a simple organism, *E. coli*, is shown in Fig. 8. On the vertical axis, the degree of reassociation from 0.0 at the top to 1.0 at the bottom is plotted. On the horizontal axis, log *Cot* is plotted. *Cot* is the abbreviation for DNA concentration

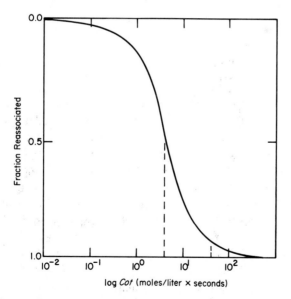

Fig. 8. Reassociation profile (*Cot* curve) for *E. coli* DNA. Renaturation of 350 base pair fragments in 0.12 *M* phosphate buffer pH 0.8 at 62°. Separation of single-stranded from double-stranded DNA at each time point by hydroxyapatite chromatography. Cot$_{1/2}$ for *E. coli* DNA under these conditions is 4.1.

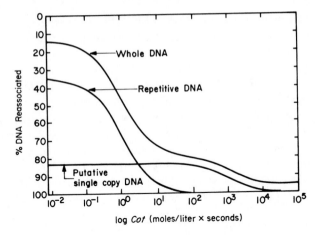

Fig. 9. Reassociation (*Cot*) curve of whole pea DNA. The DNA was sheared to a fragment length of 350 base pairs, denatured, and reassociated at 60°C in 0.12 *M* phosphate buffer pH 6.8. At varied times, samples were withdrawn and separated into single-stranded and double-stranded fractions by hydroxyapatite chromatography. Fourteen percent of DNA reassociates before measurements can begin (highly repetitive). The remainder hybridizes as two components of different sequence repetition fequencies. After R. Wilson, personal communication.

in moles nucleotide per liter $\times$ time in seconds. It is the reciprocal of the second-order reassociation rate constant for the reaction. The $Cot_{1/2}$ (half-reassociation) for *E. coli* DNA determined by hydroxyapatite chromatography under conditions, such as those outlined above, is 4.4.

Figure 9 shows the reassociation profile, or *Cot* curve (concentration of nucleotide in moles/liter $\times$ time of incubation in seconds), for pea DNA as done by hydroxyapatite chromatography. Almost 20% of the DNA has reannealed before the first measurement could be made ($Cot = 5 \times 10^{-1}$). This fraction consists of very highly repeated sequences (more than 1×10^5 copies per sequence). A similar fraction is found in the DNA's of all higher organism and is known for animals (e.g., mouse and *Drosophila*) to be centromeric, i.e., clustered at the centromere. The remainder of the pea *Cot* curve clearly consists of two components, one more rapidly reannealing and constituting about 65% of the genome and a more slowly reannealing component constituting about 17%. The *Cot* curve of Fig. 9 has further been dissected by a computer program (finger program of R. Britten, personal communication), which plots the *Cot* curves of the two components, assigning the theoretical second-order reaction rate curves to each of them. From these we can readily determine the $Cot_{1/2}$ observed for each component. The computer program also calculates the pure $Cot_{1/2}$ for each fraction, that is, the

$Cot_{1/2}$ for each component which would be found if that component were reassociated by itself.

The most slowly reannealing component contains, according to chemical determination (Birnstiel *et al.*, 1963), about 200 times as much DNA as the *E. coli* genome and reanneals about 50 times more slowly. If one assumes that the chemical determination is correct, there is no single copy DNA in the pea genome. If one assumes that the most slowly reannealing fraction is the single copy fraction, then we must conclude that the size of pea genome is about one-fourth of that measured chemically on shoot apex cells, as done by Birnstiel *et al.* There is much evidence that somatic cells of peas and other plants are largely polyploid. The present findings suggest that on the average vegetative cells of pea shoots are octaploid, a wholly reasonable state of affairs (Table VI) (R. Wilson, personal communication).

As a general conclusion it may turn out that the only sure way to measure genome size is by reassociation kinetics coupled with the assumption that the most slowly reannealing component is the single copy component. This method was earlier used to establish the genome size of the cellular slime mold, *Dictyostelium discoideum* (Firtel and Bonner, 1972). In this instance, no practical method of chemical determinations appeared to be available.

In some instances particularly in the case of polyploid species, such as the wheat varieties studied by Bendich and McCarthy (1970), the single copy portion of the genome contains many rather closely related sequences that cross-hybridize with base pair mismatch. Higher criteria (higher temperature of reassociation) help to resolve this problem.

TABLE VI

Calculations from Data of Fig. 9 of the Characteristics of Two of the Three Different Complexity Components of the Pea Genome[a]

Component	Fraction of $Cot_{1/2}$		Complexity[b]	Base pairs in genome[c] fraction (10^8)	Repetition number
	Total	Pure			
Very rapidly reannealing	0.18	—	—	2.2	—
Moderately rapidly reannealing	0.65	0.66	7.3×10^5	8.0	1.3×10^3
Slowly reannealing	0.17	195	2.1×10^8	2.1	1

[a] From R. Wilson, personal communication.
[b] Complexity is the calculated number of base pairs in different sequences.
[c] Based on the assumption that the slowly reannealing fraction is single copy.

The moderately rapidly reannealing portion of the pea genome contains about 200 times as much DNA as does the *E. coli* genome, yet the pea genome DNA reassociates 6.6 more rapidly. Therefore, it consists of repeated sequences. The average number of copies per family of these repeated sequences is 1325 [(200 × 4.4)/0.66].

IV. Concluding Remarks

Molecular biology of the plant cell nucleus has progressed rapidly in some directions and very slowly in other directions. Thus, the separation and sequencing of plant histones started as early and has progressed almost as rapidly as similar studies on mammalian histones. The determination of the number of ribosomal genes per genome was first discovered for the pea plant and is as well understood as the determination in higher animals. On the other hand, there is as yet a very incomplete understanding of the structure of the plant genome, that is, the exact number of single copy genes contained in the plant genome. There is no understanding of the way in which repetitive and single copy sequences are interspersed in the plant genome. Such an understanding is already at hand for several animal genomes. However, there is little understanding of transcription of chromatin in the nucleus of the plant cell. Although we know that messenger RNA's produced by plant nuclei contain polyadenylate at 3′ end, as do the messenger RNA's of animal cells, we still do not know the extent to which such messenger RNA's are derived from giant nuclear transcripts, such as those described for animal cells, nor do we know to what extent they are derived from shorter transcripts specifically designated as single copy genes for the production of messenger RNA's. We have no knowledge of the number of functional genes in the plant genome. When we consider the fact that genetics as a science started with Mendel's study of the pea plant, it is a sad commentary on the state of plant science today that we have such an incomplete understanding of plant nuclear metabolism and plant genetic activity. Let us hope that in the near future this deficit can be remedied.

REFERENCES

Bekhor, I., Kung, G., and Bonner, J. (1969). *J. Mol. Biol.* **39**, 351.
Bendich, A., and Anderson, R. (1974). *Proc. Nat. Acad. Sci. U.S.* **71**, 1511.
Bendich, A., and McCarthy, B. (1970). *Genetics* **65**, 545.
Birnstiel, M., Chipchase, M., and Bonner, J. (1961). *Biochem. Biophys. Res. Commun.* **6**, 161.
Birnstiel, M., Chipchase, M., and Hays, R. J. (1962). *Biochim. Biophys. Acta.* **52**, 728.

Birnstiel, M., Chipchase, M., and Hyde, B. (1963). *Biochim. Biophys. Acta* **76**, 454.
Birnstiel, M., Wallace, H., Sirlin, J., and Fitchberg, M. (1966). *Nat. Cancer Inst., Monogr.* **23**, 431.
Birnstiel, M., Speirs, J., Purdom, I., and Jones, P. K. (1968). *Nature (London)* **219**, 454.
Bonner, J., and Garrard, W. T. (1974). *Life Sci.* **14**, 209.
Bonner, J., and Huang, R. C. (1963). *J. Mol. Biol.* **6**, 169.
Bonner, J., Chalkley, R., Dahmus, M., Fambrough, D., Fujimura, F., Huang, R. C., Huberman, J., Jensen, R., Marushige, K., Ohlenbusch, H., Olivera, B., and Widholm, J. (1968a). *In* "Methods in Enzymology" (L. Grossman and K. Moldave, eds.), Vol. 12, Part B, pp. 3–65. Academic Press, New York.
Bonner, J., Dahmus, M., Fambrough, D., Huang, R. C., Marushige, K., and Tuan, D. Y. (1968b). *Science* **159**, 47.
Bonner, J., Garrard, W. T., Gottesfeld, J., Holmes, D. S., Sevall, J. S., and Wilkes, M. (1973). *Cold Spring Harbor Symp. Quant. Biol.* **38**, 303.
Britten, R., and Kohne, D. (1968). *Science* **161**, 529.
Brown, D., and Weber, C. (1968). *J. Mol. Biol.* **34**, 661.
Burton, K. (1959). *Biochem. J.* **62**, 315.
Cairns, J. (1963). *J. Mol. Biol.* **6**, 208.
Chipchase, M., and Birnstiel, M. (1963). *Proc. Nat. Acad. Sci. U.S.* **49**, 692.
DeLange, R., Smith, E., Fambrough, D., and Bonner, J. (1968). *Proc. Nat. Acad. Sci. U.S.* **61**, 7.
Dische, J., and Schwartz, K. (1937). *Mikrochim. Acta* **2**, 13.
Fambrough, D., and Bonner, J. (1966). *Biochemistry* **5**, 2563.
Fambrough, D., and Bonner, J. (1968). *J. Biol. Chem.* **243**, 4434.
Fambrough, D., and Bonner, J. (1969). *Biochim. Biophys. Acta* **175**, 113.
Fambrough, D., Fujimura, F., and Bonner, J. (1968). *Biochemistry* **7**, 575.
Filner, P. (1965). *Exp. Cell Res.* **39**, 33.
Firtel, R., and Bonner, J. (1972). *J. Mol. Biol.* **66**, 339.
Holmes, D. S., and Bonner, J. (1973). *Biochemistry* **12**, 2330.
Holmes, D. S., and Bonner, J. (1974a). *Biochemistry* **13**, 849.
Holmes, D. S., and Bonner, J. (1974b). *Proc. Nat. Acad. Sci. U.S.* **71**, 1108.
Horgen, P., and Key, J. (1973). *Biochim. Biophys. Acta* **294**, 227.
Huang, R. C., and Bonner, J. (1962). *Proc. Nat. Acad. Sci. U.S.* **48**, 1216.
Huang, R. C., Maheshwari, N., and Bonner, J. (1960). *Biochem. Biophys. Res. Commun.* **3**, 689.
Ingle, J., Pearson, G., and Sinclair, J. (1973). *Nature (London), New Biol.* **242**, 193.
Johnson, J., St. John, T., and Bonner, J. (1974). *J. Biol. Chem.* **378**, 424.
Kavenoff, R., and Zimm, B. (1973). *Chromosoma* **41**, 1.
Kedes, L., and Birnstiel, M. (1971). *Nature (London), New Biol.* **230**, 165.
Kedes, L., and Gross, P. R. (1969). *J. Mol. Biol.* **42**, 559.
Kornberg, A. (1957). *In* "The Chemical Basis of Heredity" (W. D. McElroy and B. Glass, eds.), p. 579. Johns Hopkins Press, Baltimore, Maryland.
Lehman, R., Bessman, M., Sims, E., and Kornberg, A. (1958). *J. Biol. Chem.* **223**, 163.
Leighton, T., Dill, B., Stock, J., and Phillips, C. (1971). *Proc. Nat. Acad. Sci. U.S.* **68**, 677.
Li, H. J., and Bonner, J. (1971). *Biochemistry* **10**, 1461.
Mans, R. (1974). *Plant Physiol.* Abstr., p. 26.
Marmur, J. (1961). *J. Mol. Biol.* **3**, 208.

Matsuda, K., and Siegel, A. (1967). *Proc. Nat. Acad. Sci. U.S.* **55,** 673.

Meselson, M., and Stahl, F. (1958). *Proc. Nat. Acad. Sci. U.S.* **44,** 461.

Mondol, H., Mandal, R., and Biswas, B. H. (1970). *Biochem. Biophys. Res. Commun.* **40,** 1194.

Mondol, H., Mandal, R., and Biswas, B. H. (1972). *J. Biochem.* **25,** 463.

Mory, Y., Chen, D., and Sarid, S. (1974). *Plant Physiol.* **53,** 377.

Nomura, M. (1973). *Science* **179,** 864.

Patthy, L., Smith, E. L., and Johnson, J. (1973). *J. Biol. Chem.* **248,** 6834.

Rho, J., and Chipchase, M. (1962). *J. Cell Biol.* **14,** 183.

Robbins, E., and Brown, T. (1967). *Proc. Nat. Acad. Sci. U.S.* **57** :409.

Schmidt, G., and Tannhauser, S. (1945). *J. Biol. Chem.* **161,** 38.

Schwimmer, J., and Bonner, J. (1965). *Biochim. Biophys. Acta* **108,** 67.

Scott, N., and Ingle, J. (1973). *Plant Physiol.* **51,** 677.

Steiner, G., Mullmix, K., and Bogorad, L. (1970). *Proc. Nat. Acad. Sci.* **68,** 338.

Sueoka, N., and Cheng, T. (1962). *J. Mol. Biol.* **4,** 161.

Thornburg, W., and Siegel, A. (1973). *Biochemistry* **12,** 2750.

Tobin, E., and Klein, A. (1974). *Plant Physiol.* Abstr., p. 37.

Ts'o, P. O. P., and Sato, C. (1959). *Exp. Cell Res.* **12,** 224.

Verma, D. P., Nash, D., and Schulman, H. (1974). *Plant Physiol.* Abstr., p. 9.

Watson, J., and Crick, F. (1953). *Nature (London)* **171,** 737 and 964.

Wetmur, J., and Davidson, N. D. (1968). *J. Mol. Biol.* **31,** 319.

4

Cell Membranes

A. A. BENSON AND ALICE TANG JOKELA

I. Introduction

The plant cell membrane was first recognized when Nägeli and Cramer (1855) observed the impermeability of algal and moss cell surfaces to several pigments. They found the surface layer of the cell discontinuous and distinguishable from the cytoplasm and called this layer the *plasmamembran*. Based on evidence using micromanipulation and dark field microscopy, Plowe (1931) concluded that the surface structure of the onion root tip cell is different from that of the cytoplasm. She suggested that the plasmalemma is both protective and elastic. Electron microscopy of plant cells has revealed a wide variety of membrane-bound organelles, such as nuclei, chloroplasts, mitochondria, vacuoles, and Golgi bodies (Fig. 1). The name "plasmalemma" is now used to designate the cell's outer membrane, while the term "cell membrane" is more generally used for all the membrane systems of the cell.

The extensive membrane systems in plant cells constitute significant

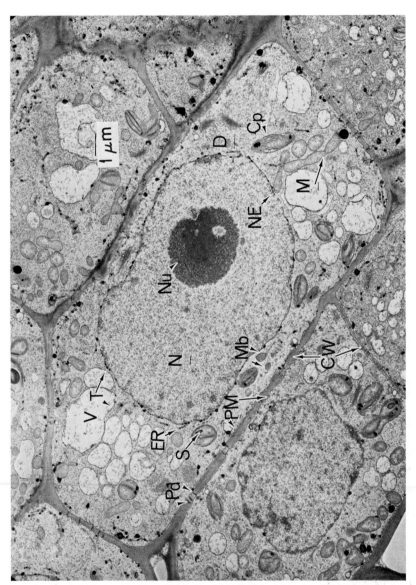

Fig. 1. Typical plant cell fixed with glutaraldehyde and OsO_4 and stained with uranyl and lead salts (2760×). Nucleus (N), nucleolus (Nu), nuclear envelope (NE), mitochondria (M), chloroplast (Cp), cell wall (CW), plasmalemma (PM), plasmodesmata (Pd), starch (S), endoplasmic reticulum (ER), vacuole (V), tonoplast (T), microbody (Mb), dictyosomes (D). (Electron micrograph courtesy of M. C. Ledbetter and K. R. Porter.)

portions of their cellular material. They provide microenvironments essential for the characteristic chemical functions they perform. Thus membranes are now viewed not only as impermeable structural entities but also as specific, selective, adaptable, and active mediators that regulate and maintain major chemical functions of the plant cell. In addition to their permeability functions, membranes perform many chemical functions for the cell. They provide for the highly efficient coordination and sequence of the cells' reactions. Most of these are incompatible with a situation in which enzyme molecules and substrates are randomly distributed within the cell. Sir Rudolph Peters (1956, 1969) described the system of membrane-directed structure as the "cytoskeleton." Recent studies have revealed many multistep metabolic reactions are indeed associated with cell membranes. Cell membranes often exhibit the property of contractility, a phenomenon associated with their functions in active transport and adaptation to changes in environment.

New physical techniques have complemented the classic anatomical, physiological, and organelle fractionation methodology. These include spectrometric measurements, such as optical rotatory dispersion (ORD) and circular dichroism (CD) (Lenard and Singer, 1966), nuclear magnetic resonance (NMR) (Cherry et al., 1971) and electron paramagnetic resonance (EPR) (Jost et al., 1973), calorimetric techniques including differential thermal analysis (Steim, 1970), fluorescent probe studies (Tasaki et al., 1968), combined enzymatic treatment and physicochemical studies (Fleischer et al., 1967), and thermodynamic considerations (Singer, 1971). Interdisciplinary advances have begun to lift the veil from both molecular architecture and its functional relationships in cell membranes.

II. Composition of Cell Membranes

Isolation and purification of membrane systems are prerequisites for studying their chemical components and metabolic activities. These in turn will indicate the membrane's role in the cell. Unfortunately, the isolation of plant cell membranes other than those from the chloroplast has been less fruitful than that of mammalian cell membranes. The strong cell walls surrounding plant cells have impeded fractionation. With improved enzymatic methods for removing cellulose walls, however, more gentle and selective fractionation procedures are becoming possible.

A. Lipids

Plant cell membranes, like those of animal and bacteria, include lipids and lipoproteins as major molecular species. Lipids account for

20 to 50% of the dry weight of cell membrane preparations. Such membrane lipids are amphiphatic molecules—they interact with both the common liposolvents and with water. As a result they form biphasic or micellar structures. They make up the lipid bilayer component of the membrane and interact with its protein and lipoprotein structures. Phospholipids, glycolipids, sulfolipids, and polyisoprenoids are the four major classes of lipids found in plant cell membranes. Triglycerides, the predominant energy storage lipids, are not important in cell membranes. Unlike animal cell membranes, plant membranes do not contain cholesterol. The lipids had been viewed as compounds providing merely structural and permeability barrier properties of cell membranes. Evidence has been presented indicating that a variety of specific membrane lipids now appear essential for many enzymatic activities (Jurtshuk *et al.*, 1961; Livne and Racker, 1969; Rothfield *et al.*, 1969). The widely different degrees of extractability of the plant membrane lipids indicate the heterogeneous nature of membrane structure.

1. Phospholipids

Plant phospholipids are glycerol phosphatides containing two long-chain fatty acids. The first hydroxyl group of the diglyceride glycerol is bound to choline, ethanolamine, serine, glycerol, or inositol through a phosphodiester linkage. Of these, the zwitterionic phosphatidylcholine (lecithin), phosphatidylethanolamine (cephalin), and phosphatidylserine are predominant in mitochondria and salt-translocating organelles, while the anionic phosphatidylinositol and phosphatidylglycerol are important in photosynthetic tissues (Fig. 2). Diphosphatidylglycerol (cardiolipin) occurs in high concentrations in both plant and animal mitochondria.

The common fatty acid components of the phospholipid of plant membranes are palmitic (hexadecanoic) acid (16:0), oleic (Δ^9-octadecenoic) (18:1) acid, linoleic ($\Delta^{9,12}$-octadecadienoic) (18:2) acid, and α-linolenic ($\Delta^{9,12,15}$-octadecatrienoic) (18:3) acid. Certain fatty acids are associated with specific phospholipids and with specific position on the glycerol moiety. Wheeldon (1960) reported that 75% of the fatty acids of phosphatidylethanolamine isolated from cabbage leaves are saturated, while only 30% of the fatty acids are saturated in phosphatidylcholine isolated from the same source. Sastry and Kates (1964a) found almost all the saturated fatty acids to be in the α-position in phosphatidylcholine of runner bean leaves. The most dramatic example of fatty acid specificity was observed by Haverkate and van Deenen (1965). Phosphatidylglycerol of chloroplasts possessed up to 35% of Δ^3-*trans*-hexadecenoic acid (16:1) (Fig. 3), an acid not found in any other lipid components

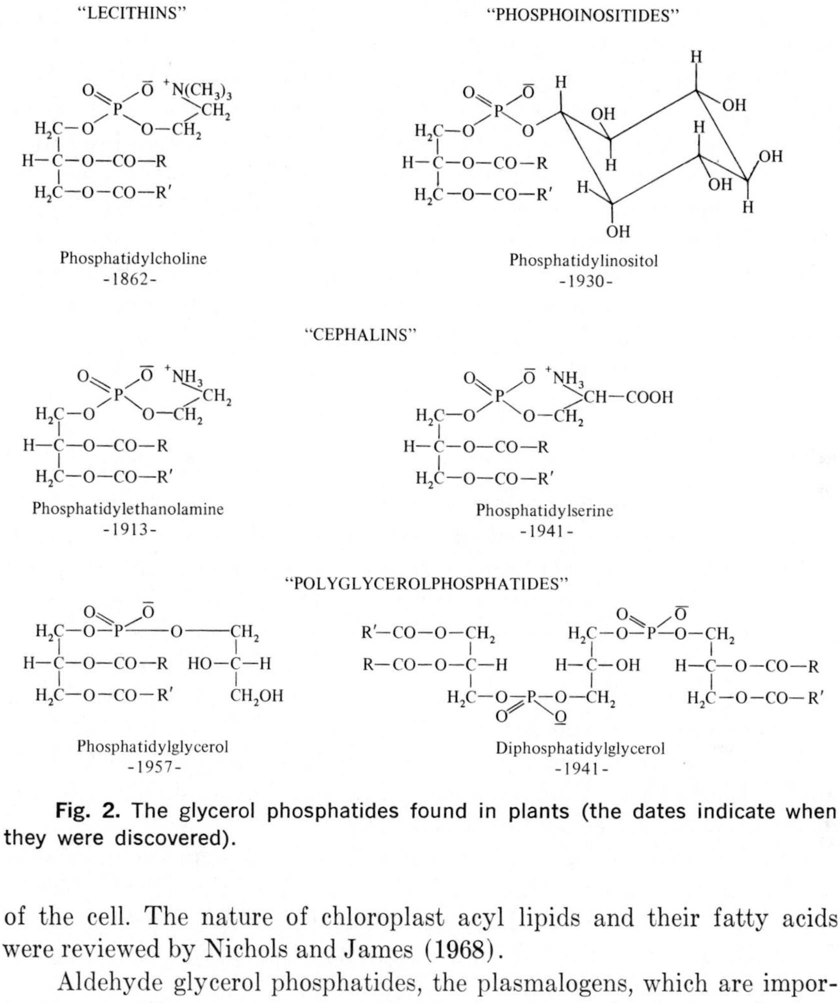

"LECITHINS"　　　　　　　　　　　"PHOSPHOINOSITIDES"

Phosphatidylcholine
-1862-

Phosphatidylinositol
-1930-

"CEPHALINS"

Phosphatidylethanolamine
-1913-

Phosphatidylserine
-1941-

"POLYGLYCEROLPHOSPHATIDES"

Phosphatidylglycerol
-1957-

Diphosphatidylglycerol
-1941-

Fig. 2. The glycerol phosphatides found in plants (the dates indicate when they were discovered).

of the cell. The nature of chloroplast acyl lipids and their fatty acids were reviewed by Nichols and James (1968).

Aldehyde glycerol phosphatides, the plasmalogens, which are important mammalian membrane components, do not occur in plant membranes. Neither do the closely related glycerol ether phosphatides. Lysophospholipids are often observed in plant lipid extracts. They may result from

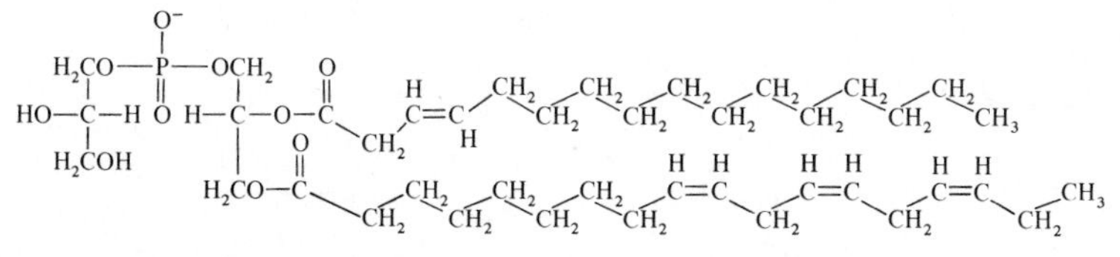

Fig. 3. Phosphatidylglycerol of chloroplasts. Note the Δ^3-*trans*-hexadecenoic acid in the 2 position.

hydrolysis of one of the two fatty esters by lipases activated during grinding and preparation of the tissue.

Phospholipids are degraded by three types of lipase. Lysolipids are the result of phospholipase A action in removing the fatty acid in the 2-position of glycerol. Phospholipase C, most active in bacterial preparations, yields diglyceride and phosphoryl esters of choline, for example. Phospholipase D is an enzyme in leaf vascular tissues which is active after grinding or during slow solvent extractions. It was found in the midrib tissue of Romaine lettuce leaves by Yang *et al.* (1967), and its action as a transphosphatidylating agent was established. By this mechanism, new phosphatides containing the fatty acids of phosphatidylcholine may be prepared.

2. GLYCOLIPIDS

Glycolipids are the predominant amphipathic lipids of plant membranes. They comprise 80% of the lipids of spinach lamellar lipoprotein (Wintermans, 1960). The major glycolipids are monogalactosyl and digalactosyl diglycerides, 1-[β-D-galactopyranosyl]-2,3-diacyl-D-glycerol, and 1-[α-D-galactopyranosyl-(1 → 6)-β-D-galactopyranosyl]-2,3-diacyl-D-glycerol (Fig. 4). The galactosyl diglycerides are synthesized by reaction of uridine diphosphate galactose (UDP-Gal) with diglycerides in chloroplasts (Ferrari and Benson, 1961; Neufeld and Hall, 1964). Digalactosyl diglyceride is formed by galactosylation of the monogalactolipids, although the reaction appeared to proceed at a different site and by an independent mechanism (Mudd, 1967).

Biosynthesis of lamellar galactolipids is a function of galactosyl transferase systems operating in the envelope of the chloroplast (Douce, 1974). The membrane-associated enzyme transfers the galactosyl group from UDP-Gal to diglyceride, mono-, di-, and trigalactolipid. Contiguity

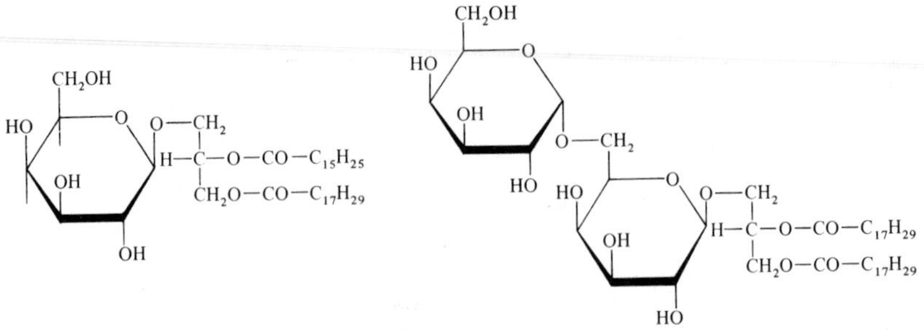

Fig. 4. The galactolipids found in plants.

of the envelope and the lamellar membrane system must facilitate migration of galactolipids to the lamellae during their formation. The biosynthesis of glycolipids, unlike that of phospholipids, was inhibited when the chloroplast protein synthesis was inhibited (Bishop and Smillie, 1970). The phospholipids are synthesized in the cytoplasm.

Fatty acid specificity was observed in glycolipids in which the monogalactolipid may have up to 25% hexadecatrienoic (16:3) acid, while the digalactosyl diglycerides contain almost exclusively linolenic acid (18:3). Bishop *et al.* (1971) reported a threefold higher galactolipid content in "bundle sheath" than in "mesophyll" chloroplasts of maize and sorghum. Since "mesophyll" chloroplasts possess grana formed from appressed lamellar membranes, it is possible that galactolipids are essential components of only one side of lamellar membrane surfaces. Galactolipase, which removes the fatty acyl groups from both monogalactolipid and digalactolipid, is uniquely active in runner bean (*Phaseolus multiflorus*) leaves (Sastry and Kates, 1964b). In isolated systems or in damaged cells, free fatty acids liberated by galactolipase affect the chloroplast's structure and impair its electron transport function.

3. SULFOLIPID

The plant sulfolipid is unique in that the sulfur atom is directly linked to the carbon-6 of the sugar. It is a sulfonic acid rather than the usual sulfate ester as is the case in animal sulfatides. The structure of plant sulfolipid was elucidated by Benson *et al.* (1959) as 6-sulfo-α-D-quinovopyranosyl-(1 → 1')-2',3'-diacyl-D-glycerol (Fig. 5). The normally slow turnover of plant sulfolipid is accelerated in low sulfate media (Shibuya *et al.*, 1965; Miyachi and Miyachi, 1966). Little is known of its biosynthesis. Sulfolipid is much more effective than any phospholipid in stabilizing the chloroplast CF_1 coupling factor against its cold inactivation (Livne and Racker, 1969). In spite of its high concentration in green plants, the relation of chloroplast metabolism to sulfonic acid biosynthesis and degradation is not yet understood.

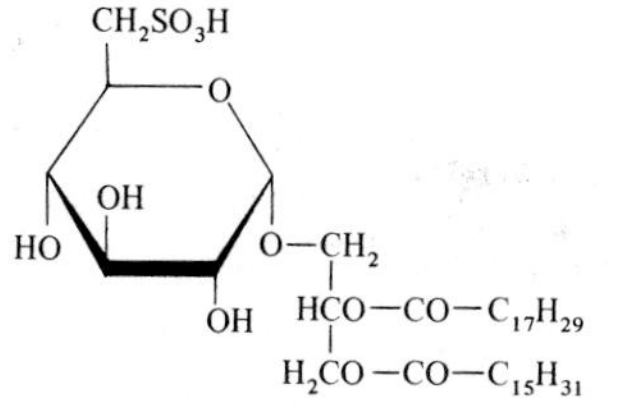

Fig. 5. The plant sulfolipid.

A sulfolipase capable of removing the two fatty acyl groups has been found in *Scenedesmus* (Yagi and Benson, 1962). The final product, sulfoquinovosyl glycerol is the major soluble sulfur-containing component of *Chlorella* and is a stable component of plant extracts. The sulfolipid of alfalfa contains both saturated palmitic acid (16:0) and unsaturated linolenic acid (18:3). An unusual C_{25} isoprenoid acid is the major fatty acid component of the sulfolipid of several cold-tolerant early blooming plants (Kuiper and Stuiver, 1972).

4. OTHER LIPIDS

Chlorophylls, carotenoids, and plastoquinones in chloroplast membrane and ubiquinones in mitochondrial membranes are some of the other important lipid components. Detailed information on these lipids will be presented in Chapters 6 and 25 of this book.

B. Proteins

Proteins constitute the largest fraction by weight of most cell membranes. They are distinguished as *peripheral* and *integral proteins* (Singer and Nicolson, 1972; Guidotti, 1972). *Peripheral proteins* are weakly bound and may be dissociated from membranes by relatively mild treatment, such as extraction with salt solutions or chelating agents Their amino acid composition resembles that of all the other "soluble" proteins. They usually contain little lipid; they are molecularly dispersed in aqueous buffers after once having been solubilized. *Peripheral proteins* like cytochrome *c* or phosphoglycolic acid phosphatase having cationic surface regions appear to associate with membranes by electrostatic interactions.

The bulk of the membrane proteins are integral proteins. They may be solubilized by detergents or some protein denaturants. The intractability of the membrane protein in aqueous solution and the dependence of its properties upon preparation procedures has complicated the interpretation of observations. In the case of plant membranes, even gross analysis of the protein has been difficult because of lack of pure, isolated membrane material. Green and associates (Richardson *et al.*, 1963) proposed that a "structural protein" forms a major portion of the animal mitochondrial integral membrane protein. This "structural protein" was by definition the insoluble residue left after exhaustive extraction of enzymes from the membranes. It was postulated to exist in all other cell membrane systems (Lenaz *et al.*, 1968). However, recent physical and chemical information indicates that there is little evidence for the existence of a special type of protein unique to membranes. Rather, a broad variety of proteins seem to exist in different cell membranes. Only about 20 membrane proteins have so far been studied in any detail (Guidotti, 1972).

A unique homogeneous lipid-free membrane surrounds gas vacuoles in some blue-green algae, such as *Microcystis aeruginosa,* and in *Halobacterium halobium.* Jones and Jost (1970) characterized the membrane of *M. aeruginosa* vacuoles as a homogeneous monolayer of 15,000 MW protein which resembled "membrane structural protein" in many ways. It aggregated readily. It contained no cysteine or S–S cross links. It contained 50% hydrophobic amino acids. In the electron micrographs, the membrane appeared as a monolayer of protein subunits. While not a lipoprotein, this homogeneous protein membrane provides a working model for hydrophobic association of membrane proteins in a monolayer. One of the basic features that membrane proteins logically must have is the capacity to interact with lipids to form stable lipoprotein structures. The properties of proteins are determined by their amino acid composition and sequence, and it appears that the integral membrane proteins are indeed structurally adapted for interaction with certain membrane lipids. They often have an excess and in at least one case, 69% of nonpolar amino acids. Membrane proteins have very little disulfide cross-linking and, therefore, may be better able to accommodate added lipid components after synthesis in aqueous media.

III. Structure of Cell Membranes

A. Interaction of Lipids and Proteins

It is obvious that understanding the interaction of lipids and proteins in plant cell membranes will be basic to interpretation of data on membrane structure and function. The following interactions are now considered by most investigators to be of major significance.

1. HYDROPHOBIC INTERACTIONS

When a protein molecule folds in aqueous solution, the nonpolar groups tend to be folded inside, away from contact with water, with a decrease of free energy accompanying each buried nonpolar group. This negative free energy results from the positive entropy gain as the ordered water molecules surrounding hydrophobic nonpolar groups in the unfolded state of the protein molecule are converted to disordered liquid water. The combined free energy change for a macromolecular protein, where many nonpolar groups occur, assumes a large negative value, and the equilibrium lies far on the side of the folded protein in water. This is now considered the major force stabilizing the native proteins in aqueous solution. Crystallographic studies confirm that the majority of nonpolar amino acids are located in the interior of protein molecules in the cases where three-dimensional structure of the proteins are known. In an exactly analogous way, hydrophobic interaction contributes to the bilayer or micelle formation by polar lipids. The importance of hydrophobic interaction in cell membrane structure has been emphasized by Singer (1971, 1974).

In nonaqueous solvents (Singer, 1962), the entropy changes are much smaller, and the hydrophobic interactions are of less consequence. They result in unfolding and extensive conformational change of the proteins.

2. Hydrogen Bonds

Hydrogen bonds formed between the carbonyl oxygen and the amide hydrogen in peptide chains are a major stabilizing force in proteins in the dry state. They can form when interatomic distances are under 2 Å. In aqueous media, any exposed protein hydrogen bonds are quickly replaced by hydrogen bonds with water. Hydrogen bonds, therefore, do not contribute much to stabilization of protein molecules, except those formed in the anhydrous interior of the macromolecule (Klotz and Franzen, 1962). Hydrogen bonds between water molecules surrounding the protein, however, exert pressure on enclosed hydrophobic molecules. It is as if the proteins were clenched within a shell of water whose hydrogen bonding leads to internal pressures estimated at several thousand atmospheres.

3. Electrostatic Interactions

Membrane lipoproteins possess many ionic groups, many of them ionized at physiological pH. The electrostatic force between either similarly or oppositely charged groups is, therefore, expected to play a role in membrane lipid–protein interactions. The force involved is considerable, being 4–5 kcal/mole between two charges at a distance of 5 Å. The additive effect of interaction with the many ammonium groups in polylysine in inhibiting photophosphorylation is striking when compared to that of an equivalent concentration of ammonium ion (Dilley et al., 1968). Association of polycationic histones (Brand et al., 1972) and peripheral proteins with membranes is a related phenomenon. When electrostatic interactions contribute to the stability of a macromolecular system, it can be expected that changes in ionic strength of the solution should alter the stability of the system.

B. Molecular Architecture of Membranes

As methods for membrane study developed, there evolved a series of molecular models for describing cell membrane structure: The lipid bilayer model, the lipoprotein subunit model, and the lipid bilayer-globular protein or fluid mosaic model.

1. THE DAVSON–DANIELLI–ROBERTSON LIPID BILAYER MODEL

This model (Fig. 6A) (Danielli and Davson, 1935; Robertson, 1964) appears to provide the lamellar matrix for biological membranes. It has been the subject of extensive experimental study and is consistent with many of the physical properties of membranes. The Davson–Danielli–Robertson model drew support from the following evidence: (1) Chemical analysis of isolated membranes indicate lipids and proteins to be generally present in amounts compatible with the model. (2) Membrane lipids form stable bilayers (myelin figures). Many of the physical properties of such bilayers (e.g., conductance, thermal phase transitions, birefringence, x-ray diffraction) are similar to those of natural membranes. Differential thermal analysis, which records heats of transition derived from phase changes in artificial bilayers and in isolated membrane systems, are strikingly similar (Steim, 1970). The methods, however, could be insensitive to 20% of specific lipoprotein components within a membrane bilayer system. (3) Membranes generally are more permeable to small hydrophobic molecules than to water-soluble molecules. (4) Lipid bilayers interact with certain proteins. The fixed, stained, and sectioned artificial membrane thus formed yields an electromicrographic image very similar

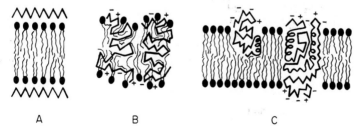

A   B   C

Fig. 6. Diagrammatic representations of membrane structure: (A) according to Danielli and Davson (1935); (B) lipoprotein monolayer model (Benson, 1966); (C) lipid bilayer globular protein model (Singer, 1971).

to those of the natural membranes. The consistent appearance of two parallel dark lines in electron micrographs of heavy metal stained sections was taken as strong supporting evidence for the lipid bilayer model, even though they were separated by 80 Å instead of the 40 Å width of a bilayer. (5) Electron micrographs of freeze-etch replicas of myelin and of parts of many mammalian membranes are consistent with the bilayer model (Branton, 1967). Hydrophobic interactions, dependent upon presence of liquid water, diminish in the frozen state, and bilayer membranes readily cleave between the two lipid layers at the interior of the bilayer. In an elegant series of experiments Branton (1966) experimentally demonstrated that lipid bilayer membranes are weakened in frozen tissues and that they cleave in the hydrophobic region between the two lipid layers. A metallized replica of the fractured bilayer revealed the smooth surface of the hydrocarbon moieties of the lipid monolayer.

2. The Lipoprotein Subunit Model

The main feature of this model (Fig. 6B) (Benson, 1966; Green and Oda, 1961) is that the membrane proteins are intimately associated with amphipathic lipids by hydrophobic interactions. The lipid molecules are arranged in a bilayer but are also intercalated into the hydrophobic regions of the proteins via interaction with the hydrocarbon chains of lipids. The hydrophilic polar groups of the lipids are at the membrane surface in contact with water. The presence of repeating subunits or lipid mosaic and the specificity of interaction between lipid and protein structure were implied in the model. The high degree of fatty acid specificity observed in biological membranes cannot be easily explained. Appropriate fluidity would be the only requirement for simple lipid bilayer formation (Benson, 1974). The demonstration of a direct relationship between α-linolenic acid (18:3) content and oxygen producing capability in chloroplast (Irwin and Bloch, 1963) and the requirement of specific lipid molecules for activity of many membrane enzymes (Rothfield et al., 1969) indicated intimate involvement of lipids in membrane function. These relationships are only compatible with the lipid bilayer model if one considers the membrane as a combination of functional lipoprotein supported in a lipid bilayer.

3. The Lipid Bilayer–Globular Protein Model

Models presented by Lenard and Singer (1966), Sjöstrand (1971), Wallach and Zahler (1966), and Singer (1974) incorporated amphipathic proteins or lipoproteins in the lipid bilayer structure of Davson and

Danielli (Figs. 6C and 7). When the lipoprotein–lipid bilayer membrane is frozen in water, the hydrophobic interactions between hydrocarbon chains and protein and between the hydrocarbon chains themselves are weakened, allowing cleavage to occur along the hydrophobic interface. In Fig. 7B the "split" membrane is shown as a smooth monolayer surface which contains the "naked" lipoprotein components. The surface, as revealed by electron microscopy of a metallic replica included proteins lying in a lipid monolayer. The lipid monolayer is seen from the exposed interior, the hydrophobic hydrocarbon groups of the amphipathic lipids. The replica of the opposing side of the membrane also showed protein molecules, but of different size and shape. This indicates asymmetry of the membrane and is supported by a considerable body of chemical, enzymatic, and independent ultrastructural evidence (Singer, 1974). Such asymmetry allows important conclusions to be drawn regarding membrane structure and function. The many functions of membranes, such as transport, electrical potentials, antigenic properties, require asymmetry.

In functional cell membranes amphipathic proteins or lipoproteins are relatively free to migrate in the lipid bilayer. Such proteins have been observed by Nicolson and Singer (1971) on the outer surfaces of red cell membranes using ferritin-stained conconavalin A and other

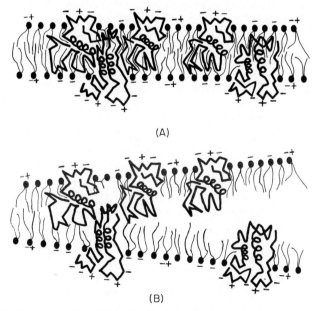

(A)

(B)

Fig. 7. Composite lipoprotein–lipid bilayer membrane model before (A) and after (B) cleavage in freeze-etch preparation.

hemagglutinins to detect specific polysaccharide binding sites (Ji and Nicolson, 1974). All the mammalian membranes examined so far were asymmetrical. They, therefore, possess components that are unable to revolve or migrate from one side of the membrane to the other. The observations of rapid rates of diffusion of antigenic sites on the surface of human–mouse heterokaryons (Frye and Edidin, 1970) indicated that glycoproteins of the membrane migrate freely in the lipid bilayer. A significant difference between plant membranes, which have been examined and those bacterial and animal membranes upon which the model is developed is the apparent lack or scarcity of glycoproteins in plant membranes. Erythrocyte membrane glycoproteins have been important factors in development of the lipid mosaic model. Their role in plant membranes appears to have been assumed by the glycolipids. Those in mammalian cells, the cerebrosides and gangliosides, may perform similar functions. In membranes where asymmetric lipoproteins arranged on each surface of the bilayer are possible, one must interpret data and theories with both caution and imagination. The essential point to remember is that membrane function is the result of genetically controlled protein synthesis. The "personality" of a membrane lies in the amino acid sequences of its integral parts. A model for a membrane like this is shown in Fig. 7A. This model is consistent with ultrastructural information as well as with physical and physiological properties of biological membranes. However, the diversity of membrane function and composition must be reflected in variety in membrane structure. A single rigidly defined model would be difficult to reconcile with all biological membrane structures.

C. Biosynthesis and Assembly of Cell Membranes

How a cell assembles complex supramolecular structures, such as its membranes and organelles, is marvelous but not yet fully understood.

Engelman *et al.* (1967) solubilized plasmalemma of *Mycoplasma laidlawii* in detergent. The reaggregated membrane upon removal of detergent showed typical "unit membrane" morphology. The reaggregated membrane also had an amino acid composition, lipid to protein ratio, and buoyant density very similar to those of the original membrane. The possible existence of a protein subunit that is soluble in water and which can be assembled into membranes by addition of lipid molecules in living cells would offer an alternative way for membrane biosynthesis. The membrane components could then be synthesized at sites different from the site of assembly. The slow incorporation of [³H]leucine into the plasma membrane of rat liver cells and its continuous incorporation hours after inhibition of protein synthesis in the cell suggested that such a soluble precursor protein, synthesized prior to its incorporation into plasma membrane, may exist (Ray *et al.*, 1968).

Membrane reconstitution experiments have had limited success. Since amphipathic lipids form stable micelles in water, there is no way to transfer lipid from micelles to protein acceptor in water. In nonaqueous solvents (Singer, 1962) or in the dry or frozen state, however, it has been possible to reintroduce membrane lipids into membrane protein with recovery of limited biological function. In each case the lipid is free to enter the appropriate interstices of the protein where it is locked in place by careful restoration of the aqueous medium. Chloroplast lamellar electron transport has been restored, in part, by condensing lipid and protein by such techniques. Biosynthesis of membrane, performed in suitable microenvironments within the cell, may utilize such processes. The asymmetry and organization of integral protein components of membranes probably require greater control than are possible in simple macroscale experiments. The protein subunits of the red cell membrane, for example, are associated with well-defined asymmetry. Although the associated membrane subunits are free to migrate over the membrane, they are not free to rotate or to move independently of their associated subunit at the other side of the membrane. To reconstitute molecular organizations at this level would be difficult. Membrane biogenesis is reviewed by Morré (1975).

IV. Properties of Specific Cell Membranes

A. Plasmalemma

The plasmalemma of plant cells encloses the "living" part of the cell. It appears under the electron microscope as a typical unit membrane approximately 100 Å thick. The freeze-etch replica micrograph (Fig. 8) reveals its granular construction. Globular particles of 80 to 130 Å diameter were observed on both the outer and inner surfaces of the plasmalemma of pea and onion root tips (Northcote and Lewis, 1968). Particles 80 Å in diameter are randomly distributed on the plasmalemma of *Chlorella*, and it was suggested that they are synthetic units for microfibrils of the cell wall (Staehelin, 1966). Larger particles of 150 to 180 Å in diameter are found on the surface of the yeast plasmalemma. The high carbohydrate reported for yeast protoplast membrane (Matile, 1970) is presumably due to the presence of such articles and indicates that they are tightly bound to the plasmalemma. Tight "extraneous coats" on cell surfaces complicate the isolation of plant membranes. They appear to form structural and functional complexes with the plasmalemma.

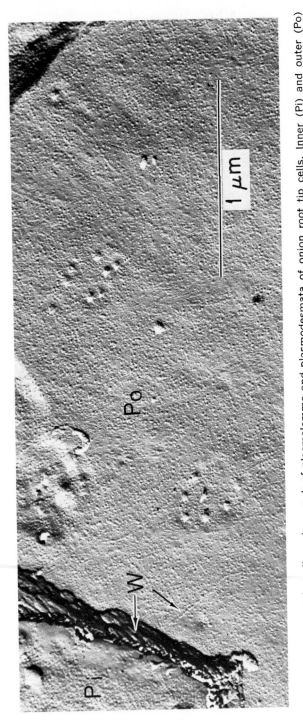

Fig. 8. Freeze-etch replica micrograph of plasmalemma and plasmodesmata of onion root tip cells. Inner (Pi) and outer (Po) plasmalemma and cell wall (W) are shown in the picture. The particles on the outer plasmalemma surface are arranged in row (arrows). Pit fields composed of about eight plasmodesmata are present. (Courtesy of D. H. Northcote and R. Lewis, Department of Biochemistry, University of Cambridge.)

The solute content of a plant cell is regulated primarily by its plasmalemma; organelle membranes further modify compositions within the microcompartments they enclose. Two basic mechanisms are recognized which regulate the cellular solute compositions: (1) passive permeability and (2) active transport. Passive permeability is a process of diffusion promoted by differences in concentrations and impeded by surface barriers. It is well known that low molecular weight compounds have different coefficients of permeability through plant cell membranes. Water is most permeable and appears to move as single molecules, like a gas, through the membrane. Lipoidal compounds are more permeable than polar compounds. These differences were the basis of Overton's (1900) classical "lipid solubility theory" of membrane permeability. Collander (1949) postulated that pores in the plasmalemma exist which are lined with lipid molecules. Although these would be consistent with contemporary concepts of membrane structure (Singer, 1974), real evidence is lacking. (2) What is now known as "active transport" has long been studied in plant systems because of its central importance in plant nutrition (Steward and Sutcliffe, 1959). The term is used to describe the movement of ions or molecules through cell boundaries against concentration gradients at the expense of metabolic energy. Membrane active transport studies have approached the problem through either kinetic analysis of ion transport or by characterization of the carrier molecules and enzyme systems involved in the transport. Epstein (1966) interpreted Na^+ and K^+ absorption as the result of two independent mechanisms operating at high and low concentration ranges. Laties (1969) attributed the second mechanism to selection by vacuolar membrane. Nissen (1974) has interpreted the process as a single multiphasic mechanism in each membrane operative over the entire concentration range. The plasmalemma appears to control the rate of uptake at low external salt concentrations, while the tonoplast membrane may become rate limiting at the higher concentrations.

In bacterial systems transport activities are performed by membrane proteins whose synthesis can be under genetic control. Galactoside permease "carrier protein" from *E. coli* cells was demonstrated by characterization of a small lipoprotein located in the plasmalemma (Fox and Kennedy, 1965). Slayman and Tatum (1965) had reported a cation transport mutant of *Neurospora* which was genetically deficient in a membrane protein. Similar systems have been reported for bacterial and animal cells (Pardee, 1968). Ion carrying proteins, ionophores, must function in plant membranes. Valinomycin, a cyclic polypeptide that facilitates K^+ transport, has served as a model and stimulated research. Its structure with a "hole" appropriate for a potassium ion is related to that of the

β-helix form of proteins or polypeptides. The β-helix structure differs from α-helix structure in that each turn of the helix involves six instead of three amino acid residues. It is stabilized by hydrogen bonding only when the peptide is constructed with alternating L- and D- (or glycine) amino acids (Urry and Ohnishi, 1974). Natural ionophores or protein segments functioning as such may assume such conformations. Na$^+$, K$^+$-activated membrane bound ATPase is well documented in animal and bacterial cell membranes (Skou, 1965) and is associated with ion transport processes. This enzyme has been reported from sugar beets, bean roots, and mangroves and presumably is located in the plasmalemma (Kylin and Gee, 1970). However, it has not been found in all plant systems investigated (Bonting and Caravaggio, 1966; Leggett, 1968). In the lipid bilayer–globular protein membrane model discussed in Section II,B, ATPase together with other enzymes of cell membranes were envisioned as a subunit of the membrane structure (Sjöstrand, 1971). Ion-stimulated conformational change of the enzyme was credited with the translocation of the specific ions. Such conformational change is considered driven by specific membrane ATPase activity. The obligatory involvement of associated peripheral membrane protein that binds the substrate at the site of the translocating integral protein(s) has been envisaged by Singer (1974).

Another interesting mechanism of membrane transport involves the ingestion of relatively large volumes of liquid through invagination of the plasmalemma and subsequent pinching off of the vesicles within the cell. This process is called pinocytosis and is often observed in metabolically active plant cells (Fig. 9). In some cases, large quantities of fluid are absorbed during extension growth of plant cells, a significant amount of it through pinocytosis (Weiling, 1961).

The plasmalemma also performs important functions in resorption, excretion, and secretion. The latter function was clearly demonstrated by the work of Northcote and Pickett-Heaps (1966) where radioisotope labeled carbohydrates are seen accumulated in Golgi bodies, which later fuse with the plasmalemma to deposit the synthesized cell wall material outside. Similarly, Thomson et al. (1969) observed salt concentration within microvesicles of salt gland cells of *Tamarix* and mangrove species. The vesicles were observed to fuse with the plasmalemma to place the concentrated salt solution outside of the leaf. In the mangrove *Aegialitis* salt excretion can exceed 20 mg per leaf per day.

Plasmodesmata are specialized structures of the plasmalemma which have been demonstrated in both sections and freeze-etch electron micrographs (Fig. 8) (Robards, 1968). They are canals that penetrate through cell walls. They are lined by plasmalemma and are traversed by a mem-

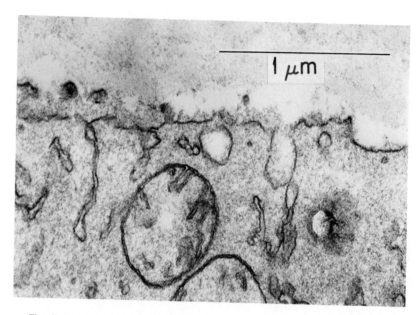

Fig. 9. Pinocytosis in the cells of the root tip (*Ricinus communis*). (Courtesy of A. Frey-Wyssling, Swiss Federal Institute of Technology.)

branous tubule (desmotubule) about 200 Å in diameter. Plasmodesmata appear to provide for intercellular movement of substances as well as to equilibrate membrane potentials. Virtually all cells of the higher plants are interconnected by these structures.

B. Plastid Membranes

Both chloroplast and mitochondria are bounded by two membranes. The inner membrane elaborates and invaginates during development within the lumen or stroma of the plastid and exhibits coordinated structure made up of molecules of various enzymes, structural proteins, and cofactors which perform photosynthetic electron transport and enzymatic reactions. Other storage or synthetic organelles in plant cells, such as leucoplasts and amyloplasts, are also called plastids. The reason is that they are also bounded by two membranes, although the development and invagination of their inner membranes are usually much less extensive.

Douce *et al.* (1973) and Poincelot (1973, 1974) isolated the chloroplast outer envelope and examined its composition and studied its properties. Douce (1974) discovered its unique capability for enzymatic transfer of galactosyl moieties from uridine diphosphate galactose to diglyceride, monogalactosyl diglyceride and digalactosyl diglyceride and thereby implicated the chloroplast envelope in synthesis of the two major lamellar lipid components. Contiguity of the lamellar membranes with the inner membrane of the envelope suggests a "lamellar flow" mechanism for transfer to galactolipids from their site of synthesis to the grana. Being a fluid mosaic membrane, such flow is consistent with observations in other systems. Another envelope-specific enzyme system recognized by Douce *et al.* (1973) was a Mg^{2+}-specific ATPase by which he characterized the membrane system.

The chloroplast envelope has the light yellow color of violaxanthin. Envelopes isolated from illuminated chloroplasts had much more orange zeaxanthin. Jeffrey *et al.* (1974) characterized the pigment transformations as a function of the violaxanthin–zeaxanthin (epoxide–olefin) transition. This reversible chloroplast carotenoid cycle is revealed most markedly in the chloroplast envelope. Structure and function of chloroplast membrane are also discussed in Chapter 7 and those of mitochondria in Chapter 5.

C. Golgi Membrane

Golgi bodies consist characteristically of a series of folded and interconnected membranes that are concentrically curved (Fig. 10). Small vesicles are usually observed along the periphery of the Golgi bodies. They have been isolated from animal cells as well as from plant cells (Morré and Mollenhauer, 1974). Those from animal sources that have been analyzed have phospholipid contents considerably higher than other cell membrane fractions. Small amounts of RNA also are involved. The ontogenic relationship between Golgi membrane and plasmalemma is clearly indicated, as the Golgi membrane becomes the new plasmalemma during cell wall synthesis of plant cells. Golgi membrane has also been shown to fuse with existing plasmalemma during cell secretion (see Section III,A).

D. Nuclear Membrane and Endoplasmic Reticulum

The nucleus of a plant cell is surrounded by a double membrane. Pores about 1180 Å in diameter extend through cytoplasm and nucleoplasm. They carry particles of 130 Å in diameter on both sides (Northcote and Lewis, 1968), which can be seen distributed over the nuclear membrane both in freeze-etch replica and in stained section electron micrographs. The lumen between the double membrane of the nucleus is continuous with that of the endoplasmic reticulum. In plant cells, the endoplasmic reticulum is approximately 70Å thick, and both its inside and outside surfaces include particles 90 to 130 Å in diameter revealed

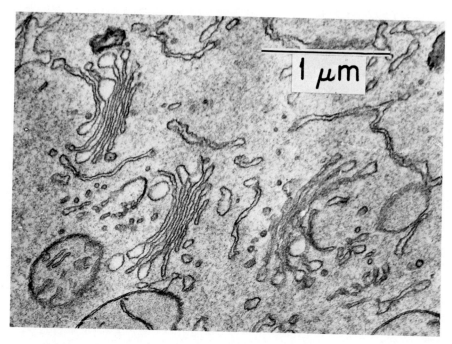

Fig. 10. Golgi bundles in the root meristem of (*Ricinus communis*). (Courtesy of A. Frey-Wyssling, Swiss Federal Institute of Technology.)

by freeze-etch electron micrographs (Fig. 11). The pattern of distribution of endoplasmic reticulum changes throughout growth and differentiation of the cell, suggesting their dynamic role in cell development. Translocation of cellular compounds and assistance in protein and lipid biosynthesis are the major functions postulated for the endoplasmic reticulum. It was suggested that the smooth endoplasmic reticulum in wheat root tip cells is concerned with the processes of transport and aggregation of the microtubular subunits (Burgess and Northcote, 1968).

E. Vacuolar Membrane

A mature living plant cell is characterized by the presence of large vacuoles. Stained electron micrographs indicate that the vacuolar envelope has a thickness of approximately 100 Å with an apparent asymmetry similar to that of the plasmalemma. Freeze-etch preparations reveal characteristic granular appearance. Small particles averaging about 85 Å in diameter are seen on both the concave and convex surfaces of the vacuolar membrane (Fig. 11) (Branton, 1966).

The origin of the vacuole membrane has been suggested as either from endoplasmic reticulum (Buvat, 1962) or from Golgi membrane

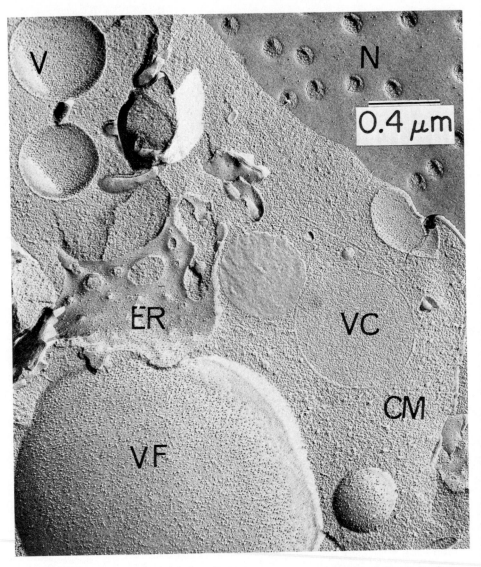

Fig. 11. Part of an onion root tip cell showing particles on the vacuolar membrane, concave and convex surfaces (VC and VF). The endoplasmic reticulum (ER) and the nuclear membrane (N) can also be seen. CM, cell membrane; V, vacuole. (Courtesy of D. Branton, Department of Botany, Harvard University.)

(Marinos, 1963; Morré, 1975). With techniques for vacuole isolation now available (Wagner and Siegelman, 1975; Leigh and Branton, 1975) the chemistry and physiology of these important membranes can be investigated. The function of vacuoles is discussed in Chapter 8.

GENERAL REFERENCES

Bolis, L., and Pethica, B. A., eds. (1968). "Membrane Models and the Formation of Biological Membranes." North-Holland Publ., Amsterdam.

Bolis, L., Katchalski, A., Keynel, R. D., Lowenstein, W. R., and Pethica, B. A. (1970). "Permeability and Function of Biological Membranes." North-Holland Publ., Amsterdam.

Branton, D. (1969). *Annu. Rev. Plant Physiol.* **20**, 209.

Dallner, G., Siekvitz, P., and Palade, G. E. (1966). *J. Cell Biol.* **30**, 73.

Davson, H., and Danielli, J. F. (1945). "The Permeability of Natural Membranes." Cambridge Univ. Press, London and New York.

Frey-Wyssling, A., and Muhlethaler, K. (1964). "Ultrastructural Plant Cytology." Elsevier, Amsterdam.

Järnefelt, J., ed. (1968). "Regulatory Function of Biological Membrane." Elsevier, Amsterdam.

Ji, T. H., and Benson, A. A. (1968). *Biochim. Biophys. Acta* **150**, 686.

Ledbetter, M. C., and Porter, K. R. (1970). "Introduction to the Fine Structure of Plant Cells." Springer-Verlag, Berlin and New York.

Moor, H., and Mühlethaler, K. (1963). *J. Cell Biol.* **17**, 609.

Nystrom, R. A. (1973). "Membrane Physiology." Prentice-Hall, Englewood Cliffs, New Jersey.

Rothfield, L. I., ed. (1971). "Structure and Function of Biological Membranes." Academic Press, New York.

Singer, S. J. (1974). *Ann. Rev. Biochem.* **43**, 805.

Singer, S. J., and Nicolson, G. L. (1972). *Science* **175**, 720.

van Deenen, L. L. M. (1968a). *In* "Membrane Models and the Formation of Biological Membranes" (L. Bolis and B. A. Pethica, eds.), pp. 98–104. North-Holland Publ., Amsterdam.

van Deenen, L. L. M. (1968b). *In* "Regulatory Function of Biological Membranes" (J. Järnefelt, ed.), pp. 72–86. Elsevier, Amsterdam.

Villaneuva, J. R., and Ponz, F., eds. (1970). "Membranes: Structure and Function." Academic Press, New York.

REFERENCES

Benson, A. A. (1966). *J. Amer. Oil Chem. Soc.* **43**, 265.

Benson, A. A. (1974). *Symp. Soc. Develop. Biol.* **30**, 153.

Benson, A. A., Daniel, H., and Wiser, R. (1959). *Proc. Nat. Acad. Sci. U.S.* **45**, 1582.

Bishop, D. G., and Smillie, R. M. (1970). *Arch. Biochem. Biophys.* **139**, 179.

Bishop, D. G., Anderson, K. S., and Smillie, R. M. (1971). *Biochim. Biophys. Acta* **231**, 412.

Bonting, S. L., and Caravaggio, L. L. (1966). *Biochim. Biophys. Acta* **112**, 519.

Brand, J., Baszynski, T., Crane, F. L., and Krogmann, D. W. (1972). *J. Biol. Chem.* **247**, 2814.

Branton, D. (1966). *Proc. Nat. Acad. Sci. U.S.* **55**, 1048.

Branton, D. (1967). *Exp. Cell Res.* **45**, 703.

Burgess, J., and Northcote, D. H. (1968). *Planta* **80**, 1.

Buvat, R. (1962). *Proc. Int. Congr. Electron Microsc., 5th, 1962* Article W1.

Cherry, J. B., Hsu, K., and Chapman, D. (1971). *Biochem. Biophys. Res. Commun.* **43**, 351.

Collander, R. (1949). *Physiol. Plant.* 3, 300.

Danielli, J. F., and Davson, H. (1935). *J. Cell. Comp. Physiol.* 5, 495.

Dilley, R. A. (1968). *Biochemistry* 7, 338.

Douce, R. (1974). *Science* 183, 852.

Douce, R., Holtz, R. B., and Benson, A. A. (1973). *J. Biol. Chem.* 248, 7215.

Engelman, D. M., Terry, T. M., and Morowitz, H. J. (1967). *Biochim. Biophys. Acta* 135, 381.

Epstein, E. (1966). *Nature (London)* 212, 1324.

Ferrari, R. A., and Benson, A. A. (1961). *Arch. Biochem. Biophys.* 93, 185.

Fleischer, S., Fleischer, B., and Stoeckenius, W. (1967). *J. Cell Biol.* 32, 193.

Fox, C. F., and Kennedy, E. P. (1965). *Proc. Nat. Acad. Sci. U.S.* 54, 891.

Frye, L. D., and Edidin, M. (1970). *J. Cell Sci.* 7, 319.

Green, D. E., and Oda, T. (1961). *J. Biochem. (Tokyo)* 69, 742.

Guidotti, G. (1972). *Annu. Rev. Biochem.* 41, 804.

Haverkate, F., and van Deenen, L. L. M. (1965). *Biochim. Biophys. Acta* 106, 78.

Irwin, J., and Bloch, K. (1963). *Biochem. Z.* 338, 496.

Jeffrey, S. W., Douce, R., and Benson, A. A. (1974). *Proc. Nat. Acad. Sci. U.S.* 71, 807.

Ji, T. H., and Nicolson, G. L. (1974). *Proc. Nat. Acad. Sci. U.S.* 71, 2212.

Jones, D. D., and Jost, M. (1970). *Arch. Mikrobiol.* 70, 43.

Jost, P., Griffith, O. H., Capaldi, R. A., and Vanderkooi, G. (1973). *Proc. Nat. Acad. Sci. U.S.* 70, 480.

Jurtshuk, P., Jr., Sekuzu, I., and Green, D. E. (1961). *Biochem. Biophys. Res. Commun.* 6, 76.

Klotz, I. M., and Franzen, J. S. (1962). *J. Amer. Chem. Soc.* 84, 3461.

Kuiper, P. J. C., and Stuiver, B. (1972). *Plant Physiol.* 49, 307–309.

Kylin, A., and Gee, R. (1970). *Plant Physiol.* 45, 169.

Laties, G. G. (1969). *Annu. Rev. Plant Physiol.* 20, 89.

Leggett, J. E. (1968). *Annu. Rev. Plant Physiol.* 19, 333.

Leigh, R. A., and Branton, D. (1975). *Plant Physiol. Suppl.* 56, (2), 52.

Lenard, J., and Singer, S. J. (1966). *Proc. Nat. Acad. Sci. U.S.* 56, 1828.

Lenaz, G., Haard, N. F., Silman, H. I., and Green, D. E. (1968). *Arch. Biochem. Biophys.* 128, 293.

Livne, A., and Racker, E. (1969). *J. Biol. Chem.* 244, 1332.

Marinos, N. G. (1963). *J. Ultrastruct. Res.* 9, 177.

Matile, P. (1970). *In* "Membranes: Structure and Function" (J. R. Villanueva and F. Ponz, eds.), pp. 39–51. Academic Press, New York.

Miyachi, S., and Miyachi, S. (1966). *Plant Physiol.* 41, 479.

Morré, D. J. (1975). *Ann. Rev. Plant Physiol.* 26, 441.

Morré, D. J., and Mollenhauer, H. H. (1974). *In* "Dynamic Aspects of Plant Ultrastructure" (A. W. Robards, ed.), pp. 84–137, McGraw-Hill, New York.

Mudd, J. B. (1967). *Annu. Rev. Plant Physiol.* 18, 229.

Nägeli, C., and Cramer, C. (1885). "Pflanzenphysiologische Untersuchungen," p. 21. Schulthess, Zurich.

Neufeld, E. F., and Hall, C. W. (1964). *Biochem. Biophys. Res. Commun.* 14, 503.

Nichols, B. W., and James, A. T. (1968). *In* "Plant Cell Organelles" (J. B. Pridham, ed.), pp. 163–177. Academic Press, New York.

Nicolson, G. L., and Singer, S. J. (1971). *Proc. Nat. Acad. Sci. U.S.* 68, 942.

Nissen, P. (1974). *Annu. Rev. Plant Physiol.* 25, 53.

Northcote, D. H., and Lewis, D. (1968). *J. Cell Sci.* 3, 199.

Northcote, D. H., and Pickett-Heaps, J. D. (1966). *Biochem. J.* **98,** 159.

Overton, E. (1900). *Jahrb. Wiss. Bot.* **34,** 669.

Pardee, A. B. (1968). *Science* **162,** 632.

Peters, R. (1956). *Nature (London)* **177,** 426.

Peters, R. (1969). *Proc. Roy. Soc., Ser. B* **173,** 11.

Plowe, J. Q. (1931). *Protoplasma* **12,** 196.

Poincelot, R. P. (1973). *Arch. Biochem. Biophys.* **159,** 134.

Poincelot, R. P., and Day, P. R. (1974). *Plant Physiol.* **54,** 780.

Ray, T. K., Lieberman, I., and Lansing, A. I. (1968). *Biochem. Biophys. Res. Commun.* **31,** 540.

Richardson, S. H., Hultin, H. O., and Green, D. E. (1963). *Proc. Nat. Acad. Sci. U.S.* **50,** 821.

Robards, A. W. (1968). *Planta* **82,** 200.

Robertson, J. D. (1964). *In* "Cellular Membranes in Development" (M. Locke, ed.), pp. 1–24. Academic Press, New York.

Rothfield, L., Weisen, M., and Endo, A. (1969). *J. Gen. Physiol.* **54,** 27s.

Sastry, P. S., and Kates, M. (1964a). *Biochemistry* **3,** 1271.

Sastry, P. S., and Kates, M. (1964b). *Biochemistry* **3,** 1280.

Shibuya, I., Maruo, B., and Benson, A. A. (1965). *Plant Physiol.* **40,** 1251.

Singer, S. J. (1962). *Advan. Protein Chem.* **17,** 1.

Singer, S. J. (1971). *In* "Structure and Function of Biological Membranes" (L. I. Rothfield, ed.), pp. 145–222. Academic Press, New York.

Singer, S. J. (1974). *Annu. Rev. Biochem.* **43,** 805.

Singer, S. J., and Nicolson, G. L. (1972). *Science* **175,** 720.

Sjöstrand, F. S. (1971). *In* "Cell Membrane: Biological and Pathological Aspects" (G. W. Richter, ed.), pp. 1–25. Williams & Wilkins, Baltimore, Maryland.

Skou, J. C. (1965). *Physiol. Rev.* **45,** 596.

Slayman, C. W., and Tatum, E. L. (1965). *Biochim. Biophys. Acta* **109,** 184.

Staehelin, A. (1966). *Z. Zellforsch. Mikrosk. Anat.* **74,** 325.

Steim, J. M. (1970). *In* "Liquid Crystals and Ordered Fluids," (J. F. Johnson and R. S. Porter, eds.), pp. 1–13. Plenum, New York.

Steward, F. C., and Sutcliffe, J. F. (1959). *In* "Plant Physiology" (F. L. Steward, ed.), Vol. 2, pp. 253–274. Academic Press, New York.

Tasaki, I., Watanabe, A., Sandlin, R., and Carnay, L. (1968). *Proc. Nat. Acad. Sci. U.S.* **61,** 883.

Thomson, W. W., Berry, W. L., and Liu, L. L. (1969). *Proc. Nat. Acad. Sci. U.S.* **63,** 310.

Urry, D. W., and Ohnishi, T. (1974). *In* "Peptides, Polypeptides, and Proteins." Proceedings of the Rehovoth Symposium, May 1974. (E. R. Blout, S. A. Bovey, M. Goodman, and N. Lotan, eds.), p. 230. John Wiley New York.

Wagner, G. J., and Siegelman, H. W. (1975). *Science* **190,** 1298.

Wallach, D. F. M., and Zahler, P. H. (1966). *Proc. Nat. Acad. Sci. U.S.* **56,** 1552.

Weeldon, L. W. (1960). *J. Lipid Res.* **1,** 439.

Weiling, F. (1961). *Naturwissenschaften* **48,** 411.

Wintermans, J. F. G. M. (1960). *Biochim. Biophys. Acta* **44,** 49.

Yagi, T., and Benson, A. A. (1962). *Biochim. Biophys. Acta* **57,** 601.

Yang, S. F., Freer, S., and Benson, A. A. (1967). *J. Biol. Chem.* **242,** 477.

5

Microbodies

R. W. BREIDENBACH

I. Introduction

For many years biologists have recognized that macromolecules involved in specific metabolic processes are often associated with physically distinguishable subcellular entities. Microscopists early observed mitochondria, chloroplasts, and nuclei, and biologists from various disciplines established the principal aspects of their metabolic and physiological functions over the ensuing years. For many years, these organelles received intensive study because of their size, their relative stability during isolation, and the central importance of the processes they mediate, i.e., respiration, photosynthesis, and chromosome replication. However, with the advent of electron microscopy, gentler cell disruption procedures, and more sophisticated methods for separating subcellular complexes, a whole new array of organelles and subcellular associations has been discovered. These recent discoveries have stimulated interest in the role of physical compartmentation as a general means of metabolic regulation.

We now recognize the importance of physical compartments in separating those metabolic pathways having common intermediates, so as to separate the reactions that would otherwise compete for these intermediates. Second, we realize that physical compartments may serve to maintain locally high concentrations of reactants in a given sequence of steps in a metabolic pathway. Third, physical compartments may restrict certain toxic or reactive metabolites from general access to the cell contents. This evolution of our recognition and understanding of compartmentation and subcellular organization leads one to wonder what more fragile and less conspicuous associations remain to be elucidated and what contribution these ordered states make to metabolic regulation.

This chapter, however, is restricted to describing our present understanding of microbodies, one class of such organelles, and employing them as a model of metabolic regulation by compartmentation.

II. General Description and Nomenclature

A. General Description

Most workers have adopted "microbody" as the generic term for a diverse but related group of organelles. These organelles vary in important respects from tissue to tissue and species to species, but possess various combinations of characteristics typical for the group (Table I). They are typically 0.5 to 1.5 μm in diameter and have an amorphous granular or slightly fibrillar matrix bounded by a single membrane. They sometimes contain a more electron-dense globoid or crystalloid structure. The structure present, however, is variably dependent on species and in some yet unknown way on the physiological state of the cells they oc-

TABLE I

Common Characteristics of Microbodies

1. Single bounding membrane
2. Diameter 0.5–1.5 μm
3. Amorphous granular or fibrillar matrix
4. Electron-dense globoid or crystalloid
5. Frequent close association with the endoplasmic reticulum
6. Sedimentation coefficient ($s_{20,w}$) of about 5×10^4
7. Buoyant density of 1.22–1.25 gm cm^{-3}
8. Positive histochemical test for catalase with diaminobenzidine
9. Presence of oxidases generating H_2O_2
10. No enzyme latency

cupy. In many species or tissues, they frequently exhibit a close association with the endoplasmic reticulum.

B. Historical Perspective

Organelles with these cytological characteristics were first termed microbodies by Rhodin (1954), who described them in the proximal tubule cells of mouse kidney. Gänsler and Rouiller (1956) observed microbodies in the cortical cells of mammalian liver, in which they have since been extensively studied, both cytologically and biochemically.

Mollenhauer *et al.* (1966) first drew attention to the cytological similarities between mammalian microbodies and organelles observed in a wide variety of plant species and tissues. Since then, Newcomb and others (see Vigil, 1973, for a complete review) have contributed to the cytological characterization of plant microbodies in both leaves and storage tissues of fatty seeds.

Biochemical characterization of organelles from rat liver, shown to be identical to microbodies seen *in situ*, began with work of De Duve and his colleagues (see De Duve, 1969a,b, for review). They demonstrated that microbodies from rat liver contained a number of oxidases (see Table II) transferring hydrogen atoms from a donor to molecular oxygen to form hydrogen peroxide. The isolated organelles would peroxidatically oxidize compounds such as nitrite, certain phenols, lower aliphatic alcohols, formaldehyde, and formic acid, leading De Duve to introduce the term peroxisome to identify these properties with microbodies. However, a peroxidatic role for peroxisomes *in vivo* has been questioned (Thurman and Chance, 1969). Their peroxidatic activity is not due to a true peroxidase but is ascribable to catalase. As Thurman and Chance have pointed out, this enzyme, while being able to transfer hydrogen from peroxidizable substrates to H_2O_2, forming $2 H_2O$ and the oxidized substrate, favors the catalatic reaction where H_2O_2 acts as both the acceptor and the donor of hydrogen atoms to form H_2O and O_2. The kinetics of catalase dictate that a peroxidizable substrate must be present in a concentration about 3×10^4 times that of H_2O_2 for the peroxidatic reaction to equal the catalatic rate. Thus the significance of the peroxidatic role for peroxisomes remains an open question. Nevertheless, H_2O_2 is an important reactant or product of reactions catalyzed in the peroxisome, and, in this sense, peroxisome is an appropriate term.

In 1967, Breidenbach and Beevers discovered that in fatty seed storage tissue, enzymes of the glyoxylate cycle were associated with a subcellular particle distinct from mitochondria (Breidenbach and Beevers, 1967; Breidenbach *et al.*, 1968). They adopted the term glyoxysome to

TABLE II

Enzyme Associated with Peroxisomes from Liver and Kidney of Various Chordates

Possible function in peroxisomes	Enzyme	Reaction	References
1. H_2O_2 elimination	Catalase	$H_2O_2 \rightarrow H_2O + \frac{1}{2}O_2$	
2. Purine catabolism	Xanthine dehydrogenase[a]	Xanthine + NADH $\rightleftharpoons$ NAD + urate	Scott et al., 1969
3. Purine catabolism	Urate oxidase	Urate + $O_2 \rightarrow$ allantoin + H_2O_2	De Duve and Baudhin, 1966
4. Purine catabolism	Allantoinase[b]	Allantoin + $H_2O \rightleftharpoons$ allantoic acid	Scott et al., 1969
5. L-α-OH acid catabolism?	L-α-Hydroxyacid oxidase	L-α-Hydroxyacid + $O_2 \rightarrow$ ketoacid + H_2O_2	De Duve and Baudhin et al., 1966
6. D-Amino acid catabolism?	D-Amino acid oxidase	D-Amino acid + $O_2 \rightarrow$ ketoacid + $NH_3 + H_2O_2$	De Duve and Baudhin, 1966
7. NADH oxidation?	NADH-glyoxylate reductase	NADH + glyoxylate $\rightleftharpoons$ NAD + glycolate	Vandor and Tolbert, 1970
8. Purine cycle?	Glutamate-glyoxylate transferase	Glutamate + glyoxylate $\rightleftharpoons$ α-ketoglutarate + glycine	Vandor and Tolbert, 1970
9. ?	NADP-isocitrate dehydrogenase	NADP + isocitrate $\rightleftharpoons$ NADPH + α-ketoglu-terate + CO_2	Leighton et al., 1968

[a] Demonstrated only in avian species.
[b] Demonstrated only in amphibians.

indicate the functional significance of the compartmentation of this phase of the gluconeogenic process in the storage tissues of fatty seeds. It was later established that these organelles bore many properties that were common to animal microbodies or peroxisomes. However, the term glyoxysome was retained to designate this as a subclass of microbodies and to emphasize their role in the compartmentation of glyoxylate cycle activity (Fig. 1).

Shortly, thereafter, Tolbert and his co-workers (1968, 1969, 1970; Kisaki and Tolbert, 1969; Tolbert and Yamazaki, 1969; Yamazaki and Tolbert, 1970) reported biochemical characterization of yet another subclass of microbodies found in the chlorophyllous tissues of higher plants. *These tissues lack the glyoxylate bypass enzymes and therefore are incapable of gluconeogenesis from fat.* The leaf microbodies exhibit flavin-linked oxidase and catalatic activities as well as other characteristic properties, and Tolbert chose to designate them as peroxisomes to emphasize their similarity to mammalian microbodies. While these similarities are indeed apparent, leaf microbodies possess properties distinguishing them from mammalian peroxisomes and defining for them a metabolic role in the glycolate pathway of photosynthetic tissues (Fig. 3).

Organelles with distinct cytological characteristics of microbodies have also been observed in a wide array of nonchlorophyllous tissues of higher plants incapable of gluconeogenesis from fats. As Tolbert (1971a) pointed out in his discussion of nomenclature, little if anything is known about the biochemical properties or metabolic functions of these organelles in many of the tissues where they have been observed, and until they have been characterized they should be referred to only as microbodies.

Huang and Beevers (1971) have shown that catalase, glycolate oxidase, and urate oxidase from a variety of those sorts of tissue sediment on sucrose gradients to the density range typical for microbodies.

C. Comments on Nomenclature

As suggested by the discussion above, all of the terms applied to these organelles have their shortcomings. Their limitations stem in part from the diversity of the group and in part from our limited knowledge of the organelles. De Duve pointed out that even the term microbody is probably too inclusive, since it has been applied in some instances to organelles lacking almost all of the general characteristics cited above. Moreover, organelles have been described in higher plants that display many cytological similarities to microbodies but have no proven biochemical relationship to microbodies as defined and may in fact not be microbodies at all.

III. Isolation of Microbodies from Plants

A few comments are appropriate on the isolation of microbodies. Microbodies were first isolated from plant tissues by pelleting them by differential centrifugation under conditions of g force and time comparable to those commonly used to sediment mitochondria, i.e., after the bulk of the nuclei, plastids, and cellular debris were sedimented at 300 g for 10 minutes, a pellet containing mostly microbodies and mitochondria was obtained by sedimenting at 10,000 g for 10 to 15 minutes. Microbodies were then further purified by resuspending and sedimenting on sucrose gradients [30 to 60% (w/w)]. The first requirements for successful isolation are gentle disruption of the cells into a suitable isotonic medium at low ionic strength and rapid but gentle processing until isolation is achieved. Abrupt changes in tonicity or an increase in ionic strength during homogenization or resuspension will disrupt the particles, releasing constituent enzymes into a soluble state. For some purposes, breakage has been minimized, and resolution of intact microbodies has been maximized by overlayering the initial extract directly onto a gradient in either a zonal rotor or swinging-bucket rotor without pelleting and resuspending.

Another note of caution concerns the construction of gradients. Many types of gradients have been used successfully in isolating microbodies. These include continuous gradients with different slopes or combinations of slopes over different regions and discontinuous or step gradients. The latter are useful, but they can lead to artifacts and misinterpretations if they are used for preparations from tissues that have not been studied previously with continuous gradients.

More detailed information has been compiled on isolation procedures and microbody enzyme assays, for which see Tolbert (1971b), Beevers *et al.* (1974), and Beevers and Breidenbach (1974).

IV. Distribution of Microbodies

Microbodies are widely distributed in the plant kingdom (Table III). Organelles with the cytological characteristics of microbodies have been observed in gymnosperms, angiosperms, pteridophytes, and thallophytes. Biochemical and cytochemical evidence supporting these observations has been reported for all but pteridophytes.

Microbodies are observed in a wide range of tissues, e.g., roots, leaves, stems, cotyledons, endosperm, gametophytes, fruits, rhizoids, nectaries, and pollen (Mollenhauer *et al.*, 1966; Huang and Beevers, 1971; Ruis, 1971). Biochemical and cytochemical characterization, however, has been largely restricted to leaf and seed storage tissues.

Leaf peroxisomes are found in the chlorophyllous tissues of both C_3 and C_4 species (Frederick and Newcomb, 1971). In C_4 species they occur in both bundle sheath and mesophyll cells, but more abundantly in the former. They are also found in the nonchlorophyllous tissue of variegated leaves (Gruber *et al.*, 1972).

Glyoxysomes are found in the storage tissues of fatty seeds. Depending on the species, the storage tissue may consist of endosperm, cotyledon or gametophyte tissues. In seeds with carbohydrate as a major storage material (e.g., corn and barley), they have been found in the scutellum

TABLE III

Distribution of Microbodies in the Plant Kingdom[a]

Species	How characterized[b]	References
Algae		
Chlorella pyrenoidosa,	1, 2, 3	Gergis, 1971; Codd and Schmid,
C. vulgaris		1971, 1972
Polytomella caeca	3	Gerhardt, 1971
Euglena gracilis	1, 2, 3	Graves *et al.*, 1971a,b, 1972; Brody
		and White, 1973
Micrasterias fimbriata (Ralfs)	1, 2	Tourte, 1972
Klebsormidium flaccidum	1, 2	Stewart *et al.*, 1972
Fungi		
Ascomycetes		
Saccharomyces cerevisiae	1, 2	Avers and Federman, 1968; Hoff-
		mann *et al.*, 1970; Todd and
		Vigil, 1972
Neurospora crassa	3	Flavell and Woodward, 1971; Kobr
		et al., 1969
Sclerotinia sclerotiorum	1	Maxwell *et al.*, 1970, 1972
Phycomycetes		
Phycomyces blakesleeanus	1	Thornton and Thimann, 1964
Fungi imperfecti		
Botrytis cinerea	1, 2	Pitt, 1969
Bryophytes		
Anthoceros sp.	1	Mollenhauer *et al.*, 1966
Pteriodophytes		
Pteridium aquilinum	1	Mollenhauer *et al.*, 1966
Equisetum sp.	1	Mollenhauer *et al.*, 1966
Gymnosperms		
Pinus ponderosa	1, 3	Ching, 1970
Angiosperms		
Many tissues in numerous	1, 2, 3	
monocots and dicots, in-		
cluding both C_3 and C_4		
species (see text)		

[a] This list is illustrative, not comprehensive.

[b] 1, cytologically; 2, cytochemically; 3, biochemically.

(Longo and Longo, 1970a,b) and aleurone tissue (Jones, 1972). These tissues are only a small fraction of the total seed tissues of these species, but contain high proportions of lipid.

It is interesting to compare the wide distribution in plants with the more limited distribution in animals (see Hruban and Rechcigle, 1969, for review). Microbodies have been observed cytologically and characterized biochemically in birds, amphibians, fishes, and mammals, all members of the chordates—the most highly evolved group of animals. In these organisms, however, they are restricted to kidney and liver tissues. Some protozoa (e.g., *Tetrahymena pyriformes*, *Acanthamoeba* spp.) that can utilize acetate as a carbon source also possess microbodies. Evidence is lacking, however, as to their occurrence in other animal phyla.

These observations and the fact that a number of the enzymes found in animal peroxisomes may be present only in very small amounts or even absent from certain mammalian species or individuals within a species led De Duve (1969ab; De Duve and Baudhuin, 1966) to suggest that peroxisomes might be considered a "fossil organelle" that is dispensible in modern organisms. However, he also pointed out, "One cannot overlook the fact that peroxisomes have persisted throughout evolution. . . . It is difficult not to assume that selective pressure has favored their retention and therefore they perform some function wherever found." This seems particularly clear in higher plants, where, as we shall see, microbodies appear to perform a function in at least two different but highly important specialized processes.

V. Functional Roles

A. Role of Glyoxysomes in Gluconeogenesis from Fats in Higher Plants

Most tissues of higher plants cannot utilize acetate, or long-chain fatty acids catabolized to acetate, as a source of carbon for gluconeogenesis. These tissues have no pathway for transforming acetate into precursors of carbohydrates. In the storage tissues of fatty seeds, however, as in certain microorganisms that can use acetate as a sole source of carbon for growth, a pathway is present that permits conversion of acetate into gluconeogenic four-carbon dicarboxylic acids (Kornberg and Krebs, 1957; Kornberg and Beevers, 1957). This ability stems from reactions catabolized by two enzymes, isocitrate lyase (EC 4.1.3.1) and malate synthase (EC 4.1.3.2), referred to as the *glyoxylate bypass* enzymes. These enzymes, together with certain enzymes found in all tissues as part of the tricarboxylic acid (TCA) cycle, make up the *glyoxylate cycle* (Fig. 1 and Table IV).

As can be seen, both the TCA cycle and the glyoxylate cycle use the same reactions to produce isocitrate from acetyl-CoA and oxalacetate but beyond that point they diverge. The TCA cycle thereafter leads

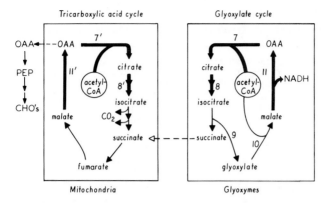

Fig. 1. Association of glyoxylate cycle and tricarboxylic acid cycle enzymes with glyoxysomes and mitochondria in fat-storing seed tissues. Heavy lines indicate reactions common to both organelles and both cycles. Numbers refer to enzymes listed in Table IV. OAA, oxalacetate; PEP, phosphoenol pyruvate; CHO's, carbohydrates.

to the formation of succinate and two CO_2 molecules via successive decarboxylations, whereas the glyoxylate cycle cleaves isocitrate to form one 4-carbon dicarboxylic acid molecule (i.e., succinate) and the 2-carbon fragment glyoxylate. Glyoxylate can then condense with a second acetyl-CoA molecule to form a second 4-carbon dicarboxylic acid molecule (i.e., malate). In this fashion the four entering carbons of two acetyl-CoA molecules are conserved to form one 4-carbon dicarboxylic acid.

Both of the glyoxylate bypass enzymes, isocitrate lyase and malate synthase, as well as all of the essential enzymes of the TCA cycle are localized in the glyoxysomes in fat-storing seed tissues (Breidenbach *et al.*, 1968; Cooper and Beevers, 1969a). These include enzymes also present in the mitochondria and functioning in the TCA cycle. They include condensing enzyme (EC 4.1.3.7) and aconitase (EC 4.2.1.3), catalyzing the reactions leading from acetyl-CoA to isocitrate, and malate dehydrogenase (EC 1.1.1.37) which converts malate to oxalacetate, the initial acceptor molecule for acetyl-CoA. Succinate cannot be metabolized by glyoxysomes and must be exported to the mitochondria to be oxidized to malate and oxaloacetate, with the concomitant conservation of energy. Oxalacetate formed from succinate is presumably the direct precursor of triose in a decarboxylative reaction catalyzed by phosphoenolpyruvate carboxylase (PEP carboxylase). This reaction essentially completes the process of gluconeogenesis.

Further evidence of the role of glyoxysomes in compartmentalizing the gluconeogenic process in fat-storing seed tissues emerges from the

TABLE IV

Enzymes Associated with Glyoxysomes Functioning in Gluconeogenesis from Fats

Function in glyoxysomes	Enzyme number in Fig. 1	Reaction	References
Hydrolysis of triglycerides	1. pH 9 lipase	Triglyceride + H_2O → glycerol + free fatty acid	Muto and Beevers, 1974
Fatty acid oxidation	2. Acyl-CoA synthetase	Fatty acid + ATP + CoA → Acyl-CoA + AMP + P $\sim$ P	Cooper and Beevers, 1969b; Cooper, 1971
	3. Acyl-CoA dehydrogenase	Acyl-CoA + acceptor → 2,3-dehydroacyl-CoA + reduced acceptor	Cooper and Beevers, 1969b; Hutton and Stumpf, 1969
	4. Enoyl-CoA hydratase	2,3-Transenoyl-CoA → H_2O → 3-hydroxyacyl CoA	Hutton and Stumpf, 1969
	5. 3-Hydroxyacyl-CoA dehydrogenase	3-Hydroxyacyl CoA + NAD → 3-ketoacyl CoA + NADH	Hutton and Stumpf, 1969
	6. 3-Ketoacyl-CoA thiolase	3-Ketoacyl-CoA + CoA → acyl-CoA + acetyl-CoA	Hutton and Stumpf, 1969
Glyoxylate cycle	7. Citrate synthase	Acetyl-CoA + oxalacetate → citrate + CoA	Breidenbach et al., 1968
	8. Aconitate hydratase	Citrate → cis-aconitate + H_2O cis-aconate + H_2O → isocitrate	Cooper and Beevers, 1969a
Glyoxylate bypass	9. Isocitrate lyase	Isocitrate → succinate + glyoxylate	Breidenbach et al., 1968
	10. Malate synthase	Glyoxylate + acetyl-CoA → malate + CoA	Breidenbach et al., 1968
Glyoxylate cycle	11. Malate dehydrogenase	Malate + NAD → oxalacetate + NADH	Breidenbach et al., 1968
H_2O_2 elimination	12. Catalase	$2 H_2O_2$ → $2 H_2O + O_2$	Breidenbach et al., 1968

findings that castor bean glyoxysomes possess a lipase Muto and Beevers, 1974) and the complete β-oxidation sequence leading to acetyl-CoA (Cooper and Beevers, 1969b; Hutton and Stumpf, 1969; Cooper, 1971) (Fig. 2 and Table IV).

In addition to the enzymes listed in Table IV that are directly implicated in gluconeogenesis, there are a number of enzymes reported to be associated with glyoxysomes but having no clearly perceived relationship to the gluconeogenic process itself. Those enzymes are listed in Table V.

Glycolate oxidase might be postulated to operate in concert with

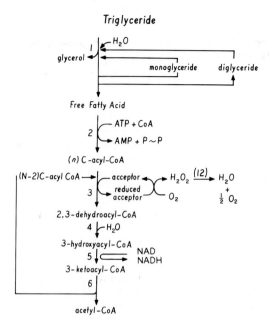

Fig. 2. Reactions leading from triglycerides to acetyl-CoA found to be housed in glyoxysomes in fat-storing seed tissues. Numbers refer to enzymes listed in Table IV.

NADH-glyoxylate reductase (Table V) to oxidize NADH formed by oxidation of malate and 3-hydroxyacyl-CoA within the glyoxysome. The level of activity of the reductase is very low, however, and it is actually more active toward hydroxypyruvate than glyoxylate, casting some doubt on the probability of a role for these two enzymes as a terminal oxidase (Lord and Beevers, 1972).

It could be postulated also that glycolate oxidase activity is just one catalytic activity in a spectrum of activities toward a various α-hydroxyacids, some of which might be products of β-oxidation of certain fatty acids. The activity of this enzyme from castor beans, such as the well-characterized enzyme from leaves (Frigerio and Harbury, 1958), is restricted to oxidation of glycolate, L-lactate, and DL-α-hydroxybutyrate (R. Lee and R. W. Breidenbach, unpublished), making this role unlikely also.

Allantoinase and urate oxidase, enzymes catalyzing two of the steps in the pathway catabolizing purines to glyoxylate and urea, are both associated with glyoxysomes (St. Angelo and Ory, 1970; Theimer and Beevers, 1971), suggesting a role for glyoxysomes in compartmenting this pathway. Attempts have failed to detect allantoicase, the enzyme cata-

TABLE V

Enzymes Reported in Glyoxysomes That Have No Perceivable Direct Role in Gluconeogenesis from Fats

Possible function in glyoxysomes	Enzyme	Reaction	References
?	Glycolate oxidase	Glycolate + $O_2 \rightarrow$ glyoxylate + H_2O_2	Breidenbach et al., 1968
?	Glyoxylate reductase (hydroxypyruvate reductase)	Glyoxylate or hydroxypyruvate + NADH $\leftrightarrow$ glycolate or D-glycerate	Lord and Beevers, 1972
Purine catabolism	Urate oxidase	Urate + $O_2 \rightarrow$ allantoin + H_2O_2	Theimer and Beevers, 1971
	Allantoinase	Allantoin + $H_2O \leftrightarrow$ allantoic acid	St. Angelo and Ory, 1970; Theimer and Beevers, 1971
?	Glutamate-oxalacetate transaminase	Glutamate + oxalacetate $\leftrightarrow \alpha$-ketoglutarate + aspartate	Cooper and Beevers, 1969a
Aromatic amino acid catabolism	L-Histidine ammonia lyase	L-Histidine $\rightarrow$ urocanate + NH_3	Ruis and Kindl, 1970
	L-Phenylalanine ammonia lyase	L-Phenylalanine $\rightarrow$ trans-cinnamate acid + NH_3	Ruis and Kindl, 1971
	L-Tyrosine ammonia lyase	L-Tyrosine $\rightarrow$ p-coumaric acid + NH_3	Ruis and Kindl, 1970
		p-Coumaric $\rightarrow$ p-hydroxybenzoic acid	Kindl and Ruis, 1971
Amino acid catabolism	D-Amino acid oxidase	D-Amino acid $\rightarrow$ keto acid + NH_3	Beevers, 1971
	L-Amino acid oxidase	L-Amino acid $\rightarrow$ keto acid + NH_3	Beevers, 1971

lyzing the final hydrolysis to glyoxylate and urea, in the particles. Together with the failure to find evidence for xanthine oxidase in glyoxysomal preparations, the absence of allantoicase (Theimer and Beevers, 1971) confounds any interpretation of the role of glyoxysomes in purine catabolism.

Enzymes involved in the metabolism of aromatic amino acids have also been reported as constituents of castor bean glyoxysomes (Ruis and Kindl, 1970; Kindl and Ruis, 1971). These enzymes showed their highest specific activity in the glyoxysomal fraction, but by far the largest total amount of activity of these enzymes was found in the soluble fraction, leaving the role of glyoxysomes in aromatic amino acid metabolism unestablished.

B. Role of Leaf Peroxisomes

Microbodies are abundant in the leaf tissues of higher plants. They have received much attention in the last six years, stimulated by interest in their involvement in the metabolism of glycolate and related compounds implicated in the process of photorespiration. Glycolate has long been recognized as an early and important product of photosynthesis. Under conditions of high pO_2 and low pCO_2, net photosynthesis is inhibited (Warburg, 1920), and a significant portion of the photosynthetically fixed CO_2 flows through pathways having glycolate as an intermediate. These organelles were linked to photorespiration by findings of Tolbert et al. (1968) that some of the enzymes of the glycolate pathway (Fig. 3 and Table VI) are associated with leaf peroxisomes.

As currently viewed, phosphoglycolate is the product of an oxidative reaction catalyzed by ribulose-1,5-diphosphate carboxylase (EC 4.1.39). This oxidative reaction is an alternative to the carboxylative reaction (Bassham and Kirk, 1962; Whittingham et al., 1967) that condenses CO_2 with ribulose 1,5-diphosphate and cleaves the transient intermediate into two 3-phosphoglycerate molecules. In the alternative reaction, O_2 competes with CO_2 at the binding site. If O_2 is bound instead of CO_2, the ribulose 1,5-diphosphate is cleaved to phosphoglycolate and 3-phospho-glycerate (PGA) instead of two 3-PGA (Bowes and Ogren, 1971; Ogren and Bowes, 1971; Andrews et al., 1973; Lorimer et al., 1973; Laing et al., 1974). As visualized by Tolbert (Randall et al., 1971), the phosphate is hydrolyzed from phosphoglycolate by a specific phosphatase (phosphoglycolate phosphatase, EC 3.1.3.18) in a process that transports gly-

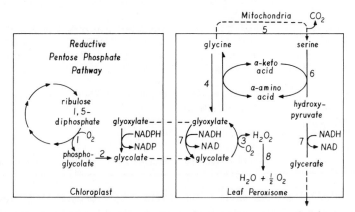

Fig. 3. Reactions involved in glycolate metabolism in photorespiring tissues of higher plants. Numbers refer to enzymes listed in Table VI.

TABLE VI

Enzymes Associated with Leaf Peroxisomes

Function in glycolate metabolism in leaves	No. in Fig. 3	Enzyme	Reaction	References
Formation of phosphoglycolate	1	Ribulose-diphosphate oxygenase	Ribulose 1,5-diphosphate $+ O_2 \rightarrow$ phosphoglycolate $+$ 3-phosphoglycerate	Bowes and Ogren, 1972; Andrews et al., 1973; Lorimer et al., 1973
Formation of glycolate?	2	Phosphoglycolate phosphatase	Phosphoglycolate $+$ $H_2O \leftrightarrow$ glycolate $+$ Pi	Randall et al., 1971
Glycolate oxidation photorespiration?	3	Glycolate oxidase	See Table IV	Tolbert et al., 1968, 1969
Formation of glycine for "1C metabolism", conversion of triose, or biosynthetic needs	4	Glutamate-glyoxylate transaminase	Glutamate $+$ glyoxylate $\leftrightarrow \alpha$-ketoglutarate $+$ glycine	Kisaki and Tolbert, 1969; Yamazaki and Tolbert, 1970
Formation of serine for "1C metabolism" conversion to triose or for biosynthetic needs	5	Serine hydroxymethyltransferase	2 glycine $\rightarrow$ serine $+$ $CO_2 + NH_3$	Kisaki and Tolbert, 1969
Triose formation	6	Serine-α-keto acid transaminase	Serine $+$ glyoxylate $\leftrightarrow$ glycine $+$ hydroxypyruvate	Yamazaki and Tolbert, 1970
Triose formation oxidation of NADH	7	NADH-glyoxylate reductase (hydroxypyruvate reductase)	Glyoxylate $+$ NADH $\leftrightarrow$ glycolate $+$ NAD $\rightarrow$ or hydroxypyruvate $+$ NADH $\leftrightarrow$ D-glycerate $+$ NAD	Tolbert et al., 1968, 1969, 1970; Yamazaki and Tolbert, 1970
Oxidation of NADPH excess reducing power	8	NADPH-glyoxylate reductase	Glyoxylate $+$ NADPH $\leftrightarrow$ glycolate $+$ NADPH	Tolbert et al., 1970
H_2O_2 elimination	8	Catalase	See Table IV	Tolbert et al., 1968
?	—	L-Malate:NAD oxidoreductase	L-Malate $+$ NAD $\leftrightarrow$ oxalacetate $+$ NADH	Yamazaki and Tolbert, 1969

colate outside the chloroplast outer envelope. Glycolate is then presumed to be transported to the leaf peroxisomes, where it is oxidized by glycolate oxidase (EC 1.1.31) to glyoxylate and then transaminated to glycine by glutamate-glyoxylate aminotransferase (EC 2.6.14). The next reaction in the glycolate pathway cannot be demonstrated in peroxisomes but does occur in mitochondria. In this reaction, two glycine molecules react to form L-serine and CO_2 (serine hydroxymethyl transferase, EC 2.1.2.1). L-Serine can then participate in another transamination reaction to yield hydroxypyruvate (serine–pyruvate aminotransferase), and hydroxypyruvate can be reduced to D-glycerate (glyoxylate reductase, EC1.1.1.26). Both of the latter reactions are catalyzed by enzymes found in leaf peroxisomes.

These reactions can conserve three of every four carbons diverted to glycolate by forming triose, or they can serve as major synthetic pathways leading to glycine and serine. Serine may be of particular interest as a point of entry into one-carbon metabolism.

The relative contributions of the competitative O_2 inhibition of CO_2 fixation and the photorespiratory loss of CO_2 are still under investigation (Ludwig and Canvin, 1971; Laing et al., 1974), but there is little question that leaf peroxisomes have a role in the latter.

As with glyoxysomes from fat-storing seed tissues, there are enzymes that have no apparent role in the function of leaf peroxisomes as it is now understood. Furthermore, microbodies that are indistinguishable in their enzymatic properties from peroxisomes from chlorophyllous tissues are present in comparable numbers in achlorophyllous (nonphotosynthetic) leaf tissues (Gruber et al., 1972). This is in contrast to the restriction of glyoxysomes and the glyoxylate bypass enzymes to fat-storing tissues.

On the other hand, leaf peroxisomes are not as abundant in mesophyll cells of C_4 species as they are in the bundle sheath cells (Tolbert et al., 1969; Frederick and Newcomb, 1971; Huang and Beevers, 1972). This is consistent with the lower photorespiratory rates of these plants and strengthens the argument for a functional relationship between leaf peroxisomes, photosynthesis, and photorespiration.

The enzymes found in leaf peroxisome preparations that have no readily understood role in the compartmented metabolic role proposed for the organelle are urate oxidase and NADP-isocitrate dehydrogenase. These enzymes are common to microbodies from many sources, and in no case is there a clear understanding of their function as a constituent of these organelles.

C. Role of Microbodies in Nonphotosynthetic Tissues

Microbodies have been observed cytologically in a wide variety of tissues other than fat-storing seed tissues and green leaves or cotyledons (see Table III). As pointed out earlier, microbodies with the same enzyme complement as leaf peroxisomes are found in nongreen regions of leaves, where they may perform a function similar to that of the peroxisomes even though the microbodies are not directly associated with photosynthesis.

Huang and Beevers (1971) have characterized the microbodies isolated from a variety of tissues, such as potato tubers, flower petals, and roots. These microbodies exhibit catalase activity, urate oxidase activity, and glycolate oxidase activity, but lack glyoxylate reductase and transaminase activities typical of leaf peroxisomes. Enzymes of the glyoxylate cycle are also absent. Huang and Beevers characterized these as "nonspecialized microbodies," more similar to the peroxisomes of rat liver. At least at this time, it is not possible to satisfactorily assign a function to these microbodies. Perhaps, as De Duve has suggested for animal peroxisomes, microbodies from these tissues are derived by progressive loss of enzymes from an ancestral organelle and no longer have a functional role. Alternatively, however, it may be that they, like leaf peroxisomes and glyoxysomes, have evolved a specialized function that we cannot yet understand. Perhaps we can hope that the interrelationships between microbodies as well as their particular functions in various tissues will yield to further study.

D. Role of Microbodies in Anapleurotic Reactions in Microorganisms

As pointed out previously, the glyoxylate bypass enzymes are found in a variety of microorganisms (including algae, fungi, bacteria, and protozoa) when they are grown on acetate as a carbon source. Microbodies have been observed cytologically in many of them, i.e., yeast (Avers and Federman, 1968), *Chlorella* (Gergis, 1971), *Microthamnion* (Watson and Arnott, 1973), *Polytomella caeca* and *Chlorogonum elongatum* (Berger and Gerhardt, 1971), *Euglena* (Graves *et al.*, 1971a), *Sclerotinia sclerotium* (Maxwell *et al.*, 1970, 1972), and *Klebsormidium* (Stewart *et al.*, 1972). A positive diaminobenzidine reaction (a cytochemical test for catalase) has been obtained for some species (Todd and Vigil, 1972; Stewart *et al.*, 1972; Gerhardt and Berger, 1971), further substantiating the identity of these structures as microbodies. The obvious question as to their biochemical constitution and functional role in the carbon metabolism of these organisms has also been asked by a number of workers (Perlman and Mahler, 1970; Kobr *et al.*, 1969; Begin-Heick, 1973; Muller *et al.*, 1968; Brody and White, 1972, 1973; Codd and Schmid, 1972; Graves *et al.*, 1971b, 1972). Results at present are ambiguous. Some of the enzymes of the glyoxylate cycle are reported to be particle-bound, while others are not. A complete complement of the enzymes necessary for an operative glyoxylate cycle has not been found in microbodies from any of the microorganisms studied. The reason, at least in part, may be the greater difficulty in obtaining intact organelles from these organisms, or it may reflect differences in regulatory mechanisms between unicellular and multicellular organisms, since microorganisms are adaptive to nutritional source and rapidly dividing, while the nondividing storage tissues of higher plants undergo programed developmental changes.

VI. Biogenesis of Microbodies

What is the genesis of microbodies? This question can have several answers, differing with the particular biological system investigated.

In populations of dividing cells, the characteristic number of microbodies must be generated in each of the derivatives of a cell division. A number of ultrastructural studies and some biochemical studies have been made of microbodies in microorganisms, such as algae and fungi (see Avers, 1971, for review), but very little understanding of microbody biogenesis has resulted. Brody and White (1972, 1973) recently investigated the environmental regulation of microbody enzymes of dark-grown, greening, and light-grown *Euglena gracilis*, but nothing has been done with synchronously dividing cultures of microorganisms. Neither has there been any intensive study of microbody biogenesis and ontogeny in meristematic tissues of higher plants.

Understanding of microbody biogenesis is somewhat better for nondividing cell populations that exhibit dramatic increases in various microbody enzymes during tissue development and differentiation. One example of such system is fat-storing seed tissues that mobilize their stored reserves for export to the seedling axis. In some cases (e.g., castor bean endosperm), these tissues degenerate when their reserves are depleted. In other cases (e.g., squash cotyledons), the tissues persist after their reserves are depleted, and they become green and leaflike or nearly leaflike. Another example of this type of system is the development and greening of true leaves. All three types of nondividing cell populations have been studied extensively (Gerhardt and Beevers, 1970; Longo and Longo, 1970a,b; Gruber *et al.*, 1970, 1973; Trelease *et al.*, 1971; Vigil, 1970; Feirabend and Beevers, 1972a,b; Kagawa *et al.*, 1973; Kagawa and Beevers, 1975; Gruber *et al.*, 1973).

Organelle profiles are difficult to discern in unimbibed seed tissues, but indistinct profiles that could be glyoxysomes can be observed in electron micrographs of unimbibed squash cotyledons (Lott, 1968). Profiles, unquestionably those of microbodies, can be seen very early during germination of sunflower, squash, and castor bean.

Microbody enzyme activities are all very low, if detectable at all, in unimbibed endosperm of castor bean (Gerhardt and Beevers, 1970), peanut cotyledons, and maize scutella (Longo and Longo, 1970a,b). The glyoxylate bypass enzymes are at or below the limits of detection, even after a single day of germination.

Protein sedimenting to the buoyant density characteristic of microbodies is observed at the earliest stages of germination in castor bean endosperm (Gerhardt and Beevers, 1970) and maize scutellum (Longo and Longo, 1970a,b). Whether this protein is actually microbody protein itself, however, is uncertain. In contrast, very little protein is present in the microbody region of gradients from peanut cotyledons during the early stages of germination (Breidenbach et al., 1966). In all of these tissues, the amount of microbody protein as well as the *specific activity* of the microbody enzymes increases to a maximum at later stages of germination. It is known also that the increased levels of isocitrate lyase and malate synthase are due to *de novo* synthesis of these two proteins (Longo, 1968). In tissues, such as castor bean endosperm and corn scutellum, which degenerate when the stored reserves are depleted, enzyme activities and microbody protein both decline together.

Cotyledons that will persist and green follow a similar pattern when germinated in the dark (Gruber et al., 1970; Trelease et al., 1971; Kagawa et al., 1973; Kagawa and Beevers, 1975). As the reserves are depleted, however, low levels of leaf peroxisome enzymes, i.e., glycolate oxidase and hydroxypyruvate reductase, slowly increase. If the seedlings are exposed to light at any point during germination, the decline of glyoxylate bypass enzymes is hastened, and tremendous increases in leaf peroxisome enzymes begin (Kagawa et al., 1973; Kagawa and Beevers, 1975).

The development of leaf peroxisomes in the young expanding leaves of wheat seedlings (Fierabend and Beevers, 1972a,b) and bean seedlings (Gruber et al., 1973) has been studied under various light regimes. In contrast to the cotyledons described above, these tissues are never capable of gluconeogenesis from fats and never exhibit glyoxylate bypass enzyme activities or contain glyoxysomes. Detectable levels of glycolate oxidase, glyoxylate (or hydroxypyruvate) reductase, and catalase can be demonstrated in these expanding leaves after 2 to 3 days of germination. If the seedlings are kept in the dark, the level of these leaf peroxisome enzymes increases slowly. Just as in greening cotyledons, whenever the leaves are exposed to light, glycolate oxidase and the hydroxypyruvate reductase activities begin to increase at a tremendously greater rate.

Cytological studies of bean leaves showed that microbody profiles are already present at 3 days, the earliest stages examined (Gruber *et al.*, 1973). During expansion and greening, however, their average size appears to increase. Unfortunately, none of the data reported from those studies establish whether the number of profiles increases during this period. The relationship to other organelles and membrane systems does alter during leaf cell development, as discussed in more detail below.

The behavior of catalase in both persistent cotyledons and developing leaves merits special consideration. During the germination of fat-storing cotyledons that persist and green, catalase reaches its maximum activity coincident with the maxima for the glyoxylate bypass enzymes, and then immediately begins to decline. The bypass enzymes disappear completely, however, whereas catalase declines more slowly and stabilizes at a level that is apparently characteristic of the level of catalase in green tissue. In developing leaves, catalase developmental patterns also do not coincide with those of the other leaf peroxisome enzymes. In both wheat and bean leaves, the activity of catalase increases only slightly faster in the light than in the dark, while the rate of increase of other peroxisomal enzymes is strongly influenced by light.

Assuming that enzyme activities accurately reflect amounts of enzyme protein, any model for microbody biogenesis must account for differences in the relative amounts of the various microbody enzymes at different stages of tissue development.

Changes in the enzyme composition of the microbody enzymes for the population could be explained by addition of new protein to a generally stable population of microbodies, or, alternatively, by constant turnover of the population. No definitive evidence indicating turnover of microbody proteins has been provided for plants. Results were inconclusive from attempts to study catalase turnover with isotopically labeled δ-aminolevulenic acid (Radin and Breidenbach, 1971).

Turnover of rat liver peroxisome protein has been studied by De Duve and co-workers (Leighton *et al.*, 1969; Poole *et al.*, 1969; Poole, 1969). Depending upon how turnover was determined, the range of half-life values obtained was 1.5 to 3.5 days. Those workers discuss the evidence for and against several models for microbody biogenesis in rat liver. However, simple extrapolation of their turnover results to plant systems is of questionable value at best.

Some evidence is available to aid our understanding of the biogenesis of the membrane of microbodies in plants. Beevers and co-workers (Lord *et al.*, 1972, 1973; Kagawa *et al.*, 1973; Moore, 1974) studied membrane synthesis in germinating castor bean endosperm. By pulse labeling with [^{14}C]choline, they showed that labeled lecithin appeared first in the endoplasmic reticulum and then in other membrane systems. The kinetics of labeling are consistent with the idea that lecithin is synthesized and incorporated into the endoplasmic reticulum and then transferred to glyoxysomes and mitochondria. The intracellular localization of the enzymes of the biosynthetic pathway for lecithin, phosphorylcholine-glyceride transferase (EC 2.7.8.2), was clearly associated with the endoplasmic reticulum.

Cytological studies in plants have frequently pointed to the close association of plant microbodies with the endoplasmic reticulum (Mollenhaur et al., 1966; Frederick et al., 1968; Vigil, 1970). Pointed out in some instances have been what appear to be connections between the endoplasmic reticulum (ER) and the microbody membrane (Frederick et al., 1968; Vigil, 1970). The endoplasmic reticulum in juxtapostion with the microbody membrane is smooth, but transition to rough ER usually occurs near the association. Similar observations have been made for animal systems. From developmental studies on bean leaves, Gruber et al. (1973) pointed out that the associations between microbodies and swollen regions of the cisternae of the smooth endoplasmic reticulum are much more prominent in nonvacuolate cells of young leaves than in the developed vacuolate cells of expanded green leaves. Thus, both the biochemical and the cytological evidence is consistent with the idea that the membrane of microbodies may be derived from the endoplasmic reticulum. However, it does not exclude simple transport of membrane lipids alone. Evidence further supporting the idea that microbody membranes originate as endoplasmic reticulum is found in the similarities between the membrane-associated enzyme activities of the endoplasmic reticulum and microbodies reported by Donaldson et al. (1972).

Evidence is meager on the mechanism for protein inclusion in the organelle. The synthesis of protein on the polysomes association with rough endoplasmic reticulum and its encapsulation by the smooth membranes presents an attractive hypothesis, but, as pointed out earlier, it is not known whether there is a continuous addition or exchange of material with existing microbodies or, on the other hand, whether new microbodies with different constituent compositions continually replace older microbodies.

Another question related to biogenesis that has been explored is whether microbodies might possess their own DNA and have some degree of autonomy. Ching (1970) reported in vitro protein synthesis by isolated glyoxysomes from Pinus cotyledons. Also, a small amount of RNA has been found to be tenaciously associated with glyoxysomes from castor bean endosperm (Gerhardt and Beevers, 1969). Attempts have failed, however, to demonstrate *in vitro* protein synthesis in preparations of castor bean glyoxysomes (H. Beevers, personal communication).

A circular DNA species with a buoyant density identical to that of nuclear DNA has been reported in yeast by Clarke-Walker (1972, 1973). This DNA species appears to be associated with a particulate fraction that persists in respiratory-deficient petite mutants (e.g., mutants lacking competent mitochondria as such, as well as the buoyant density DNA species characteristic of yeast mitochondria). Clarke-Walker suggested that this DNA species might be of peroxisomal origin. However, no cytological or biochemical evidence has been presented to identify the particles in question as microbodies. Structures with characteristic mitochondrial enzyme markers can be isolated from neutral petites and would contribute to the particulate preparation.

Myers and Cantino (1971) clearly identified a DNA species associated with organelles, termed α particles, found in the motile zoospores of the water mold *Blastocladiella emersonii*. These organelles, although much smaller than most peroxisomes or glyoxysomes, have some morphological similarities. No biochemical evidence yet resolves their relationship, if any, to microbodies.

On the other hand, Douglass et al. (1973) were unable to identify any unique DNA species associated with glyoxysomes from castor bean endosperm. Purified mitochondria and proplastids clearly showed enrichment of species with distinct, buoyant densities. The DNA found in glyoxysome preparations, on the other hand, showed variable proportions of DNA with buoyant densities characteristic of nuclear, proplastid, and mitochondrial DNA and no others. The component with nuclear DNA buoyant density hybridized with RNA synthesized *in vitro* on authentic nuclear DNA as well as did the template DNA itself.

From the foregoing it is clear that much remains unknown about the mode of origin of microbodies.

GENERAL REFERENCE

Hogg, J. F., ed. (1969). "The Nature and Function of Peroxisomes (Microbodies, Glyoxysomes)," Ann. N.Y. Acad. Sci., Vol. 168, N.Y. Acad. Sci., New York.

REFERENCES

Andrews, T. J., Lorimer, G. H., and Tolbert, N. E. (1973). *Biochemistry* 12, 11.
Avers, C. J. (1971). *Sub-Cell. Biochem.* 1, 25
Avers, C. J., and Federman, M. (1968). *J. Cell Biol.* 37, 555.
Bassham, J. A., and Kirk, M. (1962). *Biochem. Biophys. Res. Commun.* 9, 376.

Beevers, H. (1971). *Proc. Int. Congr. Biochem., 8th.*

Beevers, H., and Breidenbach, R. W. (1974). *In* "Methods in Enzymology" (S. Fleischer and L. Packer), Vol. 31, Part A, p. 565. Academic Press, New York.

Beevers, H., Theimer, R. R., and Gerhardt, R. (1974). *In* "Biochemische Cytologie der Pflanzenzelle" (G. Jacobi, ed.), p. 127. Thieme, Stutgart.

Begin-Heick, N. (1973). *Biochem. J.* **134**, 607.

Berger, C., and Gerhardt, B. (1971). *Planta* **96**, 326.

Bowes, G., and Ogren, W. L. (1972). *J. Biol. Chem.* **247**, 2171.

Breidenbach, R. W., and Beevers, H. (1967). *Biochem. Biophys. Res. Commun.* **27**, 462.

Breidenbach, R. W., Castelfranco, P. A., and Peterson, C. A. (1966). *Plant Physiol.* **41**, 803.

Breidenbach, R. W., Kahn, A., and Beevers, H. (1968). *Plant Physiol.* **43**, 705.

Brody, M., and White, J. E. (1972). *FEBS (Fed. Eur. Biochem. Soc.) Lett.* **23**, 149.

Brody, M., and White, J. E. (1973). *Develop. Biol.* **31**, 348.

Ching, T. M. (1970). *Plant Physiol.* **46**, 475.

Clarke-Walker, G. D. (1972). *Prec. Nat. Acad. Sci. U.S.* **69**, 388.

Clarke-Walker, G. D. (1973). *Eur. J. Biochem.* **32**, 263.

Codd, G. A., and Schmid, G. H. (1971). *Planta* **99**, 230.

Codd, G. A., and Schmid, G. H. (1972). *Arch. Mikrobiol.* **81**, 264.

Cooper, T. G. (1971). *J. Biol. Chem.* **246**, 3451.

Cooper, T. G., and Beevers, H. (1969a). *J. Biol. Chem.* **244**, 3507.

Cooper, T. G., and Beevers, H. (1969b). *J. Biol. Chem.* **244**, 3514.

De Duve, C. (1969a). *Proc. Roy. Soc., Ser. B* **173**, 71.

De Duve, C. (1969b). *Ann. N.Y. Acad. Sci.* **168**, 369.

De Duve, C., and Baudhuin, P. (1966). *Physiol. Rev.* **46**, 323.

Donaldson, R. P., Tolbert, N. E., and Schnarrenberger, C. (1972). *Arch. Biochem. Biophys.* **152**, 199.

Douglass, S. A., Criddle, R. S., and Breidenbach, R. W. (1973). *Plant Physiol.* **51**, 902.

Feierabend, J., and Beevers, H. (1972a). *Plant Physiol.* **49**, 28.

Feierabend, J., and Beevers, H. (1972b). *Plant Physiol.* **49**, 33.

Flavell, R. B., and Woodward, D. O. (1971). *J. Bacteriol.* **105**, 200.

Frederick, S. E., and Newcomb, E. H. (1971). *Planta* **96**, 152.

Frederick, S. E., Newcomb, E. H., Vigil, E. L., and Wergin, W. P. (1968). *Planta* **81**, 229.

Frigerio, N. A., and Harbury, H. A. (1958). *J. Biol. Chem.* **231**, 135.

Gänsler, H., and Rouiller, C. (1956). *Schweiz. Z. Allg. Pathol. Bacteriol.* **19**, 217.

Gergis, M. S. (1971). *Planta* **101**, 180.

Gerhardt, B. (1971). *Arch. Mikrobiol.* **80**, 205.

Gerhardt, B., and Beevers, H. (1969). *Plant Physiol.* **44**, 1475.

Gerhardt, B., and Beevers, H. (1970). *J. Cell Biol.* **44**, 94.

Gerhardt, B., and Berger, C. (1971). *Planta* **100**, 155.

Graves, L. B., Hanzely, L., and Trelease, R. N. (1971a). *Protoplasma* **72**, 141.

Graves, L. B., Trelease, R. N., and Becker, W. M. (1971b). *Biochem. Biophys. Res. Commun.* **44**, 280.

Graves, L. B., Trelease, R. N., Grell, A., and Becker, W. M. (1972). *J. Protozool.* **19**, 527.

Gruber, P. J., Trelease, R. N., Becker, W. M., and Newcomb, E. H. (1970). *Planta* **93**, 269.

Gruber, P. J., Becker, W. M., and Newcomb, E. H. (1972). *Planta* **105**, 114.

Gruber, P. J., Becker, W. M., and Newcomb, E. H. (1973). *J. Cell. Biol.* **56**, 500.

Hoffmann, H.-P., Szabo, A. S., and Avers, C. J. (1970). *J. Bacteriol.* **104**, 581.

Hruban, Z., and Rechcigle, M. (1969). *Int. Rev. Cytol., Suppl.* **1**, 1–296.

Huang, A. H. C., and Beevers, H. (1971). *Plant Physiol.* **48**, 637.

Huang, A. H. C., and Beevers, H. (1972). *Plant Physiol.* **50**, 242.

Hutton, D., and Stumpf, P. K. (1969). *Plant Physiol.* **44**, 508.

Jones, R. L. (1972). *Planta* **103**, 95.

Kagawa, T., and Beevers, H. (1975). *Plant Physiol.* **55**, 258.

Kagawa, T., McGregor, D. I., and Beevers, H. (1973). *Plant Physiol.* **51**, 66.

Kindl, H., and Ruis, H. (1971). *Phytochemistry* **10**, 2633.

Kisaki, T., and Tolbert, N. E. (1969). *Plant Physiol.* **44**, 242.

Kobr, M. J., Vanderhaeghe, F., and Combepine, G. (1969). *Biochem. Biophys. Res. Commun.* **37**, 640.

Kornberg, H. L., and Beevers, H. (1957). *Biochim. Biophys. Acta* **26**, 531.

Kornberg, H. L., and Krebs, H. A. (1957). *Nature (London)* **179**, 988.

Laing, W. A., Ogren, W. L., and Hageman, R. H. (1974). *Plant Physiol.* **54**, 678.

Leighton, F., Poole, B., Beaufay, H., Baudhuin, P., Coffey, W., Fowler, S., and De Duve, C. (1968). *J. Cell Biol.* **37**, 482.

Leighton, F., Poole, B., Lazarow, P. B., and De Duve, C. (1969). *J. Cell Biol.* **41**, 521.

Longo, C. P. (1968). *Plant Physiol.* **43**, 660.

Longo, C. P., and Longo, G. P. (1970a). *Plant Physiol.* **45**, 249.

Longo, C. P., and Longo, G. P. (1970b). *Plant Physiol.* **46**, 599.

Lord, J. M., and Beevers, H. (1972). *Plant Physiol.* **49**, 249.

Lord, J. M., Kagawa, T., and Beevers, H. (1972). *Proc. Nat. Acad. Sci. U.S.* **69**, 2429.

Lord, J. M., Kagawa, T., Moore, T. S., and Beevers, H. (1973). *J. Cell Biol.* **57**, 659.

Lorimer, G. H., Andrews, T. J., and Tolbert, N. E. (1973). *Biochemistry* **12**, 18.

Lott, J. (1968). Ph.D. Thesis, University of California, Davis.

Ludwig, L. J., and Canvin, D. T. (1971). *Plant Physiol.* **48**, 712.

Maxwell, D. P., Williams, P. H., and Maxwell, M. D. (1970). *Can. J. Bot.* **48**, 1689.

Maxwell, D. P., Williams, P. H., and Maxwell, M. D. (1972). *Can. J. Bot.* **50**, 1743.

Myers, R. B., and Cantino, E. C. (1971). *Arch. Mikrobiol.* **78**, 252.

Mollenhauer, H. H., Morré, D. J., and Kelly, A. G. (1966). *Protoplasma* **62**, 44

Moore, T. S. (1974). *Plant Physiol.* **54**, 164.

Muller, M., Hogg, J. F., and De Duve, C. (1968). *J. Biol. Chem.* **243**, 5385.

Muto, S., and Beevers, H. (1974). *Plant Physiol.* **54**, 23.

Ogren, W. L., and Bowes, G. (1971). *Nature (London), New Biol.* **230**, 159.

Perlman, P. S., and Mahler, H. R. (1970). *Arch. Biochem. Biophys.* **136**, 245.

Pitt, D. (1969). *J. Histochem. Cytochem.* **17**, 613.

Poole, B. (1969). *Ann. N.Y. Acad. Sci.* **168**, 229.

Poole, B., Leighton, F., and De Duve, C. (1969). *J. Cell Biol.* **41**, 535.

Radin, J. W., and Breidenbach, R. W. (1971). *Plant Physiol.* **47**, Suppl., 168.

Randall, D. D., Tolbert, N. E., and Gremel, D. (1971). *Plant Physiol.* **48**, 480.

Rhodin, J. (1954). "Aktiebolaget Godvil." Monograph. Karolinska Institute, Stockholm.

Ruis, H. (1971). *Hoppe-Seyler's Z. Physiol. Chem.* **352**, 1105.

St. Angelo, A. J., and Ory, R. L. (1970). *Biochem. Biophys. Res. Commun.* **40**, 290.

Scott, P. J., Visentin, L. P., and Allen, J. M. (1969). *Ann. N.Y. Acad. Sci.* **168**, 244.

Stewart, K. D., Floyd, G. L., Mattox, K. R., and Davis, M. E. (1972). *J. Cell Biol* **54**, 431.

Theimer, R. R., and Beevers, H. (1971). *Plant Physiol.* **47**, 246.

Thornton, R. M., and Thimann, K. V. (1964). *J. Cell. Biol.* **20**, 345.

Thurman, R. G., and Chance, B. (1969). *Ann. N.Y. Acad. Sci.* **168**, 348.

Todd, M. M., and Vigil, E. L. (1972). *J. Histochem. Cytochem.* **20**, 344.

Tolbert, N. E. (1971a). *Ann. Rev. Plant Physiol.* **22**, 45.

Tolbert, N. E. (1971b). *In* "Methods in Enzymology" (A. San Pietro, ed.), Vol. 23, Part A, p. 665. Academic Press, New York.

Tolbert, N. E., and Yamazaki, R. K. (1969). *Ann. N.Y. Acad. Sci.* **168**, 325.

Tolbert, N. E., Oeser, A., Kisakin, T., Hageman, R. H., and Yamazaki, R. K. (1968). *J. Biol. Chem.* **243**, 517.

Tolbert, N. E., Oeser, H., Yamazaki, R. K., Hageman, R. H., and Kisaki, T. (1969). *Plant Physiol.* **44**, 135.

Tolbert, N. E., Yamazaki, R. K., and Oeser, A. (1970). *J. Biol. Chem.* **245**, 5129.

Tourte, M. (1972). *Planta* **105**, 20.

Trelease, R. N., Becker, W. M., Gruber, P. J., and Newcomb, E. H. (1971). *Plant Physiol.* **48**, 461.

Vandor, S. L., and Tolbert, N. E. (1970). *Biochim. Biophys. Acta* **215**, 449.

Vigil, E. L. (1970). *J. Cell Biol.* **46**, 435.

Vigil, E. L. (1973). *Sub-Cell. Biochem.* **2**, 237.

Warburg, O. (1920). *Biochem. Z.* **103**, 188.

Watson, M. W., and Arnott, H. J. (1973). *J. Phycol.* **9**, 15.

Whittingham, C. P., Coombs, J., and Marker, A. H. F. (1967). *In* "The Biochemistry of Chloroplasts" (T. W. Goodwin, ed.), Vol. 2, p. 155. Academic Press, New York.

Yamazaki, R. K., and Tolbert, N. E. (1969). *Biochem. Biophys. Acta* **178**, 4.

Yamazaki, R. K., and Tolbert, N. E. (1970). *J. Biol. Chem.* **245**, 5137.

6

The Chloroplast

RODERIC B. PARK

I. Introduction

A major difference between cells of green plants and those of animals is the inclusion in most plant cells of a unique set of organelles called plastids. Plastids may assume a variety of morphological forms and functional roles. For example, in the potato tuber, we find the starch-containing amyloplast and in red tomatoes the lycopene-containing chromoplast. These plastids and others are structures arising from a common proplastid precursor.

All the plastids are involved in pathways of metabolism peculiar to plants, but the chloroplast is of particular interest for two reasons. First, there is now excellent evidence that the entire photosynthetic process is embodied within this organelle. Second, evidence accumulated for the past 70 years shows that chloroplasts possess a remarkable degree of genetic and metabolic autonomy. In this sense chloroplasts appear to be an organism within an organism and are now widely regarded as endosymbionts in green plants. Thus the chloroplast is both the primary

115

transducer for the biosphere by which electromagnetic energy is converted to chemical energy, and is also of great interest as a semiautonomous biological entity within itself.

II. Physiological Studies of the Photosynthetic Process

By the early nineteenth century it was established that green plant photosynthesis involved carbon dioxide fixation into sugar with evolution of oxygen. The experiments of Ingenhousz (1779) had shown that photosynthesis required illumination of the chlorophyll-containing portions of the plant. The following quantitative work of de Saussure (1804) and others enabled chemists to write the expression for green plant photosynthesis according to the familiar photosynthetic equation:

$$6\ CO_2 + 6\ H_2O \xrightarrow[\text{chlorophyll}]{h\nu} C_6H_{12}O_6 + 6\ O_2 \tag{1}$$

Plant physiologists realized that this expression was a gross oversimplification of the photosynthetic mechanism. Undoubtedly many different reactions contributed to the overall stoichiometry of photosynthesis. The problem, then, was what experiment to do with the green plant in order to demonstrate the component reactions that yield the overall process summarized in Eq. (1). Blackman (1905) and Blackman and Matthaei (1905) approached this problem by varying some of the parameters in Eq. (1) and observing the effect on the overall photosynthetic rate. They varied carbon dioxide concentration, light intensity, and temperature and observed the photosynthetic rate as a function of these variables. These findings are summarized in Fig. 1. Under conditions of excess light and rate-limiting concentrations of carbon dioxide, the photosynthetic rate was temperature dependent. This experiment indicated that the carbon dioxide fixation or dark reactions of photosynthesis were normal, temperature-dependent reactions. Under excess CO_2 and limited light the photosynthetic rate was temperature independent. This indicated that the light-mediated reactions of photosynthesis were temperature-independent or photochemical reactions. That photochemical reactions are temperature-independent is a common photographic observation, namely, that the camera aperture and speed is dependent only on light intensity and is independent of temperature. Warburg (1925) correctly interpreted the results of Blackman's experiments as signifying that, to a first approximation, photosynthesis consisted of two broad classes of reactions: light reactions and Blackman,

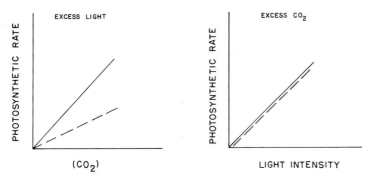

Fig. 1. Photosynthetic rate of whole plants as a function of carbon dioxide concentration and light intensity at high temperature (solid line) and low temperature (dashed line). From data of Blackman (1905) and Blackman and Matthaei (1905).

or as they are now known, dark reactions. This physiological concept of two sets of reactions in photosynthesis is still valid today and, as will be shown later, has a morphological basis within the chloroplast. Twenty-five years after Blackman's work, Emerson and Arnold (1932a) showed, using *Chlorella* cells, that the light and dark reactions of photosynthesis could be separated in time. This experiment consisted of exposing the cells to a brief flash of light (3 mseconds) followed by varying lengths of dark period. The results of this experiment are presented in Fig. 2. These experiments showed that, for the light to be used efficiently, the

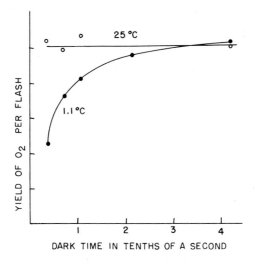

Fig. 2. Yield of O_2 per flash of light (3 mseconds) as a function of the length of dark period following the flash. From the data of Emerson and Arnold (1932a).

dark period had to be many times longer than the light period. Emerson interpreted this experiment as follows. During the light flash certain photochemical reactions took place which supplied energy for the later photosynthetic reactions; during the subsequent dark period, CO_2 was fixed and the activated photochemical system returned to its initial state. If the dark period was too short, the photochemical system was not completely regenerated, and the absorption of light by the photosynthetic system was consequently inefficient. It is now known that the light and dark reactions of photosynthesis as well as being physiologically separable in time are also spatially separated from one another in the chloroplast.

III. Localization of Light and Dark Reactions within the Chloroplast; Association with Specific Chloroplast Structures

A. Historical Evidence

The physiological experiments discussed in Section II say nothing about the localization of photosynthesis within the green plant cell. The experiments leading to the localization of both light and dark reactions of photosynthesis in the chloroplast span a period of 71 years from 1894 to 1965. A brief review of chloroplast structure is useful before discussing these experiments. The chloroplast in higher plant cells, as viewed by the light microscope, is usually a saucer-shaped body about 4–10 μm in diameter and 1–2 μm thick (see Fig. 3). The green color of the chloroplast is due to the abundance of chlorophyll within the chloroplast. When the chloroplast is seen from above in planar view by light microscopy, the green background is interrupted by a number of dark bodies about 0.4 μm in diameter, termed grana (see Fig. 3). The grana areas of the chloroplast apparently contain so much chlorophyll that they are totally absorbing and therefore appear black to the eye. The chloroplast is not a static organelle within the cell but can change shape as well as move. Such changes in shape were recorded with the light microscope over 80 years ago and are summarized in a review by Heitz (1936). Heitz noted that chloroplasts in many plants formed "pseudopodia" that could extend into the cell cytoplasm and might eventually become separated from the chloroplast. His drawing of this phenomenon is reproduced in Fig. 4. Wildman and co-workers (1962) have continued these studies using time lapse photography and phase optics to record these shape changes. The significance of the release and uptake of materials by the chloroplast is still a matter of speculation.

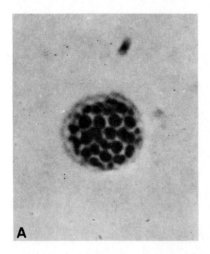
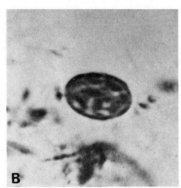

Fig. 3. Top view (A) and side view (B) of a chloroplast as seen by light microscopy. The dark regions within the chloroplasts are grana. From data of Heitz (1936).

Although it was first observed by Ingenhousz (1779) that only the green parts of plants are capable of photosynthesis, evidence of the direct involvement of the chloroplast comes from the four experiments discussed below.

Engelmann (1894) constructed a microscope condenser that enabled him to irradiate portions of photosynthetic cells with a small beam of light. With an appropriate assay for photosynthesis, he could illuminate different parts of cells and then tell which portion of the cell acted as the light receptor for photosynthesis. The assay he used consisted of bacteria that were motile only under aerobic conditions and moved toward areas of increasing oxygen concentration. These bacteria were mixed with the photosynthetic cells of the alga *Spirogyra,* and Engelmann observed that only when the chloroplast was illuminated did oxygen evolution occur. The original figure illustrating this experiment is reproduced in Fig. 5. Englemann also observed from the migration of the bacteria toward the chloroplast that the site of oxygen evolution was the chloroplast itself. This experiment showed that photosynthetic light absorption and oxygen evolution both occurred within the chloroplast. These results then indicate that at least the light reactions of photosynthesis leading to O_2 evolution occur within the chloroplast.

Further support for localization of the light reaction within the chloroplast came from the experiments of Hill (1937). Hill found that isolated chloroplasts, when illuminated, performed oxygen evolution if

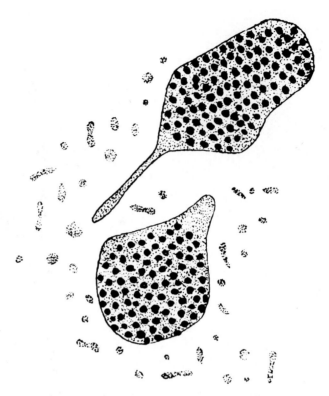

Fig. 4. Drawing of pseudopodia formation by chloroplasts. From Heitz (1936).

a suitable electron acceptor was present. The electron acceptor used was a ferric salt that was reduced to the ferrous form during the reaction. This reaction can be formulated as shown in Eq. (2).

$$\left.\begin{array}{c} ? \\ \\ O_2 \end{array}\right) \oplus \boxed{\begin{array}{c} \text{Light} \\ \downarrow \\ \underset{\longrightarrow}{e^-} \\ \hline \text{Chloroplast} \end{array}} \ominus \left(\begin{array}{c} Fe^{3+} \\ \\ Fe^{2+} \end{array}\right. \tag{2}$$

Some electron donor was oxidized to yield oxygen gas with the concomitant reduction of Fe^{3+}. The identity of the electron donor was established by Ruben et al. (1941), who showed with ^{18}O labeling experiments that oxygen evolved in photosynthesis has the same ^{18}O to ^{16}O ratio as the oxygen of the water in which the cells are suspended. Thus, the Hill reaction can be formulated as shown in Eq. (3).

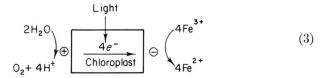

$$(3)$$

It is seen that the Hill reaction differs from photosynthesis in that iron rather than CO_2 is the terminal electron acceptor. Hill was unsuccessful in his attempts to use CO_2 as the Hill oxidant. That the enzyme machinery for carbon dioxide fixation also exists within the chloroplast was not shown until almost thirty years later, by Arnon *et al.* (1954) with the aid of $^{14}CO_2$, and a greater array of biochemical cofactors than was available to Hill. Though Arnon's group qualitatively demonstrated the occurrence of photosynthetic CO_2 fixation by isolated chloroplasts, the rates were low seldom exceeding 2–5 μmoles CO_2/mg chlorophyll/hour. Intact spinach leaves, on the other hand, possess photosynthetic rates as high as 245 μmoles CO_2/mg chlorophyll/hour.

Microscopic observation of chloroplasts isolated by the Arnon procedure shows that their limiting membranes are generally broken,

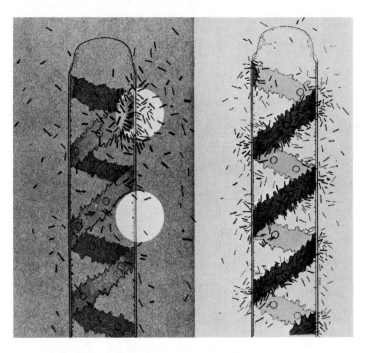

Fig. 5. Localization of photosynthetic O_2 production in *Spirogyra*. Reproduced from Engelmann (1894).

whereas milder grinding procedures leave the limiting membrane intact. These two types of chloroplasts are readily distinguished in the light microscope using phase optics (see Fig. 6). Intact chloroplasts are highly refractile, appearing very bright, whereas the broken chloroplasts appear dark. Using the terminology of Spencer and Unt (1965), the intact chloroplasts are called class I and the broken chloroplasts are called class II. Improved isolation procedures allowed Jensen and Bassham (1966) to prepare class I chloroplasts that in proper incubation medium retained CO_2 fixation rates as high as 155 μmoles CO_2/mg chlorophyll/ hour, a rate similar to intact plants. Thus the Arnon experiments show qualitatively and the Jensen and Bassham experiments show quantitatively that the entire photosynthetic apparatus is associated with

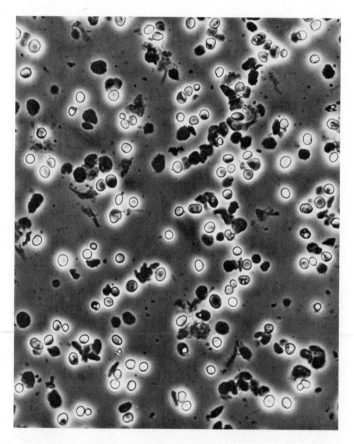

Fig. 6. Isolated spinach chloroplasts. Light colored chloroplasts have intact outer membranes and are called class I. Darker chloroplasts have broken outer membranes and are called class II.

isolated chloroplasts. These experiments are the basis for the now generally accepted belief that both the dark and light reactions of photosynthesis occur within the chloroplasts.

It is interesting to note that not all higher green plants are suitable starting material for chloroplast isolation. Plant biochemists in search of complex enzymatic systems, as a rule use a domesticated plant that is edible! These plants have been selected for thousands of years for the exclusion of oxalic acid, tannins, alkaloids, and other noxious tasting substances which are also toxic to enzymes. For example, spinach and lettuce, which are eaten raw, are excellent sources of chloroplasts, whereas tobacco, which has not been selected for its edible qualities, contains considerable quantities of phenolic compounds that cause difficulties during chloroplast isolation.

B. Structure as Related to Function in Chloroplasts

In recent years, the light and dark reactions of photosynthesis have been associated with specific chloroplast structures. Before considering this association, a brief discussion of chloroplast ultrastructure as revealed by the electron microscope is necessary. Figures 7 and 8 present typical thin sections of a higher plant (*Spinacia oleracea*) mesophyll chloroplast and a green alga (*Chlorella pyrenoidosa*) chloroplast. These oxygen-evolving photosynthetic systems are characterized by the presence of a lamellar system surrounded by an embedding matrix. The stacked membranes in spinach are called grana lamellae or small thylakoids, and the single membrane sacs are called stroma lamellae or large thylakoids. *Chlorella* and *Spinacia* have organized plastids that are separated from the cell cytoplasm by two membranes. The matrix surrounding the lamellae is referred to as the stroma portion of the chloroplast. Additionally, the chloroplast of *Chlorella* is seen to contain pyrenoids. The function of these bodies is unknown. Since starch grains are often present in the region surrounding the pyrenoid, it has been postulated that the pyrenoid may be involved in starch formation.

Experiments with the higher plant chloroplasts have shown that the light reactions and associated electron transport reactions of photosynthesis are located in the lamellae, and the CO_2 fixation or dark reactions of photosynthesis are located in the stroma region of the chloroplast (Park and Pon, 1961). An experiment that illustrates this finding is reported in Table I and Fig. 9. Isolated spinach chloroplasts were sonically ruptured and centrifuged at high speed. Under these conditions the chlorophyll-containing lamellae sediment, while the colorless stroma material (about 50% of the total chloroplast protein) remains suspended.

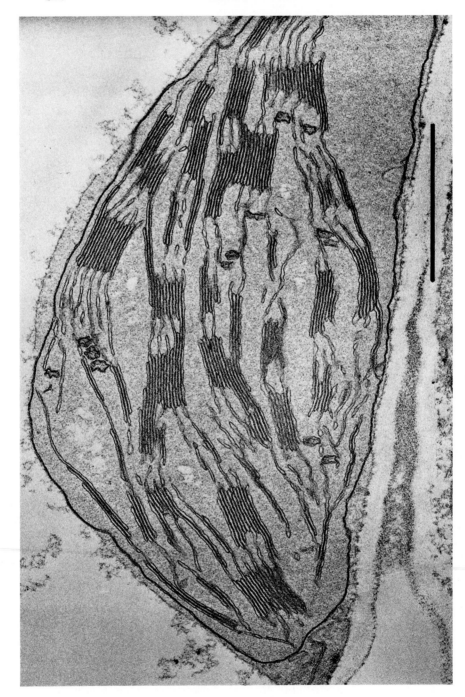

Fig. 7. Electron micrograph of a cross section through the spongy mesophyll of a mature spinach leaf. Scale equals 1 μm.

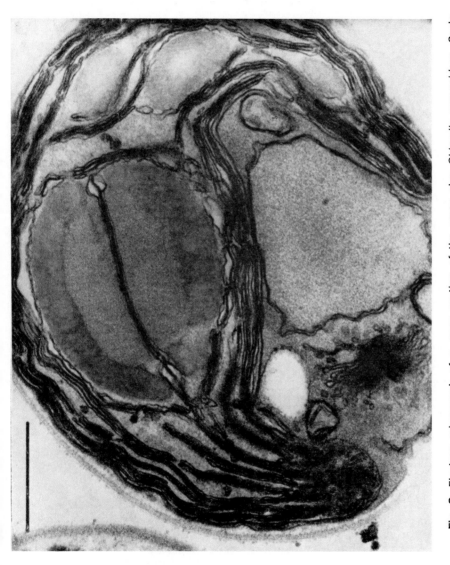

Fig. 8. Electron micrograph of a cross section of the green alga *Chlorella pyrenoidosa*. Scale equals 0.5 μm.

TABLE I

Fixation of $^{14}CO_2$ by Isolated Chloroplasts and Chloroplast Fractions[a]

	cpm of $^{14}CO_2$ fixed during 30 minutes	
Fraction	Light	Dark
1. Total chloroplast sonicate	2,100,000	40,000
2. Lamellae precipitated from equal volume of sonicate	18,000	4,200
3. Colorless supernatant proteins left after precipitation of lamellae in	30,000	53,000
4. Lamellae plus supernatant	3,000,000	22,000

[a] From Park and Pon (1961).

Fig. 9. A comparison of the products of $^{14}CO_2$ fixation by chloroplast lamellae (A), chloroplast stroma (B), and a mixture of the two (C). From the data of Park and Pon (1961).

The separated lamellae perform the Hill reaction and photosynthetic phosphorylation, while the stroma supernatant contains more than 95% of the CO_2 fixation enzyme of photosynthesis ribulosediphosphate carboxylase. The relative CO_2 fixation capacities of these two fractions by themselves and after recombination are shown in Table I. Two-dimensional radioautographs of the CO_2 fixation pattern of illuminated stroma alone, illuminated lamellae alone, and the illuminated recombined system are shown in Fig. 9. The chromatograms show that $^{14}CO_2$ fixation by the stroma material alone produces only trace amounts of phosphoglyceric acid, the first stable carbon product in photosynthesis. When the lamellar fraction alone is incubated with $^{14}CO_2$, a small amount of phosphoglyceric acid and its reduction products occur. However, when the two fractions are mixed together, a sixtyfold increase in CO_2 fixation takes place. The lamellae provide the reducing agent and ATP necessary to drive the photosynthetic carbon cycle, which is located in the stroma. The assignment of the light reactions and associated electron transport reactions to the chlorophyll-containing lamellar system and the assignment of the dark or carbon cycle reactions of photosynthesis to the embedding matrix are presumed to hold for all eukaryotic photosynthetic systems. Since the CO_2 fixation and carbon cycle reactions of the stroma enzymes are considered in detail in Chapter 24, they will not be considered further here.

The internal membrane system of the chloroplast carries out light-dependent oxidation of water and reduction of the soluble cofactor ferredoxin. The transport of electrons from water to ferredoxin involves two light reactions, one leading to oxidation of water and concomitant reduction of an intermediate chain of electron carriers. This light reaction is called photosystem II (PS II). A second light reaction, photosystem I (PS I), leads to oxidation of the intermediate chain of electron carriers and concomitant reduction of ferredoxin. A simplified diagram of these reactions is presented in Fig. 10. This formulation of electron transport has been called the Z scheme. As Arnon and co-workers (1954) have shown, phosphorylation accompanies flow of electrons through the intermediate electron transport chain and may be of either the cyclic or noncyclic type (see Chapter 25).

Now that we have established that the light reactions and electron transport reactions of photosynthesis reside within the internal membrane system of the chloroplast, further questions remain. One of the most obvious concerns the uniformity of function and chemical composition throughout the membrane system. The cross-sectional picture shown in Fig. 7 shows that the internal membranes are made up of regions in which the membranes are unappressed in contact with the stroma and

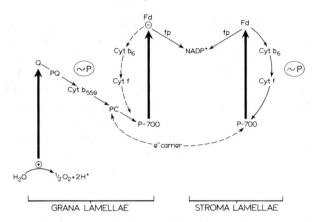

Fig. 10. A schematic presentation of photosynthetic electron transport in grana and stroma lamellae from Park and Sane (1971). Fd, ferredoxin; PC, plastocyanin; fp, flavoprotein; Cyt, cytochrome; PQ, plastoquinone; Q, quencher. The ATP ($\sim$P) and NADPH produced are utilized in CO_2 reduction by stroma enzymes.

of regions in which they are appressed to form grana stacks. The unappressed membranes are called stroma lamellae, and the appressed membranes grana lamellae. Serial sections of chloroplasts have been used to reconstruct the lamellar system, thus giving a three-dimensional view of the relation of stroma lamellae to grana lamellae. Such a reconstruction by Wehrmeyer (1964) is shown in Fig. 11. Subsequent studies by Paolillo (1970) have shown that several stroma lamellae can be connected to one small thylakoid. There is increasing evidence that the stroma lamellae carry out only PS I and cyclic phosphorylation, while the grana carry out both photosystems and cyclic as well as noncyclic photophosphorylation. This evidence is best appreciated by following the

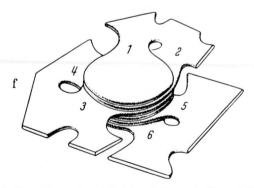

Fig. 11. A three-dimensional interpretation of the relationship of grana lamellae to stroma lamellae from serial sections by Wehrmeyer (1964).

course of an experiment performed by Sane *et al.* in 1970. First, chloroplasts are isolated in a high salt medium, which tends to destroy the limiting membrane yielding class II chloroplasts that release the soluble stroma material while maintaining the lamellar structure. The lamellae after isolation and washing in this medium appear as shown in Fig. 12. It is seen that the grana stacks persist and are interconnected by stroma lamellae. The membrane preparation is then passed through a needle valve (French) press at low pressure, and the grana stacks are sheared away from the interconnecting stroma lamella as shown in Fig. 13. By fractional centrifugation, it is then possible to separate the larger grana stacks from the smaller stroma lamellae, and their properties can be compared (Tables II and III). It is seen that stroma lamellae account for about 15% of the total membrane material, possess only PS I, have a high chlorophyll a to chlorophyll b ratio, high P-700 content, and are depleted in cytochrome b_{559}. Grana lamellae, on the other hand, possess both photoreactions and a complete Z scheme, and less P-700. Stroma lamellae also are depleted in chlorophyll compared with grana lamellae by about 40% on either a total protein or total lipid basis. It is somewhat misleading to compare the two columns in Table III directly, since a large portion of the stroma lamellar protein is made up of the components at 56 and 52kD (see Fig. 16) which are subunits of the coupling factor

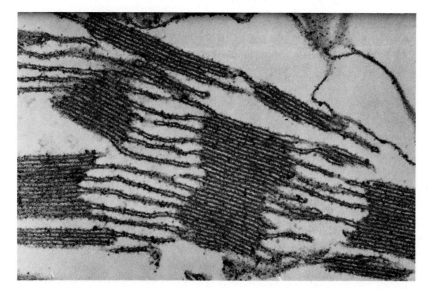

Fig. 12. Class II chloroplasts isolated in high salt medium prior to treatment with a needle valve press.

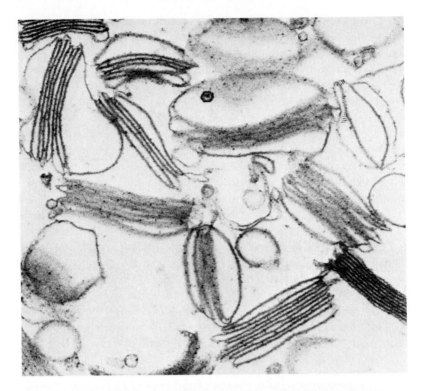

Fig. 13. Class II chloroplasts after passage through a needle valve press.

(CF$_1$) for photosynthetic phosphorylation and ribulosediphosphate carboxylase.

That the membrane material in the high chlorophyll a to chlorophyll b fraction actually arises from stroma lamellae can be shown by a comparison of the freeze fracture images of the grana and stroma regions of a class I chloroplast in Fig. 14. In the freeze fracturing process, chloroplasts in their aqueous suspending medium are rapidly frozen and the frozen sample is then fractured under a vacuum. The exposed fractured surface is then replicated with a mixture of platinum and carbon. The replica is recovered, cleaned and observed in the electron microscope. A large number of experiments have shown that the fracture plane associated with membranes actually occurs in a plane within the membrane. Thus the particle covered faces in Fig. 14 are actually internal fracture planes within the membrane. Membrane surfaces can also be observed by freezing chloroplasts in a dilute medium and subliming ice from the specimen following fracturing. The type of structure seen following such an experiment and its interpretation is given in Fig. 15.

TABLE II

Photochemical Activities of Fractions Separated by Differential Centrifugation of a French Press Homogenate[a, b]

Fraction	DCIP reduction (PS 2)	NADP+ reduction from ascorbate (PS 1)
Original chloroplasts	174	172
French press homogenate	52	87
1K	100	62
10K	74	75
40K	—	87
160K	0	169

[a] From Sane et al. (1971).
[b] Values are expressed as micromoles reduced per milligram chlorophyll per hour.

The particles on the exterior A′ surface of the membrane may be either the coupling factor for phosphorylation or the enzyme ribulosediphosphate carboxylase, the CO_2 fixing enzyme in most plants. Whereas grana in Fig. 14 contain the normal B and C faces, stroma lamellae are seen

TABLE III

Relative Compositions of Stroma Lamellae and Grana Lamellae[a,b]

	Stroma lamellae	Grana lamellae
Total chloropyhll	278	401
Chlorophyll a	238	281
Chlorophyll b	40	130
P-700	2.5	0.6
β-Carotene	21	17
Lutein	10	29
Violaxanthin	15	20
Neoxanthin	8	16
Phospholipid	76	66
Monogalactosyl diglyceride	231	214
Digalactosyl diglyceride	172	185
Sulfolipid	65	59
Cyt b (total)	1.0	3.4
Cyt f	0.5	0.7
Manganese	0.3	3.2

[a] From data of Allen et al. (1972) and Trosper and Allen (1972).
[b] Values in micromoles of component per gram of membrane protein.

Fig. 14. A class I chloroplast after freeze fracture. Whereas the grana contain large and small particles, the interconnecting stroma lamella contains only
small particles.

to contain only small particles. Observation of the separated fractions
gives the same result, the centrifugally separated grana show by freeze
fracture the normal B and C faces, whereas the high chlorophyll a to
chlorophyll b fraction shows the typical small particle composition of the
stroma lamellae.

Some differences in grana and stroma lamellae reside in their peptide
compositions, as revealed by SDS gel electrophoresis. In Fig. 16, scans of
the peptide compositions of grana and stroma lamellae are compared. It is
obvious that stroma lamellae are enriched in components at 56 and 52kD
in comparison with grana lamellae. These two components comprise the
major subunits of the coupling factor for photosynthetic phosphorylation
(56 and 53kD) and a major component (52kD) of the CO_2 fixation en-

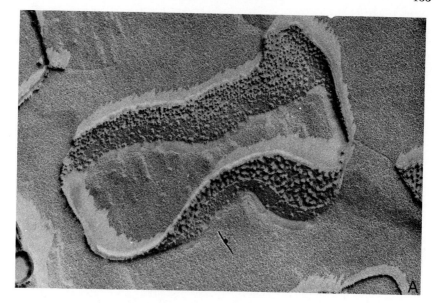

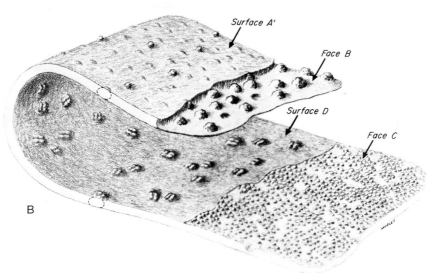

Fig. 15. The relationship of grana lamellae fracture faces and surfaces as presented by Park and Pfeifhofer (1969) and by Park and Sane (1971). (A) is a deep-etched grana lamella, and (B) is its interpretation.

zyme of photosynthesis. The differences between these two membranes are primarily quantitative and not qualitative. Thus the addition of photosystem II is not correlated with the appearance of major new pep-

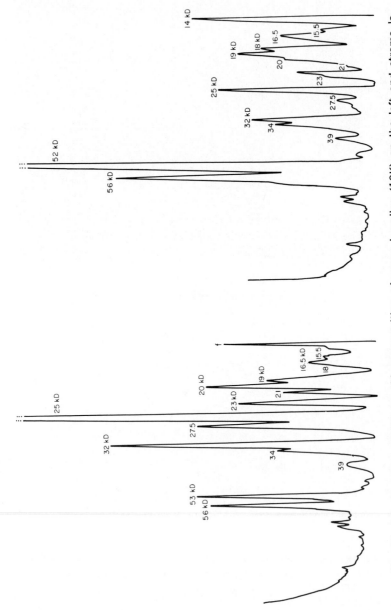

Fig. 16. A comparison of the peptide composition of grana lamellae (10K) on the left and stroma lamellae (160K) on the right. Courtesy of Dr. F. Henriques.

tides in grana lamellae, but is correlated with a major increase in the 25kD component which is believed to be associated with the light harvesting pigments for photosystems I and II. The major peptides seen in Fig. 16 play important structural roles in these membranes whereas those components known to be involved in electron transport are only a minor portion of the total peptides and are probably buried under the larger peaks. It is apparent that the small changes observed in both the lipid and peptide composition of grana and stroma lamellae account for the large functional differences between these two membranes.

Grana stacks have the interesting property of unfolding in low ionic strength medium (Izawa and Good, 1966). Both photosystems are retained in this unfolded condition, indicating that there may be conditions under which single membranes *in vivo* with the appearance of stroma lamellae have both photosystems. With this introduction, it is intriguing to look briefly at the structure of chloroplasts in the mesophyll and bundle sheath cells of sugarcane, a C_4 plant (see Chapter 24). This is one of a large group of plants with low photorespiration, low CO_2 compensation point, and very high rates of CO_2 fixation at very high light intensities. The bundle sheath cells in this instance contain almost no grana (see Fig. 17), and the question exists, do these membranes possess only PS I or are they unfolded grana possessing both PS I and PS II? The question is being actively pursued by many investigators, but the verdict is not yet in. The problems of obtaining pure active preparations of each plastid type are not trivial, and the answers to this question may not be clear until another edition of "Plant Biochemistry" is published.

IV. The Photosynthetic Unit, Its Physiological and Morphological Expression

It was mentioned earlier that Emerson and Arnold (1932a) showed that the light and dark reactions of photosynthesis could be separated in time. They further showed (Emerson and Arnold, 1932b) that the light reaction mechanism was saturated by a brief flash of light (10^{-5} second), and that up to 100 mseconds of dark period was then required to use up the products of the light reaction. When they increased the flash intensity to the point that CO_2 fixation had reached a maximum level during the dark period, they assumed that the energy conversion sites were saturated. It was then possible to divide the number of chlorophylls in the system by the number of carbon dioxide molecules fixed during the dark period to find the number of chlorophyll molecules involved in the reduction of one CO_2 molecule. This number was called the chlorophyll

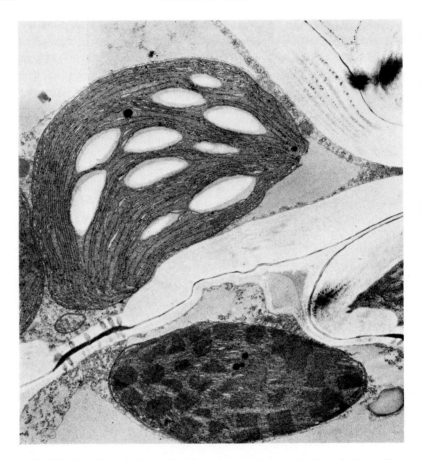

Fig. 17. Variation of chloroplast internal membrane structure in two adjacent cells (bundle sheath cell and mesophyll cell) in sugarcane leaf. Courtesy of Professor W. M. Laetsch.

unit; as determined in these experiments, it was 2500 chlorophylls per CO_2 molecule fixed. The name chlorophyll unit was subsequently changed to photosynthetic unit. The more recent determinations of this number by Kok (1956) again give a value of 2500 chlorophylls per CO_2 fixed, or per oxygen evolved. However, since the fixation of one CO_2 molecule requires approximately 10 quanta (see Chapter 25), Kok assumes that 10 flashes of light are required to yield O_2 or reduction of one CO_2. Thus, each individual unit would contain $\frac{1}{10}$ of 2500, or 250 molecules. It is the number 200–300 chlorophylls per photosynthetic unit that is generally encountered in the literature.

The photosynthetic unit then is a physiological unit of function defined by experiments with intact plants. Is there a morphological expression of the physiological photosynthetic unit? Thomas *et al.* (1953) attempted to answer this question by isolating spinach chloroplast lamellae and finding how small a fragment would still perform the Hill reaction. Their work indicated that lamellar fragments as small as 100 Å in diameter were still active in Hill reaction. These systems were, however, impure. Experiments showing the localization of the light reactions of photosynthesis and the photosynthetic pigments in chloroplast lamellae predict that a morphological expression of the photosynthetic unit should be contained in the chloroplast lamellar structure. In thin sections, such as the chloroplast lamellae shown in Figs. 7 and 8, the chloroplast lamellae appear uniform. However, shadow-cast or freeze fractured preparations of isolated chloroplast lamellae reveal a repeating structure within the grana regions of this membrane. This structure was first observed by Steinmann (1952). Work at the University of California by Park and Biggins (1964) suggests that these repeating structures may be the morphological expression of the physiological photosynthetic unit as formulated by Emerson and Arnold (1932b). For this reason, they termed these units quantasomes.

Quantasomes exist in spinach lamellae in at least three types of packing; the most organized type of packing is shown in Fig. 18. This extended array of quantasomes allows a more accurate determination of quantasome dimensions than was previously possible. These particles are 180 × 160 Å and are 100 Å thick. The crystalline packing depicted in Fig. 18 is the least common quantasome packing arrangement, but the easiest from which to get good dimensions. This is the same subunit that is seen on the D surface of the grana lamellae in Fig. 15, where the quantasomes are randomly arrayed. Since arrays of B face internal particles have the same periodicity as the D surface particles, we assume that a B face particle underlies each D surface unit and comprises part of the quantasome. Unfortunately, there are still no experiments that unambiguously demonstrate either the function or chemical composition of this membrane subunit. However, recent observations from Thornber's laboratory (Thornber 1975) indicate that the 25kD component of grana lamellae (Fig. 16) is a major component of the light harvesting complex which in turn would make it a major component of the photosynthetic unit. Recent work by Henriques in our laboratory has shown that the great increase in the 25kD component which accompanies greening in lettuce leaves is accompanied by a corresponding increase in the appearance of the large particles on the B fracture face. Thus, there is some evidence that the B face particle may be related to a morphological expression of the light gathering photosynthetic unit.

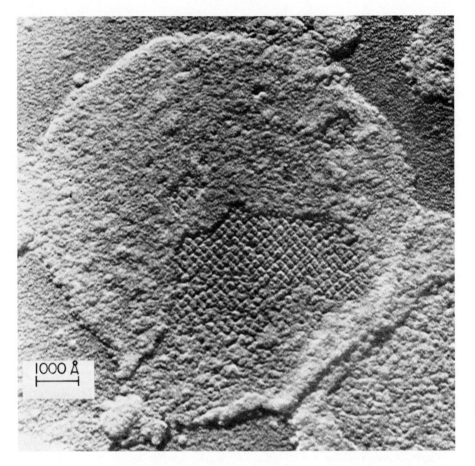

Fig. 18. Paracrystalline quantasome array in spinach chloroplast small thylakoid which is air dried and shadowed. The array corresponds to packed D surface subunits (see Fig. 15) each of which is underlain by a large B face particle and some small C face particles.

V. The General Biology of Chloroplasts

Chloroplasts are initially interesting because of their photosynthetic function. However, from a broader point of view, photosynthesis is but one of the interesting biological facets of chloroplasts. Recent work has shown that chloroplasts synthesize protein and that chloroplasts undergo shape and conformational changes during electron transport (Packer, 1963), which is related to ion transport (Schwartz, 1971). The genetic autonomy of chloroplasts as related to their nucleic acid content, methods of chloroplast reproduction, and the metabolic capacities of chloroplasts other than photosynthesis are all prospectively exciting fields in chloroplast study. A few of these topics are considered here.

A. The Genetic Autonomy of Chloroplasts

The general genetic observations concerning chloroplast development have been summarized by Sinnott and Dunn (1939). "Particularly important among these are certain traits in plants involving chloroplast development and constituting the so-called 'albomaculatus' types of leaf variegation, in which the normal green tissue is irregularly spotted with patches of paler green or white, a type of variation intensively studied by Correns and others. These may be small or may include entire leaves or branches. This character occurs in a wide variety of plants, and its inheritance has been determined in more than 20 genera. Flowers and wholly green branches produce seed which grow into normal plants; flowers on variegated branches yield offspring that have variegated foliage, and flowers from branches wholly white give progeny without chlorophyll; but, in every case the source of the pollen has no influence on the offspring. Inheritance is wholly maternal. Variegation seems to be determined by agencies localized in the cytoplasm rather than in the chromosomes. A satisfactory explanation of the mechanism of inheritance for such a trait is available, however, since variegation is evidently the result of differences in chloroplast development and since the primordia of these bodies, from which the plastids of the whole plant are ultimately derived, are present in the cytoplasm of the egg."

That chloroplast precursors, the proplastids, exist in eggs is shown in Fig. 19, which is an electron micrograph of the cotton egg. The maternal inheritance of chloroplast then is most easily explained if the chloroplast genetic material is contained not in the cell nucleus, but within the maternal cytoplasm. Another line of evidence supporting the genetic autonomy of chloroplasts comes from experiments involving streptomycin inhibition of chloroplast formation. Provasoli et al. (1951) showed that *Euglena gracilis* grown in the presence of 40 μg of streptomycin per liter was permanently bleached after several generations. These cells never regained their chloroplasts. The cell thus lost its autotrophic character and became dependent on added chemical energy sources. The nucleus of the bleached cell was unable to direct the synthesis of new chloroplasts. Again these results are explained if streptomycin interferred with chloroplast reproduction so that the new cells contained no chloroplast genome with their self-contained genetic sys-

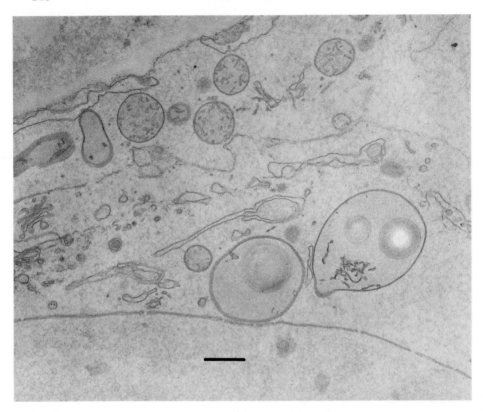

Fig. 19. Proplastids in the egg of cotton. The proplastids are just above the nuclear membrane in the lower portion of the micrograph. Calibration line equals 1 μm. Courtesy of Professor William Jensen.

tem. In this sense, the *Euglena* was cured of its chloroplast infection. Recent evidence shows that the relationship of the chloroplast to the cytoplasm is highly complex and that both nuclear and chloroplast DNA contribute to the coding of chloroplast components (Kung *et al.*, 1972).

B. Chloroplast Reproduction

Another biologically interesting aspect of chloroplasts concerns chloroplast reproduction. There appear to be two general methods for chloroplast reproduction. The first involves the division of the mature chloroplasts, and the second involves the division of a chloroplast precursor, the proplastid, from which the mature chloroplast arises. The division of mature chloroplasts is commonly seen not only in many algae, such as *Chlorella* (see Fig. 20), but in higher plants as well. Division of chloroplasts often accompanies cell division and enlargement in leaves. Plastids also arise from organelles called proplastids, which in turn undergo division. An interesting facet of chloroplast biochemistry concerns the development of chloroplasts from proplastids. The proplastid generally has few or no internal chlorophyll-containing lamellae. Under normal conditions the internal lamellae originate from an invagination of the inner of the double membranes that surround the plastid (Manton, 1962).

Chloroplast development from proplastids takes a different morphological path if chlorophyll synthesis is inhibited or mineral deficiency, such as manganese deficiency, occurs. Inhibition of chlorophyll synthesis may be caused by etiolation, mutant blocks, or streptomycin. Under these conditions, the proplastid rather than forming a lamella gives rise to a highly reticulate body within the chloroplast, referred to as a prolamellar body. Upon resumption of chlorophyll synthesis, the prolamellar body gives rise to the normal chlorophyll-containing chloroplast lamellae. This is illustrated in Fig. 21 from the data of Muhlethaler and Frey-Wyssling (1959).

C. The Endosymbiont Theory

During the past 10 years, evidence has rapidly accumulated which shows that both chloroplasts and mitochondria share many properties in common with bacteria. They contain their own DNA, which like prokaryotic DNA is circular, as well as 70 S ribosomes characteristic of bacteria. Isolated chloroplasts are also capable of protein synthesis and as mentioned earlier undergo "fission" in many cells. This has led to the theory that chloroplasts are actually prokaryotic derivatives living symbiotically within a eukaryotic cell. Presumably, such a prokaryotic organism could have originally been pinocytotically ingested by the eukaryotic cell. Thus, the outer of the two membranes around a chloroplast could be regarded as the host exclusion membrane. The inner membrane, on the other hand, can be regarded as the prokaryotic cell membrane that proliferates to form the internal membrane system.

Intriguing evidence for the endosymbiont theory comes from observations carried out in Muscatine's laboratory (Trench *et al.*, 1969). They observed that certain marine nudibranchs when feeding on siphonaceous algae ingest the plant cell cytoplasm and assimilate chloroplasts into the digestive cells of their guts presumably by a pinocytotic mechanism. These chloroplasts persist in the animal cells and subsequent experiments have shown that in some animals the chloroplasts not only

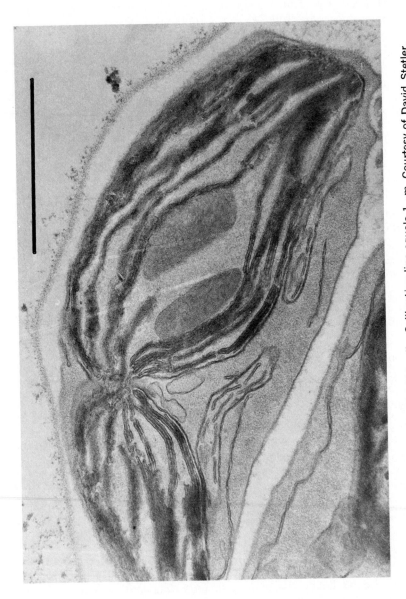

Fig. 20. Chloroplast division in *Chlorella*. Calibration line equals 1 μm. Courtesy of David Stetler.

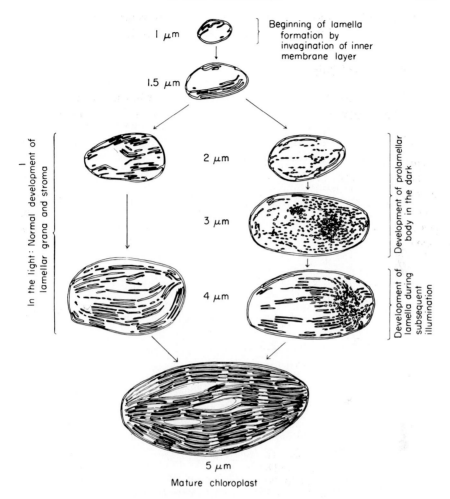

Fig. 21. Plastid development in light and dark. Adapted from data of Muhlethaler and Frey-Wyssling (1959). The development of lamellae from the prolamellar body is now thought to take place by membrane growth rather than the accretion of vesicles dispersed from the prolamellar body. It is apparent that the earlier fixative procedures may have caused the vesicle dispersion.

persist for the life of the animal but that in light the animal actually evolves oxygen! This uptake of chloroplasts has several fascinating implications. First, it may serve as a model for an evolutionary event in which a prokaryotic O_2 evolving organism formed a symbiotic relationship with a heterotrophic organism to form the first eukaryotic autotrophic plant. Second, it suggests that the interactions between chloroplast and cytoplasm may not be as complicated as previously thought. It is remarkable that biological organisms so remotely related as green algae and mollusks would form an endosymbiotic relationship.

D. The Future of Chloroplast Studies

Attention is already shifting from photosynthesis, the most spectacular property of chloroplasts, to deeper questions of chloroplast biology. How is chloroplast development controlled? Will it ever be possible to culture chloroplasts *in vitro?* To what extent can animal cells other than those of certain nudibranchs be successfully infected with chloroplasts? How is the chloroplast evolutionarily related to blue-green algae, the only prokaryotic photosynthetic group that evolves oxygen? The answers to these questions and further exploration of chloroplast photosynthetic function will challenge investigators for many years to come.

GENERAL REFERENCES

Kirk, J. T. O., and Tilney-Bassett, R. A. E. (1967). "The Plastids." Freeman, San Francisco, California.
Margulis, L. (1971). *Sci. Amer.* **225**, No. 2, 48.

REFERENCES

Allen, C. F., Good, P., Trosper, T., and Park, R. B. (1972). *Biochem. Biophys. Res. Commun.* **48**, 907.
Arnon, D. I., Allen, M. B., and Whatley, F. R. (1954). *Nature (London)* **174**, 394.
Blackman, F. F. (1905). *Ann. Bot. (London)* **19**, 281.
Blackman, F. F., and Matthaei, G. L. C. (1905). *Proc. Roy. Soc., Ser. B* **76**, 402.
de Saussure, T. (1804). "Recherches chimique sur la végétation." Nyon, Paris
Emerson, R., and Arnold, A. (1932a). *J. Gen. Physiol.* **15**, 391.
Emerson, R., and Arnold, A. (1932b). *J. Gen. Physiol.* **16**, 191.
Engelmann, T. W. (1894). *Arch. Gesamte Physiol. Menschen Tiere* **57**, 375.
Heitz, E. (1936). *Planta* **26**, 134.
Hill, R. (1937). *Nature (London)* **139**, 881.
Ingenhousz, J. (1779). "Experiments upon Vegetables, Discovering Their Great Power of Purifying the Common Air in the Sunshine, and of Injuring It in the Shade and at Night." Emsley, London.
Izawa, S., and Good, N. E. (1966). *Plant Physiol.* **41**, 544.
Jensen, R. G., and Bassham, J. A. (1966). *Proc. Nat. Acad. Sci. U.S.* **56**, 1095.
Kok, B. (1956). *Biochim. Biophys. Acta* **21**, 245.
Kung, S. D., Thornber, J. P., and Wildman, S. G. (1972). *FEBS Lett.* **24**, 185.
Lyttleton, J. W. (1962). *Exp. Cell Res.* **26**, 312.
Manton, I. (1962). *J. Exp. Bot.* **13**, 325.
Muhlethaler, K., and Frey-Wyssling, A. (1959). *J. Biophys. Biochem. Cytol.* **6**, 507.
Packer, L. (1963). *Biochim. Biophys. Acta* **75**, 12.

Paolillo, D. J. (1970). *J. Cell Sci.* **6**, 243.

Park, R. B., and Biggins, J. (1964). *Science* **144**, 1009.

Park, R. B., and Pfeifhofer, A. O. (1969). *J. Cell Sci.* **5**, 299.

Park, R. B., and Pon, N. G. (1961). *J. Mol. Biol.* **3**, 1.

Park, R. B., and Sane, P. V. (1971). *Annu. Rev. Plant Physiol.* **22**, 395.

Provasoli, L., Hutner, S. H., and Pinter, I. J. (1951). *Cold Spring Harbor Symp. Quant. Biol.* **16**, 113.

Ruben, S., Randall, M., Kamen, M. D., and Hyde, J. L. (1941). *J. Amer. Chem. Soc.* **63**, 877.

Sane, P. V., Goodchild, D. J., and Park, R. B. (1970). *Biochim. Biophys. Acta* **216**, 162.

Schwartz, M. (1971). *Annu. Rev. Plant Physiol.* **22**, 469.

Sinnott, E. W., and Dunn, L. C. (1939). "Principles of Genetics," p. 246. McGraw-Hill, New York.

Spencer, D., and Unt, H. (1965). *Aust. J. Biol. Sci.* **18**, 197.

Steinmann, E. (1952). *Exp. Cell Res.* **3**, 367.

Thomas, J. B., Blaauw, O. H., and Duysens, L. N. M. (1953). *Biochim. Biophys. Acta* **10**, 230.

Thornber, P. J. (1975). *Annu. Rev. Plant Physiol.* **26**, 127.

Trench, R. K., Greene, R. W., and Bystrom, B. G. (1969). *J. Cell Biol.* **42**, 404.

Trosper, T., and Allen, C. F. (1973). *Plant Physiol.* **51**, 584.

Warburg, O. (1925). *Biochem. Z.* **166**, 386.

Wehrmeyer, W. (1964). *Planta* **62**, 272.

Wildman, S. G., Hongladarom, T., and Honda, S. I. (1962). *Science* **138**, 434.

7

Plant Microtubules

PETER K. HEPLER

I. Introduction

Microtubules constitute a class of structurally similar organelles that occur in virtually all eukaryotic cells, both plant and animal, and they are now recognized as important cellular elements in motility and morphogenesis. Since their discovery as the prominent component of the axoneme of cilia and flagella, where they are constituted in the familiar $9 + 2$ pattern, microtubules have received special attention as important structural and motile elements in the mitotic spindle apparatus and as cytoskeletal structures, particularly in animal cells.

Their occurrence throughout plants is also widespread, and in addition to the above-mentioned associations, microtubules form the fibrous component of the cytokinetic apparatus—the phragmoplast—and they frequently reside in the cortical cytoplasm where they appear to participate in controlling cell wall pattern and the orientation of the cellulose microfibrils. Their widespread occurrence in plants, especially their disposition within the cortical cytoplasm, only became apparent following the introduction of glutaraldehyde as a superior fixative for electron microscopy. Although to date the investigation of microtubules from animal systems dominates the field, interest in their particular roles in plants has been generated in an increasing number of laboratories. In our understanding of microtubule biochemistry, however, we are still largely dependent upon results from animal systems, but it would seem reasonable that microtubule structure and molecular function is similar in both plants and animals.

This chapter will discuss microtubule composition, structure, and function as they relate to processes in higher plants. Particular emphasis will be directed to their role in cell division and in the control of cell wall organization. The topic of plant microtubules has been reviewed by Newcomb (1969) and more recently by Hepler and Palevitz (1974) and by Pickett-Heaps (1975). Microtubule structure and biochemistry has been discussed in detail in several recent reviews (Olmsted and Borisy, 1973; Roberts, 1974; Stephens, 1971; Wilson and Bryan, 1974). The role of microtubules in chromosome motion has also received considerable attention (Bajer and Molè-Bajer, 1971b, 1972; Forer, 1969; Luykx, 1970; Nicklas, 1971). Reviews on the extensive investigations of microtubules in animal systems have been presented by Bardele (1973), Slautterback (1963), Porter (1966), Roberts (1974), and Tilney (1971b).

II. Structure of Microtubules

Electron microscopy of thin sections shows that microtubules are circular in cross section, and measure about 240 Å in diameter (Ledbetter and Porter, 1963; Porter, 1966; Newcomb, 1969). They have a darkly stained outer cortex and a more lightly stained inner core. Although the core is often referred to as being "hollow," we can only say with confidence that its contents, following preparation for electron microcopy, fail to accept any of the commonly used electron stains. Microtubules often have a halo around the outside which is only slightly stained (Ledbetter and Porter, 1963). Longitudinal sections reveal that microtubules are long structures of undetermined length, sometimes as long as a few microns, and that they may be straight or curved, but not in sharply kinked configurations. They give the impression of having rigidity. High magnification of the cross section of the microtubule reveals that its cortex is composed of 13 protofilament rows in which the individual subunits are about 40 to 50 Å in diameter (Ledbetter and Porter, 1964; Warner and Satir, 1973). Associated with the outer surface of the microtubule are fine filaments about 20–50 Å wide and from 50–400 Å long referred to as arms or cross bridges, which may serve to link tubules to adjacent elements or to nearby membranous components (Bannister and Tatchell, 1968; Burgess, 1970a; Fuge and Müller, 1972; Grimstone and Cleveland, 1965; Hepler and Jackson, 1968; Hepler et al., 1970; Krishan and Buck, 1965; McIntosh, 1973, 1974; McIntosh et al., 1973; McIntosh and Porter,

1967; Mooseker and Tilney, 1973; Robison, 1966; Roth *et al.*, 1970; Tilney and Byers, 1969; Tucker, 1968, 1970; Wilson, 1969). Earlier work on the fine structure of the flagella outer doublet microtubules showed that the A subfiber always contained a pair of arms projecting from its surface. Biochemical studies by Gibbons (1963) showed upon isolation and characterization that these arms were a 14 S protein, dynein, possessing ATPase activity. A similar high molecular weight protein possessing ATPase activity has been isolated from nonflagellar sources, including neurotubules (Burns and Pollard, 1974; Gaskin *et al.*, 1974) and the axostyle of *Saccinobacculus* (Mooseker and Tilney, 1973).

Microtubule arms and cross bridges are especially evident in highly organized systems of microtubules. Examples are the pharyngeal basket of *Nassula* (Tucker, 1968, 1970), the axoneme of the protozoan *Saccinobacculus* (Grimstone and Cleveland, 1965), and the axopodia of the *Echinosphaerium* (Roth *et al.*, 1970; Tilney and Byers, 1969). It is interesting to note, in *Echinosphaerium*, at least, that there are two distinct sizes of cross bridges: short links about 70 Å long connect microtubules with a row, and longer links about 300 Å connect microtubules between adjacent rows. One of the problems in the study of the cross bridge has been the difficulty of resolving such small structures in the electron microscope. In the highly ordered systems at least, they are comparatively clear. As the systems of microtubules become less ordered, it seems that the cross bridges become more and more difficult to discern. Even in the interlocking microtubule coils of the axopodia of *Echinosphaerium*, the cross bridges are rather faint and often one has to rely on rather extensive staining procedures in order to see them (Tilney and Byers, 1969).

An increasing number of reports has indicated the presence of cross bridges in the spindle apparatus of a wide variety of organisms, including both plant and animal (Fuge and Müller, 1972; Hepler and Jackson, 1968; Hepler *et al.*, 1970). Microtubule arms are also observed attached to membranous components, including the nuclear envelope (Franke, 1971a), endoplasmic reticulum (Franke, 1971b), plasma membrane (Cronshaw, 1967; Kiermayer, 1968; Ledbetter and Porter, 1963; Robards, 1969), and between microtubules and vesicular components in the cytoplasm (Hepler *et al.*, 1970; Roth *et al.*, 1970).

Although microtubule cross bridges and arms project from the walls of microtubules at varying angles, their position on the surface of the tubule appears to be defined by the presence of nearby links. Roth and co-workers (1970), in a detailed ultrastructure study of the microtubules in the axopodia of *Echinosphaerium*, find that if a link or arm is bound to the surface of the microtubule, there is a preference for the next similar link to be bound to the opposite surface of the microtubule. In *Echino-*

sphaerium, which has two different kinds of links, they find, in addition, that the binding of one type of subunit causes the next cross bridge to be of the opposite type and to be nearby, approximately one or two subunits away. The results of Roth *et al.* (1970) suggest that allosterism is involved in the placement of microtubule arms and cross bridges. The binding of a cross bridge at one particular site, they argue, affects a region on the microtubule surface, making it either susceptible or nonsusceptible to the presence of other kinds of links.

Further evidence for the presence of defined binding sites for bridges is provided in a recent study by McIntosh (1974) wherein particular attention has been given to the periodicity of arms or bridges along the length of the microtubule. It is recognized that in the outer doublets of flagella the dynein arms exhibit a 240 Å period along the tubule; however, in less ordered systems bridges do not occur with such regularity, and initial impressions suggest that a regular period is lacking. Using combined analyses of direct measurements and optical diffraction of embedded and negatively stained material, McIntosh (1974) presents evidence supporting a periodic arrangement of bridges. The data are consistent with the idea that one bridge binding site occurs on each 80 Å tubulin dimer, but that only a portion of these sites is filled. Thus bridges may occur at intervals of 80, 160 Å, etc., with 160 and 240 Å being observed frequently.

The occurrence of arms and cross bridges on the surface of microtubules raises the interesting question of how these structures may function. Many workers would agree that they act as stabilizing elements, and in support of this contention, it is not uncommon to see the more prominent bridges among stable arrays of microtubules. Tucker (1970) finds in *Nassula* that the position of the microtubules seems to be determined before the cross bridges become evident. He argues that pattern formation of microtubules is under the control of initiating sites and that the microtubule cross bridges simply serve as stabilizing factors. Tilney and Byers (1969) and, Tilney (1971a) however, argue that the cross bridges themselves may be the important factors in determining the precise array of microtubules in the axopods of the *Echinospaerium* and in the axoneme of *Raphidiophrys.* In their studies of microtubules, which are reassembling following treatments with cold temperature, Tilney and Byers (1969) note that the patterns seem to be governed by the presence of the two kinds of cross bridges. Thus they argue that the cross bridge–tubule combination defines and determines the pattern.

A third function for the cross bridge might be in motility. It has been suggested by McIntosh and Porter (1967) that the cross bridges between the microtubules in the helix around the elongating spermatid nucleus of the chicken are mechanochemical units capable of causing adjacent microtubules to slide (Bannister and Tatchell, 1968). The concept of microtubule sliding being caused by mechanochemically active cross bridges has been developed into a model of mitosis (McIntosh *et al.*, 1969) and will be treated later in this chapter.

III. Composition and Assembly of Microtubules

Remarkable progress has been made in the isolation and characterization of tubulin, the protein subunit of microtubules (for reviews, see Olmsted and Borisy, 1973; Stephens, 1971; Wilson and Bryan, 1974). Tubulin from a wide variety of sources consists of a dimer of 120,000 MW which upon reduction and alkylation yields two monomeric subunits each about 55,000 MW. Detailed analysis of tubulin reveals that there are equimolar amounts of two slightly different monomers, α-tubulin and β-tubulin, and some current evidence suggests that the dimer is a heterodimer of the α and β subunits (Ludueña *et al.*, 1974). The dimer binds two moles of guanidine nucleotide; one mole of GDP is tightly bound while one mole of GTP is loosely bound and readily exchangeable. Each dimer also binds the antimitotic agents colchicine, vinblastine, and podophyllotoxin (Wilson and Bryan, 1974). The binding of colchicine prevents the dimers from polymerizing and has been used widely in biochemical and cytological studies to disrupt microtubule structure and hence their function. The discovery by Weisenberg (1972) that *in vitro* polymerization of microtubules can be caused primarily by lowering the calcium ion concentration in the solution has marked a significant advancement in studies of tubule assembly and has provided important insight for our thinking about *in vivo* control of tubule formation and function.

With the exception of studies on the flagellar tubules of the green alga *Chlamydomonas* (Olmsted *et al.*, 1971; Witman *et al.*, 1972a,b), the biochemistry of plant microtubules has received only little attention. A colchicine binding protein of 120,000 MW has been isolated from vascular tissue of *Heracleum* (Hart and Sabnis, 1973). In yeasts, a protein of 110,000 MW, which binds the colchicine derivative, Colcemid, ten times more effectively than colchicine has been isolated and partially characterized (Haber *et al.*, 1972). Although plant microtubules may be similar in most or all respects to those of animal origin, there is reason to suspect differences. Of particular pertinence is the observation that mitoses in plants generally require much higher concentrations of colchicine to achieve inhibition than they do in animal cells. Even in *Haemanthus* endosperm, where the cells lack an enclosing wall and can be exposed directly to the drug, the lower limit for inhibition is around 1×10^{-4} M (Hepler and Jackson, 1969), a concentration which is 100- to 1000-fold higher than that needed to inhibit dividing HeLa cells (Taylor, 1965a). Plant microtubules thus may differ markedly from animal microtubules in their affinity for colchicine, and possibly they may possess differences in their biochemical properties. It cannot be argued that plant spindle tubules are of the stable, ciliary type, since they are sensitive to cold and require careful fixation with glutaraldehyde to be preserved. The biochemical properties of plant microtubules deserve attention and could probably be easily analyzed in liquid endosperm tissue such as coconut milk. It would be especially intriguing to investigate the properties of microtubules from plants in the genera of *Colchicum, Vinca,* and *Podophyllum,* from which the principal antimitotic agents are derived.

Despite considerable progress on the structure and biochemistry of microtubule protein, much less is known about the cellular factors that control the assembly of the subunits into intact microtubules. Not only are the microtubules polymerized at specific times during cell division, but they are assembled in specific places within the cell. Thus, how the cell controls microtubule assembly is a question of prime importance and one that relates directly to microtubule function and to any cellular process, such as mitosis, which requires microtubules. The pioneer work by Inoué and co-workers on spindle fibers, using the polarizing light microscope (Inoué, 1952, 1953, 1964; Inoué and Bajer, 1961; Inoué and Sato, 1967) has demonstrated two important concepts about the birefringent spindle elements (microtubules). The first is that the spindle fibers behave as if they were composed of subunits that were in an equilibrium between an assembled and a disassembled state. By following the increase or decrease in birefringence, it has been possible to determine changes in spindle fiber assembly during the mitotic cycle (Inoué and Bajer, 1961). Alterations of the assembly equilibrium can also be brought about experimentally by several different agents (Inoué, 1964). Cold temperature, high hydrostatic pressure, and colchicine, for example, destroy birefringence, while warm temperatures and heavy water allow the birefringence to reform or even enhance it significantly over control conditions. Microscopic studies on dividing cells and other systems, notably the axopodia of the protozoan *Echinosphaerium,* have convincingly demonstrated that the loss of birefringence, e.g., by cold temperature, high pressure, and colchicine (see Tilney, 1971b, for review), can be directly correlated with the depolymerization of microtubules. Taken together, the results lead Inoué and Sato (1967) to postulate that the birefringent elements are composed of subunits that are bonded hydrophobically, in a process involving the displacement of water.

The second major point that emerges from the study using polarized light is that certain points or regions within the cell are the organizing areas of birefringent material (Inoué, 1964). These are called initiating sites or, more recently, microtubule-organizing centers (MTOC) (Pickett-Heaps, 1969d). In dividing endosperm cells of *Haemanthus*, for example, birefringent elements appear to grow from the kinetochore during prometaphase and from the phragmaplast during mid-anaphase to telophase (Inoué, 1964). Experimental studies have confirmed that these two zones are initiating sites for the birefringent fibers. Using a UV microbeam, Inoué (1964) selectively destroyed the kinetochore and followed the regrowth of the spindle fibers. Focusing the microbeam directly on the kinetochore destroyed the birefringence and the ability for it to reform, whereas focusing the microbeam on the region of the spindle fiber distal from the kinetochore caused only a local reduction in birefringence that would quickly grow back. Similar studies showed that the midzone of the phragmoplast was also a region where assembly of birefringent fibers was being initiated.

In recent years several ultrastructural studies have drawn attention to the presence of a flocculent and moderately stained material that characterizes sites of microtubule assembly (for reviews, see Pickett-Heaps, 1969d; Porter, 1966; Tilney, 1971b; Hepler and Palevitz, 1974). In plant cells, for example, this moderately stained material is particularly prominent at kinetochores (Bajer and Molè-Bajer, 1969; Braselton and Bowen, 1971; Harris and Bajer, 1965; Hepler and Jackson, 1969; Pickett-Heaps and Fowke, 1970) and is also observed in the midzone of the phragmoplast (Hepler and Jackson, 1968). In lower plants, especially in fungi and some algae, an amorphous material may be a major component of the spindle pole bodies from which the tubules of the mitotic apparatus emanate (Hepler and Palevitz, 1974; Pickett-Heaps, 1969d, 1975). In other cells, primarily of animal origin, the flocculent material can be observed at the spindle poles surrounding the centrioles, associated with satellites near basal bodies (Szöllosi, 1964; Tilney and Goddard, 1970) and in the midbody of cells in telophase (Byers and Abramson, 1968; Paveletz, 1967; Robbins and Gonatas, 1964). It is especially evident in the formation of centrioles and basal bodies that the characteristic triplet tubule configuration is derived from a structure that initially appears as a blob of amorphous, flocculent material usually in association with a maturing centriole (Fulton, 1971; Kalnins and Porter, 1969).

Experiments designed to test the roles of the microtubule-organizing centers in the formation of tubules have been performed by Tilney and Goddard (1970). In sea urchin blastula following treatment with cold, they find that upon rewarming, new microtubules form in association with the satellite bodies that contain this moderately stained flocculent material. While numerous studies support the idea that the moderately stained, flocculent substance is associated with initiation of microtubules, almost nothing is known about its composition or function.

The control, *in vivo*, of microtubule assembly has been given considerable attention by Rosenbaum *et al.* (1969) in their study on flagellar regeneration in the alga, *Chlamydomonas*. Following amputation of the two flagella normally possessed by the alga, new ones rapidly regenerate in a process requiring protein synthesis. If, however, only one flagellum is removed the intact one will shorten while the new one elongates. When both flagella are the same length they usually elongate at the same rate. Several experiments using colchicine to block microtubule assembly and cycloheximide to inhibit protein synthesis indicate that both assembly and disassembly occur independently of protein synthesis. The ultimate length of the flagella depends upon the pool size of subunits and is thus governed by the equilibrium between the assembled and disassembled states. Autoradiographic studies further suggest that new microtubule subunits are added by tip growth at the distal end of the flagellum.

IV. The Spindle Apparatus: Mitosis and Cytokinesis

The existence of a fibrous component in the spindle apparatus was conclusively demonstrated in studies of living plant and animal cells using the polarizing light microscope (Inoué, 1953). Birefringent elements that are oriented parallel to the axis of the spindle apparatus occur in close association with the chromosomes during prometaphase to anaphase, and with the developing cell plate during mid-anaphase to telophase (Inoué and Bajer, 1961). While most workers agree that the birefringent elements can be equated to microtubules, there are some who challenge this view. It has been shown, for example, that microtubules make up less than 50% of the mass of the spindle apparatus (Forer, 1969). Microtubules can be removed from the isolated mitotic apparatus without destroying overall spindle structure (Borisy and Taylor, 1967; Bibring and Baxandall, 1968). Furthermore, fixations that retain microtubules reduce or perhaps even destroy birefringence (Inoué and Sato, 1967; Forer, 1969). Finally, the UV microbeam studies of Forer (1965, 1966), which indicated, on the one hand, that a chromosome may move in anaphase even when its kinetochore fibers have a region of reduced birefringence, while, on the other hand, a chromosome may not move in spite of the fact that its microbeamed kinetochore fibers lacked a reduced zone of birefringence, support his contention that microtubules cannot be both the birefringent material and the elements that cause chromosome separation.

In support of the equivalence of microtubules and birefringence we note that microtubules are the principal oriented component in the spindle apparatus and that they are present in *Haemanthus* endosperm cells (see Figs. 2–4), for example, in numbers exceeding 10,000 (P. K. Hepler, unpublished observations). Microtubules also show the same transformations during the mitotic cycle as does the birefringent material (Hepler and Jackson, 1968), and both are destroyed by agents such as colchicine and cold temperature. In tissues and isolated cells where microtubules have been destroyed but birefringence remains, ultrastructure studies reveal that there are aligned rows of ribosomes still present (Goldman and Rebhun, 1969; Kane and Forer, 1965), and it is postulated that these rows give rise to the residual birefringence (Goldman and Rebhun, 1969). It is probable that the alignment of the ribosome rows was established by the microtubules before the latter were disrupted (Goldman and Rebhun, 1969; Rebhun and Sawada, 1969). Finally, the UV microbeam studies are inconclusive, since we do not know the full extent of their effects. While microtubules are known to be broken and disrupted by the microbeam (Bajer and Molè-Bajer, 1971a), it is highly likely that a few intact ones remain, and these could be sufficient to cause chromosomes motion. Although it is unknown whether microtubules are the force generating component of the spindle apparatus, the fact that colchicine arrests cells in metaphase strongly suggests that microtubules are necessary for chromosome separation to occur.

In higher plants the dividing cell, which has been most extensively studied at the ultrastructural level, is the endosperm cell of the African Blood Lily, *Haemanthus katherinae* (Bajer, 1968a,b,c; Bajer and Molè-Bajer, 1969, 1971a, 1972; Harris and Bajer, 1965; Hepler and Jackson 1968, 1969; Molè-Bajer, 1969). These studies, especially together with analyses of some other kinds of cells, enable us to describe mitosis with particular emphasis on the transformations and orientations of microtubules during karyokinesis and cytokinesis. In early prophase, microtubules appear in the clear zone next to the nucleus (Bajer and Molè-Bajer, 1969; Esau and Gill, 1969; Evert and Deshpande, 1970; Sakai, 1969) (Fig. 1). They are randomly oriented, but as prophase progresses they become oriented parallel to one another in a process which involves first, the mutual alignment of small groups of microtubules (Fig. 1). Subsequently, these groups become aligned in such a way that a continuous sheet or mantle of microtubules is formed, encasing the late prophase nucleus (Bajer and Molè-Bajer, 1969). In some cells, for example, in tobacco, a polar aggregation of microtubules is apparent at prophase

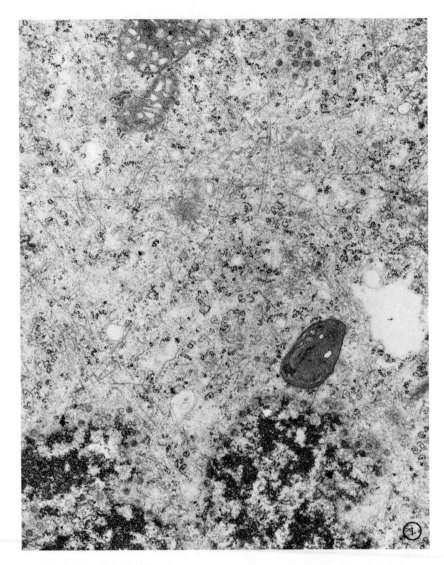

Fig. 1. Prophase in *Haemanthus* endosperm. An electron micrograph from a section that is tangential to the surface of the nuclear envelope and passes into the adjacent clear zone. Microtubules are abundant. Although the individual elements may be aligned in small groups, there is no overall orientation of the microtubules. Nuclear pores are observed in face view and numerous profiles of membrane-bound polysomes are evident. ×16,800 (P. K. Hepler and W. T. Jackson, unpublished).

(Esau and Gill, 1969). In either case, however, it appears that the microtubules that will eventually form the continuous spindle appear outside of the nuclear envelope before prometaphase.

At prometaphase the nuclear envelope breaks in a process which may involve the active participation of the tubules of the continuous spindle (Bajer and Molè-Bajer, 1969). Following breakdown of the nuclear envelope microtubules for the first time are observed among the chromosomes (Fig. 2). The continuous spindle microtubules invade the nuclear region from the outside concomitantly with the appearance, presumably by growth, of chromosomal microtubules at the kinetochores. Although it is suggested that the continuous tubules make contact with the chromosomes and directly become kinetochore microtubules (Bajer and Molè-Bajer, 1969), the evidence is not compelling. It is very difficult from electron micrographs to decide whether a microtubule has struck and become attached to a chromosome or whether a microtubule has grown from the kinetochore itself. The latter explanation is more consistent with the fact that the kinetochore is a site for microtubule initiation and with the accepted concept of an equilibrium between the assembled and disassembled states acting in the formation of microtubules from subunits (Inoué and Sato, 1967). As the transition from prometaphase to metaphase proceeds, the continuous tubules progressively intermingle with the chromosomes and with the tubules from the kinetochores (Bajer and Molè-Bajer, 1969; Figs. 3 and 4). It is during this stage that the chromosomes are becoming aligned at the metaphase plate.

During anaphase the sister chromatids separate and move to the opposite poles. Although there is a marked reduction in the number of microtubules, some of the interzone tubules always remain, and these form the beginnings of the phragmoplast in mid- to late anaphase (Bajer, 1968a; Hepler and Jackson, 1968; Lambert and Bajer, 1972) (Fig. 5). In *Haemanthus* endosperm cells the phragmoplast usually forms at the edge of the spindle and subsequently grows inward (Bajer, 1968a). When the entire midsection of the spindle has been consolidated, then the phragmoplast undergoes its normal outward or centrifugal growth (Bajer, 1968a; Hepler and Jackson, 1968).

One must be cautious about generalizing on cell plate formation from observations on *Haemanthus* alone. In many different kinds of cells, including root tip, leaf mesophyll, moss protonema (Cronshaw and Esau, 1968; Esau and Gill, 1969; Hepler and Newcomb, 1967; Ledbetter and Porter, 1963; Lehmann and Schulz, 1969; Roth *et al.*, 1966; Whaley *et al.*, 1966), the phragmoplast normally forms simultaneously along the mid-$\frac{1}{2}$ or -$\frac{2}{3}$ of the equatorial zone between the daughter nuclei and grows only centrifugally; it does not grow centripetally as in *Haemanthus*. It

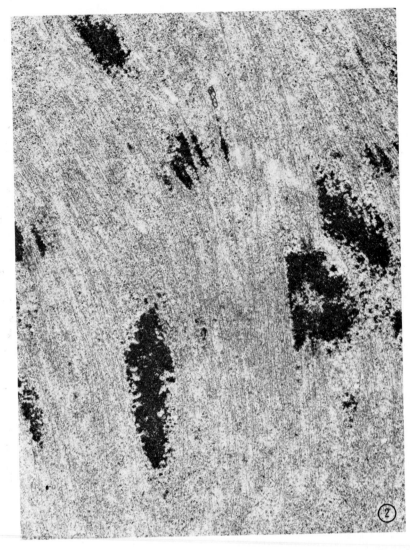

Fig. 2. Prometaphase in *Haemanthus* endosperm. The nuclear envelope has broken and the sheath of parallel continuous tubules is invading the region of the chromosomes. Microtubules are observed emanating from kinetochores (right-hand side of micrograph). ×5400 (P. K. Hepler and W. T. Jackson, unpublished).

is well to keep in mind that the *Haemanthus* cells have been removed from the embryo sack and flattened down considerably (less than 10 μm) on a microscope slide before examination. This treatment, especially the

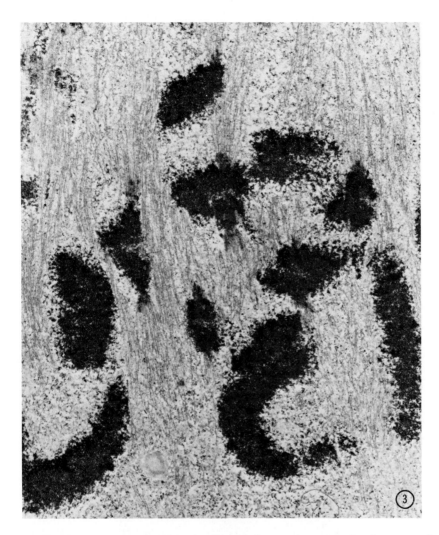

Fig. 3. Late prometaphase in *Haemanthus* endosperm. Continuous and kinetochore tubules intermingle as the chromosomes become aligned on the metaphase plate. Several pairs of chromosomes with kinetochores are evident. ×5400 (Hepler *et al.*, 1970).

flattening, may lead to distortions in the formation of the cell plate. Another aspect of cell plate formation in *Haemanthus* which differs markedly from the same process in other plant cells is the ability of endosperm cells to form a phragmoplast in the absence of a preceding mitosis. Bajer (1968a) has even reported the presence and formation of a phrag-

moplast in isolated anucleated droplets of cytoplasm. Uncoupling of cell plate formation from nuclear division may be stimulated by the artificial culturing techniques, or it may reflect a normal process in *Haemanthus* endosperm, since it is known that during normal development the cells first undergo free nuclear division and only later form cell walls.

With the lateral centrifugal growth of the cell plate, there is an extensive buildup of microtubules (Hepler and Jackson, 1968) (Figs. 6 and 7) in accord with what is observed in polarized light microscope studies (Inoué and Bajer, 1961). Microtubules occur in bundles, within which the individual elements on one side of the plate interdigitate with those on the other side (Fig. 7). Specifically, where the microtubules appear to interdigitate an electron-dense material, the presumed microtubule initiating substance is observed (Hepler and Jackson, 1968). Vesicles that are derived possibly from the Golgi apparatus (Bajer, 1968a), or possibly from the endoplasmic reticulum (Hepler and Jackson, 1968), appear to flow along the microtubules into the midzone of the phragmoplast, where they subsequently fuse to form the cell plate (Fig. 7). As portions of the cell plate are completed, the microtubules again depolymerize (Hepler and Jackson, 1968).

V. The Spindle Pole: Planes of Division

Retracing our steps back to the beginning of mitosis, one is confronted with the problem of how the dividing cell defines its poles and thus establishes its axis of division. In endosperm cells of *Haemanthus*, the problem would not appear to be acute, since there is no preferential axis of division. However, many plant cells are known to divide in predictable and often predetermined planes, as in the normal divisions of elongating root tip cells and, perhaps more strikingly, in the asymmetric divisions of the epidermal cells of monocots giving rise to the stomatal complex, where, in addition, there are characteristic nuclear migrations preceding mitosis (Bünning, 1957). The phragmoplast itself may control the plane of division in certain cells, for example, in the periclinal divisions of cambial initials where the cell plate continues growth when separated temporally and spatially from nuclear division (Esau and Gill, 1965; Evert and Deshpande, 1970) and in dividing guard mother cells of *Allium* where the entire spindle–phragmoplast structure including

Fig. 4. A higher magnification view of a portion of Fig. 3 showing the relationship between continuous and kinetochore tubules. The kinetochores are characterized by a fan-shaped bundle of microtubules. Continuous tubules run along the right-hand side of the micrograph. ×18,750 (Hepler et al., 1970).

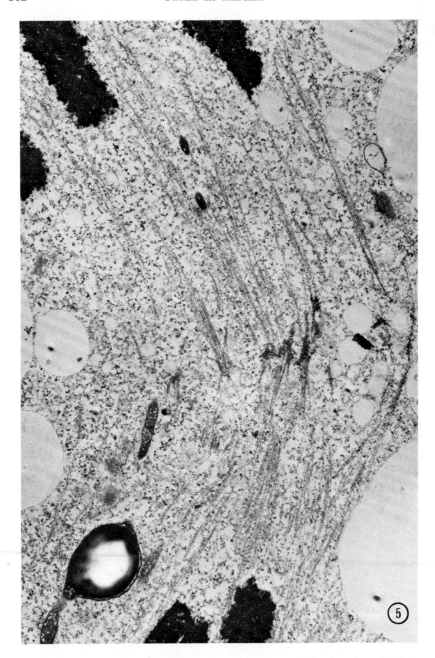

Fig. 5. Late anaphase in *Haemanthus* endosperm. A few bundles of tubules remain in the interzone at anaphase, and these become the beginnings of the phragmoplast. Note the densely staining accumulations at the ends of the microtubule bundles. ×7000 (P. K. Hepler and W. T. Jackson, unpublished).

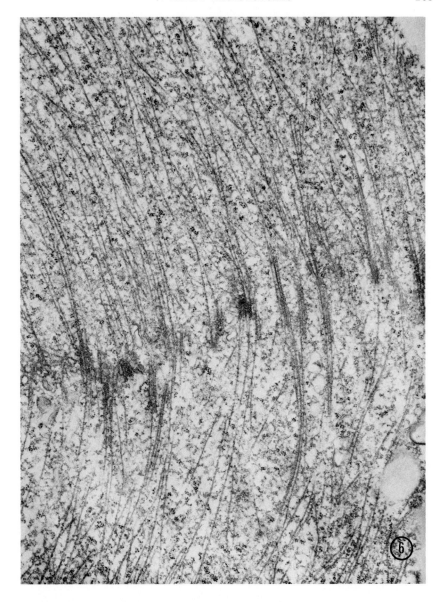

Fig. 6. The phragmoplast in *Haemanthus* endosperm. During the centrifugal growth of the cell plate, the phragmoplast is composed of bundles of microtubules. Densely staining material can often be observed in the midzone of these bundles. ×20,000 (Hepler and Jackson, 1968).

Fig. 7. The phragmoplast in *Haemanthus* endosperm. Microtubules on both sides of the cell plate appear to overlap. Vesicles from a cytoplasm rich in endoplasmic reticulum appear to flow into the plate. ×31,200 (Hepler and Jackson, 1968).

nuclei rotates during anaphase–telophase from an oblique orientation to one yielding the proper longitudinal plane of the cell plate (Palevitz and Hepler, 1974). In many other cells, however, the positioning of the spindle apparatus at the beginning of mitosis appears to define and control the subsequent plane of the cell plate. In these cells, one's attention is drawn to the spindle pole, its nature, organization, and function.

In spite of numerous attempts to elucidate the nature of the spindle pole in cells of higher plants, its structure, composition, and function continue to perplex us (Bajer and Allen, 1966a; Inoué and Bajer, 1961; Hepler and Palevitz, 1974). In *Haemanthus*, for example, the pole may cover a broad area up to 50 μm wide within which microtubules apparently terminate. While other cytoplasmic organelles, such as plastids and mitochondria, may be aggregated in the polar regions, there is no structure that, because of its association with the pole or with the spindle microtubule, is implicated in functioning as a spindle organizer (Bajer and Molè-Bajer, 1969). Most higher plant cells fit the above description, although a few notable exceptions exist. In studies of pollen mother cells during meiosis, aggregations of endoplasmic reticulum have been observed in the spindle poles (Sakai, 1969; Wilson, 1970). In addition, the spindle poles are quite defined and the microtubules are disposed in focused arrays. Groups of microtubules are also observed among elements of the endoplasmic reticulum at the poles of the spindle in other dividing plant cells (Cutter and Hung, 1972; Esau and Gill, 1969; Hanzely and Schjeide, 1973), while a massive accumulation of the endoplasmic reticulum with microtubules is noted in centrifuged root tip cells (Burgess and Northcote, 1968).

The situation in lower plants is quite different, since clearly defined centrosomes have been observed in some dividing cells (for reviews, see Pickett-Heaps, 1969d, 1975; Hepler and Palevitz, 1974). Centrioles, spindle pole bodies, and related structures have been observed at the spindle pole in several different fungi and algae. In some lower plants, including the nonflowering vascular plants, it is especially noteworthy that the cells giving rise to the motile sperm undergo a transition in their spindle structure from an anastral to an astral type (Lepper, 1956; Sharp, 1912). A centriole or centriole equivalent arises *de novo* and appears to participate in the formation and organization of the spindle apparatus (Moser and Kreitner, 1970). In the water fern, *Marsilea*, Sharp in 1914 reported that the centriole equivalent, called a blepharoplast, arises during the second of four spermatogenous divisions and resides at the poles of the spindle apparatus for the final divisions. Electron microscopic investigations of the ninth (last) division in microspores of *Marsilea* reveal that the blepharoplast forms as a double structure and during prophase migrates to the opposite poles of the spindle apparatus, where it remains throughout division (P. K. Hepler, unpublished observations). Subsequently, during a differentiation phase, the blepharoplast gives rise to 100–150 centrioles with the characteristic nine triplet tubule cylinder wall, and these become the basal bodies for an equivalent number of flagella of the motile sperm (Mizukami and Gall, 1966). While the blepharoplast possesses a unique structure that differs from that of a mature centriole, its ability to act as the focus for microtubules (P. K. Hepler, unpublished observations) and to subsequently give rise to basal bodies, clearly mark it as an important organelle in microtubule organization (Mizukami and Gall, 1966).

The spindle pole in plants exhibits a wide range in morphology from a centriole, blepharoplast, or spindle pole body to no apparent structure at all. To date, however, it has been difficult to assess the role of the centrosome when present, other than to acknowledge that it represents a defined MTOC. The presence of a discrete centrosome at the spindle pole may only be fortuitous (Pickett-Heaps, 1969d); it may be a sufficient but not necessary condition for microtubule organization. It is possible that elements of the ER observed at the spindle pole play an important role in microtubule organization; for example, the ER may function like the sarcoplasmic reticulum of muscle and release or sequester calcium ions under the proper signal. By altering the concentration of calcium in local regions of the cell, the ER might effectively control microtubule polymerization and depolymerization. Even if the ER were to function in this manner, we still would not understand how it became positioned in the first place to regulate and define the axis of the spindle.

The discovery by Pickett-Heaps and Northcote (1966a,b) of a band of microtubules encircling the preprophase nuclei gave much added stimulus to our thinking on the questions of planes of division, nuclear position, and spindle alignment in dividing plant cells. In symmetrically dividing root tip cells and in the asymmetrically dividing guard mother cells and subsidiary cells of the stomata complex, it was found that during preprophase there was invariably a dense band of microtubules positioned in a plane that "indicated" the plane of the new cell plate (Pickett-Heaps and Northcote, 1966a). Although this basic observation has been repeated by others (Burgess, 1969, 1970c; Burgess and Northcote, 1967, 1968; Cronshaw and Esau, 1968; Deysson and Benbadis, 1968; Newcomb, 1969), it is important to note that the preprophase bands have not been observed in the endosperm of *Haemanthus* (Bajer and Molè-Bajer, 1971; Hepler, unpublished observations), in asymmetrically dividing pollen cells (Burgess, 1970b; Heslop-Harrison, 1968), or in dividing cambial cells (Evert and Deshpande, 1970), and thus it does not occur universally. The notion that the preprophase band might be involved in positioning of the nuclei prior to division, has been put forth by Burgess and Northcote (1967) and Burgess (1970c). Their observations supported the notion that the preprophase band was more usually aligned with the equator of the spindle apparatus, rather than with the plane of the new cell plate. However, subsequent studies by Pickett-Heaps (1969a) showed that the preprophase band became apparent only after nuclei in asymmetrically dividing cells had already undergone their polarized movements. That the nuclei may be stabilized in their position by the preprophase band is supported by observations showing that nuclei possessing preprophase bands were less sensitive to centrifugation than those that did not (Pickett-Heaps, 1969b). Experiments using the drug caffeine to block the formation of the phragmoplasts and thus cause the development of binucleate cells have shown that, in spite of this severe disruption, the preprophase band, when it appears, always occurs in the plane anticipating the plane of the cell plate (Pickett-Heaps, 1969c).

During division, as the cell progresses from preprophase to prophase, the band of microtubules breaks down. Thus, although it may have some role in orienting the spindle (Burgess and Northcote, 1967), it is difficult to understand the mechanism, since the band is absent at a time when spindle orientation is occurring. Pickett-Heaps (1975) currently favors the view that the preprophase band is the result and not the cause of polarization within the cell. He has suggested, furthermore, that the microtubules of the band may migrate intact and become directly incorporated into the spindle apparatus (Pickett-Heaps, 1969b). The evidence for this is at best suggestive. It would seem more likely that the band breaks down and spindle tubules form anew.

It may be possible to reconcile several observations about the preprophase band and to assign a function to it if we consider that it is, in part, the product of a cortical (plasmalemma associated?) MTOC. The appearance of the tubules themselves at preprophase may be somewhat misleading and only reflect the high levels of tubulin subunits just prior to mitosis, which, in the absence of active nucleative centers like the poles or the kinetochores, simply polymerize as a function of an equilibrium driven reaction onto the hypothetical cortical MTOC. With the beginning of prophase–prometaphase, other nucleating centers compete for the available subunits and the preprophase band breaks down. However, the cortical MTOC may remain, and during anaphase when spindle tubules are depolymerizing it might once again effectively compete for subunits and catalyze tubule formation, possibly in such a way as to influence the direction of growth of the centrifugally expanding phragmoplast and cell plate. It has been suggested that such a cortical membrane associated MTOC could participate in the rotation of the spindle–phragmoplast structure in dividing guard mother cells of *Allium*, possibly by generating shearing forces through tubules between itself and the surface of the spindle (Palevitz and Hepler, 1974).

VI. Mechanism of Mitosis

Ever since the discovery of mitosis, there has been a continuing interest on the nature of its mechanism. This has been the subject of books and numerous articles (Bajer and Molè-Bajer, 1972; Mazia, 1961; Schrader, 1953; Wilson, 1928); for a penetrating discussion of the mechanism of mitosis the reader is directed to a review by Nicklas (1971). Since the discovery of microtubules as the spindle fibers common to virtually all dividing eukaryotic cells, much attention has been focused specifically on how they might work to produce chromosome motion (Bajer and Molè-Bajer, 1971b, 1972; Dietz, 1969; Forer, 1969; Inoué and Sato, 1967; McIntosh et al., 1969; Nicklas, 1971; Subirana, 1968). But despite our increasingly detailed understanding of microtubule composition and structure the basic motile mechanism has not yet been elucidated. With the development of new techniques, for example, chromosome micromanipulation (Nicklas and Staehly, 1967), the use of laser and UV microbeams (Forer, 1965, 1966; Inoué, 1964) to selectively destroy small portions of the spindle, and the quantitative evaluation of microtubule distribution in the spindle (Brinkley and Cartwright, 1971; McIntosh and Landis, 1971), the means are becoming available wherein the different models for mitosis can be tested experimentally. Coupled with the intense current interest in several different laboratories, there can be good reason for expecting considerable progress in clarifying the mechanism of mitosis in the next few years.

The dynamic equilibrium between the intact microtubule and its protein subunits has been the basis for one of the more enduring models of mitosis (Dietz, 1969; Inoué and Sato, 1967). The underlying concept suggests that the disassembly of microtubules produces the force that moves chromosomes. Upon removal of subunits, the microtubule will shorten, and if the distal end of the kinetochore tubule is firmly embedded in the pole, the chromosome will be pulled toward the pole. Calculations show that the system can produce the required force (Nicklas, 1971). Also, in at least one example, it has been shown through the use of colchicine that the artificial depolymerization of the spindle fibers can actually cause the chromosomes to migrate (Inoué, 1952). There are, however, significant problems with this theory. First, there is, in higher plants, no apparent structure at the spindle pole in which the tubules are embedded or to which they are anchored. Second, this model requires that the material that is moved be linked to the microtubules at a kinetochore or some other such attachment, yet there is considerable evidence for the motion of particles and unattached acentric fragments in the spindle (Bajer, 1958, 1967; Freed and Lebowitz, 1970; Rebhun, 1963, 1964;

Taylor, 1965b). Third, it is difficult to explain how depolymerization would move unattached vesicles to the central plane of the phragmoplast during cytokinesis, at a time when some fragments and materials are eliminated from the midzone at anaphase–telophase (Bajer, 1965; Bajer and Allen, 1966b). Fourth, the work of Forer (1966) indicates that the mechanism of mitosis has two components, which cannot be readily accounted for by the equilibrium model. Thus, while the concept of microtubule shortening as a mechanism for chromosome motion fails to explain many features of division, it must be maintained, nonetheless, that the lengthening or growth of microtubules through the addition of new subunits, under control of special initiating sites, plays an important role in mitosis, for example, in pole separation and in keeping telophase nuclei apart. Furthermore, if the kinetochore tubules are constantly turning over and new subunits are being inserted at the proximal or kinetochore end, the constant movement of particles or states to the pole even during metaphase observed by Allen et al. (1969) can be explained.

A provocative paper by Östergren et al. (1960) suggests that there is a shearing force between the spindle fibers and the matrix. This concept has been developed in light of recent findings about microtubules by Subirana (1968), who specifically postulates that actin-like microtubules, which are polarized through the development of shearing forces at their surface, might "swim" through a myosin-like matrix. The postulate readily accounts for alignment of chromosomes at the metaphase plate, since oppositely directed kinetochore tubules would exert equal force on the still joined sister chromatids. Following a split of the kinetochore these same tubule forces would then move the chromosomes to the poles. Fiber shortening is a result and not the cause of anaphase motion. This model would certainly be more attractive if the microtubule proteins were actin-like, but recent studies indicate that they are not. Subirana (1968) suggests that the continuous tubules switch polarity at the midzone of the cell and are so disposed as to move particles from the midzone to the pole, thus not accounting for the motion of vesicles into the cell plate at telophase. Particle saltation would appear to be governed by a hydrodynamic driving force resulting from the tubule–matrix shear, and within a small region, or half-spindle, would be undirectional. However, the fact that particles can move from pole to pole (Freed and Lebowitz, 1970; Rebhun, 1963) and that nearby particles can move in opposite directions suggests that motion is independent of the surrounding matrix.

A mitosis model by McIntosh et al. (1969) suggests, in contrast to the microtubule–matrix idea put forth by Östergren et al. (1960) and Subirana (1968), that the microtubules themselves interact and can slide past one another. They postulate that the microtubule arms or cross bridges, which have been observed in the spindle apparatus, are mechano-chemical elements possibly containing an ATPase. The arms are thought to bind to other elements, such as a neighboring tubule or a vesicle, and to cause that element to move relative to the microtubule surface. Micro-tubule polarity is an important part of this model, since it is suggested that tubules that are parallel can bind but cannot produce force between one another, whereas tubules that are antiparallel can bind and produce force and thus slide. The polarity is thought to operate in such a way that a microtubule slides in the direction of its growth, i.e., away from its site of initiation. The congression of the chromosomes to the meta-phase plate can be accounted for by the combined factors of microtubule growth from the kinetochore and the interaction through bridges of the kinetochore tubules with both parallel and antiparallel continuous tu-bules. The metaphase plate would thus be the null point between two equal but oppositely directed and acting groups of kinetochore tubules.

With the splitting of the kinetochore, the sister chromatids are pulled to their respective poles in this model (McIntosh et al., 1969), in a process involving sliding between antiparallel kinetochore and continuous tu-bules. As the kinetochore tubules move, the continuous tubules would be pulled from the pole where they were initiated and slide toward the opposite pole. During middle and late anaphase, sliding could continue between antiparallel continuous tubules in the interzone. A point would be reached where there would be too few bridges for further movement, and this would create the overlap zone observed in phragmoplast tubules. Two important observations about the phragmoplast can be ex-plained: (1) The initiating ends of the continuous tubules are reposi-tioned from the poles to the midzone where they can direct the new tubule growth that is known to occur (Inoué and Bajer, 1961; Hepler and Jack-son, 1968), and (2) the polarity of these repositioned continuous tubules is such that vesicles, which may be moved by bridge action, would be directed from both sides of the interzone to the midplane or cell plate. The model also has the attractive feature of accounting for particle salta-tion as the result of bridge action between a vesicle and adjacent tubule.

Support for this model (McIntosh et al., 1969) comes from several lines of work, mainly those studies showing that cross-bridged microtu-bules can and do slide relative to one another. In the ciliate, Stentor, there are rows of cross-bridged microtubules that have been shown to slide over one another during elongation and contraction (Bannister and Tatchell, 1968). In the elongation of the chicken spermatid nucleus, it has been shown (McIntosh and Porter, 1967) that the microtubules of the helix surrounding the nucleus slide during development, and more recently in movement of the axostyle of Saccinobacculus, ultrastructural studies reveal that cross-bridged rows of microtubules slide (McIntosh, 1973). That sliding occurs in cilia and flagella seems evident from the work of Satir (1967) and also from the studies of Summers and Gibbons (1971). In membraneless preparations of cilia, which had been treated lightly with trypsin and subsequently supplied exogenous ATP, Summers and Gibbons (1971) observed that the cilium elongated to approximately six times its normal length. The most reasonable explanation seems to be that the outer doublet tubules slide over one another to cause this increased elongation. Evidence that sliding occurs between spindle tubules, while not so compelling as in those examples mentioned above, is nevertheless implicated from the experiments that show that weak concentrations of colchicine that will inhibit late anaphase spindle motions do not block the early anaphase separation of chromosomes up to a distance equal to the length of the original metaphase spindle (Oppenheim et al., 1973). Oppenheim et al. (1973) suggest that colchicine in low doses will not affect existing structures but will prevent growth of new tubules and elongation of the existing ones, and thus it seems reasonable that the early anaphase separation of chromosomes that are unaffected by the drug may be mediated by a sliding phenomenon.

The model by McIntosh et al. (1969) predicts that during anaphase chromosomes on one side of the interzone should not pass to the other side of the interzone due to the opposite polarity of the microtubules. The experiments of Nicklas et al. (1970) and Nicklas and Koch (1972), however, show that a chromosome during late anaphase can move across the interzone, contrary to the expectations of the model. These observations and the well-known phenomenon of particle elimination toward the pole have prompted Nicklas (1971) to suggest that there are two sets of continuous or "interpolar" microtubules. These sets overlap in the midzone and are oppositely polarized such that bridge action always moves material to the poles. The interpolar microtubules are, in addition, stationary and provide a "scaffolding" on which bridges between them and antiparallel kinetochore tubules can act. In this model (Nicklas, 1971), however, all material, chromosomes, particles, etc., are moved poleward, and therefore it fails to account for the migration of vesicles to the cell plate during late anaphase and telophase (Bajer, 1965; Bajer and Allen, 1966b). The opposing movement of different particles in the interzone might be accounted for in the McIntosh et al. (1969) model by assuming that cell plate vesicles are brought into the midzone by bridge action, while the larger particles and acentric fragments are eliminated toward the pole by phragmoplast tubule growth.

A microtubule sliding mechanism for chromosome motion (McIntosh et al., 1969) further predicts a specific redistribution of microtubules in the metaphase to anaphase transition. Two laboratories have undertaken the arduous task of quantitatively determining microtubule distribution during various phases of cell division in cultured mammalian cells (Brinkley and Cartwright, 1971; McIntosh and Landis, 1971). If all microtubules are repositioned as a result of sliding then it is expected that during anaphase–telophase the midzone between the separating chromosomes would be composed of the overlapping ends of antiparallel tubules. Direct counts should yield a ratio of 2:1 between the overlapped and nonoverlapped regions. The data, however, indicate a ratio of 1.5:1 and, therefore, it seems likely that some microtubules are continuous through the midzone. These results may be accounted in part by the failure of some microtubules to slide or by the growth of new elements from the midzone.

A mitotic model put forth by Bajer (1973) uses the energy derived from lateral interaction between microtubules to cause chromosome motion. Observations on *Haemanthus* endosperm cells show that microtubules fan out from the kinetochore and in addition reveal the presence of segments of skewed, i.e., nonparallel, tubules throughout the spindle. Bajer (1973), therefore, suggests that a progressive "zipping" occurs between nonparallel tubules of a kinetochore and the surrounding polar region. The lateral interaction between these nonparallel elements forces the kinetochore tubule to become parallel with the skewed nonkinetochore element. Through a series of these events, like a sailboat tacking into the wind, the chromosome moves to the pole. This model does not require sliding but may use cross bridges as the units that produce the lateral association of microtubules. Since the force is generated between the chromosome and the pole to which it moves, the "zipper" model (Bajer, 1973) does not readily account for the observations from UV microbeam studies (Forer, 1969), which indicate that movement in both halves of the spindle may be coupled. Based on our current understanding of spindle structure, there is no reason, especially in *Haemanthus* endosperm, which lacks a defined spindle role, to suspect that nonkinetochore fragments would be sufficiently well anchored such that they would attract and thus pull a kinetochore tubule toward them. The converse seems equally plausible, wherein a nonkinetochore fragment would be attracted and pulled toward the kinetochore tubule. The latter condition would not produce movement of the chromosome to the pole.

Although microtubules have dominated our thinking about the mitotic mechanisms, no experiment has proved that they are the force generating structures. Microtubules might be a scaffolding and only transmit the force produced by other molecules, such as an actomyosin complex (Forer, 1974; Luykx, 1970; Rebhun, 1972). The recent discovery of a filamentous material in the spindle apparatus of crane fly spermatocytes and other cells that can bind heavy meromyosin (Behnke *et al.*, 1971; Forer and Behnke, 1972; Gawadi, 1971) supports this idea, although much additional work is now needed to follow up the preliminary reports.

The mechanism of mitosis continues both to intrigue and elude us. No one model explains with satisfaction all the facts and observations, but the development of new and testable models helps in providing an important framework for further experiments. One of the most exciting recent developments, for example, has been the production of lysed cells with partially functional mitotic apparatus (Cande *et al.*, 1974; Inoué *et al.*, 1974; Rebhun *et al.*, 1974). The hope in these studies is to remove unnecessary components and produce a mitotic spindle in which chromosome motion can be started and stopped experimentally through the addition of macromolecules and cofactors, such as tubulin, dynein, and ATP. While the electron microscope has done much to clarify the structure of the spindle apparatus, in particular the spindle microtubule, we must be alert to the presence of new structures or heretofore unrecognized relationships between the well-known structures. Microtubule appears to play a central role in defining the structure of the spindle apparatus, in moving the chromosomes during karyokinesis, and in establishing the cellplate during cytokinesis, but the important question of how they work remains unanswered.

VII. Cortical Microtubules

The discovery by Ledbetter and Porter (1963) of microtubules in the cortex of plant cells which are oriented parallel to the direction of the cellulose microfibrils of the adjacent cell walls gave new impetus to our thinking about the mechanism by which the cell controls the formation of its wall. Prior to the observations of Ledbetter and Porter, a system of cortical cytoplasmic filaments had been predicted by Green (1962), who found in a study of *Nitella* that colchicine caused the young, growing internode cells to become spherical. Analysis of the cells in the polarized light microscope revealed that the wall was composed of randomly oriented cellulose microfibrils. Green (1962, 1963) postulated that a cytoplasmic element similar to the colchicine-sensitive fibers of the spindle apparatus was present in the cortex of *Nitella* cells and was involved in controlling the orientation of the cellulose microfibrils of the adjacent layers of the cell wall.

In discussing the role of microtubules in cell wall formation, it is important to separate the processes of synthesis and deposition of the wall microfibrils from that process which determines their orientation. It seems highly probable that microtubules do not affect the synthesis of the cellulose microfibrils, since cell wall formation occurs in the absence of microtubules, for example, in the tips of tip-growing cells (Green, 1969) such as pollen tubes (Rosen, 1968), root hairs (Newcomb and Bonnett, 1965), or fungal hyphae (Grove and Bracker, 1970). Some algae that develop from wall-less swarmers loose their microtubules at the time when the cell wall is being formed (Gawlik and Millington, 1969; Millington and Gawlik, 1970; Preston and Goodman, 1968). Finally, there are several reports showing that even in the presence of colchicine, cellulose microfibrils continue to be deposited in the wall (Brennan, 1971; Green, 1962, 1963; Hepler and Fosket, 1971; Marx-Figini, 1971; Pickett-Heaps, 1967a).

Numerous reports, including studies of material prepared by the freeze-etch procedure (Northcote, 1969; Northcote and Lewis, 1968), as well as by fixation with glutaraldehyde, show a strong correlation between the orientation of the microtubules in the cytoplasm and the cellulose microfibrils in the wall (for reviews, see Hepler and Palevitz, 1974; Newcomb, 1969; O'Brien, 1972; Pickett-Heaps, 1975). This has been observed in cells forming primary and secondary walls and includes examples of widely different species from angiosperms to algae. In some types of cells it is particularly striking that the microtubules are localized specifically over highly patterned wall thickenings and oriented parallel to the underlying cellulose microfibrils, as in the developing bands of secondary wall (Cronshaw, 1967; Cronshaw and Bouck, 1965; Esau et al., 1966; Hepler and Fosket, 1971; Hepler and Newcomb, 1964; Pickett-Heaps, 1966, 1967a; Wooding and Northcote, 1964; see Roberts, 1969, for review) and in secondary xylem where they are found around the rim of bordered pits (Robards and Humpherson, 1967). Similarly, in developing stomatal complexes, microtubules are specifically associated with the wall thickening that surrounds the pore (Kaufman et al., 1970; Singh and Srivastava, 1972).

In spite of the evidence that confirms the mutual orientation between cellulose and the microtubules, there have been several reports that have emphasized that microtubules are not always observed aligned with the underlying cellulose microfibrils (Chafe and Wardrop, 1970; Newcomb and Bonnett, 1965; O'Brien, 1972; Pickett-Heaps, 1968a; Robards, 1968; Srivastava, 1966; Srivastava and O'Brien, 1966a,b). One would expect that some microtubules, for example, in an elongating root tip cell, would deviate from a strict transverse alignment, since some of the cellulose microfibrils also do so. Images of skewed microtubules might also reflect a transient state wherein an element had just been formed but had not yet become aligned. In view of the fact that some of the reports purporting to show a lack of correlation between wall microfibrils and the cytoplasmic microtubules are based on studies that, by the admission of the authors, are insufficient to permit a detailed account of cellulose orientation (Srivastava and O'Brien, 1966a), the matter cannot be considered resolved.

Two studies, however, show a clear nonalignment of the microtubules with the underlying microfibrils of the cell wall, and these deserve special mention. The first study by Newcomb and Bonnett (1965) involved an analysis of the microtubule–microfibril orientation in rapidly growing root hairs. They observed that the microtubules in the root hair were predominantly axially aligned, and that they extended to within 3 or 4 μm from the cell tip. The cellulose microfibrils, however, were randomly oriented near the tip and were found aligned in an axial direction about 25 μm back from the tip. Thus, there was a zone approximately 20 μm in length wherein tubules of an axial orientation were localized in the cortical cytoplasm next to a cell wall that was composed of randomly oriented cellulose microfibrils. Although these observations argue against the relationship of microtubules and microfibrils, the following points must be taken into consideration. As pointed out by Newcomb and Bonnett (1965), the root hair grows at approximately 100 μm/hour, and thus the 20 μm zone would be replaced about every 12 to 15 minutes. It seems possible that the microtubules that are present in that 20 μm zone are becoming aligned in the axial direction in anticipation of their subsequent role in determining cellulose microfibril orientation. It would not be unreasonable to find a 12 to 15 minute lag period before the appearance of the layer of longitudinally oriented microfibrils. One should also bear in mind that the axially oriented fibrils may extend closer than 25 μm to the tip but simply not be resolved, since they are too few in number.

The question of cytoplasm–cell wall relationship and the role of microtubules has been considered by Chafe and Wardrop (1970). In collenchyma cells of *Apium*, they found that microtubules may either be parallel to or at marked angles to the microfibrils of the wall. They noted that the wall itself is composed of successive layers of cellulose microfibrils and that adjacent layers possess a different orientation. Chafe and Wardrop (1970) argue that microtubules that are aligned at marked angles to the underlying cellulose microfibrils may be in the process of initiating a new layer of microfibrils. In their structural analysis, they compared the relationship of the membrane particles discussed by Preston and Goodman (1968) and the microtubules to the developing cell wall, concluding that the microtubules are the cytoplasmic element most likely involved in the orientation of the cellulose microfibrils.

The theory that microtubules participate in the orientation of cellulose microfibrils is strengthened by several studies using the drug colchicine (Brennan, 1971; Green, 1962; Hepler and Fosket, 1971; Pickett-Heaps, 1967a). The earlier work of Green (1962) had shown that colchicine-treated *Nitella* cells became spherical and possessed walls composed of randomly oriented cellulose microfibrils. Root tip cells are also known to swell in the presence of colchicine, and a recent electron microscopic study confirms that the cortical microtubules have been destroyed (Brennan, 1971). Although cellulose microfibril orientation has not been examined in detail, it is argued that the lack of microtubules leads to a disordered deposition of cellulose microfibrils which weakens the ability of the wall to withstand the stresses imposed by turgor pressure, and thus permits the cells to swell (Brennan, 1971). Ethylene, a plant growth hormone, also causes cells to swell and, in addition, is reported to disrupt microfibril orientation (Apelbaum and Burg, 1971). While it mimics colchicine in some respects, it is unknown whether ethylene affects microtubule structure or organization.

More extensive studies have been performed on the effect of colchicine on secondary wall deposition in tracheary elements (Barlow, 1969; Hepler and Fosket, 1971; Pickett-Heaps, 1967a; Roberts and Baba, 1968). While secondary cell wall is continually deposited in the presence of the drug, the wall thickenings become markedly deformed (compare Fig. 8 with Figs. 9 and 10 and compare Figs. 11 and 12 with Figs. 13 and 14). Rather than being deposited in well-defined ridges, the secondary wall becomes smeared out over the primary wall (Figs. 9, 10, 13, and 14). Ultrastructural examination reveals that the microfibrils in the presence of the drug are deposited in swirls, sometimes at right angles to those deposited under normal conditions (Hepler and Fosket, 1971). These studies strongly indicate that microtubules participate in determining the orientation of the cellulose microfibrils in the wall.

In addition to affecting the orientation of the cellulose microfibrils, microtubules also control the position or pattern of the secondary wall bands in differentiating xylem elements (Hepler and Fosket, 1971). Pickett-Heaps (1966) has shown that cortical microtubules appear in groups along the wall prior to the formation of the secondary bands. These groups seem to determine where the thickening will be deposited. Removal of the microtubules with colchicine in wound vessel elements of *Coleus* at the inception of differentiation causes the secondary wall to be spread over the entire primary wall surface (Hepler and Fosket, 1971) (Fig. 13). Microtubules thus are involved in both determination and maintenance of the banded pattern of secondary wall in xylem elements.

Although microtubules are important elements in determining the

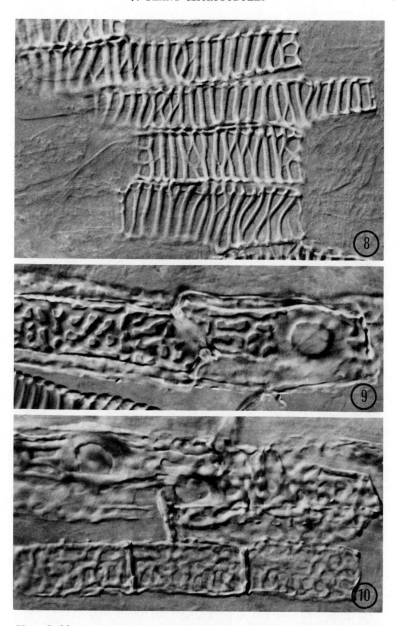

Figs. 8–10. Macerated and cleared wound vessel elements of *Coleus* examined in the Nomarski differential interference contrast microscope. Figure 8, untreated control. Note discrete, reticulate banding of the secondary wall. Figures 9 and 10, treated with colchicine. The drug causes the wall to become smeared. Note, however, that circular perforation plates are still evident. ×1200 (Hepler and Fosket, 1971).

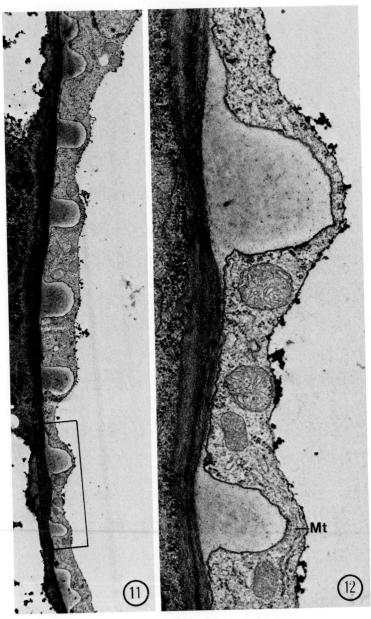

Figs. 11 and 12. Wound vessel elements of *Coleus;* untreated controls. Secondary wall is deposited in discrete bands along the primary wall. Microtubules (Mt) (Fig. 12) specifically overlie the thickenings of secondary wall. Outlined area in Fig. 11, shown at higher magnification in Fig. 12. Figure 11, ×5950; Fig. 12, ×29,750 (Hepler and Fosket, 1971).

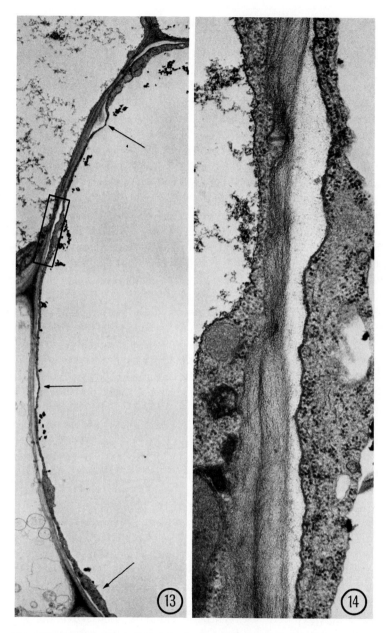

Figs. 13 and 14. Wound vessel elements of *Coleus* treated with colchicine. Secondary wall is smeared over the primary wall. Microtubules are absent. Outlined area in Fig. 13 is shown at higher magnification in Fig. 14. Figure 13, ×5100; Fig. 14, ×44,200 (Hepler and Fosket, 1971).

pattern in the formation of the secondary wall, other factors also participate. The secondary thickenings of adjacent xylem elements of *Coleus* are deposited directly opposite each other across the intervening primary wall, even in the presence of colchicine, indicating that the process is independent of microtubules. However, in the initiation of a new element that lacks a neighboring xylem member, microtubules seem to be the sole cytoplasmic organelle responsible for pattern formation (Hepler and Fosket, 1971).

How microtubules act in the orientation of cellulose microfibrils is an important question on which some speculation but few facts exist. Ledbetter and Porter (1963) suggested that the microfibril orientation was determined by the streaming pattern within the cell and that perhaps microtubules caused the streaming. Streaming as a process in the orientation of microfibrils is also a part of the model by Marx-Figini and Schulz (1966). With the exception of the bundles of microtubules that are observed in the endoplasm of the coenocytic alga, *Caulerpa* (Sabnis and Jacobs, 1967), however, there is no good correlation between streaming and the microtubules of the cell. Studies of *Nitella* show that streaming is neither caused by microtubules nor is it the orienting mechanism for cellulose microfibrils. Streaming in *Nitella* occurs only in the endoplasm, and is separated from the wall surface and the plasmalemma by a stationary layer of cytoplasm, the ectoplasm. Furthermore, the flow is oriented in the axial direction of a cell, while the microtubules (Pickett-Heaps, 1967b) and the cellulose microfibrils (Green, 1963) are oriented transversely around the cell. Studies using colchicine show that streaming is not blocked (Wessells *et al.*, 1971). Streaming is probably caused by microfilaments that are observed at the interface between the endoplasm and the ectoplasm (Nagai and Rebhun, 1966), and which are similar to muscle actin since they bind rabbit muscle heavy meromyosin (Palevitz *et al.*, 1974).

Rather than causing a streaming of the entire cytoplasm, it has been suggested that microtubules are involved in directing vesicles (for example, the Golgi vesicles), which may contain enzymes and wall precursors along defined channels (Maitra and De, 1971; Northcote, 1969; Pickett-Heaps, 1968c; Robards, 1968). To explain how the cellulose microfibrils become oriented, Northcote (1969) suggests that the flow of materials from the cytoplasm may detach the plasmalemma particles, presumed to have cellulose synthetic capacity, and move them in the direction of flow, thus imposing an orientation to the microfibrils as they are formed.

Another possibility is that microtubules give rise to orientation through the production of shearing forces developed at their surface. By interacting directly with the plasmalemma through cross bridges (Cronshaw, 1967; Kiermayer, 1968; Ledbetter and Porter, 1964; Robards, 1968, 1969) the microtubule, in a manner similar to that postulated by McIntosh *et al.* (1969), might develop shearing forces within the membrane which would generate a directed flow. It is conceivable that an oriented flow of macromolecules within the membrane would align enzyme complexes or precursor material, such as a growing cellulose chain, and thus govern the resulting orientation of the microfibril (Hepler and Fosket, 1971; Hepler and Palevitz, 1974).

While progress has been made in our thinking about microtubule function in cell wall formation, there are important questions that have not been answered. Of prime importance to our understanding of cellular morphogenesis is the mechanism by which the cell positions the microtubules in the first place. Green and King (1966) and more recently Green *et al.* (1970) suggest that in *Nitella* microtubules become aligned in young internode cells during a phase of marked lateral expansion. The strain of the lateral expansion would thus align the microtubules, and, once aligned, they would remain so, even in the face of tremendous deformation perpendicular to their axis. While this idea is attractive, in particular for the *Nitella* system, it cannot explain, for example, in regenerating xylem elements of *Coleus,* how the microtubule bands are patterned when there is little or no growth and presumably no significant strain. It is especially intriguing to those studying pattern formation in tracheary elements to speculate how microtubules become aligned in anticipation of subsequent wall depositions. It seems reasonable to suspect that membrane–microtubule interactions play an important role.

VIII. Conclusion and Prospect

Although our knowledge of the structure and composition of microtubules and of their relationship to many cell processes has increased markedly since their discovery as common cellular elements about 13 years ago, important questions about them remain unanswered. The mechanism of microtubule action and the cellular control of assembly have yet to be deciphered. The dividing cell will continue to be the object of intense investigation, and with the availability of elegant single cell systems, such as the endosperm of *Haemanthus,* and the possibility of experimentally modifying these cells, our understanding of the role of microtubules in chromosome and particle motion will be significantly enhanced. Perhaps even more revealing to an elucidation of the molecular mechanism of microtubule action will be the analysis of those systems, such as cilia, flagella, and axostyles of Pyrosnymphidae, and the axopods of Heliozoa, possessing highly ordered arrays of microtubules and in which outward movement can be related to the detailed internal structure and, in particular, to the tubule–tubule interactions. Because of the high degree of regularity in tubule spacing and in cross bridge periodicity, the opportunity exists for detecting through analysis by optical diffraction and related techniques subtle changes in microtubule structure and cross bridge conformation. While the cross bridge is becoming the focus of much current attention, further theories on its function would benefit greatly from a knowledge of its composition and possible enzyme activity, as has been achieved for the dynein arms of cilia.

The control of microtubule formation and position in the cell are questions that also relate intimately to the control of cell division and to the even less well understood processes of cellular morphogenesis. In dividing cells at least, certain regions have been experimentally identified as microtubule-initiating zones, although their composition and mechanism of activation and function at specific times throughout the cell cycle remain obscure. It is even more perplexing how cortical microtubules become organized, since here we have no evidence even for initiating sites. It would seem reasonable, therefore, to look toward the plasmalemma as the cellular component on which factors participating in the control of microtubule assembly and spatial organization may reside.

ACKNOWLEDGMENTS

I thank my colleagues at Stanford University for helpful discussions and criticism during the preparation of this chapter. Supported by Grants GB-25152 and BMS 74-15245 from the National Science Foundation.

REFERENCES

Allen, R. D., Bajer, A., and LaFountain, J. (1969). *J. Cell Biol.* **43**, 4a.
Apelbaum, A., and Burg, S. P. (1971). *Plant Physiol.* **48**, 648–652.
Bajer, A. (1958). *Chromosoma* **9**, 319–331.
Bajer, A. (1965). *Exp. Cell Res.* **37**, 376–398.
Bajer, A. (1967). *J. Cell Biol.* **33**, 713–720.
Bajer, A. (1968a). *Chromosoma* **24**, 383–417.
Bajer, A. (1968b). *Chromosoma* **25**, 249–281.
Bajer, A. (1968c). *Symp. Soc. Exp. Biol.* **22**, 287–310.
Bajer, A. (1973). *Cytobios* **8**, 139–160.
Bajer, A., and Allen, R. D. (1966a). *Science,* **151**, 572–574.
Bajer, A., and Allen, R. D. (1966b). *J. Cell Sci.* **1**, 455–462.
Bajer, A., and Molè-Bajer, J. (1969). *Chromosoma* **27**, 448–484.
Bajer, A., and Molè-Bajer, J. (1971a). *Proc. Int. Congr. Electron Microsc., 7th, 1970* pp. 267–268.

Bajer, A., and Molè-Bajer, J. (1971b), *Advan. Cell Mol. Biol.* **1**, 213–266.

Bajer, A., and Molè-Bajer, J. (1972). *Int. Rev. Cytol.* **34**, Suppl. 1–271.

Bannister, L. H., and Tatchell, E. C. (1968). *J. Cell Sci.* **3**, 295–308.

Bardele, C. F. (1973). *Cytobiol.* **7**, 442–488.

Barlow, P. W. (1969). *Protoplasma* **68**, 79–83.

Behnke, O., Forer, A., and Emmerson, J. (1971). *Nature (London)* **234**, 408–410.

Bibring, T., and Baxandall, J. (1968). *Science* **161**, 377–379.

Borisy, G. G., and Taylor, E. W. (1967). *J. Cell Biol.* **34**, 535.

Braselton, J. P., and Bowen, C. C. (1971). *Caryologia* **24**, 49–58.

Brennan, J. R. (1971). *Phytomorphology* **20**, 309–315.

Brinkley, B. R., and Cartwright, J., Jr. (1971). *J. Cell Biol.* **50**, 416–431.

Bünning, E. (1957). *Protoplasmatologia* **8**, 1–86.

Burgess, J. (1969). *Planta* **87**, 259–270.

Burgess, J. (1970a). *Planta* **92**, 25–28.

Burgess, J. (1970b). *Protoplasma* **69**, 253–264.

Burgess, J. (1970c). *Protoplasma* **71**, 77–89.

Burgess, J., and Northcote, D. H. (1967). *Planta* **75**, 319–326.

Burgess, J., and Northcote, D. H. (1968). *Planta* **80**, 1–14.

Burns, R. G., and Pollard, T. D. (1974). *FEBS (Fed. Eur. Biochem. Soc.) Lett.* **40**, 274–280.

Byers, B., and Abramson, D. H. (1968). *Protoplasma* **66**, 413–435.

Cande, W. Z., Snyder, J., Smith, D., Summers, K., and McIntosh, J. R. (1974). *Proc. Nat. Acad. Sci. U.S.* **71**, 1559–1563.

Chafe, S. C., and Wardrop, A. B. (1970). *Planta* **92**, 13–24.

Cronshaw, J. (1967). *Planta* **72**, 78–90.

Cronshaw, J., and Bouck, G. B. (1965). *J. Cell Biol.* **24**, 415–431.

Cronshaw, J., and Esau, K. (1968). *Protoplasma* **65**, 1–24.

Cutter, E. G., and Hung, C.-Y. (1972). *J. Cell Sci.* **11**, 723–737.

Deysson, G., and Benbadis, M. (1968). *C. R. Soc. Biol.* **162**, 601–604.

Dietz, R. (1969). *Naturwissenschaften* **56**, 237–248.

Esau, K., and Gill, R. H. (1965). *Planta* **67**, 168–181.

Esau, K., and Gill, R. H. (1969). *Can. J. Bot.* **47**, 581–591.

Esau, K., Cheadle, V. I., and Gill, R. H. (1966). *Amer. J. Bot.* **53**, 765–771.

Evert, R. F., and Deshpande, B. P. (1970). *Amer. J. Bot.* **57**, 942–961.

Forer, A. (1965). *J. Cell Biol.* **25**, 95–117.

Forer, A. (1966). *Chromosoma* **19**, 44–98.

Forer, A. (1969). *In* "Handbook of Molecular Cytology" (A. Lima-de-Farina, ed.), pp. 553–601. North-Holland Publ., Amsterdam.

Forer, A. (1974). *In* "Cell Cycle Controls" (G. M. Padilla, I. L. Cameron, and A. M. Zimmerman, eds.), pp. 319–336. Academic Press, New York.

Forer, A., and Behnke, O. (1972). *Chromosoma* **39**, 145–173.

Franke, W. W. (1971a). *Z. Naturforsch. B* **26**, 626–627.

Franke, W. W. (1971b). *Exp. Cell Res.* **66**, 486–489.

Freed, J. J., and Lebowitz, M. M. (1970). *J. Cell Biol.* **45**, 334–354.

Fuge, H., and Müller, W. (1972). *Exp. Cell Res.* **71**, 241–245.

Fulton, C. (1971). *In* "Origin and Continuity of Cell Organelles" (J. Reinert and H. Ursprung, eds.), p. 170. Springer-Verlag, Berlin and New York.

Gaskin, F., Kramer, S. B., Cantor, C. R., Adelstein, R., and Shelanski, M. L. (1974). *FEBS (Fed. Eur. Biochem. Soc.) Lett.* **40**, 281–286.

Gawadi, N. (1971). *Nature (London)* **234**, 410.

Gawlik, S. R., and Millington, W. F. (1969). *Amer. J. Bot.* **56**, 1084–1093.

Gibbons, I. R. (1963). *Proc. Nat. Acad. Sci. U.S.* **50**, 1002–1010.

Goldman, R. D., and Rebhun, L. I. (1969). *J. Cell Sci.* **4**, 179–209.

Green, P. B. (1962). *Science* **138**, 1404–1405.

Green, P. B. (1963). *Symp. Soc. Study Develop. Growth* **21**, 203–234.

Green, P. B. (1969). *Annu. Rev. Plant Physiol.* **20**, 365–394.

Green, P. B., and King, A. (1966). *Aust. J. Biol. Sci.* **19**, 421–437.

Green, P. B., Erikson, R. O., and Richmond, P. A. (1970). *Ann. N.Y. Acad. Sci.* **175**, 712–731.

Grimstone, A. V., and Cleveland, L. R. (1965). *J. Cell Biol.* **24**, 387–400.

Grove, S. N., and Bracker, C. E. (1970). *J. Bacteriol.* **104**, 989–1009.

Haber, J. E., Peloquin, J. G., Halvorson, H. O., and Borisy, G. G. (1972). *J. Cell Biol.* **55**, 355–367.

Hanzely, L., and Schjeide, O. A. (1973). *Cytobios* **7**, 147–162.

Harris, P., and Bajer, A. (1965). *Chromosoma* **16**, 624–636.

Hart, J. W., and Sabnis, D. D. (1973). *Planta* **109**, 147–152.

Hepler, P. K., and Fosket, D. E. (1971). *Protoplasma* **72**, 213–236.

Hepler, P. K., and Jackson, W. T. (1968). *J. Cell Biol.* **38**, 437–446.

Hepler, P. K., and Jackson, W. T. (1969). *J. Cell Sci.* **5**, 727–743.

Hepler, P. K., and Newcomb, E. H. (1964). *J. Cell Biol.* **20**, 529–533.

Hepler, P. K., and Newcomb, E. H. (1967). *J. Ultrastruct. Res.* **19**, 498–513.

Hepler, P. K., and Palevitz, B. A. (1974). *Annu. Rev. Plant Physiol.* **25**, 309–362.

Hepler, P. K., McIntosh, J. R., and Cleland, S. (1970). *J. Cell Biol.* **45**, 438–444.

Heslop-Harrison, J. (1968). *J. Cell Sci.* **3**, 457–466.

Inoué, S. (1952). *Exp. Cell Res., Suppl.* **2**, 305–314.

Inoué, S. (1953). *Chromosoma* **5**, 487–500.

Inoué, S. (1964). *In* "Primitive Motile Systems in Cell Biology" (R. D. Allen and N. Kamiya, eds.), pp. 549–598. Academic Press, New York.

Inoué, S., and Bajer, A. (1961). *Chromosoma* **12**, 48–63.

Inoué, S., and Sato, H. (1967). *J. Gen. Physiol. Suppl.* **50**, 259–292.

Inoué, S., Borisy, G. G., and Kiehart, D. P. (1974). *J. Cell Biol.* **62**, 175–184.

Kalnins, V. I., and Porter, K. R. (1969). *Z. Zellforsch. Mikrosk. Anat.* **100**, 1–30.

Kane, R. E., and Forer, A. (1965). *J. Cell Biol.* **25**, 31–39.

Kaufman, P. B., Petering, L. B., Yocum, C. S., and Basic, D. (1970). *Amer. J. Bot.* **57**, 33–49.

Kiermayer, O. (1968). *Planta* **83**, 223–236.

Krishan, A., and Buck, R. C. (1965). *J. Cell Biol.* **24**, 433–443.

Lambert, A., and Bajer, A. (1972). *Chromosoma* **39**, 101–144.

Ledbetter, M. C., and Porter, K. R. (1963). *J. Cell Biol.* **19**, 239–250.

Ledbetter, M. C., and Porter, K. R. (1964). *Science* **144**, 872–874.

Lehmann, H., and Schulz, D. (1969). *Planta* **85**, 313–325.

Lepper, R. J. (1956). *Bot. Rev.* **22**, 375–417.

Ludueña, R., Wilson, L., and Shooter, E. M. (1974). *J. Cell Biol.* **63**, 202a.

Luykx, P. (1970). *Int. Rev. Cytol., Suppl.* **2**, 1–173.

McIntosh, J. R. (1973). *J. Cell Biol.* **56**, 324–339.

McIntosh, J. R. (1974). *J. Cell Biol.* **61**, 166–187.

McIntosh, J. R., and Landis, S. C. (1971). *J. Cell Biol.* **49**, 468–497.

McIntosh, J. R., and Porter, K. R. (1967). *J. Cell Biol.* **35**, 153–173.

McIntosh, J. R., Hepler, P. K., and Van Wie, D. G. (1969). *Nature (London)* **224**, 659.

McIntosh, J. R., Ogata, E. S., and Landis, S. C. (1973). *J. Cell Biol.* **56**, 304–323.

Maitra, S. C., and De, D. N. (1971). *J. Ultrastruct. Res.* **34**, 15–22.

Marx-Figini, M. (1971). *Biochim. Biophys. Acta* **237**, 75–77.

Marx-Figini, M., and Schulz, G. V. (1966). *Biochim. Biophys. Acta* **112**, 81–101.

Mazia, D. (1961). *In* "The Cell" (J. Brachet and A. E. Mirsky, eds.), Vol. 3, pp. 77–412. Academic Press, New York.

Millington, W. F., and Gawlik, S. R. (1970). *Amer. J. Bot.* **57**, 552–561.

Mizukami, I., and Gall, J. (1966). *J. Cell Biol.* **29**, 97–111.

Molè-Bajer, J. (1969). *Chromosoma* **26**,427–448.

Mooseker, M. S., and Tilney, L. G. (1973). *J. Cell Biol.* **56**, 13–26.

Moser, J. W., and Krietner, G. L. (1970). *J. Cell Biol.* **44**, 454–458.

Nagai, R., and Rebhun, L. I. (1966). *J. Ultrastruct. Res.* **14**, 571–589.

Newcomb, E. H. (1969). *Annu. Rev. Plant Phys.* **20**, 253–288.

Newcomb, E. H., and Bonnett, H. T. (1965). *J. Cell Biol.* **27**, 575–589.

Nicklas, R. B. (1971). *Advan. Cell Biol.* **2**, 225–297.

Nicklas, R. B., and Koch, C. A. (1972). *Chromosoma* **39**, 1–26.

Nicklas, R. B., and Staehly, C. A. (1967). *Chromosma* **21**, 1–16.

Nicklas, R. B., Koch, C. A., and Marek, L. F. (1970). *J. Cell Biol.* **47**, 148a.

Northcote, D. H. (1969). *Proc. Roy. Soc., Ser. B* **173**, 21–30.

Northcote, D. H., and Lewis, D. R. (1968). *J. Cell Sci.* **3**, 199–206.

O'Brien, T. P. (1972). *Bot. Rev.* **38**, 87–118.

Olmsted, J. B, and Borisy, G. G. (1973). *Annu. Rev. Biochem.* **42**, 507–540.

Olmsted, J. B., Witman, G. B., Carlson, K., and Rosenbaum, J. L. (1971). *Proc. Nat. Acad. Sci. U.S.* **68**, 2273–2277.

Oppenheim, D. S., Hauschka, B. T., and McIntosh, J. R. (1973). *Exp. Cell Res.* **79**, 95–105.

Östergren, G., Molè-Bajer, J., and Bajer, A. (1960). *Ann. N.Y. Acad. Sci.* **90**, 381–408.

Palevitz, B. A., and Hepler, P. K. (1974). *Chromosoma* **46**, 297–326.

Palevitz, B. A., Ash, J. F., and Hepler, P. K. (1974). *Proc. Nat. Acad. Sci. U.S.* **71**, 363–366.

Paveletz, N. (1967). *Naturwissenschaften* **20**, 533–535.

Pickett-Heaps, J. D. (1966). *Planta* **71**, 1–14.

Pickett-Heaps, J. D. (1967a). *Develop. Biol.* **15**, 206–236.

Pickett-Heaps, J. D. (1967b). *Aust. J. Biol. Sci.* **20**, 539–551.

Pickett-Heaps, J. D. (1968a). *Aust. J. Biol. Sci.* **21**, 255–274.

Pickett-Heaps, J. D. (1968c). *Protoplasma* **65**, 181–205.

Pickett-Heaps, J. D. (1969a). *Aust. J. Biol. Sci.* **22**, 375–391.

Pickett-Heaps, J. D. (1969b). *J. Ultrastruct. Res.* **27**, 24–44.

Pickett-Heaps, J. D. (1969c). *J. Cell Sci.* **4**, 397–420.

Pickett-Heaps, J. D. (1969d). *Cytobios* **1**, 257–280.

Pickett-Heaps, J. D. (1975). *In* "Dynamic Aspects of Plant Ultrastructure" (A. W. Robards, ed.), pp. 219–225. McGraw-Hill, New York.

Pickett-Heaps, J. D., and Fowke, L. C. (1970). *Aust. J. Biol. Sci.* **23**, 71–92.

Pickett-Heaps, J. D., and Northcote. D. H. (1966a). *J. Cell Sci.* **1**, 109–120.

Pickett-Heaps, J. D., and Northcote. D. H. (1966b). *J. Cell Sci.* **1**, 121–128.

Porter, K. R. (1966). *Principles Biomol. Organ., Ciba Found. Symp., 1965* pp. 308–334.

Preston, R. D., and Goodman, R. N. (1968). *J. Roy. Microsc. Soc.* [3] **88**, 513–527.

Rebhun, L. I. (1963). *In* "The Cell in Mitosis" (L. Levine, ed.), pp. 67–106. Academic Press, New York.

Rebhun, L. I. (1964). *In* "Primitive Motile Systems in Cell Biology" (R. D. Allen and N. Kamiya, eds.), pp. 503–525. Academic Press, New York.

Rebhun, L. I. (1972). *Int. Rev. Cytol.* **32**, 93–137.

Rebhun, L. I., and Sawada, N. (1969). *Protoplasma* **68**, 1–22.

Rebhun, L. I., Rosenbaum, J., Lefebvre, P., and Smith, G. (1974). *Nature (London)* **249**, 113–115.

Robards, A. W. (1968). *Protoplasma* **65**, 449–464.

Robards, A. W. (1969). *Planta* **88**, 376–379.

Robards, A. W., and Humpherson, P. G. (1967). *Planta* **77**, 233–238.

Robbins, E., and Gonatas, N. K. (1964). *J. Cell Biol.* **21**, 429–463.

Roberts, L. W. (1969). *Bot. Rev.* **35**, 201–250.

Roberts, K. (1974). *Prog. Biophys. Molec. Biol.* **28**, 371–420.

Roberts, L. W., and Baba, S. (1968). *Plant Cell Physiol.* **9**, 315–321.

Robison, W. G. (1966). *J. Cell Biol.* **29**, 251–265.

Rosen, W. G. (1968). *Annu. Rev. Plant Physiol.* **19**, 435–462.

Rosenbaum, J. L., Moulder, J. E., and Ringo, D. L. (1969). *J. Cell Biol.* **41**, 600–619.

Roth, L. E., and Shigenaka, Y. (1970). *J. Ultrastruct. Res.* **31**, 356–374.

Roth, L. E., Wilson, H. J., and Chakraborty, J. (1966). *J. Ultrastruct. Res.* **14**, 460–483.

Roth, L. E., Pihlaja, D. J., and Shigenaka, Y. (1970). *J. Ultrastruct. Res.* **30**, 7–37.

Sabnis, D. D., and Jacobs, W. P. (1967). *J. Cell Sci.* **2**, 465–472.

Sakai, A. (1969). *Cytologia* **34**, 57–70.

Satir, P. (1967). *J. Gen. Physiol.* **50**, Suppl., 241.

Schrader, F. (1953). "mitosis, The Movements of Chromosomes in Cell Division." Columbia Univ. Press, New York.

Sharp, L. W. (1912). *Bot. Gaz. (Chicago)* **54**, 89–119.

Sharp, L. W. (1914). *Bot. Gaz. (Chicago)* **58**, 419–431.

Singh, A. P., and Srivastava, L. M. (1972). *Protoplasma* **76**, 61–82.

Slautterback, D. B. (1963). *J. Cell Biol.* **18**, 367–388.

Srivastava, L. M. (1966). *J. Cell Biol.* **31**, 79–93.

Srivastava, L. M., and O'Brien, T. P. (1966a). *Protoplasma* **61**, 257–276.

Srivastava, L. M., and O'Brien, T. P. (1966b). *Protoplasma* **61**, 277–293.

Stephens, R. E. (1971). *In* "Biological Macromolecules" (S. N. Timasheff and G. D. Fasman, eds.), Vol. 4, pp. 355–391. Dekker, New York.

Subirana, J. A. (1968). *J. Theor. Biol.* **20**, 177–123.

Summers, K. E., and Gibbons, I. R. (1971). *Proc. Nat. Acad. Sci. U.S.* **68**, 3092–3096.

Szöllosi, D. (1964). *J. Cell Biol.* **21**, 465–479.

Taylor, E. W. (1965a). *J. Cell Biol.* **25**, 145–160.

Taylor, E. W. (1965b). *Proc. Int. Congr. Rheol., 4th, 1963* Part 4, pp. 175–191.

Tilney, G. (1971a). *J. Cell Biol.* **51**, 837–854.

Tilney, L. G.((1971b). *In* "Origin and Continuity of Cell Organelles" (J. Reinert and H. Ursprung, eds.), pp. 222–260. Springer-Verlag, Berlin and New York.

Tilney, L. G., and Byers, B. (1969). *J. Cell Biol.* **43**, 148–165.

Tilney, L. G., and Goddard, J. (1970). *J. Cell Biol.* **46**, 564–575.

Tucker, J. B. (1968). *J. Cell Sci.* **3**, 493–514.

Tucker, J. B. (1970). *J. Cell Sci.* **7**, 793–821.

Warner, F. D., and Satir, P. (1973). *J. Cell Sci.* **12**, 313–326.

Weisenberg, R. C. (1972). *Science* **117**, 1104–1105.

Wessells, N. K., Spooner, B. S., Ash, J. F., Bradley, M. O., Ludueña, M. A., Taylor, E. L., Wrenn, J. T., and Yamada, K. M. (1971). *Science* **171**, 135.

Whaley, W. G., Dauwalder, M., and Kephart, J. (1966). *J. Ultrastruct. Res.* **15**, 169–180.

Wilson, E. B. (1928). "The Cell in Development and Heredity, 3rd ed. Macmillan, New York.

Wilson, H. J. (1969). *J. Cell Biol.* **40**, 854–859.

Wilson, H. J. (1970). *Planta* **94**, 184–190.

Wilson, L., and Bryan, J. (1974). *Advan. Cell Mol. Biol.* **3**, 21–72.

Witman, G. B., Carlson, K., Berliner, J., and Rosenbaum, J. L. (1972a). *J. Cell Biol.* **54**, 507–539.

Witman, G. B., Carlson, K., and Rosenbaum, J. L. (1972b). *J. Cell Biol.* **54**, 540–555.

Wooding, F. B. P., and Northcote, D. H. (1964). *J. Cell Biol.* **23**, 327–337.

8

Vacuoles

Ph. MATILE

I. Introduction

Plant cells are subdivided into several morphologically distinct compartments that are separated from one another by membranes. The *vacuome* (all the vacuoles present in a cell) is certainly one of the most conspicuous of these compartments. It is by far the most voluminous compartment in parenchyma cells. It is distinguished by its physical properties. Early plant cytologists (see Küster, 1951) conceived of vacuoles as watery regions separated by semipermeable membranes from the jellylike, colloidal cytoplasm. This membrane is known as the *tonoplast* because it is held in tension by the osmotic properties of the vacuolar fluid, the *cell sap.*

Vacuoles are ubiquitous in the plant kingdom, yet they vary widely in size, shape, number per cell, and even color. They are always surrounded by a simple unit membrane, a property shared with other organelles (such as peroxisomes, or constituents of the Golgi complex).

The identification of an organelle as a vacuole requires certification by several criteria. Classic French cytology used the accumulation of basic vital stains, such as neutral red, for the identification of vacuoles (see Dangeard, 1956). This criterion recognizes that the vacuole often accumulates organic acids and phenolic compounds capable of binding basic stains. It also includes all of the structures that are ontogenetically related to the minute organelles of meristematic cells that stain with neutral red, such as the specialized vacuoles that develop into the aleurone grains (protein bodies) in seeds. The spherosome, an oil-containing vacuole-like organelle has no affinity for neutral red and would consequently not be classified as a typical vacuole. Nevertheless, it shall be considered here because it bears other features typical of vacuoles.

Vacuoles have in the past been considered to represent sites of temporary or final deposition of metabolic waste products. The term waste product means that the substance has no presently known function. It is generally recognized also that the organic compounds and the inorganic solutes of the cell sap, principally in the vacuole, are responsible for the production of the osmotic pressure that is important to the water relations of plant tissue (turgor, rigidity, extension growth, cold resistance). As fixed nitrogen is a factor that is always scarce in nutrition, it has been argued that vacuoles represent a means of saving nitrogen-rich cytoplasm.

In all of these and similar functional interpretations, the vacuome plays a passive role. Apart from possibly actively accumulating substances by the tonoplast, the picture is that of a watery enclave in the cytoplasm. Conversely, the concept of an active participation in plant metabolism can also be imagined because of the presence of various digestive enzymes in vacuoles. Protein synthesis takes place outside of the vacuome in the cytoplasm. Protein breakdown, which may also play an important role in biochemical differentiation, involves enzymes stored at least in part in the vacuome. Turnover of cytoplasmic constituents involves both synthetic and lytic processes. The coexistence of these two different processes in a single individual cell would require specific cell compartments.

II. Biochemistry of Vacuoles

A. Experimental Approaches

Data on the biochemistry of vacuoles obtained from direct analysis are scarce if compared with the abundant knowledge available in the case of other organelles. This fact reflects the difficulties of the experimental approach. Isolation, the prerequisite for any decent investigation, is restricted to those cells and tissues whose vacuoles can be liberated at least partially and obtained in pure form. Methods for preparing the large central vacuoles of parenchyma cells are not available, and the size and fragility of these structures imply that it may be difficult to work out an appropriate isolation technique. In any case, the polymorphism of vacuoles will perhaps never allow a quantitative routine analysis of the vacuome in a given tissue.

Direct analysis of the vacuolar fluid has been performed only in a few cases of giant cells, such as those of *Nitella* and *Valonia*, whose saps can be aspirated through inserted syringes. The corresponding results concerning the accumulation of inorganic solutes are reviewed by Steward and Sutcliffe (1959).

1. ISOLATION OF VACUOLES

Conventional techniques of tissue extraction will normally result in the disruption or bursting of the central vacuole of parenchyma cells. Hence, the isolation of vacuoles is restricted either to meristems, whose small vacuoles are not destroyed upon grinding or blending of the tissue, or to selected cell types that can be suitably prepared for the purpose.

From the viewpoint of convenience the laticifers seem to represent an attractive system accessible by simple tapping. In certain species (e.g., *Hevea brasiliensis* or the Papaveracae) the vacuoles of laticifers are numerous and small. They can be easily isolated from the latex by differential or density gradient centrifugation (Pujarniscle, 1968; Matile et al., 1970). Although these vacuoles seem to represent typical plant vacuoles with regard to many biochemical properties, they have the disadvantage of originating from an extremely specialized tissue.

Difficulties of tissue homogenization with regard to the release of intact vacuoles are chiefly caused by the mechanical resistance of plant cell walls. Therefore, the most efficient techniques involve the conversion of cells into osmotically stabilized naked protoplasts that can subsequently be gently lysed under suitable conditions. Yeast cells are most useful for this purpose. Such cells are readily converted into protoplasts upon exposure to cell wall degrading enzymes (e.g., snail gut enzyme) and their vacuoles endure the osmotic shock needed to lyse the protoplasts. The procedure for isolating vacuoles from *Saccharomyces cerevisiae* (Matile and Wiemken, 1967) is illustrated in Fig. 1. Metabolic lysis, i.e., bursting of protoplasts in the presence of a metabolized sugar, has also been used for preparing yeast vacuoles (Indge, 1968a). In the case of *Neurospora* it is possible to obtain intact vacuoles by osmotically shocking conidia whose cell walls have been partially degraded by the action of hydrolases (Matile, 1971). Osmotic lysis of yeast proto-

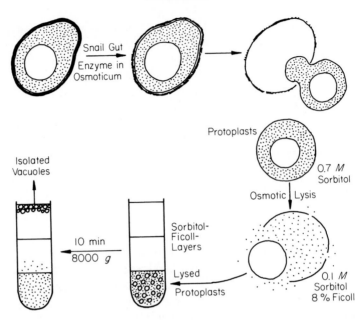

Fig. 1. Isolation of vacuoles from yeast protoplasts. Method of Matile and Wiemken (1967).

plasts yields vacuoles that are leaky with regard to micromolecules. If protoplasts are lysed under isotonic conditions, e.g., by gentle mechanical disintegration (see Wiemken and Nurse, 1973b) or by use of polycations (Dürr *et al.*, 1975), the micromolecular content of the vacuoles is preserved. Cocking (1960) has noticed the occurrence of free vacuoles in preparations of higher plant protoplasts. This observation suggests that the method of isolating vacuoles from protoplasts could possibly be modified for higher plant cells. A more convenient source of higher plant vacuoles is meristematic tissue. By grinding root tips in the presence of quartz sand and in a suitable medium, small vacuoles up to about 2 µm in diameter can be extracted and isolated by combined differential and density gradient centrifugation (Matile, 1968a). There are, however, other problems in dealing with root meristems. The cells of the tissue are heterogeneous and so are the physical properties of their vacuoles (size, density). The yield of isolated vacuoles depends (capriciously) on the initial grinding.

Aleurone grains have been isolated from a variety of reserve tissues using various techniques of extraction and centrifugation (see, e.g., Matile, 1968b; Ory and Henningsen, 1969; Morris *et al.*, 1970). The extraction of ungerminated seeds may result in damage to the tonoplast resulting in the loss of its water-soluble constituents. Furthermore, aleurone grains can be difficult to separate from starch grains. Yatsu and Jacks (1968) have avoided the latter difficulty by selecting an oil storing species, cotton. They also use glycerol as a nonaqueous extraction medium. This prevents the washing out of water-soluble substances from the vacuole.

In contrast to all other types of vacuoles, spherosomes are stable and can be readily extracted from any tissue. Separation from other cellular components is easily achieved by floating in aqueous media because the high content of lipid spherosomes causes them to have densities lower than unity (see, e.g., Jacks *et al.*, 1967; Ching, 1968; Matile and Spichiger, 1968; Mollenhauer and Totten, 1971b).

2. SEQUENTIAL EXTRACTION OF SUBCELLULAR COMPONENTS

The isolation of other cell components is generally associated with the bursting of vacuoles. It is, therefore, impossible to account quantitatively for the catalog of vacuolar components and enzyme activities. However, the isolation of vacuoles is restricted to a few suitable objects that furthermore must be kept under special conditions (e.g., protoplast formation) prior to liberation of vacuoles. A fruitful experimental approach to the dynamic properties of the vacuome requires methods that would allow the rapid estimation of vacuolar contents independent of the condition of the cell or tissue. Such an approach is in fact the stepwise extraction of yeast cells based on the rapid loss of semipermeability of the plasmalemma after treatment with positively charged proteins at low ionic strength (Schlenk *et al.*, 1970). If the exposure of yeast cells to basic proteins is carried out in an isoosmotic solution the vacuoles remain intact and the small molecules of the cytoplasm can be readily extracted. The vacuoles can then be ruptured osmotically and extracted by exposing the cells to water (Fig. 2). This technique of stepwise extraction has proved to be most useful in investigations on the compartmentation of small molecules. The localization of amino

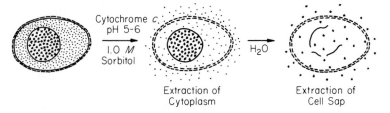

Fig. 2. Stepwise extraction of yeast cells. Modified after Schlenk *et al.* (1970).

acid pools in yeast cells has been elucidated by employing an advanced technique of stepwise extraction (Wiemken and Nurse, 1973a,b). A similar susceptibility to basic proteins of protoplasts prepared from *Avena* coleoptiles (Ruesink, 1971) would suggest that the technique may be successfully applied to higher plant cells as well.

B. Vacuolar Enzymes

In preparations obtained by conventional homogenization of plant tissues a large fraction, sometimes all of the activity of hydrolytic enzymes, is soluble. One cannot conclude from this result that these enzymes were originally present in the cytoplasmic matrix that contributes to the soluble cell fraction. Hydrolases, such as proteases, exopeptidases, nucleases, and other esterases, would interfere with the metabolic machinery and are therefore locked up in a distinct compartment—the lysosome. The discovery of the lysosome came about by the fact that rat liver homogenates contain sedimentable particles that release hydrolases when lysed (see de Duve, 1969). Plant tissue lysosomes were not so readily detected. However, the first attempt to isolate root meristem vacuoles yielded evidence that these contained hydrolytic enzymes (Matile, 1966). The lysosomal nature of plant vacuoles is now well documented. It appears from Table I that the complement of vacuolar hydrolases includes enzymes capable of hydrolyzing proteins, nucleic acids, phosphate and other esters, and glycosidic bonds. The list of vacuolar hydrolases is certainly incomplete, a circumstance that is partially due to lack of knowledge of plant hydrolases. A morphological study on senescence in the corolla of *Ipomoea* has shown that the cellular digestive processes may ultimately lead to the complete disappearance of structured cytoplasmic material (Matile and Winkenbach, 1971). This finding indicates the existence in this tissue of a hydrolytic enzyme system capable of eventually breaking down all cytoplasmic macromolecules. Isolated yeast vacuoles contain all of the proteolytic enzymes (endo- and exopeptidases and possibly dipeptidases) necessary for completely hydrolyzing the octapeptide angiotensin II blocked at both the terminal amino and carboxyl groups (Wiemken, 1969).

A complete analysis of the compartmentation of hydrolases is difficult to achieve because all vacuoles present in a given material are never completely recovered in any isolate. Furthermore, many hydrolases have a dual or multiple localization, e.g., acid phosphatase whose presence has been demonstrated in vacuoles, dictyosomes, endoplasmic reticulum, and the extracellular space (Fig. 3) in a variety of objects

TABLE I

Hydrolases Detected in Isolated Vacuoles

Source	Structure isolated	Hydrolases	Reference
Zea mays	Small vacuoles from root meristem	Endo- and exopeptidases, RNase, DNase, phosphatase, phosphodiesterase, acetylesterases, β-amylase, α- and β-glucosidase, β-galactosidase	Matile (1966, 1968a, unpublished results)
Asplenium fontanum	Small vacuoles from meristematic fronds	Protease, RNase, DNase, phosphatase, phosphodiesterase, β-galactosidase	Coulomb (1971)
Solanum tuberosum	Small vacuoles from young dark grown shoots	RNase, phosphatase, phosphodiesterase, acetylesterase	Pitt and Galpin (1973)
Hevea brasiliensis	Vacuoles (lutoides) from latex	Endopeptidase, RNase, DNase, phosphatase, phosphodiesterase, β-glucosidase, β-galactosidase, β-N-acetylglucosaminidase	Pujarniscle (1968)
Chelidonium majus	Vacuoles from latex	Endopeptidase, RNase, phosphatase	Matile *et al.* (1970)
Saccharomyces cerevisiae	Vacuoles from lysed protoplasts	See Table II	
Coprinus lagopus	Vacuoles from vegetative hyphae and fruiting bodies	Endopeptidases, RNase, phosphatase, β-glucosidase, chitinase	Iten and Matile (1970)
Neurospora crassa	Vacuoles from macroconidia	Endopeptidases, aminopeptidase, RNase, phosphatases, phosphodiesterase, invertase	Matile (1971)
Acetabularia mediterranea	Stratified subcells prepared from stalks	RNase, phosphatase	Lüscher and Matile (1974)
Nitella axilliformis	Sap from internodal central vacuole	Phosphatase, carboxypeptidase	Doi *et al.* (1975)

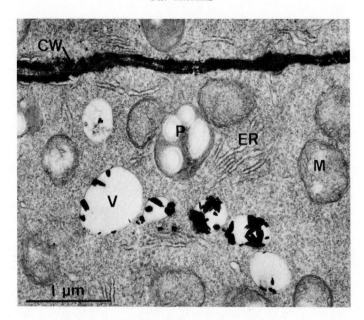

Fig. 3. Cytochemical demonstration of acid phosphatase activity in small vacuoles (V) and cell walls (CW). Mitochondria (M), plastids (P), endoplasmic reticulum (ER), and the cytoplasmic matrix are devoid of reaction product. Root meristem of *Cucumis sativus*. Courtesy of N. Poux.

(e.g., Poux, 1970). In *Saccharomyces cerevisiae* only one out of four aminopeptidase isozymes is localized in vacuoles. In addition, it is induced only under culture conditions that cause biochemical differentiation of cells (Matile *et al.*, 1971).

To demonstrate that a particular enzyme is vacuolar, it is required to show that not only is the activity present in the subcellular preparation, but, in addition, that it is absent from other compartments and that it is releasable by lysis of the compartment in which it is contained (Table II). Whether or not this latency of a vacuolar enzyme can be shown depends on the stability of the isolated vacuoles under the conditions of the assay (see, e.g., Pujarniscle, 1969; Matile, 1971).

Yeast cells contain inhibitor proteins specific for proteases A and B as well as for carboxypeptidase. As these inhibitors are located in the cytoplasm, it appears that the vacuolar proteolytic enzymes are fully active in the living cells (Lenney *et al.*, 1974).

Unspecific acid phosphatase has gained an important position among the hydrolases not only because it is usually active and can easily be assayed but also because it can be demonstrated cytochemically at the level of electron microscopy. Its vacuolar localization (Fig. 3)

TABLE II

Enzymes Present or Absent in Vacuoles Isolated from Protoplasts of *Saccharomyces cerevisiae*[a]

Enzyme	Ratio of specific activities isolated vacuoles/lysed protoplasts
Mitochondrial and soluble cytoplasmic enzymes	
Cytochrome oxidase	<0.01
Ethanol dehydrogenase	<0.01
Hydrolases with exclusive intracellular localization	
Protease A	20.7
Protease B	40.2
Carboxypeptidase	24.0
RNase	19.2
α-Mannosidase[b]	20.0
Acetylesterase	20.5
Intra- and extracellularly localized hydrolases	
Leucyl aminopeptidase	7.7
Invertase	20.1
β-Glucosidase[c]	27.8
Acid phosphatase (Co^{2+})[d]	15
Alkaline phosphatase(Mg^{2+})[d]	<40
Specific hydrolases absent from vacuoles	
α-Glucosidase[d]	<0.01
α-Glycerophosphatase	<0.01
Oxidoreductases probably localized in tonoplast	
NADH:cytochrome c oxidoreductase	0.016
NADH:DIP oxidoreductase	0.638

[a] Data unless otherwise stated compiled from Matile and Wiemken (1967) and from Wiemken and Nurse (1973b). *Saccharomyces cerevisiae* strain LBG 1022.

[b] Van der Wilden *et al.* (1973).

[c] Cortat *et al.* (1972).

[d] Derepressed cells.

has indeed been demonstrated repeatedly using this technique (see, e.g., Poux, 1970; Berjak, 1972). Cytochemical work has also led to the observation that acid phosphatase represents a conspicuous enzyme of aleurone grains (see, e.g., Poux, 1965). As shown in Table III these specialized vacuoles are not only filled with reserve proteins but also with proteases capable of mobilizing these same proteins. The same remark applies to the presence of phytase activity and phytate in aleurone grains. The spectrum of hydrolases detected in aleurone vacuoles from cotyledons of germinating pea seeds suggests that the digestive ca-

TABLE III

Hydrolase Activities Present in Isolated Aleurone Grains and Spherosomes

Source	Enzyme	Reference
Aleurone grains		
Ungerminated cotton seeds	Endopeptidase, phosphatase	Yatsu and Jacks (1968)
Cotyledons of germinating pea seed	Endopeptidase, RNase, phosphatase	Matile (1968b)
	Acetylesterase, α-glucosidase, β-amylase	
Ungerminated barley seeds	Endopeptidase, phytase	Ory and Henningsen (1969)
Ungerminated hempseeds	Endopeptidase	St. Angelo et al. (1969)
Broad bean seeds	Endopeptidase, phosphatase, phytase	Morris et al. (1970)
Mung bean seeds	Caseolytic activity, carboxypeptidase, α-mannosidase, N-acetylglucosaminidase	Harris and Chrispeels (1975)
Sorghum bicolor seeds	Protease, RNase, phosphatase, pyrophosphatase, phytase, α- and β-glucosidase, β-galactosidase	Adams and Novellie (1975)
Spherosomes		
Castor bean endosperm	Lipase	Ory et al. (1968)
Tobacco endosperm	Endopeptidase, RNase, DNase, phosphatase, acetylesterase, lipase	Matile and Spichiger (1968)
Douglas fir seeds	Lipase	Ching (1968)
Maize scutellum	Lipase	P. Matile (unpublished)

pacity of these organelles may not be restricted to the reserves stored in the vacuolar compartment (Matile, 1968b).

A storage organelle analogous to the aleurone vacuole is the spherosome, known to be the site of triglyceride accumulation. Although spherosomes have some cytological properties which distinguish them from true vacuoles (see Section IV), they nevertheless share the lysosomal nature. At least one type of hydrolase, lipase, seems to be associated with spherosomes from some oleaginous tissues (Table III). Enzyme cytochemistry at the light microscope level has yielded evidence for the localization of acid phosphatase and other hydrolases in spherosomes of nonoily tissues (see Matile, 1969). The corresponding activities have not been detected, however, in isolated spherosomes (Yatsu et al., 1971). Tobacco endosperm seems to represent an exception among oleaginous tissue. Its spherosomes were found to contain those hydrolases typical of plant vacuoles (Matile and Spichiger, 1968). Lipase activity has not been detected in spherosomes isolated from peanuts (Jacks *et al.*, 1967).

C. Substances Deposited in Vacuoles

Using cytochemical tests and vital stains, the early plant cytologists produced a vast amount of knowledge about the chemical peculiarities of vacuolar contents. Inorganic salts, sugars, organic acids, amino acids, amides, lipids, mucilages, gums, tannins, anthocyanins, and flavones have been detected in vacuoles. The corresponding reviews (e.g., Küster, 1951) provide abundant information on these matters. This early work, while informative, does not satisfy today's desire for precise and quantitative data.

1. INORGANIC SUBSTANCES

Approaches to the problem of accumulation of inorganic solutes in vacuoles have been largely indirect. The corresponding evaluations are mostly based on the kinetics of radioisotope losses from labeled tissues. It is generally assumed that the pools of ions that exchange very slowly with the external medium are localized in the cell sap. Moreover, it has been shown that only a fraction of the total pool of phosphate and of nitrate is metabolized. In the case of orthophosphate, indirect evidence favoring the vacuolar localization of the passive pool has been presented (Ullrich *et al.*, 1965). Only the coenocytic vesicles of *Valonia* and other giant cells, which are large enough to allow the insertion of micropipettes or microelectrodes, have lent themselves for studies on either the chemical composition of the cell sap or the ion transporting capacity of the tonoplast (see Steward and Sutcliffe, 1959; Gutknecht and Dainty, 1968).

Because the permeability properties of tonoplasts may be altered and solutes released upon the isolation of vacuoles, it is doubtful whether cell fractionation will be useful in the study of intracellular distribution of inorganic ions. Ribaillier *et al.* (1971) have presented quantitative estimates on the distribution of some inorganic ions between vacuoles and cytoplasm (serum) in *Hevea* latex. The corresponding concentration ratios are different for different ion species. Mg^{2+} was found to be accumulated 90-fold, Ca^{2+} 6-fold, and Cu^{2+} 2-fold in the vacuole. In contrast, equal concentrations of K^+ were found in vacuole and cytoplasm. It should be noted that in the case of *Hevea*, the separation of vacuoles can be performed by centrifugation of the whole latex (that is, the natural "medium" of the vacuoles); the cytoplasm is not replaced by an artificial medium as is normally the case in cell fractionation work. The estimated distribution of inorganic ions may, therefore, really correspond with the *in vivo* situation. Acid-soluble inorganic phosphates represent further constituents that are accumulated in the vacuolar fraction of *Hevea* latex (Ribaillier *et al.*, 1971). A similar result has been found by Indge (1968b) in vacuoles prepared from yeast protoplasts. About 40% of the acid-soluble phosphorus present in the protoplast was associated with a crude preparation of vacuoles. This vacuolar phosphorus was present largely as polyphosphates (volutin) and is probably responsible for the absorption of neutral red by yeast vacuoles (Indge, 1968a). Of the total K^+ pool of yeast protoplasts, only 20% was recovered from the vacuolar fraction (Indge, 1968b).

2. METABOLIC INTERMEDIATES

Many of the intermediates of plant metabolism are formed in excess. Well known is the accumulation of tricarboxylic acid cycle intermediates and other acids responsible for the acidity of most extracts from plant material. The sap squeezed from lemons has pH values as low as 2.5, indicating that inactivation of cytoplasmic enzymes would occur if the acids were stored in the cytoplasm. It is therefore necessary to postulate that acids formed in excess (or inhibitors, such as malonic acid, present in many species) are stored in cellular regions that are remote from centers of metabolism. The existence of corresponding discrete pools of intermediates that are not in equilibrium with the metabolized pools has been demonstrated repeatedly (see Beevers *et al.*, 1966). It is generally assumed that these storage pools are localized in the vacuoles. This assumption is either based on the well-known acidity of cell saps demonstrated, e.g., by the color of anthocyanins, natural indicators present in certain cell saps, or by vital indicator stains such as bromophenol blue. Another indication is given by the observation that the relative sizes of storage pools of organic acids are larger in the highly vacuolated proximal segments of rootlets as compared with the less vacuolated meristematic tips (MacLennan *et al.*, 1963). Acid accumulation has also been demonstrated directly in vacuoles of *Hevea* latex. This compartment contains citric acid at a concentration up to 24 times higher than the cytoplasm. In contrast, malic acid is present at about equal concentrations in vacuoles and cytoplasm of *Hevea* latex (Ribaillier *et al.*, 1971). There are storage pools of amino acids in plant cells just as is the case with organic acids. Thus, amino acids have been detected in vacuoles prepared from yeast protoplasts. These contain 23% of the total amino acid pool (Indge, 1968b). Stepwise extraction of *Candida utilis* cells by Wiemken and Nurse (1973a,b) revealed that an astonishingly small fraction, only 5 to 10% of total amino acid, is in the cytoplasm. A large vacuolar pool accounted for at least 90% of the total amino acid pool and predominantly contained the nitrogen-rich species of amino acids. If the cells were supplied with inorganic nitrogen, this pool was filled with endogenously produced amino acids. If, however, certain amino acids were supplied as the sole nitrogen source, they were accumulated preferentially and accounted for 50% of the vacuolar pool.

In yeasts the accumulation of purines in vacuoles is conspicuous if the cells are supplied with purines as a source of nitrogen. The food yeast, *Candida utilis*, is able to utilize various purines that are slowly metabolized but held at high intracellular concentrations. The vacuolar localization of these storage pools has been clearly demonstrated in the case of uric acid and isoguanin. The low water solubility of these compounds causes them to form large crystals in the vacuoles (Roush, 1961) (Fig. 4). The vacuolar deposition of purines has also been demonstrated by UV microscopy and by employing the technique of stepwise cell extraction. If *Saccharomyces cerevisiae* is grown in a medium containing methionine, *S*-adenosylmethionine is synthesized in excess and deposited in the vacuoles (Svihla *et al.*, 1963). The cells are unable to

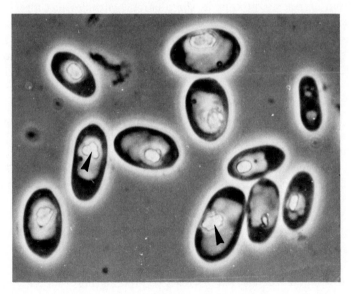

Fig. 4. Crystals of uric acid (arrowheads) in vacuoles of *Candida utilis* cells grown on uric acid as source of nitrogen. ×5000. Courtesy of A. Wiemken.

accumulate this compound if it is added to the culture medium. In contrast, other biological sulfonium compounds, S-methyl-L-methionine and dimethyl-3-propiothetin, are readily taken up by cells if supplied under aerobic conditions. Their intracellular distribution differs, however, considerably from that of S-adenosylmethionine. Dimethyl-3-propiothetin is located predominantly in the cytoplasm, whereas nearly equal amounts of S-methyl-L-methionine are found in vacuoles and cytoplasm (Schlenk et al., 1970). These observations suggest that there are specific mechanisms responsible for governing the intracellular distribution of each different compound.

3. RESERVE SUBSTANCES

The existence of discrete pools of sugars in plant cells has been demonstrated indirectly in various ways. Direct evidence for the localization of storage pools in the vacuoles is not available. In *Hevea* latex sugars seem to be present mainly in the cytoplasm (Ribaillier, 1971). It is logical to assume, however, that the large quantities of sucrose stored in sugarcane internodes and in similar organs are present in the cell sap. In contrast to these examples of sugar-accumulating species, most plants convert large quantities of reserves into osmotically inactive macromolecules (protein, starch, glycogen, etc.) or neutral fat. The deposition of reserve protein in the aleurone vacuoles has already been mentioned. Similar vacuolar distribution has been established for specific storage proteins in a variety of species. By use of specific fluorescent antibodies Graham and Gunning (1970) have demonstrated the exclusive localization of legumin and vicilin in the aleurone grains of bean cotyledon cells. This result is in agreement with analytical work on isolated *Vicia faba* aleurone grains (Morris et al., 1970). Protein bodies isolated from soybean meal contained 70% of the total protein, and the major reserve protein, glycinin, was identified in the isolate. In addition, the preparation contained phytic acid (myoinositol hexaphosphate), the conspicuous phosphate reserve of seeds (Tombs, 1967). Cell fractionation work (Lui and Altschul, 1967) as well as cytochemical tests (Poux, 1965) have shown that the so-called globoids (conspicuous inclusions in aleurone grains of many species) consist of precipitated phosphates. Alternatively phytate may not be segregated into distinct globoids but rather may be associated with the reserve protein (Tronier et al., 1971).

The vacuolar deposition of reserve proteins may not be restricted to seeds. In leaves of Solanaceae (tomato, potato), a protein characterized by its inhibitory action on chymotrypsin and other proteinases is

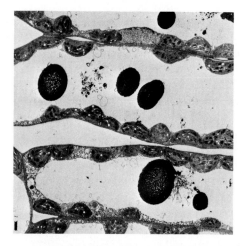

Fig. 5. Palisade mesophyll of a tomato leaf whose juice contained 565 μg chymotrypsin inhibitor protein per ml. Note the large electron-dense deposits in the vacuoles. Glutaraldehyde–osmium fixation. Courtesy of L. K. Shumway and C. A. Ryan.

synthesized in considerable amounts under certain environmental conditions. The occurrence of this chymotrypsin inhibitor protein in leaves has been found to be correlated with the presence of conspicuous electron-dense bodies in the vacuoles (Fig. 5) (Ryan and Shumway, 1971).

A structure analogous to the protein vacuoles is represented by the spherosomes. These oil vacuoles are ubiquitous in plant cells. In oil seeds they may occupy a large fraction of the total volume of reserve cells (Frey-Wyssling *et al.*, 1963). Over 90% of the total lipid is contained in the spherosomes isolated from corresponding reserve tissues (Jacks *et al.*, 1967; Matile and Spichiger, 1968).

4. ALKALOIDS, PHENOLIC COMPOUNDS, AND POLYTERPENES

The hydrophilic nature of alkaloids suggests that these compounds could be dissolved and accumulated in vacuoles. However, the example of nicotine, which is localized predominantly in the cell walls and in plastids of *Nicotiana rustica* leaf petioles (Müller *et al.*, 1971), indicates that this is not necessarily the case. In *Macleaya cordata* tissue cultures, the diffuse yellow color of alkaloid-containing idioblasts is due to the presence of sanguinarine. Neumann and Müller (1967), employing cytochemical and autoradiographical techniques at the electron microscopic level, have been able to demonstrate the vacuolar localization of alkaloids in these cells. If hexachloroplatinic acid is added to the fixation medium, large electron-dense clumps of precipitated alkaloid are formed in the vacuoles. In another Papaveraceae, *Chelidonium majus*, similar bodies are present in the vacuoles of laticifers (Matile *et al.*, 1970). The latex of this species is yellow, and so are the vacuoles isolated from the latex. Sanguinarine, barberine, and several other alkaloids are present in the isolate in a concentrated form. They are probably bound to phenols, which are also present in the cell sap (Jans, 1974).

Phenolic compounds represent another group of metabolic products deposited in vacuoles. This localization is clear even by light microscope examination of cells containing flavonoid pigments. It can also easily be demonstrated by use of ferric ion as a cytochemical reagent in the case of colorless phenols. The iron–phenol complex is colored. Tannin inclusions in vacuoles have also been demonstrated by the electron microscope (Chafe and Durzan, 1973; Baur and Wilkinshaw, 1974).

The vacuolar localization of the last class of excretion products, isoprenoids, is based on electron microscopic observations. In species of the genus *Euphorbia*, the rubber particles first occur in cytoplasmic vesicles (Marty, 1968) and are ultimately transferred into the large central vacuole (Schnepf, 1964; Schulze *et al.*, 1967; Marty, 1971). Conversely, in species such as *Hevea, Papaver,* and *Taraxacum,* the rubber granules are deposited in the cytoplasmic matrix (Schulze *et al.*, 1967). Apparently, a special compartmentation is not required in the case of the metabolically inert polyterpenes. Other isoprenoids are excreted into the extracellular space.

III. Functions of Vacuoles

A. Lysosomal Function

1. Autophagy: Turnover and Differentiation

In plant tissues, many macromolecules essential for vital functions are subjected to turnover. Whether a particular species of protein or nucleic acid is metabolically labile is often difficult to determine as is the exact turnover rate. It is generally taken for granted that nuclear DNA is metabolically stable. Protein and cytoplasmic nucleic acids do, however, turn over, as demonstrated by the gradual disappearance of isotope labels after a period of isotope incorporation (the so called pulse-chase experiment). The study on synthesis and turnover of a selected enzyme protein, nitrate reductase, carried out by Zielke and Filner (1971), included both density and radioisotope labeling of proteins in cultured tobacco cells under various conditions. This study exemplifies the difficulties encountered in the quantitative analysis of simultaneous synthesis and degradation of protein in individual cells. Nitrate reductase is a short-lived enzyme (i.e., it turns over rapidly). There are most probably vast differences in turnover rates of different species of protein. In any case, if turnover occurs under steady-state conditions (constant enzyme content), synthesis and degradation are in equilibrium. It is difficult to understand the reason for turnover. If, however, certain species of enzymes are broken down and are simultaneously replaced by other enzymes, cells change their metabolic abilities. This constitutes biochemical differentiation. Such differentiation is in response to changing environmental conditions. The regulation of metabolic activity is, therefore, determined, on the one hand, by control of enzyme synthesis (induction, repression), and, on the other hand, by control of enzyme degradation. A good example of biochemical differentiation is the secretion of hydrolytic enzymes which is induced by gibberellic acid in barley aleurone layers. These enzymes are synthesized at the expense of reserve proteins mobilized in the aleurone vacuoles. This has been demonstrated by density labeling of α-amylase and other induced enzymes in the presence of $H_2^{18}O$, which upon proteolysis is incorporated into amino acids (see Filner et al., 1969).

In the context of vacuoles the most interesting aspect of enzyme turnover concerns the necessity of compartmentation of lytic and synthetic processes. The localization of the cellular digestive enzymes in vacuoles suggests the involvement of the vacuome in the degradative part of turnover reactions. An important consequence of this compartmentation concerns interaction between vacuoles and cytoplasm. Cellular digestive processes seem to require not only lysosomal enzyme activities in the vacuoles but also the transport of cytoplasmic material into the vacuome. The corresponding process, *autophagy*, appears to involve an active participation of the vacuolar membranes.

An autophagic activity of vacuoles has been suspected because of the presence in the cell sap of material of cytoplasmic origin. Remnants of mitochondria and other membraneous material, ribosomes and unidentified material, have been encountered in vacuoles of a variety of plant cells (e.g., Poux, 1963; Malkoff and Buetow, 1964; Thornton, 1968; Coulomb and Buvat, 1968; Zandonella, 1970; Matile and Winkenbach, 1971; Fineran, 1971). The incorporation of cytoplasmic macromolecules and structures into the vacuome appears to involve increased amount of membranes. Invaginations of the tonoplast resulting in the

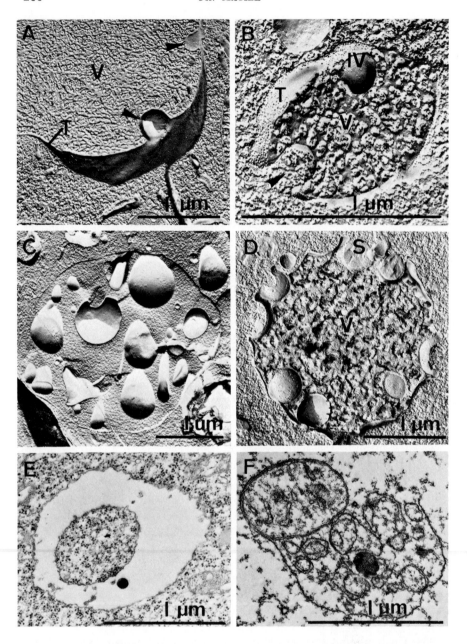

Fig. 6. Autophagic activity of vacuoles (V). (A) Invaginating (arrowheads) tonoplast (T) of a vacuole in a root meristem cell of *Zea mays*. (B) Cross-fractured invagination (arrowhead) and an intravacuolar vesicle (IV) in a small meristematic vacuole. *Zea mays*. (A) and (B) from Matile and Moor (1968). (C) Autophagic

formation of intravacuolar vesicles, which contain cytoplasmic material, have been observed in corn root meristem cells (Matile and Moor, 1968) and other objects (see, e.g., Fineran, 1971; Mesquita, 1972) using freeze-etching as well as ultrathin sectioning techniques (Fig. 6).

In this process preexisting vacuoles are involved in the sequestration of cytoplasmic material. A second type of autophagy is characterized by the wrapping of large portions of cytoplasm in endoplasmic reticulum, followed by the disorganization and decay of the sequestered material (Fig. 7) (Buvat, 1968; Marty, 1970; Villiers, 1972; Mesquita, 1972; Cresti et al., 1972). Apparently, there are only minor differences between these two types of autophagy. Vacuolization and autophagic activity may occur simultaneously, or they may be temporally separated. In both cases, the strict compartmentation of the vacuolar hydrolases and their involvement in the degradation of the sequestered cytoplasm is evident (Fig. 8).

A problem still to be solved is that of the specificity of autophagy. On morphological grounds, it is difficult to evaluate the biochemical significance of the process. Fineran (1971) has presented micrographs indicating the specific sequestration of certain cytoplasmic structures, and Villiers (1971) has been able to demonstrate specificity of autophagic sequestration of proplastids in *Fraxinus* seeds that had been subjected to prolonged dormancy. Apart from such morphologically discernible selectivity, the sequestration normally seems to comprise an average portion of cytoplasm. These facts do not help to explain the different turnover rates of different species of macromolecules. It must be assumed that additional mechanisms of autophagy remain to be discovered.

A functional significance of intracellular digestion in vacuoles is suggested by studies on biochemical differentiation in yeast. In *Saccharomyces cerevisiae*, a facultative anaerobe, which, in addition, is characterized by a repressor effect of glucose on the synthesis of respiratory enzymes, the phenomenon of biochemical differentiation is particularly conspicuous. If the organism is cultured on a glucose medium, respiration is repressed and ethanol is produced until the gradual disappearance of glucose from the medium ultimately results in derepression of respiration. At this stage the culture enters a lag phase, during which the dif-

---

vacuole containing numerous intravacuolar vesicles in a hymenial cell of the senescing gill of *Coprinus lagopus*. From Iten and Matile (1970). (D) Uptake of spherosomes (S) into a vacuole of *Saccharomyces cerevisiae*. Courtesy of H. Moor. (E) Intravacuolar vesicle containing a portion of cytoplasmic matrix with ribosomes. *Avena sativa* root tip. Courtesy of B. A. Fineran. (F) Ruptured intravacuolar vesicle containing a mitochondrion and membraneous material in a large vacuole of an *Avena* root meristem cell. Courtesy of B. A. Fineran.

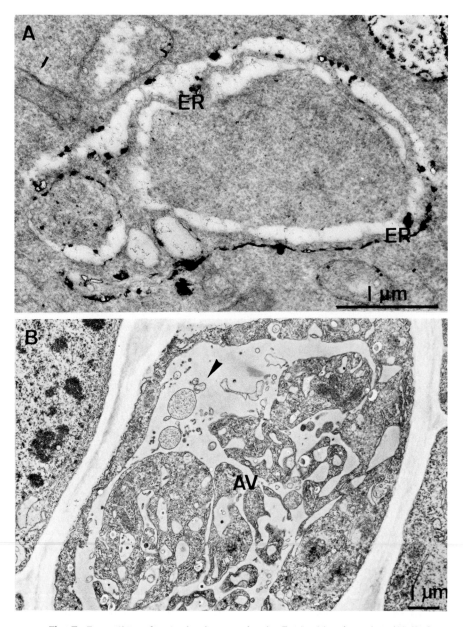

Fig. 7. Formation of autophagic vacuoles in *Euphorbia characias*. (A) Endoplasmic reticulum (ER) wrapping up portions of cytoplasmic matrix. Electron-dense material represents reaction product of acid phosphatase in the ER cisterna. Root meristem. (B) Autophagic vacuole (AV) in a meristematic laticifer. Beginning decay (arrowhead) of the sequestered cytoplasmic material. Courtesy of F. Marty.

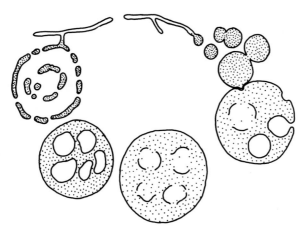

Fig. 8. Diagram summarizing the two forms of autophagic activity illustrated in the Figs. 6 and 7.

ferentiation from fermentative to oxidative metabolism takes place. It is this lag phase separating the two exponential phases of diauxic growth (fermentative on glucose to oxidative on ethanol) which is characterized by the severalfold increase of proteolytic and other hydrolase activities (Wiemken, 1969; Matile *et al.*, 1970). In contrast, these activities are comparatively low in exponentially growing cells that are fully adapted to the environmental conditions. Stationary phase cells that are deprived of exogeneous nutrients are characterized by autophagic metabolism, and, correspondingly, the activities of the vacuolar enzymes are high in these cells. In fact, Halvorson (1960) has demonstrated that protein is synthesized in stationary yeast cells at the expense of existing proteins. Turnover rates as high as 7% of the total protein per hour have been estimated in stationary yeast cells (Fukuhara, 1967).

2. SENESCENCE AND AUTOLYSIS

Senescence may be thought of as the final phase of development. It represents an unbalanced biochemical differentiation that ultimately results in cell death. The decreasing contents of protein, nucleic acids, and other cellular constituents in senescing tissues points to the involvement of lytic processes. Protein synthesis continues in senescing organs, but synthesis counterbalances hydrolysis only partially. The observed gradual decrease of total protein content is, therefore, due to changes in the relative rates of synthesis and degradation. The coexistence of these opposed processes in senescing tissues indicates that the compartmentation of the cellular digestion in the vacuome is maintained.

In broad bean cotyledons about 80% of the RNA is gradually

broken down during the first 30 days of germination. The significance of cellular compartmentation of ribonuclease and ribosomes appears from the very rapid cleavage of ribosomal RNA when the same cotyledons were homogenized and incubated for a few hours (Payne and Boulter, 1974). The disruption of the membranes upon homongenization obviously results in an uncontrolled catabolism of RNA.

Morphological observations suggest the importance of autophagic activity as indicated by the presence of sequestered cytoplasmic material in vacuoles of senescent cells (Matile and Winkenbach, 1971). The decay and death of individual cells appear to be unsynchronized within a leaf and even within a tissue (Ragetli et al., 1970; Matile and Winkenbach, 1971). Nevertheless, senescence of plant organs seems to be as rigorously timed as are other ontogenetic processes. This circumstance indicates the existence of a precise regulation. In fact, hormonal control of senescence has been detected in many cases. The most conspicuous effects are those caused by cytokinins. If cytokinins are applied to excised organs that are on the road to senescence, the appearance of the typical phenomena of aging (yellowing of green leaves, decreasing protein content, etc.) are delayed. It appears from these investigations (reviewed by Kende, 1971) that cytokinins may inhibit proteolysis rather than stimulate protein synthesis. This could be due either to reduced synthesis of lysosomal enzymes or to slower transport of cytoplasmic material into the vacuome. Indeed, the activity of proteases and RNase is lower in leaves treated with cytokinin as compared with untreated controls (see Kende, 1971). Since degradation of cytoplasmic constituents must, however, always be preceded by the membrane activities associated with autophagy, it is likely that cytokinins also control these processes.

In the course of development of the *Ipomoea* corolla, its DNA content is practically constant after cell division is completed in the bud. DNA is not degraded before the onset of wilting, i.e., in the early afternoon of the day of anthesis. In contrast, RNA begins to decrease on the day preceding anthesis. During wilting, the content of DNA is reduced by about 70% indicating the breakdown of a considerable fraction of nuclear DNA (Matile and Winkenbach, 1971). Since nuclei are never encountered in vacuoles among the autophagocytized material, the disappearance of DNA must be due to cellular autolysis. In contrast to autophagy, autolysis represents an uncontrolled digestive process. It is initiated by the breakdown of the tonoplasts resulting in the mixing of vacuolar hydrolases with the cytoplasm. Although rupture of the tonoplast has been considered in relation to cell senescence (Treffry et al., 1967; Ragetli et al., 1970; Mittelhäuser and van Steveninck, 1971), the recognition of its lethal consequence depends on whether the lysosomal nature of the vacuole is considered (Berjak and Villiers, 1970; Matile and Winkenbach, 1971; Berjak, 1972). In flax seedling cotyledons ruptured tonoplasts and autolysis have been observed as soon as 6 hours after treatment with the herbicide Paraquat (Harris and Dodge, 1972). Whether in normal senescence the abolishment of compartmentation of digestive enzymes is the cause of cell death or vice versa is unknown. In any case, autolysis is irreversible as indicated by the digestion of nuclei (Matile and Winkenbach, 1971).

B. Accumulation and Mobilization

1. METABOLIC INTERMEDIATES

The food yeast *Candida utilis* is able to utilize a wide variety of amino acids. If arginine, glycine, or ornithine is supplied as the sole source of nitrogen, each is utilized rapidly. Simultaneously, accumulation occurs to such an extent that amino acids represent the most abundant class of cellular small molecules. Isotope kinetic analysis demonstrates that the amino acids newly absorbed from the medium do not mix freely with the total amino acid pool present in the cells. Radioactive arginine, for example, is rapidly processed, its radioactivity appearing much faster in arginine derivatives than predicted on the assumption that the labeled arginine taken up from the medium is diluted by the total unlabeled arginine pool. Hence, the existence of at least two discrete arginine pools must be postulated: a small precursor pool subject to arginine metabolism and a large storage pool (Wiemken and Nurse, 1973a,b). The same situation holds for other amino acids as well. The existence of an inactive pool of amino acids has also been demonstrated in higher plants (Hellebust and Bidwell, 1963). Precursor pools turn over rapidly, i.e., are rapidly saturated by newly absorbed amino acids, whereas storage pools are only slowly diluted in isotope kinetic studies.

Sequential extraction of *Candida utilis* has revealed that the large storage pool of amino acids is to be found in the vacuoles (Wiemken and Nurse, 1973a). Only 5–10% of the total arginine is present in the cytoplasmic precursor pool. The large vacuolar deposit seems to function as a source of nutrients in situations in which yeast cells turn to endogeneous metabolism. This occurs not only in starving yeast cells but also in the course of the cell cycle. Wiemken *et al.* (1970), using synchronously growing cultures of *Saccharomyces cerevisiae*, have observed that culture dry mass remains practically constant during the budding phase, while the total protein content increases exponentially. The utilization of amino acids stored in the vacuome during budding has, in fact, also been demonstrated by the technique of sequential cell extraction. By taking advantage of the considerable cell density changes that occur in the course of the cell cycle, populations of exponentially growing cells can be fractionated according to their position in the cell cycle. This is done by centrifugation in density gradients of osmotically inactive media (Wiemken et al., 1970). Initial budding cells, characterized by endogenous metabolism, contain much lower amounts of stored amino acids than do late budding and double cells (Nurse and Wiemken, 1974). The vacuolar amino acid pool is utilized during initiation of budding and is gradually built up again as budding progresses. The changes of size of the vacuolar storage pools are, in turn, most probably responsible for the density changes of the cells. Initial budding cells containing a small storage pool are much heavier than double cells containing a large storage pool. Morphologically, the mobilization of stored amino acids is correlated with a conspicuous shrinkage and fragmentation of the vacuoles at the onset of budding. The refilling of this pool is, in turn, correlated with swelling and fusion of small vacuoles so that only one or few large vacuoles are present after completion of the budding cycle (Fig. 9) (Wiemken et al., 1970).

In higher plant tissues, radioisotope studies on the metabolism of organic acids have yielded conclusive evidence for the existence of separate pools (see Beevers et al., 1966). MacLennan et al. (1963) distinguished discrete pools accessible or inaccessible to metabolizing enzymes. In tissues that accumulate large quantities of certain acids, only a small fraction may be present in the metabolic pool, for example, in *Bryophyllum* leaves only 2% of the total isocitric acid pool is turned over, whereas 98% is removed from metabolism. The formation of storage pools is species specific; in addition, it is reversible, as indicated by the diurnal

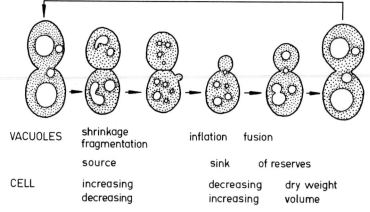

VACUOLES	shrinkage fragmentation		inflation	fusion
	source		sink	of reserves
CELL	increasing decreasing		decreasing increasing	dry weight volume

Fig. 9. Dynamism of the yeast vacuome (adapted from Wiemken *et al.*, 1970).

changes of acid accumulation in leaves of plants with Crassulacean acid metabolism. As mentioned above, there are cogent reasons to assume that storage pools of organic acids are located in the vacuoles.

Vacuolar storage pools of metabolic intermediates represent an important device for maintaining homeostasis within the cytoplasm. As pointed out by Wiemken and Nurse (1973b) plant cells are more or less directly exposed to environmental changes; in this situation the vacuole with its nutritious cell sap represents a large internal environment that buffers the cytoplasm against environmental changes.

2. POSSIBLE MECHANISMS OF SOLUTE ACCUMULATION

Since membrane transport is a fashionable topic in contemporary cell biology, one is tempted to speculate that permeases localized in the tonoplast are responsible for solute accumulation in vacuoles. Indeed, that transport of inorganic ions into vacuoles is permease mediated has been concluded from extensive studies in perfused *Valonia* cells (see Gutknecht and Dainty, 1968) and also from compartment analysis based on the kinetics of isotope exchange in various tissues. Active transport across the tonoplast would, in fact, appear to be required for the accumulation of higher solute concentrations (activities) in the cell sap than in the cytoplasmic matrix.

Isolated vacuoles of *Candida utilis* did not show active transport of *S*-adenosylmethionine, uric acid, and several amino acids (see Section II,C,2) which they accumulated *in vivo* (Nakamura and Schlenk, 1974). Boller *et al.* (1975) characterized a specific transport system in isolated vacuoles of *Saccharomyces cerevisiae* which catalyzes the exchange of arginine added to the medium with arginine present in the vacuoles. Although the presence of this carrier in the tonoplast suggests the existence of controlled interactions between vacuolar storage pools and cytoplasmic pools of arginine, it does not explain the accumulation of arginine within the vacuoles.

Charged solutes may be trapped by nondiffusable counter-ions present in any particular compartment. An example of this type of accumulation is illustrated by the behavior of cationic vital stains. Among the vacuolar anions responsible for binding these stains, organic acids, flavones, tannins, and other phenolic compounds have been recognized. Donnan equilibria and precipitation of salts are responsible for accumulation, and the corresponding phenomena have been widely used in the cytochemical diagnosis of cell saps (see e.g., Kinzel and Pischinger, 1962). The example of alkaloid accumulation in the vacuole of *Chelidonium* latex demonstrates a strikingly analogous behavior of these biological cations.

Isolated vacuoles are capable of rapidly absorbing sanguinarine (one of the colored alkaloids present in the vacuoles) dissolved in the medium (Fig. 10A). This process does not depend on an energy source, although the alkaloid is transported against a concentration gradient. The absorption is, however, accompanied by the release of other alkaloids into the medium, indicating that ion exchange takes place in the cell sap. The vacuolar anions responsible for trapping sanguinarine are most probably phenols. The limited capacity of isolated vacuoles to absorb sanguinarine from the medium (Fig. 10B) may be determined by the amount of exchangeable ions present in the cell sap. It is interesting to note that the isolated vacuole is lysed as soon as it can no longer remove the alkaloid from the medium. Sanguinarine is destructive to biological membranes. These facts demonstrates the importance of the vacuolar deposition of sanguinarine (Matile *et al.*, 1970; Jans, 1973).

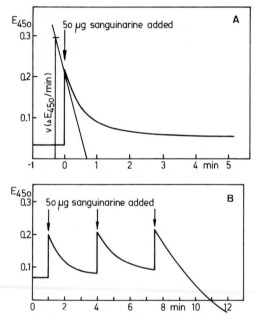

Fig. 10. Absorption of the alkaloid sanguinarine in vacuoles isolated from *Chelidonium* latex. Both cuvettes of a double-beam spectrophotometer were filled with a suspension of vacuoles (turbidity nearly compensated). The changes in optical density at 450 nm are due to the changes of sanguinarine concentration in the medium. Upon the repetition of the experiment illustrated in (A), the third addition of alkaloid results in rapid bursting of vacuoles (B). The turbidity of the reference is no longer compensated; alkaloids are completely soluble at the end of the experiment. (From Matile *et al.*, 1970.)

The formation of crystals of calcium oxalate in vacuoles (Schötz *et al.*, 1970) represents another example of accumulation which could in in part be explained without the participation of permease-mediated transport. The vacuolar deposition of oxalic acid would merely require the presence in the cell sap of calcium ions. Likewise, the movement of the barely water-soluble uric acid into the vacuoles of *Candida utilis* (Fig. 4) could be caused by the comparatively low pH and, hence, by the ionic composition of the cell sap. A common feature of all of these possible mechanisms for accumulation is their dependency on properties of the cell which were established before accumulation began.

Tannins, which can be detected cytochemically, appear in the endoplasmic reticulum (ER) cisterna. After the formation of vacuoles (see Section IV), they can be detected in the ER-derived provacuoles and finally in the large vacuoles (Chafe and Durzan, 1973; Baur and Walkinshaw, 1974). It has been observed that dictyosome vesicles are autophagocytized by vacuoles (see, e.g., Matile and Moor, 1968; Berjak and Villiers, 1970; Fineran, 1971). The significance of this phenomenon is not completely clear. However, it would seem that vacuoles are made in part from dictyosome residues. It would seem likely that organic compounds, such as acidic polysaccharides and phenols, are deposited in vacuoles via Golgi-mediated internal secretion. These compounds could in turn be responsible for the accumulation of appropriate cations. Likewise, ion transport into vacuoles associated with ER-derived vesicles containing (proteinaceous?) counterions has been proposed (see Costerton and Macrobbie, 1970). Lastly, intracellular digestion of macromolecules after autophagy may result in the generation of micromolecules in the vacuome.

A common feature of vacuolar deposition of solutes is the reversibility of accumulation. The diurnal variation of organic acid accumulation in leaves of the Crassulaceae or the cyclic change in amino acid deposition in yeast vacuoles show that there is metabolic regulation of solute accumulation. The smallness of the soluble cytoplasmic pool of amino acids in yeast cells suggests that only catalytic amounts of metabolic intermediates may be present in the cytoplasmic matrix. It should be emphasized that this large difference between storage and turnover pools has been found in exponentially growing cells, which are characterized by intensive amino acid metabolism.

3. RESERVES

Apart from starch, which is formed in amyloplasts, the reserves accumulated in seeds are located in modified vacuoles—reserve proteins in aleurone grains and triglycerides in spherosomes. Corresponding synthetic capacities associated with these organelles seem to represent a prerequisite of accumulation. Results concerning the association of lipid-synthesizing activity with spherosomes are, however, contradictory (Semadeni, 1966; Jacks et al., 1967). One possible site of synthesis of reserve proteins has been established by employing pulse chase labeling and autoradiographic techniques. Bailey et al. (1970) have demonstrated in cultured slices of broad bean cotyledons that labeled leucine incorporated into globulin is initially associated with rough endoplasmic reticulum. In the course of the chase, the label gradually disappeared from the reticulum and moved into the protein bodies. The involvement of the ER in the synthesis of reserve proteins is indicated by the association of developing aleurone grains with masses of rough ER (Öpik, 1968; Briartry et al., 1969). Since direct connections between ER filaments and protein bodies have not been observed, the mechanism of transport of reserve proteins into their final compartment is not yet clear. If the newly synthesized proteins were first released into the cytoplasmic matix, it would be necessary to postulate a specific mechanism of transport across the tonoplast of aleurone vacuoles. A more probable sequence is the initial release of the reserve proteins into the ER cisterna, followed by the production of vesicles filled with reserve protein which subsequently discharge their content into aleurone vacuoles. A mechanism of this kind is indicated by electron micrographs showing the origin and development of protein granules in maize endosperm (Khoo and Wolf, 1970). In this tissue vesicles pinched off from the ER have membranes still associated with polysomes. This observation indicates that protein synthesis may continue as these vesicles develop either directly into protein bodies or fuse with existing aleurone vacuoles. Upon the maturation of seeds the gradual desiccation results in the precipitation of phytate, which forms more or less spherical enclaves, globoids, in a proteinaceous matrix (see e.g., Engelman, 1966; Jones, 1969). In species containing seed globulins a second type of inclusion, the cristalloid, appears within the aleurone grain (Fig. 11) (see, e.g., Poux, 1965; Lott and Vollmer, 1973).

As mentioned above the aleurone grains of ungerminated seeds contain hydrolases that are most probably involved in the mobilization of reserves. It is unlikely that these enzymes are deposited in the aleurone vacuoles simultaneously with their prospective substrates. Rather they are probably synthesized after deposition of the reserves and the end of seed development. In any case, the presence of hydrolases in dormant seeds allows the immediate onset of reserve mobilization upon induction of seed germination. Morphologically, this process is characterized by the gradual disappearance of protein matrix, crystalloids, and globoids com-

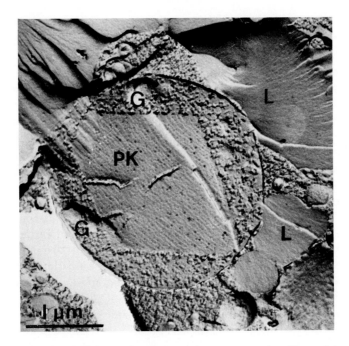

Fig. 11. Freeze-etching showing an aleurone vacuole with protein crystal (PK) and globoids (G) in an endosperm cell of *Ricinus communis*. L, Lipid droplets (spherosomes). Courtesy of A. M. Schwarzenbach.

bined with a considerable swelling of aleurone vacuoles (see, e.g., Briarty *et al.*, 1970; Jones and Price, 1970). In barley aleurone cells, Paleg and Hyde (1964) have observed a drastic stimulation of these morphological changes by treatment with gibberellic acid. This finding suggests that the contact between reserves and hydrolases may be subjected to hormonal regulation. Moreover, it appears that mobilization is correlated with synthetic activities that are also dependent on hormonal induction (see Filner *et al.*, 1969). Protein synthesis is most probably a common capacity of reserve cells, and the confinement of the unspecific lytic processes to the vacuome appears as a prerequisite for an organized process of reserve mobilization.

C. Turgor

Water relations, turgor pressure, and (in turn) the rigidity of plant tissues are related to the accumulation of osmotically active substances in the vacuoles. Moreover, the importance of turgor in cell extension is demonstrated by the effect of osmotically active substances on growth rate. If turgor is shifted by submerging *Nitella* internodes in metabolically inert osmotically active solutes, the elongation rate is immediately reduced. Green *et al.* (1971) have shown that elongation takes place only if the turgor exceeds the yielding threshold of the wall. However, resumption of growth after a fall in turgor has been found to be connected with an increased extensibility of the cell wall rather than with increased turgor, which, in Green's experiments, was measured directly by means of a micromanometer inserted into the vacuole. Hence, growth is a function of turgor, but the regulation of cell elongation is not primarily exercized by the osmotic properties of the cell sap.

Naked protoplasts are ideal objects for separately investigating the involvement of turgor in cell extension; these protoplasts should respond to growth hormone by rupturing if an increase in internal osmotic pressure is induced. Bursting of the protoplasts induced by auxins has indeed been observed by Cocking (see Power and Cocking, 1970). This effect was not observed when protoplasts from the classic object of the study of hormonal regulation of growth, *Avena* coleoptiles, were tested (Ruesink and Thimann, 1965). A recent reinvestigation has shown, however, that auxin-induced bursting can be produced in *Avena* protoplasts if they are kept under osmotic conditions corresponding to incipient plasmolysis (Hall and Cocking, 1974). It is interesting to note that the dependency of bursting response upon the concentration of auxin closely follows the concentration dependency of extension growth in the respective organs. It seems, therefore, that turgor pressure represents one of the prerequisites of growth that are subjected to hormonal regulation.

There are other phenomena that indicate the existence of turgor-regulating mechanisms in plant cells. A particularly illustrative example is represented by the guard cells of stomata whose reversible change of shape is responsible for stomata aperture opening and closing. Guyot and Humbert (1970) have observed that in the dark, when the stoma are closed, the vacuome consists of numerous tiny vacuoles. Upon induction of opening, these vacuoles inflate and coalesce into a single large vacuole. This behavior points to a reversible deposition of osmotically active substances in the vacuoles. It also reminds one of the behavior of vacuoles observed in the course of the budding cycle in yeast (Fig. 9). In this case, the cyclic inflation and shrinkage of the vacuome seems to be related with alternating deposition and withdrawal of soluble reserves that the refilling of the vacuolar pools during the budding process is, in addition, necessary for the turgor-dependent extrusion of the bud. The ability of yeast cells to regulate the turgor is seen in the response to turgor shifts caused by osmotically active substances. The initial plasmolysis is overcome by the increase of internal osmotic pressure of the cells by use of cellular material and not by the uptake of solute from the culture medium (Lillejoj and Ottolenghi, 1966).

IV. Ontogeny of the Vacuome

The vacuoles of yeast cells represent persistent organelles. At the onset of budding, the large vacuoles of the mother cells are fragmented into numerous small vacuoles. In the course of budding, these organelles are distributed between mother and daughter cell. After completion of the budding cycle one or few large vacuoles have developed by gradual inflation and fusion of existing vacuoles (Fig. 9) (Wiemken *et al.*, 1970). As a consequence, compounds that were accumulated in the vacuome of the mother cell are evenly distributed between mother and daughter cell.

This cyclic process of vacuole formation contrasts with the unidirectional process that takes place in higher plants. The youngest cells of meristems seem to be completely deprived of vacuoles, the vacuome being gradually formed as these cells develop into parenchymatous cells. The initial vacuolar components are of submicroscopic dimension. Vesicles, 0.1 to 0.3 μm in diameter are pinched off from the rough surfaced endoplasmic reticulum (Matile and Moor, 1968; Mesquita, 1969; Berjak, 1972) (Fig. 12A and B). Upon inflation and fusion of these provacuoles, they develop into vacuoles of microscopic dimensions. In root meristems stained vitally with neutral red, these small vacuoles appear as an extended reticular system. As the process of vacuolation continues, the number of vacuoles is reduced by extensive fusion (Fig. 12C and D) until at the end of development coalescence into a single large vacuole occurs.

The inflation of the vacuome, which takes place during the expansion of meristem cells, must be accompanied by the accumulation of osmotically active substances in the cell sap. These may in part be produced through autophagy and intracellular digestion of cytoplasmic material, which can in fact be observed in the early phase of development of vacuoles (Matile and Moor, 1968; Fineran, 1971). Conversely, the incorporation of dictyosome vesicles into vacuoles (Matile and Moor, 1968; Fineran, 1971; Berjak, 1972) may be involved in the process of accumulation. Vacuolation and autophagy may also be joined together in a single process (Fig. 7). In this case, filaments of the endoplasmic reticulum are responsible for the sequestration of large portions of the cytoplasm. The tonoplasts therefore appears to originate from the ER. Although this membrane is probably differentiated during the development of vacuoles, some enzymes typical of the ER appear to persist in the tonoplast (Matile, 1968a). Even in yeast vacuoles whose development is not connected with

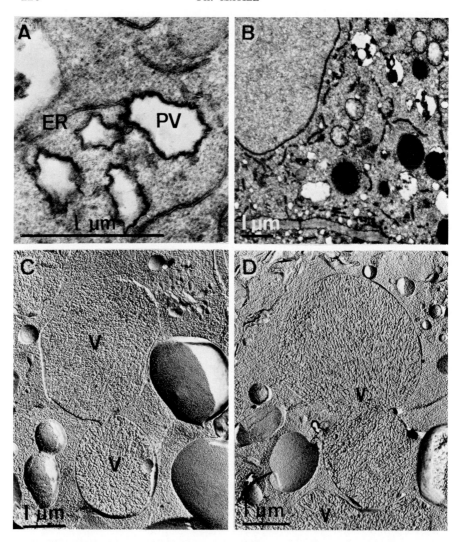

Fig. 12. Vacuolation in root meristems. (A) Vesiculation of the endoplasmic reticulum (ER). Reaction product of acid phosphatase is present in the provacuoles (PV). Root meristem of *Lepidium sativum.* Courtesy of P. Berjak. (B) Acid phosphatase activity associated with developing vacuoles. Root cap initial of *Zea mays.* Courtesy of P. Berjak. (C) and (D) Fusion of vacuoles (V) in meristematic root cells of *Zea mays.* From Matile and Moor (1968).

the ER, these enzymes have been detected (see Table II) (Matile and Wiemken, 1967). In yeast the tonoplast seems to be associated with a small population of polysomes (Wiemken, 1969) suggesting that the synthesis of lysosomal enzymes might proceed in this membrane. Whether in yeast or higher plant vacuoles, how the regulation of lysosomal enzyme activity is in fact exercised through protein synthesis in the tonoplast is not yet known. In higher plant cells another possibility, the fusion of existing vacuoles with primary lysosomes (ER-derived provacuoles?), must be considered (see Matile, 1974).

The origin of aleurone vacuoles from the ER has already been mentioned. These protein bodies may develop directly from provacuoles (Khoo and Wolf, 1970). Conversely, at early stages of development they may be indistinguishable from vacuoles, which are subsequently filled with reserves (e.g., Buttrose, 1963; Engleman, 1966). In cotyledons of legumes, the initial cell expansion is associated with the formation of large vacuoles, which upon subdivision differentiate into protein bodies (Öpik, 1968). The vacuolar nature of aleurone grains derives furthermore from the cytological changes in germinating seeds. After mobilization of the reserves they eventually coalesce into a single central vacuole. In species whose cotyledons develop into green leaves, this vacuole eventually represents the central vacuole of the parenchyma cells (e.g., Treffry et al., 1967).

Spherosomes are characterized by a surrounding membrane which, in contrast to the normal triple-layered membrane structure, consists of only a single electron-dense contour. This membrane anomaly is explained by the ontogeny of spherosomes. These organelles originate from small provacuole-like vesicles pinched off from the ER (Frey-Wyssling et al., 1963); recently, Schwarzenbach (1971) has been able to demonstrate that the lipids accumulated in spherosomes are deposited in the hydrophobic central layer of the membrane of these prospherosomes. As lipid accumulation proceeds this layer increases in volume and separates the outer and inner electron-dense membrane leaflets. The anomaly of the spherosomal membrane is possibly responsible for the fate of these organelles upon lipid mobilization. Fusion between spherosomes and vacuoles has never been observed. In yeast they are autophagocytized selectively under certain conditions (Fig. 6D), and their disappearance in the vacuole indicates the presence of lipase in this compartment. In oleaginous tissues of seeds they gradually decrease in volume during germination. The transformation of spherosomes into flattened saccules, observed in cotyledons of germinating legume seeds (Mollenhauer and Totten, 1971a), suggests that in this case the triglycerides may be mobilized by the action of spherosomal lipase.

GENERAL REFERENCES

Dangeard, P. (1956). "Le vacuome de la cellule végétale. Protoplasmatologia," Vol. III D, p. 1. Springer-Verlag, Berlin and New York.

de Duve, C. (1969). *In* "Lysosomes in Biology and Pathology" (J. T. Dingle and H. B. Fell, eds.), Vol. 1, pp. 3–40. North-Holland Publ., Amsterdam.

Küster, E. (1951). "Die Pflanzenzelle," 2nd ed. pp. 473–526. Fischer, Jena.

Matile, P. (1969). *In* "Lysosomes in Biology and Pathology" (J. T. Dingle and H. B. Fell, eds.) Vol. 1, pp. 406–430. North-Holland Publ., Amsterdam.

Matile, P. (1974). *In* "Dynamic Aspects of Plant Ultrastructure" (A. W. Robards, ed.), pp. 178–218. McGraw-Hill, New York.

Matile, P. (1975). "The Lytic Compartment of Plant Cells." Cell Biology Monographs, Vol. 1. Springer-Verlag, Berlin and New York.

REFERENCES

Adams, C. A., and Novellie, L. (1975). *Plant Physiol.* **55**, 7.

Bailey, C. J., Cobb, A., and Boulter, A. (1970). *Planta* **95**, 103.

Baur, P. S., and Wilkinshaw, C. H. (1974). *Can. J. Bot.* **52**, 615.

Beevers, H., Stiller, M. L., and Butt, V. S. (1966). *In* "Plant Physiology" (F. C. Steward, ed.) Vol. 4 B, pp. 119–242. Academic Press, New York.

Berjak, P. (1972). *Ann. Bot. (London.)* [N.S.] **36**, 73.

Berjak, P., and Villiers, T. A. (1970). *New Phytol.* **69**, 929.

Boller, T., Dürr, M., and Wiemken, A. (1975). *Eur. J. Biochem.* **54**, 81.

Briarty, L. G., Coult, D. A., and Boulter, D. (1969). *J. Exp. Bot.* **20**, 358.

Briarty, L. G., Coult, D. A., and Boulter, D. (1970). *J. Exp. Bot.* **21**, 513.

Buttrose, M. S. (1963). *Aust. J. Biol. Sci.* **16**, 768.

Buvat, R. (1968). *C. R. Acad. Sci.* **267**, 296.

Chafe, S. C., and Durzan, D. J. (1973). *Planta* **113**, 251.

Ching, T. M. (1968). *Lipids* **3**, 482.

Cocking, E. C. (1960). *Nature (London)* **187**, 962.

Cortat, M., Matile, P., and Wiemken, A. (1972). *Arch. Mikrobiol.* **82**, 189.

Costerton, J. W. F., and MacRobbie, E. A. C. (1970). *J. Exp. Bot.* **21**, 535.

Coulomb, C., and Buvat, R. (1968). *C. R. Acad. Sci.* **267**, 843.

Coulomb, P. (1971). *J. Microsc. (Paris)* **11**, 299.

Cresti, M., Pacini, E., and Sarfatti, G. (1972). *J. Submicrosc. Cytol.* **4**, 33.

Doi, E. Ohtsuri, C., and Matoba, T. (1975). *Plant Sci. Lett.* **4**, 243.

Dürr, M., Boller, T., and Wiemken, A. (1975). *Arch. Microbiol.* **105**, 319.

Engleman, E. M. (1966). *Amer. J. Bot.* **53**, 231.

Filner, P., Wray, J. L., and Varner, J. E. (1969). *Science* **165**, 358.

Fineran, B. A. (1971). *Protoplasma* **72**, 1.

Frey-Wyssling, A., Grieshaber, E., and Mühlethaler, K. (1963). *J. Ultrastruct. Res.* **8**, 506.

Fukuhara, H. (1967). *Biochim. Biophys. Acta* **134**, 143.

Graham, T. A., and Gunning, B. E. S. (1970). *Nature (London)* **228**, 81.

Green, P. B., Erickson, R. O., and Buggy, J. (1971). *Plant Physiol.* **47**, 423.

Gutnecht, J., and Dainty, J. (1968). *Oceanogr. Mar. Biol.* **6**, 163.

Guyot, M., and Humbert, C. (1970). *C. R. Acad. Sci.* **270**, 2787.

Hall, M. D., and Cocking, E. C. (1974). *Protoplasma* **79**, 225.

Halvorson, H. O. (1960). *Advan. Enzymol.* **22**, 99.

Harris, N., and Dodge, A. D. (1972). *Planta* **104**, 201.

Harris, N., and Chrispeels, M. J. (1975). Plant Physiol. **56**, 292.

Hellebust, J. A., and Bidwell, R. G. S. (1963). *Can. J. Bot.* **41**, 985.

Indge, K. J. (1968a). *J. Gen. Microbiol.* **51**, 441.

Indge, K. J. (1968b). *J. Gen. Microbiol.* **51**, 447.

Iten, W., and Matile, P. (1970). *J. Gen. Microbiol.* **61**, 301.

Jacks, T. J., Yatsu, L. Y., and Altschul, A. M. (1967). *Plant Physiol.* **42**, 585.

Jans, B. (1973). *Ber. Schweiz. Bot. Ges.* **83**, 306.

Jones, R. L. (1969). *Planta* **85**, 359.

Jones, R. L., and Price, J. M. (1970). *Planta* **94**, 191.

Kende, H. (1971). *Int. Rev. Cytol.* **31**, 301.

Khoo, U., and Wolf, M. J. (1970). *Amer. J. Bot.* **57**, 1042.

Kinzel, H., and Pischinger, I. (1962). *Protoplasma* **55**, 550.

Lillehoj, E. B., and Ottolenghi, P. (1966). *Abh. Deut. Akad. Wiss. Berlin, Kl. Med.* **6**, 145.

Lenney, J. T., Matile, P., Wiemken, A., Schellenberg, H., and Meyer, J. (1974). *Biochem. Biophys. Res. Commun.* **60**, 1378.

Lott, J. N. A., and Vollmer, C. M. (1973). *Protoplasma* **78**, 255.

Lui, N. S. T., and Altschul, A. M. (1967). *Arch. Biochem. Biophys.* **121**, 678.

Lüscher, A., and Matile, P. (1974). *Planta* **118**, 323.

MacLennan, D. H., Beevers, H., and Harley, J. L. (1963). *Biochem. J.* **89**, 316.

Malkoff, D. B., and Buetow, D. E. (1964). *Exp. Cell Res.* **35**, 58.

Marty, F. (1968). *C. R. Acad. Sci.* **267**, 299.

Marty, F. (1970). *C. R. Acad. Sci.* **271**, 2301.

Marty, F. (1971). *C. R. Acad. Sci.* **272**, 399.

Matile, P. (1966). *Z. Naturforsch. B* **21**, 871.

Matile, P. (1968a). *Planta* **79**, 181.

Matile, P. (1968b). *Z. Pflanzenphysiol.* **58**, 365.

Matile, P. (1971). *Cytobiologie* **3**, 324.

Matile, P., and Moor H. (1968). *Planta* **80**, 159.

Matile, P., and Spichiger, J. (1968). *Z. Pflanzenphysiol.* **58**, 277.

Matile, P., and Wiemken, A. (1967). *Arch. Mikrobiol.* **56**, 148.

Matile, P., and Winkenbach, F. (1971). *J. Exp. Bot.* **22**, 759.

Matile, P., Jans, B., and Rickenbacher, R. (1970). *Biochem. Physiol. Pflanz.* **161**, 447.

Matile, P., Wiemken, A., and Guyer, W. (1971). *Planta* **96**, 43.

Mesquita, J. F. (1969). *J. Ultrastruct. Res.* **26**, 242.

Mesquita, J. F. (1972). *Cytologia* **37**, 95.

Mittelheuser, C. J., and Van Steveninck, R. M. (1971). *Protoplasma* **73**, 239.

Mollenhauer, H. H., and Totten, C. (1971a). *J. Cell Biol.* **48**, 395.

Mollenhauer, H. T., and Totten, C. (1971b). *J. Cell Biol.* **48**, 533.

Morris, G. F. I., Thurman, D. A., and Boulter, D. (1970). *Phytochemistry* **9**, 1707.

Müller, E., Nelles, A., and Neumann, D. (1971). *Biochem. Physiol. Pflanz.* **162**, 272.

Neumann, D., and Müller, E. (1967). *Flora (Jena) Abt. A* **158**, 479.

Nurse, P., and Wiemken, A. (1974). *J. Bacteriol.* **117**, 1108.

Öpik, H. (1968). *J. Exp. Bot.* **19**, 64.

Ory, R. L., and Heningsen, K. W. (1969). *Plant Physiol.* **44**, 1488.

Ory, R. L., Yatsu, L. Y., and Kircher, H. W. (1968). *Arch. Biochem. Biophys.* **123**, 255.

Paleg, L., and Hyde, B. (1964). *Plant Physiol.* **39**, 673.

Payne, P. I., and Boulter, D. (1974). *Planta* **117**, 251.

Pitt, D., and Galpin, M. (1973). *Planta* **109**, 233.

Poux, N. (1963). *C. R. Acad. Sci.* **257**, 736.

Poux, N. (1965). *J. Microsc. (Paris)* **4**, 771.

Poux, N. (1970). *J. Microsc. (Paris)* **9**, 407.

Power, J. B., and Cocking, E. C. (1970). *J. Exp. Bot.* **21**, 64.

Pujarniscle, S. (1968). *Physiol. Veg.* **6**, 27.

Pujarniscle, S. (1969). *Physiol. Veg.* **7**, 391.

Ragetli, H. W. J., Weintraub, M., and Lo, E. (1970). *Can. J. Bot.* **48**, 1913.

Ribaillier, D., Jacob, J. L., and d'Auzac, J. (1971). *Physiol. Veg.* **9**, 423.

Roush, A. H. (1961). *Nature (London)* **190**, 449.

Ruesink, A. W. (1971). *Plant Physiol.* **47**, 192.

Ruesink, A. W., and Thimann, K. V. (1965). *Proc. Nat. Acad. Sci. U.S.* **54**, 56.

Ryan, C. A., and Shumway, L. K. (1970). *In* "Proteinase Inhibitors." de Gruyter, Berlin.

St. Angelo, A. J., Ory, R. L., and Hansen, H. J. (1969). *Phytochemistry* **8**, 1135.

Schlenk, F., Dainko, J. L., and Svihla, G. (1970). *Arch. Biochem. Biophys.* **140**, 228.

Schnepf, E. (1964). *Protoplasma* **58**, 193.

Schötz, F., Diers, L., and Bathelt, H. (1970). *Z. Pflanzenphysiol.* **63**, 91.

Schulze, C., Schnepf, E., and Mothes, K. (1967). *Flora (Jena), Abt. A* **158**, 458.

Schwarzenbach, A. M. (1971). *Cytobiologie* **4**, 145.

Semadeni, E. G. (1966). *Planta* **72**, 91.

Steward, F. C., and Sutcliffe, J. F. (1959). *In* "Plant Physiology" (F. C. Steward, ed.) Vol. 2, pp. 253–478. Academic Press, New York.

Svihla, G., Dainko, J. L., and Schlenk, F. (1963). *J. Bacteriol.* **85**, 399.

Thornton, R. M. (1968). *J. Ultrastruct. Res.* **21**, 269.

Tombs, M. P. (1967). *Plant Physiol.* **42**, 797.

Treffry, T., Klein, S., and Abrahamsen, M. (1967). *Aust. J. Biol. Sci.* **20**, 859.

Tronier, B., Ory, R. L., and Heningsen, K. W. (1971). *Phytochemistry* **10**, 1207.

Ullrich, W., Urbach, W., Santarius, A., and Heber, U. (1965). *Z. Naturforsch. B* **20**, 905.

Van der Wilden, W., Matile, P., Schellenberg, M., Meyer, J., and Wiemken, A. (1973). *Z. Naturforsch. C* **28**, 416.

Villiers, T. A. (1971). *Nature (London), New Biol.* **233**, 57.

Villiers, T. A. (1972). *New Phytol.* **71**, 145.

Wiemken, A. (1969). Thesis No. 4340. Swiss Federal Institute of Technology, Zürich.

Wiemken, A., and Nurse, P. (1973a). *Planta* **109**, 293.

Wiemken, A., and Nurse, P. (1973b). *Proc. Int. Spec. Symp. Yeasts, 3rd, 1973,* Part II, p. 331.

Wiemken, A., Matile, P., and Moor, H. (1970). *Arch. Mikrobiol.* **70**, 89.

Yatsu, L. Y., and Jacks, T. J. (1968). *Arch. Biochem. Biophys.* **124**, 466.

Yatsu, L. Y., Jacks, T. J., and Hensarling, T. P. (1971). *Plant Physiol.* **48**, 675.

Zandonella, P. (1970). *C. R. Acad. Sci.* **271**, 70.

Zielke, H. R., and Filner, P. (1971). *J. Biol. Chem.* **246**, 1772.

9

The Primary Cell Wall

PETER ALBERSHEIM

I. Introduction

The cell walls of plants are fundamentally involved in many aspects of plant biology, including the morphology, growth, and development of plant cells and the interactions between plant hosts and their pathogens. Plant cell walls are semirigid structures surrounding the cytoplasmic membrane of the cells. In a plant tissue, the cell wall of each cell merges with the walls of adjacent cells, giving the tissue physical coherence and strength (Esau, 1960). The morphology of a given plant tissue is thus determined by the morphology of the cell walls within it.

In growing plant tissues, however, the cell walls are not simply rigid, static shells. The cell walls must expand as the cells grow, and new components must be inserted into the existing wall structure (Cleland, 1971). Moreover, the walls must change in size, shape, and chemical composition as the cells of the tissue differentiate. The differentiation of plant cell walls is one of the most obvious and important manifestations of differentiation within the tissue (Albersheim, 1965).

Cell walls function as the skin as well as the skeleton of plants. They form a barrier that protects the cells from invasion by viral, bacterial, and fungal pathogens (Albersheim, 1965). Many pathogens attack the cell walls of the host plant by secreting enzymes that degrade components of the walls: the nature, specificity, and sequence of induction of these pathogen-secreted enzymes are related to the molecular structure of the plant cell wall matrix (Albersheim *et al.*, 1969; Cooper and Wood, 1973; English *et al.*, 1971).

Owing to the multifaceted biological importance of the plant cell wall, the structure of the wall has long been the object of intensive study. The cell wall is conveniently considered to be of two types, a thin primary wall and a thicker secondary wall (Albersheim, 1965). The primary wall is that part of the wall laid down by young, undifferentiated cells that are still growing. The primary cell wall is transformed into a secondary wall when the cell stops growing. The primary cell walls of a variety of higher plants appear to have many features in common and may, in fact, have very similar structures. This is not true of secondary walls, where the composition and ultrastructure vary considerably from one cell type to another. This chapter will be concerned with the structure and function of primary cell walls only.

Cellulose, hemicellulose, pectic polysaccharide, structural protein, and lignin have been identified as the major components of the plant cell wall. These components have been discussed in several recent reviews (Albersheim, 1965; Aspinall, 1970; Cleland, 1971; Frey-Wyssling, 1969;

Lamport, 1970; Mühlethaler, 1967; Northcote, 1972; Timell, 1964, 1965; Whistler and Richards, 1970). Lignin is a characteristic component of secondary walls (Albersheim, 1965) and will therefore not be discussed further.

II. The Noncellulosic Structural Components of Primary Cell Walls

A. The Attributes of Suspension-Cultured Cells for the Study of Cell Wall Structure

There are several reasons for selecting the cell walls isolated from suspension-cultured cells for detailed study. The most important is that *cultured cells can be grown as a homogeneous tissue possessing primary, but no secondary, walls.* Intact plant tissues contain a variety of cell types and a mixture of primary and secondary walls. The problems associated with cell wall structural analysis would be magnified by having to deal with more than one type of primary cell wall and by the presence of secondary cell wall material, since methods for separating secondary wall polymers from the primary wall are not available.

Suspension-cultured cells possess another attribute that has proved particularly advantageous for structural studies. *These cells secrete into their culture medium polysaccharides that are similar in composition to the noncellulosic polysaccharides of the cell wall* (Becker et al., 1964). It has been suggested (Becker et al., 1964) and now confirmed (Bauer et al., 1973; Burke et al., 1974; Wilder and Albersheim, 1973) that some of these soluble polymers are structurally related to cell wall polysaccharides. Therefore, these extracellular polymers offer a convenient source of material for developing techniques to study the wall polymers.

B. The Composition of the Walls of Suspension-Cultured Sycamore Cells

The composition of the cell walls isolated from suspension-cultured sycamore cells is presented in Table I (Talmadge et al., 1973). The amounts of each of the monosaccharide constituents of the noncellulosic polysaccharides of the wall were measured as their alditol acetate derivatives by gas chromatographic analysis; these sugars account for 63% of the wall. The remainder of the cell wall consists of cellulose (23%) and protein (10%). The composition of different walls is constant with the exception of the content of the noncellulosic glucose, which accounts for from 8 to 14% of the wall preparations. This variation results from the presence in the wall preparations of varying amounts of starch. Evidence has been obtained that the starch, rather than being a structural

TABLE I

Composition of the Cell Walls Isolated from Suspension-Cultured Sycamore Cells[a]

Components	Weight % composition	Mole % composition of total carbohydrate
Rhamnose	3.1	3.9
Fucose	1.3	1.7
Arabinose	21.0	28.2
Xylose	7.6	10.2
Mannose	0.3	0.3
Galactose	12.8	14.5
Glucose (noncellulosic)	3.7	4.2
Glucose (cellulosic)	23	24
Galacturonic acid	13.4	13.2
Protein (total)	10	
Hydroxyproline	2	

[a] Sycamore walls were hydrolyzed and the neutral sugars and uronic acids determined by gas chromatography (Talmadge et al., 1973). Cellulose was determined by the method of Updegraff (1969). Protein was estimated by multiplying the nitrogen content by 6.25. Nitrogen analyses were performed by the Kjeldahl procedure at the Pediatrics Microanalytical Laboratories, University of Colorado Medical Center. Hydroxyproline was determined by the method of Kivirikko and Liesmaa (1959) as described in Talmadge et al. (1973). Sycamore wall preparations contain a contaminant of starch representing about 10% by weight which may be removed by treatment with amylase. The glucose derived from starch has been excluded from the data presented in this table.

polymer of the wall, is a contaminant of the isolation procedure (Talmadge et al., 1973). The data presented in this chapter are calculated on the basis of cell walls that do not contain starch.

Analysis of the structure of the cell wall of suspension-cultured sycamore cells has relied heavily on methylation analysis to identify, and to determine quantitatively, the sugar linkages that constitute the total cell wall as well as the individual wall polymers. Methylation analysis of the isolated but unfractionated cell wall is important, since it permitted a quantitative summary of all but the minor sugar linkages present in the total wall. Subsequent studies on wall fractions obtained by the action of purified hydrolytic enzymes have been made quantitative by comparing the amounts of the sugar linkages found in each fraction to the amounts present in the total wall. These results demonstrate that the sycamore cell wall is composed of five major structural components: an arabinogalactan, cellulose, hydroxyproline-rich glycoprotein, a rhamnogalacturonan, and a xyloglucan. The arabinogalactan is the least well characterized of the wall polysaccharides; the arabinogalactan probably represents more than one polysaccharide, but is likely to consist of an arabinan attached to a galactan.

Fractionation of the cell wall following methylation was effected by differential solubility of wall polymers (Talmadge et al., 1973). Methylated cell walls were separated into chloroform–methanol-soluble and -insoluble fractions. The glycosyl linkage patterns indicated that the chloroform–methanol-soluble polymers were the noncellulosic wall polysaccharides: the arabinogalactan, the rhamnogalacturonan, and the xyloglucan. The chloroform–methanol–insoluble polymers were primarily incompletely methylated cellulose and methylated tetraarabinosides attached to the hydroxyproline-rich protein. This particular fractionation of methylated wall polymers is characteristic of sycamore cells, but is not a general property of all primary cell walls.

The detection, in the protein containing chloroform–methanol-insoluble wall fraction of sycamore cell walls, of terminal, 2-linked, and 3-linked arabinosyl residues in the proportion of 1:2:1 provided evidence that these arabinosyl residues represent tetraarabinosides that are glycosidically attached to the hydroxyproline residues of the hydroxyproline-rich glycoprotein (Talmadge et al., 1973). The tetraarabinosides represent 20% of the chloroform–methanol–insoluble wall fraction and approximately 9% of the total cell wall. The tetraarabinoside content of the wall (9%) can be accounted for in toto if one of these tetrasaccharides is attached to every hydroxyproline residue of the wall (2%, see Table I). These data on the glycoprotein component of sycamore walls are in very good agreement with the results reported by Lamport for the tetraarabinosides attached to hydroxyproline in cell walls of suspension-cultured tomato cells (Lamport, 1965, 1970). There is considerable evidence that such tetraarabinosides are attached to hydroxyproline residues in the cell walls of many other plants (Lamport and Miller, 1971), including the walls of suspension-cultured sycamore cells (Heath and Northcote, 1971). The results of Talmadge et al. (1973) also agree with those of Karr (1972) who has reported that apparently identical arabinosyl-(1 → 2)-arabinosyl-(1 → 2)-arabinosyl-(1 → 3)-arabinosyl-(1 → 4)-proline units are present in the cytoplasmic proteins of sycamore cells. Karr (1972) suggests and Chrispeels (1969) and Dashek (1970) have presented evidence that the cytoplasmic hydroxyproline-rich glycoprotein is a precursor of the cell wall glycoprotein.

The amounts of each of the wall polysaccharides except the rhamnogalacturonan have been estimated using the glycosyl linkage data obtained by methylation analysis (Talmadge *et al.*, 1973). For example, 4,6-linked glucosyl residues represent 12.7% of the sugars in the fraction of the wall soluble in chloroform–methanol following methylation. Since this fraction represents 55% of the total wall, this glucosyl residue represents approximately 7.0% of the total cell wall. It has been determined that 4,6-linked glucosyl residues represent 32% of the purified xyloglucan polymer (Bauer *et al.*, 1973). Therefore, if all of the 4,6-linked glucosyl residues in the cell wall are derived from xyloglucan, then 21% of the total cell wall is xyloglucan. A similar result is obtained if one of the other glycosyl residues present in xyloglucan, for example, terminal xylose or terminal fucose, is used as the basis for this calculation. The amounts of arabinogalactan, cellulose, and hydroxyproline tetraarabinosides were estimated using similar calculations. The amount of rhamnogalacturonan in the wall was estimated by assuming that this polymer is composed entirely of rhamnose and galacturonic acid and that this is the only polymer within the wall in which these residues occur. Thus, the rhamnogalacturonan content was determined from the amounts of rhamnose and galacturonic acid found in the cell wall (Table I). The polymer composition of the wall is given in Table II. Essentially the entire cell wall is accounted for by the sum of the amounts of the five polymeric components.

Lamport (1965) has used chemical extraction procedures to determine that cell walls isolated from suspension-cultured sycamore cells contain 36% pectin, 34% hemicellulose, and 27% cellulose. He reports that

TABLE II

Calculated Polymer Composition of Sycamore Cell Walls[a]

Wall component	Cell walls (%)
Arabinogalactan	20
Cellulose	23
Protein	10
Rhamnogalacturonan	16
Tetraarabinosides (attached to hydroxyproline)	9
Xyloglucan	21
	99

[a] Protein was estimated by multiplying the nitrogen content by 6.25. Nitrogen analyses were performed by the Kjeldahl procedure at the Pediatrics Microanalytical Laboratories, University of Colorado Medical Center. The amount of the other components was estimated as described in the text and by Talmadge *et al.* (1973).

most of the wall protein is in the hemicellulose fraction. Using the values presented in Table II, and including both protein and xyloglucan in the hemicellulose fraction, the corresponding values of Talmadge *et al.* (1973) would be 34% pectin, 38% hemicellulose, and 26% cellulose. Roelofsen (1959) has noted that primary cell walls are typically one-third cellulose, one-third hemicellulose, and one-third pectin plus protein. Thus, the values reported by Talmadge *et al.* (1973) are in good agreement with the typical primary wall values given by Roelofsen (1959). *These components are found in all primary cell walls.*

C. The Pectic Polymers of Sycamore Cell Walls

1. THE ENDOPOLYGALACTURONASE PRODUCTS

The structural studies of sycamore cell walls were made possible by the availability of purified hydrolytic enzymes that are capable of degrading isolated cell walls. The most important of these enzymes is an endopolygalacturonase obtained from the culture medium of *Colletotrichum lindemuthianum* (English *et al.*, 1972). Among the fungal enzymes studied, the endopolygalacturonase is uniquely capable of initiating the degradation of isolated plant cell walls. Isolated cell walls, which have been subjected to the action of the endopolygalacturonase, have greatly increased susceptibility to degradation by a purified endoglucanase (Bauer *et al.*, 1973; Keegstra *et al.*, 1973) and by pronase (Keegstra *et al.*, 1973).

The structures of the sycamore wall pectic polymers—the neutral arabinogalactan and acidic rhamnogalacturonan—have been partially determined from the analysis of those portions of these two polymers which are released from sycamore walls by the action of the endopolygalacturonase.

This enzyme specifically degrades the homogalacturonan regions of the cell wall pectic polysaccharides and releases 16% of the wall as soluble products. The soluble material represents about 50% of the total pectic polysaccharides of the wall (Table II) and includes 75% of the galacturonic acid present in the wall (Table I). The wall material released by this enzyme consists of approximately equal amounts of short oligomers of galacturonic acid and of polymeric material composed of 26% galacturonic acid and 74% neutral sugars.

The short oligomers released by the endopolygalacturonase consist of an approximately equimolar mixture of tri-, di-, and monogalacturonic acid. These same products are found when citrus polygalacturonic acid is treated with this endopolygalacturonase (English *et al.*, 1972). This indicates that this enzyme has the same mode of action on the cell wall pectic polysaccharides as on soluble polygalacturonic acid, and that the enzyme is incapable of hydrolyzing oligomers containing fewer than four residues.

2. The Structure of the Rhamnogalacturonan—A Pectic Polymer

Talmadge *et al.* (1973) have presented a tentative structure for the rhamnogalacturonan (the acidic pectic polymer) of sycamore walls (Fig. 1). This structure is based on the analyses of the fractions obtained from endopolygalacturonase treatment of cell walls and on the analysis of the galacturonosyl-containing oligomers obtained by partial acid hydrolysis of isolated cell walls. *The rhamnogalacturonan is not a straight chain molecule. The zigzagged shape results from the presence of 2-linked rhamnosyl residues in an otherwise linear* α-*(1→4)-linked galacturonan chain.* When an arabinogalactan chain is attached to carbon-4 of a rhamnosyl residue, the rhamnose forms a Y-shaped branch point. These observations came as a result of building CPK space filling models of

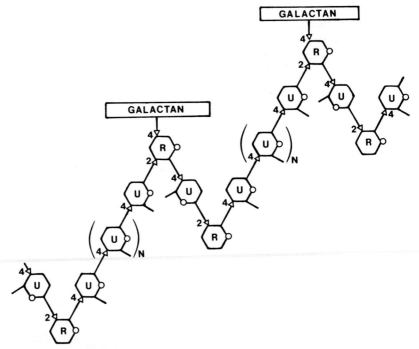

Fig. 1. A proposed structure for the rhamnogalacturonan. The structure is based on evidence presented in the text and in Talmadge *et al.* (1973). The sugar residues in the figure are designated as R = rhamnose and U = galacturonic acid.

these structures. A similar observation was made by Simmons in studying the O-antigen polysaccharides of *Shigella flexneri*. He noted that the $(1 \rightarrow 2)$-rhamnosyl linkages of that oligosaccharide caused a "buckling" of the otherwise linear polymer (Simmons, 1971). Rees and Wight (1971), using model-building computations, came to the same conclusions concerning the structure of rhamnogalacturonans.

The rhamnosyl residues are not uniformly distributed in the chain, but probably occur as rhamnosyl-$(1 \rightarrow 4)$-galacturonosyl-$(1 \rightarrow 2)$-rhamnosyl units. This sequence appears to alternate in the polymer with a homogalacturonan sequence containing approximately 12 residues of 4-linked galacturonic acid. These two sequences give rise to the two major fractions obtained upon endopolygalacturonase treatment of sycamore cell walls. The oligogalacturonosides arise from endopolygalacturonase hydrolysis of the homogalacturonan sequences of the rhamnogalacturonan polymer, whereas the neutral rich polymers arise from the rhamnose-rich regions of the rhamnogalacturonan polymer.

Degradation of the galacturonan chain by the endopolygalacturonase demonstrated that the glycosidic linkages of the galacturonosyl residues are in the α configuration (English *et al.*, 1972). Talmadge *et al.* (1973) have obtained chemical evidence that the rhamnosyl residues are in the β configuration. Diborane reduction of the methylated oligogalacturonosides released by the endopolygalacturonase substantiated the 4-linkage of the galacturonosyl residues in the galacturonan chains.

The linear galacturonan portion of the rhamnogalacturonan (Fig. 1) gives rise to larger oligogalacturonosides when intact cell walls are subjected to partial acid hydrolysis (Talmadge *et al.*, 1973). These oligogalacturonosides, which represent about 5% of the cell wall, have rhamnosyl residues at their reducing ends. The presence of rhamnose in these oligomers provides evidence that the two regions in the proposed structure are linked in an alternating sequence in the same polymer. The rough estimate of between 10 to 14 galacturonosyl residues in the galacturonan region was obtained from gel filtration chromatography of the oligogalacturonosides isolated following partial acid hydrolysis of cell walls.

Several lines of evidence have been obtained for the occurrence of rhamnosyl residues in the galacturonan chains. As discussed above, relatively large oligogalacturonosides containing rhamnose residues at their reducing ends have been isolated following partial acid hydrolysis of intact cell walls. In addition, the most abundant aldobiuronic acid isolated from partially hydrolyzed cell walls was identified by methylation analysis and diborane reduction as galacturonosyl-$(1 \rightarrow 2)$-rhamnose (Talmadge *et al.*, 1973). This same aldobiuronic acid has also been reported to occur in the pectic polysaccharides obtained from a number of different plant sources, including sycamore cell walls (Aspinall *et al.*, 1967b, 1968; Barrett and Northcote, 1965; Rees and Wight, 1969; Stoddart *et al.*, 1967).

As indicated above, the rhamnose-rich sequence in Fig. 1 gives rise to neutral sugar-rich polymers when cell walls are treated with endopolygalacturonase. The neutral sugar-rich polymers are not further degraded by the endopolygalacturonase. The known specificity of this enzyme suggests, therefore, that there are no more than two galacturonosyl residues between neighboring rhamnosyl residues. The rhamnose content of each neutral sugar-rich region is accounted for by two rhamnose residues, since the rhamnose of this region is equivalent to approximately 14% of the linear galacturonan region, which has been estimated to contain 12 galacturonic acid residues. The fact that the predominant aldouronide isolated from partially acid hydrolyzed cell walls is galacturonosyl-$(1\rightarrow2)$-rhamnose rather than galacturonosyl-$(1\rightarrow4)$-galacturonosyl-$(1\rightarrow2)$-rhamnose indicates that in the region rich in neutral sugar, there is only one galacturonosyl residue between neighboring rhamnosyl residues. These results, along with the fact that the region rich in neutral sugar contains galacturonic acid and rhamnose in the molar ratio of 2:1, suggest that the rhamnogalacturonan portion of the neutral sugar-rich polymer has an average structure of galacturonosyl-$(1\rightarrow2)$-rhamnosyl-$(1\rightarrow4)$-galacturonosyl-$(1\rightarrow2)$-rhamnosyl-$(1\rightarrow4)$-galacturonosyl-$(1\rightarrow4)$-galacturonosyl.

DEAE–Sephadex ion exchange chromatography demonstrates that the neutral sugars of the neutral sugar-rich endopolygalacturonase product are covalently attached to galacturonosyl residues (Talmadge *et al.*, 1973). Most of these neutral sugar residues, which represent 76% of this fraction, are part of the arabinogalactan. The methylation analyses of the neutral sugar-rich fraction and of isolated, intact cell walls indicate that approximately 50% of the rhamnosyl residues are branched, having a substituent at carbon-4 as well as at carbon-2. As this is the major, if not the only, branch point of the rhamnogalacturonan chain, this 2,4-linked rhamnosyl represents the point of attachment of at least some of the neutral side chains.

3. The Structure of the Arabinogalactan—A Pectic Polymer

As indicated earlier, most of the arabinogalactan that is released by endopolygalacturonase treatment is part of neutral sugar-rich but nevertheless acidic polymers. An estimate of the average number of glycosyl residues in each arabinogalactan can be made by dividing the total amount of arabinose and galactose in the neutral sugar-rich fraction by the amount of branched (2,4-linked) rhamnosyl residues. This calculation indicates that the arabinogalactan consists on the average of 6 galactosyl residues and 7 arabinosyl residues (Talmadge *et al.*, 1973). The number of arabinosyl residues in a single chain may be greater if, as appears likely, some galactan chains do not have arabinan chains attached. Methylation analysis of the neutral sugar-rich polymers indicates that, of the 6 galactosyl residues in an average side chain, three galactosyl residues are 4-linked, one is terminal, and one is 6-linked. In addition, 3-linked and branched (2,4-, 3,6- and 4,6-linked) galactosyl residues occur with a frequency of less than once in every chain. The linkages of the arabinosyl residues also reflect the complex nature of the arabinogalactan. An average chain contains three terminal arabinosyl residues, two branched (3,5-linked) arabinosyl residues, and two 5-linked arabinosyl residues. There are also smaller amounts of 3-linked, 2,5-linked, and double branched (2,3,5-linked) arabinosyl residues that occur with a frequency of less than once in every chain.

The structure of the arabinogalactan has been further examined by subjecting the neutral sugar-rich polymers to weak acid hydrolysis (Talmadge *et al.*, 1973). Following the hydrolysis, the sample was fractionated on DEAE–Sephadex into neutral and acidic components, which were then separately chromatographed on a Bio-Gel P-2 column. The results from this analysis indicate that about 85% of the arabinosyl residues were released by hydrolysis from the acidic rhamnogalacturonan fragment, while 75% of the galactosyl residues remained attached to the acidic fragment. This preferential cleavage of arabinose was accompanied by only minor changes in the linkages of the sugar residues remaining in the acidic fraction. *These results suggest that the galactosyl residues are present as a linear chain that is attached at its reducing end to the rhamnogalacturonan main chain, and that the arabinosyl residues are in the form of a branched chain attached either to the galactosyl backbone or directly to the rhamnogalacturonan main chain.*

The neutral sugar-rich polymers released by endopolygalacturonase contain, in addition to the rhamnogalacturonan fragment and the arabinogalactan, small amounts of xyloglucan (Bauer *et al.*, 1973). The xyloglucan of the endopolygalacturonase-liberated polymers fractionates as an acidic polymer on DEAE–Sephadex, indicating a covalent attachment between the xyloglucan wall component and the pectic polysaccharides (Talmadge *et al.*, 1973). After weak acid hydrolysis of these polymers, over 70% of the xyloglucan still fractionates as an acidic polymer on DEAE–Sephadex. This indicates that the xyloglucan is probably attached to the galactan backbone of the arabinogalactan; this result is further evidence that acid labile arabinosyl residues are not interspersed in the galactan chain. *As most arabinogalactan chains are covalently attached to the rhamnogalacturonan, the arabinogalactan serves as a cross-link between the xyloglucan and rhamnogalacturonan components of the wall.*

Based on the results of the weak acid hydrolysis experiments and on the methylation analyses, a tentative structure has been proposed by Talmadge *et al.* (1973) for the arabinogalactan (Fig. 2). The model assumes that about two in three galactan chains have arabinan chains attached, but, as stated above, the arabinan chains could be attached directly to the rhamnogalacturonan. It should be emphasized that, although it is consistent with the available data, the proposed structure is very tentative.

The minor amounts of 3- and 3,6-linked galactosyl residues and the minor amounts of 3- and 2,5-linked arabinosyl residues present in the neutral sugar-rich fraction probably arise from the presence of a small amount of an arabinogalactan that is distinctly different from the arabinogalactan that cross-links the xyloglucan and rhamnogalacturonan. Evidence is presented in Section III,B that this second type of arabinogalactan is present as a minor constituent in sycamore cell walls and that this polymer may be important in the attachment of the pectic polymers to cell wall protein (Keegstra *et al.*, 1973).

The arabinogalactan that interconnects the xyloglucan and rhamnogalacturonan is the least characterized of the five major structural polymers that constitute the sycamore cell wall. Enzymes, which can specifically degrade this arabinogalactan, have been purified and will be used for further structural characterization. Preliminary results using a partially purified *endo*-β-1,4-galactanase (J. M. Labavitch, M. McNeil, and P. Albersheim, unpublished results) show that treatment of walls with this enzyme permits urea to extract xyloglucan molecules from the wall; xyloglucan could not be extracted from walls that were not pretreated with the endogalactanase. This result provides further evidence that at least a portion of the xyloglucan molecules are connected to the wall matrix through galactan chains.

The models proposed for the two sycamore pectic polymers (Figs. 1 and 2) are generally consistent with previous structural studies on pectic polysaccharides. Rhamnogalacturonans appear to be a common feature of all pectic polysaccharides. Pectic arabinogalactans containing a β-1,4-linked galactan backbone have been isolated from soybean seed (Aspinall *et al.*, 1967a). Talmadge *et al.* (1973) have obtained evidence that suggests that the galactosyl residues of the arabinogalactan of sycamore cell walls are in the β configuration. The highly branched arabinan

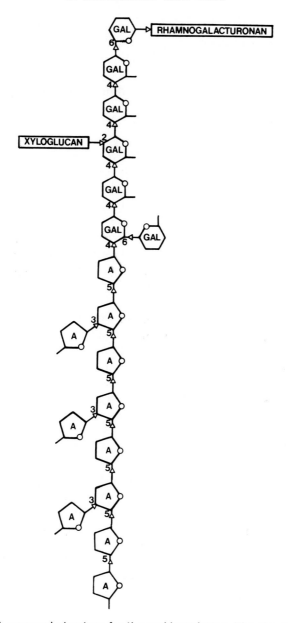

Fig. 2. A proposed structure for the arabinogalactan. The structure is based on evidence presented in the text and in Talmadge *et al.* (1973). The sugar residues in the figure are designated as A = arabinose and GAL = galactose.

region of sycamore pectic polysaccharides contains similar linkages to those present in the pectic arabinans isolated from soybean, lemon peel, and mustard seed (Aspinall and Cottrell, 1971).

D. The Hemicellulose of Suspension-Cultured Sycamore Cells

1. Sycamore Extracellular Xyloglucan

Sycamore cells, when grown in suspension culture, secrete polysaccharides into their culture medium (Becker et al., 1964). *The fractionation of deesterified sycamore extracellular polysaccharides (SEPS) on DEAE-Sephadex yields three major fractions: a neutral fraction containing xyloglucan, a weakly acidic fraction containing arabinogalactan, and a strongly acidic fraction containing rhamnogalacturonan* (Bauer et al., 1973). These three fractions correspond in general to the three major fractions obtained by Aspinall et al. (1969) using a different fractionation procedure. The exact composition and the relative amounts of the three major fractions of SEPS vary considerably depending on the media used to culture the cells.

Aspinall et al. (1969) have shown that SEPS xyloglucan has the property of binding strongly, but noncovalently, to cellulose, and they have used this property in their purification of xyloglucan. The first indication that arabinose and galactose might be present in the xyloglucan polymer came from the finding that significant and reproducible amounts of these two sugars were found in the SEPS material which bound to cellulose, and which could be partially eluted from cellulose with 8 M urea or 1 N sodium hydroxide (Bauer et al., 1973).

Additional evidence for the presence of arabinose and galactose in SEPS xyloglucan was obtained by gel filtration chromatography of cellulose-purified SEPS xyloglucan on agarose Bio-Gel A-1.5m (Bauer et al., 1973). The sugar compositions of polymers from three different regions of the xyloglucan peak were quite similar, indicating that the polymers from these three regions were structurally homogeneous, but differed in size or degree of aggregation. Methylation analysis of SEPS xyloglucan confirmed the principal findings of Aspinall et al. (1969). In addition, this analysis showed the presence of residues of terminal arabinose, terminal and 2-linked galactose, and 2,4,6-linked glucose (Bauer et al., 1973).

2. Sycamore Wall Xyloglucan

Sycamore cell walls contain a xyloglucan polysaccharide that is essentially identical to that isolated from SEPS (Bauer et al., 1973). A consideration of the methods used to extract xyloglucan from the cell walls leads to the conclusion that this polysaccharide is covalently linked to the pectic polysaccharides of the walls and is noncovalently bound to the cellulose fibers.

The extraction of untreated sycamore walls with 8 M urea solubilizes small amounts of a 4-linked glucan, which is probably cellulose, but does not solubilize any other polysaccharide or sugar in appreciable amounts (Bauer *et al.*, 1973). Since 8 M urea is widely used for the disruption of hydrogen bonds, and since the cellulose fibers of the cell wall are believed to be held together by interchain hydrogen bonds (Frey-Wyssling, 1969), the solubilization of some cellulose by 8 M urea might be expected. From the fact that other sugars and polysaccharides are not released from untreated walls by 8 M urea, it appears that urea does not cause the cleavage of covalent bonds in the wall matrix.

The extraction of endopolygalacturonase-pretreated walls with 8 M urea releases a wall fraction that contains xyloglucan (Bauer *et al.*, 1973). The extraction of untreated walls with urea releases only the 4-linked glucan discussed above. The subsequent treatment of these walls with the endopolygalacturonase does not release the xyloglucan. Thus, the effect of urea appears to be chemically reversible when the urea is removed. *This supports the view that the disruption of noncovalent (hydrogen) bonds is involved in the release of xyloglucan from endopolygalacturonase-pretreated walls by urea.* Aspinall et al. (1969) have shown that SEPS xyloglucan binds noncovalently to cellulose, and that the cellulose-bound xyloglucan can be partially solubilized by extraction with 8 M urea. It seems likely that the release of the xyloglucan-containing fraction from endopolygalacturonase-pretreated walls by urea involves the disruption of hydrogen bonds between the xyloglucan and the cellulose fibers of the walls. *Moreover, since urea releases xyloglucan-containing polymers from endopolygalacturonase-pretreated walls, but not from untreated walls, it appears that, in the native cell wall, the xyloglucan polysaccharides are covalently attached to the wall matrix by galacturonosylic bonds.*

Similar results are obtained when cell walls are treated with *endo-β-1,4-glucanases* (Bauer *et al.*, 1973). The endoglucanase from *Trichoderma viride*, acting on untreated sycamore walls, releases only about 1% of the untreated walls as soluble products. The same enzyme, however, solubilizes 10–15% of the walls if the walls have been pretreated with endopolygalacturonase. The endoglucanase-liberated material contains the sugars, in the appropriate linkages, which are characteristic of xyloglucan. It appears that those polysaccharides, which are removed from untreated walls by the action of endopolygalacturonase, are able to block the access of *T. viride* endoglucanase to xyloglucan in the *untreated* walls.

The method of isolating the xyloglucan fragments produced by endoglucanase treatment of endopolygalacturonase-pretreated walls provides no direct evidence that the polysaccharides removed from endopolygalacturonase-pretreated walls by endoglucanase were bound noncovalently to these walls. However, *T. viride* endoglucanase does hydrolyze xyloglucan to oligosaccharide fragments, and it has been shown that these oligosaccharide fragments do not bind appreciably in aqueous solution to cellulose (Bauer *et al.*, 1973). Moreover, it has been shown that *T. viride* endoglucanase is able to hydrolyze cellulose-bound SEPS xyloglucan, releasing the xyloglucan (fragments) from the cellulose. Thus, if the xyloglucan polymers were attached to the endopolygalacturonase-pretreated walls by noncovalent associations between xyloglucan and cellulose fibers, then their enzymic release from the walls by endoglucanase would be expected.

Although the heptasaccharide and nonasaccharide fragments produced by endoglucanase hydrolysis of xyloglucan do not bind to cellulose in water, these oligosaccharides do bind to cellulose at low temperatures in 60–70% ethanol or acetate (Valent and Albersheim, 1974). This is further evidence that xyloglucan chains are hydrogen-bonded to cellulose.

There is considerable evidence that sycamore wall xyloglucan is the same as that isolated from SEPS. The fractions released from endopolygalacturonase-pretreated walls by urea, endoglucanase, or dilute base clearly contain the sugars and sugar linkages which are characteristic of SEPS xyloglucan (Bauer *et al.*, 1973). Moreover, the polysaccharides, which are solubilized by urea and base, show the same ability as SEPS xyloglucan to bind noncovalently to cellulose and to form colored complexes with iodine. In addition, endoglucanase releases fragments of xyloglucan from endopolygalacturonase-pretreated walls which give the same basic elution pattern when fractionated on Bio-Gel P-2 as the elution pattern obtained by fractionation of endoglucanase-pretreated SEPS xyloglucan. This correspondence between the elution patterns of xyloglucan fragments from the two sources is extremely unlikely unless the wall-derived xyloglucan and SEPS xyloglucan have the same basic structure. A comparison of the sugar and sugar linkage compositions of these fragments showed that the xyloglucans from the two sources are virtually identical.

3. The Structure of Xyloglucan

Studies of the composition and properties of fragments produced by dilute acid hydrolysis of xyloglucan have demonstrated that the cellulose-binding property of xyloglucan cannot be due to the presence of discrete, cellulose-like regions in the polymer that contain only residues of unbranched glucose (Bauer *et al.*, 1973). Even relatively small fragments of xyloglucan retain their ability to bind to cellulose. It has been shown also that the rhamnosyl, fucosyl, and arabinosyl residues of the polymer are not required for cellulose binding, and that the xylosyl and galactosyl residues of the fragments do not prevent the binding to cellulose (Bauer *et al.*, 1973).

Considerable information about the structure of xyloglucan has been obtained by analysis of the fragments produced by endoglucanase hydrolysis. The xyloglucan fragments produced by *T. viride* endoglucanase hydrolysis appear as discrete peaks when fractionated on Bio-Gel P-2 (Bauer *et al.*, 1973). Proposed structures for the three principal endoglucanase-derived xyloglucan oligosaccharides are presented in Fig. 3. Oligosaccharide A represents at least 12% of the sycamore xyloglucan, oligosaccharide B represents 28%, and oligosaccharide C 35%. Other xyloglucan fragments, closely related to those presented in Fig. 3, have also been characterized (Bauer *et al.*, 1973). A comparison of the experimentally determined sugar linkage compositions of the xyloglucan oligosaccharides and the sugar linkage compositions for the oligosaccharide structures presented in Fig. 3 is given in Table III. In addition to methylation analysis, which yielded the data of Table III, the oligosaccharide structures presented in Fig. 3 are based on sugar composition analysis, comparison of the gel filtration elution volumes with the elution volumes of known oligosaccharides, identification of the sugar residue at the reducing end of the oligosaccharide, and knowledge of the specificity of the enzyme (*endo*-β-1,4-glucanase) that liberated the oligosaccharides.

The structure proposed for oligosaccharide A (Fig. 3) is more complex and therefore more difficult to establish than the structures of oligosaccharides B and C. The distinguishing feature of oligosaccharide A is the presence of the doubly branched 2,4,6-linked glucose (Bauer *et al.*, 1973). There is approximately one residue of this component for every 20 residues in the oligosaccharide. The oligosaccharide cannot be much larger than 20 residues or it would have voided the Bio-Gel P-2 column despite its highly branched structure. If the structures of the more prevalent heptasaccharide B and nonasaccharide C are used in the biosynthesis of the oligosaccharide A, then a structure for oligosaccharide A which is consistent with its approximate size and sugar linkage composition may be generated by the combination of one heptasaccharide unit, one nonasaccharide unit, one residue of 4-linked glucose, and one residue of 2,4,6-linked glucose with arabinose and fucosylgalactosylxylose side chains.

Any structure proposed for oligosaccharide A should account for the fact that the glucosylic bonds of the two internal residues of 4-linked glucose (which are indicated by arrows in the proposed structure shown in Fig. 3) are *not* hydrolyzed by the *T. viride* endoglucanase, even though this enzyme produces oligosaccharides B and C by hydrolyzing the glucosylic bonds between residues of 4-linked glucose and 4,6-linked glucose.

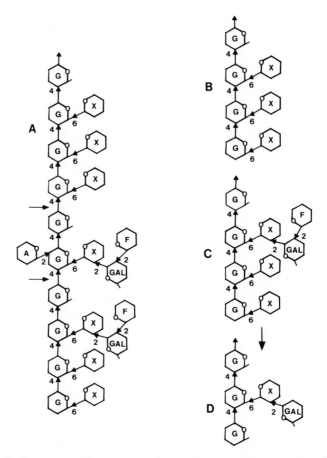

Fig. 3. The proposed structure of endoglucanase-derived xyloglucan oligosaccharides. The significance of these oligosaccharides is discussed in the text. The structures are based on the data presented in Table III, in Bauer *et al.* (1973), and in the text. For simplicity, the anomeric configurations of the glycosidic linkages are not indicated. The resistance to endoglucanase hydrolysis of the glycosidic linkages marked by arrows, in the structure proposed for the oligosaccharide of Peak A, is discussed in the text. The sugar residues in the figure are designated as A = arabinose, F = fucose, G = glucose, GAL = galactose, and X = xylose.

It has been suggested that the *T. viride* endoglucanase cannot hydrolyze the two indicated glucosylic bonds in oligosaccharide A because the side chain at the 2-position of the doubly branched, 2,4,6-linked glucosyl residue sterically hinders the enzyme (Bauer *et al.*, 1973).

The location of the fucosylgalactose side chain of the nonasaccharide (Fig. 3, oligosaccharide C) was recently ascertained by subjecting this

TABLE III

Actual and Calculated Mole Percent Sugar Linkage Compositions of Endoglucanase-Derived Xyloglucan Oligosaccharides[a]

Sugar and linkage	Bio-Gel P-2 peak					
	A		B		C	
	Expt.	Calc.	Expt.	Calc.	Expt.	Calc.
Fucose						
T-Fuc	5.7	7.1	0	0	11.4	11.1
Arabinose						
T-Ara	2.7	2.4	0	0	0	0
Xylose						
T-Xyl	26.1	26.2	41.3	42.9	20.4	22.2
2-Xyl	6.8	7.1	0.7	0	11.3	11.1
Galactose						
T-Gal	2.7	2.4	0.5	0	2.3	0
2-Gal	6.0	7.1	0	0	9.4	11.1
Glucose						
4-Glc	13.8	14.3	12.7	14.3	12.8	11.1
6-Glc	5.1	4.8	13.4	14.3	10.3	11.1
4,6-Glc	23.0	23.9	27.3	28.6	21.1	22.2
2,4,6-Glc	4.8	4.8	0	0	0	0

[a] The experimentally determined sugar linkage compositions of the three endoglucanase-derived xyloglucan oligosaccharides are compared with the sugar linkage compositions calculated for the proposed structures of these oligosaccharides shown in Fig. 3 (Bauer et al., 1973). The sugar linkage compositions are presented in this table as mole percent of the total carbohydrate. The glycosidic linkages are indicated by numerical prefixes; thus, 2-Gal means that sugars are glycosidically linked to the second carbon of the galactosyl residues. Terminal residues are indicated by T- (e.g., T-Fuc).

oligosaccharide to enzyme hydrolysis by a combination of an α-fucosidase, an α-xylosidase, and a β-glucosidase. The product of this treatment was isolated and clearly shown to be the pentasaccharide D illustrated in Fig. 3 (P. Albersheim, B. Valent, and M. McNeil, unpublished results). This pentasaccharide could only arise from a nonasaccharide having the fucosylgalactosyl side chain on the first xylosyl residue as indicated in oligosaccharide C of Fig. 3.

4. A COMPARISON OF SYCAMORE XYLOGLUCAN AND "AMYLOIDS"

As previously noted by Aspinall et al. (1969), the xyloglucan from SEPS is structurally similar to the so-called seed "amyloids." These "amyloids," which are also xyloglucans, are found in the seeds of a wide variety of plants, and derive their name from the fact that they form colored complexes with iodine, as does starch (amylose) (Kooiman, 1960, 1961, 1967; Siddiqui and Wood, 1971). Sycamore xyloglucan also forms colored complexes with iodine. It has been suggested (Gould *et al.*, 1971) that the failure of Aspinall *et al.* (1969) to detect formation of an iodine complex with SEPS xyloglucan may have been due to the presence of protein in their preparation. The chemical basis of xyloglucan–iodine complex formation is not known, although Gould *et al.* (1971) have recently suggested that formation of the complex between iodine and the "amyloid" from white mustard seeds involves the interaction of iodine molecules or iodide ions within the interstices between *aggregated* xyloglucan chains. This suggestion is consistent with the observation of Bauer *et al.* (1973) that the oligosaccharides produced by endoglucanase hydrolysis of SEPS xyloglucan do not form colored iodine complexes. Since the endoglucanase-derived xyloglucan oligosaccharides have little or no ability to bind to cellulose, it seems unlikely that they would be able to associate strongly enough with each other to provide interchain "holes" for complex formation with iodine. Bauer *et al.* (1973) have also observed that SEPS xyloglucan bound to cellulose is able to form colored iodine complexes, which, in the model of Gould *et al.* (1971), would indicate that suitable "holes" for iodine complexing are formed when chains of xyloglucan associate with chains of cellulose.

There is a striking similarity between the structure proposed by Bauer *et al.* (1973) for sycamore xyloglucan and that proposed in the elegant study by Kooiman (1961) for the "amyloid" obtained from the seeds of *Tamarindus indicus*. Kooiman has isolated and characterized the oligosaccharides produced by endoglucanase hydrolysis of the tamarind "amyloid." Three such oligosaccharides were obtained in significant amounts. One of these oligosaccharides was shown to have a structure identical to heptasaccharide B (Fig. 3) of sycamore xyloglucan. The other two oligosaccharides were found by Kooiman to contain either one or two residues of terminal galactose linked to the 2-position of xylosyl residues in the heptasaccharide unit. Using an enzyme preparation called "luizym," Kooiman (1961) was also able to hydrolyze the tamarind xyloglucan so that almost all of the xylosyl residues of the polymer were recovered in the disaccharide 6-*O*-α-D-xylopyranosyl-D-glucopyranose. The essentially quantitative isolation of this disaccharide clearly demonstrates that all of the glucosyl residues in the polymer are present in a cellulose-like glucan backbone, and that all of the xylosyl residues occur as monoxylosyl side chains linked to the 6 position of glucosyl residues in the glucan backbone.

Preliminary investigations by Talmadge *et al.* (1973) of the ano-meric configuration of the glycosylic linkages in sycamore xyloglucan, using the chromium trixoide–acetic acid oxidation method of Hoffman *et al.* (1972), suggest that the fucosyl residues of this polymer are linked in the α-configuration, and that all of the other glycosylic linkages exist in the β-configuration. The α-configuration of the fucosyl residues was confirmed by hydrolysis with an α-fucosidase (P. Albersheim, B. Valent, and M. McNeil, unpublished results). Kooiman (1961) has reported that the galactosyl and glucosyl residues of the tamarind xyloglucan are gly-cosylically linked in the β-configuration. However, Kooiman has pre-sented strong evidence that the xylosyl residues are linked α-(1→6) to the glucosyl residues, whereas the evidence of Talmadge *et al.* (1973) in favor of the β-(1→6) linkage for the xylosyl residues in sycamore xyloglucan is not conclusive. It has now been confirmed by the action of an α-xylosidase that the xylosyl residues are linked in the α-configura-tion (P. Albersheim, B. Valent, and M. McNeil, unpublished results).

The molecular weight of SEPS xyloglucan has been estimated by hypoidite oxidation to be approximately 7600, which corresponds to about 56 sugar residues in the polymer (Bauer *et al.*, 1973). About 25–30 of these residues would be glucosyl residues in the β-(1→4)-linked back-bone of the polymer. The molecular weights of the "amyloid" xyloglucans from *T. indicus* and *Annona muricata* L. were estimated by Kooiman (1967), using a similar hypoiodite method, and found to be approximately 10,750 and 11,500, respectively. These molecular weights correspond to polymers containing 70 and 74 sugar residues. Using a periodate oxidation method, however, Kooiman (1967) found the molecular weight of the *A. muricata* xyloglucan to be only 8800, corresponding to a polymer of 56 residues. (The average molecular weights of the sugar residues in these three xyloglucans are different because their sugar compositions are different; thus, the sycamore and *A. muricata* xyloglucans can have different molecular weights but the same number of residues.) Since the methods used for molecular weight determinations yield only rough esti-mates of the average size of polymers this large, these three xyloglucans can be considered to have comparable molecular weights.

In view of the structural similarities between sycamore xyloglucan (Bauer *et al.*, 1973) and the seed "amyloid" xyloglucans (Kooiman, 1961, 1967; Siddiqui and Wood, 1971), it will be of considerable interest to ascertain whether or not the "amyloid" xyloglucans are incorporated into the structure of the primary cell walls of the germinating seedlings. The fate of the "amyloid" xyloglucan from white mustard seeds has been studied by Gould *et al.* (1971). These investigators found that, before germination, two xyloglucan fractions could be obtained from the cotyledons: a soluble xyloglucan that could be extracted with hot EDTA solutions and an insoluble xyloglucan that required further extraction with aqueous alkali or lithium thiocyanate. After germination, the soluble xyloglucan fraction was not detected, but the insoluble xyloglucan was still present. The disappearance of the soluble xyloglucan after germination led Gould *et al.* (1971) to the conclusion that the soluble xyloglucan is a storage polysaccharide that is metabolized upon germination. It seems likely, however, that the difference between the soluble and insoluble xyloglucans is the binding of the latter to cellulose, and that quite possibly the disappearance of the soluble xyloglucan after germination involves the binding of this fraction to cellulose or its incorporation into newly synthesized cell walls.

III. The Connections between the Structural Components of a Primary Cell Wall

A. The Covalent Connection between the Xyloglucan and the Arabinogalactan of the Walls of Suspension-Cultured Sycamore Cells

There are several wall fractions that provide evidence for the presence of a covalent connection between the hemicellulosic xyloglucan and the pectic arabinogalactan. The arabinogalactan polymers released by endopolygalacturonase (Section II,C,3) provide evidence for this interconnection. Endopolygalacturonase is able to release arabinogalactan chains, for these chains are held in the wall by covalent attachment to the rhamnogalacturonan (Talmadge *et al.*, 1973); the rhamnogalacturonan is the substrate of the endopolygalacturonase. Some of the arabinogalactan chains released by endopolygalacturonase have, in addition to fragments of the rhamnogalacturonan, small xyloglucan fragments covalently attached (Talmadge *et al.*, 1973).

The best evidence for the existence of an interconnection between xyloglucan and the pectic polymers, however, comes from an analysis of the fragments solubilized by endoglucanase treatment of endopolygalacturonase-pretreated walls (Keegstra *et al.*, 1973). The neutral fraction of the carbohydrate solubilized by endoglucanase consists of xyloglucan fragments (Bauer *et al.*, 1973). However, approximately 40% of the total material released by endoglucanase is acidic. These acidic polymers contain the sugars and sugar linkages characteristic of the acidic arabinogalactan (Fig. 2) released by endopolygalacturonase (Talmadge *et al.*, 1973) and also contain the sugars and sugar linkages characteristic of xyloglucan (Fig. 3 and Bauer *et al.*, 1973). Since these acidic polymers are solubilized by the action of endoglucanase, which splits only β-1,4-

glucosidic bonds, the pectic fragments must be connected, prior to endo-glucanase treatment, to the wall matrix by the β-1,4-linked glucosyl residues of the xyloglucan chains. This conclusion is substantiated by the cochromatography of the xyloglucan fragments with the acidic arabino-galactan on DEAE–Sephadex (Bauer et al., 1973). Since xyloglucan, by itself, is a neutral polysaccharide, the cochromatography of xyloglucan with the acidic arabinogalactan indicates that this xyloglucan is covalently linked to the acidic arabinogalactan.

The covalent attachment of xyloglucan to the acidic arabinogalactan provides an explanation for the observation that the release of xyloglucan from the walls by urea requires the prior hydrolysis of galacturonosylic bonds by endopolygalacturonase (Section II,D,2), and also provides an explanation for the fact that exhaustive endopolygalacturonase treatment releases from the wall only about 65% of the arabinogalactan side chains of the rhamnogalacturonan polymers (Talmadge et al., 1973). This agrees with the finding that 30–35% of the xyloglucan polymers extracted with base are covalently linked to the acidic arabinogalactan. Since xyloglucan is covalently linked in the native cell wall to the arabinogalactan side chains of the rhamnogalacturonan, it is the expected result that xyloglucan cannot be released from the wall by urea unless either the arabino-galactan or rhamnogalacturonan chains are broken. As mentioned above, endopolygalacturonase treatment of the walls does permit urea to extract xyloglucan, and, more recently, we have found that an endo-β-1,4-galactanase also permits urea to extract xyloglucan (J. M. Labavitch, M. McNeil, and P. Albersheim, unpublished results). Furthermore, in agreement with the observed results, it is expected that those arabinogalactan side chains, which are not attached to xyloglucan, will be released from the wall by exhaustive endopolygalacturonase treatment (the neutral sugar-rich fraction), whereas those arabinogalactan side chains that are attached to xyloglucan will remain in the endopolygalacturonase-pre-treated walls until the noncovalent binding of xyloglucan to cellulose is disrupted by the action of urea, base, or endoglucanase.

Urea and base solubilize xyloglucan from endopolygalacturonase-pretreated walls (Section II,D,2). A portion of the xyloglucan polymers solubilized in this manner is covalently attached to pectic polysaccharides. The only difference between these polymers and the acidic xyloglucan fragments released by endoglucanase is that the urea- and base-extracted polymers contain a larger amount of xyloglucan attached to each pectic fragment (Bauer et al., 1973). This is the expected result, since the xyloglucan polymers had not been subjected to degradation with endoglucanase prior to the urea or base extraction.

The question of where and how the xyloglucan and pectic polymers are interconnected cannot be deduced from the data presented. If the xyloglucan chains were attached directly to the rhamnogalacturonan, it is likely that some of the acidic xyloglucan fragments would lack arabinogalactan chains. This is true since the arabinogalactan chains are known to be attached by their reducing ends to the rhamnogalacturonan (Talmadge *et al.*, 1973). The presence of arabinogalactan in all of the fractions that contain xyloglucan suggests that the xyloglucan polymers are glycosidically attached to the arabinogalactan, and through the arabinogalactan chains to the rhamnogalacturonan. The evidence obtained by weak acid hydrolysis also indicates that the xyloglucan is covalently linked to the rhamnogalacturonan through the galactan portion of the arabinogalactan (Talmadge *et al.*, 1973). In addition, some arabinogalactan is released from endopolygalacturonase-pretreated walls by endoglucanase, urea, and base. This result is best explained by a covalent attachment between xyloglucan and arabinogalactan.

Since the xyloglucan is attached to the cellulose fibers by hydrogen bonds (Bauer *et al.*, 1973; Valent and Albersheim, 1974), the reducing ends of the xyloglucan chains are free to attach glycosidically to the arabinogalactan chains. If this model is correct, then a galacturonosyl residue, liberated by the action of endopolygalacturonase, would be the anticipated reducing end of the acidic polymers released by endoglucanase from endopolygalacturonase-pretreated cell walls. If the hypothesis that the xyloglucan is attached to the pectic fragments is wrong, and it is the pectic fragments that are glycosidically attached to the xyloglucan chains, then one would expect to find glucosyl residues at the reducing ends of the acidic polymers released by endoglucanase. Keegstra *et al.* (1973), using sodium [^{3}H]borohydride to radiolabel the sugar residue on the reducing end of these polymers, obtained evidence that the polymers terminate in galacturonic acid. This result supports the hypothesis that the reducing end of the xyloglucan is attached to the pectic polymer, and indicates that the alternative hypothesis is not true. However, the exact position of the attachment of the xyloglucan to the pectic polysaccharide remains to be determined.

B. Evidence for a Covalent Linkage between the Polysaccharides and Structural Protein of the Walls of Suspension-Cultured Sycamore Cells

In considering the question of whether wall polysaccharides are covalently linked to structural protein, the most obvious possibility is the attachment of wall polysaccharides to the oligoarabinosides known to be attached to the hydroxyl groups of almost all of the many hydroxyproline residues of the sycamore cell wall (Lamport, 1969). Hydroxyproline accounts for 2% (w/w) of the wall and represents over 20% of the amino acid residues of the sycamore cell wall. Lamport's evidence suggests that there is only one type of structural protein in the cell wall. Lamport has established that any carbohydrate that is connected to the oligoarabinosides must be attached by an alkali-labile bond (1969). Since glycosidic bonds are stable to the alkali treatment used by Lamport, the glycosidic attachment of wall polysaccharide to oligoarabinosides is ruled out. The galactose, which has been reported to be directly or indirectly attached to these arabinosyl oligosaccharides, may, in fact, be glycosidically attached to the hydroxyl groups of the serine residues of this protein. Serine has been shown to be present in each enzymically released glycopeptide fragment that contains galactose (Lamport, 1969). This hypothesis is supported by results from this laboratory which indicate that most of the serine residues of the cell wall protein contain glycosidically linked sugars (D. Burke and P. Albersheim, unpublished results). This was established by treating cell walls with mild alkali, which causes the β elimination of any glycosyl residue linked to serine and results in dehydration of serine to give dehydroalanine. Serine residues with unsubstituted hydroxyl groups are not affected by this treatment. A dramatic decrease in the serine content of the cell wall is observed after alkali treatment, while the serine content of a model peptide is not reduced. Thus, a covalent linkage to serine seemed a likely attachment point between the polysaccharides and protein of plant cell walls. Recently, Lamport *et al.* (1973) isolated from tomato cell walls a galactose-containing peptide and provided strong evidence that the galactose is attached to the serine residue of this hydroxyproline-rich peptide. This result contradicted an earlier report (Heath and Northcote, 1971) that cast doubt on the possibility that serine residues in the hydroxyproline-rich protein have galactosyl units glycosidically attached. Thus, the available evidence is in favor of the wall polysaccharide being connected to the hydroxyproline-containing protein through the serine residues of this protein.

The glycoprotein of SEPS (sycamore extracellular polysaccharides) (Bauer *et al.*, 1973) could present an interesting model for the connection between the polysaccharides and the structural protein of the cell wall. The glycoprotein of SEPS binds to DEAE–Sephadex and was purified from the DEAE–Sephadex eluent by chromatography at pH 2 on a column of SE-Sephadex C-50 (Keegstra *et al.*, 1973). Polysaccharides that have no cationic properties at pH 2 came directly through the column, while those that contain positively charged groups (presumably only those molecules containing covalently bound protein) were absorbed. The

column was then eluted with a linear gradient of sodium chloride to remove glycoproteins. An arabinogalactan eluted from the SE-Sephadex
column, simultaneously in the linear salt gradient, with a hydroxyproline-
containing protein. This suggests that the arabinogalactan and the protein are parts of the same glycoprotein molecule. Methylation analysis
of this polymer indicated that the arabinogalactan is a highly branched
structure containing predominantly 3,6-linked galactosyl residues as
branch points with a single arabinosyl residue as the most prevalent side
chain (Keegstra *et al.*, 1973). These results are similar to those reported
by Aspinall *et al.* (1969) for this arabinogalactan and are also similar
to arabinogalactans isolated from coniferous woods (Timell, 1965) and
from plant gums (Aspinall, 1969). The SEPS arabinogalactan differs
from those in wood in that it contains rhamnose and a higher percentage
of arabinose. An arabinogalactan isolated from maple (*Acer saccharum*)
sap (Adams and Bishop, 1960) is similar to the SEPS (*Acer pseudoplatanus*) arabinogalactan in all respects.

One possible structure for the arabinogalactan from SEPS is shown
in Fig. 4. This structure has two features that suggest that it is an interesting cell wall component. The first is the attachment of this polysaccharide to a hydroxyproline-containing protein, a known component of
plant cell walls (Lamport, 1965, 1969). This suggests that this arabinogalactan is a wall component, and that it may be a connecting point
between wall polysaccharides and wall protein. The second striking feature of this polymer is the presence of a terminal rhamnosyl residue. Since
all of the rhamnose in the cell wall is accounted for by that covalently
linked in the rhamnogalacturonan (Talmadge *et al.*, 1973), the rhamnosyl
residue in this arabinogalactan might act as a primer to which a rhamnogalacturonan can be attached in the cell wall. *This finding leads to the
hypothesis that the rhamnogalacturonan is connected to the hydroxyproline-rich wall protein through a highly branched arabinogalactan.* This
hypothesis is further supported by the ability of the arabinogalactan to
act as a cation and thereby bind to SE-Sephadex and also by the simultaneous elution of the arabinogalactan and hydroxyproline-containing protein from both DEAE–Sephadex and SE-Sephadex during gradient elution. The fact that SEPS xyloglucan is identical to the xyloglucan of
sycamore cell walls makes it reasonable to use the arabinogalactan of
SEPS as a model for a similar polymer within the cell wall. Since the
arabinogalactan of SEPS has a high proportion of 3,6-linked galactosyl
residues, it can be readily distinguished from the arabinogalactan side
chains of the pectic polymers in which the galactosyl residues are predominantly 4-linked and unbranched.

One approach in searching for a connection between the rhamno-

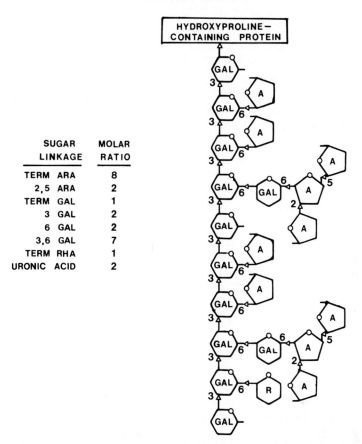

SUGAR LINKAGE	MOLAR RATIO
TERM ARA	8
2,5 ARA	2
TERM GAL	1
3 GAL	2
6 GAL	2
3,6 GAL	7
TERM RHA	1
URONIC ACID	2

Fig. 4. Molar ratio of the glycosyl derivatives present in SEPS arabinogalactan and a proposed structure for this polysaccharide. The molar ratios were determined by Keegstra *et al.* (1973). Since uronic acids are not recovered from the methylation analysis under the conditions used, it was not possible to tell how these are linked in this structure. The structure shown is not unique to the data, but it is consistent with and accounts for the data available. The sugar residues in the figure are designated as A = arabinose, GAL = galactose, and R = rhamnose.

galacturonan and the structural protein of the wall is to use a protease to hydrolyze the wall protein in order to release glycopeptides for further study. However, if the carbohydrate portion of the glycoprotein is attached through its *nonreducing* end to other wall polysaccharides, it would be necessary to use walls that have been pretreated with polysaccharide degrading enzymes before the glycopeptides are released. This possibility was tested by Keegstra *et al.* (1973) who found that less than 0.3% of the wall carbohydrate is released from untreated walls by the

action of pronase. On the other hand, when endopolygalacturonase-pretreated walls are used, more than 2% of the wall carbohydrate is solubilized by pronase, and almost 4% of the wall carbohydrate is solubilized
if endopolygalacturonase–endoglucanase-pretreated walls are subsequently treated with pronase. *These results indicate that the carbohydrate of the cell wall glycoprotein is attached to other wall polysaccharides.* These wall polysaccharides must be partially degraded before the
protease can solubilize a significant fraction of the wall carbohydrate.

A portion of the material released by pronase from endopolygalacturonase–endoglucanase-pretreated cell walls was purified and found to
consist of 95% carbohydrate and 5% protein (Keegstra *et al.,* 1973).
Approximately 12% of this protein is hydroxyproline. The carbohydrate
of this fraction is very similar to both the neutral sugar-rich pectic fragments released by endopolygalacturonase (Section II,C,1) (Talmadge *et
al.,* 1973) and to the pectic fragments in the acidic endoglucanase products (Section II,D,2) (Bauer *et al.,* 1973). This fraction of the pronase
products, therefore, consists predominantly of pectic fragments. *Since the
pectic fragments were solubilized by the action of a protease, this finding
constitutes evidence for a linkage between the pectic polysaccharides and
the protein.*

The question of how the pectic polysaccharides are attached to the
wall protein cannot be definitively determined from the data available.
However, the data are consistent with the hypothesis that the pectic polysaccharides are connected to the wall protein by short arabinogalactan
chains similar to those attached to the hydroxyproline-containing protein
of SEPS (Fig. 4). The pronase-released fragments of the wall, like the
glycoprotein in SEPS, include hydroxyproline-containing peptides and
have a higher proportion of 3- and 3,6-linked galactosyl residues than
do the pectic fragments isolated from the wall by treatment with either
endopolygalacturonase (Talmadge *et al.,* 1973) or endoglucanase (Bauer
et al., 1973). These results are consistent with the idea that the pronase
fragments are rich in the arabinogalactan chains associated with the reducing ends of the pectic polysaccharides, that is, those chains that appear to connect the rhamnogalacturonan and the hydroxyproline-rich
protein. In summary, the evidence that the pectic polysaccharides are
connected, in some manner, to the hydroxyproline-rich protein is highly
suggestive but not conclusive.

C. The Bonding of Xyloglucan to Cellulose

Crystalline cellulose fibers make up an important part of the framework of the cell walls of all higher plants. Electron microscopy and x-ray
diffraction have led to a rather detailed description of the structure of

this wall component (Frey-Wyssling, 1969; Roelofsen, 1965; Wilson, 1964). The linear glucan molecules of cellulose are bound together by hydrogen bonds. The bonding between approximately 40 glucan chains results in 35 Å diameter thread-like fibers. In secondary cell walls, these very elongated elementary fibers are aggregated into 150 and 250 Å diameter ropelike structures (Frey-Wyssling, 1969).

The structure of xyloglucan consists basically of a cellulose-like β-(1→4)-linked glucan backbone with frequent xylosyl side chains attached to the 6-position of the glucosyl residues in the backbone (Bauer et al., 1973). On the basis of x-ray data and model studies, it is believed that cellulose fibers are held together by hydrogen bonds between oxygens of alternating glycosidic bonds in one glucan chain and the primary hydroxyl groups at position 6 of glucosyl residues in another chain (Frey-Wyssling, 1969). Thus, when xyloglucan binds to cellulose, every second glycosidic oxygen of the glucan chain of xyloglucan is available to act as an acceptor for hydrogen bond formation with the hydrogen of a primary hydroxyl group at position 6 of a glucosyl residue of a given cellulose chain. However, only about one-fourth of the glucosyl residues in the xyloglucan polymer have primary hydroxyl groups available at position 6 to act as donors for interchain hydrogen bonding. Thus, the bonding between a xyloglucan chain and a cellulose chain would be weaker than between two cellulose chains.

The fucosyl-(1→2)-galactosyl-(1→2)-xylose side chains of sycamore xyloglucan probably play an important role in preventing further lateral associations, giving a monolayer of xyloglucan on the surface of the cellulose fiber. Xyloglucan structures built with CPK models show that the 1→2 linkages of the fucosylgalactosylxylose side chains cause them to curl over either the top or the bottom face of the β-(1→4)-linked glucan backbone. If the bottom face of the glucan backbone of xyloglucan is hydrogen bonded to a cellulose fiber, then the trisaccharide side chains would hinder access of other β-(1→4)-linked glucan chains to the top face of the xyloglucan backbone.

It has been calculated that there is approximately enough xyloglucan in sycamore cell walls to encapsulate all of the cellulose fibers in these walls with a monolayer of xyloglucan (Bauer et al., 1973). This rough calculation is based on the relative amounts of cellulose and xyloglucan in the sycamore wall, on the approximate proportion of cellulose chains that are exposed at the surface of an elementary fiber (estimated to be about 50%), and on the assumption that xyloglucan can bind to all of the cellulose chains that are at the surface of a cellulose fiber. The observations that isolated cell walls will not bind any more xyloglucan, that xyloglucans are present in the extracellular medium of cell suspension cultures, and that xyloglucans are found in the soluble fraction of plant

tissue homogenates (J. M. Labavitch and P. Albersheim, unpublished results) all support the hypothesis that the cellulose fibers are completely
coated with xyloglucan.

Xyloglucan is not the only plant cell wall polysaccharide with a
structure suited to the formation of interchain hydrogen bonds with cellulose. Polysaccharides appearing in the classical hemicellulose fraction of
secondary plant cell walls (xylans, mannans, glucomannans, and galacto-
glucomannans) and the dominant hemicellulose of monocot primary cell
walls (arabinoxylan) also have structures that are well suited for hydrogen bonding to cellulose chains. Several such polysaccharides have, in
fact, been reported to bind to cellulose *in vitro* (Blake and Richards,
1971; Grant *et al.*, 1969; M. McNeil, P. Albersheim, L. Taiz, and R.
Jones, unpublished results). It is possible that these polysaccharides, and
the xyloglucans, may belong to a single class of functionally related polymers that have in common the ability to bind noncovalently to cellulose.
Bauer *et al.* (1973) have suggested that the structural function of the
hemicellulosic polysaccharides is to interconnect the cellulose fibers and
the pectic polysaccharides of the wall, and that this function is based
on the ability of the hemicellulosic polysaccharides to bind noncovalently
to cellulose and to bind covalently, through glycosylic bonds at their reducing ends, to the pectic polysaccharides.

The partial and nonspecific cleavage of chemical bonds in the plant
cell wall by the aqueous alkali used to obtain the classical hemicellulose
fraction of the wall has made it difficult to recognize the polysaccharides
appearing in this classical fraction as discrete, but interconnected, polymers, and has made it difficult to correlate the appearance of these polysaccharides in this fraction with their place and structural function in
the native cell wall. Bauer *et al.* (1973) have suggested, therefore, that
the term "hemicellulose" be redefined to include only those plant cell
wall polysaccharides that are found to bind noncovalently to cellulose.
This operational definition is based on a chemical property of the polysaccharides which is relatively easy to measure and which is clearly related to the proposed biological function of these polymers.

IV. A Tentative Molecular Structure of the Walls of Suspension-Cultured Sycamore Cells

A. The Model

The results presented here can be summarized most readily in terms
of a model of the sycamore cell wall (Fig. 5) (Keegstra *et al.*, 1973).
Although other structures are possible, the one presented is consistent

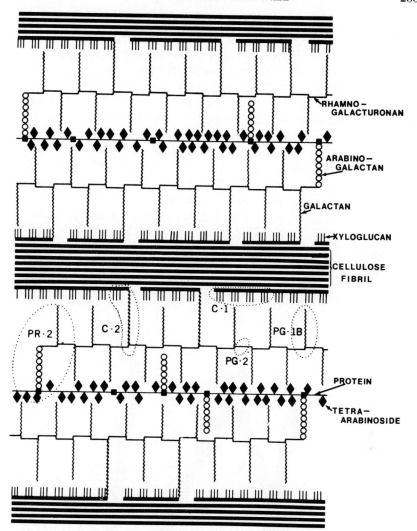

Fig. 5. Tentative structure of sycamore cell walls. The structure presented is based on the data described in the text and in Keegstra et al. (1973). This model is not intended to be quantitative, but an effort was made to present the wall components in approximately proper proportions. The distance between cellulose fibers is expanded to allow room to present the interconnecting structure. A discussion of this model is included in the text. The circled areas are representative wall fractions released by the degradative enzymes. They are PG-1B and PG-2 released by endopolygalacturonase (Talmadge et al., 1973), C-1 and C-2 released by endoglucanase (Bauer et al., 1973), and PR-2 released by pronase (Keegstra et al., 1973).

with all of the data obtained. The model utilizes the fact that xyloglucan
has been shown to bind tightly to purified cellulose (Aspinall *et al.*, 1969)
as well as to the cellulose of the cell wall (Bauer *et al.*, 1973). Several
lines of experimental evidence discussed earlier in this chapter indicate
that the reducing ends of the xyloglucans are attached to the arabinogalactan side chains of the rhamnogalacturonan (Fig. 5). In contrast to
the model presented in Fig. 5, which shows the pectic polysaccharides
covalently attached to the protein, it is possible to construct a coherent,
cross-linked structure of the cell wall using only the linkages between
cellulose and xyloglucan and between xyloglucan and the pectic polysaccharides. In such a model (Fig. 6) a single pectic polysaccharide is attached through xyloglucan chains to more than one cellulose fiber; and
a single cellulose fiber is attached through xyloglucan chains to more than
one pectic polysaccharide. This arrangement would result in a cross-linking of the cellulose fibers and is consistent with all of the data obtained
except data suggesting a connection between protein and polysaccharide.
The hypothetical structure presented in Fig. 5 suggests that the pectic
polysaccharides are attached to the serine residues of the wall protein

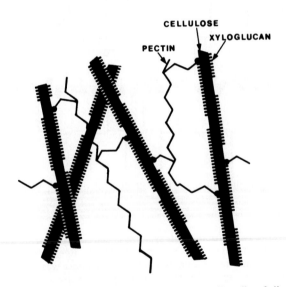

Fig. 6. A highly simplified model of the primary cell walls of dicots. Cellulose
fibers are coated with molecules of hemicellulose (xyloglucan). The xyloglucan
chains are strongly attached to the cellulose fibers via numerous hydrogen
bonds. Some of the xyloglucan chains are covalently attached through their reducing ends to a pectic polysaccharide. This sequence of interconnections, repeated many times, effectively cross-links the cellulose fibers.

through an arabinogalactan chain (Fig. 4) although the evidence for this interconnection is weak. The structure of the wall is likely to contain the cross-linking features of both models.

The structure presented in Fig. 5 allows an understanding of the wall fragments that each enzyme releases. An example of that portion of the wall solubilized by each enzyme is circled and labeled. These fractions are summarized as follows. The endopolygalacturonase attacks the galacturonosyl linkages of the main pectic chain releasing tri-, di-, and mono-galacturonic acid (PG-2 in Fig. 5) as well as arabinogalactan side chains attached to acidic fragments of the main chain (PG-1B in Fig. 5) (Talmadge et al., 1973). After the pectic polysaccharide has been partially degraded by the endopolygalacturonase, endoglucanase more readily degrades xyloglucan, releasing neutral oligosaccharides (C-1 in Fig. 5) as well as pectic fragments that had been held insoluble by their connection with xyloglucan (C-2 in Fig. 5) (Bauer et al., 1973). Pronase, which cannot release carbohydrate from untreated walls, is able to release pectic fragments after endopolygalacturonase pretreatment, and larger amounts of carbohydrate are released by pronase after a combination of endopolygalacturonase and endoglucanase treatment (PR-2 in Fig. 5).

The primary cell wall of sycamore cells can be considered as a single macromolecule (Figs. 5 and 6). The rhamnogalacturonan, arabinogalactan, and xyloglucan, and, in all likelihood, the hydroxyproline-rich protein, are interconnected by covalent bonds, while the many hydrogen bonds that interconnect cellulose and xyloglucan make this connection as strong as a covalent bond. It has been suggested (Lamport, 1970) that the plant cell wall contains a protein–glycan network analogous to the peptidoglycan network of bacterial cell walls (Ghuysen, 1968). The results of Keegstra et al. (1973) support this analogy as they find that the structural component of the sycamore cell wall is composed of well-defined, interconnected polymers in the form of a large "bag-shaped" molecule (Weidel and Pelzer, 1964).

An important aspect of the model presented is that it provides a framework for interpreting results already obtained. It is rather difficult to compare the structures of the wall components described by Keegstra et al. (1973) with the data in the literature because of the wide variety of wall preparatory procedures used as well as the heterogeneity of chemically extracted fractions (Dever et al., 1968; Ray, 1963; Stoddart et al., 1967). Most of the preparatory procedures that have been used result in the presence of water-soluble polymers in the wall preparations. While these polymers may be interesting in their own right, their presence confuses the study of the structural portion of the wall. In addition, the chemical extraction procedures that have been used to solubilize classical wall fractions cause a wide variety of effects. For example, the acid solutions that generally have been used to extract the pectic polymers (Lamport, 1970) result in the hydrolysis of bonds such as arabinosyl or rhamnosyl glycosides (Talmadge et al., 1973). On the other hand, the strong alkali used to extract hemicellulose simultaneously results in transelimination of uronic acids (Albersheim, 1959; Neukom and Deuel, 1958) and β-elimination of serine glycosides (Spiro, 1970).

Despite the difficulties described above, there are important findings that have been reported in the literature that are consistent with the results reported here. For example, the relative amounts of the wall accounted for by pectin, hemicellulose, cellulose, and protein as determined by methylation analysis and enzymic fractionation agree closely with the values obtained by chemical fractionation (Section II,B). And many specific features of the sycamore cell wall are found in the cell walls of other plants. Hydroxyproline-containing proteins with their associated oligoarabinosides are widespread in the plant kingdom (Lamport, 1970; Lamport and Miller, 1971). Kooiman (1960) has demonstrated that xyloglucans are present in the cell walls of the cotyledons or endosperm of a wide variety of plants. Moreover, there is even a report which provides some evidence of a connection between the xyloglucan and pectic polysaccharides of the cell walls of mustard cotyledons; this report describes a pectic polysaccharide that has been purified to a state that ". . . if not homogeneous, consists of a family of related species . . ." (Rees and Wight, 1969). Methylation analysis was used to demonstrate that their preparation contained xyloglucan as well as the pectic polymers. Although they considered the xyloglucan to be a contaminant, we interpret their data as evidence in support of a covalent linkage between these wall components.

An interesting observation concerning the structure of plant cell walls has been reported by Grant et al. (1969). They have isolated a soluble mucilage particle from mustard seedlings and have speculated that this particle may represent a structural unit of the cell wall. The particle consists of a cellulose fiber encapsulated by other polysaccharides. The composition of the encapsulating polysaccharides suggests that they are xyloglucan and pectic polymers. Thus, the "cell wall unit" of mustard seedlings may be similar to the structure of the cell walls of the distantly related sycamore tree.

The evidence presented above and to be presented in the following section strongly sustains the hypothesis that the interrelationship between the structural components of the primary cell walls of all higher plants is comparable.

B. Evidence That the Structural Model of the Sycamore Cell Wall Is Applicable to the Walls of Other Cells

The first solid evidence that the model of the walls of sycamore cells is applicable to the walls of other dicots came from a comparison of the hemicellulose of bean and of sycamore (Wilder and Albersheim, 1973). *The walls of Red Kidney bean (Phaseolus vulgaris) suspension-cultured cells have a hemicellulose (xyloglucan) which is extraordinarily similar to that of sycamore cells.* The similarity of these polymers is demonstrated by their susceptibility to the same hydrolytic enzyme, an endoglucanase, and by the fact that enzymic hydrolysis of each yields an almost identical mixture of oligosaccharide fragments. The best evidence, however, of this similarity is obtained by comparing the partially methylated alditol acetates derived from the xyloglucans of each of these plants. The

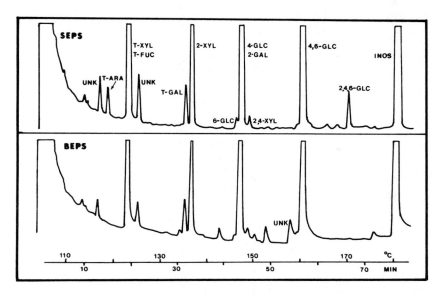

Fig. 7. Comparison of the gas chromatograms of the alditol acetate derivatives of permethylated xyloglucans isolated from the extracellular polysaccharides of suspension-cultured sycamore (SEPS) and bean (BEPS) cells. The chromatograms were obtained from Bauer *et al.* (1973) and Wilder and Albersheim (1973). The initial peak in each chromatogram is due to the acetic anhydride used as a solvent. The abbreviations used are ARA = arabinose, FUC = fucose, GAL = galactose, GLC = glucose, MAN = mannose, XYL = xylose, UNK = unknown. The glycosidic linkages to each sugar derivative are indicated by numerical prefixes: thus, 4,6-GLC indicates that sugars are glycosidically linked in the polysaccharide to the 4 and 6 carbons of the glucosyl residues. Terminal residues are indicated by T- (e.g., T-XYL). The inositol hexaacetate used as a standard is indicated by INOS.

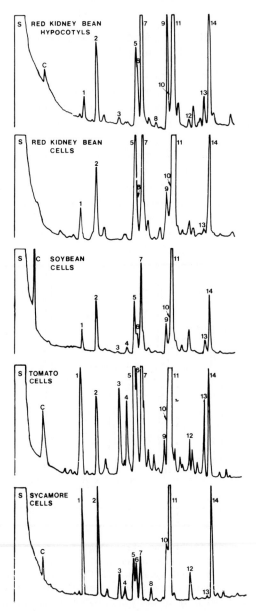

Fig. 8. Comparison of the gas chromatograms of the alditol acetates of the chloroform–methanol-soluble fraction of permethylated walls (Talmadge *et al.*, 1973) isolated from 8-day-old Red Kidney bean hypocotyls and from cell suspension-cultures of Red Kidney bean, soybean, tomato, and sycamore. The chromatograms were obtained as described in Talmadge *et al.* (1973). The initial peak in each chromatogram, labeled S, is due to the acetic anhydride used

partially methylated alditol acetates of bean and sycamore xyloglucans are very similar (Fig. 7).

The manner of attachment of xyloglucans within these walls is also similar as shown by combining chemical, enzymic, and liquid chromatographic methods for isolating the xyloglucans and xyloglucan fragments from the bean and sycamore cell walls (Bauer et al., 1973; Wilder and Albersheim, 1973).

Further evidence that the sycamore cell wall model is a general one was obtained by comparing the partially methylated alditol acetates derived from a variety of plant cell walls with those derivatives obtained from sycamore and bean. Wolfgang D. Bauer synthesized these derivatives from the walls of cultured tomato cells (*Lycopersicon esculentum*), Mina Fisher made the derivatives from the walls of suspension-cultured soybean (*Glycine max*) cells, and Bernard Nusbaum obtained this information for both suspension-cultured Red Kidney bean (*P. vulgaris*) cells and for 8-day-old hypocotyls of Red Kidney beans. The gas chromatograms of the partially methylated alditol acetates obtained from these different plants are compared in Fig. 8. It is evident that all the cell walls are composed of the same sugars linked in the same way. Quantitative examination of the peaks shows that the proportion of the different glycosyl linkages in the walls varies somewhat, but the major components are the same. These data suggest that all of these cell walls are composed of similar polymers. *This information and the fact that these various plant cell walls are degraded in the same manner by the same enzymes that degrade the sycamore walls and that this enzymic degradation yields similar chromatographic fractions has convinced us that the suspension-cultured cells of such diverse plants as beans, tomatoes, and sycamore trees, have, architecturally, very similar walls.*

The fact that the walls of the hypocotyls of Red Kidney beans yield partially methylated alditol acetates similar to those obtained from suspension-cultured bean cells (Fig. 8) is evidence that the suspension-cultured cells are representative of primary walls found in the intact plant. Apparently, secondary wall polysaccharides yield relatively small

---

as a solvent. The peak labeled C is an unknown contaminant present in one of the reagents. The identities of the peaks are: 1, terminal arabinose; 2, terminal xylose and terminal fucose; 3, 2-linked arabinose and 2-linked rhamnose; 4, 3-linked arabinose; 5, 5-linked arabinose; 6, terminal galactose; 7, 2- and 4-linked xylose and an unknown component; 8, 2,4-linked rhamnose; 9, 4-linked mannose; 10, predominantly 4-linked galactose; 11, predominantly 4-linked glucose; 12, 6-linked galactose and 2,3,5-linked arabinose; 13, 4,6-linked galactose; 14, 4,6-linked glucose.

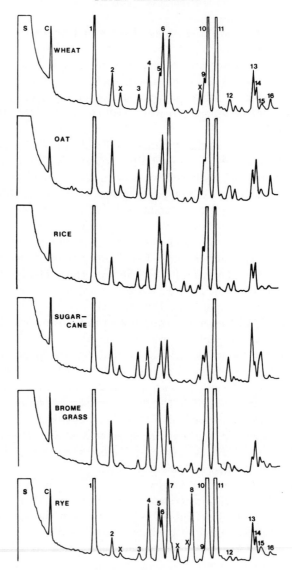

Fig. 9. Comparison of the gas chromatograms of the alditol acetates obtained from the permethylated walls of six suspension-cultured grasses. The chromatograms were obtained as described (Talmadge *et al.*, 1973; Burke *et al.*, 1974). The peaks are 1, terminal arabinose; 2, terminal xylose; 3, 2-linked arabinose; 4, 3-linked arabinose; 5, 5-linked arabinose; 6, terminal galactose; 7, 2- and 4-linked xylose; 8, 3-linked glucose; 9, 3- and 4-linked galactose; 10, 2,5-linked arabinose and 4-linked glucose; 11, 3,4-linked xylose; 12, 2,3,5-linked arabinose and 6-linked galactose; 13, 2,3,4-linked xylose; 14, 4,6-linked glucose; 15, 4,6-linked galactose; 16, 3,6-galactose. S, peak due to acetic anhydride used as solvent; C, an unknown contaminant present in one of the reagents.

amounts of partially methylated alditol acetates when treated in the same manner that is successful in synthesizing these derivatives from the polysaccharides of primary cell walls. D. H. Northcote and co-workers (personal communication) have analyzed the sugar composition of walls of cambial cells excised from sycamore trees and found the composition to be identical to that we reported for suspension-cultured sycamore cells. T. E. Timell and B. W. Simson (personal communication) have been investigating the cell walls of cambial cells removed from the stems of the aspen tree and have found polymers very similar to those obtained from suspension-cultured sycamore cells. Thus, we conclude that *the model of the walls of sycamore cells grown in suspension-culture acts as a model for the walls of other suspension-cultured cells, and is also a valid model for the primary walls of cells in intact plants.*

Our laboratory became interested in investigating whether suspension-cultured monocot cell walls could be compared in structure to the dicot cell walls. Burke *et al.* (1974) surveyed the cell walls of six different suspension-cultured grasses. The chromatograms shown in Fig. 9 compare the fingerprint patterns for the partially methylated alditol acetates obtained from the cell walls of these different grasses. The great similarity in the glycosyl linkages in these walls is immediately apparent. *It is interesting that although the wheat and rice cultures were derived from root tissue, the oat and brome grass from embryo tissue, the sugarcane from internode tissue, and the rye grass from endosperm, the walls of all of these tissue cultures are very similar.* However, only the cells derived from endosperm have, in significant amounts, the partially methylated alditol acetate indicative of a β-1,3-glucan. The derivative from this polymer is number 8 in the endosperm-derived rye cell walls (Fig. 9, bottom). We believe the β-1,3-glucan is a storage polysaccharide present in the endosperm cells and is not a structural component of endosperm cell walls.

The proportion of the various sugar derivatives in the walls of the suspension-cultured grass cells is not the same, but all of the walls are based on the same sugars linked in the same manner. These partially methylated alditol acetate patterns are very different from those obtained from suspension-cultured dicots (Fig. 8). *Yet, available evidence enables one to predict that the monocot cell walls will be based on an architectural plan similar to that of the dicots.* It is probable that the monocot cell walls will have cellulose fibers interconnected via hemicellulose and pectic polymers. The dominant hemicellulose of the primary cell walls of monocots (arabinoxylan) is not the same as the hemicellulose in the dicots (xyloglucan), but both monocots and dicots do contain, as a major primary wall constituent, a hemicellulose that hydrogen bonds to cellulose (Bauer *et al.*, 1973; McNeil *et al.*, 1975), and both monocots and dicots contain pectic polymers (Talmadge *et al.*, 1973; Bauer *et al.*, 1973).

V. The Nonstructural Components of Primary Cell Walls

A. The Water-Soluble Polysaccharides

It is difficult to be sure that nonstructural macromolecules associated with the cell wall fraction of tissue homogenates are, in fact, associated with the wall prior to cellular disruption. An interesting class of such polymers is the pectic-like polysaccharides that remain associated with the cell wall in 500 mM phosphate but which can subsequently be extracted from such walls by distilled water. Jones *et al.* (1972) have found that these polymers are equivalent to about 5% of the walls isolated from tomato stems, while A. Anderson and P. Albersheim (unpublished results) have found that similar polysaccharides are equivalent to as much as 15% of the walls isolated from bean hypocotyls. These polysaccharides may be similar to the galacturonic acid-containing polymers that are secreted into the culture medium by suspension-cultured cells (see, for example, Becker *et al.*, 1964). The polysaccharides extracted by water from isolated cell walls efficiently induce pathogenic fungi to secrete endopolygalacturonase (Jones *et al.*, 1972). More recently, J. M. Labavitch, M. McNeil, and P. Albersheim (unpublished results) have isolated xyloglucans and arabinogalactans of wall-like structure from the soluble fraction of pea stem homogenates. Pulse-labeling experiments (Labavitch and Ray, 1974) suggest that at least a portion of the water-soluble xyloglucan fraction was, at one time, a structural component of the cell wall.

B. Cell Wall Enzymes

Many enzymes have been reported to be associated with the cell wall, but, again, evidence for such association which relies on the presence of the enzymes in the wall fraction of tissue homogenates must be viewed with skepticism. Lamport (1970) and Cleland (1971) have evaluated much of the literature concerning cell wall enzymes. Since then, Keegstra and Albersheim (1970) have shown that β-glucosidase, and α- and β-galactosidase are associated with the cell walls of suspension-cultured sycamore cells. These enzymes can be assayed while they are attached to the walls of the intact sycamore cells, and the enzymes can be extracted from the walls without disrupting the cells. Klis (1971) has shown that an α-glucosidase is associated with the cell walls of intact *Convolvulus* callus and that a number of other enzymes are associated with isolated cell wall fragments (Klis *et al.*, 1974). Nevins (1970) has suggested that a number of glycosidases are present in the walls of bean hypocotyls, and Dowler *et al.* (1974) have provided evidence for the presence of several glycosidases in the cell walls of *Avena* coleoptiles. Thus, it is very likely that some polysaccharide-degrading enzymes are associated with the cell walls of intact tissues.

C. The Cell Wall Proteins Which Inhibit Polygalacturonases Secreted by Plant Pathogens

A unique class of recently discovered cell wall macromolecules consists of proteins that specifically and efficiently inhibit polygalacturonases secreted by plant pathogens (Albersheim and Anderson, 1971). All microbial plant pathogens secrete enzymes capable of degrading the polysaccharides of plant cell walls (Albersheim *et al.*, 1969), but in each case that has been examined the only enzyme capable of initiating wall degradation is a pectic-degrading enzyme. The cell walls of Red Kidney bean hypocotyls, of tomato stems, and of sycamore cells grown in suspension culture contain proteins that have the ability to inhibit, completely and specifically, the activities of the early polygalacturonases secreted by *Colletotrichum lindemuthianum*, *Fusarium oxysporum* f. sp. *lycopersici*, *Sclerotium rolfsii*, and *Helminthosporium maydis* (Albersheim and Anderson, 1971; Anderson and Albersheim, 1972; Fisher *et al.*, 1973; Jones *et al.*, 1972). The activities of other polysaccharide-degrading enzymes secreted by these pathogens are unaffected by plant extracts containing the polygalacturonase inhibitors. However, inhibition of the endopolygalacturonase is sufficient to prevent significant wall degradation by an enzyme mixture that contains an endopolygalacturonase as well as other polysaccharide-degrading enzymes.

The possible role of these polygalacturonase inhibitors in protecting the plant against infection is apparent. Another role for these or similar inhibitors could be in controlling the rate of cell growth by regulating wall-loosening, and similar inhibitors could participate in tissue differentiation by controlling the rate at which primary cell walls are transformed into secondary cell walls. Control of wall differentiation by such inhibitors could take the following form. It was noted in Section III,C that the fucosylgalactose side chains of xyloglucan would prevent lateral association of the elementary cellulose fibers present in primary cell walls. There is considerable evidence that the cellulose fibers of primary cell walls aggregate to form thicker fibers in secondary cell walls (Frey-

Wyssling, 1969). The cleavage from xyloglucan of the fucosyl and galactosyl residues by exoglycanases within the cell wall could initiate this fundamental differentiation. Therefore, inhibitors of the exoglycanases would control the rate of the differentiation. This fairytale-like story is enacted in even more elaborate form by differentiating yeast cells (Cabib and Farkas, 1971; Cabib and Ulane, 1973).

VI. What Does All This Tell Us about the Mechanism of Cell Wall Extension?

A. Some Observations Concerning the Mechanisms Underlying Elongation Growth

Many, if not most, of the workers in plant cell wall research have as a goal an elucidation of the mechanism underlying control of cell wall extension. Cleland (1971) has lucidly summarized the current thinking about wall extension. It is generally agreed that addition of auxin to tissues deficient in this hormone quickly causes the primary cell walls of the tissue to be loosened or weakened, such that the rate of cell extension is increased. Perhaps not as widely held, but nevertheless accepted by this writer, is the view that auxin initiates wall-loosening so quickly that *de novo* protein synthesis and *de novo* polysaccharide synthesis cannot participate in this *initiation*. Thus, our laboratory has examined the structure of primary cell walls with the idea that initiation of wall extension probably results from a rearrangement or alteration of the existing wall structure.

Hormone stimulation of elongation growth results from a *temporary* weakening or relaxation of the wall (Cleland, 1971). There is considerable evidence that pH 5 catalyzes the relaxation of the wall in a manner similar to that catalyzed by hormones (Adams *et al.*, 1973; Bonner, 1934; Evans, 1967; Montague *et al.*, 1973; Nitsch and Nitsch, 1956; Rayle, 1973; Rayle and Cleland, 1970). There is also evidence that the hormones activate ion pumps within the cell membrane and that these ion pumps lower the pH of the wall (Cleland, 1971, 1973; Fisher and Albersheim, 1974; Hager *et al.*, 1971; Marrè *et al.*, 1973a,b; Rayle, 1973). An hypothesis based on these considerations would suggest that the direct action of the hormones is on the cell membrane, and that the reactions within the cell wall, which permit elongation growth, take place more efficiently at pH 5 than at pH 7. As far as we have been able to determine, there are no bonds within the cell wall which would be nonenzymically degraded at pH 5 but which are stable at pH 7. We had suggested (Keegstra *et al.*, 1973) that the rate at which the hydrogen bonds between xyloglucans and the cellulose fibers are made and broken might be affected by such a pH change. Our more recent evidence suggests that this is not probable (Valent and Albersheim, 1974). However, if elongation growth is catalyzed by enzymes, then the pH 5 effect makes sense. It is an attractive idea to consider a wall-loosening enzyme that is more effective at pH 5 than at pH 7, although it is not necessary to limit one's search for a wall-loosening enzyme to assays at pH 5.

Rayle *et al.* (1970) and Rayle and Cleland (1972) have shown that the cell walls of freeze-thawed coleoptile sections are weaker when buffered at pH 5 than at pH 7. The pH 5 weakening does not take place if the coleoptile sections have been subjected to treatments that would denature enzymes. This supports the idea that the wall-loosening process is mediated by enzymes that remain active after freezing.

I would be surprised only if I learned that enzymes do not participate directly in catalyzing the growth of plant cells. A critical catalytic function for wall enzymes in plants is supported by evidence that wall enzymes play such a role in growth of bacterial cells (Fiedler and Glaser, 1973).

B. The Properties to Be Expected of an Enzyme Capable of Catalyzing Elongation Growth

This is an interesting question. The role of such an enzyme is to break bonds that interconnect the cellulose fibers within the polysaccharide matrix. The ideal enzyme could not only break bonds but could cross-connect new polysaccharide partners so that during growth the wall maintains its strength. Even though new wall polymers are being synthesized during growth, the existing polymers cannot be continually weakened, for, if they were, an extended cell wall would be weaker per unit length than an unextended wall. If growth were catalyzed by the breaking of bonds without resynthesis, then every time a wall doubles in length, half of the wall would consist of old, relatively degraded polymers and half of new, relatively undegraded polymers. This is true since the wall maintains about the same thickness while elongating, that is, it maintains the same mass per unit length. *As the walls of cells that have elongated manyfold are about as strong as the walls of unelongated cells, there must be no net breakage of bonds during growth, even though it is clear that bonds must be broken, at least temporarily, to permit the cellulose fibers to move past each other. Any attempt to explain wall growth must account for this maintenance of wall strength.*

The fact that a wall-loosening enzyme will be interacting with polysaccharides rather than with small oligosaccharides dictates that the enzyme will be a glycanase rather than a glycosylase. The latter, which can hydrolyze di- and trisaccharides as well as synthetic glycosides, such as the p-nitrophenol derivatives, has little or no activity on polysaccharides (Barras *et al.*, 1969; Reese and Mandels, 1963; Reese *et al.*, 1968). Thus, although glycosylases and even glycanases have been reported to be present in cell walls (Keegstra and Albersheim, 1970; Kivilaan *et al.*, 1971; Klis, 1971; Klis *et al.*, 1974; Nevins, 1970; Spencer and MacLachlan, 1972; Strauss, 1962), it makes little sense to imply that glycosylases are likely to be responsible for cell wall loosening (Johnson *et al.*, 1974).

Exoglycanases split the glycosyl linkages of mono- or disaccharides at the terminal nonreducing ends of polysaccharides. Endoglycanases attack at internal positions of polysaccharides. Only endoglycanases would be effective in breaking the connection between two polysaccharide chains.

A very interesting property of endoglycanases, or, if we follow the suggestions of Hehre *et al.* (1973), endoglycosylases, is that these enzymes not only catalyze the breaking of glycosylic bonds, but they also catalyze the synthesis of such bonds. *Glycosylases are not hydrolases, but rather are transglycosylases that, in some cases, catalyze hydrolysis by transferring a glycosylic bond from a sugar to water.* Even the amylases, which are considered classical examples of hydrolases because of their ability to convert starch into maltose, are true transglycosylases. Hehre *et al.* (1971) have dramatically demonstrated this by showing that the α-amylases from a variety of bacteria efficiently catalyze the synthesis of higher oligosaccharides from disaccharides or disaccharide analogs.

The transglycosylation capacity of a rather large number of enzymes has been well documented. For example, enzymes are known which synthesize polyfructans from sucrose, and other enzymes synthesize polyglucans from this disaccharide (Dedonder, 1972; Hassid, 1970). Of even greater importance for our purposes was the demonstration that enzymes can transfer the glycosylic bond from a sugar to an amino acid residue of the enzyme itself. Voet and Abeles (1970) were the first to isolate a glycosyl enzyme intermediate. This enzyme, sucrose phosphorylase, transfers the glucosylic bond of sucrose to itself. Of course, stable glycosylic bonds between oligosaccharides and peptides are very common, being widely found in nature as glycoproteins and mucopolysaccharides.

I should like to suggest that the enzyme in the plant cell wall which catalyzes cell growth is likely to be an endotransglycosylase that transfers a portion of a polysaccharide to itself. This reaction can be used to weaken the interconnection between two cellulose fibers (Fig. 10). Since such reactions are reversible, if the polysaccharide–enzyme should

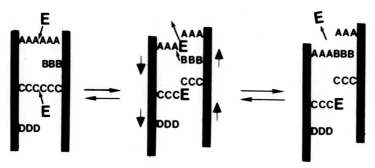

**A, B, C, D = POLYSACCHARIDES   CROSS – LINKING
CELLULOSE   FIBERS**

E = TRANSGLYCOSYLASE

Fig. 10. A hypothetical schematic representation of how a transglycosylase could temporarily loosen the interconnections between cellulose fibers with the result that the fibers can, under the impetus of turgor pressure, move relative to one another. In this hypothetical scheme to explain cell elongation, it is possible to envision that hormones activate an ion pump in the plasma membrane which results in a lowering of the pH of the wall from 7 to 5 and, further, that the transglycosylase more efficiently relaxes the wall at pH 5 than at pH 7.

find another polysaccharide partner of the proper sort, brought to it by slippage of the cellulose fibers within the wall, the enzyme can transfer the polysaccharide from itself to the new partner. Thus, we can envision two cellulose fibers moving past each other by making and breaking bonds between the interconnecting polysaccharides. Such a mechanism has been suggested for slippage of peptidoglycan chains past each other in bacteria (Fiedler and Glaser, 1973).

It is reasonable that such a transglycosylase be more active at pH 5 than at pH 7; it is even possible that the equilibrium between polysaccharide attached to enzyme and polysaccharide attached to polysaccharide favors the attachment of polysaccharide to enzyme at pH 5, while it favors the interconnection between the two polysaccharides at pH 7. This would result in a wall that is weaker at pH 5 than at pH 7.

C. How Can We Look for a Transglycosylase Which Catalyzes the Loosening of Primary Cell Walls?

This is the goal of our research. The problem is to identify the glycosylic linkage that controls the rate of elongation growth, then to isolate a polysaccharide fragment containing this linkage, and, finally, to extract an enzyme from the cell wall which can catalyze the making and breaking of this interconnection. This is a tall order, but one clearly worth the effort. If we are successful, we believe we will have identified the linkage in all dicot primary cell walls which regulates the rate of growth of these cells. It is likely, too, that analogous reactions take place in monocot cell walls and that the determination of this reaction in dicots will assist us in identifying the reaction in monocots. In pursuing this research, we have an additional benefit of continuing to learn more about cell wall structure.

One could approach this problem in a random fashion, that is, by looking at the cell wall structure and picking out likely points of attack, then isolating these structural entities, and, finally, attempting to extract from the wall an enzyme that will split the wall fragment. This random approach is not as dubious as one might expect, since the wall is made up of a large number of repeating units with only a few sites as probable locations for a transglycosylase to be effective in loosening the wall.

Our approach to this problem may have been simplified by the findings of Labavitch and Ray at Stanford University (1974). These workers tried to find out whether one or more components of the cell wall turn over more rapidly as a result of auxin-stimulated elongation growth. Their experiment involved incubating pea stem sections in radioactive glucose for sufficient time to label the wall uniformly. The sections were then placed in an excess of unlabeled glucose in order to chase out any radiolabeled metabolites from the precursor pools of the cell wall. The stem sections were then divided into three equal lots. One was harvested immediately, a second was placed in unlabeled glucose in the absence of auxin, and the third was placed in unlabeled glucose in the presence of auxin. The conclusion from their experiment was that the auxin-stimulated pea stem sections, which grew more rapidly than those which were placed in the absence of the hormone, did, in fact, have a wall component that was removed at a more rapid rate than in the controls. This wall component was found in the soluble fraction of the homogenate. The component that turns over more rapidly was determined to be a polysaccharide containing xylose and glucose. In earlier experiments, which are related to those described above demonstrating that auxin stimulates the turnover of a xylose- and glucose-containing polymer in a dicot, Loescher and Nevins (1972) reported that auxin stimulates the turnover of a noncellulosic glucan in a monocot.

More recently, as a postdoctoral fellow in our laboratory, John Labavitch has demonstrated that this xylose- and glucose-containing polysaccharide of peas is, in fact, a xyloglucan like that reported in beans and sycamore (Bauer *et al.*, 1973; Wilder and Albersheim, 1973). Thus, auxin, when stimulating elongation growth of pea stems, stimulates the removal of existing xyloglucan molecules from the cell wall as well as the insertion of new xyloglucan molecules into the wall. The remaining polymers of the wall are not stimulated by auxin to turn over more quickly. Therefore, the interconnection between xyloglucan and the neutral pectic polymers, to which the xyloglucan is covalently attached, is likely to be synthesized and degraded during elongation growth. These experiments may have pinpointed the site within the cell wall which is degraded and synthesized by the wall-loosening enzyme.

Michael McNeil has been able to isolate a fragment of the wall which contains the xyloglucan attached to the pectic galactan. We are now in the process of investigating the detailed structure of this wall fragment and trying to determine whether the bond between these molecules, the xyloglucan and the galactan, is, in fact, broken and resynthesized during elongation growth. The procedure is to find out whether the radioactivity of the xyloglucan portion of this wall fragment disappears more quickly than the radioactivity of the galactan portion of this fragment during auxin-stimulated growth. If the radioactivity in the xyloglucan portion of the molecule decreases more rapidly than in the galactan portion when the fragment is isolated from radiolabeled tissue that had been subsequently treated with auxin, we will be satisfied that this interconnection is being broken and resynthesized during the growth process. If we become convinced of this, we will then look for an enzyme in the wall that can catalyze the separation of the xyloglucan and galactan portions of the isolated wall fragment, and, perhaps, attach the xyloglucan portion of the fragment to the enzyme.

ACKNOWLEDGMENTS

The author is indebted beyond measure to Wolfgang D. Bauer, Kenneth Keegstra, and Kenneth W. Talmadge as well as to the other members of this laboratory who have made this chapter possible. They are scientists who love the pursuit of knowledge. This work was supported in part by the Energy Research and Development Administration (ERDA) (11-1)-1426.

REFERENCES

Adams, G. A., and Bishop, C. T. (1960). *Can. J. Chem.* **28**, 2380.
Adams, P. A., Kaufman, P. B., and Ikuma, H. (1973). *Plant Physiol.* **51**, 1102.
Albersheim, P. (1959). *Biochem. Biophys. Res. Commun.* **1**, 253.
Albersheim, P. (1965). *In* "Plant Biochemistry" (J. Bonner and J. E. Varner, eds.), 2nd ed., pp. 151–186. Academic Press, New York.
Albersheim, P., and Anderson, A. (1971). *Proc. Nat. Acad. Sci. U.S.* **68**, 1815.
Albersheim, P., Jones, T. M., and English, P. D. (1969). *Annu. Rev. Phytopathol.* **7**, 171.

Anderson, A., and Albersheim, P. (1972). *Physiol. Plant Pathol.* **2**, 339.

Aspinall, G. O. (1969). *Advan. Carbohyd.-Chem. Biochem.* **24**, 333.

Aspinall, G. O. (1970). *In* "The Carbohydrates" (W. Pigman and D. Horton, eds.), 2nd ed., Vol. 2B, pp. 515–536. Academic Press, New York.

Aspinall, G. O., and Cottrell, I. W. (1971). *Can. J. Chem.* **49**, 1019.

Aspinall, G. O., Begbie, R., Hamilton, A., and Whyte, J. N. C. (1967a). *J. Chem. Soc., C* p. 1065.

Aspinall, G. O., Cottrell, I. W., Egan, S. V., Morrison, I. M., and Whyte, J. N. C. (1967b). *J. Chem. Soc., C* p. 1071.

Aspinall, G. O., Gestetner, B., Molloy, J. A., and Uddin, M. (1968). *J. Chem. Soc., C* p. 2554.

Aspinall, G. O., Molloy, J. A., and Craig, J. W. T. (1969). *Can. J. Biochem.* **47**, 1063.

Barras, D. R., Moore, A. E., and Stone, B. A. (1969). *Advan. Chem. Ser.* **95**, 105–138.

Barrett, A. J., and Northcote, D. H. (1965). *Biochem. J.* **94**, 617.

Bauer, W. D., Talmadge, K. W., Keegstra, K., and Albersheim, P. (1973). *Plant Physiol.* **51**, 174.

Becker, G. E., Hui, P. A., and Albersheim, P. (1964). *Plant Physiol.* **39**, 913.

Blake, J. D., and Richards, G. N. (1971). *Carbohyd. Res.* **17**, 253.

Bonner, J. (1934). *Protoplasma* **21**, 406.

Burke, D., Kaufman, P. B., McNeil, M., and Albersheim, P. (1974). *Plant Physiol.* **54**, 109.

Cabib, E., and Farkas, V. (1971). *Proc. Nat. Acad. Sci. U.S.* **68**, 2052.

Cabib, E., and Ulane, R. (1973). *Biochem. Biophys. Res. Commun.* **50**, 186.

Chrispeels, M. (1969). *Plant Physiol.* **44**, 1187.

Cleland, R. (1971). *Annu. Rev. Plant Physiol.* **22**, 197.

Cleland, R. (1973). *Proc. Nat. Acad. Sci. U.S.* **70**, 3092.

Cooper, R. M., and Wood, R. K. S. (1973). *Nature (London)* **246**, 309.

Dashek, W. V. (1970). *Plant Physiol.* **46**, 831.

Dedonder, R. (1972). *In* "Biochemistry of the Glycosidic Linkage: An Integrated View" (R. Piras and H. G. Pontis, eds.), pp. 21–78. Academic Press, New York.

Dever, J. E., Jr., Bandurski, R. S., and Kivilaan, A. (1968). *Plant Physiol.* **43**, 50.

Dowler, M. J., Rayle, D. L., Cande, W. Z., Ray, P. M., Durans, H., and Zenk, M. H. (1974). *Plant Physiol.* **53**, 229.

English, P. D., Jurale, J. B., and Albersheim, P. (1971). *Plant Physiol.* **47**, 1.

English, P. D., Maglothin, A., Keegstra, K., and Albersheim, P. (1972). *Plant Physiol.* **49**, 293.

Esau, K. (1960). "Plant Anatomy." Wiley, New York.

Evans, M. (1967). Ph.D. Thesis, University of California, Santa Cruz.

Fielder, F., and Glaser, L. (1973). *Biochim. Biophys. Acta* **300**, 467.

Fisher, M. L., and Albersheim, P. (1974). *Plant Physiol.* **53**, 464.

Fisher, M. L., Anderson, A., and Albersheim, P. (1973). *Plant Physiol.* **51**, 489.

Frey-Wyssling, A. (1969). *Fortsch. Chem. Org. Naturst.* **27**, 1.

Ghuysen, J. M. (1968). *Bacteriol. Rev.* **32**, 425.

Gould, S. E. B., Rees, D. A., and Wight, N. J. (1971). *Biochem. J.* **124**, 47.

Grant, G. T., McNab, C., Rees, D. A., and Skerrett, R. J. (1969). *Chem. Commun.* p. 805.

Hager, A., Menzel, H., and Krauss, A. (1971). *Planta* **100**, 47.

Hassid, W. Z. (1970). *In* "The Carbohydrates" (W. Pigman, D. Horton, and A. Herp, eds.), 2nd ed., Vol. 2A, pp. 301–373. Academic Press, New York.

Heath, M. F., and Northcote, D. H. (1971). *Biochem. J.* **125**, 953.

Hehre, E. J., Genghof, D. S., and Okada, G. (1971). *Arch. Biochem. Biophys.* **142**, 382.

Hehre, E. J., Okada, G., and Genghof, D. S. (1973). *Advan. Chem. Ser.* **117**, 309–333.

Hoffman, J., Lindberg, B., and Svensson, S. (1972). *Acta Chem. Scand.* **26**, 661.

Johnson, K. D., Daniels, D., Dowler, M. J., and Rayle, D. L. (1974). *Plant Physiol.* **53**, 224.

Jones, T. M., Anderson, A., and Albersheim, P. (1972). *Physiol. Plant Pathol.* **2**, 153.

Karr, A. L., Jr. (1972). *Plant Physiol.* **50**, 275.

Keegstra, K., and Albersheim, P. (1970). *Plant Physiol.* **45**, 675.

Keegstra, K., Talmadge, K. W., Bauer, W. D., and Albersheim, P. (1973). *Plant Physiol.* **51**, 188.

Kivilaan, A., Bandurski, R. S., and Schulze, A. (1971). *Plant Physiol.* **48**, 389.

Kivirikko, K. I., and Liesmaa, M. (1959). *Scand. J. Clin. Lab. Invest.* **11**, 128.

Klis, F. M. (1971). *Physiol. Plant.* **25**, 253.

Klis, F. M., Dalhuizen, R., and Sol, K. (1974). *Phytochemistry* **13**, 55.

Kooiman, P. (1960). *Acta. Bot. Neer.* **9**, 208.

Kooiman, P. (1961). *Rec. Trav. Chim. Pays-Bas* **80**, 849.

Kooiman, P. (1967). *Phytochemistry* **6**, 1665.

Labavitch, J. M., and Ray, P. M. (1974). *Plant Physiol.* **53**, 699.

Lamport, D. T. A. (1965). *Advan. Bot. Res.* **2**, 151.

Lamport, D. T. A. (1969). *Biochemistry* **8**, 1155.

Lamport, D. T. A. (1970). *Annu. Rev. Plant Physiol.* **21**, 235.

Lamport, D. T. A., and Miller, D. H. (1971). *Plant Physiol.* **48**, 454.

Lamport, D. T. A., Katona, L., and Roerig, S. (1973). *Biochem. J.* **133**, 125.

Loescher, W., and Nevins, D. J. (1972). *Plant Physiol.* **50**, 556

McNeil, M., Albersheim, P., Taiz, L., and Jones, R. (1975). *Plant Physiol.* **55**, 64.

Marrè, E., Lado, P., Caldogno, R. R., and Colombo, R. (1973a). *Plant Sci. Lett.* **1**, 179.

Marrè, E., Lado, P., Caldogno, R. R., and Colombo, R. (1973b). *Plant Sci. Lett.* **1**, 185.

Montague, M. J., Ikuma, H., and Kaufman, P. B. (1973). *Plant Physiol.* **51**, 1026.

Mühlethaler, K. (1967). *Annu. Rev. Plant Physiol.* **18**, 1.

Neukom, H., and Deuel, H. (1958). *Chem. Ind. (London)* p. 683.

Nevins, D. J. (1970). *Plant Physiol.* **46**, 458.

Nitsch, J. P., and Nitsch, C. (1956). *Plant Physiol.* **31**, 94.

Northcote, D. H. (1972). *Annu. Rev. Plant Physiol.* **23**, 113.

Ray, P. M. (1963). *Biochem. J.* **89**, 144.

Rayle, D. L. (1973). *Planta* **114**, 63.

Rayle, D. L., and Cleland, R. (1970). *Plant Physiol.* **46**, 250.

Rayle, D. L., and Cleland, R. (1972). *Planta* **104**, 282.

Rayle, D. L., Haughton, P. M., and Cleland, R. (1970). *Proc. Nat. Acad. Sci. U.S.* **67**, 1814.

Rees, D. A. and Wight, N. J. (1969). *Biochem. J.* **115**, 431.

Rees, D. A., and Wight, A. W. (1971). *J. Chem. Soc., B* p. 1366.

Reese, E. T., and Mandels, M. (1963). *In* "Advances in Enzymic Hydrolysis of Cellulose and Related Materials" (E. T. Reese, ed.), pp. 197–234. Macmillan, New York.

Reese, E. T., Maguire, A. H., and Parrish, F. W. (1968). *Can. J. Biochem.* **46**, 25.

Roelofsen, P. A. (1959). "The Plant Cell Wall," p. 128. Borntraeger, Berlin.

Roelofsen, P. A. (1965). *Advan. Bot. Res.* **2**, 29.

Siddiqui, I. R., and Wood, P. J. (1971). *Carbohyd. Res.* **17**, 97.

Simmons, D. A. R. (1971). *Eur. J. Biochem.* **18**, 53.

Spencer, F. S., and Maclachan, G. A. (1972). *Plant Physiol.* **49**, 58.

Spiro, R. G. (1970). *Annu. Rev. Biochem.* **39**, 599.

Stoddart, R. W., Barrett, A. J., and Northcote, D. H. (1967). *Biochem. J.* **102**, 194.

Straus, J. (1962). *Plant Physiol.* **37**, 342.

Talmadge, K. W., Keegstra, K., Bauer, W. D., and Albersheim, P. (1973). *Plant Physiol.* **51**, 158.

Timell, T. E. (1964). *Advan. Carbohyd. Chem.* **19**, 247.

Timell, T. E. (1965). *Advan. Carbohyd. Chem.* **20**, 409.

Updegraff, D. M. (1969). *Anal. Biochem.* **32**, 420.

Valent, B., and Albersheim, P. (1974). *Plant Physiol.* **54**, 105.

Voet, J. G., and Abeles, R. H. (1970). *J. Biol. Chem.* **245**, 1020.

Weidel, W., and Pelzer, H. (1964). *Advan. Enzymol.* **26**, 193.

Whistler, R. L., and Richards, E. L. (1970). *In* "The Carbohydrates" (W. Pigman and D. Horton, eds.), 2nd ed., Vol. 2A, pp. 447–469. Academic Press, New York.

Wilder, B., and Albersheim, P. (1973). *Plant Physiol.* **51**, 889.

Wilson, K. (1964). *Int. Rev. Cytol.* **17**, 1.

II

Basic Metabolism

10

Regulation of Enzyme Activity in Metabolic Pathways

JACK PREISS AND TSUNE KOSUGE

I. Introduction

Metabolism in the cell is concerned with the production of energy and the formation of low molecular weight substances, such as amino acids, lipids, and carbohydrates, which are eventually utilized for synthesis of the various macromolecules and the structural components required for growth. Interplayed with these activities are the processes of active transport of metabolites and inorganic ions as well as those processes that convert metabolic energy into mechanical energy. It is apparent, therefore, that the orderly growth and maintenance of a cell requires the development of intricate and sensitive mechanisms to coordinate and integrate all the various biosynthetic and degradative processes therein.

This chapter is concerned mainly with the description and the physiological significance of the various mechanisms used in control of enzyme activity by the metabolites present in the cell. Initially, various phenomena associated with regulation of metabolic systems will be discussed, and later in the chapter a number of studies on the control of metabolic pathways in plants will be reviewed. It must be emphasized, however, that regulation of a metabolic system is rarely effected by a single mechanism. Usually controls on both enzyme synthesis (or degradation) and on enzyme activity are integrated to produce a more efficient modulation. Other phenomena, such as compartmentation of a metabolic system in a subcellular organelle, may also occur along with the other modes of regulation. For information on control manifested by compartmentation the reader is referred to other chapters and in references listed at the end of this chapter. The reader is also referred to Chapter 16. These contain information on control of enzyme synthesis and degradation in plants. For information on the regulation of protein synthesis in animals and in bacteria and on the phenomena associated with control of bacterial protein synthesis, such as feedback repression, substrate induction, and catabolite repression, the reader is referred to references at the end of this chapter.

II. Modulation of Enzyme Activity by Small Metabolites

A. Feedback or End Product Inhibition

Umbarger (1956) and Yates and Pardee (1956) showed that the end product of a metabolic pathway was able to inhibit the first enzyme unique to the pathway. Thus, isoleucine inhibited threonine deaminase [reaction (1)], an enzyme in the isoleucine biosynthetic pathway (Umbarger, 1956), and cytidine triphosphate (CTP) inhibited the enzyme unique to the pyrimidine biosynthetic pathway, aspartate transcarbamylase [reaction (2)] (Yates and Pardee, 1956).

$$\text{Threonine} \rightarrow \text{ammonia} + \alpha\text{-ketobutyrate} \tag{1}$$
$$\text{Aspartate} + \text{carbamyl-P} \rightarrow P_i + \text{carbamyl aspartate} \tag{2}$$

These investigators recognized the importance of these findings in terms of metabolic control, and Umbarger, comparing the modulation by isoleucine with the electronic circuit of a vacuum tube, coined the term, "feedback inhibition." Since then there have been numerous examples of metabolic pathways being regulated by feedback inhibition (Stadtman, 1966, 1970; Atkinson, 1966, 1969).

Feedback inhibition is an effective means of control, as regulation of the first step in a pathway modulates the entire metabolic flux through the pathway. As the concentration of the end product builds up, it would tend to lower the rate of its own formation by inhibiting the first enzyme in the sequence. In this type of control the undesired accumulation of the intermediate metabolites is also prevented. When the concentration of the end product is sufficiently lowered by metabolic utilization, inhibition is released and synthesis commences.

Generally the end product inhibitor has no structural similarity to the substrate of the enzyme it is regulating. Thus isoleucine does not resemble threonine, the substrate for threonine deaminase, and CTP certainly does not resemble aspartate, the substrate for aspartate transcarbamylase. This led Gerhart and Pardee (1962) and Monod et al. (1963) to suggest that regulatory enzymes contain binding sites for the inhibitor molecule that are distinct from the substrate site. As these metabolic inhibitors did not resemble the substrates of the regulatory enzyme, they were called *allosteric* inhibitors (Monod et al., 1963). Since enzymes are highly specific in their substrate requirements, Monod et al. (1963) also proposed that the allosteric effectors of regulatory enzymes exerted their effects by binding at specific sites (allosteric sites) on the enzyme, thereby causing changes in the conformational state which modified the catalytic site. Thus, an inhibitor could either decrease the affinity for the substrates (increase K_m) at the catalytic site, or affect the catalytic efficiency of the enzyme (decrease V_{max}), or do both.

B. Evidence That the Allosteric Site Is Separate or Distinct from the Catalytic Sites

Evidence for distinct sites for the allosteric inhibitor may be found by kinetic studies. If a noncompetitive type, or mixed type, inhibition relationship exists between inhibitor and substrate this indicates separate sites for each. If competitive type inhibition (reversal of inhibition by substrate) is seen, however, this does not necessarily mean that the inhibitor is competing for the same site as the substrate. The inhibitor may still bind at a site distinct from the substrate, but in this case binding of the inhibitor by the enzyme results in a conformational change of the protein structure so that the substrate site is able to bind the substrate weakly or not at all. This phenomenon results in competitive type inhibition, even though substrate and inhibitor bind at different sites. Other methods for showing that the substrate and regulatory sites are not equivalent have been used. Selective desensitization of various enzymes to allosteric effectors has been achieved, showing that the activity of the allosteric sites can be abolished without the catalytic activity being affected. Thus, CTP inhibition of aspartate transcarbamylase disappears when the enzyme is heated for 5 minutes at 55°C (Gerhart and Pardee, 1962), although the enzyme activity is undiminished. Mercurous ions (Hg^{2+}) at low concentration (10^{-5} M) prevent histidine inhibition of the ATP–phosphoribosyl pyrophosphate (PRPP) pyrophosphorylase without altering the enzyme activity (Martin, 1962). Papain (proteolytic) digestion destroys the adenosine monophosphate (AMP) inhibition of mammalian fructose diphosphatase with very little effect on enzyme activity (Taketa and Pogell, 1965). Certain regulatory enzymes (e.g., threonine deaminase) have been isolated from mutant bacteria and have been shown to have lost their allosteric site (Cohen, 1965, 1969). These "mutant" enzymes still retain their normal catalytic activity. The isolation of these mutants have provided important information on the physiological importance of feedback inhibition as well as providing very strong suggestive evidence for the nonidentity of the allosteric and catalytic sites. All these types of experiments suggest that the allosteric site is separate from the catalytic site. It has been shown that aspartate transcarbamylase can be dissociated into dissimilar subunits (Gerhart and Schachman, 1965). The subunits containing the binding sites for the in-

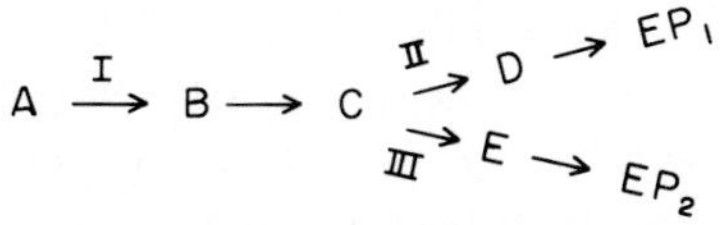

$$A \xrightarrow{\text{I}} B \longrightarrow C \underset{\text{III}}{\overset{\text{II}}{\rightrightarrows}} \begin{array}{l} D \rightarrow EP_1 \\ E \rightarrow EP_2 \end{array}$$

Fig. 1. Synthesis of end products 1 and 2 (EP_1 and EP_2) by a branched pathway. A, B, C, D, and E are the metabolites in the pathway. Enzyme I catalyzes the first step in the pathway, and enzymes II and III catalyze the first steps after the branch point.

hibitor CTP can be physically separated from the substrate binding sub-units, thus providing direct proof of nonidentity of the sites (Gerhart and Schachman, 1965).

C. Feedback Regulation Phenomena Associated with Branched Biosynthetic Pathways

In a metabolic pathway only concerned with the synthesis of one end product, there is no problem with respect to control since feedback inhibition may be a sufficient mechanism. However, regulation of a branched pathway, where a number of end products are synthesized from a common precursor and a number of enzymatic steps for the synthesis of the end products are shared, presents a number of problems in regulation. As shown in Fig. 1, feedback regulation of a branched biosynthetic pathway by the end products EP_1 and EP_2 may lead to a situation where an excess of either could lead to a decrease in the rate of synthesis of B from A, which would lead to a decrease in the formation of both end products. However, studies of the various branched pathways have revealed a number of mechanisms to resolve these complications.

1. SPECIFIC FEEDBACK INHIBITION OF MULTIPLE ENZYMES

The first step in the pathway common to both the synthesis of EP_1 and EP_2 may be catalyzed by two distinct enzymes, one of which is inhibited by EP_1 and the other by EP_2 (Fig. 2). Therefore, a mechanism

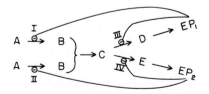

Fig. 2. Regulation of a branched pathway by multiplicity of the first enzyme in the pathway (enzymes I and II). The nomenclature is the same as indicated in Fig. 1. The lines connect the end products with the steps they are inhibiting.

is provided where excessive supply of either EP_1 or EP_2 causes inhibition of B formation. The maximal decrease in B formation that can be obtained by either EP_1 or EP_2 is restricted to that portion of the total activity that is catalyzed by the specific enzyme in question. Thus, if EP_1 totally inhibits enzyme I, enzyme II still continues to catalyze synthesis of B, which can still be utilized for synthesis of E as well as D. This mechanism, therefore, requires additional control phenomena to be completely effective. Thus, EP_1 would also prevent the ultimate conversion of C to D by also inhibiting enzyme III, and EP_2 would inhibit enzyme IV therefore preventing C from entering into the branched part of the pathway leading to EP_2.

This type of control mechanism is found in *Escherichia coli* for the branched pathway leading from aspartate to the synthesis of the amino acids, lysine, methionine, threonine, and isoleucine (Cohen, 1965, 1969). Similarly, in *E. coli* the first common step in the biosynthesis of the three aromatic amino acids, tryptophan, tyrosine, and phenylalanine, is catalyzed by three separate enzymes, each of which is subject to feedback control by one of the three amino acids (Smith *et al.*, 1962; Brown and Doy, 1963, 1966; Doy and Brown, 1965).

The formation of multiple enzymes capable of catalyzing similar reactions also occurs in situations where the enzymatic step is involved both in the biosynthesis and degradation of a metabolite. For example, in *E. coli* the formation of α-ketobutyrate from threonine is the first unique step toward isoleucine biosynthesis. It is also the first step in the degradation of threonine when the organism utilizes the amino acid as a carbon source under anaerobic conditions. The threonine deaminase used for biosynthetic purposes is feedback inhibited by isoleucine, while the deaminase induced during threonine degradation is not inhibited by isoleucine, but appears to be activated by 5'-adenylate (Umbarger, 1969). Since threonine degradation in the anaerobically growing organism is associated with energy production, the finding that it is activated by AMP can be rationalized in terms of regulation by the energy requirements of the cell during growth (see Section V).

2. SEQUENTIAL FEEDBACK INHIBITION

Regulation of a branched pathway sometimes may not occur by inhibition of the first step in the pathway by the ultimate end products. As seen in Fig. 3, the last common metabolite in the pathway, C, will feedback inhibit enzyme I. The end products EP_1 and EP_2 will feedback inhibit enzymes II and III, respectively, the first divergent steps in the pathways to the end products. Thus, accumulation of EP_1 causes inhibi-

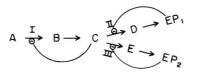

Fig. 3. Regulation of a branched pathway by sequential feedback inhibition. The nomenclature is the same as in Figs. 1 and 2.

tion of enzyme II, preventing conversion of C to D. Compound C will then be utilized for the synthesis of EP₂. When the latter becomes present in excess also, it will inhibit enzyme III, which in turn causes accumulation of C. Accumulation of C shuts down the whole pathway by inhibition of enzyme I. Examples of this sequential pattern of feedback regulation are seen in the biosynthesis of the aromatic amino acids in *Bacillus subtilis* (Nester and Jensen, 1966) and in the biosynthesis of threonine and isoleucine in *Rhodopseudomonas spheroides* (Datta, 1966, 1969; Datta and Prakash, 1966).

3. Concerted (Multivalent) Feedback Inhibition

In this type of regulation, the first step in the pathway A → B is inhibited by neither of the ultimate end products alone, but when both are present simultaneously they act in concert to inhibit enzyme I (Fig. 4). As in sequential feedback inhibition secondary controls by EP₁ and EP₂ are usually observed on the enzymatic steps II and III. Thus a situation may arise where EP₁ accumulates and inhibits its own synthesis by inhibiting enzyme II. C is thus made available for synthesis of EP₂. As EP₂ accumulates, both end products cause inhibition of the whole pathway by inhibiting enzyme I. The aspartokinases of *Rhodopseudomonas capsulata* (Datta and Gest, 1964) and of *Bacillus polymyxa* (Paulus and Gray, 1964) have been found subject to concerted feedback inhibition by threonine plus lysine. Threonine and lysine alone are incapable of inhibition of the aspartokinases from the above microorganisms.

A variation of concerted feedback inhibition called synergistic feedback inhibition is observed in some systems where the end products EP₁

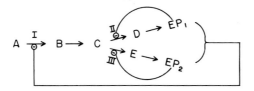

Fig. 4. Regulation of a branched pathway by concerted feedback inhibition. The nomenclature is the same as in Figs. 1 and 2.

and EP_2 each can inhibit the reaction $A \rightarrow B$, but very poorly. However, the presence of the two end products together inhibit much more effectively than the sum of the inhibition for the end products acting independently. Thus, if inhibitor D inhibited step I 20% at a certain concentration, and inhibitor E also inhibited 20% at a given concentration, the sum of the inhibition when both inhibitors were present should be 36%. In certain allosteric systems, the inhibition observed in the presence of the two "weak" inhibitors is greater, perhaps being over 90%. Examples of this are the mammalian and bacterial glutamine phosphoribosylpyrophosphate amidotransferases [reaction (3)], the first enzyme in purine nucleotide biosynthesis (Casky et al., 1964; Wyngaarden, 1972).

Glutamine + phosphoribosyl pyrophosphate →
$$\text{phosphoribosylamine} + \text{glutamate} + PP_i \quad (3)$$

This enzyme is inhibited by both nucleotides of 6-hydroxypurines, guanosine monophosphate (GMP), and 6-aminopurines, adenosine monophosphate (AMP). The extent of inhibition exerted by mixtures containing both 6-aminopurines and 6-hydroxypurines is greater than the sum of the fractional inhibition obtained with each nucleotide independently for the pigeon liver (Casky et al., 1964) and the Aerobacter aerogenes (Nierlich and Magasanik, 1965) enzymes. The synergistic nature of the inhibitions by 6-amino- and 6-hydroxypurine ribonucleotides on the first step of the purine biosynthetic pathway permits an effective curtailment of biosynthesis when both types of inhibitors are in excess simultaneously, but allows for a more moderate control when only one kind of purine is present in excess.

Synergistic feedback inhibition is also seen for the glutamine synthetase of Bacillus licheniformis (Hubbard and Stadtman, 1967). The enzyme is only slightly effected by low concentrations of AMP, histidine, or glutamine. However, a combination of either AMP plus histidine or of glutamine plus histidine causes almost complete inhibition of the activity.

4. CUMULATIVE FEEDBACK INHIBITION

This type of inhibition was observed for the E. coli glutamine synthetase [reaction (4)].

$$NH_3 + \text{glutamate} + ATP \rightarrow \text{glutamine} + ADP + P_i \quad (4)$$

The amide nitrogen of glutamine is utilized in E. coli for synthesis of tryptophan, AMP, CMP, glucosamine 6-phosphate, histidine, and carbamyl phosphate. In addition, the amide nitrogen can be utilized for the synthesis of various amino acids by transamination reaction. Individually the above six compounds, as well as glycine and alanine, can only partially inhibit the above enzyme, even at high concentrations. These inhibitors act independently of each other; therefore, when they are present simultaneously their total inhibitory effect is cumulative (Woolfolk and Stadtman, 1964, 1967). At concentrations where tryptophan inhibits 16%, CTP 14%, AMP 41%, and carbamyl phosphate 13%, together their cumulative inhibition is 63%. Alone no compound gives appreciable inhibition, together the inhibition is significant. When all 8 inhibitors are combined, the *cumulative* inhibition can reach 93%. With this mechanism, the enzyme activity is diminished progressively in response to excessive production of each end product. Glutamine synthetase may not be considered an enzyme in a branched pathway. However, since the product of its reaction, glutamine, participates in the biosynthesis of many metabolites, the cumulative type inhibition observed in the *E. coli* system appears to be an ideal type of mechanism to modulate its activity. It should be stressed, however, that this regulation is also in concert with the feedback type of control observed for the unique enzymatic steps present in the various branched pathways. Thus as each end product accumulates in excess, it curtails only that amount of the total glutamine synthetase activity which presumably is required for its formation as well as feedback inhibiting reactions in its metabolic pathway.

III. Regulation of Enzyme Activity by Chemical Modification

A. Phosphorylation and Adenylylation Reactions

In bacteria and in mammals a number of enzymes have been shown to be regulated by enzyme-catalyzed covalent attachment of specific groups. These processes may lead to changes in the primary or quaternary structures of the enzyme, thus rendering the enzyme into a more inactive or more active form. As indicated in Table I the inactive and active forms of phosphorylase, phosphorylase *b* kinase, glycogen synthetase, and pyruvate dehydrogenase are interconvertible with each other by a phosphorylation–dephosphorylation mechanism. The amino acid functional group modified is a serine hydroxyl group (Fischer *et al.*, 1971; Larner and Villar-Palasi, 1971). Similarly, adenylylation of glutamine synthetase alters its cation requirement from Mg^{2+} to Mn^{2+} and converts the enzyme to a form that is more susceptible to cumulative feedback inhibition (Shapiro and Stadtman, 1968). Reconversion to the active forms occurs by phosphorolysis of the phosphodiester bond by a specific enzyme (An-

TABLE I

Enzymes Regulated by Modification Reactions

Enzyme	Origin	Modification reactions	Physiological effect	References[a]
Glycogen phosphorylase	Mammals, fungi	(1) 2 Phos b + 4 ATP → Phos a + 4 ADP	Phos a = active	1–5
		(2) Phos a + H_2O → 4 P_i + 2 Phos b	Phos b = inactive	
Glycogen synthetase (GS)	Mammals, fungi	(1) GS I + ATP → GS D + ADP	GS D = inactive	5–11
		(2) GS D + H_2O → GS I + P_i	GS I = active	
Phosphorylase b kinase (Phos b K)	Mammals	(1) Phos b K_I + ATP → Phos b K_a + ADP	Phos b K_I = inactive	1–5, 13–16
		(2) Phos b K_a + H_2O → Phos b K_I + P_i	Phos b K_a = active	
Pyruvate dehydrogenase (PDH)	Mammals	(1) PDH A + ATP → PDH I + ADP	PDH A = active	17–20
		(2) PDH I + H_2O → P_i + PDH A	PDH I = inactive	
Glutamine synthetase (Glut S)	E. coli	(1) 12 ATP + Glut S → (AMP)$_{12}$-Glut S + 12 PP_i	(AMP)$_{12}$-Glut S = inactive	21–38
		(2) (AMP)$_{12}$-Glut S + 12 P_i → 12 ADP + Glut S	Glut S = active	

[a] Key to references:

1. Fischer et al., 1971.
2. Fischer et al., 1970.
3. Krebs and Fischer, 1962.
4. Graves and Wang, 1972.
5. Soderling and Park, 1974.
6. Traut and Lipmann, 1963.
7. Friedmann and Larner, 1963.
8. Larner, 1967.
9. Larner and Villar-Palasi, 1971.
10. Hers et al., 1970a.
11. Hers et al., 1970b.
12. Stahman and Hers, 1973.
13. Heilmeyer et al., 1970.
14. Haschke et al., 1970.
15. Brostrom et al., 1971.
16. Walsh et al., 1971.
17. Linn et al., 1969.
18. Wieland and Jagow-Westermann, 1969.
19. Wieland and Siess, 1970.
20. Reed et al., 1973.
21. Shapiro and Stadtman, 1970.
22. Kingdon et al., 1967.
23. Shapiro, 1969.
24. Wulff et al., 1967.
25. Anderson et al., 1970.
26. Hennig et al., 1970.
27. Anderson and Stadtman, 1971.
28. Heilmeyer et al., 1969.
29. Ebner et al., 1970.
30. Brown et al., 1971.
31. Holzer, 1969.
32. Holzer and Duntze, 1971.
33. Stadtman, 1970.
34. Stadtman and Ginsburg, 1974.
35. Ginsburg and Stadtman, 1973.
36. Segal et al., 1974.
37. Mangum et al., 1973.
38. Adler et al., 1975.

derson and Stadtman, 1970). These reactions are enzyme-catalyzed and in turn these modifying enzymes appear to be regulated either by hormone action or by allosteric action in order to prevent their simultaneous antagonistic function. For example, the enzyme that catalyzes the adenylnylation of *E. coli* glutamine synthetase is inactive when the deadenylnylation process has been activated because of various allosteric inhibitions.

The effect of chemical modification can drastically alter the catalytic properties of these enzymes with respect to V_{max} and the apparent affinities for substrates and allosteric effector molecules. For example, muscle phosphorylase exists in two forms, phosphorylase *a* and phosphorylase *b* (Table I). Phosphorylase *b*, believed to be the physiologically inactive form, is highly dependent on the presence of its allosteric effector AMP for activity, while phosphorylase *a* exhibits 67% of its total activity in the absence of 5'-adenylate. Muscle glycogen synthetase also has an inactive form, the D form, and an active form, the I form. The D form is dependent on the presence of the activator glucose 6-phosphate (glucose 6-P) for activity, while the I form is fully active, or independent of the presence of glucose 6-P for activity. The K_m for UDP-glucose is much higher for the D form than for the I form. It is of interest that the protein kinase that phosphorylates glycogen synthetase I (active form) to convert it to the inactive D form also phosphorylates inactive phosphorylase *b* kinase to transform it into the active form, suggesting the glycogen synthesis and degradation in muscle is coordinated to the point where one is "switched-on" as the other is "turned off" (Soderling *et al.*, 1970).

Enzyme-catalyzed chemical modification usually occurs along with other regulatory effects (allosteric control, repression, or derepression of protein synthesis), and is a mechanism that amplifies these effects. For further details, the reader is referred to a number of excellent reviews listed in Table I, on the regulation of these enzymes and others by chemical modification.

It should be pointed out that the analogous enzymes in plants, starch phosphorylase, starch synthetase, pyruvate dehydrogenase, and glutamine synthetase, do not appear to be regulated by chemical modification. However, this type of regulation may be very important in carbon metabolism during photosynthesis, especially in the inactivation and activation of certain enzymes during light–dark transitions. A number of studies suggest that certain enzymes of the CO_2 fixation pathways in plant leaves and in green algae may be activated in the light and inactivated in the dark (see Chapter 6). It is implied or suggested that the activation of these enzymes may occur by modification by photoreductants or related metabolites arising from the photosynthetic process. This would be an elegant type of regulation consistent with the function of these enzyme activities in CO_2 fixation occurring during photosynthesis.

Rat liver xanthine oxidase occurs in two interconvertible forms, and it is believed the mechanism of interconversion is reduction and oxidation of sulfhydryl groups (Della Corte and Stirpe, 1968; Stirpe and Della Corte, 1969, 1970).

B. Proteolysis

Another important mechanism by which the activity of enzymes may be controlled is modification by limited proteolysis. Conversion of pepsinogen to pepsin (Herriot, 1938) and of trypsinogen to trypsin (Northrup *et al.*, 1948) are obvious examples. Enzyme activation by limited proteolysis is also important in the regulatory processes concerned with coagulation of the blood (Davie *et al.*, 1969).

This type of regulation may occur in certain plant systems, but at present little is known about the prevalence of this potential regulatory mode in plants.

An obvious disadvantage of limited proteolysis as a mode of regulation compared to the phosphorylation–dephosphorylation and adenylation–deadenylation mechanisms is apparent. Limited proteolysis is an irreversible step and would be unsatisfactory when interconversion of inactive and active forms of enzymes is required for the "switching on or off" of a metabolic process.

IV. Regulation of "Futile Cycles" between Enzymes Involved in Gluconeogenesis and Glycolysis (Scrutton and Utter, 1968)

In glycolysis, the phosphofructokinase reaction catalyzes the formation of fructose 1,6-diphosphate (fructose 1,6-diP) from fructose 6-phosphate (fructose 6-P) [reaction(5)]. This reaction is physiologically irreversible.

$$\text{Fructose 6-phosphate} + \text{ATP} \rightarrow \text{fructose 1,6-diphosphate} + \text{ADP} \tag{5}$$
$$\underline{\text{Fructose 1,6-diphosphate} \rightarrow \text{fructose 6-phosphate} + \text{P}_i} \tag{6}$$
$$\text{ATP} \rightarrow \text{ADP} + \text{P}_i$$

Fructose 1,6-diP can be hydrolyzed to fructose 6-P during gluconeogenesis in a reaction catalyzed by fructose diphosphatase [reaction (6)]. It is apparent that the combined action of these two enzymes will function as an ATPase if their activities are not controlled.

Similarly, the combined reactions of pyruvate kinase, a glycolytic enzyme, and pyruvate carboxylase and phosphoenolpyruvate carboxykinase, gluconeogenic enzymes, will also give a futile cycle leading to loss of energy [reactions (7)–(9)].

$$\text{Phosphoenol pyruvate} + \text{ADP} \rightarrow \text{pyruvate} + \text{ATP} \tag{7}$$
$$\text{Pyruvate} + CO_2 + \text{ATP} \rightarrow \text{ADP} + P_i + \text{oxalacetate} \tag{8}$$
$$\text{Oxalacetate} + \text{ATP} \rightarrow CO_2 + \text{ADP} + \text{phosphoenol pyruvate} \tag{9}$$

$$\overline{\text{ATP} \rightarrow \text{ADP} + P_i}$$

In order to prevent the wasteful cycling between fructose 6-P and fructose 1,6-diP and between phosphoenol pyruvate and pyruvate, certain regulatory controls must be present.

In mammalian systems, the tissues that conduct both gluconeogenesis and glycolysis are liver and kidney. In these tissues, phosphofructokinase is activated by AMP, and fructose 1,6-diP and is inhibited by ATP (Stadtman, 1966; Atkinson, 1966, 1969; Scrutton and Utter, 1968). Some of these allosteric effectors have opposite effects on fructose diphosphatase; AMP and fructose diphosphate inhibit the phosphatase. In conjunction with these effects, citrate, 3-phosphoglycerate, and phosphoenol pyruvate are effective inhibitors of phosphofructokinase (Krzanowski and Matschinsky, 1969), while 3-phosphoglycerate has been reported to be an effective activator of the fructose diphosphatase activity, which is inhibited by ATP (Pogell et al., 1971; Taketa et al., 1971). Thus, under conditions where AMP and FDP are high, glycolysis would be stimulated (high phosphofructokinase activity, low or negligible fructose-1,6-diphosphatase) and gluconeogenesis would be suppressed. Conversely, low AMP and fructose 1,6-diP concentrations are conditions where glycolysis is inhibited and the gluconeogenic process proceeds. Fructose 6-P is also known to antagonize the inhibitions of both ATP and citrate (Passonneau and Lowry, 1964). Lowered levels of fructose 6-P, which also occur with lowered rates of glycolysis, would also allow the available citrate and ATP concentrations to inhibit fructokinase activity (Passonneau and Lowry, 1964). Accumulation of 3-phosphoglycerate during gluconeogenesis (Exton and Park, 1969) would also inhibit phosphofructokinase activity and stimulate fructose-1,6-diphosphatase activity. Therefore, allosteric phenomena affecting both enzymes acting in the fructose 6-P–fructose 1,6-diP cycle ensure that the two enzymes are not active at the same time.

There are a number of other enzymatic steps that occur in metabolism to catalyze opposing reactions. As indicated previously, coupling of the pyruvate carboxylase, phosphoenolpyruvate carboxykinase, and pyruvate kinase reactions would also lead to a "futile cycle." Similarly, the coupling of liver hexokinase (glucokinase) and liver microsomal glucose-

6-phosphatase reactions would also lead to a loss of energy. Coordinate regulation of these reactions entail allosteric phenomena as well as other regulatory elements, such as compartmentation and control of enzyme synthesis and activation by hormone action (Scrutton and Utter, 1968). It is probable that the same regulatory factors are also prevalent in co-ordinating the activity of directly opposed enzymatic steps in plants. A very good discussion of regulation of futile cycle pathways in mammalian tissues is discussed by Newsholme and Start (1973). Experiments attempting to quantitate the extent of cycling in the P-fructokinase–fructose-1,6-diphosphatase futile cycle are discussed in a review by Clark and Lardy (1975).

V. Control of Metabolism by Adenylate Energy Charge

Krebs (1964) suggested that glycolysis and gluconeogenesis are regulated by the level of 5′-AMP in the cell or by the ratio of 5′-AMP to ATP. This was based on the observations that 5′-AMP stimulated phosphofructokinase activity and inhibited fructose diphosphatase activity in mammalian systems. Atkinson (1968a,b, 1969, 1970, 1971), Atkinson and Walton (1967), Shen et al. (1968), and Klungsøyr et al. (1968) have recognized that many enzymes have as allosteric effectors either ATP, ADP, or AMP and therefore would respond to either the ATP/AMP or the ATP/ADP ratios in the cell. Atkinson has suggested that the ratio of AMP, ADP, and ATP should be buffered against wide fluctuations in order for the cell to remain operational. Thus, production of energy in the form of ATP must be regulated so that it is equal to the rate of utilization. Since the adenylate pool of ATP plus ADP plus AMP is constant, the utilization of ATP should result in the formation of either ADP (in phosphate transfer or cleavage reactions) or AMP (in pyrophosphate transfer or cleavage reactions). These three nucleotides are in equilibrium with each other via the adenylate kinase reaction [reaction (10)].

$$ATP + AMP \rightleftharpoons 2 \, ADP \tag{10}$$

Atkinson (Atkinson and Walton, 1967) proposed the term "energy charge" to define the energy state of the adenylate system as

$$\text{Energy charge} = \frac{ATP + \frac{1}{2} \, ADP}{ATP + ADP + AMP} \tag{11}$$

The energy charge of a cell would be 1.0 if the adenine nucleotides are completely in the form of ATP and 0 if they are all AMP. The energy charge value can also be considered as being one-half the number of anhydride-bound phosphate groups available in the cell per adenosine moiety. In studying the effect of energy charge on a number of enzymes from mammalian and microbial sources, Atkinson's group found that enzymes involved in generating ATP [e.g., NAD-linked isocitrate dehydrogenase (Shen *et al.*, 1968) and phosphofructokinase (Atkinson, 1968a)] were active at low energy charge values and became less active at high energy charge. Enzymes utilizing ATP for biosynthetic purposes [PRPP synthetase (Atkinson, 1968a), and mammalian ATP-dependent citrate cleavage enzyme (Atkinson and Walton, 1967), and ADP-glucose pyrophosphorylase (Shen and Atkinson, 1970)] were less active at low energy charge and highly active at high energy charge. The response curves of the two different classes of enzymes are shown in Fig. 5. The enzymes involved in metabolic sequences synthesizing ATP and the enzymes involved in ATP utilization were most sensitive at the energy charge value of 0.8; i.e., the steepest part of the curves were at the value 0.8. The consistency of response observed in these studies *in vitro* suggested strongly that the regulatory enzymes have been designed by selection to participate in maintaining energy charge *in vivo* at values between 0.8 and 0.9. It has been shown that the energy charge values in a great number of living cells range between 0.8 and 0.95 (Chapman *et al.*, 1971). These results suggested that the regulatory response of both the ATP

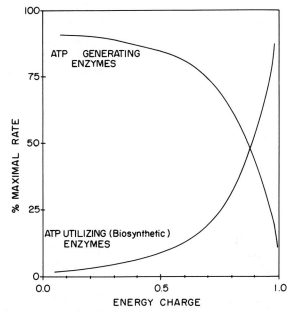

Fig. 5. An ideal response of some regulatory enzymes to energy charge. From Atkinson (1968a).

generating and utilizing enzymes were designed to keep the energy level of the cell constant. What is remarkable is that most enzymes tested in this way show a very sensitive response in the region of energy charge of 0.8.

Atkinson's hypothesis provides a convenient way to determine the regulatory effect of the adenylate effector molecules on enzyme activity *in vitro* in what may perhaps be a closer approximation of the conditions *in vivo* where the adenine nucleotides are in equilibrium due to the adenylate kinase reaction. It also allows one to determine the extent of interaction of other regulatory effects on enzymes (due to substrates or effectors) with those due to the energy charge giving one perhaps a clearer insight into the physiological significance of these interactions. For example, the effect of citrate, an allosteric inhibitor of phosphofructokinase on the energy charge response of the enzyme was studied (Shen *et al.*, 1968), and it was shown to increase the steepness of the enzyme response to energy charge under conditions where fructose 6-P concentrations overcame ATP inhibition. Thus, under conditions where fructose 6-P would ordinarily overcome the ATP inhibition of the enzyme, citrate, a negative modifier, increased the sensitivity of the enzyme to ATP inhibition and therefore its response to energy charge.

The response of phosphofructokinase to energy charge and the modulation of its energy charge response to both fructose 6-P and citrate are consistent with the role the enzyme plays in an amphibolic sequence. Glycolysis is not only an ATP generating sequence but it also supplies the carbon skeletons for biosynthesis of many cellular constituents, such as amino acids, purines, and pyrimidines. As indicated by Atkinson (1969), if phosphofructokinase was only regulated by energy charge, then under conditions where rapid growth may occur and energy charge is high, inhibition of phosphofructokinase would limit the supply of carbon for cellular synthesis. Thus, at high energy charge, the inhibition of phosphofructokinase can be relieved by the high concentrations of the substrate fructose 6-P, which would accumulate in an actively growing organism. Accumulation of citrate, an intermediate in the amphibolic tricarboxylic acid cycle, would be a signal that sufficient metabolites are present for biosynthesis. The excess concentration of citrate plus high energy charge would then inhibit the phosphofructokinase activity significantly.

The allosteric effectors of some other regulatory enzymes involved in biosynthetic pathways or in ATP utilizing pathways can also interact with the energy charge responses of these enzymes. Higher concentrations of the feedback inhibitor lysine decrease the activity of the *E. coli* aspartokinase at a constant energy charge, while greater activity is observed at lower concentrations of lysine (Klungsyer *et al.*, 1968). The lysine inhibition effects may also be counteracted to some extent by an increase in the energy charge. Thus, the two requirements for biosynthetic activity would appear to be both high energy charge and low concentrations of feedback inhibitor.

It is of interest that Pradet and Bomsel (1968), Bomsel and Pradet (1968), and Pradet (1969) have independently developed the energy charge hypothesis, although not in as great detail as Atkinson's group, in studying various plant systems. However, as will be described later, the plant phosphofructokinases are not affected by either AMP or ADP. They are inhibited by ATP and citrate, and these inhibitions are partially relieved by phosphate. Phosphate is a potent inhibitor of the plant and algal ADP-glucose pyrophosphorylases. Therefore, the levels of inorganic phosphate as well as the adenine nucleotides may have to be taken into account in trying to correlate energy charge levels with regulatory processes in plants.

Pacold and Anderson (1975) have shown that chloroplast and cytoplasmic pea leaf 3-phosphoglycerate (3-P-glycerate) kinase activities *in vitro* are regulated by energy charge. AMP inhibits the enzyme activity in both directions: formation of 3-P-glycerate (ATP generating reaction) and formation of 1,3-di-P-glycerate (ATP utilizing reaction). ATP inhibits the enzyme in the direction of 3-P-glycerate formation. The region of greatest sensitivity to energy charge was between $0.8 \rightarrow 1.0$ for both reactions. At high energy charge ($\sim$0.95), 3-P-glycerate kinase activity would be increased greatly in the direction toward 1,3-di-P-glycerate formation and, therefore, in the cell toward carbohydrate synthesis. At low energy charge ($\sim$0.7), the enzyme activity would be predominantly in the direction of 3-P-glycerate formation and toward glycolysis. Thus the direction of the 3-P-glycerate kinase reaction may be regulated during photosynthesis and in the dark. This was suggested by the results of Heber and Santarius (1970), who found that energy charge in *Elodea* chloroplasts and cytoplasm is high in the light (0.95 and 0.94, respectively) and low in the dark (0.67 and 0.76, respectively).

VI. Kinetic Properties of Allosteric Enzymes

A. Methods for Analyzing Kinetic Parameters of Allosteric Enzymes

For many regulatory enzymes, a plot of substrate concentration versus velocity yields a curve that does not obey the Michaelis–Menten equation, $V = V_{max}S/(K_m + S)$. Umbarger (1956), in studying the *E. coli* threonine deaminase showed that the substrate saturation curve was sigmoid in shape. This type of curve has been observed for many regula-

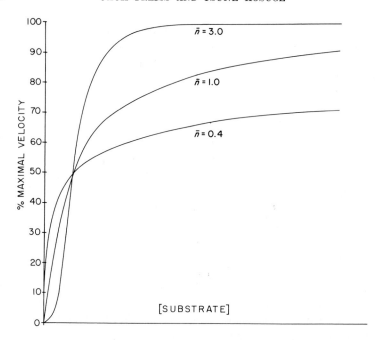

Fig. 6. Theoretical substrate saturation curves. A curve showing no co-operativity (noninteracting binding sites $\bar{n} = 1$), a curve showing positive co-operativity ($\bar{n} = 3$), and a curve showing negative cooperativity ($\bar{n} = 0.4$).

tory enzymes, and a theoretical curve showing sigmoidal-type kinetics is shown in Fig. 6 and compared with the normal hyperbolic curve. These complex kinetics have usually been interpreted as suggesting the presence of multiple binding sites on an enzyme interacting in a cooperative manner, so that the binding of one substrate molecule facilitates the binding of additional substrate molecules. Thus, a sigmoidal curve signifies positive cooperativity among substrate binding sites. Sigmoidal curves have also been observed for allosteric activators and inhibitor molecules.

The differences between the sigmoidal curve and normal hyperbolic curve can also be seen in double reciprocal plots of substrate concentration versus velocity (Fig. 7). A straight line is obtained for the curve obeying Michaelis–Menten kinetics, while the curve showing positive cooperative effects is concave upward. As indicated by Koshland *et al.* (1966) and by Taketa and Pogell (1965), the ratio of the concentration of substrate giving 90% of maximal velocity to the substrate concentration giving 10% of maximal velocity, defined as R_s (Koshland *et al.*, 1966), is equal to 81 for hyperbolic curves (seen in Fig. 1) while the same ratio for sigmoidal curves would be much less than 81.

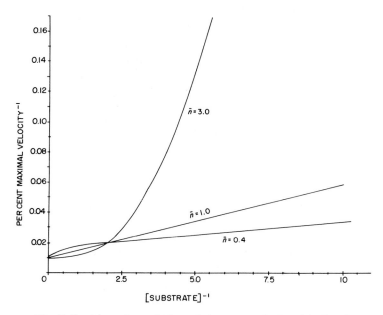

Fig. 7. Double-reciprocal plots of the curves depicted in Fig. 6.

Another anomalous curve not obeying Michaelis–Menten kinetics is also seen in Figs. 6 and 7. Although it appears to be qualitatively similar to the hyperbolic curve it has an R_s value greater than 81 and is concave downward in the double reciprocal plot. This type of curve has been interpreted to suggest *negative* cooperativity between substrate sites; i.e., the binding of the first substrate molecule hinders the binding, or causes a decrease in the affinity for the binding, of the next substrate molecule. This type of kinetic phenomenon has been shown to occur for a number of enzymes, among those being rabbit muscle glyceraldehyde-3-P dehydrogenase (Conway and Koshland, 1968), ADP-glucose pyrophosphorylase (Gentner and Preiss, 1967, 1968; Govons *et al.*, 1973), aspartate transcarbamylase, CTP synthetase (Levitzki and Koshland, 1969), and phosphoenolpyruvate carboxylase (Corwin and Fanning, 1968). Positive and negative cooperativity phenomena can be observed in the kinetics for the above enzymes.

The curves giving positive and negative cooperativity are usually analyzed by what is called a Hill plot (Fig. 8). Hill (1913) noted the sigmoidal nature of the O_2 saturation curve for hemoglobin and analyzed it in terms of

$$y = Kx^n/(1 + Kx^n)$$

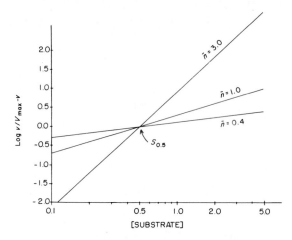

Fig. 8. Hill plots of the curves depicted in Fig. 6.

where y is the fraction of saturation of hemoglobin by O_2, x is the partial pressure of O_2, n is the number of O_2 molecules bound per hemoglobin molecule, and K is a constant. This equation can be applied to the study of enzyme-catalyzed reactions in the form

$$\log [v/(V_{max} - v)] = n \log (S) - \log K$$

where v is the reaction velocity, V_{max} is the maximal reaction velocity, K is a constant, S is the substrate concentration, n is the number of sites at which a substrate may bind, and its value may be obtained from the slope when $\log [v/(V_{max} - v)]$ is plotted against $\log S$.

Atkinson (1966) and Atkinson *et al.* (1965) have clarified the physical significance to be assigned to the slope, n, of a plot of $\log [v/(V_{max} - v)]$ versus $\log S$ by deriving the Hill equation using Michaelis–Menten assumptions and assuming that concentrations of enzyme–substrate complexes containing fewer than n molecules (ES_{n-1}, Es_{n-2}, etc. $\simeq 0$) are negligibly small, i.e., infinite cooperation among all the substrate binding sites. If cooperativity is very strong, n would approximate the actual number of substrate binding sites. However, this is rarely the case as n values are usually the combined result of the number of binding sites and the strength of interaction between the binding sites. The n values in the Hill plot, therefore, are usually well below the actual number of binding sites. For example, for hemoglobin, which contains four binding sites, the Hill coefficient is 2.8 (Wyman, 1948). The Hill coefficient will indicate, however, the minimal number of binding sites for a substrate or effector molecule giving a positive cooperative saturation curve. If there is no interaction between binding sites then n will have a value of 1. Since the symbol n is usually reserved for meaning actual number of binding sites, the symbol $\bar{n}$ will be used for the Hill slope. $\bar{n}$ has also been referred to in the literature as the Hill coefficient, as the interaction coefficient (Changeux, 1961), and as the apparent order of reaction (Atkinson et al., 1965).

Figure 8 shows Hill plots of the three different curves seen in Fig. 6. For the curve exhibiting positive cooperativity $\bar{n} = 3.0$, and $\bar{n}$ for the hyperbolic curve is 1.0. The curve showing negative cooperativity has an $\bar{n}$ value of less than 1 (0.4), which is consistent with the interpretation that the interaction of the binding sites leads to a progressive decrease in affinity for the substrate.

It should be emphasized that the Hill plots shown in Fig. 8 are plots for ideal curves where maximum cooperativity occurs. In actual cases, the Hill plots may be only linear over a short range, usually about the midpoint where $v/(V_{max} - v) = 1$, and deviate from linearity at the extremes.

As indicated earlier, the effect of allosteric modifiers is either to decrease or increase the enzyme's apparent affinity for the substrate or to increase or decrease V_{max}. The allosteric modifiers also may change the shape of the substrate saturation curves, i.e., either increase or decrease the Hill slope. However, these effects are not observed in all systems (Atkinson, 1966).

The concentration of substrate required for 50% of maximal velocity is obtained from the midpoint of the Hill curve where $\log [v/(V_{max} - v)] = 0$ or where $v/(V_{max} - v) = 1$. In the case where the plot of substrate versus velocity is hyperbolic and $\bar{n} = 1$, this value is equivalent to the K_m value of the Michaelis–Menten equation. In order to distinguish the value for half saturating concentration of substrate attained from curves exhibiting cooperative kinetics from the Michaelis–Menten parameter, Koshland et al. (1966) have suggested the nomenclature $S_{0.5}$. Extending this to the case of allosteric effectors, one could use the terms $A_{0.5}$ and $I_{0.5}$ to define the concentration of activator required for half-maximal activation and the concentration of inhibitor required for 50% inhibition, respectively. The kinetic parameters of an inhibition curve can be analyzed by a Hill plot with the following equation (Taketa and Pogell, 1965),

$$\log [v/(V_0 - v)] = \log K_I - \bar{n} \log I$$

where v is the velocity of reaction in presence of inhibitor, V_0 is velocity in absence of inhibitor, K_I is a constant, $\bar{n}$ the Hill coefficient, and I is inhibitor concentration.

The nature of the sigmoid curve as well as its steepness is of considerable practical significance with respect to the physiological function of the allosteric enzyme. In this curve there is a threshold concentration below which enzyme activity is relatively insensitive to changes in concentrations of substrate or effectors, but above which relatively large changes in enzyme activity are elicited by slight changes in effector concentration. This situation ensures extreme sensitivity of the enzyme activity to very narrow, selected ranges of substrate and effector concentrations.

The curve exhibiting negative cooperativity may have evolved for the opposite reason. It would dampen variations in reaction rate due to fluctuations of substrate or effector concentration. This may be important in the case of cyclic AMP phosphodiesterase (Russell *et al.*, 1972). Increased concentrations of cyclic AMP would not increase its degradative rate because of the negative cooperativity phenomena, thus enabling the increased formation of cyclic AMP due to hormone effects to function physiologically for longer periods of time. In the case of a feedback inhibitor, negative cooperativity would prevent total inhibition of the enzyme by the inhibitor, and this would be significant if the product of the enzyme reaction is also a substrate for a metabolic pathway different from the one synthesizing the inhibitor end product.

Other mathematical analyses of saturation curves and substrate and effector binding curves of regulatory enzymes have been discussed in a number of reviews and experimental papers (Koshland, 1969, 1970; Kirtley and Koshland, 1967; Haber and Koshland, 1967; Cornish-Bowden and Koshland, 1970; Monod *et al.*, 1965; Rubin and Changeux, 1966). These analyses were developed for the formulation of various models to explain cooperative binding. Some are discussed briefly below.

B. Proposed Models for Regulatory Enzymes

1. MONOD–WYMAN–CHANGEUX MODEL (SYMMETRY OR CONCERTED MODEL) (1965)

This theory postulates the following.

1. All allosteric enzymes are polymers consisting of two or more identical subunits.

2. These subunits exist in at least two different conformations, R and T in equilibrium with each other.

3. Each of the identical subunits possesses a single catalytic site, specific for the substrate, and a separate allosteric site for each allosteric effector (positive or negative).

4. For each conformational state the catalytic sites and allosteric sites have equal affinities for their respective ligands.

5. The subunits are linked to each other by noncovalent bonds in such a way that they are all equivalent, and, therefore, the polymer has at least one axis of symmetry. This symmetry is conserved when the subunits undergo transition from one state to the other. The transition from one state to the other $(R \leftrightarrow T)$ involves simultaneous changes in all the subunits of the polymer. Therefore, in a tetramer (a polymer consisting of four subunits), if one subunit is transformed into R from T, then the other three subunits undergo the same transformation simultaneously.

If the R and T states have different affinities for the effector molecules and substrates, binding of these molecules will cause a change in the equilibrium between the R and T states and will cause the displacement of the equilibrium in favor of that conformational state that has the greater affinity for the effector molecule or substrate. The change in one subunit conformation will cause, in concerted fashion, a change of all the subunits into the favored conformational state. The facilitation of the subsequent binding of the other molecules to the other subunits will then occur. This manner of binding can, therefore, explain the sigmoid-shaped velocity versus effector molecule or substrate concentration curves. Monod *et al.* (1965) have developed the mathematical formulation of this model.

2. KOSHLAND MODEL (SEQUENTIAL MODEL) (Koshland *et al.*, 1966; Koshland, 1969, 1970)

Many of the postulates of the Monod–Wyman–Changeux model are consistent with the Koshland model. There is one important difference; there is no simultaneous transformation to a conformational state of all the subunits on binding of substrate or effector molecule. Rather, there is a sequential transformation from one state to another. Thus, a tetramer will undergo the following:

$$\text{TTTT} + \text{S} \xrightarrow{\text{S}} \text{RTTT} \xrightarrow{\text{S}} \text{RRTT} \xrightarrow{\text{S}} \text{RRRT} \rightarrow \text{RRRR}$$

Conformational change occurs by "induced fit" (Koshland, 1958, 1959, 1963) of the substrate or effector molecules. The conformation change induced by the binding causes new interactions between the subunits, which results in a change in the affinity of the unoccupied binding sites which is the basis for the observed cooperative effects. A mathematical formulation stressing that subunit interaction causes the cooperative effects observed, and including molecular parameters describing the energy of the conformational change, the strength of subunit interactions and the affinity of the effector or substrate molecule to the defined conformational states makes the model very general and applicable to most allosteric enzymes (Koshland, 1969, 1970; Kirtley and Koshland, 1967; Haber and Koshland, 1967; Cornish-Bowden and Koshland, 1970). It should be pointed out that the sequential model can predict negative cooperativity by having selective interactions among different subunits. This type of kinetic phenomena cannot be predicted by the concerted model.

Both the symmetry and sequential models stress the importance of subunit conformations to explain cooperative binding. However, other theories have been advanced to explain the kinetic phenomena associated with regulatory enzymes. A number of investigators have postulated that sigmoidal curves can be obtained from two-substrate enzymes containing only one catalytic site if these reaction mechanisms involve alternate pathways where substrate binding or release are rate-limiting steps (Ferdinand, 1967; Rabin, 1967; Sweeney and Fisher, 1968). The association and dissociation of subunits can also lead to sigmoidal curves for binding of substrates and effector molecules. The reader is referred to Frieden (1967) and Nichols et al. (1967) for more detailed information on the various models, as they are beyond the scope of this chapter.

It is probable that no one present theory or model will account for the observed kinetics of all the regulatory enzymes studied.

VII. Regulation of Enzymes Involved in Carbon Metabolism during Photosynthesis

Since the regulation of the C_4 pathway and the C_3 pathway are covered in Chapter 24, the discussion of the regulation of the enzymes involved in those pathways will not be covered here. The discussion of regulation of enzyme activity will be limited to processes not covered elsewhere (see also Chapters 17, 19, and 24) and where regulatory-type phenomena have been clearly demonstrated.

A. Regulation of Ribulose-5-P Kinase

The discovery that 5'-adenylate (AMP) inhibited the ribulose-5-P kinase of several chemosynthetic and photosynthetic organisms suggested that this enzyme is under control of the energy charge state in the cell

(Johnson and Peck, 1965; Johnson, 1966; Gale and Beck, 1966; McFadden and Tu, 1967; Mayeux and Johnson, 1967; MacElroy *et al.*, 1968; Abd-El-Al and Schlegel, 1974). AMP inhibited by competing with the substrate ATP for the phosphoribulose kinase of *Thiobacillus ferroxidans* (Gale and Beck, 1966). For the *Thiobacillus thioparus* phosphoribulokinase, concentration versus reaction rate curves were sigmoidal for both ATP and AMP, suggesting cooperativity between the substrate sites as well as between the inhibitor binding sites (MacElroy *et al.*, 1968). Reduced nicotinamide adenine dinucleotide (NADH) was also reported to stimulate the ATP-dependent CO_2 fixation catalyzed by extracts of *Hydrogenomonas facilis* (MacElroy *et al.*, 1969), *Hydrogenomonas eutropha* H16 Abd-El-Al and Schlegel, 1974), and *Rhodopseudomonas spheroides* (Rindt and Ohmann, 1969). The effect of NADH appeared to be on the ribulose-5-P kinase and not on the ribulose-1,5-diP carboxylase or phosphoribulose isomerase. For the three organisms, AMP inhibited or reversed the activation caused by NADH.

It, therefore, appears that CO_2 fixation in the above organism occurs only at high energy charge or when the ratio of ATP to AMP is high. Carbon dioxide fixation would occur in *Hydrogenomonas* only if a reducing atmosphere is present for the production of NADH. The above results would also provide a control mechanism for CO_2 fixation in *R. spheroides* when the organism changes from dark aerobic metabolism to anaerobic light metabolism. Anaerobic light metabolism results in the production of NADH, thus activating the ribulose-5-P kinase. On transition to dark aerobic metabolism, the concentration of NADH decreases thus lowering the ribulose-5-P kinase activity. NADH levels have been shown to rapidly decrease in another closely related photosynthetic organism, *Rhodospirillum rubrum*, during the change from anaerobic light to aerobic dark conditions (Jackson and Crofts, 1968). The inhibition by AMP would ensure that ribulose 5-P kinase activity and, therefore, CO_2 fixation occur only when the energy level of the cell is high.

B. Regulation of Citrate Synthase in Photosynthetic Bacteria

Both *Rhodospirillum rubrum* and *Rhodopseudomonas spheroides* are capable of autotrophic growth in the light on $CO_2 + H_2$. They can also grow in the dark, aerobically, or in the light anaerobically on tricarboxylic acid (TCA) or TCA-related metabolites. Under autotrophic or heterotrophic conditions of growth, both bacteria contain a relatively high concentration of citrate synthase (Anderson and Fuller, 1967; Eidels and Preiss, 1970a,b). The specific activity of the citrate synthase in both organisms is about twofold greater in cells grown aerobically in the dark than in cells grown anaerobically in the light with malate, acetate, or pyruvate as carbon sources. These observations would be consistent with a more active TCA cycle in the dark than in the light.

Since an important function of the TCA cycle is to generate reducing power, the activity of the TCA cycle would be expected to decrease in the light, where reducing units could be obtained from photochemical processes in these photosynthetic bacteria. Although the amount of citrate synthase is relatively low in dark-grown cells, it still showed the highest activity of any of the enzymes investigated in the cell-free extracts of *Rhodopseudomonas capsulata* (Eidels and Preiss, 1970b). Because of this, the regulation of the citrate synthase activity by metabolites is of interest.

The citrate synthase of both *R. rubrum* and *R. capsulata* were found to be inhibited strongly by NADH (Eidels and Preiss, 1970a). AMP reversed this inhibition. The enzyme studied in most detail was the *R. capsulata* enzyme. At 0.1 mM, NADH increased the $S_{0.5}$ of acetyl-CoA from 0.19 to 0.58 mM and converted the acetyl-CoA curve from a hyperbolic form to a sigmoidal form. The K_m for oxalacetate was also increased from 20 to 55 μM by 0.1 mM NADH. The addition of 0.46 mM AMP completely overcame the inhibition by 0.1 mM NADH and restored the $S_{0.5}$ value of acetyl-CoA as well as the hyperbolic shape of its saturation curve. AMP by itself (in the absence of NADH) had no effect on $V_{\max}$ or the $S_{0.5}$ (K_m) values of the substrates. High concentrations of acetyl-CoA (>1.5 mM) also completely overcame the inhibition by 0.1 mM NADH. NADH inhibition was noncompetitive with oxalacetate.

The concentration of NADH required to elicit 50% inhibition ($I_{0.5}$) at 0.12 and 0.34 mM acetyl-CoA were 26 and 74 μM, respectively. At 0.18 mM acetyl-CoA, the concentrations of AMP required to reverse 50% of the inhibition caused by 0.1, 0.49, and 0.98 mM NADH, were 4.3, 26, and 51 μM, respectively. The AMP saturation curves were sigmoidal with $\bar{n}$ values between 1.5 and 1.9.

Similar results have recently been observed for the citrate synthase from *Rhodopseudomonas spheroides* (Borriss and Ohmann, 1972). The enzyme from *R. capsulata, R. spheroides*, and *R. rubrum* is, therefore, most sensitive to fluctuations of both AMP and NADH concentrations. Under conditions of low "energy charge" and low reducing power, the activity of citrate synthase would be maximal, resulting in an active TCA cycle. An increase in reduced pyridine nucleotides caused by substrate oxidation in the dark or by photosynthesis in the light with the concomitant increase in energy charge caused by an increase of ATP production would result in the inhibition of citrate synthase of these photosynthetic bacteria and ultimately to a decrease in the activity of the TCA cycle.

It was shown that in *R. rubrum* under dark aerobic conditions, the total diphosphopyridine nucleotide pool was present as NAD+ (Jackson and Crofts, 1968), while under light anaerobic conditions 70% of the NAD+ + NADH pool was in the form of NADH. These results are also consistent with a lower citrate synthase activity in the light anaerobic conditions.

The finding of NADH inhibition of citrate synthase and its reversal by AMP in *R. capsulata* is also consistent with the observation of Weitzman and Jones (1968). They reported that citrate synthases from gramnegative organisms was inhibited by NADH. In this group, those organisms (strict aerobes) that do not metabolize glucose via glycolysis but utilize an alternate pathway [e.g., Entner–Doudoroff pathway (Entner and Doudoroff, 1952)] were very sensitive to reversal of inhibition by AMP, while those organisms (facultative anaerobes) that metabolize glucose via glycolysis were not sensitive to this reversal. *R. capsulata*, which can metabolize glucose via the Entner–Doudoroff pathway but is not a strict aerobe, would be included in the latter group. Weitzman and Jones (1968) concluded that organisms, in which AMP (or ADP) acting as a "low energy signal," activates the key glycolytic enzymes (such as phosphofructokinase or pyruvate kinase) would not require a similar low energy signal at the level of their citrate synthase. An the other hand, organisms in which the Embden–Meyerhof pathway is absent or in which there is no regulation of its key enzymes, would require a low energy signal to control the TCA cycle at the level of entry to the cycle, namely, citrate synthase. The latter appears to be the case with *R. capsulata*.

It is of interest to compare the regulatory properties of the citrate synthase from *R. rubrum*, *R. spheroides*, and *R. capsulata* with the regulatory properties of the ribulose-5-P kinase from the closely related organism of *Rhodopseudomonas spheroides* (Rindt and Ohmann, 1969). The kinase is activated by NADH and inhibited by AMP. Thus, under conditions where the ribulose-5-P kinase is active (e.g., photosynthesis), citrate synthase and the TCA cycle activities would be low. The reciprocal relationship between synthase and ribulose-5-P kinase in photosynthetic bacteria with respect to their modulation by the energy charge and the "reducing state" of the cell could be responsible for the regulation and interaction between two of the carbon pathways present in these organisms. It should be pointed out that the mode of regulation of citrate synthase present in mitochondria of higher plants and animals is different and this will be discussed later.

VIII. Regulation of the Biosynthesis of Starch in Photosynthetic Tissues

It is generally agreed that biosynthesis of the α-1,4-glucosyl linkage of starch occurs according to reactions (12) (Espada, 1962), (13) (Munch-Peterson *et al.*, 1953), and (14).

$$\text{ATP} + \alpha\text{-glucose 1-P} \rightarrow \text{ADP-glucose} + \text{PP}_i \qquad (12)$$
$$\text{UTP} + \alpha\text{-glucose 1-P} \rightarrow \text{UDP-glucose} + \text{PP}_i \qquad (13)$$
$$\text{ADP-glucose (UDP-glucose)} + \alpha\text{-1,4-glucan primer} \rightarrow$$
$$\alpha\text{-1,4 glucosylglucan} + \text{ADP (UDP)} \quad (14)$$

ADP-glucose is more effective as a glucosyl donor than UDP-glucose in reaction (3) (Recondo and Leloir, 1961). In fact, it has been shown that the leaf chloroplast glucosyltransferases are specific for ADP-glucose and deoxy-ADP-glucose (Doi *et al.*, 1964; Frydman and Cardini, 1964; Ghosh and Preiss, 1965a). Other sugar nucleotides are virtually inactive. It appears that regulation of α-1,4-glucan synthesis (Preiss, 1969, 1973) occurs at the level of ADP-glucose synthesis [reaction (12)]. The find-ing that control occurs at the site of ADP-glucose synthesis is consistent with the concept that regulation of a biosynthetic pathway occurs at the first unique step of the pathway. In plants and in bacteria, the unique step is ADP-glucose pyrophosphorylase.

A. ADP-Glucose Pyrophosphorylases of Higher Plants and Green Algae

1. LEAF ADP-GLUCOSE PYROPHOSPHORYLASES

All ADP-glucose pyrophosphorylases from leaves of higher plants and from green algae are activated by 3-phosphoglycerate (3-P-glycerate) and are inhibited by orthophosphate (Ghosh and Preiss, 1965b, 1966; Preiss *et al.*, 1967; Sanwal *et al.*, 1968; MacDonald and Strobel, 1970; Sanwal and Preiss, 1967; Ribereau-Gayon and Preiss, 1971; Fur-long and Preiss, 1969a). Other glycolytic intermediates, such as phos-phoenol pyruvate (PEP), fructose 1,6-diP, and fructose 6-P, activate to lesser extents, and at much higher concentrations of the lesser ac-tivators fructose 6-P is most active giving about 20 to 60% of the activation noted for 3-P-glycerate. The enzyme of this class studied in the greatest detail is that obtained from spinach leaf (Ghosh and Preiss, 1965, 1966; Preiss *et al.*, 1967; Ribereau-Gayon and Preiss, 1971). 3-P-glycerate decreases the K_m for ADP-glucose from 0.93 to 0.15 mM, the K_m of pyrophosphate from 0.50 to 0.04 mM, and the K_m of ATP from

0.45 to 0.04 mM. The K_m value of glucose 1–P is 0.07 mM in the absence of 3-P-glycerate and is decreased to 0.04 mM in its presence. All substrate saturation curves are hyperbolic in the presence or absence of 3-P-glycerate. The $MgCl_2$ saturation curve is sigmoidal in the presence or absence of the activator, and its $S_{0.5}$ (1.6 mM) is not changed by 3-P-glycerate.

The stimulation of the spinach leaf enzyme by 3-P-glycerate is dependent on pH because of the different pH optima of the activated and unactivated reaction. Stimulation of ADP-glucose synthesis can vary from 9- to 80-fold. The 3-P-glycerate activation curve is hyperbolic in shape at pH 7.0 and 7.5, but becomes progressively sigmoidal as the pH increases. At pH 8.5 the Hill constant $\bar{n}$, is 1.8. ADP-glucose pyrophosphorylase was isolated from ten other leaf sources and studied. The specificity of the activation did not change whether the enzyme was from a plant fixing CO_2 via the Calvin–Benson pathway or the Hatch–Slack pathway (Sanwal et al., 1968; Furlong and Preiss, 1969a). The concentration of 3-P-glycerate required for 50% of maximal activation varied from 7 μM for the barley enzyme to 370 μM for the sorghum leaf enzyme.

Inorganic phosphate proved to be an effective inhibitor of ADP-glucose synthesis for all the leaf and algal enzymes. ADP-glucose synthesis catalyzed by the spinach leaf enzyme is inhibited 50% by 22 μM P_i in the absence of activator at pH 7.5 (Preiss et al., 1967). In the presence of 1 mM 3-P-glycerate, 50% inhibition of ADP-glucose synthesis requires 1.3 mM phosphate. Thus, the activator increases $I_{0.5}$ about 450-fold. The 3-P-glycerate saturation curve, normally hyperbolic, becomes sigmoidal in the presence of the inhibitor, phosphate. P_i at 0.5 mM increases the $A_{0.5}$ of 3-P-glycerate from 20 to 230 μM and increases $\bar{n}$ from 1.0 to 1.9. At 0.75 mM P_i, the $\bar{n}$ and $A_{0.5}$ values of 3-P-glycerate are increased to 2.5 and 300 μM, respectively. Conversely, 3-P-glycerate increases the sigmoidicity of the P_i inhibition curve. The Hill interaction coefficient for P_i in the absence of 3-P-glycerate is 1.2 and is increased to 2.9 in its presence (1 mM). Thus, the inhibitor, phosphate, causes an increase in the interaction among activator sites and the presence of the activator, 3-P-glycerate increases the interaction among the inhibitor sites. Phosphate is a noncompetitive or mixed inhibitor with respect to the substrates, ADP-glucose, PP_i, ATP, and glucose 1-P (Ghosh and Preiss, 1966).

The concentration of phosphate required for 50% inhibition of the other leaf ADP-glucose pyrophosphorylases varies from 20 μM for the barley enzyme to 200 μM for the sorghum leaf enzyme (Sanwal et al., 1968). In all cases phosphate inhibition was reversed or overcome by the activator 3-P-glycerate.

2. *Chlorella pyrenoidosa* ADP-Glucose Pyrophosphorylase

The ADP-glucose pyrophosphorylase from this green alga is very similar in properties to the leaf enzymes (Sanwal and Preiss, 1967). The V_{max} of synthesis and pyrophosphorolysis are increased 18- and 7-fold, respectively, by 3-P-glycerate at pH 8.5, the optimum for both the activated and unactivated reaction. The algal enzyme is also inhibited by phosphate. The $I_{0.5}$ value is 0.18 mM and the Hill $\bar{n}$ is 1.3 in the absence of 3-P-glycerate. In the presence of 2 mM 3-P$_i$-glycerate, $I_{0.5}$ and $\bar{n}$ are increased to 1.0 and 1.6 mM, respectively. Conversely, the $A_{0.5}$ of 3-P-glycerate is 0.4 mM, and the Hill interaction coefficient $\bar{n}$ is 1.0 in the absence of P$_i$. In the presence of 0.1 mM P$_i$, $\bar{n}$ is increased to 1.3 and $A_{0.5}$ to 0.5 mM, while in the presence of 0.5 mM P$_i$, $A_{0.5}$ is 0.72 mM and $\bar{n}$ is 1.6. These results are similar to those observed for the leaf enzymes, in that greater interaction is seen between the inhibitor sites when activator is present and between the activator sites when inhibitor is present.

The ATP and ADP-glucose saturation curves for the *Chlorella pyrenoidosa* enzyme are sigmoidal in the presence or absence of the activator 3-P-glycerate. This result differs from that for the spinach leaf enzyme. 3-P-glycerate only decreases the $S_{0.5}$ values about 1.6-fold; ATP from 0.8 to 0.5 mM, and ADP-glucose from 2.8 to 1.8 mM.

The ADP-glucose pyrophosphorylases of *Chlorella vulgaris*, *Scenedesmus obliquus*, and *Chlamydomonas reinhardii* are also activated by 3-P-glycerate and inhibited by orthophosphate (Sanwal and Preiss, 1967).

B. Physiological Significance of 3-Phosphoglycerate Activation and Phosphate Inhibition of the Leaf and Algal ADP-Glucose Pyrophosphorylases

Because of the great sensitivity of the leaf ADP-glucose pyrophosphorylases to 3-P-glycerate, the primary CO_2 fixation product of photosynthesis, and P$_i$ it is suggested that they play a significant role in the regulation of starch biosynthesis. The level of P$_i$ has been shown to decrease in leaves during photosynthesis because of photophosphorylation, and glycolytic intermediates are known to increase in the chloroplast in light. This situation would, therefore, contribute to conditions necessary for optimal starch synthesis via the increased rate of formation of ADP-glucose. In the light, the levels of ATP and reduced pyridine nucleotides are also increased, leading to the formation of sugar phosphates from 3-P-glycerate. In the dark, there is an increase in phosphate concentration with concomitant decreases in the levels of glycolytic intermediates, ATP, and reduced pyridine nucleotides. This would lead to inhibition of ADP-glucose synthesis and therefore starch synthesis. In order to confirm this hypothesis, knowledge of the concentrations of the various effector molecules at the actual site of the ADP-glucose pyrophosphorylase is necessary. However, no information of this sort is available, and at present it would be difficult to obtain.

At best, the results obtained by workers on the concentrations of the glycolytic intermediates (Heber, 1967), phosphate, and ATP (Santarius and Heber, 1965; Heber and Santarius, 1965) in the chloroplast qualitatively support the hypothesis of regulation of starch synthesis by 3-P-glycerate, other glycolytic intermediates, and phosphate levels.

Heber and Santarius (1965) and Santarius and Heber (1965) have shown that the concentration of P_i in the chloroplasts of spinach leaf in the dark is about 5 to 10 mM and decreases about 30 to 50% in the light. At these concentrations 3-P-glycerate can partially reverse the inhibition by phosphate. In a kinetic experiment, it was shown that at 5 mM 3-P-glycerate, the rate of ADP-glucose synthesis was increased fivefold when the phosphate concentration was decreased from 10 to 7.5 mM. A 23-fold increase was observed when the phosphate concentration was decreased to 5 mM (Sanwal et al., 1968). Thus, under these conditions a decrease of phosphate concentration of only 30 to 50% in the chloroplast may cause a significant acceleration of ADP-glucose synthesis and, therefore, of starch synthesis.

Recently Kanazawa et al. (1972) have shown that both starch and ADP-glucose synthesis occur in Chlorella pyrenoidosa cells in the light. Starch synthesis ceases abruptly, and the ADP-glucose level drops below detectable limits when the light is turned off. UDP-glucose levels do not change perceptibly in the light to dark transition. ADP-glucose is not detectable at any time later in the dark despite the high steady state level of ATP and hexose phosphates. Kanazawa et al. (1972) indicate that this observation provides strong support for the postulated regulatory role of ADP-glucose pyrophosphorylase in starch synthesis in vivo. Thus the allosteric effects exerted by 3-P-glycerate and P_i appear to be physiologically important. Since the level of 3-P-glycerate does not appreciably change in the dark to light transition (Heber, 1967; Kanazawa et al., 1972) while the phosphate levels appear to increase in the dark and decrease in the light (Santarius and Heber, 1965; Heber and Santarius, 1965), it is possible that the variation of this effector molecule is the most important control element.

MacDonald and Strobel (1970) reported that wheat leaves infected with the fungus, Puccinia striiformis, accumulated more starch than non-infected leaves. They could correlate starch accumulation with the inverse of the variation observed in P_i levels in diseased leaves during the infection process. They indicated that their data suggested that in diseased leaves the variations in the level of P_i and, to a lesser extent, variations in the level of activators of the wheat leaf ADP-glucose pyrophosphorylase (3-P-glycerate, fructose-diP, etc.) regulated the rate of starch synthesis via control of the activity of ADP-glucose pyrophosphorylase.

C. Regulation of Glycogen Synthesis in Photosynthetic Bacteria

The reactions involved in glycogen synthesis in bacteria are the same as those described for starch synthesis in leaves of higher plants and green algae. The mode of regulation of starch biosynthesis observed in plants is also the same for bacterial glycogen synthesis (Preiss, 1969, 1973), namely, regulation occurs at the level of ADP-glucose synthesis. Whereas ADP-glucose synthesis in higher plants is activated by 3-P-glycerate, the primary CO_2 fixation product of the Calvin–Benson cycle and the first glycolytic intermediate formed in CO_2 assimilation in the Hatch-Slack pathway, the *Rhodospirillum rubrum* ADP-glucose pyrophosphorylase is activated by pyruvate (Furlong and Preiss, 1969a,b). The only other metabolite found to activate the enzyme is α-ketobutyrate. Another distinct property of the *R. rubrum* enzyme is that it is not inhibited by either P_i, 5'-AMP, or ADP. No inhibitor of physiological importance has been found for this enzyme. Pyruvate increases the maximal velocity of pyrophosphorolysis and synthesis of ADP-glucose about twofold. It also decreases the $S_{0.5}$ values for ATP (from 3.4 to 0.36 mM with 5 mM pyruvate) and for ADP-glucose (from 2.0 to 0.38 mM with 25 mM pyruvate). The decreases in the K_m values for pyrophosphate and α-glucose 1-P are only about 1.5- to 2.0-fold, however. Pyruvate also decreases the $S_{0.5}$ value for $MgCl_2$ about 1.5- to 2.0-fold and shifts the pH optimum of ADP-glucose synthesis from 8.5 to 7.5.

Rhodospirillum rubrum is capable of growth under a number of heterotrophic conditions as well as under autotrophic conditions. ADP-glucose pyrophosphorylase activity is seen whether the cells are grown aerobically in the dark with malate or anaerobically in the light with malate, acetate, acetate $+$ CO_2, or CO_2 $+$ H_2. The activator specificity of the pyrophosphorylase does not change with cells grown under different conditions (Furlong and Preiss, 1969a).

Thus, pyruvate alone is important in the regulation of glycogen synthesis in *R. rubrum*. This is consistent with the observations made by Stanier *et al.* (1959). These investigators showed that incubation of starved cells of *R. rubrum* in the light with succinate, malate, or pyruvate caused accumulation of glycogen. If the cells were incubated with acetate or butyrate, the reserve polymer that accumulated was poly-β-hydroxybutyrate and only small amounts of glycogen accumulated. The accumulated poly-β-hydroxybutyrate was utilized if CO_2 was made available to cells, and under these conditions glycogen accumulated. Glycogen was also formed if R. rubrum was incubated with CO_2 plus acetate or CO_2 plus H_2. The pattern of labeling of glycogen by [1-^{14}C]- and [2-^{14}C]succinate suggested that the hexose units of the polysaccharide were formed by conversion of the succinate to pyruvate, and subsequent hexose synthesis through a reversal of the glycolytic sequence. Thus, Stanier et al. (1959) concluded that compounds (such as succinate, malate, or glutamate) which led to formation of pyruvate resulted in glycogen formation. Other studies (Kikuchi et al., 1963) also suggested that in R. rubrum grown in the light under anaerobic conditions, dicarboxylic acids liberate CO_2 mainly at the levels of malate and oxalacetate to yield pyruvate.

Since incubation of R. rubrum with acetate gave little glycogen, but incubation of the cells with acetate + CO_2 did give rise to significant amounts of glycogen, Stanier et al. (1959) suggested that CO_2 may play an essential role in the formation of C_3 compounds from acetate by R. rubrum. In this respect, Cutinelli et al. (1951) have shown that CO_2 is an important carbon source during photosynthetic growth of R. rubrum with acetate. The incorporation of CO_2 specifically into the carboxyl group of alanine and the incorporation of the carboxyl and methyl groups of acetate into the α- and β-carbon atoms of alanine, respectively, suggested to these investigators the formation of pyruvate by addition of CO_2 to an acetyl derivative. Buchanan et al. (1967) have demonstrated the formation of pyruvate from CO_2 and acetyl-CoA [reaction (15)] in cell-free extracts of R. rubrum that had been grown on CO_2 and H_2. This enzymatic reaction required reduced ferredoxin (FdH_2).

$$\text{Acetyl-CoA} + CO_2 + FdH_2 \rightarrow Fd + CoA + \text{pyruvate} \qquad (15)$$

Thus mechanisms for the synthesis of pyruvate, the allosteric activator of ADP-glucose synthesis in R. rubrum, are available in this photosynthetic organism grown under various nutritional conditions that give rise to accumulation of glycogen. In this respect, the demonstration of the following reaction (16) in R. rubrum by Buchanan and Evans (1965) is pertinent to glycogen synthesis in this organism.

$$\text{Pyruvate} + \text{ATP} \rightarrow \text{phosphoenol pyruvate} + \text{AMP} + P_i \qquad (16)$$

This unique reaction is catalyzed by phosphoenolpyruvate synthase and is distinct from pyruvate kinase. Two energy-rich bonds of ATP are cleaved to give rise to phosphoenol pyruvate (PEP) plus AMP plus P_i and to allow the equilibrium of the reaction lie in favor of PEP formation. Because of this reaction pyruvate may be considered the first glycolytic intermediate in gluconeogenesis in *R. rubrum*.

The central position that pyruvate plays in carbon metabolism is thus reflected in its function as the sole activator for ADP-glucose synthesis in that organism.

Rhodopseudomonas capsulata, another photosynthetic anaerobe, has an ADP-glucose pyrophosphorylase activated by pyruvate and fructose 6-P (Eidels *et al.*, 1970). In contrast to *R. rubrum*, *R. capsulata* is able to grow on glucose as well as on various TCA cycle intermediates. The glucose is catabolized via the Entner–Doudoroff pathway (Entner and Doudoroff, 1952). The presence of the two activators for the *R. capsulata* pyrophosphorylase may be rationalized in that *R. capsulata* utilizes a pathway for glucose degradation as well as being able to grow heterotrophically or autotrophically in the light.

It appears that the nature of the activator for ADP-glucose synthesis in a particular organism may be correlated with the mode of carbon assimilation of the organism. Other bacteria that contain ADP-glucose pyrophosphorylases that are activated by fructose 6-P also catabolize glucose via the Entner–Doudoroff pathway. Bacteria, using the glycolytic pathway for glucose catabolism, contain an ADP-glucose pyrophosphorylase that is activated by fructose diP (Preiss, 1969, 1973). Evidence that the activation seen for *Escherichia coli* ADP-glucose pyrophosphorylase *in vitro* is physiologically important in the regulation of glycogen synthesis has been obtained by isolation of mutants of *E. coli* accumulating about three to five times as much glycogen as the parent strain and containing an ADP-glucose pyrophosphorylase having stronger affinities than the parent enzyme for its activator metabolite, fructose diphosphate, and a weaker affinity for the inhibitor, 5′-adenylate (Preiss, 1969, 1973).

IX. Regulation in Higher Plant Nonchlorophyllous Tissue

A. Regulation of Glycolysis, Gluconeogenesis, and Tricarboxylic Acid Pathways

Since ATP is synthesized via photochemical processes in plants, the energy producing pathways of glycolysis and the tricarboxylic acid cycle would not be expected to be functional in the chloroplast. Indeed, chloroplasts do not contain the full complement of either glycolytic or TCA cycle enzymes (Laties, 1950; Heber *et al.*, 1967; Latzko and Gibbs,

1968). However, these latter pathways would be operative in the non-photosynthetic tissues of plants, as they synthesize their ATP via oxidation phosphorylation. Since the primary function of these two pathways is to provide energy, it would not be surprising to find that the activities of these pathways are regulated by the energy level of the cell. These pathways are also concerned with supplying organic intermediates for biosynthetic reactions, and thus one would also expect to observe regulatory phenomena concerned with this function superimposed on the controlling factors concerned with regulating the cell's energy level. A few examples of the available information on plant enzymes possibly involved in the regulation of glycolysis, TCA cycle activity, and gluconeogenesis will be presented.

1. PHOSPHOFRUCTOKINASE (PFK)

Generally, the mammalian phosphofructokinases are inhibited by ATP and citrate and by phosphoenol pyruvate and 3-P-glycerate (Stadtman, 1966; Atkinson, 1966, 1969; Krzanowski and Matschinsky, 1969). AMP reverses the inhibition caused by ATP and citrate. Fructose 6-P, orthophosphate, ADP and 3′,5′-cyclic AMP also interact with the inhibitors in the various systems and usually overcome the inhibition caused by ATP and citrate. The yeast and *Escherichia coli* enzymes are very similar to the mammalian enzymes with some variation. The *E. coli* PFK is not inhibited by citrate, and ADP rather than AMP reverses the ATP inhibition (Blangy *et al.*, 1968). Some reviews on the PFK's of animals and microorganisms are noted (Mansour, 1972; Bloxham and Lardy, 1973; Ramaiah, 1974).

The first report on the regulatory properties of a plant PFK (Lowry and Passonneau, 1964) indicated that the enzymes from parsley leaves and avocado fruit were inhibited by ATP. This inhibition was relieved by increasing concentrations of fructose 6-P for both enzymes. Phosphate decreased the ATP inhibition for the parsley enzyme but not for the avocado enzyme. AMP, ADP, and 3′,5′-cyclic AMP had no effect on these enzymes. PFK from carrots (Dennis and Coultate, 1966) was inhibited both by citrate and ATP, the two inhibitors working synergistically with each other. Phosphate and fructose 6-P could relieve the inhibition by ATP and citrate, and AMP or ADP were inhibitory rather than relieving the inhibition by ATP. Essentially the same results were obtained for the enzyme from brussels sprouts leaves (Dennis and Coultate, 1967b). It was also noticed that the enzyme from the most immature tissues showed the greatest regulatory control and that from mature and senescent leaves the least.

Phosphoenol pyruvate at very low concentrations (50% inhibition at 20 μM) inhibited the pea seed PKF (Kelly and Turner, 1969a,b, 1970). This inhibition was relieved by fructose 6-P and inorganic phosphate. ATP and citrate also inhibited this enzyme, and phosphate stimulated activity in the presence of the inhibitors. Other inhibitors of the pea seed enzyme were 3-P-glycerate, 2-P-glycerate, and 6-P-gluconate. The inhibition caused by these compounds were relieved by high concentrations of fructose 6-P (Kelly and Turner, 1970).

The corn scutellum PFK had similar properties *in vitro* to other plant phosphofructokinases (Garrad and Humphreys, 1968). However, since the levels of citrate, ATP, or phosphate remained unchanged when tissue slices of corn scutellum incubated in 0.1 M fructose solution had increased their glycolytic rate 4- to 7-fold, the authors of this study suggested that PFK may not be regulated by the level of these metabolites. The levels of fructose 6-P concentration did double, however, and it is quite possible that in the conditions present in the scutellum cells a simple doubling of the fructose 6-P levels would be able to produce a 4- to 7-fold increase in the PFK activity by reversal of the ATP inhibition. Variation in the levels of phosphoenol pyruvate, a potential inhibitor, was not reported, and compartmentation of effector molecules may also be involved.

Of interest is the purification and characterization of the PFK from the appendices of *Sauromatum*. Johnson and Meeuse (1972) studied the properties of this enzyme in an attempt to correlate its mode of regulation with its possible role in the respiratory "flare-up" manifested by the tissue during flowering. The enzyme properties were similar to other plant phosphofructokinases. ADP and AMP were inhibitors, and not activators; ATP and citrate also inhibited but their inhibitions could be partially reversed by fructose 6-P, Mg^{2+}, or phosphate. The experiments suggest, but do not prove, that PFK activity is high during the time when respiration is high because of low ADP and ATP concentrations in the *Sauromatum* appendix on the first day of flowering. It is important also to note that the high temperature and alkaline pH optima of the enzyme are consistent with the cellular environment noted on the first day of flowering. The enzyme also appears to be absolutely dependent on the concentration of NH_4^+. Ammonia is released in significant quantities on the first day of flowering by the appendices.

2. Pyruvate Kinase

The regulatory properties of pyruvate kinase from many mammalian tissues and microorganisms have been described and reviewed (Seubert and Schoner, 1971). Generally phosphoenol pyruvate, the substrate, shows sigmoidal saturation curves, and the enzyme may be activated by fructose-diP or AMP and inhibited by ATP. These are essentially the properties of the pyruvate kinases of several *Euglena gracilis* strains (Ohmann, 1969; Vaccaro and Zeldin, 1974). ATP, citrate, and Ca^{2+} were found to be inhibitors of the *Euglena* enzymes. Recently, Duggleby and Dennis (1973a,b) have shown that the pyruvate kinase from cotton seeds is activated by AMP and inhibited by citrate and ATP but not activated by fructose-diP. These results are consistent with pyruvate kinase being regulated by energy charge and feedback inhibited by tricarboxylic acid activity.

Germinating cotton seed is a gluconeogenic tissue. Thus it is reasonable to assume that TCA cycle flux toward sucrose and hexose biosynthesis in this tissue would necessitate a shut down of pyruvate kinase activity to allow carbon flow to proceed to hexose without great loss via pyruvate kinase activity. The regulatory effects described may be those effective in shutting down pyruvate kinase activity.

3. NAD-Linked Isocitrate Dehydrogenase (ICDH)

It has been shown that ICDH in yeast and fungi is activated by AMP and that its mammalian counterpart is stimulated by ADP (Stadtman, 1966; Atkinson, 1966, 1969). No NAD-linked ICDH has been found in bacteria. The enzyme in mammals and fungi appears to be controlled by the energy charge of the cell. Two plant NAD-linked isocitrate dehydrogenases have been studied with respect to regulatory properties. Both the pea mitochondrion (Cox and Davies, 1967; 1969) and the *Brassica napus* L. enzymes (Coultate and Dennis, 1969; Dennis and Coultate, 1967a) were unresponsive to either AMP or ATP. The *Brassica* enzyme was inhibited by ATP, but the inhibition appeared to be due to chelation of Mg^{2+}, which was required for the reaction. Both enzymes were inhibited by high concentrations of inorganic anions and were activated by citrate. With both enzymes a plot of isocitrate concentration versus velocity gave sigmoidal curves. The *Brassica* ICDH was very sensitive to NADH inhibition (Dennis and Coultate, 1967a). It is possible that the principal control of ICDH is via the NAD/NADH ratio, which should reflect the energy charge of the mitochondria through the operation of the electron transport chain.

4. Fructose Diphosphatase (FDPase)

Fructose diphosphatase is considered to be an important enzyme in gluconeogenesis (Stadtman, 1966; Atkinson, 1966). In mammals and in

E. coli (Fraenkel *et al.*, 1966) it was found to be inhibited by 5'-adenylate. This observation is consistent with the concept that high energy charge would stimulate gluconeogenesis and low energy charge would inhibit this process (see Section IV). The alkaline fructose diphosphatases from castor bean endosperm (Scala *et al.*, 1968a,b) and wheat embryo (Bianchetti and Sartirana, 1967) are also inhibited by 5'-adenylate. It should be noted that both of these plant tissues are gluconeogenic in their metabolism. The endosperm of the germinating castor bean utilizes acetate derived from aliphatic acids for the biosynthesis of sucrose (Canvin and Beevers, 1961; Kornberg and Beevers, 1967). The wheat embryo enzyme appears when the embryo is grown on glycerol or ethanol as a carbon source, but it is not present when growth takes place in the presence of glucose (Bianchetti and Sartirana, 1967, 1968). Two FDPases have been noted in castor bean (Scala *et al.*, 1968b, 1969). One is present in the ungerminated castor bean and is AMP sensitive, while the other, which is not AMP sensitive, appears after 3 days germination and seems to be induced by gibberellic acid.

5. Citrate Synthase from Plant Mitochondria and Glyoxysomes

The citrate synthase of mammalian tissues have been shown to be inhibited by ATP (Hathaway and Atkinson, 1965; Jaangard *et al.*, 1968). It is believed this property signifies regulation of the mitochondrial enzyme by energy charge. High energy charge would, therefore, inhibit citrate synthase, the first enzyme in the ATP-generating pathway, the tricarboxylic acid cycle.

The citrate synthases from mitochondria of lemon fruit (Bogin and Wallace, 1966) and castor bean (Axelrod and Beevers, 1972) are also inhibited by ATP. For both enzymes ATP inhibition is competitive with acetyl-CoA. For the lemon fruit enzyme the K_m of oxalacetate was found not to be affected. Thus, the citrate synthases from plants may also be under regulatory control. The germinated castor bean also contains another subcellular organelle, the glyoxysome, whose function is to convert the acetyl-CoA arising from fatty acid breakdown to succinate via the glyoxalate pathway (Breidenbach and Beevers, 1967). The glyoxysome also contains a citrate synthase whose function is to convert the acetyl-CoA to citrate, which is eventually converted to succinate by aconitase and isocitrate lyase. This citrate synthase is not inhibited by ATP.

Thus, the same tissue contains two citrate synthases, with two different functions. The one involved in ATP production is subject to control by energy charge while that involved in gluconeogenesis is not.

6. Regulation of Sucrose Biosynthesis in Nonchlorophyllous Tissue

The biosynthesis of sucrose may occur by two different pathways in chlorophyllous and nonchlorophyllous tissue (Cardini *et al.*, 1955; Leloir and Cardini, 1955; Mendicino, 1960; Haq and Hassid, 1965; Hawker, 1967; Slabnik *et al.*, 1968): by transfer of glucose from UDP-glucose (or from ADP-glucose) to fructose [reaction (17) below: sucrose synthase]; or by transfer of glucose from UDP-glucose to fructose 6-P to yield sucrose phosphate [reaction (18) below: sucrose-P synthase] which is then hydrolyzed to sucrose by phosphatase action (Hawker, 1966; Hawker and Hatch, 1966) [reaction (19)].

$$\text{ADP-glucose or UDP-glucose} + \text{fructose} \rightarrow \text{sucrose} + \text{ADP or UDP} \quad (17)$$
$$\text{UDP-glucose} + \text{fructose 6-P} \rightarrow \text{sucrose-P} + \text{UDP} \quad (18)$$
$$\text{Sucrose-P} \rightarrow P_i + \text{sucrose} \quad (19)$$

It is believed that the route of sucrose formation during photosynthesis from CO_2 is reactions (18) and (19). This conclusion is based on experiments (Gibbs *et al.*, 1967; Putman and Hassid, 1954) showing that equal labeling of the glucose and fructose moieties of sucrose occur before there is significant incorporation of $^{14}CO_2$ into the neutral sugar fraction during photosynthetic CO_2 assimilation. At the same time, both the UDP-glucose and fructose 6-P pools are highly labeled. Equal labeling of the glucose and fructose would not be expected if reaction (17) is the predominant mechanism for sucrose formation. Recent kinetic studies with wheat germ sucrose-P synthase indicated that both substrates, fructose 6-P and UDP-glucose, exhibit sigmoidal rate versus concentration curves, suggesting that synthase activity was modulated by the substrates (Preiss and Greenberg, 1969). The sigmoidal shape of the substrate saturation curves would provide a more sensitive response to the fluctuation of substrate concentrations. Although these results have been obtained with a sucrose-P synthase from nonphotosynthetic tissue, it is tempting to speculate that the same allosteric properties may be associated with the leaf sucrose-P synthase (Bird *et al.*, 1974). Bird *et al.* (1974) have demonstrated that the leaf sucrose-P synthase is a cytoplasmic enzyme and is not associated with the choroplast. The rapid rate of sucrose synthesis in leaves during photosynthesis could then be explained by the increase of both UDP-glucose and fructose 6-P owing to CO_2 fixation (Gibbs *et al.*, 1967; Bassham and Jensen, 1967).

The sucrose synthase of *Phaseolus aureus* seedlings has been purified to apparent homogeneity (Delmer, 1972). At present no regulatory phenomena have been found for this enzyme. It is believed, though not proved, that the physiological importance of this enzyme is for the conversion of sucrose to the sugar nucleotide via the reversal of reaction (17) (Slabnik *et al.*, 1968; Murata *et al.*, 1964; De Fekete and Cardini, 1964). The formation of sugar nucleotide from sucrose may be important in endosperm or reserve tissues in the conversion of sucrose to starch.

7. ADP-GLUCOSE PYROPHOSPHORYLASES OF NONCHLOROPHYLLOUS PLANT TISSUE

The ADP-glucose pyrophosphorylases occurring in nonphotosynthetic plant tissues, maize endosperm and embryo (Dickinson and Preiss, 1969a,b), wheat germ, etiolated peas and mung bean seedlings, potato tuber, carrot roots, and avocado mesocarp (Preiss *et al.*, 1967) are qualitatively similar to the leaf enzymes in that they are activated by 3-phosphoglycerate (3-P-glycerate). The stimulation by 3-P-glycerate is 1.5- to 10-fold for these enzymes. The enzyme representative of this group and studied in most detail is the one isolated from maize endosperm (Dickinson and Preiss, 1969a). Activation of ADP-glucose synthesis by 3-P-glycerate is 1.5- to 2-fold at pH 7.9 and 3- to 4-fold at pH 6.7. However, the $A_{0.5}$ value for 3-P-glycerate is very high (2.2 mM) compared with that of the leaf enzymes. Fructose 6-P also stimulates about 3-fold with an $A_{0.5}$ of 4.0 mM. Phosphate, at 3 mM, causes 50% inhibition in the absence of activator. In the presence of 10 mM 3-P-glycerate, the $I_{0.5}$ is 10 mM. Phosphate also changed the 3-P-glycerate saturation curve from a hyperbolic form to a slightly sigmoidal curve.

As with other ADP-glucose pyrophosphorylases, the activator 3-P-glycerate increases the apparent affinity of the enzyme for the substrates. Both the $MgCl_2$ and ATP saturation curves are sigmoidal in the absence of 3-P-glycerate, but in the presence of 3-P-glycerate the ATP saturation curve becomes hyperbolic. The saturation curve for glucose 1-P is hyperbolic in the absence of 3-P-glycerate and also in its presence.

Maize tissue also has an ADP-glucose pyrophosphorylase in the embryo, which is distinct from the endosperm enzyme (Dickinson and Preiss, 1969b; Preiss *et al.*, 1971). The embryo enzyme is more stable at 60°C and is more sensitive to inhibition by P_i ($I_{0.5} = 0.32$ mM). Furthermore, 3-P-glycerate has no effect on the phosphate inhibition, changing neither $I_{0.5}$ nor the hyperbolic shape of the curve. The 3-P-glycerate saturation curve is sigmoidal, with $A_{0.5}$ being 4.2 mM and $\bar{n} = 1.9$. Stimulation of V_{max} is threefold. In contrast to all other ADP-glucose pyrophosphorylases studied, the activator *decreases* the apparent affinity for the substrates, ATP and glucose 1-P.

The variation in properties of the maize enzymes as compared to the leaf enzymes may reflect differences between leaf and endosperm cells with respect to intracellular levels of metabolites. However, the most important phenomenon regulating starch biosynthesis in endosperm may be regulation of synthesis of the starch biosynthetic enzymes, ADP-glucose pyrophosphorylase and ADP-glucose:α-glucan-4-glucosyltransferase (Tsai et al., 1970; Ozbun et al., 1973).

Starch-deficient maize mutants shrunken-2 and brittle-2, which have only about 10 to 12% of the ADP-glucose pyrophosphorylase activity observed in the normal maize endosperm, synthesize only 25 to 30% as much starch as normal maize (Dickinson and Preiss, 1969b; Tsai and Nelson, 1966). These data would suggest that the major portion, if not all, of the starch synthesized in the normal endosperm is via the ADP-glucose pathway.

8. REGULATORY PROPERTIES OF ENZYMES INVOLVED IN BIOSYNTHESIS OF SUGAR NUCLEOTIDES

The biosynthesis of cell wall components containing various sugars are usually catalyzed by various transglycosylases that utilize as their substrates sugar nucleotides (for more information on this subject, see Nikaido and Hassid, 1971, and Chapters 10 and 13). A number of enzymes involved in the biosynthesis of sugar nucleotides utilized in synthesis of cell walls have been shown to be regulated by feedback inhibition. Generally, the first unique enzyme in the pathway of the synthesis of a sugar nucleotide is inhibited by the product of the pathway.

One such example is the L-glutamine-fructose-6-P amidotransferase from *Phaseolus aureus* seeds (Vessal and Hassid, 1972), the first enzyme in the pathway for the biosynthesis of uridine diphosphate–N-acetyl-D-glucosamine (UDP-gNAc) which catalyzes the following irreversible reaction (Pogell and Gryder, 1957; Ghosh et al., 1960; Mayer et al., 1968):

$$\text{D-Fructose 6-P} + \text{L-glutamine} \rightarrow \text{D-glucosamine 6-P} + \text{L-glutamate} \quad (20)$$

The enzyme was competitively inhibited with respect to D-fructose 6-P by UDP-gNAc with a K_i value of 13 μM. The K_i value of 13 μM reported for the mung bean enzyme is in good agreement with the value reported for the HeLa cell (Kornfeld, 1967) and rat liver (Kornfeld et al., 1964) enzymes.

The mung bean enzyme could be completely desensitized to UDP-gNAc inhibition in 24 hours if it was stored in the presence of fructose 6-P and in the absence of glutamine. The enzyme retained some but not all of its activity. Both the catalytic and regulatory activities could be completely retained if the enzyme was stored with L-glutamine. The above data suggest that the inhibitor site and catalytic sites are distinct from each other.

An interesting possible control phenomenon has been reported for the UDP-glucose pyrophosphorylase from *Lilium longiflorium* pollen (Hopper and Dickinson, 1972) This enzyme is abundant in plant cells capable of rapid growth and cell wall formation and is important for the synthesis of UDP-glucose [a glucosyl donor for synthesis of sucrose, steryl glucosides (Alpaslan and Mudd, 1970), callose (Feingold *et al.*, 1958), hemicellulose, and β-1,3- and 1,4-glucans (Hassid, 1969; Ray *et al.*, 1969; Ordin and Hall, 1968; Peaud-Penol and Axelos, 1970; Tsai and Hassid, 1971)]. UDP-glucose is also utilized for the synthesis of UDP-galactose, UDP-glucuronate, UDP-galacturonate, UDP-xylose, and UDP-L-arabinose (Nikaido and Hassid, 1971; Hassid, 1967, 1969). The ungerminated pollen of *L. longiflorium* contains a UDP-glucose pyrophosphorylase that is inhibited by UDP-glucose, UDP-glucuronate, UDP-xylose, and UDP-galactose (Hopper and Dickinson, 1971; Dickinson *et al.*, 1973). The inhibition is noncompetitive with the substrate α-glucose 1-P and mixed competitive and noncompetitive with UTP. The apparent inhibition constants (K_i) are UDP-glucose, 0.13 mM; UDP-glucuronate, 0.75 mM; UDP-galacturonate, 0.93 mM; UDP-xylose, 1.6 mM; and UDP-galactose, 4.8 mM. In various combinations tested these inhibitors showed simple additive inhibition when present at low individual concentrations and in the presence of low concentrations of UTP and glucose 1-P.

These kinetic studies suggested the possibility that the above uridine diphosphate sugars and uridine diphosphate sugar acids may participate *in vivo* in a cumulative feedback manner at the UDP-glucose pyrophosphorylase level. Thus, the flow of carbon into the various pathways leading to cell wall polysaccharide synthesis is modulated by the respective sugar nucleotides. Such modulation could coordinate the rate of carbon flow into these pathways with the rate of usage of the various UDP-sugars. In view of the relative concentrations of the above respective sugar nucleotides found in the cell and their apparent inhibition constants for UDP-glucose pyrophosphorylase, it would appear that UDP-glucose, the product of the enzyme, is the most important inhibitor.

UDP-glucose has also been reported to be an inhibitor competitive with UTP for the UDP-glucose pyrophosphorylases from rat liver (Kornfeld, 1965), erythrocyte, heart, and mung bean (Tsuboi *et al.*, 1969) in contrast to the pollen enzyme.

There are two known routes in plants for the production of the cell wall substituents, pentoses and uronides. Figure 9 shows that glucuronosyl and pentosyl nucleotides may be formed via the UDP-glucose pyro-

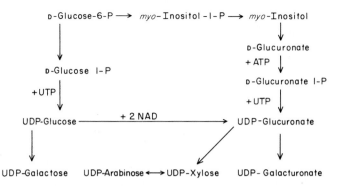

Fig. 9. Synthesis of sugar nucleotide cell wall precursors via the UDP-glucose dehydrogenase and the inositol pathways.

phosphorylase plus UDP-glucose dehydrogenase pathway (Strominger and Mapson, 1957) or via the inositol pathway (Loewus, 1969, 1971, 1973).

Some possible regulatory controls in these pathways have been recognized at other steps in addition to the UDP-glucose pyrophosphorylase reaction by Dickinson *et al.* (1973) and by Davies and Dickinson (1972). UDP-glucose dehydrogenase activity from *Lilium longiflorium* pollen was found to be feedback inhibited by UDP-glucuronate, UDP-galacturonate, and most potently by UDP-xylose. The normally hyperbolic UDP-glucose saturation curve was changed to a sigmoid form in the presence of UDP-xylose.

Preliminary experiments suggest that glucuronate 1-P and UDP-glucuronate may be potent inhibitors of the lilly pollen glucuronate kinase (Dickinson, *et al.*, 1973), thus suggesting feedback regulation of sugar nucleotide synthesis via the inositol pathway at the level of the glucuronate kinase. Both UDP-glucose dehydrogenase and glucuronkinase are the committed steps in their respective pathways (Fig. 9) leading to the uronic acids and pentoses required for cell wall synthesis.

X. Regulation of Nitrogen Metabolism

Most of the work on regulatory enzymes in nitrogen metabolism deals with those associated with amino acid biosynthesis. Since amino acid biosynthesis and its regulation is discussed in Chapter 17, we will cite here selected examples of enzyme regulation in tetrapyrrole, pyrimidine, and purine biosynthesis in photosynthetic organisms.

A. Regulation of Tetrapyrrole Synthesis

1. SUCCINYL-CoA SYNTHASE

In bacteria and mammals it is well documented that this enzyme fulfills an important function by providing succinyl-CoA that is utilized for the synthesis of δ-aminolevulinic acid (ALA). Wider and Tigier (1970) reported that the synthase from soybean callus tissue was inhibited by hemin and that ATP provided a homotropic, cooperative effect on the enzyme as deduced by an $\bar{n}$ value of 1.7 for that substrate. They proposed that such interactions of the enzyme with ATP and hemin help regulate tetrapyrrole synthesis in soybean callus tissue. Although ALA synthase has been reported to occur in tobacco callus tissue, its occurrence in other plant tissues has not been conclusively demonstrated. Consequently the general involvement of succinyl-CoA synthetase in tetrapyrrole synthesis in plants remains to be established.

2. AMINOLEVULINIC ACID (ALA) SYNTHASE FROM PHOTOSYNTHETIC BACTERIA AND PLANTS

The formation of δ-aminolevulinic acid from the condensation of glycine and succinyl-CoA represents the first committed step in the synthesis of tetrapyrroles in many organisms. ALA synthase, which catalyzes this reaction, has been purified from mammals and bacteria including photosynthetic bacteria. It is generally accepted that this enzyme represents a key point in the regulation of tetrapyrrole biosynthesis in these organisms (Lascelles, 1968; Burnham, 1969; Granick and Sassa, 1971).

Gibson et al. (1961) first reported that heme functioned as a negative effector of the enzyme from photosynthetic organisms. Subsequently, Burnham and Lascelles (1963) showed that the enzyme from R. spheroides was inhibited by hemin and heme but not by bacteriochlorophyll and bacteriochlorophyllide.

The R. spheroides enzyme has been purified 1300-fold (Warnick and Burnham, 1971). Its molecular weight was estimated to be 57,000 and the K_m values for glycine and succinyl-CoA were, respectively, 10 mM and 25 μM. Plots of rate as a function of substrate concentration were hyperbolic for both succinyl-CoA and glycine. Hemin was a potent inhibitor of the purified enzyme and reduced enzyme activity over 50% at 5 μM. Although significant inhibition occurred at 1 μM hemin, complete inhibition was not achieved by higher hemin concentrations (Burnham, 1969). Nevertheless, the sensitivity of ALA synthase to inhibition by very low concentrations of hemin is in keeping with its proposed function of regulating tetrapyrrole synthesis (Lascelles and Hatch, 1969; Burnham and Lascelles, 1963). Porra *et al.* (1972) have examined the nature of hemin inhibition of ALA synthase and propose that inhibition is caused by formation of a coordination complex between the enzyme and the iron of the hemin.

Mechanisms other than feedback inhibition by hemin appear to help regulate activity of the enzyme in *R. spheroides*. Cell-free preparations of the bacterium contain a low molecular weight, light sensitive inhibitor of ALA synthase that is distinct from hemin, protoporphyrin, and bacteria chlorophyll (Tuboi *et al.*, 1969). Such preparations also yielded a low molecular weight activator of the enzyme (Neuberger *et al.*, 1973). Both active and inactive forms of the synthase occur in extracts of the bacterium. Treatments, such as dialysis and gel filtration, that remove small molecular compounds convert the inactive form to an active form. In addition two forms of the synthase have been separated and purified to homogeneity (Tuboi *et al.*, 1970a; Fanica-Gaignier and Clement-Metral, 1973a). They appear to have similar molecular weights and both are repressible by high O_2 tensions. However, the synthesis of one form (F-I) is induced by low O_2 tensions, and induction of the other form (F-II) requires both low O_2 tension and illumination. Only the F-I form exists in active and inactive forms (Tuboi *et al.*, 1970b). Claims have been made that the F-I enzyme is a soluble, cytoplasmic form associated with dark respiratory metabolism and that the F-II form, located in the chromatophore, is associated with photometabolism (Fanica-Gaignier and Clement-Metral, 1973c). Such phenomenon are undoubtedly associated with mechanisms that help regulate tetrapyrrole synthesis in *R. spheroides*, and their precise functions should become clear with additional investigation.

Correlations have been sought between tetrapyrrole synthesis and the adenylate pool in photosynthetic bacteria. When conditions were invoked in cultures of *R. spheroides* to cause an initial lag in the synthesis of bacteriochlorophyll, a sharp drop in intracellular ATP content preceded onset of bacteriochlorophyll synthesis. A relationship between increased rate of bacteriochlorophyll synthesis and reduced intracellular ATP was seen, but no correlation was observed between energy charge and bacteriochlorophyll synthesis (Fanica-Gaignier *et al.*, 1971). Interestingly enough, the negative correlation between reduced intracellular ATP and onset of chlorophyll synthesis was not observed in *R. rubrum* (Oelze and Kamen, 1971). In an extension of these studies with *R. spheroides*, it was reported that 1 m*M* ATP inhibited ALA synthase 72–80%, whereas, at the same concentration, ADP inhibited only 10–14% and AMP was without effect. It has been suggested that ATP inhibits by acting on a thiol group at the active center of the enzyme (Fanica-Gaig-nier and Clement-Metral, 1973b). It is not known if ATP is specific in its effect, since other nucleoside triphosphates were not tested. It was nevertheless proposed that in *R. spheroides* ATP imparts a regulatory function on photopigment production at the step catalyzed by ALA syn-thase (Fanica-Gaignier and Clement-Metral, 1971). Presumably, the sig-nificance of regulating bacteriochlorophyll production by ATP is associ-ated with the role that the photopigment has in the production of ATP by photophosphorylation. However, the existence of such a regulatory system in *R. spheroides* requires more evidence that is presently available.

In plants, although there is good evidence that ALA is a precursor for tetrapyrroles there is little information on the manner in which it is produced. Unlike the bacterial systems, glycine and succinate are poor precursors of ALA in several plants, and the enzymatic formation of ALA from succinyl-CoA and glycine reported in extracts of tobacco callus tis-sue (Wider *et al.*, 1971) has not been conclusively demonstrated in other plant tissues (Beale and Castelfranco, 1974). The observation that gluta-mate and related 5-carbon compounds are effective precursors of ALA in bean, barley, and cucumber lends support to the proposal that ALA synthesis in plants occurs by a pathway different from that which occurs in bacteria and mammals (Beale and Castelfranco, 1974).

3. ALA Dehydratase from Photosynthetic Bacteria and Plants

The conversion of ALA to porphobilinogen which is catalyzed by ALA dehydratase may represent a second point of control of pyrrole syn-thesis in *R. spheroides*. This is supported by the observation that the enzyme from this bacterium was inhibited by hemin and protoporphyrin and activated by Li^+, NH_4^+, Rb^+, and K^+ (Nandi *et al.*, 1968). In the presence of K^+, the enzyme formed an equilibrium mixture of monomer, dimer, and trimer species, all enzymatically active (Nandi and Shemin, 1968). Mn^{2+} and Mg^{2+} at low concentration activated, but these ions be-came inhibitory at high concentrations. Na^+ also activated the enzyme, but functioned in a manner distinct from that of the other four mono-valent cations since it did not promote association of the monomer. In contrast, the dehydratase from *R. capsulata* did not require metal cations, was not inhibited by hemin, and showed no homotropic cooperative effects with the substrate as did the *R. spheroides* enzyme (Nandi *et al.*, 1968; Nandi and Shemin, 1973).

The enzyme also has been found in several plants. However, in con-trast to the *R. spheroides* enzyme, the partially purified ALA dehydratase from tobacco was not inhibited by hemin, protoporphyrin IX, or cytochrome c (Shetty and Miller, 1969a). Mg^{2+} was essential for activity, and Mn^{2+}, in the absence of Mg^{2+}, stimulated at low concentrations but inhibited at high concentrations. At pH 7.4, the tobacco enzyme was virtually inactive in Tris hydrochloride buffer and was slightly active in Tris buffer at pH 8 (Shetty and Miller, 1969b). On the other hand, the enzyme from wheat leaves was active in Tris buffer at pH 7.5, not activated by orthophosphate, and completely inhibited by 3.3 mM ATP (Nandi and Waygood, 1967). The suggestion by Nandi and Waywood that the ATP effect might be associated with a regulatory function deserves further investigation. The ATP effect may in part be associated with its capacity to bind divalent cations, which are needed for optimal dehydratase activity since metal chelators, such as EDTA and pyrophosphate, also were strongly inhibitory.

4. EVIDENCE FOR *in Vivo* REGULATION OF TETRAPYRROLE SYNTHESIS IN PHOTOSYNTHETIC BACTERIA

Control *in vivo* of porphyrin synthesis by the regulatory enzymes described above was studied in *R. spheroides* by Burnham and Lascelles (1963) who found that, in cell suspensions, exogenous heme reduced synthesis of porphyrin when glycine and α-ketoglutarate were used as precursors. However, when ALA was used, heme did not inhibit porphyrin synthesis. Such results suggest that regulation of tetrapyrrole synthesis by heme occurred *in vivo* at the step catalyzed by ALA synthase but not at the step catalyzed by ALA dehydratase. These results, therefore, do not support the view that regulatory properties reported for ALA dehydratase (Nandi *et al.*, 1968; Nandi and Shemin, 1968) are involved in controlling tetrapyrrole synthesis *R. spheroides*. Further studies may reveal a regulatory function of the latter enzyme in this bacterium.

5. INTERACTION OF METABOLIC SEQUENCES IN TETRAPYRROLE AND AMINO ACID SYNTHESIS

In intermediary metabolism, there is no sequence of reactions that can function independently of other reaction sequences. Activity of a given pathway is affected by the activity of pathways with which it interacts. Interaction of metabolic sequences, therefore, represents an important mechanism that helps regulate metabolism. Of the many such interactions that occur in the plant, the interaction between amino acid metabolism and tetrapyrrole synthesis may be used as an example. Methionine is required for the synthesis of chlorophyll and is involved in the methylation of magnesium protoporphyrin to yield its monomethyl ester (Gibson *et al.*, 1963; Granick and Sassa, 1971). Since methionine is produced by a pathway common to the synthesis of the aspartate family of amino acids (see Chapter 18), metabolites that regulate synthesis of this group of amino acids likewise can affect the synthesis of chlorophyll. That such regulatory interactions apparently occur *in vivo* is suggested by the work of Gibson and his colleagues (Gibson *et al.*, 1962a,b) who found that addition of threonine to suspensions of *R. spheroides* inhibited bacteriochlorophyll synthesis and caused accumulation of coproporphyrin. Addition of either homocysteine or methionine restored bacteriochlorophyll synthesis. Apparently threonine had inhibited the synthesis of methionine and made it unavailable for the methylation of magnesium protoporphyrin. *In vitro*, threonine was a potent inhibitor of homoserine dehydrogenase, one of the enzymes that regulates production of methionine and other members of the aspartate family of amino acids. Consequently, the threonine effect was related to inhibition of homoserine dehydrogenase (Gibson *et al.*, 1962b).

B. Pyrimidine Biosynthesis

1. Carbamylphosphate Synthase from Plants

The reaction catalyzed by this enzyme carries out an important function of bringing CO_2 and amino nitrogen into organic combination. Carbamyl phosphate, the product of the reaction catalyzed by the enzyme, is a precursor for the synthesis of arginine and the pyrimidines. Depending upon the source of the enzyme, glutamine, ammonia, or both serve as the amino donor. O'Neal and Naylor (1968, 1969) partially purified the glutamine-dependent enzyme from pea seedlings and found that it was inhibited by various nucleotides, including AMP, ADP, UMP, UDP, UTP, and GTP. In common with the enzyme from certain nonphotosynthetic bacteria, UMP was the most potent inhibitor of the plant enzyme. Ornithine partially reversed the inhibition of the pea enzyme by UMP. In contrast, ornithine not only completely reversed UMP inhibition, it activated the enzyme from *E. coli* and *S. typhimurium*. Moreover, in contrast to the effects on the pea enzyme, AMP activated the *E. coli* enzyme (Abd-El-Al and Ingraham, 1969; Anderson and Meister, 1966).

2. Aspartate Transcarbamylase from Plants

The enzyme catalyzes the conversion of carbamyl phosphate and aspartate to yield carbamyl aspartate and represents the first enzyme in the pathway for pyrimidine synthesis. Work on this regulatory enzyme from *E. coli* helped demonstrate the concept of metabolic control by feedback inhibition (Gerhart and Pardee, 1962). A similar regulatory enzyme was therefore sought and subsequently demonstrated in higher plants (Neumann and Jones, 1962). Investigations on the enzyme from lettuce seedlings revealed that it was inhibited by CMP, UMP, UDP, and UTP; purine nucleotides were not inhibitory. UMP was the most effective inhibitor of the several compounds tested, inhibiting 90% at 5 mM. Such properties have been reported for the enzyme from other plant sources (Neumann and Jones, 1964; Johnson *et al.*, 1973). Succinate was also an inhibitor of the lettuce and cowpea enzyme (Neumann and Jones, 1962; Johnson *et al.*, 1973). In photosynthetic as well as other organisms, the reactions catalyzed by this enzyme and carbamylphosphate synthase represent points of control of pyrimidine synthesis by regulatory enzymes (Fig. 10).

3. COMPENSATORY CONTROL IN PYRIMIDINE AND ARGININE BIOSYNTHESIS

Effective mechanisms are needed to regulate the interacting pathways for arginine and pyrimidine biosynthesis (Fig. 10) particularly when cellular demands for amino acids and the pyrimidines differ. An interesting type of compensatory control of carbamylphosphate synthase is proposed to function in such situations in bacteria (Stadtman, 1970). For example, if the demand for pyrimidine nucleotides is low and requirement for arginine is high, the nucleotides will accumulate and inhibit the enzyme. Reduced synthesis of carbamyl phosphate will occur and will limit the synthesis of arginine. However, the limited availability of carbamyl phosphate leads to an accumulation of ornithine, which will re-

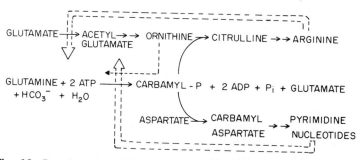

Fig. 10. Regulation of carbamylphosphate synthase by compensatory control through antagonism of end product inhibition. This scheme is adopted from Stadtman (1970) who also describes other types of compensatory control mechanisms.

verse pyrimidine nucleotide inhibition and restore carbamyl phosphate production. Production of the pyrimidine nucleotides will still remain curtailed, since these metabolites feedback inhibit aspartate transcarbamylase. Thus carbamyl phosphate synthesized under such conditions will be utilized mainly for the synthesis of arginine. The synthesis of arginine nevertheless remains under control, since accumulation of the amino acid will feedback inhibit N-acetylglutamate synthetase and limit ornithine production.

Such a proposed mechanism may function in plants, since the regulatory properties of the plant carbamylphosphate synthase and aspartate transcarbamylase resemble those of the bacterial enzymes. Moreover, it has recently been reported that the conversion of glutamate to acetyl glutamate in radish and *Chlorella* is feedback inhibited by both arginine and ornithine (Morris and Thompson, 1971; C. F. Morris and J. F. Thompson, personal communication). Thus, to represent the probable situation that exists in plants, the scheme shown for bacteria in Fig. 10 would include ornithine as a feedback inhibitor of acetyl glutamate synthesis. The plant system would seem to be more sensitive to modulation by ornithine, since the amino acid not only relieves pyrimidine nucleotide inhibition of carbamyl phosphate synthesis, it also feedback inhibits acetyl glutamate synthesis.

C. Purine Biosynthesis

1. Adenine Phosphoriboslytransferase from Plants

The enzyme that catalyzes the reaction between purine and 5-phosphoribosyl pyrophosphate to yield purine nucleotide and pyrophosphate may function in a secondary pathway for purine nucleotide synthesis in the plant. Presumably, in some bacteria purines that are freed from nucleotides by hydrolytic cleavage are salvaged by this system for reuse in nucleic acid synthesis. The enzyme from soy bean callus tissue was inhibited 92% by 0.62 mM AMP and activated 21% by 60 μM ATP (Nicholls and Murray, 1968). These properties resemble those reported for the enzyme from Ehrlich ascites tumor cells (Murray, 1967). Such regulatory properties may indicate a significant, as yet undefined, role of the enzyme in relation to purine metabolism in the plant.

We have already described some of the interesting regulatory properties of the glutamine phosphoribosylpyrophosphate amidotransferase, which catalyzes the first committed step in the principal pathway of purine synthesis in microorganisms (Gots, 1971; Wyngaarden, 1972). Information is presently lacking on the occurrence of a similar enzyme in plants. If such an enzyme does exist in plants, its role in regulating purine biosynthesis likely would be a significant one.

XI. Summary

Sections VII through X have provided a number of examples of regulatory controls of enzyme activity *in vitro*. Many of these cases seem to illustrate regulation of enzyme activity by energy charge (ATP, ADP, AMP, and P_i), by reduced pyridine nucleotides, feedback regulation by end products of pathways, activation by important metabolites in pathways, etc. The results presented are consistent with the notions of the functions of these enzymes in their metabolic pathways. However, there is very little direct evidence that the *in vitro* studies are relevant to the actual *in vivo* situation or if they have physiological importance. With plant systems, only in a few cases have attempts been made to correlate metabolite change with change in enzyme activity. Thus, the direction of research efforts in the future may be to develop the various techniques required for such demonstration. Perhaps a system where mutants of various organisms or plant cells (cell culture?) having enzymes with altered regulatory properties can be developed in the future. An important problem in developing such systems would be concerned with detection of such mutants. Mutant strains of the unicellular green alga *Chlamydomonas reinhardii* have aided in the understanding of the sequence of the photosynthetic electron transport system (Levine, 1968). One mutant of the above alga which lacks the capacity for photosynthetic CO_2 fixation was found to lack ribulose-1,5-diP carboxylase activity (Levine and Togasaki, 1965).

GENERAL REFERENCES

References on Regulation by Compartmentation

Chapell, J. B., and Robinson, B. H. (1967). *Biochem. Soc. Symp.* **27**, 123.

Fritz, I. B. (1968). *In* "Cellular Compartmentalisation and Control of Fatty Acid Metabolism" (F. C. Gran, ed.), p. 40. Academic Press, New York.

Lardy, H. A., Paetkan, V., and Walter, P. (1965). *Proc. Nat. Acad. Sci. U.S.* **53**, 1410.

Lehninger, A. L. (1964). "The Mitochondrion." Benjamin, New York.

Lowenstein, J. M. (1967). *Biochem. Soc. Symp.* **27**, 61.

Srere, P. A. (1967). *Biochem. Soc. Symp.* **27**, 11.

Srere, P. A., and Mosbach, K. (1974). *Annu. Rev. Microbiol.* **28**, 61.

References on Control of Protein Synthesis in Animals

Arias, I., Doyle, D., and Shimke, R. T. (1969). *J. Biol. Chem.* **244**, 3303.

Goldberg, A. L., and Dice, J. F. (1974). *Annu. Rev. Biochem.* **43**, 835.

Shimke, R. T. (1964). *J. Biol. Chem.* **239**, 3808.
Shimke, R. T. (1967). *Curr. Top. Cell. Regul.* **1**, 77–124.

References on Regulation of Bacterial Protein Synthesis
1. Feedback Repression
Ames, B. N., and Garry, B. (1959). *Proc. Nat. Acad. Sci. U.S.* **45**, 1453.
Ames, B. N., and Martin, R. G. (1964). *Annu. Rev. Biochem.* **33**, 235.
Englesberg, E. (1971). *In* "Metabolic Pathways" (H. J. Vogel, ed.), 3rd ed., Vol. 5, pp. 257–296. Academic Press, New York.
Freundlich, M., Burns, R. V., and Umbarger, H. E. (1962). *Proc. Nat. Acad. Sci. U.S.* **48**, 1804.
Vogel, H. J., and Vogel, R. H. (1967). *Annu. Rev. Biochem.* **36**, 519.

2. Substrate Induction
Canovas, J. L., and Stanier, R. Y. (1967). *Eur. J. Biochem.* **1**, 289.
Cohen, M., and Monod, J. (1953). *Symp. Soc. Gen. Microbiol.* **2**, 132.
Jacob, F. (1966). *Science* **152**, 1470.
Monod, J. (1966). *Science* **154**, 475.
Stanier, R. Y., Hegeman, G. D., and Ornston, L. N. (1963). *Collog. Int. Cent. Nat. Rech. Sci.* p. 228.
Vogel, H. J. (1957). *In* "The Chemical Basis of Heredity" (W. D. McElroy and B. Glass, eds.), p. 276. Johns Hopkins Press, Baltimore, Maryland.

3. Catabolite Repression
DeCrombrugghe, B., Varmus, H. E., Perlman, R. L., and Pastan, I. (1969). *J. Biol. Chem.* **244**, 5828.
Magasanik, B. (1961). *Cold Spring Harbor Symp. Quant. Biol.* **26**, 249.
Pastan, I., and Perlman, R. L. (1968). *Proc. Nat. Acad. Sci. U.S.* **61**, 1336.
Perlman, R. L., and Pastan, I. (1968). *J. Biol. Chem.* **243**, 5420.
Rickenberg, H. V. (1974). *Annu. Rev. Microbiol.* **28**, 353.
Ullman, A., and Monod, J. (1968). *FEBS (Fed. Eur. Biochem. Soc.) Lett.* **2**, 57.
Zubay, G., and Chambers, D. A. (1971). *In* "Metabolic Pathways" (H. J. Vogel, ed.), 3rd ed., Vol. 5, pp. 297–347. Academic Press, New York.

REFERENCES

Abd-El-Al, A., and Ingraham, J. L. (1969). *J. Biol. Chem.* **244**, 4033.
Abd-El-Al, A. T. H., and Schlegel, H. G. (1974). *Biochem. J.* **139**, 481.
Adler, S. P., Purich, D., and Stadtman, E. R. (1975). *J. Biol. Chem.* **250**, 6264.
Alpaslan, O., and Mudd, J. B. (1970). *Plant Physiol.* **45**, 255.
Anderson, L. E., and Fuller, R. C. (1967). *Plant Physiol.* **42**, 4917.
Anderson, P. M., and Meister, A. (1966). *Biochemistry* **5**, 3164.
Anderson, W. B., and Stadtman, E. R. (1970). *Biochem. Biophys. Res. Commun.* **41**, 704.
Anderson, W. B., and Stadtman, E. R. (1971). *Arch. Biochem. Biophys.* **143**, 482.
Anderson, W. B., Hennig, S. B., Ginsburg, A., and Stadtman, E. R. (1970). *Proc. Nat. Acad. Sci. U.S.* **67**, 1417.
Atkinson, D. E. (1966). *Annu. Rev. Biochem.* **35**, 85.
Atkinson, D. E. (1968a). *Biochem. Soc. Symp.* **27**, 23.
Atkinson, D. E. (1968b). *Biochemistry* **7**, 4030.
Atkinson, D. E. (1969). *Annu. Rev. Microbiol.* **23**, 47.
Atkinson, D. E. (1970). *In* "The Enzymes" (P. D. Boyer, ed.), 3rd ed., Vol. 1, pp. 461–489. Academic Press, New York.

Atkinson, D. E. (1971). *In* "Metabolic Pathways" (H. J. Vogel, ed.), 3rd ed., Vol. 5, pp. 1–21. Academic Press, New York.

Atkinson, D. E., and Walton, G. M. (1967). *J. Biol. Chem.* **242**, 3239.

Atkinson, D. E., Hathaway, J. A., and Smith, E. C. (1965). *J. Biol. Chem.* **240**, 2682.

Axelrod, B., and Beevers, H. J. (1972). *Biochim. Biophys. Acta* **256**, 175.

Bassham, J. A., and Jensen, R. G. (1967). *In* "Harvesting the Sun" (A. S. San Pietro, F. A. Greer, and T. J. Army, eds.), pp. 79–100. Academic Press, New York.

Beale, S. I., and Castelfranco, P. A. (1974). *Plant Physiol.* **23**, 297.

Bianchetti, R., and Sartirana, M. L. (1967). *Biochem. Biophys. Res. Commun.* **27**, 378.

Bianchetti, R., and Sartirana, M. L. (1968). *Life Sci.* **7**, 121.

Bird, I. F., Cornelius, M. J., Keys, A. J., and Whittingham, C. P. (1974). *Phytochemistry* **13**, 59.

Blangy, D., Buc, H., and Monod, J. (1968). *J. Mol. Biol.* **31**, 13.

Bloxham, D. P., and Lardy, H. A. (1973). *In* "The Enzymes" (P. D. Boyer, ed.), 3rd ed., Vol. 8, pp. 240–278. Academic Press, New York.

Bogin, E., and Wallace, A. (1966). *Biochim. Biophys. Acta* **128**, 190.

Bomsel, J. L., and Pradet, A. (1968). *Biochim. Biophys. Acta* **162**, 230.

Borriss, R., and Ohmann, E. (1972). *Biochem. Physiol. Pflanzen.* **163**, 328.

Breidenbach, R. W., and Beevers, H. J. (1967). *Biochem. Biophys. Res. Commun.* **27**, 462.

Brostrom, C. O., and Hunkeler. F. L., and Krebs, E. G. (1971). *J. Biol. Chem.* **246**, 1961.

Brown, K. D., and Doy, C. H. (1963). *Biochim. Biophys. Acta* **77**, 170.

Brown, K. D., and Doy, C. H. (1966). *Biochim. Biophys. Acta* **118**, 157.

Brown, M. S., Segal, A., and Stadtman, E. R. (1971). *Proc. Nat. Acad. Sci. U.S.* **68**, 2949.

Buchanan, B. B., and Evans, M. C. W. (1965). *Biochem. Biophys. Res. Commun.* **22**, 484.

Buchanan, B. B., Evans, M. C. W., and Arnon, D. I. (1967). *Arch. Mikrobiol.* **59**, 32.

Burnham, B. F. (1969). *In* "Metabolic Pathways" (D. M. Greenberg, ed.), 3rd ed., Vol. 3, pp. 403–537. Academic Press, New York.

Burnham, B. F., and Lascelles, J. (1963). *Biochem. J.* **87**, 462.

Canvin, D. T., and Beevers, H. J. (1961). *J. Biol. Chem.* **236**, 988.

Cardini, C. E., Leloir, L. F., and Chiriboga, J. (1955). *J. Biol. Chem.* **214**, 149.

Casky, C. T., Ashton, D. M., and Wyngaarden, J. B. (1964). *J. Biol. Chem.* **239**, 2570.

Changeux, J. P. (1961). *Cold Spring Harbor Symp. Quant. Biol.* **26**, 313.

Chapman, A. G., Fall, L., and Atkinson, D. E. (1971). *J. Bacteriol.* **108**, 1072.

Clark, M. G., and Lardy, H. A. (1975). *In* "Biochemistry of Carbohydrates" (W. J. Whelan, ed.), MTP Int. Biochem. Ser. 1, Vol. 5, pp. 224–266. Buttersworth, London and Raven Press, Baltimore, Maryland.

Cohen, G. N. (1965). *Annu. Rev. Microbiol.* **19**, 419.

Cohen, G. N. (1969). *Curr. Top. Cell. Regul.* **1**, 183.

Conway, A, and Koshland, D. E. (1968). *Biochemistry* **7**, 4011.

Cornish-Bowden, A., and Koshland, D. E., Jr. (1970). *J. Biol. Chem.* **245**, 6241.

Corwin, L. M., and Fanning, G. R. (1968). *J. Biol. Chem.* **243**, 3517.

Coultate, T. P., and Dennis, D. T. (1969). *Eur. J. Biochem.* **7**, 153.

Cox, G. F., and Davies, D. D. (1967). *Biochem. J.* **105**, 729.

Cox, G. F., and Davies, D. D. (1969). *Biochem. J.* **113**, 813.

Cutinelli, C., Ehrensvard, G., Reio, L., Saluste, E., and Stjernholm, R. (1951). *Ark. Kemi* **3**, 315.

Datta, P. (1966). *J. Biol. Chem.* **241**, 5836.

Datta, P. (1969). *Science* **165**, 556.

Datta, P., and Gest, H. (1964). *Nature (London)* **203**, 1259.

Datta, P., and Prakash, L. (1966). *J. Biol. Chem.* **241**, 5827.

Davie, O. E. W., Hougie, C., and Lundblad, R. L. (1969). *In* "Recent Advances in Blood Coagulation" (L. Pollen, ed.), p. 13. Churchill, London.

Davies, M. D., and Dickinson, D. B. (1972). *Arch. Biochem. Biophys.* **152**, 53.

De Fekete, M. A. R., and Cardini, C. E. (1964). *Arch. Biochem. Biophys.* **104**, 173.

Della Corte, E., and Stirpe, F. (1968). *Biochem. J.* **108**, 349.

Delmer, D. P. (1972). *J. Biol. Chem.* **247**, 3822.

Dennis, D. T., and Coultate, T. P. (1966). *Biochem. Biophys. Res. Commun.* **25**, 187.

Dennis, D. T., and Coultate, T. P. (1967a). *Life Sci.* **6**, 2353.

Dennis, D. T., and Coultate, T. P. (1967b). *Biochim. Biophys. Acta* **146**, 129.

Dickinson, D. B., and Preiss, J. (1969a). *Arch. Biochem. Biophys.* **130**, 119.

Dickinson, D. B., and Preiss, J. (1969b). *Plant Physiol.* **44**, 1058.

Dickinson, D. B., Hopper, J. E., and Davies, M. D. (1973). *In* "Biogenesis of Plant Cell Wall Polysaccharides" (F. Loewus, ed.), p. 29. Academic Press, New York.

Doi, A., Doi, K., and Nikuni, Z. (1964). *Biochim. Biophys. Acta* **92**, 628.

Doy, C. H., and Brown, K. D. (1965). *Biochim. Biophys. Acta* **104**, 377.

Duggleby, R. G., and Dennis, D. T. (1973a). *Plant Physiol.* **52**, 312.

Duggleby, R. G., and Dennis, D. T. (1973b). *Arch. Biochem. Biophys.* **155**, 270.

Ebner, E., Wolf, D., Gancedo, C., Elsasser, S., and Holzer, H. (1970). *Eur. J. Biochem.* **14**, 535.

Eidels, L., and Preiss, J. (1970a). *J. Biol. Chem.* **245**, 2937.

Eidels, L., and Preiss, J. (1970b). *Arch. Biochem. Biophys.* **140**, 75.

Eidels, L., Edelmann, P. L., and Preiss, J. (1970). *Arch. Biochem. Biophys.* **140**, 60.

Entner, N., and Doudoroff, M. (1952). *J. Biol. Chem.* **196**, 853.

Espada, J. (1962). *J. Biol. Chem.* **237**, 3577.

Exton, J. H., and Park, C. R. (1969). *J. Biol. Chem.* **244**, 1424.

Fanica-Gaignier, M., and Clement-Metral, J. D. (1971). *Biochem. Biophys. Res. Commun.* **44**, 192.

Fanica-Gaignier, M., and Clement-Metral, J. (1973a). *Eur. J. Biochem.* **40**, 13.

Fanica-Gaignier, M., and Clement-Metral, J. (1973b). *Eur. J. Biochem.* **40**, 19.

Fanica-Gaignier, M., and Clement-Metral, J. (1973c). *Biochem. Biophys. Res. Commun.* **55**, 610.

Fanica-Gaignier, M., Clement-Metral, J. D., and Kamen, M. D. (1971). *Biochim. Biophys. Acta* **226**, 135.

Frieden, C. (1967) *J. Biol. Chem.* **242**, 4045.

Feingold, D. S., Neufeld, E. F., and Hassid, W. Z. (1958). *J. Biol. Chem.* **233**, 783.

Ferdinand, W. (1967). *Biochem. J.* **98**, 278.

Fischer, E. H., Pocker, A., and Saari, J. C. (1970). *Essays Biochem.* **6**, 23–68.

Fischer, E. H., Heilmeyer, L. M. G., and Haschke, R. H. (1971). *Curr. Top. Cell. Regul.* **4**, 211–251.

Fraenkel, D. G., Pontremoli, S., and Horecker, B. L. (1966). *Arch. Biochem. Biophys.* **114**, 4.

Frieden, C. (1967). *J. Biol. Chem.* **242**, 4045.

Friedman, D. I., and Larner, J. (1963). *Biochemistry* **2**, 669.
Frydman, R. B., and Cardini, C. E. (1964). *Biochem. Biophys. Res. Commun.* **17**, 407.
Furlong, C. E., and Preiss, J. (1969a). *Progr. Photosyn. Res.* **3**, 1604–1617.
Furlong, C. E., and Preiss, J. (1969b). *J. Biol. Chem.* **241**, 2539.
Gale, N. L., and Beck, J. V. (1966). *Biochem. Biophys. Res. Commun.* **24**, 792.
Garrad, L. A., and Humphreys, T. E. (1968). *Phytochemistry* **7**, 1949.
Gentner, N., and Preiss, J. (1967). *Biochem. Biophys. Res. Commun.* **27**, 417.
Gentner, N., and Preiss, J. (1968). *J. Biol. Chem.* **243**, 5882.
Gerhart, J. C., and Pardee, A. B. (1962). *J. Biol. Chem.* **237**, 891.
Gerhart, J. C., and Schachman, H. K. (1965). *Biochemistry* **4**, 1054.
Ghosh, H. P., and Preiss, J. (1965a). *Biochemistry* **4**, 1354.
Ghosh, H. P., and Preiss, J. (1965b). *J. Biol. Chem.* **240**, 960.
Ghosh, H. P., and Preiss, J. (1966). *J. Biol. Chem.* **241**, 4491.
Ghosh, S., Blumenthal, H. J., Davidson, E., and Roseman, S. (1960). *J. Biol. Chem.* **235**, 1265.
Gibbs, M., Latzko, E., Everson, R. G., and Cockburn, W. (1967). *In* "Harvesting the Sun" (A. San Pietro, F. A. Green, and T. J. Army, eds.), pp. 111–130. Academic Press, New York.
Gibson, K. D., Matthew, M., Neuberger, A., and Tait, G. H. (1961). *Nature (London)* **192**, 204.
Gibson, K. D., Neuberger, A., and Tait, G. H. (1962a). *Biochem. J.* **83**, 550.
Gibson, K. D., Neuberger, A., and Tait, G. H. (1962b). *Biochem. J.* **84**, 483.
Gibson, K. D., Neuberger, A., and Tait, G. H. (1963). *Biochem. J.* **88**, 325.
Ginsburg, A., and Stadtman, E. R. (1973). *In* "The Enzymes of Glutamine Metabolism" (E. R. Stadtman and S. Prusiner, eds.), p. 9. Academic Press, New York.
Gots, J. S. (1971). *In* "Metabolic Pathways" (H. J. Vogel, ed.), 3rd ed., Vol. 5, pp. 225–255. Academic Press, New York.
Govons, S., Gentner, N., Greenberg, E., and Preiss, J. (1973). *J. Biol. Chem.* **248**, 1731.
Granick, S., and Sassa, S. (1971). *In* "Metabolic Pathways" (H. J. Vogel, ed.), 3rd ed., Vol. 5, pp. 77–141. Academic Press, New York.
Graves, D. J., and Wang, J. H. (1972). *In* "The Enzymes" (P. D. Boyer, ed.), 3rd ed., Vol. 7, pp. 435–482. Academic Press, New York.
Haber, J. E., and Koshland, E. E. (1967). *Proc. Nat. Acad. Sci. U.S.* **58**, 2087.
Haq, S., and Hassid, W. Z. (1965). *Plant Physiol.* **40**, 591.
Haschke, R. H., Heilmeyer, L. M. G., Meyer, F., and Fischer, E. H. (1970). *J. Biol. Chem.* **245**, 6657.
Hassid, W. Z. (1967). *Annu. Rev. Plant Physiol.* **18**, 253.
Hassid, W. Z. (1969). *Science* **165**, 137.
Hathaway, J. A., and Atkinson, D. E. (1965). *Biochem. Biophys. Res. Commun.* **20**, 661.
Hawker, J. S. (1966). *Phytochemistry* **5**, 1191.
Hawker, J. S. (1967). *Biochem. J.* **105**, 943.
Hawker, J. S., and Hatch, M. D. (1966). *Biochem. J.* **99**, 102.
Heber, U. W. (1967). *In* "The Biochemistry of Chloroplasts" (T. W. Goodwin, ed.), Vol. 2, pp. 71–78. Academic Press, New York.
Heber, U. W., and Santarius, K. A. (1965). *Biochim. Biophys. Acta* **109**, 390.
Heber, U. W., and Santarius, K. A. (1970). *Z. Naturforsch. B* **25**, 718.
Heber, U. W., Hallier, U. W., and Hudson, M. A. (1967). *Z. Naturforsch. B* **22**, 1200.

Heilmeyer, L., Battig, F., and Holzer, H. (1969). *Eur. J. Biochem.* **9**, 259.

Heilmeyer, L. M. G., Meyer, F., Haschke, R. H., and Fischer, E. H. (1970). *J. Biol. Chem.* **245**, 6649.

Hennig, S. B,, Anderson, W. F., and Ginsburg, A. (1970). *Proc. Nat. Acad. Sci. U.S.* **67**, 1761.

Herriott, R. M. (1938). *J. Gen. Physiol.* **21**, 501.

Hers, H. G., DeWulf, H., and Stalmans, W. (1970a). *FEBS (Fed. Eur. Biochem. Soc.) Lett.* **12**, 73.

Hers, H. G., DeWulf, H., Stalmans, W., and Van den Burghe, G. (1970b). *Advan. Enzyme Regul.* **8**, 171.

Hill, A. J. (1913). *Biochem. J.* **7**, 471.

Holzer, H. (1969). *Advan. Enzymol.* **32**, 297.

Holzer, H., and Duntze, W. (1971). *Annu. Rev. Biochem.* **40**, 345.

Hopper, J. E., and Dickinson, D. B. (1972). *Arch. Biochem. Biophys.* **148**, 523.

Hubbard, J. S., and Stadtman, E. R. (1967). *J. Bacteriol.* **94**, 1016.

Jaangard, N. O., Unkeless, J., and Atkinson, D. E. (1968). *Biochim. Biophys. Acta* **151**, 225.

Jackson, J. B., and Crofts, A. R. (1968). *Biochem. Biophys. Res. Commun.* **32**, 908.

Johnson, E. J. (1966). *Arch. Biochem. Biophys.* **114**, 178.

Johnson, E. J., and Peck, H. D., Jr. (1965). *J. Bacteriol.* **89**, 1041.

Johnson, L. B., Niblett, C. L., and Shively, O. D. (1973). *Plant Physiol.* **51**, 318.

Johnson, T. F. and Meeuse, B. J. D. (1972). *Proc., Kon. Ned. Akad. Wetensch., Ser. C* **75**, 1.

Kanazawa, T., Kanazawa, K., Kirk, M. R., and Bassham, J. A. (1972). *Biochim. Biophys. Acta* **256**, 656.

Kelly, G. T., and Turner, J. F. (1969a). *Biochem. Biophys. Res. Commun.* **30**, 195.

Kelly, G. T., and Turner, J. F. (1969b). *Biochem. J.* **115**, 481.

Kelly, G. T., and Turner, J. F. (1970). *Biochim. Biophys. Acta* **208**, 360.

Kikuchi, G., Tsuiki, S., Muto, A., and Yamada, H. (1963). *In* "Microalgae and Photosynthetic Bacteria" (Japanese Society of Plant Physiologists, eds.), p. 547. Univ. of Tokyo Press, Tokyo.

Kingdon, H. S., Shapiro, B. M., and Stadtman, E. R. (1967). *Proc. Nat. Acad. Sci. U.S.* **58**, 1703.

Kirtley, M. E., and Koshland, D. E. (1967). *J. Biol. Chem.* **242**, 4192.

Klungsøyr, L., Hageman, J. H., Fall, L., and Atkinson, D. E. (1968). *Biochemistry* **7**, 4035.

Kornberg, H. L., and Beevers, H. J. (1967). *Nature (London)* **180**, 35.

Kornfeld, R. (1967). *J. Biol. Chem.* **242**, 3135.

Kornfeld, S. (1965). *Fed. Proc., Fed. Amer. Soc. Exp. Biol.* **24**, 536.

Kornfeld, S., Kornfeld, R., Neufeld, E. G., and O'Brien, P. J. (1964). *Proc. Nat. Acad. Sci. U.S.* **52**, 371.

Koshland, D. E., Jr. (1958). *Proc. Nat. Acad. Sci. U.S.* **44**, 98.

Koshland, D. E., Jr. (1959). *J. Cell. Comp. Physiol.* **54**, 245.

Koshland, D. E., Jr. (1963). *Cold Spring Harbor Symp. Quant. Biol.* **28**, 473.

Koshland, D. E., Jr. (1969). *Curr. Top. Cell. Regul.* **1**, 1–27.

Koshland, D. E., Jr. (1970). *In* "The Enzymes" (P. D. Boyer, ed.), 3rd ed., Vol. 1, pp. 341–396. Academic Press, New York.

Koshland, D. E., Jr., Néméthy, G., and Filmer, D. (1966). *Biochemistry* **5**, 365.

Krebs, E. G., and Fischer, E. H. (1962). *Advan. Enzymol.* **24**, 263.

Krebs, H. A. (1964). *Proc. Roy. Soc., Ser. B* **159**, 454.

Krzanowski, J., and Matshinsky, F. M. (1969). *Biochem. Biophys. Res. Commun.* **34**, 816.

Larner, J. (1967). *Ann. N.Y. Acad. Sci.* **29**, 162.

Larner, J., and Villar-Palasi, C. (1971). *Curr. Top. Cell. Regul.* **3**, 195–236.

Lascelles, J. (1968). *Biochem. Soc. Symp.* **28**, 49–59.

Lascelles, J., and Hatch, T. P. (1969). *J. Bacteriol.* **98**, 712.

Laties, G. G. (1950). *Arch. Biochem. Biophys.* **27**, 404.

Latzko, E., and Gibbs, M. (1968). *Z. Pflanzphysiol.* **59**, 184.

Lavergne, D., Bismuth, E., and Champigny, M. L. (1974). *Plant Sci. Lett.* **3**, 391.

Leloir, L. F., and Cardini, C. E. (1955). *J. Biol. Chem.* **214**, 157.

Levine, R. P. (1968). *Science* **162**, 678.

Levine, R. P., and Togasaki, R. L. (1965). *Proc. Nat. Acad. Sci. U.S.* **53**, 987.

Levitzki, A. E., and Koshland, D. E. (1969). *Proc. Nat. Acad. Sci. U.S.* **62**, 1121.

Linn, T. C., Pettit, F. H., and Reed, L. J. (1969). *Proc. Nat. Acad. Sci. U.S.* **62**, 234.

Loewus, F. (1969). *Ann. N.Y. Acad. Sci.* **165**, 577.

Loewus, F. (1971). *Annu. Rev. Plant Physiol.* **22**, 237.

Loewus, F. (1973). *In* "Biogenesis of Plant Cell Wall Polysaccharides" (F. Loewus, ed.), pp. 1–27. Academic Press, New York.

Lowry, O. H., and Passonneau, J. V. (1964). *Naunyn-Schmiedebergs Arch. Exp. Pathol. Pharmakol.* **248**, 185.

MacDonald, P. W., and Strobel, G. A. (1970). *Plant Physiol.* **46**, 126.

MacElroy, R. D., Johnson, E. J., and Johnson, M. K. (1968). *Biochem. Biophys. Res. Commun.* **30**, 678.

MacElroy, R. D., Johnson, E. J., and Johnson, M. K. (1969). *Arch. Biochem. Biophys.* **131**, 272.

McFadden, B. A., and Tu, C. C. L. (1967). *J. Bacteriol.* **93**, 886.

Mangum, J. H., Magni, G., and Stadtman, E. R. (1973). *Arch. Biochem. Biophys.* **158**, 514.

Mansour, T. E. (1972). *Curr. Top. Cell. Regul.* **5**, 1.

Martin, R. G. (1962). *J. Biol. Chem.* **237**, 257.

Mayer, F. C., Bikel, I., and Hassid, W. Z. (1968). *Plant Physiol.* **43**, 1097.

Mayeux, J. V., and Johnson, E. J. (1967). *J. Bacteriol.* **94**, 409.

Mendicino, J. (1960). *J. Biol. Chem.* **235**, 3347.

Monod, J., Changeux, J. P., and Jacob, F. (1963). *J. Mol. Biol.* **6**, 306.

Monod, J., Wyman, J., and Changeux, J. P. (1965). *J. Mol. Biol.* **12**, 88.

Morris, C. J., and Thompson, J. F. (1971). *Plant Physiol.* **47**, Suppl., 17.

Munch-Peterson, A., Kalckar, H. M., Cutolo, E., and Smith, E. E. B. (1953). *Nature (London)* **172**, 1036.

Murata, T., Sugiyama, T., and Akazawa, T. (1964). *Arch. Biochem. Biophys.* **107**, 92.

Murray, A. W. (1967). *Biochem. J.* **103**, 271.

Nandi, D. L., and Shemin, D. (1968). *J. Biol. Chem.* **243**, 1231.

Nandi, D. L., and Shemin, D. (1973). *Arch. Biochem. Biophys.* **158**, 305.

Nandi, D. L., and Waygood, E. R. (1967). *Can. J. Biochem.* **45**, 327.

Nandi, D. L., Baker-Cohen, K. F., and Shemin, D. (1968). *J. Biol. Chem.* **243**, 1224.

Nester, E. W., and Jensen, R. A. (1966). *J. Bacteriol.* **91**, 1591.

Neuberger, A., Sandy, J. D., and Tait, G. H. (1973). *Biochem. J.* **136**, 491.

Neumann, J., and Jones, M. E. (1962). *Nature (London)* **195**, 709.

Neumann, J., and Jones, M. E. (1964). *Arch. Biochem. Biophys.* **104**, 438.

Newsholme, E. A., and Start, C. (1973). "Regulation in Metabolism." Wiley, New York.

Nichol, L. W., Jackson, W. T. H., and Winzor, D. J. (1967). *Biochemistry* **6**, 2449.

Nicholls, P. B., and Murray, A. W. (1968). *Plant Physiol.* **43**, 645.

Nierlich, D. P., and Magasanik, B. (1965). *J. Biol. Chem.* **240**, 358.

Nikaido, H., and Hassid, W. Z. (1971). *Advan. Carbohyd. Chem.* **26**, 351.

Northrop, J. H., Kunitz, M., and Herriot, R. (1948). "Crystalline Enzymes," 2nd ed. Columbia Univ. Press, New York.

Oelze, J., and Kamen, M. D. (1971). *Biochim. Biophys. Acta* **234**, 137.

Ohmann, E. (1969). *Arch. Mikrobiol.* **67**, 273.

O'Neal, T. D., and Naylor, A. W. (1968). *Biochem. Biophys. Res. Commun.* **31**, 322.

O'Neal, T. D., and Naylor, A. W. (1969). *Biochem. J.* **113**, 271.

Ordin, L., and Hall, M. A. (1968). *Plant Physiol.* **43**, 473.

Ozbun, J. L., Hawker, J S., Greenberg, E., Lammel, C., Preiss, J., and Lee, E. Y. C. (1973). *Plant Physiol.* **51**, 1.

Pacold, I., and Anderson, L. E. (1973). *Biochem. Biophys. Res. Commun.* **51**, 139.

Pacold, I., and Anderson, L. E. (1975). *Plant Physiol.* **55**, 160.

Passonneau, J. V., and Lowry, O. H. (1964). *Advan. Enzymol. Regul.* **2**, 265.

Paulus, H., and Gray, E. (1964). *J. Biol. Chem.* **239**, 4008.

Peaud-Penol, C., and Axelos, M. (1970). *FEBS (Fed. Eur. Biochem. Soc.) Lett.* **8**, 224.

Pogell, B. M., and Gryder, R. M. (1957). *J. Biol. Chem.* **228**, 701.

Pogell, B. M., Taketa, K., and Sarngadharan, M. E. (1971). *J. Biol. Chem.* **246**, 1947.

Porra, R. J., Irving, E. A., and Tennick, A. M. (1972). *Arch. Biochem. Biophys.* **148**, 37.

Pradet, A. (1969). *Physiol. Veg.* **7**, 261.

Pradet, A., and Bomsel, J. L. (1968). *C.R. Acad. Sci.* **226**, 2416.

Preiss, J. (1969). *Curr. Top. Cell. Regul.* **1**, 125–160.

Preiss, J. (1973). *In* "The Enzymes" (P. D. Boyer, ed.), 3rd ed., Vol. 8, pp. 73–120. Academic Press, New York.

Preiss, J., and Greenberg, E. (1969). *Biochem. Biophys. Res. Commun.* **36**, 289.

Preiss, J., Ghosh, H. P., and Wittkop, J. (1967). *In* "The Biochemistry of Chloroplants" (T. W. Goodwin, ed.), Vol. 2, pp. 131–153. Academic Press, New York.

Preiss, J., Lammel, C., and Sabraw, A. (1971). *Plant Physiol.* **47**, 104.

Putman, E. W., and Hassid, W. Z. (1954). *J. Biol. Chem.* **207**, 885.

Rabin, B. R. (1967). *Biochem. J.* **102**, 22C.

Ramaih, A. (1974). *Curr. Top. Cell. Regul.* **8**, 297.

Ray, P. M., Shiniger, T. L., and Ray, M. M. (1969). *Proc. Nat. Acad. Sci. U.S.* **64**, 605.

Recondo, E., and Leloir, L. F. (1961). *Biochem. Biophys. Res. Commun.* **6**, 85.

Reed, L. J., Pettit, F. H., Roche, T. E., and Butterworth, P. J. (1973). *In* "Protein Phosphorylation in Control Mechanisms" (F. Huijing and E. Y. C. Lee, eds.), pp. 83–97. Academic Press.

Ribereau-Gayon, G., and Preiss, J. (1971). *In* "Methods in Enzymology" (A. San Pietro, ed.), Vol. 23, pp. 618–629. Academic Press, New York.

Rindt, K. P., and Ohmann, E. (1969). *Biochem. Biophys. Res. Commun.* **36**, 357.

Rubin, M., and Changeux, J. P. (1966). *J. Mol. Biol.* **21**, 265.

Russell, T. R., Thompson, W. J., Schneider, F. W., and Appleman, M. M. (1972). *Proc. Nat. Acad. Sci. U.S.* **69**, 1791.

Santarius, K. A., and Heber, U. (1965). *Biochim. Biophys. Acta* **102**, 39.

Sanwal, G. G., and Preiss, J. (1967). *Arch. Biochem. Biophys.* **119**, 454.

Sanwal, G. G., Greenberg, E., Hardie, J., Cameron, E. C., and Preiss, J. (1968). *Plant Physiol.* **43**, 417.

Scala, J., Patrick, C., and Macbeth, G. (1968a). *Life Sci.* **7**, 407.

Scala, J., Patrick, C., and Macbeth, G. (1968b). *Arch. Biochem. Biophys.* **127**, 576.

Scala, J., Patrick, C., and Macbeth, G. (1969). *Phytochemistry* **8**, 37.

Scrutton, M. C., and Utter, M. F. (1968). *Annu. Rev. Biochem.* **37**, 249.

Segal, A., Brown, M. S., and Stadtman, E. R. (1974). *Arch. Biochem. Biophys.* **161**, 319.

Seubert, W., and Schoner, W. (1971). *Curr. Top. Cell. Regul.* **3**, 237.

Shapiro, B. M. (1969). *Biochemistry* **8**, 659.

Shapiro, B. M., and Stadtman, E. R. (1968). *J. Biol. Chem.* **243**, 3769.

Shapiro, B. M., and Stadtman, E. R. (1970). *Annu. Rev. Microbiol.* **24**, 501.

Shen, L. C., and Atkinson, D. E. (1970). *J. Biol. Chem.* **245**, 3996.

Shen, L. C., Fall, L., Walton, G. M. and Atkinson, D. E. (1968). *Biochemistry* **7**, 404.

Shetty, A. S., and Miller, G. W. (1969a). *Biochem. J.* **114**, 331.

Shetty, A. S., and Miller, G. W. (1969b). *Biochim. Biophys. Acta* **185**, 458.

Slabnik, E., Frydman, R. B., and Cardini, C. E. (1968). *Plant Physiol.* **43**, 1063.

Smith, L. C., Ravel, J. M., Lox, S. R., and Shine, W. (1962). *J. Biol. Chem.* **237**, 3566.

Soderling, T. R., and Park, C. R. (1974). *Advan. Cyclic Nucleotide Res.* **4**, 283.

Soderling, T. R., Hickenbottom, J. P., Reinmann, E. M., Hunkeler, F. L., Walsh, D. A., and Krebs, E. G. (1970). *J. Biol. Chem.* **245**, 6317.

Stadtman, E. R. (1966). *Advan. Enzymol.* **28**, 41.

Stadtman, E. R. (1970). *In* "The Enzymes" (P. D. Boyer, ed.), 3rd ed., Vol. 1, pp. 397–459. Academic Press, New York.

Stadtman, E. R., and Ginsburg, A. (1974). *In* "The Enzymes" (P. D. Boyer, ed.), 3rd ed., Vol. 10, pp. 755–807. Academic Press, New York.

Stahman, W., and Hers, H. G. (1973). *In* "The Enzymes" (P. D. Boyer, ed.), 3rd ed., Vol. 9, pp. 310–362. Academic Press, New York.

Stanier, R. Y., Doudoroff, M., Kunisawa, R., and Contopoulou, R. (1959). *Proc. Nat. Acad. Sci. U.S.* **45**, 1246.

Stirpe, F., and Della Corte, E. (1969). *J. Biol. Chem.* **244**, 3855.

Stirpe, F., and Della Corte, E. (1970). *Biochim. Biophys. Acta* **212**, 195.

Strominger, J. L., and Mapson, L. W. (1957). *Biochem. J.* **66**, 567.

Sweeney, J. R., and Fisher, J. R. (1968). *Biochemistry* **7**, 571.

Taketa, K., and Pogell, B. M. (1965). *J. Biol. Chem.* **240**, 651.

Taketa, K., Sarngadharan, M. G., Watanabe, A., Aoe, H., and Pogell, B. M. (1971). *J. Biol. Chem.* **246**, 5676.

Traut, R. R., and Lipmann, F. (1963). *J. Biol. Chem.* **238**, 1213.

Tsai, C. M., and Hassid, W. Z. (1971). *Plant Physiol.* **47**, 740.

Tsai, C. Y., and Nelson, O. E. (1966). *Science* **151**, 341.

Tsai, C. Y., Salamini, F., and Nelson, O. E. (1970). *Plant Physiol.* **46**, 299.

Tsuboi, K. K., Fukunaga, K., and Petricciani, J. C. (1969). *J. Biol. Chem.* **244**, 1008.

Tuboi, S., Kim, H. J., and Kikuchi, G. (1969). *Arch. Biochem. Biophys.* **130**, 92.

Tuboi, S., Kim, H. J., and Kikuchi, G. (1970a). *Arch. Biochem. Biophys.* **138**, 147.

Tuboi, S., Kim, H. J., and Kikuchi, G. (1970b). *Arch. Biochem. Biophys.* **138**, 155.

Umbarger, H. E. (1956). *Science* **123**, 848.

Umbarger, H. E. (1969). *Annu. Rev. Biochem.* **38,** 323.

Vaccaro, D., and Zelden, M. H. (1974). *Plant Physiol.* **54,** 617.

Vessal, M., and Hassid, W. Z. (1972). *Plant Physiol.* **49,** 977.

Walsh, D. A., Perkins, J. P., Brostrom, C. O., Ho, E. S., and Krebs, E. G. (1971). *J. Biol. Chem.* **246,** 1968.

Warnick, G. R., and Burnham, B. F. (1971). *J. Biol. Chem.* **246,** 6880.

Weitzman, P. D. J., and Jones, D. (1968). *Nature (London)* **219,** 270.

Wider, E. A., and Tigier, H. A. (1970). *FEBS (Fed. Eur. Biochem. Soc.) Lett.* **9,** 30.

Wider, E. A., Batlle, A. M. D. C., and Tigier, H. A. (1971). *Biochim. Biophys. Acta* **235,** 511.

Wieland, O., and Jagow-Westermann, B. V. (1969). *FEBS (Fed. Eur. Biochem. Soc.) Lett.* **3,** 271.

Wieland, O., and Siess, E. (1970). *Proc. Nat. Acad. Sci. U.S.* **65,** 947.

Woolfolk, C. A., and Stadtman, E. R. (1964). *Biochem. Biophys. Res. Commun.* **17,** 313.

Woolfolk, C. A., and Stadtman, E. R. (1967). *J. Bacteriol.* **118,** 736.

Wulff, K., Mecke, D., and Holzer, H. (1967). *Biochem. Biophys. Res. Commun.* **28,** 740.

Wyman, J. (1948). *Advan. Protein Chem.* **4,** 407.

Wyngaarden, J. B. (1972). *Curr. Top. Cell. Regul.* **5,** 135–176.

Yates, R. A., and Pardee, A. B. (1956). *J. Biol. Chem.* **221,** 757.

11

Mono- and Oligosaccharides

J. E. GANDER

I. Introduction

Chemical analyses of plants show that they contain a diverse group of polysaccharides, oligosaccharides, monosaccharides, and derivatives of monosaccharides. Investigations of the sequence and regulation of the reactions through which monosaccharides pass in their conversion to polysaccharides continues to be an active field of investigation and is discussed elsewhere in this book. Investigations to date suggest that oligosaccharide and polysaccharide metabolism is frequently similar. This discussion will focus on the metabolism of oligosaccharides unique to plants

337

and on the ancillary substances and reactions resulting in the degradation or biosynthesis of oligosaccharides.

Approximately a dozen oligosaccharides are found rather widely distributed and occur as major constituents of plant tissues. Another dozen are found as minor constituents of plants (Bailey, 1965). In addition, there are well over 100 monosaccharides that are naturally occurring either as monomers or as a constituent of a glycoside. There are many C-, S-, N-, and O-glycosides in the plant kingdom in which carbohydrate is attached through glycosidic linkage to an aglycone. The diversity of this group of substances is shown by the classes of substances found as aglycones. These include sterols, flavones, phenolics, and amino acids.

Unlike organisms that are dependent upon organic substances for their nutrition, organisms that carry out photosynthesis normally synthesize nearly all of their organic substances. Carbohydrate is derived from organic acids produced by photosynthesis or from reserve sources, such as polysaccharides, oligosaccharides, or the carbon skeleton of amino acids stored as proteins. Figure 1 shows a summary of the reactions to be considered in some detail in this chapter. D-Fructose 6-phosphate, D-glucose 6-phosphate, and α-D-glucose 1-phosphate are products of either photosynthesis or polysaccharide catabolism. They are the branching points in metabolism leading to other monosaccharides or to oligosaccharides. The reactions leading to these products are epimerizations, decarboxylations, oxidation–reduction, aldol condensation or retroaldol cleavage, transglycosidation, ring contraction, and pyrophosphorolysis to name a few types.

An understanding of the function of most of the oligosaccharides in plants, with the exception of sucrose, continues to be primarily speculative and inferential. However, the suggested function for some of the oligosaccharides will be discussed and the evidence bearing on this suggestion examined.

II. Transformations of Sugar Phosphates

A. Kinases, Mutases, and Isomerases

Because plants obtain carbohydrate through photosynthesis, which results in the formation of phosphorylated sugars, the roles of kinases that catalyze the phosphorylation of hexoses and pentoses might seem superfluous at first. However, plants also contain phosphatases and carbohydrases which degrade sugar phosphates, oligosaccharides, and poly-

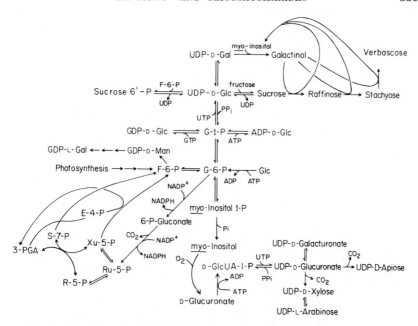

Fig. 1. Summary of reactions of mono- and oligosaccharides. The abbreviations used are UDP-D-Gal, uridine 5'-(α-D-galactopyranosyl pyrophosphate); UDP-D-Glc, uridine 5'-(α-D-glucopyranosyl pyrophosphate); ADP-D-Glc, adenosine 5'-(α-D-glucopyranosyl pyrophosphate); GDP-D-Glc, guanosine 5'-(α-D-glucopyranosyl pyrophosphate); GDP-D-Man, guanosine 5'-(α-D-mannopyranosyl pyrophosphate), GDP-L-Gal, guanosine 5'-(β-L-galactopyranosyl pyrophosphate); UDP, uridine 5'-diphosphate; UTP, uridine 5'-triphosphate; ATP, adenosine 5'-triphosphate; ADP, adenosine 5'-diphosphate; GTP, guanosine 5'-triphosphate; F-6-P, D-fructose 6-phosphate; sucrose 6'-P, sucrose 6'-phosphate; G-1-P, α-D-glucopyranosyl 1-phosphate; G-6-P, D-glucose 6-phosphate; Glc, D-glucose; PPᵢ, inorganic pyrophosphate; 6-P-gluconate, 6-phosphogluconate; Ru-5-P, D-ribulose 5-phosphate; Xu-5-P, D-xylulose 5-phosphate; R-5-P, D-ribose 5-phosphate; E-4-P, D-erythrose 4-phosphate; S-7-P, D-sedoheptulose 7-phosphate; 3-PGA, D-glyceraldehyde 3-phosphate; and *myo*-inositol 1-P, *myo*-inositol 1-phosphate.

saccharides to hexoses and pentoses. Axelrod (1965) summarized the work performed up to that time on the kinases (ATP:transphosphorylases) found in plants that phosphorylate the following sugars: D-glucose, D-galactose, D-glucuronic, D-galacturonic acids, D-xylose, and L-arabinose. The presence of these kinases provides indirect evidence that oligo- and polysaccharides which are degraded to their constituent monosaccharides are then phosphorylated. Of these kinases only wheat germ hexokinase has been highly purified (Meunier *et al.*, 1971). Its physical properties differ from yeast hexokinase and its kinetic properties have not been investigated.

D-Glucose 6-phosphate and D-fructose 6-phosphate, derived either directly from photosynthetic products or by the combined action of glyco-hydrolases (invertase on sucrose or amylase on starch) and hexokinase provide the carbon skeleton for other hexoses, pentoses, oligosaccharides, and polysaccharides (Fig. 1). This figure shows that phosphoglucomutase (α-D-glucose 1,6-diphosphate:α-D-glucose-1-phosphate phosphotransfer-ase, EC 2.7.5.1) has a major function, along with the nucleoside diphos-phate glucose pyrophosphorylases, in the formation of glucosyl donors for polysaccharide and oligosaccharide biosynthesis. Although phospho-glucomutase activity has been demonstrated in a number of plant tissues, it has not been purified to homogeneity nor has it been subjected to inten-sive kinetic analysis. Bird *et al.* (1965) have shown that *Nicotiana tabacum* chloroplasts contain enzymes (phosphoglucose isomerase, phos-phoglucomutase, and UDP-D-glucose pyrophosphorylase) necessary for the conversion of D-fructose 6-phosphate to UDP-D-glucose. They could not detect either phosphoglucose isomerase (D-glucose-6-phosphate ketol-isomerase EC 5.3.1.9) or phosphoglucomutase in whole cell preparations. Hexokinase, phosphoglucomutase, and phosphoglucose isomerase activities are found in germinating barley (Latzko and Kotze, 1965). The phos-phorylase (α-1,4-glucan:orthophosphate glucosyltransferase EC 2.4.1.1) is probably more important for the breakdown of endosperm starch and formation of hexose phosphates than the combined action of hydrolytic cleavage of starch and hexokinase-catalyzed phosphorylation of glucose. Phosphoglucomutase activity has been detected in milkweed (Zauralov, 1969), potato tubers (Verleur, 1969), *Avena* coleoptiles (Hall and Ordin, 1967), and *Citrus acida* fruit tissues (Parekh and Shah, 1971). Finding [^{14}C]D-glucose 1-phosphate in wheat root tops after administration of [^{14}C]D-glucose provides indirect evidence for phosphoglucomutase in wheat. Similar experiments have been performed for many plant tissues and similar results have been obtained. Little is known about the mecha-nism of the reaction catalyzed by phosphoglucomutase from plants.

B. Oxidative Pentose Phosphate Cycle

The oxidative pentose phosphate cycle is found in higher plants and has been investigated by Axelrod and colleagues (Axelrod, 1965), Wang and his associates (Wang *et al.*, 1962) and others. This pathway of carbon metabolism is often called the hexose monophosphate shunt because car-bon is shunted around some intermediates of the Embden–Meyerhoff–Parnas pathway. This provides an organism with versatility in providing a part of the carbon skeleton for aromatic amino acids and phenolic substances and at the same time provides a source of reducing potential (NADPH) necessary for biosynthetic reductions. Equations (1)–(3) show some of the versatility.

$$6 \text{ D-Glc 6-P} + 12 \text{ NADP}^+ \rightarrow 4 \text{ F 6-P} + 6 \text{ CO}_2$$
$$+ 2 \text{ Gly 3-P} + 12 \text{ NADPH} + 12 \text{ H}^+ \quad (1)$$
$$6 \text{ D-Glc 6-P} + 12 \text{ NADP}^+ \rightarrow 3 \text{ F 6-P} + 6 \text{ CO}_2$$
$$+ 3 \text{ E 4-P} + 12 \text{ NADPH} + 12 \text{ H}^+ \quad (2)$$
$$6 \text{ D-Glc 6-P} + 12 \text{ NADP}^+ \rightarrow 6 \text{ R 5-P} + 6 \text{ CO}_2 + 12 \text{ NADPH} + 12 \text{ H}^+ \quad (3)$$

Figure 1 also shows the pentose phosphate pathway. Equation (1) shows that when intermediates of the pathway are not used in biosynthetic reactions, 12 NADPH molecules are generated at the expense of two 3-carbon fragments. Further, D-fructose 6-phosphate may be recycled through the pentose phosphate pathway, which then provides additional NADPH. Presumably the extent of recycling is balanced with other needs of the cell. ap Rees et al. (1965) concluded that recycling of D-fructose 6-phosphate to D-glucose 6-phosphate and through the oxidative pentose phosphate cycle was restricted in pumpkin, turnip, and carrot. Wang et al. (1962) observed restricted recycling of hexose phosphates in tomato slices.

Equation (2) shows that erythrose 4-phosphate (E 4-P), a precursor of L-phenylalanine, L-tyrosine, L-tryptophan, and phenolic substances including lignin (Higuchi and Shimada, 1967), is generated from the oxidative pentose phosphate cycle. For each erythrose 4-phosphate formed, 4 NADPH molecules are generated.

Equation (3) shows that D-ribose 5-phosphate (R 5-P) is derived by the pentose phosphate cycle. D-Ribose 5-phosphate is a precursor of the ribosyl moieties of RNA's, and Fan et al. (1969) suggested it is an important precursor of DNA in plants.

Although ribose and xylulose are intermediates of the pentose phosphate cycle, the available evidence suggests that D-xylose and L-arabinose are not derived from ribose or xylulose. Ginsburg and Hassid (1956) concluded that conversion of D-glucose to D-xylose or L-arabinose did not involve the loss of the C-1 atom. Neish (1958) showed that L-arabinose was incorporated into L-arabinosyl units in plant polysaccharides without rearrangement of the carbon skeleton. In contrast, D-xylose and D-ribose are first converted to hexoses through the oxidative pentose phosphate cycle and then incorporated into polysaccharides (Ginsburg and Hassid, 1956; Neish, 1958). D-Xylose metabolism was clarified by the studies of Pubols et al. (1963) and Zahnley and Axelrod (1965). Pubols et al. (1963) obtained an enzyme from wheat germ which catalyzed the isomerization of D-xylose to D-xylulose and D-ribulose. Zahnley and Axelrod obtained from pea meal D-xylulokinase and D-ribulokinase which catalyze phosphorylation of the C-5 hydroxymethyl group. L-Arabinose appears to be metabolized in a different manner. Neufeld *et al.* (1960) obtained an enzyme from *Phaseolus aureus* that catalyzed the formation of β-L-arabinosyl 1-phosphate from ATP and L-arabinose. β-L-Arabinosyl 1-phosphate is converted to uridine 5′-(β-L-arabinosyl pyrophosphate) (UDP-L-arabinose). Feingold *et al.* (1960) were the first to demonstrate that UDP-D-xylose and UDP-L-arabinose were also derived by oxidation, decarboxylation, and epimerization of UDP-D-glucose, UDP-D-galacturonate, and UDP-D-xylose, respectively. The D-glucose 6-phosphate–*myo*-inositol–glucuronate pathway discovered by Loewus (1962) is also of importance in formation of D-xylosyl and L-arabinosyl residues.

The presence of the oxidative pentose phosphate cycle in plants is based on demonstrating (a) enzymes that catalyze the appropriate reactions, (b) intermediates of the cycle, or (c) conversion of intermediates to other intermediates and glycolytic substances (Pon, 1964). One or more of these criteria are met by a variety of plant tissues, including spinach (Axelrod and Jang, 1954), pea leaves (Axelrod and Jang, 1954; Smillie, 1962; Gibbs and Horecker, 1954), sugar beet leaves, barley leaves, tobacco leaves (Tolbert and Zill, 1954), *Sorghum vulgare* (Reed, 1961), and carrot (ap Rees and Beevers, 1960a) to name but a few of the earlier observations. Reed and Kolattukudy (1966) showed that extensive washing of red beet slices increases the fraction of glucose metabolized by the oxidative pentose phosphate pathway. They observed a large increase in 6-phosphogluconate:NADP oxidoreductase activity with increased time of washing. NADPH generated by this pathway was used for biosynthetic reductions. Kikuta *et al.* (1971) suggested that during potato callus formation the relative fraction of the glucose oxidized via the oxidative pentose phosphate cycle increased and that this has a significant role in DNA multiplication phase of callus development. Carrot callus shows similar high activity of enzymes of this cycle during development (Komamine *et al.*, 1969). The oxidative pentose phosphate pathway is functionally important in differentiating tissues during germination of peas (Dolreau, 1969), and the pathway is important during differentiation of apical pea root cells (Fowler and ap Rees, 1970). The cycle is functionally more important in organisms in the dark than in the light. 6-Phosphogluconate is not detected in spinach chloroplasts until illumination is stopped (Krause and Bassham, 1969), after which 6-phosphogluconate accumulates in chloroplasts in the dark. Oxidative pentose metabolism may provide NADPH during dark metabolism. Reductive photosynthesis provides at least part of the needed NADPH during intervals in light. The activity of D-glucose-6-phosphate:NADP oxidoreductase may be regulated by light and dark. The enzyme would be activated in the dark and inactivated by light. Isozymes of glucose-6-phosphate and 6-phosphogluconate dehydrogenases obtained from spinach chloroplasts differed from those in the cytosol (Schnarrenberger *et al.*, 1973). The data suggest that leaf cells have two oxidative pentose phosphate pathways. Muto and Uritani (1970) obtained nonlinear kinetic plots of velocity and D-glucose 6-phosphate concentration and proposed negative cooperativity between the enzyme and D-glucose 6-phosphate. It is unlikely that this property alone results in the extent of inhibition observed by Krause and Bassham (1969).

The oxidative pentose phosphate cycle, through the action of transaldolase and transketolase, provides precursors of heptoses and heptuloses. Sedoheptulose 7-phosphate is a product of both transketolase and transaldolase catalyzed reactions (Fig. 1). Rendig and McComb (1964) obtained an enzyme from alfalfa which catalyzed the synthesis of D- or L-heptuloses. The steric configuration about carbon atoms 1 through 4 was that of sedoheptulose (D-*altro*-heptulose), while that about carbon atoms 5 through 7 was dictated by the pentose precursor. Only those pentoses having a 2 S hydroxyl group (Cahn *et al.*, 1956) served as heptulose precursors. Four possible heptulose isomers resulted: D-*altro*-, D-*iodo*-, L-*gluco*-, and L-*galacto*-heptulose.

C. Nucleoside Diphosphate Sugar Pyrophosphorylases and Phosphorylases

1. UDP-D-Glucose Pyrophosphorylase

Caputto *et al.* (1950) were first to isolate a nucleoside diphosphate sugar, uridine 5′-(α-D-glucopyranosyl pyrophosphate) (UDP-D-glucose). Buchanan *et al.* (1953) were first to isolate and identify UDP-D-glucose from green plants. The isolation of this substance and many other related nucleotides, unique in either the base or sugar moiety, led to an understanding of the biosynthesis of oligo- and polysaccharides in plants. These substances are effective glycosyl donors presumably because their group transfer potential (− 7600 cal/mole) (Leloir *et al.*, 1960) at physiological pH is considerably greater than that of the glycosidic bond(s) being formed. Phosphorylation of the nucleoside diphosphate formed during glycoside (RO-sugar) synthesis [Eq. (4)] serves to pull the reaction toward completion.

$$\text{Nucleoside diphosphate sugar} + \text{ROH} \rightarrow \text{RO–sugar} + \text{nucleoside diphosphate} \quad (4)$$

UDP-D-glucose pyrophosphorylase (UTP:α-D-glucose 1-phosphate uridylyltransferase, EC 2.7.7.9) catalyzes UDP-D-glucose synthesis [Eq. (5)].

$$\alpha\text{-D-Glc 1-P} + \text{UTP} \rightleftarrows \text{UDP-D-glc} + \text{PP}_i \qquad (5)$$

α-D-Glucose 1-phosphate is derived from either D-glucose 6-phosphate or by a phosphorylase-catalyzed degradation of reserve polysaccharide. Munch-Petersen et al. (1952) first isolated UDP-D-glucose pyrophosphorylase from yeast. Burma and Mortimer (1956) showed UDP-D-glucose pyrophosphorylase activity in sugar beet leaf homogenates. Ginsburg (1958) demonstrated UDP-D-glucose pyrophosphorylase activity in mung bean, and Turner and Turner (1958) showed that pea roots, wheat shoots, and sugar cane roots contained the enzyme. Tsuboi et al. (1969) performed parallel investigations on partially purified UDP-D-glucose pyrophosphorylase from mung beans, human erythrocytes, and cardiac muscle. They observed that UDP-D-glucose was a 7- to 10-fold more potent inhibitor of the mammalian enzymes than the plant enzyme. Tovey and Roberts (1970) studied both UDP-D-glucose pyrophosphorylase and ADP-D-glucose pyrophosphorylase from wheat embryo and endosperm. The K_m for α-D-glucose 1-phosphate was 1×10^{-3} and 4.4×10^{-5} M, respectively, for the two enzymes. UDP-D-glucose pyrophosphorylase activity increases during germination. In contrast, ADP-D-glucose pyrophosphorylase activity remains constant during germination. A maximum rate of UDP-D-glucose synthesis occurs at a UTP:Mg ratio of 1:2. Gustafson and Gander (1972) obtained a highly purified (1200 μmoles UDP-D-glucose/minute/mg protein) UDP-D-glucose pyrophosphorylase from etiolated sorghum seedlings. This preparation contained protein that separated into two bands upon acrylamide gel electrophoresis. Each band contained the same specific activity. The reaction is an ordered Bi-Bi type of reaction (Cleland, 1963) and $\text{Mg} \cdot \text{UTP}^{2-}$ and $\text{Mg} \cdot \text{PP}_i^{2-}$ serve as substrates (Fig. 2). They showed that the velocity of UDP-D-glucose synthesis is a function of Mg^{2+} concentration with a UTP:Mg ratio of 1:2 giving a maximum velocity. In contrast, a maximum velocity of pyro-

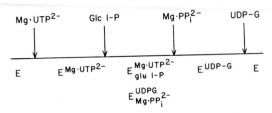

Fig. 2. Mechanism of sorghum UDP-D-glucose pyrophosphorylase catalyzed reaction. Showing the ordered Bi-Bi type of reaction (Cleland, 1963).

phosphorolysis was observed with $PP_i:Mg$ ratio of approximately 1:1. The results suggest that Mg^{2+} serves in some fashion to activate the enzyme as has been proposed by Tovey and Roberts (1970). Similar kinetics were observed when Mn^{2+} or Co^{2+} replaced Mg^{2+}. The purified sorghum enzyme showed almost absolute specificity for UTP and UDP-D-glucose. The nucleoside diphosphate glucose derivatives of thymidine, adenine, guanosine, and cytidine gave at least 1000-fold lower activity.

The kinetic parameters of sorghum UDP-D-glucose pyrophosphorylase are unlike those of the mammalian enzymes. The apparent K_m for $Mg \cdot UTP^{2-}$ (0.03 mM) is 10-fold lower than that for human erythrocyte or calf liver enzyme, and the K_i for UDP-D-glucose competing with $Mg \cdot UTP^{2-}$ is 2-fold larger than that for the mammalian enzyme. Gustafson and Gander (1972) concluded that under conditions where the ratio of UTP:Mg is approximately 1 there is little UDP-D-glucose synthesis occuring and that it is feasible that UDP-D-glucose undergoes pyrophosphorolysis if PP_i is available. The kinetics of plant UDP-D-glucose pyrophosphorylase are such that limiting Mg^{2+} concentration would favor conversion of sucrose to α-D-glucose 1-phosphate as proposed by De Fekete and Cardini (1964), Milner and Avigad (1965), and others as shown in Eqs. (6) and (7).

$$\text{Sucrose} + \text{UDP} \rightleftarrows \text{fructose} + \text{UDP-D-Glc} \qquad (6)$$
$$\text{UDP-D-Glc} + \text{PP}_i \rightleftarrows \text{UTP} + \alpha\text{-D-Glc 1-P} \qquad (7)$$

In this context, Matsumoto et al. (1968, 1969) have shown that cucumbers cultured in 200 mg NH_3 nitrogen/liter accumulates large quantities of UDP-D-glucose and that the starch content decreases. In contrast, sucrose, maltose, stachyose, fructose, and glucose concentration increases under these conditions. This investigation did not establish that UDP-D-glucose was derived from sucrose [reaction (6)] (Matsumoto et al., 1968, 1969). Apparently UDP-D-glucose was not derived from starch.

Unlike ADP-D-glucose pyrophosphorylase (Sanwal and Preiss, 1967), sorghum UDP-D-glucose pyrophosphorylase activity is not modified by 3-phosphoglycerate, D-fructose, 1,6-diphosphate, or phosphoenol pyruvate (Gustafson and Gander, 1972).

2. ADP-D-Glucose Pyrophosphorylase

ATP:α-D-glucose 1-phosphate adenylyltransferase (ADP-D-glucose pyrophosphorylase) catalyzes the synthesis of ADP-D-glucose as shown first by Espada (1962) using wheat flour. Arai and Fujisaki (1971) obtained ADP-D-glucose pyrophosphorylase activity in leaves of 20 species of plants, but found no activity in storage tissues. Preiss and colleagues

(Ghosh and Preiss, 1966; Sanwal *et al.*, 1968) showed that the enzyme is located in chloroplasts. Furthermore, ADP-D-glucose pyrophosphorylase activity is increased by 3-phosphoglycerate, D-fructose 6-phosphate, and phosphoenol pyruvate, and the activity is decreased by ADP, AMP, or orthophosphate. Sanwal and Preiss (1967) and Dickinson and Preiss (1969) reported that maximum ADP-D-glucose pyrophosphorylase activity occurs at a ATP:Mg ratio greater than 1.

The roles of UDP-D-glucose, ADP-D-glucose, and GDP-D-glucose in polysaccharide biosynthesis are discussed in Chapter 12.

3. GDP-D-GLUCOSE PYROPHOSPHORYLASE

GTP:α-D-glucose 1-phosphate guanylyltransferase (GDP-D-glucose pyrophosphorylase) was demonstrated in extracts from peas, etiolated mung bean seedlings, spinach leaves, buckwheat, mustard, and parsley (Barber and Hassid, 1964). Barber and Hassid state that separate enzymes from peas also catalyze the synthesis of appropriate nucleoside diphosphate glucosides when incubated with α-D-glucose 1-phosphate and one of the following: ATP, CTP, UTP, TTP, ITP, and dUTP.

Plant tissues contain a diverse group of nucleotide sugars. Table I shows the sugar nucleotides in European larch (*Larix decidua* Mill.) (Cumming, 1970). Table I contains substances for which there is currently no known role. The function of UDP-oligosaccharides, GDP-oligosaccharide, and ADP-galactose is unknown. UDP-oligosaccharide fraction contained two disaccharides, cellobiose and the other composed of glucose and arabinose. GDP-oligosaccharide fraction contained glucosyl and mannosyl residues. Galactose in GDP-galactose may be in the L-configuration. Su and Hassid (1962) identified GDP-L-galactose from red alga *Prophyra perforata*. Furthermore, *Phaseolus vulgaris* contains enzymes capable of converting GDP-D-mannose to GDP-L-fucose (GDP-6-deoxy-L-galactose) (Liao and Barber, 1971). In addition, GDP-mannuronic and GDP-L-guluronic acids have been isolated from the marine alga *Fucus gardneri* (Lin and Hassid, 1966).

4. UDP-D-GLUCURONIC ACID PYROPHOSPHORYLASE

Finegold *et al.* (1958) first observed UTP:α-D-glucuronic acid uridylyltransferase (UDP-D-glucuronic acid pyrophosphorylase) in *Phaseolus aureus*. With the discovery by Loewus and colleagues (1962) that D-glucuronate is a direct metabolite of 1-L-*myo*-inositol and that 1-L-*myo*-inositol is a precursor of cell wall uronic acids, the properties and role of UDP-D-glucuronic acid pyrophosphorylase took on additional impor-

TABLE I

Sugars in Nucleotides from European Larch

Base	Sugar	Percentage of total sugar for each case	
		Xylem	Cambium
Uracil	Glucose	73.2	61.0
	Galactose	9.8	7.0
	Arabinose	1.5	5.0
	Xylose	2.4	8.0
	Fructose	13.2	2.0
	Galacturonic acid	—	2.0
	Oligosaccharides[a]	—	15.0
Guanine	Glucose	46.7	49.0
	Mannose	42.0	43.0
	Galactose	11.3	7.0
	Oligosaccharide[b]	—	1.0
Adenine	Glucose	53.1	97
	Ribose[c]	3.0	3
	Galactose	5.9	trace
	Fructose	10.0	trace

[a] Two oligosaccharides; one contained only glucose, the other glucose and arabinose.

[b] Oligosaccharide contained glucose and mannose.

[c] Derived from NADH and NADPH.

tance. Roberts (1971a) purified this enzyme from barley seedlings and observed that optimum UDP-D-glucuronate synthesis occurs with a UTP:Mg ratio of about 2. In contrast, a maximum rate of pyrophosphorolysis was observed with a PP_i:Mg ratio of 1:1. UDP-D-glucuronate pyrophosphorylase has properties similar to UDP-D-glucose and ADP-D-glucose pyrophosphorylases with respect to Mg^{2+} requirement. Roberts proposed, as did Gustafson and Gander (1972), that Mg^{2+} has a role in activating the pyrophosphorylase in addition to that of binding UTP^{4-}.

5. NUCLEOSIDE DIPHOSPHATE SUGAR PHOSPHORYLASES

Carminatti and Cabib (1961) reported the presence in yeast of an enzyme that catalyzes the phosphorolysis of GDP-D-mannose to give GDP and D-mannose 1-phosphate. The preparation also catalyzed a similar phosphorolysis of UDP-D-glucose. Dankert et al. (1964) reported that wheat germ contained ADP-D-glucose phosphorylase activity. The preparation was active on ADP-D-mannose.

D. Epimerases

Leloir (1951) was the first to demonstrate the transformation of UDP-D-glucose to UDP-D-galactose in yeast extracts. This observation directed the way for the discovery of nucleoside diphosphate sugar epimerases in plants, animals, and microorganisms.

Investigations in Hassid's laboratory (Neufeld *et al.*, 1957; Feingold *et al.*, 1960; Neufeld, 1962) have shown that mung beans contain enzymes that catalyze epimerization of the C-4 hydroxyl group of UDP-D-glucose, UDP-D-galactose, UDP-D-xylose, UDP-L-arabinose, TDP-D-glucose, TDP-D-galactose, UDP-D-glucuronate, and UDP-D-galacturonate. Some of these metabolic interrelationships to monosaccharide metabolism are shown in Fig. 1.

Interest in the mechanism of epimerization of UDP-D-glucose to UDP-D-galactose is long standing. Evidence for the postulated intermediate UDP-4-keto-D-glucose was only indirect until recently. Nelsestuen and Kirkwood (1971) demonstrated that the hydrogen attached to C-4 of the hexose in UDP-6-deoxy-D-glucose is transferred to the β-position of the nicotinamide ring in NADH and that this hydrogen can be transferred to either TDP-4-keto-6-deoxy-D-glucose or UDP-4-keto-6-deoxy-D-glucose. This investigation, and that of Maitra and Ankel (1971) who isolated [4-^{3}H]glucose after treating *E. coli* UDP-D-glucose 4-epimerase with NaB^{3}H$_4$, presents convincing evidence that epimerization occurs through an oxidation reduction mechanism with UDP-4-keto sugar as an intermediate. Maitra *et al.* (1974) demonstrated that the rates of 4-epimerization of UDP-[4-^{3}H]glucose, -galactose, and -glucuronate were about one-half that observed with the nontritiated species. They suggest that a common mechanism exists for 4-epimerization of UDP-hexose, UDP-pentose, and UDP-hexuronate. Enzyme for catalyzing epimerization of UDP-D-hexose and UDP-pentose were obtained from *E. coli*, and that for UDP-D-hexuronate from blue-green alga. Partially purified UDP-L-arabinose 4-epimerase from wheat germ shows no NAD requirement, nor can one be shown by prolonged dialysis or treatment of the enzyme with charcoal or nicotinamide adenine dinucleotidase (NADase) (Fan and Feingold, 1970). If a coenzyme does not participate in the 4-epimerization this reaction would be unique among the 4-epimerases.

Interconversion of ribulose 5-phosphate and xylulose 5-phosphate, an important reaction in the oxidative pentose phosphate pathway, is a 3-epimerization. This epimerization apparently occurs via an endiol intermediate that can occur because of adjacent hydroxyl and keto groups. No coenzyme is required to facilitate this epimerization (Axelrod, 1965).

Epimerization of glucose to galactose is not restricted to the uridine nucleotide sugars. Neufeld (1962) and Katan and Avigad (1966) showed that TDP-D-glucose is converted to TDP-D-galactose by extracts of sugar beet roots. In addition, the sugar beet root extracts oxidized TPD-D-galactose to TDP-D-galacturonate.

Some sugars are derived by epimerization of more than one hydroxyl group. Roberts (1971b) observed that ^{14}C from [^{14}C]D-mannose administered to corn roots is incorporated into an alcohol-insoluble polymer that contains D-mannose, L-galactose, and L-fucose. None of the mannose was converted to D-glucose, D-galactose, D-xylose, or L-arabinose isolated from cell walls. This suggests that L-galactose and L-fucose were derived from D-mannose, possibly through GDP-D-mannose as an intermediate (see Section II,C,3). Synthesis of L-galactose and L-fucose from D-mannose require epimerizations at C-3 and C-5.

E. Glucose 6-Phosphate–*myo*-Inositol–Glucuronate Pathway

Maquenne suggested, before 1900, that *myo*-inositol was formed by cyclization of a glucose derivative (Posternak, 1965). Work by Loewus and associates (Loewus *et al.*, 1962; Loewus and Kelly, 1963; Loewus and Loewus, 1971) using parsley and Chen and Charalampous (1963, 1964, 1965a,b) using the yeast *Candida utilis* have established that D-glucose 6-phosphate is converted to D-glucuronate through 1-L-*myo*-inositol (Fig. 3). Ruis *et al.* (1967) were the first to demonstrate *myo*-inositol formation in cell-free extracts of higher plants. They incubated D-glucose 6-phosphate with cell-free preparations from 4-week-old *Phaseolus vulgaris* or *Sinapis alba* tissue and showed the formation of *myo*-inositol. The work of Chen and Charalampous (1963, 1964, 1965a,b, 1966) delineated the pathway from D-glucose 6-phosphate to *myo*-inositol with cell-free extracts from *C. utilis* and purified enzymes from this organism. Recently, M. W. Loewus and Loewus (1971, 1973) purified and

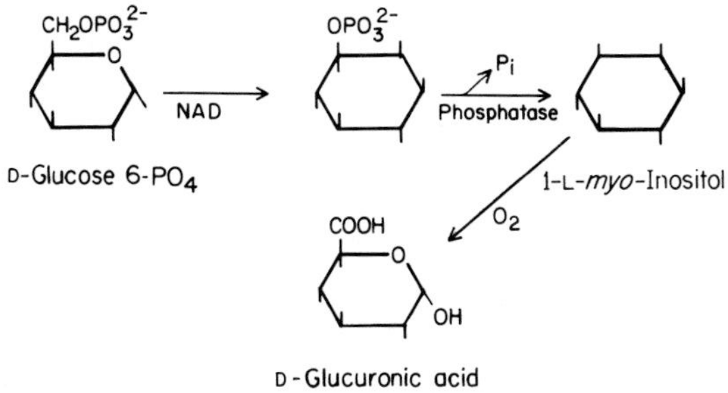

Fig. 3. D-Glucose 6-phosphate–*myo*-inositol–glucuronate pathway.

characterized D-glucose-6-phosphate cycloaldolase (NAD-dependent) from *Acer pseudoplatanus* L. (sycamore) cell cultures. Ammonium ion increases the activity of the enzyme from the sycamore as well as that obtained from *Neurospora* (Pina and Tatum, 1967) and *Candida* (Chen and Charalampous, 1965a). The mechanism of this reaction has been investigated (Barnett and Corina, 1968; Hauska and Hoffmann-Ostenhof, 1967; Pina *et al.*, 1969; Chen and Charalampous, 1967; Loewus and Loewus, 1973). The NAD⁺ requirement for cyclization of D-glucose 6-phosphate to 1-L-*myo*-inositol 1-phosphate suggests a three step reaction sequence: (1) oxidation at C-5 atom of D-glucose 6-phosphate, (2) intramolecular aldol condensation to form D-*myo*-inose-2 1-phosphate, (3) reduction of D-*myo*-inose-2 1-phosphate by enzyme-bound NADH to yield 1-L-*myo*-inositol 1-phosphate. NADH remains tightly bound to the enzyme and does not exchange with added NADH (Barnett and Corina, 1968; Chen and Charalampous, 1967). The proposed D-*xylo*-5-hexulose 6-phosphate remains enzyme bound during the course of the reaction (Sherman *et al.*, 1969). Loewus and Loewus (1973) suggest that the enzyme has the properties of a type II aldolase.

Crude cell-free preparations contain a phosphatase that hydrolyzes *myo*-inositol 1-phosphate to 1-L-*myo*-inositol (Chen and Charalampous, 1966; Eisenberg, 1967; Loewus and Loewus, 1971). The enzyme does not catalyze hydrolysis of D-glucose 6-phosphate.

Loewus *et al.* (1962) were first to note that ¹⁴C from 1-L-*myo*-inositol became incorporated into the component sugars of cell wall pectin and hemicellulose of strawberry and parsley leaves. The ¹⁴C was located in arabinose, xylose, and galacturonic acid. By contrast, hexose residues did not contain ¹⁴C. Callus tissue of *Fraxinus pennsylvanica* grown on [U-¹⁴C]inositol for 10 weeks contained little ¹⁴C in cell wall galacturonate and *myo*-inositol obtained from phosphoinositides was of high specific activity. The authors suggest that *myo*-inositol is not an important precursor of cell wall uronic acid (Jung *et al.*, 1972).

Gruhner and Hoffmann-Ostenhof (1966) isolated from barley seedlings and strawberries a *myo*-inositol oxygenase (*myo*-inositol:oxygen oxidoreductase EC 1.13.99.1). The mechanism of the reaction catalyzed by plant enzymes has not been investigated. The authors stated that it had properties similar to the yeast oxygenase (Charalampous, 1960).

The data currently available, and discussed in greater detail by Loewus (1969, 1971), suggest that the reactions from D-glucose 6-phosphate to *myo*-inositol and D-glucuronic acid provide, at least in some plant species, a direct pathway for formation of pentoses and uronic acids. It may be significant that pentoses and uronic acids, formed in relatively large quantities in plants, can be synthesized without the obligatory net reduction of NAD⁺ or NADP⁺. This pathway also provides precursors of apiose and galactinol as discussed in a latter section of this chapter.

Three-day-old wheat or bean seedlings do not incorporate a significant quantity of glucose into 1-L-*myo*-inositol (Matheson and St. Clair, 1971). Matheson and St. Clair suggest that stored phytate may serve as a reserve source of 1-L-*myo*-inositol in young tissues. This would provide more glucose for energy metabolism and hexoses for cell wall synthesis. This is a particularly interesting finding and suggests that in young seedlings phytate may provide the carbon source for pentoses and uronic acids, and that storage polysaccharides provide the carbon source for neutral hexoses and for energy metabolism.

F. Oxidation and Decarboxylation of Nucleotide Sugars

Edelman *et al.* (1955) demonstrated that carbon atom 6 of hexose was lost in the conversion of glucose to xylose isolated from wheat seedling xylan, and Ginsburg and Hassid (1956) showed that carbon atom 1 of glucose was not lost when glucose was converted to either D-xylose or L-arabinose isolated from wheat seedling hemicellulose. The labeling patterns indicated that only 15% recycling of hexose through the oxidative pentose cycle occurred. The evidence favors a pathway of metabolism in which carbon atom 6 of hexose is lost in synthesis of D-xylose and L-arabinose.

Uridine 5′-(α-D-glucopyranosyl pyrophosphate) undergoes oxidation at carbon atom 6. Solms and Hassid (1957) isolated UDP-D-glucuronic acid from mung beans. Strominger and Mapson (1957) obtained, and purified 1000-fold, UDP-D-glucose dehydrogenase (UDP-D-glucose:NAD oxidoreductase, EC 1.1.1.22) from pea seeds.

$$\text{UDP-D-Glc} + 2 \text{ NAD}^+ \rightleftarrows 2 \text{ NADH} + 2 \text{ H}^+ + \text{UDP-D-glucuronate} \qquad (8)$$

NADP⁺ cannot substitute for NAD⁺ as an electron acceptor. Although Hassid (1967) has indicated that the enzyme is widespread in the plant kingdom, there is little recorded information about the quantitative importance of this enzyme. Roberts (1971a) has shown that UDP-D-glucose dehydrogenase activity in barley seedlings is very low during the first 6 days of germination and insufficient to meet the requirements for cell wall polysaccharide biosynthesis. In contrast to UDP-D-glucose dehydrogenase activity, UDP-D-glucuronate pyrophosphorylase activity increases 15-fold during the first 6 days of germination. These studies favor the D-glucose 6-phosphate–*myo*-inositol–glucuronate pathway as the predominant one for forming UDP-D-glucuronate in young seedlings. It remains

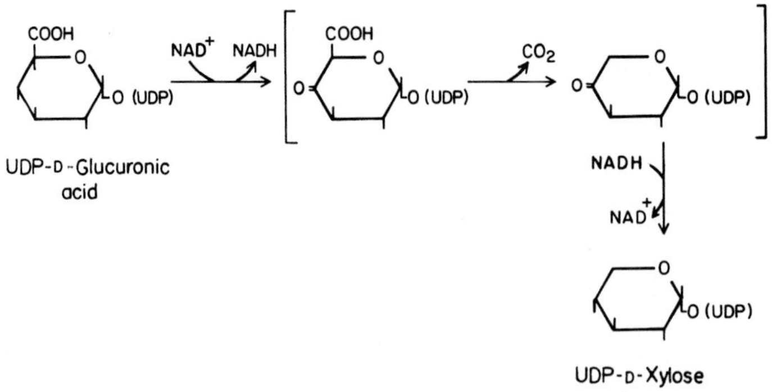

COOH
NAD⁺ NADH
UDP-D-Glucuronic acid
CO₂
NADH
NAD⁺
UDP-D-Xylose

Fig. 4. Proposed mechanism of UDP-D-glucuronate decarboxylation.

to be demonstrated that other enzymes of that pathway are active during early stages of seedling growth.

Katan and Avigad (1966) have obtained from sugar beet roots an enzyme preparation that catalyzes the conversion of TDP-D-glucose to TDP-D-galacturonate when incubated with NAD⁺. These data suggest a pathway of galacturonate formation that is independent of glucuronate formation.

Loewus and Kelly (1963) have shown that D-glucuronate is a precursor of D-xylose and L-arabinose in cell wall polysaccharides. Roberts *et al.* (1967a,b) extended this work to show that 1-L-*myo*-inositol also serves as a precursor of D-xylose in xylan and 4-*O*-methylglucuronic acid in cell wall isolated from *Zea mays* root tips. Feingold *et al.* (1960, 1964) had demonstrated previously that enzymes from higher plants catalyze the conversion of UDP-D-glucuronate to UDP-D-xylose and UDP-L-arabinose. The enzyme preparation probably contained UDP-D-xylose 4-epimerase. Ankel and Feingold (1965) isolated from wheat germ UDP-D-glucuronate carboxylase, which catalyzes conversion of UDP-D-glucuronate to UDP-D-xylose. The enzyme was purified 350-fold and a NAD⁺ requirement shown. The purified enzyme did not decarboxylate UDP-D-galacturonate. They suggested UDP-4-keto-D-glucuronate as an intermediate of the reaction (Fig. 4). Although these data are limited, they are consistent with the postulate that L-arabinose as well as D-xylose are derived from D-glucuronate (Fig. 1) (Maitra *et al.*, 1974).

III. 1-L-*myo*-Inositol Metabolism

Four of the stereoisomers of inositol along with several methyl ethers of inositol have been isolated from plants. The predominant stereoisomer is 1-L-*myo*-inositol. 1-L-*myo*-Inositol derivatives that have been identified include phosphate, indoylacetyl esters, indoylacetyl 1-L-*myo*-inositol arabinosides (Ueda *et al.*, 1970), *O*-α-D-galactopyranosyl-(1 → 2)-1-L-*myo*-inositol, phosphatidylinositol, and various complex glycolipids (Carter *et al.*, 1964; Anderson and Wolter, 1966). Imhoff and Bourdu (1970) observed that in Italian ryegrass that had received intense illumination followed by a long dark period nearly all the inositol, as well as sucrose, was in the chloroplasts.

The greatest portion of the inositol is stored as its hexaphosphate ester (phytic acid) complexed with calcium, magnesium and potassium ions (Matheson and Strother, 1969). During germination phytic acid is used as a source of glucuronate precursor as indicated previously. The activity of phytase upon phytic acid to convert it to 1-L-*myo*-inositol is modified by the concentration of orthophosphate (Sartirana and Bianchetti, 1967). Orthophosphate represses phytase synthesis in wheat (Bianchetti and Sartirana, 1967) and in mung bean (Mandal and Biswas, 1970).

A. Phytic Acid

The conversion of *myo*-inositol to phytic acid has been established by several investigators. Roberts and Loewus (1968) showed that 20% of the ^{14}C from [^{14}C]1-L-*myo*-inositol taken up was incorporated into phytic acid in the aquatic angiosperm *Wolfiella floridana*.

The biosynthesis of phytic acid is not completely understood. English *et al.* (1966) obtained an inositol kinase from germinating mung bean seeds. Inositol 1-phosphate is the product of the reaction. The authors propose that inositol 1-phosphate is an intermediate in phytic acid biosynthesis. Majumder and Biswas (1973a) present evidence that phosphorylation of inositol phosphates (inositol-P_1 to inositol-P_5) occurs via transfer of the terminal phosphoryl group of ATP to a phosphoprotein, and it serves as the direct phosphoryl donor eventually forming inositol hexaphosphate (inositol-P_6). Majumder and Biswas (1973b) found a protein inhibitor of phosphoinositol kinase which appears in the latter stages of ripening of *Phaseolus aureus* seeds. They suggest that failure to detect phosphoinositol kinase in ungerminated seeds may be due to the presence of the inhibitor rather than the absence of the kinase. They showed that ATP can be replaced partially by UTP and phosphoenol pyruvate (Majumder *et al.*, 1972). Phosphoinositol kinase and inositol kinase are different enzymes. Asada *et al.* (1969) have shown that small quantities of inositol monophosphate and large quantities of inositol hexaphosphate accumulate in wheat grains. Tanaka *et al.* (1971) have characterized the monophosphate as *myo*-inositol 2-phosphate, and they have presented evidence that it is derived by phosphorylation of 1-L-*myo*-inositol rather than degradation of phytic acid. Wheat and rice grains contain negligible quantities of inositol di-, tri-, tetra-, and penta-phosphates. They proposed that some unknown substance reacts with inositol monophosphate to form inositol monophosphate–X which serves as a substrate for further phosphorylation until all six hydroxyl groups are phosphorylated, and X along with inositol hexaphosphate are released. This hypothesis is not consistent with the observation that inositol mono- through hexaphosphates accumulate when an enzyme extract from *Lemna gibba* is incubated with ATP and 1-L-*myo*-inositol 1-phosphate (Molinari and Hoffman-Ostenhof, 1968).

Most investigators (Mori, 1969; Roberts and Loewus, 1968; Asada *et al.*, 1969) have found at least trace quantities of the various partially phosphorylated inositol species. It is not known whether these species represent intermediates in biosynthesis of, or degradation products of, phytic acid.

Little is known about the role of phytate. Biswas and Biswas (1965) purified an enzyme from mung bean that catalyzes phosphoryl transfer from phytate to GDP to form GTP. This report has not been confirmed. The occurrence of such an enzyme could provide a ready source of GTP for synthesis of protein and of GDP-sugars.

B. D-Apiose and D-Hamamelose Biosynthesis

The branched chain sugar D-apiose (3-*C*-hydroxymethyl-*aldehydo*-D-glycerotetrose) was first discovered as a component of a flavone glyco-side by Vongerichten (1901) and its structure proved by Schmidt (1930). The 5,7,4'-flavone glycoside contains D-apiofuranosyl-(1 → 2)-D-gluco-pyranosyl moieties attached by a glycosidic linkage to C-7 atom of the flavone (Hemming and Ollis, 1953) (Fig. 5). *Viburnum furcatin* contains *p*-vinylphenol-(D or L)-apiofuranosyl-(1 → 6)-β-D-glucopyranoside (Hat-

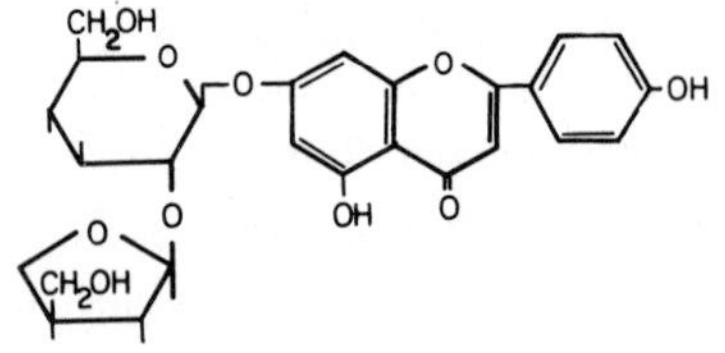

Fig. 5. Structure of 7-apigenin-*O*-β-D-apiofuranosyl-(1-2)-β-D-glucopyranoside.

tori and Imaseki, 1959). *Lemna minor* (duckweed) cell walls contain apiogalacturonans. Hart and Kindel (1970) present evidence that two apiofuranosyl residues (apiobiosyl) are attached to the galacturonan as a disaccharide. They tentatively conclude, based on proton magnetic resonance spectra and rates of hydrolysis of the apibiosyl residues, that the terminal apiosyl unit is attached β-$(1 \rightarrow 3)$ to the next apiosyl unit. The occurrence of apiose as a constituent of a polysaccharide may be very widespread (Bell, 1962), and apiose can be mistaken easily for rhamnose because they have similar chromatographic properties (Bell *et al.*, 1954).

Grisebach and Dobereiner (1964, 1966) showed that ^{14}C from [U-^{14}C]D-glucose is incorporated into apiin in parsley and also into the apiosyl moiety of apiin. The C-3' atom of apiose was derived from either C-3 or C-4 atom of glucose. Acetate or formate are not precursors of apiose and all 5 carbon atoms are derived from D-glucose. Mendicino and Picken (1965) found that apiose in both parsley and duckweed are derived from glucose with the loss of C-6 atom of glucose. They showed later that carbon atoms 1 through 5 of glucose are incorporated without randomization into apiose, and they confirmed Grisebach and Dobereiner (1964) observation that C-3' atom is derived from either C-3 or C-4 of glucose (Picken and Mendicino, 1967). Tracer studies with labeled inositol (Roberts *et al.*, 1967b) provide evidence that apiose can be derived via the D-glucose 6-phosphate–inositol–glucuronate pathway. Grisebach and Sandermann (1966) showed that D-glucuronic acid is converted to apiose. Sandermann *et al.* (1968) showed that UDP-D-glucuronate is converted to UDP-D-apiose by cell-free extracts of parsley. Sandermann and Grisebach (1968) isolated a small quantity of UDP-D-apiose from parsley. In contrast, Gustine and Kindel (1969) reported the isolation of a phosphorylated derivative of apiose which contained no uridine when UDP-D-glucuronate was incubated with a cell-free extract of duckweed. More recently Kindel and Watson (1973) have purified UDP-glucuronate cyclase 55-fold from *Lemna minor*. It catalyzes the conversion of UDP-D-glucuronate to UDP-D-apiose. Mendicino and Hanna (1970) showed that cell-free extracts of duckweed readily convert UDP-D-glucuronate to α-D-apio-D-furanosyl 1:2-cyclic phosphate. They demonstrated a requirement for NAD$^+$ and that NADP$^+$ could not substitute for NAD$^+$. UDP-pentoses are not converted to the apiose derivative. It is likely that α-D-apio-D-furanosyl 1:2-cyclic phosphate is derived from UDP-D-apiose by a nonenzymic degradation (Kindel and Watson, 1973). Baron *et al.* (1973) have purified the enzyme 1000-fold from parsley. The ratio of UDP-apiose and UDP-xylose formed from UDP-D-glucuronate was constant, and they concluded that one enzyme catalyzes both reactions. Added ammonium ion increases the quantity of

UDP-xylose formed suggesting that NH_4^+ channels the direction of synthesis (Baron et al., 1972) (see Fig. 1).

Mechanisms for apiose biosynthesis were suggested before any of the biosynthetic investigations described above were conducted. An intramolecular acyloin condensation of a 4-ketohexose to form a cyclic intermediate (Shemyakin et al., 1952) and aldol condensation between dihydroxyacetone and glycoaldehyde are inconsistent with the data (Hough and Jones, 1956). Candy et al. (1964) proposed a mechanism involving a transanular rearrangement of a hexose to form a branched chain hexose. The data are consistent with Candy's mechanism as well as the two proposed by Grisebach and Dobereiner (1966) and Mendicino and Picken (1965) (Fig. 6).

Two glycosyltransferases were obtained from parsley which catalyze the incorporation of glucose from UDP-D-glucose into apigenin followed by transfer of apiose from UDP-D-apiose with the formation of apiin (Sutter et al., 1972; Ortmann et al., 1972). Apiosyltransferase is highly specific for UDP-D-apiose as the donor, but it is not as specific for the acceptor other than it must be a β-glucoside. Parsley contains both 7-O- and 3-O-flavanol glucosides. Sutter and Grisebach (1973) separated and partially characterized a UDP-D-glucose:flavanol 3-O-glucosyltransferase from the 7-O-glucosyltransferase.

D-Hamamelose (2-C-hydroxymethyl-D-ribose) was first derived from witch hazel (Hamamelis virginiana L.) tannin by Fischer and Freudenberg (1912). They isolated the glycoside and showed that its elemental analysis corresponded to di-O-galloyl hexoside. Work by Freudenberg and colleagues and Schmidt (1929; reviewed by Shafizadeh, 1956) resulted in proof of structure of the D-hamamelose moiety and confirmation

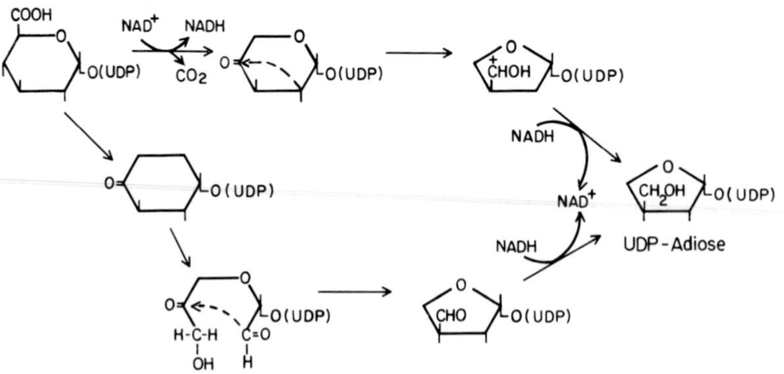

Fig. 6. Proposed mechanisms of UDP-D-apoise biosynthesis from UDP-D-glucuronic acid. The upper pathway was proposed by Grisebach and Dobereiner (1966) and the lower pathway was proposed by Picken and Mendicino (1967).

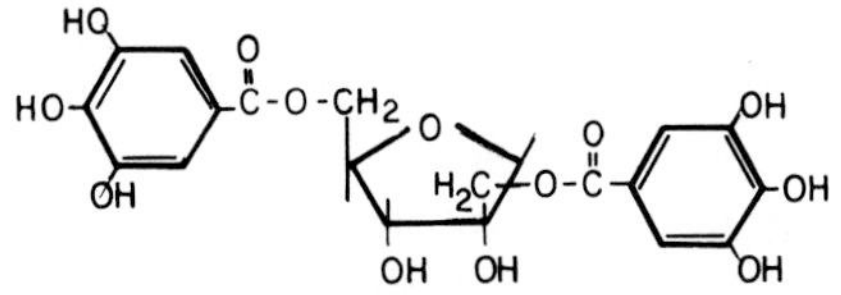

Fig. 7. Structure of di-O-galloyl-2′,5-hamamelose.

of the glycoside structure, in which galloyl units are esterified to the primary hydroxyls at C-5 and C-2′ atoms (Fig. 7).

Sellmair and co-workers (1968) and Beck (1969) isolated a hamamelitol-containing substance, clusinose, and characterized it as O-α-D-galactopyranosyl-(1 → 2)-D-hamamelitol from *Primula clusiana* Tausch. Hamamelose occurs in 75% of the 560 species (110 families) of plants investigated (Beck *et al.*, 1971). Sellmair *et al.* (1969) reported that twice as much clusinose as sucrose accumulates in *Primula* leaves. During photosynthesis in $^{14}CO_2$ clusinose is formed more slowly than sucrose. Unlike sucrose, it is not degraded during 4 days in the dark. Furthermore, the hamamelitol portion is labeled with ^{14}C much more slowly than the galactopyranosyl residue during photosynthesis in $^{14}CO_2$. This suggests that an endogenous pool of hamamelitol or hamamelitol precursors which is considerably larger than the galactose pool exists.

Eickenbusch *et al.* (1971) concluded that hamamelose biosynthesis is connected to photosynthesis. Inhibition of the Calvin cycle results in inhibition of incorporation of ^{14}C from [^{14}C]D-glucose into hamamelose. Beck *et al.* (1971) showed that [^{14}C]hamamelose diphosphate accumulates during 15 minutes of photosynthesis by spinach chloroplasts. Their evidence for hamamelose diphosphate consisted of showing that hamamelose is released upon treating sugar phosphates from the sugar diphosphate area of a chromatogram with phosphatase. They suggested that hamamelose diphosphate is synthesized in chloroplasts, transferred to the cytoplasm, and dephosphorylated by a phosphatase not present in chloroplasts. Synthesis of hamamelose 2′,5-diphosphate might result from an aldol condensation of two 3-phosphoglyceraldehyde molecules (Fig. 8).

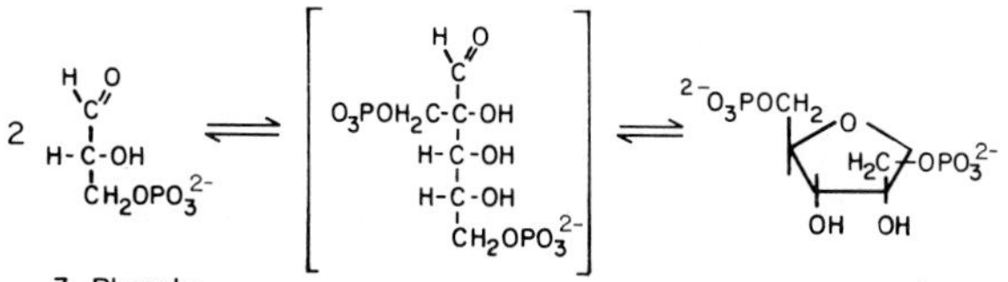

Fig. 8. Proposed mechanism of hamamelose-2′,5-diphosphate synthesis.

It is of interest to note the similarity between clusinose, galactinol, and various lichen products such as umbilicin and peltigeroside (see Section VI). In each substance, a hexose is attached to a polyol. Umbilicin and peltigeroside each contain a furanoside attached to the polyol. The role of furanosides in plants or fungi is unknown.

C. Galactinol Biosynthesis and Function

Galactinol was first isolated from sugar beets (Brown and Serro, 1953) and characterized by Kabat and associates (1953). Frydman and Neufeld (1963) showed that galactinol is synthesized by UDP-D-galactose:inositol galactosyltransferase that requires Mn^{2+} for activity. This requirement is not satisfied with Mg^{2+}. The reaction is given in Eq. (9).

$$\text{UDP-D-Gal} + myo\text{-inositol} \rightleftarrows \text{UDP}$$
$$+ O\text{-}\alpha\text{-galactopyranosyl}(1 \rightarrow 1)myo\text{-inositol} + H^+ \quad (9)$$

Tanner and Kandler (1966, 1968) confirmed this observation, and they demonstrated a physiological role for galactinol as the galactosyl donor in the biosynthesis of tri-, tetra-, and pentasaccharides; raffinose; stachyose; and verbascose was demonstrated (Senser and Kandler, 1966; Tanner et al., 1967; Lehle and Tanner, 1973). Galactinol is also an effective inhibitor of α-galactosidase activity. It is slowly hydrolyzed by α-galactosidase (Tanner, 1969). Galactinol is a common constituent in plants containing oligosaccharides of the raffinose family (Tanner, 1969). Galactinol:sucrose 6-galactosyltransferase was purified about 400-fold from Vicia faba seeds, and galactinol, p-NO_2-phenyl-α-D-galactopyranoside and raffinose, but not UDP-D-galactose, function as galactopyranosyl donors for the enzyme (Lehle and Tanner, 1973).

Tanner and Kandler (1968) suggested that the cofactor role of galactinol in biosynthesis of higher oligosaccharides may extend to other types of inositol glycosides. Little is known about the physiological function or biosynthesis of other inositol glycosides.

IV. L-Ascorbate Biosynthesis

L-Ascorbic acid biosynthesis has been investigated in both plants and animals. In animals all six carbon atoms of ascorbate are derived from D-glucose, and the sequence of carbon atoms in glucose becomes inverted during the process of L-ascorbate synthesis. L-Ascorbate biosynthesis in plants is not as clearly understood. Isherwood and Mapson (1962) and

Mapson (1967) champion the idea that the pathway of ascorbate synthesis in plants is similar to that in animals and that the uronic acids and their lactones are important intermediates. Loewus (1969, 1971) presents compelling evidence that L-ascorbate synthesis in plants is unlike that in animals. The reviews, cited above, by these investigators should be consulted for details and data.

Loewus et al. (1956, 1958) administered [1-^{14}C]- and [6-^{14}C]D-glucose to strawberries and cress seedlings. Distribution of ^{14}C in L-ascorbate subsequently isolated showed that glucose was converted to ascorbate without rearrangement or inversion of the sequence of carbon atoms. This finding is in contrast to the observation by Horowitz et al. (1952) and Horowitz and King (1953) who showed that in rats fed either [1-^{14}C]- or [6-^{14}C]D-glucose, L-ascorbate was labeled primarily in carbon atoms 6 and 1, respectively. Isherwood et al. (1954) proposed that in plants, as in the rat, the carbon skeleton of glucose became inverted during conversion of D-glucose or D-galactose to L-ascorbate. This conclusion was based on the observation that only D-glucurono-γ-lactone, L-gulono-γ-lactone, L-galactono-γ-lactone, and D-galacturonic acid methyl ester increased synthesis of L-ascorbate in cress seedlings. They proposed the following reaction sequence.

$$\text{D-Glucose} \rightarrow \text{D-glucuronic acid} \rightarrow \text{L-gluconic acid} \rightarrow \rightarrow \rightarrow \text{L-ascorbate} \qquad (10)$$

Loewus and associates (Baig et al., 1970; Finkle et al., 1960; Loewus et al., 1958; Loewus and Kelly, 1961) used L-galactono-1,4-lactone, D-glucurono-3,6-lactone, D-galacturonic acid, and L-gulono-1,4-lactone labeled in either C-6 or C-1 and showed the formation of L-ascorbate appropriately labeled as predicted by the scheme of Isherwood and Mapson. However, Loewus (1963) showed that [6-^{3}H]D-glucose is converted to [6-^{3}H]L-ascorbate in strawberries. Galacturonate isolated from pectin in the same experiment contained very little ^{3}H. These data appear to eliminate uronic acids as obligatory intermediates. In addition, an oxidation of C-1 of glucose (Loewus et al., 1956) and a change of configuration about the C-5 atom occurs in the conversion of glucose to L-ascorbate. Loewus (1969) proposed D-xylo-5-hexulosonate or its 1,4-lactone as possible intermediates and has evidence (Loewus, 1971) claiming loss of ^{3}H from [5-^{3}H]D-glucose in its conversion to L-ascorbate.

The detailed pathway for L-ascorbate synthesis from D-glucose in plants is unknown. The evidence appears to eliminate D-glucuronate, D-galacturonate, or their respective lactones as obligate intermediates. Nevertheless, plants are capable of converting uronic acids and their lactones to L-ascorbate.

V. Oligosaccharide Biosynthesis

A. Sucrose

Soon after the discovery of UDP-D-glucose by Leloir and associates they demonstrated that UDP-D-glucose served as a glucosyl donor in sucrose biosynthesis catalyzed by extracts from wheat, corn, peas, and bean germs (Leloir and Cardini, 1953, 1955). They also demonstrated that some preparations from peas formed sucrose 6'-phosphate (O-α-D-glucopyranosyl-$(1 \rightarrow 2)$-D-fructofuranosyl 6-phosphate) when D-fructose 6-phosphate was substituted for D-fructose (Leloir and Cardini, 1955). Buchanan (1953) found [^{14}C]sucrose 6'-phosphate in sugar beet leaves after 5 minute exposure in $^{14}CO_2$ and light and suggested that UDP-D-glucose was the glucosyl donor. This observation was confirmed by Bean *et al.* (1962). Bean and Hassid (1955) confirmed both observations from Leloir's laboratory. Yamaha and Cardini (1960a,b) showed that separate enzymes catalyze the synthesis of sucrose and sucrose 6'-phosphate. The properties of these enzymes are distinct (Slabnik *et al.*, 1968). Only UDP-D-glucose is an effective glucosyl donor in sucrose 6'-phosphate synthesis. In contrast, UDP-D-glucose, ADP-D-glucose, TDP-D-glucose, and GDP-D-glucose are reactants in the reversible synthesis of sucrose shown in Eq. (12).

$$\text{UDP-D-Glc} + \text{D-fructose 6-P} \rightleftarrows \text{sucrose 6'-P} + \text{UDP} + \text{H}^+ \qquad (11)$$
$$\text{UDP-D-Glc} + \text{fructose} \rightleftarrows \text{Sucrose} + \text{UDP} + \text{H}^+ \qquad (12)$$

Reactions (11) and (12) are catalyzed by sucrose-6'-phosphate synthase (UDP-D-glucose:D-fructose-6-phosphate 2-glucosyltransferase, EC 2.4.1.14) and sucrose synthase (UDP-D-glucose:D-fructose-2-glucosyltransferase, EC 2.4.1.13), respectively. Slabnik *et al.* (1968) showed that phenolic glucosides and ADP inhibit sucrose synthase but were without effect on sucrose-6'-phosphate synthase.

Reaction (12) is freely reversible with an estimated ΔG° of 1000 cal/mole (Cardini *et al.*, 1955). Mendicino (1960) reported a value of 3250 cal/mole for the ΔG° of reaction (11). Neufeld and Hassid (1963) questioned the validity of this estimate. They suggested that free energy of sucrose hydrolysis should be about -800 cal/mole less than that of sucrose 6'-phosphate. Hydrolysis of sucrose results in the formation of glucose and both fructofuranose and fructopyranose The conversion of the furanose initially formed by hydrolysis to the pyranose form pulls the overall reaction. The ratio of D-fructofuranose to D-fructopyranose at equilibrium is 1:4. Hydrolysis of sucrose 6'-phosphate results in the formation of D-glucose and D-fructofuranose 6-phosphate. Phosphate attached to carbon atom 6 locks the molecule into the furanose form. The discrepancy between the observations can be reconciled if either UDP or sucrose 6'-phosphate binds orders of magnitude more tightly to the enzyme than either UDP-D-glucose or D-fructose 6-phosphate. A valid $\Delta G°$ estimate would be difficult to obtain because of a long time interval needed to reach equilibrium. However, this is most likely not an important aspect of the physiology of sucrose formation because UDP is probably quickly converted to UTP.

Several investigations suggest that synthesis of sucrose 6'-phosphate is the primary means of forming sucrose. Putnam and Hassid (1954) infiltrated [^{14}C]glucose or [^{14}C]fructose into *Canna* leaves and showed that both hexosyl residues of sucrose were of approximately equal specific activity. In contrast, when [^{14}C]glucose was infused, the D-fructose isolated from *Canna* had little ^{14}C. These results suggest that sucrose synthesis occurs with D-fructose 6-phosphate as the glucosyl acceptor. Haq and Hassid (1965) reported that chloroplasts from sugarcane leaves contain a phosphatase that hydrolyzes both sucrose 6'-phosphate and D-fructose 6-phosphate. The chloroplasts also contained sucrose-6'-phosphate synthase. Bird *et al.* (1965) also found sucrose 6'-phosphatase in chloroplasts, but they failed to find a phosphatase that would hydrolyze fructose 6-phosphate. Hatch (1964) proposed that sucrose 6'-phosphate is synthesized in a metabolic compartment and is converted to sucrose during the process of transport to a storage compartment. Hawker and Hatch (1966) obtained a specific phosphatase from sugarcane and carrot roots that hydrolyzes sucrose 6'-phosphate but has little activity with D-fructose 6-phosphate, D-glucose 6-phosphate, D-glucose 1-phosphate, and D-fructose 1,6-diphosphate. The enzyme requires Mg^{2+}. Isherwood and Selvendran (1971) investigated the influence of changing from dark to light to dark on the quantitative relationship of phosphorylated sugars. They concluded that sucrose 6'-phosphate is most likely the first product of sucrose biosynthesis.

Priess and Greenberg (1969) carried out kinetic studies on wheat germ sucrose-6'-phosphate synthase. They obtained sigmoid kinetics with either UDP-D-glucose or D-fructose 6-phosphate as the variable substrate. They also observed that $MgCl_2$ increased the V_{max} twofold and decreased K_m of UDP-D-glucose about threefold. De Fekete (1971) purified sucrose-6'-phosphate synthase from *Vicia faba* cotyledons and demonstrated that freezing and thawing released an activator from the enzyme. The activator converted the enzyme from a form showing sigmoid kinetics to one giving hyperbolic kinetics.

In contrast to the investigations of sucrose-6'-phosphate synthase, those conducted on sucrose synthase all seem to provide indirect evidence that its physiological role is other than sucrose synthesis. Cardini and Recondo (1962) showed that sucrose synthase catalyzed the reaction of sucrose and UDP or ADP to form UDP-D-glucose or ADP-D-glucose, respectively. De Fekete and Cardini (1964), Murata et al. (1964), and Slabnik et al. (1968) suggested that the role of sucrose synthase is to convert sucrose to UDP-D-glucose. This reaction would conserve at least part of the energy of the sucrose glycosidic bond. High concentrations of sucrose in storage tissues and the furanose–pyranose equilibrium would drive this reaction toward UDP-D-glucose synthesis. Active UDP-D-glucose transglycosylases would also pull the sucrose synthase catalyzed reaction toward sucrose cleavage. The kinetics of sucrose cleavage using purified sweet potato sucrose synthase (Murata, 1971) are sigmoidal with respect to either sucrose or UDP concentration. In contrast, the kinetics of the reaction of sucrose synthesis are hyperbolic with fructose as the variable substrate. Presumably the physiologically important direction of the reaction is the one in which substrates serve to control the reaction that results in sigmoidal kinetics. Sucrose synthase was obtained in near homogenous form from rice grains (Nomura and Akazawa, 1973a). Hyperbolic kinetics were obtained for sucrose synthesis. Hill plots of data for sucrose degradation gave $\bar{n}$ values of 1.1 and 1.6 for UDP and ADP, respectively. Nomura et al. (1969) demonstrated previously that glucose from reserve starch in endosperm of germinating rice seeds is mobilized to the scutellum where it is converted to sucrose. The scutellum has an active sucrose-6'-phosphate synthase and sucrose synthase of lesser activity (Nomura and Akazawa, 1973b). Sucrose synthase from *Phaseolus aureus* seedlings was obtained by Grimes et al. (1970), and they showed that in addition to UDP-D-glucose, ADP-D-glucose, TDP-D-glucose, CDP-D-glucose, and GDP-D-glucose serve as substrates. UDP-D-glucose is the preferred substrate. Sucrose synthase activity is high in nonphotosynthetic tissue of *Phaseolus aureus* (Delmer and Albersheim, 1970). Delmer (1972a) purified the enzyme and showed that it is a tetrameric protein that catalyzes the reaction by a random mechanism. NADP+ and pyrophosphate activate the formation of UDP-D-glucose from sucrose and inhibit synthesis of sucrose (Delmer, 1972b). It is also of interest to note that the optimum pH of sucrose synthase is nearer physiological pH for sucrose degradation than for sucrose synthesis (Pressey, 1969).

Thus the available evidence suggests that the physiological role of sucrose-6'-phosphate synthase is to synthesize sucrose and that the role of sucrose synthase is to provide an available source of UDP-D-glucose, and possibly the other nucleoside diphosphate glucosides, for synthesis of glucosides. In addition, UDP-D-glucose is also a potential source of α-D-glucose 1-phosphate under conditions where UDP-D-glucose pyrophosphorylase activity favors pyrophosphorolysis (Gustafson and Gander, 1972). The net result of the action of sucrose synthase and UDP-D-glucose pyrophosphorylase links the metabolism of sucrose, the major translocatable sugar in many species, with polysaccharide and energy metabolism. This sequence of reactions conserves the energy of the glycosidic bond in sucrose and thus provides an overall efficiency that is relatively high.

Cleavage of sucrose is also achieved by invertase (O-β-D-fructofuranoside fructohydrolase EC 3.2.1.26). Tsai et al. (1970) proposed that sucrose is metabolized primarily by invertase at an early stage in developing endosperm and that at a later stage sucrose synthase becomes primarily important in sucrose cleavage in maize. In contrast, Hordeum distichum contains relatively high sucrose synthase activity in young endosperm (Baxter and Duffus, 1973). It is apparent that sucrose synthase and invertase activities need to be determined in many more species before a conclusion about the roles of these enzymes can be stated.

Particulate and soluble forms of invertase occur. Corn coleoptile invertase is attached to cell walls (Kivilaan et al., 1961). Jaynes and Nelson (1971a) reported that Zea mays L. seeds contain one bound and two soluble forms of invertase (invertase I and II). Invertase I activity reaches a maximum at 12 days after pollination. Its activity decreases during the period of greatest net starch synthesis. Invertase II is present in germinating corn embryos and in the 6-day stage endosperm. Bound invertase activity is present in the endosperm by the 6-day stage, and its activity remains constant during development. Hatch et al. (1963) described a soluble invertase from immature storage tissues. This invertase showed maximum activity at pH 5 to 5.5. Mature storage tissues contain a second soluble invertase that shows a maximum activity at pH 7.0. Lyne and ap Rees (1971) demonstrated both acid and alkaline invertase activity in pea roots. Indoleacetic acid increases and naphthyleneacetic acid decreases the activity of acid invertase in sugarcane internodal tissue (Sacher et al., 1963). Gibberellic acid and sucrose increases both soluble and wall-bound invertase in Avena stem segments. Gibberellic acid also promotes the appearance of an increase in invertase in other plant species (lentil epicotyls, sugarcane stem tissue, beet root slices, corn staminal filaments, and Jerusalem artichoke tuber) (Kaufman et al., 1973).

Pressey (1966, 1967) isolated from potato tuber a protein that inhibits invertase. Pressey and Shaw (1966) proposed that the inhibitor regulates invertase activity in the tuber. An invertase inhibitor has been demonstrated in red beets, sugar beets, and sweet potatoes (Pressey, 1968). The inhibitor is active against acid invertase, and its estimated molecular weight is 20,000 daltons. Jaynes and Nelson (1971b) reported an invertase I inactivator in corn endosperm. The inactivator appears to be a protein. Kinetics of inactivation are noncompetitive with respect to sucrose. Invertase I inactivator does not inactivate invertase II.

Schwimmer and Rorem (1960) observed that potato tubers subjected to low or high temperature stress accumulate sugars, especially sucrose. Their data appear to eliminate the possibility of an induced sucrose synthase activity resulting in response to an altered environment. It will be of interest to determine if these conditions influence the quantity of invertase inhibitor in the tuber.

B. Raffinose and Stachyose

The seeds, roots, and underground stems of many plants contain sucrose and one or more members of the raffinose family of oligosaccharides (French, 1954). These are considered as storage forms of galactosyl, glucosyl, and fructosyl residues. The structural relationship of these to one another is given in Fig. 9. Sucrose and raffinose are found in cereal grains, sugar beets, cottonseed, soybeans, and many legumes. Stachyose, along with sucrose and raffinose, is found in many, but not all, genera of Leguminosae family, and it is a dominant constituent in genera of the Labiatae (French, 1954). Senser and Kandler (1966) reported that in many plants, especially the Lamiaceae, a large percentage of the ^{14}C incorporated during photosynthesis is found in galactosides of the raffinose family. These glycosides are metabolized by plants in the dark.

The sequence of reactions by which galactose is incorporated into these oligosaccharides is unique among polymers containing hexosyl residues. Bourne et al. (1965) obtained an enzyme extract from $Vicia\ faba$ seeds that catalyzes the synthesis of raffinose from sucrose, UTP, and α-D-galactose 1-phosphate. Pridham and Hassid (1965) demonstrated

$$O\text{-}\alpha\text{-}\text{D-}Glc_p(1 \to 2)\text{-}O\text{-}\beta\text{-}\text{D-}Fruc_f$$

SUCROSE

$$O\text{-}\alpha\text{-}\text{D-}Gal_p(1 \to 6)\text{-}O\text{-}\alpha\text{-}\text{D-}Glc_p(1 \to 2)\text{-}O\text{-}\beta\text{-}\text{D-}Fruc_f$$

RAFFINOSE

$$O\text{-}\alpha\text{-}\text{D-}Gal_p(1 \to 6)\text{-}O\text{-}\alpha\text{-}\text{D-}Gal_p(1 \to 6)\text{-}O\text{-}\alpha\text{-}\text{D-}Glc_p(1 \to 2)\text{-}O\text{-}\beta\text{-}\text{D-}Fruc_f$$

STACHYOSE

Fig. 9. Structural relationship of sucrose, raffinose, and stachyose.

that galactose from UDP-D-galactose was incorporated into raffinose. However, Lehle and Tanner (1973) purified galactinol:sucrose 6-galactosyltransferase from *Vicia faba* seeds. They suggest that the crude enzyme preparation used by previous investigators catalyzed both reactions (13) and (14).

$$\text{UDP-D-Gal} + myo\text{-inositol} \rightleftarrows \text{galactinol} + \text{UDP} + \text{H}^+ \tag{13}$$
$$\text{Sucrose} + \text{galactinol} \rightleftarrows \text{raffinose} + myo\text{-inositol} \tag{14}$$
$$\text{Raffinose} + \text{galactinol} \rightleftarrows \text{stachyose} + myo\text{-inositol} \tag{15}$$
$$\text{Stachyose} + \text{galactinol} \rightleftarrows \text{verbascose} + myo\text{-inositol} \tag{16}$$

Galactinol is the galactosyl donor in the synthesis of stachyose and verbascose [Eqs. (15) and (16)] (Tanner and Kandler, 1966, 1968). *Phaseolus vulgarus* extracts catalyze stachyose formation, and *Vicia faba* extracts catalyze verbascose formation (see Fig. 1) (Tanner *et al.*, 1967). Plants containing oligosaccharides of the raffinose family also contain galactinol. Thus, Tanner (1969) concluded that galactinol has a central function in raffinose biosynthesis.

The pathway of degradation of these oligosaccharides is not known with certainty. Reid (1971) has shown that the metabolism of the raffinose family of oligosaccharides in *Trigonella foenum-graecum* L. endosperm and cotyledons results in a concomitant increase in galactose and sucrose in the endosperm and sucrose in the cotyledons. In *Vicia faba*, raffinose is hydrolyzed to galactose, and sucrose by an α-galactosidase, and galactose is phosphorylated by galactokinase (Pridham *et al.*, 1969).

In legumes galactose is stored in oligosaccharides of the raffinose family and in galactomannan polymers. The data suggest that oligosaccharides are depleted first during germination and that galactomannan is not utilized during the initial process of germination (Sioufia *et al.*, 1970). The process of hydrolysis of galactosyl residues followed by phosphorylation of galastose is inefficient as compared to the sucrose synthase type of reaction. It will be of interest to determine if appropriate plant species contain enzymes that carry out the degradation of galactofuranosyl in a way that conserves the energy of the galactosidic linkage.

C. Biosynthesis of C-, S- and O-Glycosides

A wide variety of C-, S-, and O-glycosides have been isolated from plants, and the biosynthesis of the S- and O-glycosides has been investigated extensively (Ettlinger and Kjaer, 1968; Conn and Butler, 1969; Conn, 1973; Pridham, 1965; Alston, 1968). Much of the attention has been directed toward delineating the metabolic pathways in formation of the aglycon moiety.

The physiological role, or roles, of these glycosides is largely unknown. Pridham (1960) reviewed this subject and concluded that no single function could be given to such a structurally heterogeneous group of substances. Initially these substances were considered as end products of metabolism (Staub, 1971; Miller, 1943). Other roles, such as carbohydrate reserves and stabilized nontoxic forms of phenolics also received considerable support in the older literature. More recently some of these substances have been shown to be metabolized actively. Kosuge and Conn (1961) presented evidence that 2-O-β-D-glucosyloxy-*trans*-cinnamic acid is an intermediate in coumarin biosynthesis: Furaya *et al.* (1962) that kaempferol-3-(triglucoside-*p*-coumaric acid) may be involved in regulating 3-indoylacetic acid oxidase activity; and Bough and Gander (1971) that o-β-D-glucopyranosyl-(1 → 2)-4-hydroxy-L-mandelonitrile (dhurrin) is turned over rapidly in young etiolated *Sorghum vulgare* seedlings metabolizing L-tyrosine in the dark. Abrol and Conn (1966) had previously presented similar evidence. Their evidence suggests that both the side chain and ring carbon atoms are oxidized to CO_2 (Bough and Gander, 1971). Pridham (1965) suggested that complex glycosides, such as some of the flavonoid glycosides that contain oligosaccharides attached to an aglycon, may serve as primer molecules for polysaccharide biosynthesis. Alternatively, these oligosaccharides may serve as precursor, or a storage form, of the oligosaccharide portion of glycoproteins in that particular species. Assuming that the oligosaccharide is transferred intact, comparison of the structure of the oligosaccharide in a flavonoid to that in the glycoproteins in that species could provide inferential evidence either supporting or refuting this idea.

1. C-Glycoside

In 1957 the existence of C-glycosides as natural products was first realized (Evans *et al.*, 1957) on the basis of the products obtained upon degrading vitexin (Fig. 10). Vitexin was first isolated by Perkin in 1900 who showed that it is a flavone derivative.

Wallace and Alston (1966) incubated [¹⁴C]apigenin-7β-O-glucoside

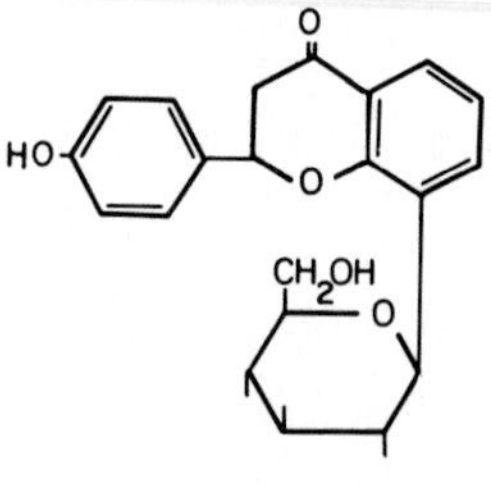

Fig. 10. Structure of vitexin, a C-glycoside.

with duckweed and obtained ^{14}C in luteolin-8β-O-glucopyranoside, but not in the C-glycosides. In a similar experiment with apigenin 8-C-glucopyranoside labeled with ^{14}C the radioactivity was not distributed to the other glucosides. They concluded that glucosylation may not be a terminal step in formation of C-glucosides. The biosynthesis of C-glycosides continues to be primarily speculative.

2. O-GLYCOSIDES

a. Cyanogenic Glycosides. Work on the biosynthesis of cyanogenic glycosides has been reviewed by Conn and Butler (1969) and Conn (1973) who along with their collaborators have made major contributions in this field.

Biosynthetic studies directed toward determining the precursor of the aglycone moiety of O-β-D-glucopyranosyl-(1 → 2)-4-hydroxy-L-mandelonitrile (dhurrin) were initiated independently in Conn's laboratory (Conn and Akazawa, 1958) and Gander's laboratory (1958). They (Koukal *et al.*, 1962; Akazawa *et al.*, 1960; Gander, 1960, 1962) established that the aglycone was derived from L-tyrosine, that tyramine was not a precursor, and that carbon atoms 2 and 3 of L-tyrosine became the nitrile and α-hydroxy carbon atoms, respectively, of 4-hydroxymandelonitrile. Uribe and Conn (1966) determined that the nitrile nitrogen of dhurrin was derived directly from the amino group of L-tyrosine, and Zilg *et al.* (1972) showed that the glucosidic linkage oxygen of the cyanogenic glucosides linamarin and lotaustralin is derived from O_2.

Mentzer (Mentzer *et al.*, 1963; Mentzer and Favre-Bonvin, 1961) proposed that 3-phenylpyruvic acid ketoxime is an intermediate in the conversion of L-phenylalanine to O-β-D-glucopyranosyl-(1 → 2)-L-mandelonitrile (prunasin) in cherry laurel. Ben-Yehoshua and Conn (1964) showed that phenylalanine is a precursor of prunasin in peaches. In addition, Mahadevan (1963) reported that 3-indole acetaldoxime was converted to indole acetonitrile in fungi and bananas. Tapper *et al.* (1967) demonstrated that α-ketoisovaleric acid oxime and isobutyraldoxime were converted to linamarin and lotaustralin in flax. Further, the oxime nitrogen remained attached to the carbon in reactions that followed in formation of linamarin and lotaustralin. Recently Tapper and Butler (1972) have shown that flax shoots accumulate an intermediate with the properties of isobutyraldoxime. Tapper *et al.* (1972) have shown that 2-hydroxyaldoximes may be precursors of the aglycone portion of linamarin and prunasin in flax and cherry laurel, respectively. Both phenylpyruvic acid ketoxime and phenylacetaldoxime were shown as possible prunasin precursors (Tapper *et al.* 1967; Hahlbrock *et al.*, 1968). Tapper and Butler (1971) had shown that the related 2-ketoxime, aldoximes, nitriles, and 2-hydroxynitriles were effective precursors of the aglycone moiety of linamarin, lotaustralin, and prunasin. They proposed a pathway for formation of cyanogenic glucosides involving N-hydroxyamino acids, aldoximes, nitriles and 2-hydroxynitriles as shown in Fig. 11.

Hahlbrock and Conn (1970) have purified UDP-D-glucose:ketone cyanohydrin β-D-glucosyltransferase from flax and demonstrated that it is highly specific for both UDP-D-glucose and acetone or butanone cyanohydrins. They had previously proposed, based on feeding experiments, that the α-hydroxynitrile was the glucosyl acceptor (Hahlbrock et al., 1968).

Farnden et al. (1973) have shown that young sorghum shoots convert p-hydroxyphenylacetaldoxime and p-hydroxyphenylacetonitrile to dhurrin. They also obtained evidence through the use of "trapping experiments" for the in vivo formation of p-hydroxyphenylacetaldoxime but not p-hydroxyphenylacetonitrile from tyrosine. Reay and Conn (1974) have partially purified a glucosyl transferase from sorghum seedlings. This enzyme catalyzes the formation of stereospecific dhurrin [p-hydroxy(S)-mandelnitrile-β-D-glucopyranoside] from UDP-D-glucose and (R,S)-p-hydroxymandelonitrile. In contrast, Zilg and Conn (1974) have shown that natural linamarin is in the (R) configuration and that when a mixture of the two enantiomers (R,S)-2-hydroxy-2-methylbutyronitrile was administered to excised flax shoots both (R)-lotaustralin and (S)-epi-lotaustralin were formed. Further, the UDP-D-glucose:ketone cyanohydrin-β-D-glucosyltransferase did not distinguish between the two epimers of (R,S)-2-hydroxy-2-methylbutyronitrile resulting in formation of both (R), lotaustralin and (S)-epi-lotaustralin.

Recently McFarlane et al. (1975) have demonstrated in vitro the complete synthesis of dhurrin from L-tyrosine, and their evidence is com-

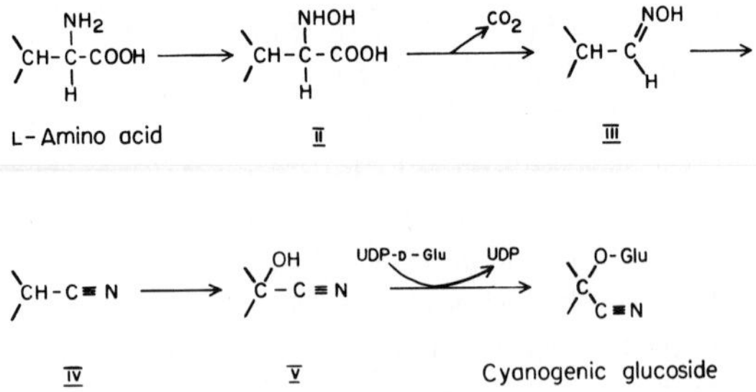

Fig. 11. Proposed pathway for the biosynthesis of a cyanogenic glucoside from and L-amino acid. Structures II through V are N-hydro-L-amino acid, an aldoxime, a nitrile, and a cyanohydrin, respectively. Figure modified from Conn and Butler (1969).

patible with the reaction scheme shown in Fig. 11. The conversion of
L-tyrosine to 4-hydroxymandelonitrile is catalyzed by particulate enzymes
and the glucosyltransferase is a soluble enzyme. A more detailed discussion of this subject is given by Conn and Butler (1969) and Conn (1973).

b. Other O-Glycosides.   UDP-D-glucose is the glucosyl donor in the glycosylation of a number of other aromatic substances. Cardini and Leloir
(1957) showed that the synthesis of arbutin (4-hydroxyphenyl β-D-gluco-
pyranoside) occurred with UDP-D-glucose as the glucosyl donor using
wheat germ extract. This work was extended by Yamaha and Cardini
(1960a,b) who showed that wheat germ extracts also contained a second
glucosyltransferase that catalyzes the formation of 4-hydroxyphenyl
β-D-gentiobioside. Trivelloni et al. (1962) reported that ADP-D-glucose
is a better glucosyl donor in this reaction than UDP-D-glucose.

Barber (1962) reported that in *Phaseolus vulgaris* TDP-D-glucose
serves as the glucosyl donor in formation of a rutin precursor, O-β-D-glu-
copyranosyl-$(1 \rightarrow 3)$-quercitin,   and   TDP-L-rhamnose   serves   as   the
rhamnosyl donor in the synthesis of rutin $[O$-α-L-rhamnosyl-$(1 \rightarrow 6)$-β-
D-glucopyranosyl-$(1 \rightarrow 3)$-quercitin] (Jacobelli *et al.*, 1958). Marsh
(1960) has obtained an enzyme from *Phaeolus vulgaris* which glucuro-
nosylates quercitin with UDP-D-glucuronate as the glycosyl donor. The
position of attachment of glucuronate has not been determined.

Glycosides are found in ester linkage to a carboxylate. Glucosyl
transfer from UDP-D-glucose to form an ester has been demonstrated
(Kosuge and Conn, 1959; Jacobelli *et al.*, 1958).

3. S-GLYCOSIDES

The thioglycosides are another class of glycosides that has been
investigated rather extensively. Their biosynthesis was reviewed by Ettlinger and Kjaer (1968) and compared to that of the cyanogenic glucosides. That review should be consulted for a more detailed description of
the work.

The aglycone portion of glucosinolates is, like the cyanogenic glucosides, derived from amino acids as proposed by Kjaer (1960). Underhill
and Chisholm (1964) showed that the C-2 atom and nitrogen of phenylalanine are incorporated intact into benzylglucosinolate (glucotropaeolin).
Evidence has been presented that suggests N-hydroxyphenylalanine and
phenylacetaldoxime are intermediates in glucotropaeolin biosynthesis
(Kindl and Underhill, 1968). In contrast, neither phenylacethydroxamic
nor phenylpyruvic acid oxime are effective precursors. Tapper and Butler
(1967) showed that isobutyraldoxime was a precursor of isopropylgluco-

sinolate in *Cochlearia officinalis* L., and that it was severalfold more effectively incorporated into the glucosinolate than either L-valine or α-ketoisovalerate oxime.

In addition to the feeding type of experiments, experiments have been conducted with cell-free preparations. Enzyme prepartions from *Sinapis albis* L., *Tropaeolum majus* L., and *Nasturtium officinale* R. Br catalyze decarboxylation of *N*-hydroxyphenylalanine to phenylacetaldoxime (Kindl and Underhill, 1968). The pathway shown in Fig. 12 was proposed for biosynthesis of the aglycone portion of glucosinolates. This pathway is consistent with data from Matsuo and Underhill (1969) who isolated [^{14}C]phenylacetothiohydroxamate from *T. majus* plants administered [1-^{14}C]phenylacetaldoxime. An enzyme has been isolated from *T. majus* which catalyzes the transfer of glucose from UDP-D-glucose to phenylacetothiohydroxamate forming desulfobenzylglucosinolate (Matsuo and Underhill, 1971). ADP-D-glucose, GDP-D-glucose, CDP-D-glucose, UDP-D-galactose, and UDP-D-xylose did not serve as glycosyl donors. Specificity toward the aglycone was not as great; however, phenylacetothiohydroxamate was the best acceptor of a number of analogues and related compounds tested.

Glucosinolate biosynthesis has also been investigated in *Seseda luteola* L. which contains glucobarbarin. Glucobarbarin and progoitrin (from thyroid) are derived from amino acids modified by single or multiple condensations with acetate. Glucobarbarin is derived from L-phenylalanine, which along with acetate results in the formation of 3-benzylmalic acid (Underhill, 1967; Underhill and Kirkland, 1972a). Underhill and Kirkland showed that C-2 atom and N of L-2-amino-4-phenylbutyric

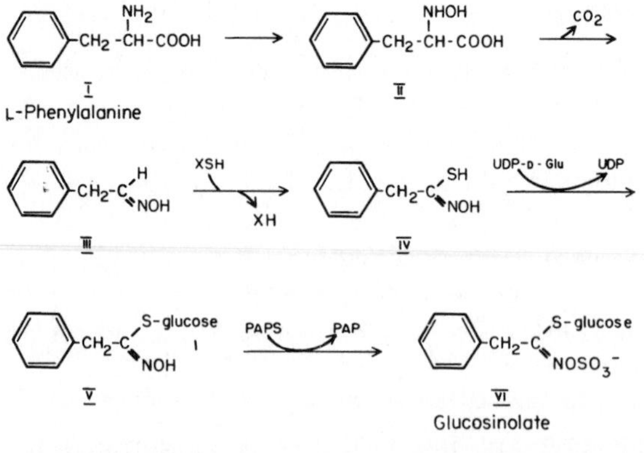

Fig. 12. Proposed pathway for formation of a glucosinolate from L-phenylalanine. Structures II through V are *N*-hydroxy-L-phenylalanine, phenylacetaldoxime, phenylacetothiohydroxamate, and desulfobenzylglucosinolate, respectively. Figure taken in part from Matsuo and Underhill (1971).

acid are incorporated intact into glucobarbarin [(S)-2-hydroxy-2-phenyl-ethyglucosinolate]. Progoitrin is derived from 2-amino-6-(methylthio)-caproic acid (Lee and Serif, 1970). Underhill and Kirkland (1972b) have isolated 2-methylpropylglucosinolate from *Conrigia orientalis* L. Andrz. and have shown that it is formed from valine and the methyl carbon of acetate via leucine.

Tryptophan serves as a precursor of indoleglucosinolates, which are widely distributed among the Crucifera family (Elliott and Stowe, 1971). Mahadevan and Stowe (1972) have isolated desthioglucobrassicin (1-S-glucosyl indolacetaldoxime) after administering 3-indoleacetal-doxime to the Crucifer woad (*Isatis tinctoria* L.). Glucobrassicin is found in cabbage and woad.

VI. Polyols, Monosaccharides, and Oligosaccharides in Lichen

Lichen is a form in which green or blue-green algae and a fungus live together symbiotically. Although an extensive investigation of the relative quantities of monosaccharides, polyols, and oligosaccharides in lichens has not been published, the available literature suggests that in addition to common hexoses, relatively large quantities of polyols accumulate. D-Ribitol, D-arabitol, *meso*-erythritol, D-mannitol, glycerol, *myo*-inositol, D-*glycero*-D-*talo*-heptitol (volemitol) and 1-deoxy-D-*glycero*-D-*talo*-heptitol have been identified from one or more lichen species (Culberson, 1969). The order Pyrenocarpeae contains volemitol and mannitol but no arabitol. In contrast, Gymnocarpeae contains arabitol and mannitol but no volemitol (Lindberg *et al.*, 1953). All of these substances are reduced products derived, at least in part, from the oxidative pentose phosphate cycle except for mannitol. These observations suggest that the symbiotic relationship results in relatively large quantities of reduced coenzymes, possibly even an imbalance of reduced over oxidized NAD⁺ or NADP⁺. This imbalance could result in the accumulation of polyols.

Mannitol accumulates in some algae species much as sucrose accumulates in vascular plants. Yamaguchi *et al.* (1969) showed that the ^{14}C labeling pattern in mannitol was similar to that of hexose phosphates derived during photosynthesis. They did not determine whether mannitol was derived by reduction of mannose or by reduction of fructose 6-phosphate to mannitol or mannitol 1-phosphate, respectively. The algal fronds contain a phosphatase that has a high degree of specificity toward mannitol 1-phosphate.

In addition to sucrose and α,α-trehalose, three unique oligosaccharides are found in lichen. Two oligosaccharides, 3-*O*-β-D-galactofura-

nosyl-D-manitol (peltigeroside) (Pueyo, 1959) and 3-O-β-D-glucopyra-
nosyl-D-mannitol were obtained from *Peltigera horizontalis*. A second
galactofuranoside, 2-O-β-D-galactofuranosyl-D-arabitol (umbilicin) has
been isolated from *Umbilicaria pustulata* L. (Lindberg *et al.*, 1952; Lindberg and Wickberg, 1962).

The biosynthetic reactions that lead to the galactofuranosides and
their physiological role is currently unknown. However, Trejo *et al.*
(1971) have isolated UDP-D-galactofuranoside from *Penicillium charlesii*
G. Smith which synthesizes a peptidophosphogalactomannan containing
galactofuranosyl residues (Preston *et al.*, 1969a,b; Gander *et al.*, 1974).

VII. Concluding Remarks

Many intriguing problems remain unsolved or at best only partially
solved. For instance, the role of phytic acid and inositol in the germinating seedling may be to provide the uronic acids and pentoses needed for
cell wall synthesis. The evidence for this is mostly circumstantial. The
biological function and degradative metabolism of the raffinose family
of oligosaccharides is mostly speculative. The role of hamamelose is completely unknown, as are the roles of branched chain sugars and sugars
in the furanosyl form. Although the biosynthesis of cyanogenic glycosides
and glucosinolates is understood reasonably well, very little is known
about their degradative metabolism and their function. The difficulty in
isolating and purifying to a state of homogeneity, enzymes from plants
is in part responsible for the lack of development of these areas of plant
biochemistry. Application of the newer, more rapid, techniques in enzyme
isolation should result in major advances in delineating metabolic pathways in plants and in determining how these pathways are controlled.

GENERAL REFERENCES

Axelrod, B., and Beevers, H. (1965). *Annu. Rev. Plant Physiol.* **7**, 267.
Bassham, J. A., and Krause, G. H. (1969). *Biochim. Biophys. Acta* **189**, 207.
Burton, K., and Krebs, H. A. (1953). *Biochem. J.* **54**, 94.
Fan, D.-F., and Feingold, D. S. (1969). *Plant Physiol.* **44**, 599.
Haynes, L. J. (1963). *Advan. Carbohyd. Chem.* **18**, 227.
Hough, L., Iyer, P. N. S., and Stacey, B. E. (1973). *Phytochemistry* **12**, 573.
Ikawa, T., Watanabe, T., and Nisizawa, K. (1972). *Plant Cell Physiol.* **13**, 1017.
Kindl, H., and Hoffmann-Ostenhof, O. (1966). *Biochem. Z.* **345**, 454.
Loewus, F. (1971). *Annu. Rev. Plant Physiol.* **22**, 337.
Krauss, M. (1969). *Ann. N.Y. Acad. Sci.* **165**, 509.
Lavintman, N., and Cardini, C. E. (1968). *Plant Physiol.* **43**, 434.
Mahadevan, S. (1973). *Annu. Rev. Plant Physiol.* **24**, 69.

Miyata, H., and Yamamoto, Y. (1969). *Plant Cell Physiol.* **10**, 875.
Turner, J. F., and Turner, D. H. (1975). *Annu. Rev. Plant Physiol.* **26**, 159.

REFERENCES

Abrol, Y. P., and Conn, E. E. (1966). *Phytochemistry* **5**, 237.
Akazawa, T., Miljanich, P., and Conn, E. E. (1960). *Plant Physiol.* **35**, 535.
Alston, R. E. (1968). *Recent Advan. Phytochem.* **1**, 305.
Anderson, L., and Wolter, K. E. (1966). *Annu. Rev. Plant Physiol.* **17**, 209.
Ankel, H., and Feingold, D. S. (1965). *Biochemistry* **4**, 2468.
ap Rees, T., and Beevers, H. (1960a). *Plant Physiol.* **35**, 830.
ap Rees, T., and Beevers, H. (1960b). *Plant Physiol.* **35**, 839.
ap Rees, T., Blanch, E., Graham, D., and Davies, D. D. (1965). *Plant Physiol.* **40**, 910.
Arai, Y., and Fujisaki, M. (1971). *Shokubutsugaku Zasshi* **84**, 76.
Asada, K., Tanaka, K., and Kasai, Z. (1969). *Ann. N.Y. Acad. Sci.* **165**, 801.
Axelrod, B. (1965). *In* "Plant Biochemistry" (J. Bonner and J. E. Varner, eds.), 2nd ed., pp. 231–257. Academic Press, New York.
Axelrod, B., and Jang, R. (1954). *J. Biol. Chem.* **209**, 847.
Baig, M. M., Kelly, S., and Loewus, F. (1970). *Plant Physiol.* **46**, 277.
Bailey, R. W. (1965). *In* "Oligosaccharides," 1st ed. p. 10. Macmillan, New York.
Barber, G. A. (1962). *Biochemistry* **1**, 463.
Barber, G. A., and Hassid, W. Z. (1964). *Biochim. Biophys. Acta* **86**, 397.
Barnett, J. E., and Corina, D. L. (1968). *Biochem. J.* **108**, 125.
Baron, D., Wellmann, E., and Grisebach, H. (1972). *Biochim. Biophys. Acta* **258**, 310.
Baron, D., Streitberger, U., and Grisebach, H. (1973). *Biochim. Biophys. Acta* **293**, 526.
Baxter, E. D., and Duffus, C. M. (1973). *Phytochemistry* **12**, 1923.
Bean, R. C., and Hassid, W. Z. (1955). *J. Amer. Chem. Soc.* **77**, 5737.
Bean, R. C., Barr, B. K., Welch, H. V., and Porter, G. G. (1962). *Arch. Biochem. Biophys.* **96**, 524.
Beck, E. (1969). *Z. Pflanzenphysiol.* **61**, 360.
Beck, E., Stransky, H., and Furbringer, M. (1971). *FEBS (Fed. Eur. Biochem. Soc.) Lett.* **13**, 229.
Bell, D. J. (1962). *Methods Carbohyd. Chem.* **1**, 260.
Bell, D. J., Hardwick, N. E., Isherwood, F. A., and Cahn, R. S. (1974). *J. Chem. Soc. (London)* p. 3702.
Ben-Yehoshua, S., and Conn, E. E. (1964). *Plant Physiol.* **39**, 331.
Bianchetti, R., and Sartirana, M. L. (1967). *Biochim. Biophys. Acta* **145**, 485.
Bird, I. F., Porter, H. K., and Stocking, C. R. (1965). *Biochim. Biophys. Acta* **100**, 366.
Biswas, S., and Biswas, B. B. (1965). *Biochim. Biophys. Acta* **108**, 710.
Bough, W. A., and Gander, J. E. (1971). *Phytochemistry* **10**, 67.
Bourne, E. J., Walter, M. W., and Pridham, J. B. (1965). *Biochem. J.* **97**, 802.
Brown, R. J., and Serro, R. F. (1953). *J. Amer. Chem. Soc.* **75**, 1040.
Buchanan, J. G. (1953). *Arch. Biochem. Biophys.* **44**, 140.
Buchanan, J. G., Lynch, V. H., Benson, A. A., Bradley, D. F., and Calvin, M. F. (1953). *J. Biol. Chem.* **203**, 935.
Burma, D. P., and Mortimer, D. C. (1956). *Arch. Biochem. Biophys.* **62**, 16.

Cahn, R. S., Ingold, C. K., and Prelog, V. (1956). *Experientia* **12**, 81.

Candy, D. J., Blumson, N. L., and Baddiley, J. (1964). *Biochem. J.* **91**, 31.

Caputto, R., Leloir, L. F., Cardini, C. E., and Paladini, A. C. (1950). *J. Biol. Chem.* **184**, 333.

Cardini, C. E., and Leloir, L. F. (1957). *Cienc. Invest.* **13**, 514.

Cardini, C. E., and Recondo, E. (1962). *Plant Cell Physiol.* **3**, 313.

Cardini, C. E., Leloir, L. F., and Chiriboga, J. (1955). *J. Biol. Chem.* **214**, 149.

Carminatti, H., and Cabib, E. (1961). *Biochem. Biophys. Acta* **53**, 417.

Carter, H. E., Brooks, S., Gigg, R. H., Strobach, D. R., and Suami, T. (1964). *J. Biol. Chem.* **239**, 743.

Charalampous, F. C. (1960). *J. Biol. Chem.* **235**, 1286.

Chen, I. W., and Charalampous, F. C. (1963). *Biochem. Biophys. Res. Commun.* **12**, 62.

Chen, I. W., and Charalampous, F. C. (1964). *J. Biol. Chem.* **239**, 1905.

Chen, I. W., and Charalampous, F. C. (1965a). *J. Biol. Chem.* **240**, 3507.

Chen, I. W., and Charalampous, F. C. (1965b). *Biochem. Biophys. Res. Commun.* **19**, 144.

Chen, I. W., and Charalampous, F. C. (1966). *J. Biol. Chem.* **241**, 2194.

Chen, I. W., and Charalampous, F. C. (1967). *Biochem. Biophys. Acta* **136**, 568.

Cleland, W. W. (1963). *Biochim. Biophys. Acta* **67**, 173.

Conn, E. E. (1973). *Biochem. Soc. Symp.* **38**, 277.

Conn, E. E., and Akazawa, T. (1958). *Fed. Proc., Fed. Amer. Soc. Exp. Biol.* **17**, 205.

Conn, E. E., and Butler, G. W. (1969). *In* "Perspectives in Phytochemistry" (J. B. Harborne and T. Swain, eds.), pp. 47–74. Academic Press, New York.

Culberson, C. F. (1969). *In* "Chemical and Botanical Guide to Lichen Products," p. 73. Univ. of North Carolina Press, Chapel Hill.

Cumming, D. F. (1970). *Biochem. J.* **116**, 189.

Dankert, M., Gonçalves, I. R., and Recondo, R. (1964). *Biochim. Biophys. Acta* **81**, 78.

De Fekete, M. A. R. (1971). *Eur. J. Biochem.* **19**, 73.

De Fekete, M. A. R., and Cardini, C. E. (1964). *Arch. Biochem. Biophys.* **104**, 173.

Delmer, D. P. (1972a). *J. Biol. Chem.* **247**, 3822.

Delmer, D. P. (1972b). *Plant Physiol.* **50**, 469.

Delmer, D. P., and Albersheim, P. (1970). *Plant Physiol.* **45**, 782.

Dickinson, D. B., and Preiss, J. (1969). *Arch. Biochem. Biophys.* **130**, 119.

Dolreau, P. (1969). *C.R. Acad. Sci., Ser. D* **269**, 1664.

Edelman, J., Ginsburg, V., and Hassid, W. Z. (1955). *J. Biol. Chem.* **213**, 843.

Eickenbusch, J. D., Sellmair, J., and Beck, E. (1971). *Z. Pflanzenphysiol.* **65**, 24.

Eisenberg, F. (1967). *J. Biol. Chem.* **242**, 1375.

Elliott, M. C., and Stowe, B. B. (1971). *Plant Physiol.* **48**, 498.

English, P. D., Dietz, M., and Albersheim, P. (1966). *Science* **151**, 198.

Espada, J. (1962). *J. Biol. Chem.* **237**, 3577.

Ettlinger, M., and Kjaer, A. (1968). *Recent Advan. Phytochem.* **2**, 59.

Evans, W. H., McGookin, A., Jurd, L., Robertson, A., and Williamson, W. R. N. (1957). *J. Chem. Soc., London* p. 3510.

Fan, D.-F., and Feingold, D. S. (1970). *Plant Physiol.* **46**, 592.

Fan, M. L., LaCroix, L. J., and Unrau, A. M. (1969). *Biochemistry* **8**, 4083.

Farnden, K. J. F., Rosen, M. A., and Liljegren, D. R. (1973). *Phytochemistry* **12**, 2673.

Feingold, D. S., Neufeld, E. F., and Hassid, W. Z. (1958). *Arch. Biochem. Biophys.* **78,** 401.

Feingold, D. S., Neufeld, E. F., and Hassid, W. Z. (1960). *J. Biol. Chem.* **235,** 910.

Feingold, D. S., Neufeld, E. F., and Hassid, W. Z. (1964). *Mod. Methods Plant Anal.* **7,** 474.

Finkle, B. J., Kelly, S., and Loewus, F. (1960). *Biochim. Biophys. Acta* **38,** 332.

Fischer, E., and Freudenberg, K. (1912). *Ber. Deut. Chem. Ges.* **45,** 2709.

Fowler, M. W., and ap Rees, T. (1970). *Biochim. Biophys. Acta* **201,** 33.

French, D. (1954). *Advan. Carbohyd. Chem.* **9,** 149.

Frydman, R. B., and Neufeld, E. F. (1963). *Biochem. Biophys. Res. Commun.* **12,** 121.

Furaya, M., Galston, A. W., and Stowe, B. B. (1962). *Nature (London)* **193,** 456.

Gander, J. E. (1958). *Fed. Proc., Fed. Amer. Soc. Exp. Biol.* **17,** 226.

Gander, J. E. (1960). *Plant Physiol.* **35,** 767.

Gander, J. E. (1962). *J. Biol. Chem.* **237,** 3232.

Gander, J. E., Jentoft, N. J., Drewes, L. R., and Rick, P. D. (1974). *J. Biol. Chem.* **249,** 2063.

Ghosh, H. P., and Preiss, J. (1966). *J. Biol. Chem.* **241,** 4491.

Gibbs, M., and Horecker, B. L. (1954). *J. Biol. Chem.* **208,** 813.

Ginsburg, V. (1958). *J. Biol. Chem.* **232,** 55.

Ginsburg, V., and Hassid, W. Z. (1956). *J. Biol. Chem.* **223,** 277.

Grimes, W. J., Jones, B. L., and Albersheim, P. (1970). *J. Biol. Chem.* **245,** 188.

Grisebach, H., and Dobereiner, U. (1964). *Biochem. Biophys. Res. Commun.* **17,** 737.

Grisebach, H., and Dobereiner, U. (1966). *Z. Naturforsch. B* **21,** 429.

Grisebach, H., and Sandermann, H. (1966). *Biochem. Z.* **346,** 322.

Gruhner, K. M., and Hoffmann-Ostenhof, O. (1966). *Hoppe-Seyler's Z. Physiol. Chem.* **347,** 278.

Gustafson, G. L., and Gander, J. E. (1972). *J. Biol. Chem.* **247,** 1387.

Gustine, D. L., and Kindel, P. K. (1969). *J. Biol. Chem.* **244,** 1382.

Hahlbrock, K., and Conn, E. E. (1970). *J. Biol. Chem.* **245,** 917.

Hahlbrock, K., Tapper, B. A., Butler, G. W., and Conn, E. E. (1968). *Arch. Biochem. Biophys.* **125,** 1013.

Hall, M. A., and Ordin, L. (1967). *Plant Physiol.* **42,** 205.

Haq, S., and Hassid, W. Z. (1965). *Plant Physiol.* **40,** 591.

Hart, D. A., and Kindel, P. K. (1970). *Biochem. J.* **116,** 569.

Hassid, W. Z. (1967). *Annu. Rev. Plant Physiol.* **18,** 253.

Hatch, M. D. (1964). *Biochem. J.* **93,** 521.

Hatch, M. D., Sacher, J. A., and Glasziou, K. T. (1963). *Plant Physiol.* **38,** 338.

Hattori, S., and Imaseki, H. (1959). *J. Amer. Chem. Soc.* **81,** 4424.

Hauska, G., and Hoffmann-Ostenhof, O. (1967). *Hoppe-Seyler's Z. Physiol. Chem.* **348,** 1558.

Hawker, J. S., and Hatch, M. D. (1966). *Biochem. J.* **99,** 102.

Hemming, R., and Ollis, W. D. (1953). *Chem. Ind. (London)* p. 85.

Higuchi, T., and Shimada, M. (1967). *Plant Cell Physiol.* **8,** 71.

Horowitz, H. H., and King, C. G. (1953). *J. Biol. Chem.* **200,** 125.

Horowitz, H. H., Doerschuk, A. P., and King, C. G. (1952). *J. Biol. Chem.* **199,** 193.

Hough, L., and Jones, J. K. N. (1956). *Advan. Carbohyd. Chem.* **11,** 185.

Imhoff, V., and Bourdu, R. (1970). *Physiol. Veg.* **8,** 649.

Isherwood, F. A., and Mapson, L. W. (1962). *Annu. Rev. Plant Physiol.* **13**, 329.

Isherwood, F. A., and Selvendran, R. R. (1971). *Phytochemistry* **10**, 579.

Isherwood, F. A., Chen, Y. T., and Mapson, L. W. (1954). *Biochem. J.* **56**, 1.

Jacobelli, G., Tabone, M. J., and Tabone, D. (1958). *Bull. Soc. Chim. Biol.* **40**, 955.

Jaynes, T. A., and Nelson, O. E. (1971a). *Plant Physiol.* **47**, 623.

Jaynes, T. A., and Nelson, O. E. (1971b). *Plant Physiol.* **47**, 629.

Jung, P., Tanner, W., and Wolter, K. (1972). *Phytochemistry* **11**, 1655.

Kabat, E. A., MacDonald, D. L., Ballou, C. E., and Fischer, H. O. L. (1953). *J. Amer. Chem. Soc.* **75**, 4507.

Katan, R., and Avigad, G. (1966). *Biochem. Biophys. Res. Commun.* **24**, 18.

Kaufman, P. B., Ghosheh, N. S., LaCroix, J. D., Soni, S. L., and Ikuma, H. (1973). *Plant Physiol.* **52**, 221.

Kikuta, Y., Akemine, T., and Tagawa, T. (1971). *Plant Cell Physiol.* **12**, 73.

Kindel, P. K., and Watson, R. R. (1973). *Biochem. J.* **133**, 227.

Kindl, H., and Underhill, E. W. (1968). *Phytochemistry* **7**, 745.

Kivilaan, A., Beaman, T. C., and Bandurski, R. S. (1961). *Plant Physiol.* **36**, 605.

Kjaer, A. (1960). *Fortschr. Chem. Org. Naturst.* **18**, 122.

Komamine, A., Morohashi, Y., and Shimokoriyama, M. (1969). *Plant Cell Physiol.* **10**, 411.

Kosuge, T., and Conn, E. E. (1959). *J. Biol. Chem.* **234**, 2133.

Koukol, J., Miljanich, P., and Conn, E. E. (1962). *J. Biol. Chem.* **237**, 3223.

Krause, G. H., and Bassham, J. R. (1969). *Biochim. Biophys. Acta* **172**, 553.

Latzko, E., and Kotze, J. P. (1965). *Z. Pflanzenphysiol.* **53**, 377.

Lee, C.-J., and Serif, G. (1970). *Biochemistry* **9**, 2068.

Lehle, L., and Tanner, W. (1973). *Eur. J. Biochem.* **38**, 103.

Leloir, L. F. (1951). *Arch. Biochem.* **33**, 186.

Leloir, L. F., and Cardini, C. E. (1953). *J. Amer. Chem. Soc.* **75**, 6084.

Leloir, L. F., and Cardini, C. E. (1955). *J. Biol. Chem.* **214**, 157.

Leloir, L. F., Cardini, C. E., and Cabib, E. (1960). *In* "Comparative Biochemistry" (M. Florkin and H. S. Mason, eds.), Vol. 2, p. 97. Academic Press, New York.

Liao, T.-H., and Barber, G. A. (1971). *Biochim. Biophys. Acta* **230**, 64.

Lin, T., and Hassid, W. Z. (1966). *J. Biol. Chem.* **241**, 3283.

Lindberg, B., and Wickberg, B. (1962). *Acta Chem. Scand.* **16**, 2240.

Lindberg, B., Wachtmeister, C. A., and Wickberg, B. (1952). *Acta Chem. Scand.* **6**, 1052.

Lindberg, B., Misiorny, A., and Wachtmeister, C. A. (1953). *Acta Chem. Scand.* **7**, 591.

Loewus, F. (1961). *Ann. N.Y. Acad. Sci.* **92**, 57.

Loewus, F. (1963). *Phytochemistry* **2**, 109.

Loewus, F. (1969). *Ann. N.Y. Acad. Sci.* **165**, 577.

Loewus, F. (1971). *Annu. Rev. Plant. Physiol.* **22**, 337.

Loewus, F., and Kelly, S. (1961). *Arch. Biochem. Biophys.* **95**, 483.

Loewus, F., and Kelly, S. (1963). *Arch. Biochem. Biophys.* **102**, 96.

Loewus, F. A., Jang, R., and Seegmiller, C. G. (1956). *J. Biol. Chem.* **222**, 649.

Loewus, F. A., Finkle, B. J., and Jang, R. (1958). *Biochem. Biophys. Acta* **30**, 629.

Loewus, F., Kelly, S., and Neufeld, E. F. (1962). *Proc. Nat. Acad. Sci. U.S.* **48**, 421.

Loewus, M. W., and Loewus, F. (1971). *Plant Physiol.* **48**, 255.

Loewus, M. W., and Loewus, F. (1973). *Plant Physiol.* **51**, 263.

Lyne, R. L., and ap Rees, T. (1971). *Phytochemistry* **10**, 2593.

McFarlane, I. J., Lees, E. M., and Conn, E. E. (1975). *J. Biol. Chem.* **250**, 4708.

Mahadevan, S. (1963). *Arch. Biochem. Biophys.* **100**, 557.

Mahadevan, S., and Stowe, B. B. (1972). *Plant Physiol.* **50**, 43.

Maitra, U. S., and Ankel, H. (1971). *Proc. Nat. Acad. Sci. U.S.* **68**, 2660.

Maitra, U. S., Gaunt. M. A., and Ankel, H. (1974). *J. Biol. Chem.* **249**, 3075.

Majumder, A. L., and Biswas, B. B. (1973a) *Phytochemistry* **12**, 315.

Majumder, A. L., and Biswas, B. B. (1973b). *Phytochemistry* **12**, 321.

Majumder, A. L., Mandal, N. C., and Biswas, B. B. (1972). *Phytochemistry* **11**, 503.

Mandal, N. C., and Biswas, B. B. (1970). *Plant Physiol.* **45**, 4.

Mapson, L. W. (1967). *In* "The Vitamins" (W. H. Sebrell, Jr., and R. S. Harris, eds.), Vol. 1, p. 386. Academic Press, New York.

Marsh, C. A. (1960). *Biochim. Bipohys. Acta* **44**, 359.

Matheson, N. K., and St. Clair, M. (1971). *Phytochemistry* **10**, 1299.

Matheson, N. K., and Strother, S. (1969). *Phytochemistry* **8**, 1349.

Matsumoto, H., Wakiuchi, N., and Takahashi, E. (1968). *Physiol. Plant.* **21**, 1210.

Matsumoto, H., Wakiuchi, N., and Takahashi, E. (1969). *Physiol. Plant.* **22**, 537.

Matsuo, M., and Underhill, E. W. (1969). *Biochem. Biophys. Res. Commun.* **36**, 18.

Matsuo, M., and Underhill, E. W. (1971). *Phytochemistry* **10**, 2279.

Mendicino, J. (1960). *J. Biol. Chem.* **235**, 3347.

Mendicino, J., and Hanna, R. (1970). *J. Biol. Chem.* **245**, 6113.

Mendicino, J., and Picken, J. M. (1965). *J. Biol. Chem.* **240**, 2797.

Mentzer, C., and Favre-Bonvin, J. (1961). *C. R. Acad. Sci.* **253**, 1072.

Mentzer, C., Favre-Bonvin, J., and Massias, M. (1963). *Bull. Soc. Chim. Biol.* **45**, 749.

Meunier, J. C., Buc, J., and Ricard, J. (1971). *FEBS (Fed. Eur. Biochem. Soc.) Lett.* **14**, 25.

Miller, L. P. (1943). *Contrib. Boyce Thompson Inst.* **13**, 185.

Milner, Y., and Avigad, G. (1965). *Nature (London)* **206**, 825.

Molinari, E., and Hoffmann-Ostenhof, O. (1968). *Hoppe-Seyler's Z. Physiol. Chem.* **349**, 1797.

Mori, H. (1969). *J. Jap. Soc. Food Nutr.* **22**, 122.

Munch-Petersen, A., Kalckar, H. M., Cutolo, E., and Smith, E. (1952). *Nature (London)* **172**, 1036.

Murata, T. (1971). *Agr. Biol. Chem.* **35**, 297.

Murata, T., Sugiyama, T., and Akazawa, T. (1964). *Arch. Biochem. Biophys.* **107**, 92.

Muto, S., and Uritani, I. (1970). *Plant Cell Physiol.* **11**, 767.

Neish, A. C. (1958). *Can. J. Biochem. Physiol.* **36**, 187.

Nelsestuen, G., and Kirkwood, S. (1971). *J. Biol. Chem.* **246**, 7533.

Neufeld, E. F. (1962). *Biochem. Biophys. Res. Commun.* **7**, 461.

Neufeld, E. F., and Hassid, W. Z. (1963). *Advan. Carbohyd. Chem.* **18**, 309.

Neufeld, E. F., Ginsburg, V., Putnam, E. W., Fanshier, D., and Hassid, W. Z. (1957). *Arch. Biochem. Biophys.* **69**, 602.

Neufeld, E. F., Feingold, D. S., and Hassid, W. Z. (1960). *J. Biol. Chem.* **235**, 906.

Nomura, T., and Akazawa, T. (1973a). *Arch. Biochem. Biophys.* **156**, 644.

Nomura, T., and Akazawa, T. (1973b). *Plant Physiol.* **51**, 979.

Nomura, T., Kono, Y., and Akazawa, T. (1969). *Plant Physiol.* **44**, 765.

Ortmann, R., Sutter, A., and Grisebach, H. (1972). *Biochim. Biophys. Acta* **289**, 293.

Parekh, L. J., and Shah, V. J. (1971). *Enzymologia* **41**, 1.

Perkin, A. (1900). *J. Chem. Soc. (Trans.) London* **77**, 416.

Picken, J. M., and Mendicino, J. (1967). *J. Biol. Chem.* **242**, 1629.

Pina, E., and Tatum, E. L. (1967). *Biochim. Biphys. Acta* **136**, 265.

Pina, E., Saldana, Y., Branner, A., and Chagoya, V. (1969). *Ann. N.Y. Acad. Sci.* **165**, 541.

Pon, N. G. (1964). *In* "Comparative Biochemistry" (M. Florkin and H. S. Mason, eds.), Vol. 7, pp. 1–92. Academic Press, New York.

Posternak, T. (1965). *In* "The Cyclitols," p. 284. Holden-Day, San Francisco, California.

Preiss, J., and Greenberg, E. (1969). *Biochem. Biophys. Res. Commun.* **36**, 289.

Pressey, R. (1966). *Arch. Biochem. Biophys.* **113**, 667.

Pressey, R. (1967). *Plant Physiol.* **42**, 1780.

Pressey, R. (1968). *Plant Physiol.* **43**, 1430.

Pressey, R. (1969). *Plant Physiol.* **44**, 759.

Pressey, R., and Shaw, R. (1966). *Plant Physiol.* **41**, 1657.

Preston, J. F., Lapis, E., Westerhouse, S., and Gander, J. E. (1969a). *Arch. Biochem. Biophys.* **134**, 316.

Preston, J. F., Lapis, E., and Gander, J. E. (1969b). *Arch. Biochem. Biophys.* **134**, 324.

Pridham, J. B. (1960). *In* "Phenolics in Plants in Health and Disease," p. 9. Pergamon, Oxford.

Pridham, J. B. (1965). *Annu. Rev. Plant Physiol.* **16**, 13.

Pridham, J. B., and Hassid, W. Z. (1965). *Plant Physiol.* **40**, 984.

Pridham, J. B., Walter, M. W., and Worth, H. G. J. (1969). *J. Exp. Bot.* **20**, 317.

Pubols, M. H., Zahnley, J. C., and Axelrod, B. (1963). *Plant Physiol.* **38**, 457.

Pueyo, G. (1959). *C.R. Acad. Sci.* **248**, 2788.

Putman, E. W., and Hassid, W. Z. (1954). *J. Biol. Chem.* **207**, 885.

Reay, P. F., and Conn, E. E. (1974). *J. Biol. Chem.* **249**, 5826.

Reed, D. J. (1961). *Plant Physiol.* **36**, xxxii.

Reed, D. J., and Kolattukudy, P. E. (1966). *Plant Physiol.* **41**, 653.

Reid, J. S. G. (1971). *Planta* **100**, 131.

Rendig, V. V., and McComb, E. A. (1964). *Plant Physiol.* **39**, 793.

Roberts, R. M. (1971a). *J. Biol. Chem.* **246**, 4995.

Roberts, R. M. (1971b). *Arch. Biochem. Biophys.* **145**, 685.

Roberts, R. M., and Loewus, F. (1968). *Plant Physiol.* **43**, 1710.

Roberts, R. M., Shah, R., and Loewus, F. (1967a). *Arch. Biochem. Biophys.* **119**, 590.

Roberts, R. M., Shah, R. H., and Loewus, F. (1967b). *Plant Physiol.* **42**, 659.

Ruis, H., Molinari, E., and Hofmann-Ostenhof, O. (1967). *Hoppe-Seyler's Z. Physiol. Chem.* **348**, 1705.

Sacher, J. A., Hatch, M. D., and Glasziou, K. T. (1963). *Physiol. Plant.* **16**, 836.

Sandermann, H., Jr., and Grisebach, H. (1968). *Biochim. Biophys. Acta* **156**, 435.

Sandermann, H., Jr., Tisue, G. T., and Grisebach, H. (1968). *Biochim. Biophys. Acta* **165**, 550.

Sanwal, G. G., and Preiss, J. (1967). *Arch. Biochem. Biophys.* **119**, 454.

Sanwal, G. G., Greenberg, E., Hardie, J. C., Cameron, E. C., and Preiss, J. (1968). *Plant Physiol.* **43**, 417.

Sartirana, M. L., and Bianchetti, R. (1967). *Physiol. Plant.* **20**, 1066.

Schmidt, O. T. (1929). *Justus Liebigs Ann. Chem.* **476**, 250.

Schmidt, O. T. (1930). *Justus Liebigs Ann. Chem.* **483**, 115.

Schnarrenberger, C., Oeser, A., and Tolbert, N. E. (1973). *Arch. Biochem. Biophys.* **154**, 438.

Schwimmer, J., and Rorem, E. S. (1960). *Nature (London)* **187**, 1113.

Sellmair, J., Beck, E., and Kandler, O. (1968). *Z. Pflanzenphysiol.* **59**, 70.

Sellmair, J., Beck, E., and Kandler, O. (1969). *Z. Pflanzenphysiol.* **61**, 338.

Senser, M., and Kandler, O. (1966). *Ber. Deut. Bot. Ges.* **79**, 210.

Shafizadeh, F. (1956). *Advan. Carbohyd. Chem.* **11**, 263.

Shemyakin, M. M., Kokhlov, A. S., and Kolosov, M. N. (1952). *Dokl. Akad. Nauk SSSR* **85**, 1301.

Sherman, W. R., Stewart, M. A., and Zinbo, M. (1969). *J. Biol. Chem.* **244**, 5703.

Sioufia, A., Percheron, F., and Courtois, J. E. (1970). *Phytochemistry* **9**, 991.

Slabnik, E., Frydman, R. B., and Cardini, C. E. (1968). *Plant Physiol.* **43**, 1063.

Smillie, R. M. (1962). *Plant Physiol.* **37**, 716.

Solms, J., and Hassid, W. Z. (1957). *J. Biol. Chem.* **228**, 357.

Staub, W. (1971). *Biochem. Z.* **82**, 48.

Strominger, J. L., and Mapson, L. W. (1957). *Biochem. J.* **66**, 567.

Su, J.-C., and Hassid, W. Z. (1962). *Biochemistry* **1**, 474.

Sutter, A., and Grisebach, H. (1973). *Biochim. Biophys. Acta* **309**, 289.

Sutter, A., Ortmann, R., and Grisebach, H. (1972). *Biochim. Biophys. Acta* **258**, 71.

Tanaka, K., Watanabe, K., Asada, K., and Kasai, Z. (1971). *Agr. Biol. Chem.* **35**, 314.

Tanner, W. (1969). *Ann. N.Y. Acad. Sci.* **165**, 726.

Tanner, W., and Kandler, O. (1966). *Plant Physiol.* **41**, 1540.

Tanner, W., and Kandler, O. (1968). *Eur. J. Biochem.* **4**, 233.

Tanner, W., Lehle, L., and Kandler, O. (1967). *Biochem. Biophys. Res. Commun.* **29**, 166.

Tapper, B. A., and Butler, G. W. (1967). *Arch. Biochem. Biophys.* **120**, 719.

Tapper, B. A., and Butler, G. W. (1971). *Biochem. J.* **124**, 935.

Tapper, B. A., and Butler, G. W. (1972). *Phytochemistry* **11**, 1041.

Tapper, B. A., Conn, E. E., and Butler, G. W. (1967). *Arch. Biochem. Biophys.* **119**, 593.

Tapper, B. A., Zilg, H., and Conn, E. E. (1972). *Phytochemistry* **11**, 1047.

Tolbert, N. E., and Zill, L. P. (1954). *Arch. Biochem. Biophys.* **50**, 392.

Tovey, K. C., and Roberts, R. M. (1970). *Plant Physiol.* **46**, 406.

Trejo, G. A., Chittenden, G. J. F., Buchanan, J. G., and Baddiley, J. (1971). *Biochem. J.* **122**, 49.

Trivelloni, J. C., Recondo, E., and Cardini, C. E. (1962). *Nature (London* **195**, 1202.

Tsai, C. Y., Salamini, F., and Nelson, O. E. (1970). *Plant Physiol.* **46**, 297.

Tsuboi, K. K., Fukunaga, K., and Petricciani, J. C. (1969). *J. Biol. Chem.* **244**, 1008.

Turner, D. H., and Turner, J. F. (1958). *Biochem. J.* **69**, 448.

Ueda, M., Ehmann, A., and Bandurski, R. (1970). *Plant Physiol.* **46**, 715.

Underhill, E. W. (1967). *Eur. J. Biochem.* **2**, 61.

Undershill E. W., and Chisholm, M. D. (1964). *Biochem. Biophys. Res. Commun.* **14**, 425.

Underhill, E. W., and Kirkland, D. E. (1972a). *Phytochemistry* **11**, 1973.

Underhill, E. W., and Kirkland, D. E. (1972b). *Phytochemistry* **11**, 2085.

Uribe, E. G., and Conn, E. E. (1966). *J. Biol. Chem.* **241**, 92.

Verleur, J. J. (1969). *Acta Bot. Neer.* **18**, 353.

Vongerichten, E. (1901). *Justus Liebigs Ann. Chem.* **318**, 121.

Wallace, J. W., and Alston, R. E. (1966). *Plant Cell Physiol.* **7**, 699.

Wang, C. H., Doyle, W. P., and Ramsey, J. C. (1962). *Plant Physiol.* **37**, 1.

Yamaguchi, T., Ikawa, T., and Nisizawa, K. (1969). *Plant Cell Physiol.* **10**, 425.

Yamaha, T., and Cardini, C. E. (1960a). *Arch. Biochem. Biophys.* **86**, 127.

Yamaha, T., and Cardini, C. E. (1960b). *Arch. Biochem. Biophys.* **86**, 133.

Zahnley, J. C., and Axelrod, B. (1965). *Plant Physiol.* **40**, 372.

Zauralov, O. A. (1969). *Fiziol. Rast.* **16**, 542.

Zilg, H., and Conn, E. E. (1974). *J. Biol. Chem.* **249**, 3112.

Zilg, H., Tapper, B. A., and Conn, E. E. (1972). *J. Biol. Chem.* **247**, 2384.

12

Polysaccharides

T. AKAZAWA

I. Starch

There are various types of reserve polysaccharide in plants. The most abundant and important is starch, which is widely distributed from lower microalgae, such as *Chlamydomonas* and *Chlorella*, to higher plants. The classical study of Sachs (1862) demonstrated the formation of the iodine-staining starch granules in light-exposed leaves. It was proposed that starch was an early end product in chloroplasts resulting from the photosynthetic reduction of CO_2. A hundred years later, Arnon and his associates (1954; Arnon, 1955; Allen *et al.*, 1955) showed that starch was produced from CO_2 and H_2O in light-exposed chloroplasts without the aid of external enzymes and substrates. Their experiments indicated that in some plants chloroplasts have a complete set of enzymatic machinery for the net synthesis of starch. The photosynthetically produced starch accumulating in chloroplasts, often called assimilation starch, is a transitory reserve carbohydrate. It disappears in the dark and is transported as sucrose through phloem to other tissues or organs for metabolic transformation. Recent studies revealed that there are two types of chloroplasts in tropical grasses, so-called C_4 plants, e.g., sugarcane and maize (Slack *et al.*, 1969; Laetsch, 1974; and also Chapters 6 and 24). Chloroplasts in bundle sheath and mesophyll cells of these plants are sharply distinguishable in their morphological structure as well as their enzymatic assembly related to the carbon metabolism. One of the characteristic features is an abundant accumulation of starch granules in bundle sheath

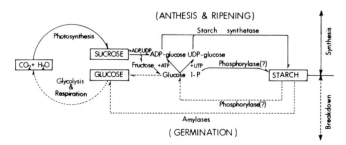

Fig. 1. Metabolic pathways of synthesis and breakdown of reserve starch in cereal seeds.

chloroplasts, as contrasted to their paucity in mesophyll chloroplasts.

The major subject for biochemical studies, however, has been starch immobilized as a reserve substance and typically present in storage organs, such as cereal seeds and potato tubers. Vigorous synthesis of reserve starch occurs in developing rice seeds from anthesis to ripening (Akazawa *et al.*, 1964) and in growing roots of sweet potato (Murata, 1970). On the other hand, during the course of germination and sprouting, reserve starch is enzymatically degraded to low molecular weight carbohydrates for further metabolic utilization. The overall enzymatic mechanisms involved in these processes are schematically shown in Fig. 1. The synthesis of reserve starch in the nonphotosynthetic organs occurs in amyloplasts, a specialized form of plastids with a function of synthesizing starch (Fig. 2). Its unique structure can be compared with that of chloroplasts accumulating starch granules during developmental stage (Fig. 3). In view of current advances in knowledge about the autonomy and continuity of plastids, clarification of the ultrastructural and biochemical uniqueness of amyloplasts can be anticipated.

A. Structure of Starch

1. Amylose and Amylopectin

It is well established that starch consists of two chemically and physically distinguishable polysaccharide fractions, amylose and amylopectin. They are separable by the fact that amylose is soluble in hot water (70°–80°C), while amylopectin is not (Meyer *et al.*, 1940). Using this conventional fractionation method, it was found that the amount of amylopectin (70–90%) in native starch is much higher than that of amylose (10–30%). However, as pointed out by Frey-Wyssling and Mühlethaler (1965), elevation of the temperature during preparation from 70°C to 90°–98°C causes an apparent increase in the amylose content. It is evi-

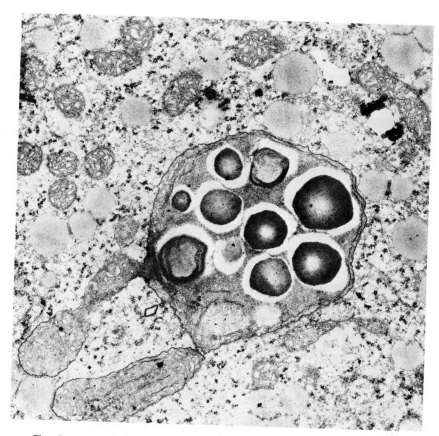

Fig. 2. Amyloplast of barley scutellum (×22,000). (Courtesy of Dr. L. K. Shumway.)

dent that a more specific method is needed for the quantitative analysis of amylose and amylopectin. Schoch (1942) developed a selective precipitation of amylose by making the crystalline complex with n-butanol. A still more refined method employing other alcohols, i.e., n-amyl alcohol and n-propyl alcohol, has been widely used for purifying starch samples of various origins (Schoch, 1945). Table I shows some representative analytical data of starch distributions in several plants (Greenwood, 1956).

Amylose is a linear polymer of $\alpha(1 \rightarrow 4)$-linked glucose with a helical conformation. The molecular weight ranges from 10,000 to 100,000. On the other hand, amylopectin is a highly branched molecule, with molecular weights ranging up to 1,000,000 and with a degree of polymerization of several thousands. In the amylopectin molecule, many chain segments

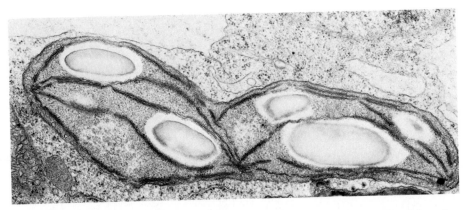

Fig. 3. Chloroplast in dividing rice callus cells (×25,200). (Courtesy of H. Miyake.)

having about 20–25 glucose units are interlinked to the main chain at $\alpha(1 \rightarrow 6)$ bonds, resulting in a highly ramified structural organization. Establishment of the primary backbone structure of amylose and amylopectin was achieved by employing chemical techniques, such as exhaustive methylation followed by hydrolysis and periodate oxidation (Pigman, 1957; Wolfrom and El Khadem, 1965). Equally important contributions in determining the primary structure of starch came from enzymatic degradation using specific carbohydrases. In contrast to the seemingly simple chemical arrangement of glucose polymerization in the amylose molecule, as envisaged from its complete hydrolysis into maltose by β-amylase, the elucidation of the branched structure of amylopectin is certainly a difficult task, and diverse molecular models have been proposed such as those by Haworth, Staudinger, and Meyer, based on the chemical analytical methods. Amylopectin is only partially attacked by β-amylase or phosphorylase, resulting in a residual end product called β-limit dextrin and ϕ-dextrin, respectively. The $\alpha(1 \rightarrow 6)$ bonds in the dextrin molecule, which are resistant to the β-amylolysis, are specifically attacked by "debranching" enzymes. One of the initial studies in this direction was carried out by Maruo and Kobayashi (1949, 1952) using isoamylase isolated from yeast. The R-enzyme isolated from broad bean was found to be similar to the yeast isoamylase (Hobson *et al.*, 1951). Still two other debranching enzymes of microbial origins, pullulanase from *Aerobacter aerogenes* (Bender and Wallenfels, 1961) and isoamylase from *Cytophage* (Gunja-Smith *et al.*, 1970), respectively, were isolated, and their basic enzymatic properties were found to be analogous to those of yeast isoamylase and R-enzyme. Whelan and his associates attempted

TABLE I

Amylose Content and Average Unit Chain Length of Amylopectin Component in Starches[a]

Starch sources	Amylose (%)	Average unit chain length in amylopectin		
		Methyla-tion	Periodate oxidation	Calculated for amylo-pectin from periodate oxidation
Arrowroot	20.5	—	27.3	22
Banana	16.8	26	26.3	22
Barley var. Pioneer	22.0	—	29.5	23
Elm tree, sapwood	21.5	26	26	20
Hevea brasiliensis, seed	20.0	—	30–31	24
Maize	24.0	27	26.5	20
Maize, hybrid "Amylomaize"	50.0	—	—	—
Maple tree, sapwood	19.0	30	29	22
Oat var. Sun II	26.0	—	27.4	20
Pea, smooth, var. Alaska	34.5	—	—	—
Pea, wrinkled, var. Steadfast	80.0	—	—	—
Potato	20.0	—	28.3	23
Potato var. Golden wonder	22.0	—	28.3	22
Rice	18.5	30	27.5	22
Sweet potato	17.8	28	28.2	23
Tapioca	16.7	—	26.2	22
Wheat	25.0	24	26.2	20

[a] From Greenwood (1956).

to explore the structure of amylopectin and glycogen using these enzymes. Results of their analytical studies on the digestion of amylopectin and its β-limit dextrin by pullulanase followed by fractionation of reaction products did not conform the Meyer's model (Lee *et al.*, 1968). Based on their additional studies using *Cytophage* isoamylase, Whelan has proposed that the ramified structure of the amylopectin molecule is more complicated than the previously accepted structure (Whelan, 1971; Lee and Whelan, 1971).

It has long been recognized that the ratio of amylose and amylopectin contents in plant starch is determined by genetic constitution. A typical example can be seen in certain varieties of maize and rice which are referred to as being waxy or glutinous (gene symbol, *wx*). In the grains of these varieties, the starch consists entirely of amylopectin

(Sprague *et al.*, 1943), in contrast to starch in the ordinary lines of non-waxy varieties (*Wx*), comprising both amylopectin and amylose (see Table I). Because of the potential industrial use of amylose, plant geneticists and breeders have attempted to develop mutants of higher amylose content. Several high amylose starches are known, such as those that occur in amylomaize and in wrinkled pea. In such cases, the amylose content ranges from 50 to 80%. However, the genetic development of high amylose mutants usually encounters difficulty because of a reduction in total starch yield. Zuber (1965) reported that the introduction of a newly discovered modifier gene into amylose extender (*ae*) or dull (*du*) mutants of maize gives a considerable increase in the amylose content (49 and 37%, respectively). During the course of these genetic experiments, several other genes, such as sugary (su_1, su_2, su^{am}), were discovered, and these gene combinations were tested for developing high amylose mutants. One can thus anticipate that comparative studies using genetic variants in relation to the interaction of constituent enzymes in the biosynthetic route as well as the clarification of their molecular entities may assist us in unveiling the secret underlying the mechanism of starch formation.

2. The Fine Structure of Starch Granules

Starch is deposited in plant cells as granular particles. It is remarkable that genetically fixed different plant species have their own specific starch granules, varying in such characteristics as size, shape, structure of the shell, and the location of the hylum, although starch molecules are identical in their chemical architecture. It is often possible to identify the plant source of starch through a microscopic examination. However, the details of fine structure or submicroscopic molecular arrangement in starch granules remain unknown, and various models have been proposed based on the molecular dimensions of amylose and amylopectin (Frey-Wyssling, 1953; Frey-Wyssling and Mühlethaler, 1965).

X-ray crystallographic analysis is a powerful tool for elucidating the fine structure of crystalline compounds having repeating lattices. Its application to structural studies of starch granules is limited, because the spherulitic texture of crystalline starch gives only diffraction rings but not spots (Zaslow, 1965; Marchessault and Sarko, 1967). Among a relatively few crystallographic studies on starches in comparison with that of other biopolymers, the classical study of Katz and Van Itallie (1930) has shown that X-ray diffraction patterns of starch granules can be classified into three different types, A, B, and C. Perhaps the greatest contribution from X-ray studies of starch has been the elucidation of the molecular structure of amylose by Rundle *et al.* (1944a), which is in fact the first helical structural model proposed for a biopolymer. They showed the helical structure of amylose, with six successive glucose rings to one revolution, 8 Å per pitch and 13 Å in diameter. Rundle *et al.* (1944b) further analyzed the structure of the complexes of amylose with iodine, alcohols, and fatty acids, and it was shown that one molecule of iodine is located in the center of each gyre of six glucose units of the amylose molecule. The blue coloration of amylose solution by iodine occurs as a consequence of the penetration of iodine molecules into the linear amylose helices. A typical experiment in determining the absorption maxima of the iodine complexes of starch are useful in estimation of the degree of polymerization (Bailey and Whelan, 1961). In iodine–amylopectin complexes, which give a violet color, only the unit chain lengths of the amylopectin branches contribute to the complex formation. On the basis of X-ray diffraction pattern of the highly crystalline amylodextrin molecules prepared from potato starch granules, Kainuma and French (1972) recently proposed the interwound (double) helical model for the B-type starch as shown in Fig. 4.

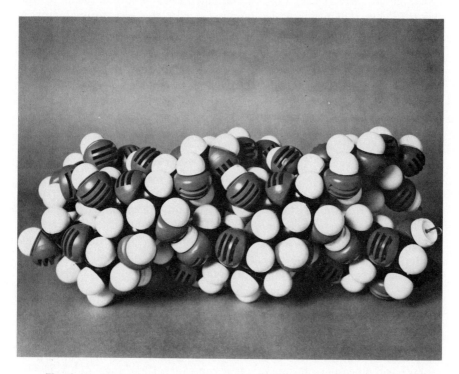

Fig. 4. Double helical model of amylose. There are 6 glucose units per turn of each strand, 10.5 Å per 3 glucose units (21 Å per 6 glucose units). (After Kainuma and French, 1972.)

The dynamic features of starch granules formation in plant cells are also suggested by the layer arrangement (shell structure) as revealed by microscopic and electron microscopic observations, although the patterns cannot be interpreted in molecular terms. As discussed in the previous edition, experiments by several workers using starch granules from different plant sources have shown that both external and internal factors influence the formation of the layered structure. Electron microscopic studies by Buttrose (1960, 1962, 1963) indicated that enzymatic rhythm may control the development of the layers. Furthermore, the radioisotope experiments by Badenhuizen and Dutton (1956) and Yoshida et al. (1958) have indicated the appositional growth of starch granules in amyloplasts. Sarko and his associates (Mencik et al., 1971) studied the anisotropic light scattering patterns of asymmetric (nonspherical) starch granules obtained from Curcuma zedoaria roots (shoti variety) with the hylum at the end, using a red continuous wave laser as a light source. By comparing the observed patterns with those predicted from the type of molecular arrangement in the granules, it was concluded that the starch polymers are arranged in a radially oriented fashion normal to the growth layer. Their conclusion basically agrees with the appositional growth history of starch granules.

B. The Enzymatic Mechanism of Starch Biosynthesis

1. PHOSPHORYLASE

The enzymatic biosynthesis of polysaccharide was first successfully achieved with phosphorylase from muscle and liver tissues by Cori and his associates (1937). The enzyme catalyzes the reversible phosphorolysis of $\alpha(1 \rightarrow 4)$-glucan molecules as illustrated in Fig. 5a. Their monumental work demonstrated the formation of α-glucan molecules (glycogen) from glucose 1-phosphate by staining blue with iodine. Experiments by Hanes (1940a,b) on starch synthesis using plant phosphorylase isolated from pea seeds and potato tubers led to a long continued belief about the important role of phosphorylase in the α-glucan biosynthesis. In the reaction, the free energy change of hydrolysis of glucose 1-phosphate ($\Delta F^\circ = -4800$ cal) (Burton and Krebs, 1953) constitutes a driving force for the synthetic reaction, as the energy level of the $\alpha(1 \rightarrow 4)$ glucosidic bond in the glucan molecule is -4300 cal. However, it must be noted that the equilibrium constant, K'_{eq}, of the α-glucan synthesizing reaction by phosphorylase will be solely determined by the ratio [inorganic phosphate] to [glucose 1-phosphate] at a constant pH (Hanes and Maskell, 1942). The fact that high cellular concentration of inorganic phosphate

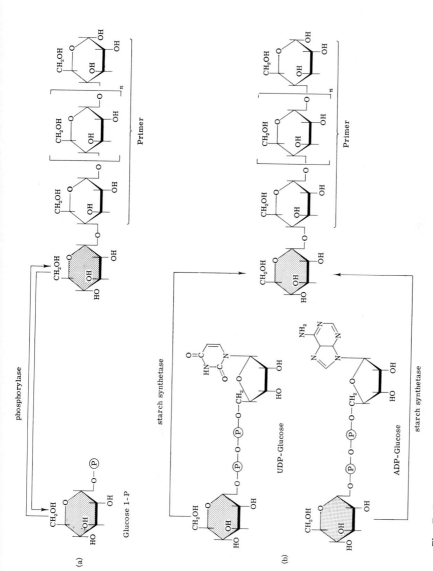

Fig. 5. Enzymatic synthesis of starch [α(1 → 4)-glucan] by (a) phosphorylase and (b) starch synthetase using glucose 1-phosphate, and either ADP-glucose or UDP-glucose, respectively, as glucosyl donor.

in animal and plant tissues greatly affects the reaction equilibrium has gradually led to the notion of a catabolic role for phosphorylase in the breakdown of glycogen and starch.

In contrast to the remarkable progress made in research on the enzymology of phosphorylases of animal origins in the last 30 years, studies on plant phosphorylases have been limited and our knowledge is still fragmentary. A comparative description of some basic structural properties of phosphorylases from plants and animals will be useful. From the initial stage of investigations by Cori's group, rabbit muscle phosphorylase was shown to exist in an active and less active form (phosphorylase a and b, respectively), and the activity of the latter form is dependent on the presence of AMP. As illustrated in Fig. 6, it was found that phosphorylase b, a dimer of molecular weight 185,000, is essentially inactive in the absence of AMP but is converted to an active tetramer of molecular weight 370,000 (phosphorylase a) (Fischer *et al.*, 1971). Phosphorylase a contains four phosphoryl serine residues, essential to the enzymatic activity. Another unique property of the glycogen phosphorylase is the tight binding of pyridoxal 5'-phosphate, which is also necessary for the catalysis, but its exact role remains unknown. The reversible dissociation of both phosphorylase a and b into monomeric subunit (molecular weight, 92,500) can be achieved by treating the enzyme molecule with SH-reagents such as *p*-mercuribenzoate. Although the liver phosphorylase has been less purified, it also exists in active and inactive forms. Plant

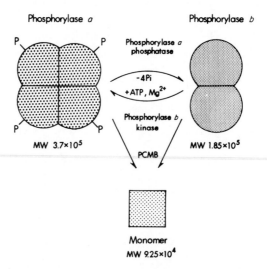

Fig. 6. Molecular interconversion of muscle glycogen phosphorylase. (Modified after Fischer *et al.*, 1971.)

phosphorylase exhibits several differences from animal phosphorylase. Lee (1960) reported that a homogeneous preparation of potato phosphorylase ($s_{20,w} = 7.9$) had a molecular weight of about 207,000. Although two molecules of pyridoxal 5′-phosphate are present per enzyme molecule, serine phosphate is absent and AMP is inert. A recent report of Kamogawa et al. (1968) on a crystalline phosphorylase from potato tuber shows $s_{20,w} = 6.5$. Their experimental results show also the presence of two molecules of pyridoxal 5′-phosphate, but the amino acid composition of the enzyme is quite similar to that of muscle phosphorylase. It was further found that the crystalline potato phosphorylase is comprised of two apparently identical polypeptides (subunits), molecular weight being 108,000 (Iwata and Fukui, 1973). On the other hand, Lee and Braun (1973) reported that the purified preparation of phosphorylase from maize seed has 7.9 $s_{w,20}$ and a molecular weight of 150,000 and that one molecule of pyridoxal 5′-phosphate is present per enzyme molecule.

Hormonal, renal, and allosteric regulation of the muscle and liver phosphorylases mediated by enzymatic reactions is tightly connected to the structural modifications of the enzyme molecules as depicted in Fig. 6, and these changes are delicately linked to the homeostatic control of carbohydrate metabolism in animals. Historically, it is important to recall that the effect of glycogenolytic hormones, such as epinephrine and glucagon, is associated with the glycogen breakdown by enhancing the glycogen phosphorylase activity (Sutherland, 1952). The absence of phosphorylase in biopsy specimens of skeletal muscle having a metabolic defect involving an inability to break down glycogen (glycogen storage disease) gave additional support for a role of phosphorylase in the carbohydrate catabolism (Mommaerts et al., 1959; Larner and Villar-Palasi, 1959). In these tissues, levels of other enzymes related to glycogen metabolism, e.g., UDP-glucose-glycogen transglycosylase (glycogen synthetase) remain normal (see below).

2. STARCH SYNTHETASE

The decisive demonstration by Leloir and his associates (Leloir and Cardini, 1957) of a new enzymatic system for glycogen biosynthesis catalyzed by glycogen synthetase opened an entirely new era of polysaccharide biochemistry. They extended their studies from glycogen to starch biosynthesis, demonstrating that essentially the same reaction mechanism operates (Leloir et al., 1961). The preparation of starch synthetase initially discovered was starch granules isolated from immature dwarf string bean seeds, and similar enzyme preparations were isolated from plant materials, such as sweet corn, potato, and other starch containing tissues (Akazawa, 1965). The general enzymatic mechanism of starch biosynthesis illustrated in Fig. 5b involves the stepwise transfer of glucose moiety from the nucleotide glucose molecule to the nonreducing end of accepting α-glucan molecule (primer) by forming the linear $\alpha(1 \rightarrow 4)$-glucan (amylose). In a thermodynamic sense, the role of UDP-glucose in the transglucosylation reaction is analogous to that of glucose 1-phosphate in the phosphorylase reaction discussed above. However, it must be stressed that the free energy change of the overall reaction of Fig. 5b is —3300 cal, favoring the reaction proceeding toward the synthesis. Moreover, the cellular concentration of inorganic phosphate does not affect the reaction equilibrium. Important progress was made by Recondo and Leloir (1961), who discovered that ADP-glucose was more efficient as a donor than was UDP-glucose. Recondo et al. (1963) and Murata et al. (1963) isolated ADP-glucose from developing maize and rice seeds, respectively.

It was considered for some time that granule-bound starch synthetase probably is the biologically active agent in starch synthesis. However, a tight association of the enzyme molecule with starch granules obviously posed an obstacle for thoroughly investigating the enzymatic properties as well as the reaction mechanism. In spite of numerous efforts to solubilize the enzyme molecule from starch particles, no one has succeeded in this attempt. On the other hand, Frydman and Cardini (1964) first demonstrated the presence of soluble starch synthetase from the seeds of sweet corn, and later several investigators reported the presence of similar soluble enzyme from various plant materials (Murata and Akazawa, 1966). These soluble enzymes exhibit absolute specificity toward ADP-glucose as the glucose donor, and the current view is that the soluble enzyme utilizing ADP-glucose plays the main role in starch synthesis. Several investigations dealing with the existence of multiple forms of soluble starch synthetase in plant tissues tell us something about the enzymatic mechanism of starch formation. Tanaka and Akazawa (1971) separated by DEAE-cellulose column at least two isozymic fractions of starch synthetase from developing rice seed extracts, each one of them utilizing the specific glucose acceptor. Fraction I enzyme shows the predominant utilization of short-chain maltooligosaccharides, whereas fraction II utilizes the longer chain α-glucan more readily. One can speculate that the chain elongation reaction in the amylose synthesis proceeds in a stepwise fashion using two different enzyme species. By means of DEAE-cellulose column chromatography, Preiss and his associates demonstrated the existence of starch synthetase isozymes in several plant tissues, e.g., spinach leaf, maize leaf, maize, and waxy maize (Ozbun et al., 1971a,b; 1972; Hawker et al., 1972, 1974). An important finding of their investigations was the "unprimed" transglucosylation reaction, which is catalyzed by one isozyme fraction in the absence of added primer. The enzyme reaction requires the presence of high concentrations of some anions and/or proteins, and the unprimed transglucosylation reaction catalyzed by mammalian liver (Krisman, 1972) and bacterial enzymes (Gahan and Conrad, 1968; Chambost et al., 1973) that transfers glucosyl residues from either UDP-glucose or ADP-glucose to the methanol-insoluble fractions was found to exhibit similar properties. With the starch synthetase from spinach leaves, one of the activating anions, citrate (0.5 M), was shown to decrease the K_m for glycogen and amylopectin by several hundredfold to as low as 0.86 μg/ml. Under these conditions a trace amount of glucan shown to be associated with the enzyme may have been acting as a primer. However the unprimed enzyme activity was not diminished even after treating the enzyme preparations with *Rhizopus* glucoamylase and α-amylase and is sharply distinguishable from α-glucan synthesis catalyzed by the crystalline potato phosphorylase (Kamogawa et al., 1968). The results, therefore, do not rule out the possibility of a *de novo* reaction, but most of the studies could be explained by the presence of an endogenous primer. Preiss and his associates hypothesize that the need for only very low concentrations of primer for optimal activity may obviate the requirement of a mechanism for *de novo* synthesis of primer. Daughter cells may acquire low amounts of primer by binding glucan from parent cells to newly formed glucan synthetase.

3. Amylopectin Biosynthesis (Enyzymatic Mechanism of Branching)

As the chemical structure of amylopectin is basically similar to glycogen, the enzymatic mechanism underlying their formation is considered to be same. The formation of $\alpha(1 \rightarrow 6)$ bonds in the glycogen molecule is known to be catalyzed by the enzyme named "branching enzyme" or amylo-$(1,4 \rightarrow 1,6)$-transglucosylase, first reported by Cori and Cori (1943). An analogous plant enzyme, Q-enzyme, was isolated from potato tuber (Haworth et al., 1944), and its purification was recently attempted for rice (Igaue, 1962) and potato (Drummond et al., 1972). Proof for the role of branching enzyme in the glycogen biosynthesis came from the structural studies on the synthetic molecules produced by joint action of glycogen synthetase and branching enzyme. It was initially reported that high molecular α-glucan particles synthesized by crystalline rabbit muscle phosphorylase and branching enzyme is undistinguishable from the native glycogen particles (Mordoh et al., 1965). However, detailed investigations demonstrated that they were different in several respects, such as iodine spectrum, sedimentation coefficient, and acid and alkali susceptibility (Mordoh et al., 1966). It was subsequently revealed that the synthetic glycogen molecules produced jointly by glycogen synthetase and branching enzyme are identical to the native glycogen (Parodi et al., 1969). A similar experiment was carried out in a plant system by Drummond et al. (1970) as glucose 1-phosphate can be converted to amylopectin type branched α-glucan molecules by joint reaction of potato phosphorylase and Q-enzyme. It was then found that the synthetic polymers were different from the native amylopectin in their unit chain profiles as examined from the Sephadex G-50 eluates of the pullulanase digests. It is naturally surmised that the biosynthetic pathway involving ADP-glucose–starch synthetase coupled with Q-enzyme is an alternative mechanism. Since the iodine spectrum of the α-glucan produced by the unprimed starch synthetase reaction reported by Preiss's group was similar to that of amylopectin (490–500 nm), which contain both α-1,4- and α-1,6-glucosidic links, it can be assumed that the branching enzyme was tightly associated with the starch synthetase isozyme fraction engaged in the de novo starch synthesis. Hawker et al. (1974) succeeded in separating these two enzyme activities in the spinach leaf preparation by affinity chromatography (ADP-hexanolamine Sepharose 4B), and addition of the purified branching enzyme was shown to greatly enhance the activity of the unprimed transglucosylation reaction. These workers postulated that the role of branching enzyme in the initiation reaction of starch synthesis is to increase the nonreducing ends in the growing glucan originated from the endogenous core molecule.

What then controls the formation of amylose and amylopectin molecules at their specific ratio in plant cells? Genetic analytical studies on the starch synthesizing enzymes using various plant mutants producing specific types of starches is a meaningful approach to understand both enzymatic processes as well as the regulatory mechanisms. In this connection, it is worthy of note that the biochemical lesion of the type 4 liver glycogen storage disease, accumulating glycogen molecules having longer inner and outer chains, is apparently a defect of branching enzyme (amylopectinosis) (Illingworth and Brown, 1964). So far no different distribution patterns of enzymes in maize varieties have been reported, and so a recent experiment by Shiefer et al. (1973) showing the existence of several multiple forms of starch synthetase in isogenic lines of maize varieties are intriguing. They found that in amylomaize (ae) activities of specific isozyme components, apparently engaged in the amylose formation, are much stronger than the waxy (wx) or sweet corn (su₁) varieties. Their results may indicate that different enzymes are involved in the biosynthesis of amylose and amylopectin.

4. SUCROSE–STARCH INTERCONVERSION

In plant cells, sucrose is the major form of transport from one tissue or organ to another, flowing into further metabolic utilization. This is typically seen in the sucrose–starch conversion processes in developing cereal seeds or in growing potato roots (see Fig. 1), and much attention has been given to its enzymatic mechanism (Akazawa, 1965; Nikaido and Hassid, 1971). The free energy change ($\Delta F°$) of hydrolysis of sucrose can be calculated as $-10,000$ cal from available thermodynamic data of the individual constituent molecules [Eq. (1)] which supports a notion of relatively high free energy potential of glucosidic bond in the sucrose molecule reported previously (Pigman, 1957; Leloir, 1964). In spite of the fact that the standard free energy change may not necessarily indi-

$$\begin{array}{ccccccc}
\text{Sucrose} & + & \text{H}_2\text{O} & \rightarrow & \text{glucose} & + & \text{fructose} \\
\text{(s)} & & \text{(l)} & & \text{(s)} & & \text{(s)} \\
-370.90\,\text{kcal} & & -56.69\,\text{kcal} & & -218.89\,\text{kcal} & & -219.33\,\text{kcal}
\end{array} \tag{1}$$

cate its physiological meaning *in vivo*, since it neglects the effect of the actual concentration of metabolites, it will be pointed out that the value is considerably higher than the generally quoted one of -7000 cal calculated from K_{eq}' of sucrose phosphorylase reaction (0.053, pH 6.6) [Eq. (2)] (Neufeld and Hassid, 1963).

$$\text{Glucose 1-phosphate} + \text{fructose} \rightleftharpoons \text{sucrose} + \text{phosphate} \tag{2}$$

Recent experiments from several laboratories support the idea that the reversal of the sucrose synthetase reaction [Eq. (3)] is the principal

$$\text{Sucrose} + \text{UDP (ADP)} \rightleftharpoons \text{UDP-glucose (ADP-glucose)} + \text{fructose} \tag{3}$$

mechanism of sucrose cleavage (Chapter 11). On the other hand, sucrose-6-phosphate synthetase coupled with sucrose-6-phosphate phosphatase, each catalyzing Eq. (4) and (5), respectively, are judged to be the principal route of the sucrose synthesis.

$$\begin{aligned}
\text{UDP-glucose (ADP-glucose)} + \text{fructose 6-phosphate} \rightarrow \\
\text{sucrose 6-phosphate} + \text{UDP (ADP)}
\end{aligned} \tag{4}$$

$$\text{Sucrose 6-phosphate} + \text{H}_2\text{O} \rightarrow \text{sucrose} + \text{phosphate} \tag{5}$$

What then is the nature of enzymatic link between sucrose cleavage and starch synthesis? The reaction of Eq. (3) proceeds more readily with UDP (UDP-glucose formation) than with ADP (ADP-glucose forma-

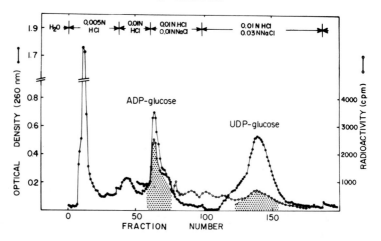

Fig. 7. Radioactive labeling of ADP-glucose and UDP-glucose in developing rice seeds fed [^{14}C]sucrose. (From Murata and Akazawa, 1966.)

tion), although starch synthetase utilizes ADP-glucose predominantly. There are apparently missing mechanism(s) between the two mutually contradictory reaction systems involved in the sucrose–starch transformation. Figure 7 is a result of experiment showing the more ready incorporation of [^{14}C]sucrose into ADP-glucose than into UDP-glucose after short-term feeding of [^{14}C]sucrose to the rice panicles (Murata and Akazawa, 1966). An important route of ADP-glucose synthesis in plant cells is that mediated by ADP-glucose pyrophosphorylase, which is allosterically controlled by several intermediary compounds of glycolytic pathway (Chapter 10), but the relationship of this reaction to the sucrose metabolism remains unknown.

C. The Enzymatic Breakdown of Starch (Amylases)

The amylases have one of the oldest histories in enzymology. At the same time, it must be pointed out from the foregoing discussion that the use of amylases has greatly contributed in understanding the molecular architecture of starch and other related reserve polysaccharides (Fischer and Stein, 1960; French, 1960; Thoma *et al.*, 1971; Whelan, 1971). Most important in the physiology of amylase is its role in hydrolyzing reserve starch in germinating seeds (Fig. 1). Classic studies on the breakdown of reserve polysaccharides in germinating starch-bearing cereal seeds indicated that the amylolytic process predominates in the endosperm tissues (James, 1953). An analytical study by Murata *et al.* (1968) on the carbohydrate metabolism of germinating rice seeds supported this mechanism.

An elegant study by Linderstrøm-Lang and Engel (1938) demonstrated the specific localization of amylase in aleurone cells of barley malt, employing their ultramicroanalytic technique of measuring enzyme activities in biological specimens. Their studies preceded by about 20 years the later work by Yomo (1960) and Paleg (1960, 1961) on the enhancement of α-amylase production in the gibberellic acid-treated barley seeds. The mechanism of enzyme formation in aleurone cells of gibberellic acid-treated barley seeds has been subsequently intensively studied by Varner and his associates (Varner and Johri, 1968 and Chapter 22), opening a new facet of hormone research.

The occurrence of multiple molecular forms of amylases in germinating barley seeds was originally reported by Frydenberg and Nielsen (1966), who attempted to examine the relationship between isozyme patterns and genetic combination. Several workers demonstrated the production of α-amylase isozymes in the cereal plants under the direct control of gibberellic acid (Momotani and Kato, 1966; Tanaka and Akazawa, 1970; Jacobsen et al., 1970). Nearly identical patterns of α-amylase isozyme detected by isoelectrofocusing on polyacrylamide gel between normally germinating rice seed and gibberellic acid-treated, embryo-free half-seed extracts strongly support a notion that gibberellic acid has an intrinsic role in the hydrolytic breakdown of reserve starch in germinating seeds (Tanaka et al., 1970). In order to test whether or not the synthesis of each individual amylase molecule is directed by a different structural gene, it is crucial to prove the production of chemically distinguishable polypeptide chains having α-amylase activities under the experimental conditions employed. This approach will be of particular importance because, in spite of much biochemical interest on the structure–activity relationship of multiple forms of plant enzymes, often there is no guarantee showing the absence of structural deformation of enzyme molecules, such as those due to the proteolytic effect during the course of enzyme isolation.

II. Inulin, Mannan, and Other Reserve Polysaccharides

In some plants photosynthetic products are stored in the form of polyfructosans. Inulin and related fructosans with $\beta(2 \rightarrow 1)$ linkages are found in the roots and tubers of the Compositae (e.g. *Dahlia*, *Helianthus*, *Cichorium*) and Campanulaceae. Another type of fructosan in which $\beta(2 \rightarrow 6)$ linkages predominate, is found in the leaf, stem and root tissue of many monocotyledonous plants, in particular Gramineae, e.g., rye, wheat, and rye glass. Inulins of *Dahlia* and Jerusalem artichoke consti-

tute more than 50% of the fresh weight of tuber tissues and have been
most thoroughly investigated. The basic structures of inulin and the grass
polyfructosans have been elucidated chiefly through the work of Hirst
and his associates in Great Britain (Hirst, 1957; Schlubach, 1958; Bacon,
1960). It is of great interest to clarify the biochemical and enzymatic
mechanisms of synthesis of specific polyfructosans as either final or temporary reserve material. Studies by Edelman and Bacon (1951a,b), Bacon
and Edelman (1951), Dedondor (1952), and Edelman and Dickerson
(1966) have been developed further by Edelman, who with Jefford (1968)
has proposed a complete scheme to explain the synthesis of inulin-
type polysaccharides in the Jerusalem artichoke. A trisaccharide,
fructosyl sucrose, is formed by a dismutation reaction between sucrose
molecules, Eq. (6), and trisaccharide then becomes donor or acceptor
in a series of transglycosylation reactions which leads to the formation
of molecules with degree of polymerization (D.P.) 30–40, Eq. (7). This
basal reaction mechanism is analogous to that of dextran synthesis in

$$\text{Sucrose (G} \sim \text{F)} + \text{sucrose (G} \sim \text{F)} \rightarrow \text{G} \sim \text{F—F} + \text{G} \qquad (6)$$

$$\text{G} \sim \text{F—F}_n + \text{G} \sim \text{F—F}_m \rightarrow \text{G} \sim \text{F—F}_{n-1} + \text{G} \sim \text{F—F}_{m+1} \qquad (7)$$

Leuconostoc mesenteroides, Eq. (8), and that of levan synthesis in *Ace-
tobactor levanicum* or *Streptococcus salivarius*, Eq. (9). The classical
studies by Hehre (1951) clearly established that these two enzymatic
reactions do not involve the phosphorylated intermediates. Therefore, this

$$n \text{ Sucrose} \rightarrow n \text{ (D-fructose)} + \text{(D-glucose)}_n \, [\alpha(1 \rightarrow 6)] \qquad (8)$$

$$n \text{ Sucrose} \rightarrow n \text{ (D-glucose)} + \text{(D-fructose)}_n \, [\beta(2 \rightarrow 6)] \qquad (9)$$

reaction mechanism of polyfructosan biosynthesis, without the expendi-
ture of external energy for activation of sucrose molecules, provides additional evidence that the glycosidic linkage in the sucrose molecule has
high energy potential. Since the structure of the grass fructosan is related
to that of the high molecular weight levan produced by bacteria, it is
often called the grass levan. However, no comparable experiments have
been reported for the biosynthesis of grass levan.

At the time of these early studies on inulin, dextran, and levan synthesis, little was known about the transglucosylation reactions involving
sugar nucleotides. When Gonzales and Pontis (1963) and Umemura *et
al.* (1967) reported the isolation of a new nucleotide sugar, UDP-fructose,
from tuber tissues of *Dahlia* and Jerusalem artichoke, respectively, an
optimistic view prevailed that this compound might be a possible intermediate in the inulin biosynthesis. However, despite intensive investiga-

tions, there is no convincing evidence in support of this hypothetical mechanism (Pontis, 1966).

Another type of reserve polysaccharide in plants is mannan (Stepanenko, 1960). Galactomannans are typically distributed in endosperm tissue of Leguminosae seeds. On the other hand, glucomannans have been typically shown to be present in a small number of tuber tissues, such as ivory nut, salep, konjak (*Amorphophallus tuberosus*), and some Orchidaceae. The chemical structures of mannans are not known definitely, and the details on their biosynthesis and breakdown remain obscure. It is conceivable that the role of reserve glucomannan in tuber tissue is analogous to that of inulin, as it disappears at the onset of sprouting. According to the analytical results of Murata (1972), sucrose comprises the major portion of soluble sugar in bulb and leaf blades of the konjak plant, with little detectable mannose, and a possible enzymatic mechanism of sucrose–mannan transformation can be expected. It will be pointed out also that recent progresses in the structural and the biosynthetic studies of membranous glucomannan in mung bean seedlings (Elbein, 1965) as well as in yeast (Kozak and Bretthauer, 1970; Smith *et al.*, 1975) will give us a clue to unveil the biochemical nature of reserve mannan.

Several other reserve polysaccharides are known to be synthesized in protozoa and various types of algae, such as brown, red, blue-green, and green algae. Paramylon, $\beta(1 \to 3)$-glucan, accumulating in *Euglena gracilis*, has been one of the most thoroughly studied compounds (Clarke and Stone, 1960; Goldemberg and Maréchal, 1963; Barras and Stone, 1968). In some fungi, for instance, *Sclerotinia sclerotiorum* (Jones, 1970) and *Schizophyllum commune* (Wessels, 1969), there is good experimental evidence available showing the utilization of wall polysaccharides to support the fungal growth. There are many other polysaccharides in nature with great potential economic and industrial use, such as the sulfated polysaccharides in some marine algae (Bourne *et al.*, 1970). The elucidation of their chemical structure as well as their biosynthetic pathway remain a challenge for future biochemical investigations.

GENERAL REFERENCES

Akazawa, T. (1965). *In* "Plant Biochemistry" (J. Bonner and J. E. Varner, eds.), 2nd ed., p. 258. Academic Press, New York.
Leloir, L. F. (1964). *Biochem. J.* **91**, 1.
Leloir, L. F. (1971). *Science* **172**, 1299.
Piras, R., and Pontis, H. G., eds. (1972). "Biochemistry of the Glycosidic Linkage: An Integrated View." Academic Press, New York.
Preiss, J. (1969). *Curr. Top. Cell. Regul.* **1**, 125–160.

Whelan, W. J., and Cameron, M. P., eds. (1964). "Control of Glycogen Metabolism." Churchill, London.
Whistler, R. L., and Paschall, E. F. eds. (1965). "Starch: Chemistry and Technology," Vol. 1. Academic Press, New York.

REFERENCES

Akazawa, T., Minamikawa, T., and Murata, T. (1964) *Plant Physiol.* **39,** 371.
Allen, M. B., Arnon, D. I., Capindale, J. B., Whatley, F. R., and Durham, L. J. (1955) *J. Amer. Chem. Soc.* **77,** 4149.
Arnon, D. I. (1955). *Science* **122,** 9.
Arnon, D. I., Allen, M. B., and Whatley, F. R. (1954). *Nature (London)* **174,** 394.
Bacon, J. S. D. (1960). *Bull. Soc. Chim. Biol.* **42,** 1441.
Bacon, J. S. D., and Edelman, J. (1951). *Biochem. J.* **48,** 114.
Badenhuizen, N. P., and Dutton, R. W. (1956). *Protoplasma* **47,** 156.
Bailey, J. M., and Whelan, W. J. (1961). *J. Biol. Chem.* **236,** 969.
Barras, D. R., and Stone, B. A. (1968). *In* "The Biology of Euglena" (D. E. Buetow, ed.), Vol. 2, p. 149. Academic Press, New York.
Bender, H., and Wallenfels, K. (1961). *Biochem. Z.* **334,** 79.
Bourne, E. J., Johnson, P. G., and Percival, E. (1970). *J. Chem. Soc., C* p. 1561.
Burton, K., and Krebs, H. A. (1953). *Biochem. J.* **54,** 94.
Buttrose, M. S. (1960). *J. Ultrastruct. Res.* **4,** 231.
Buttrose, M. S. (1962). *J. Cell Biol.* **14,** 159.
Buttrose, M. S. (1963). *Naturwissenschaften* **50,** 450.
Chambost, J. P., Favard, A., and Cattaneo, J. (1973). *Biochem. Biophys. Res. Commun.* **55,** 132.
Clarke, A. E., and Stone, B. A. (1960). *Biochim. Biophys. Acta* **44,** 163.
Cori, C. F., Colowick, S. P., and Cori, G. T. (1937). *J. Biol. Chem.* **121,** 465.
Cori, G. T., and Cori, C. F. (1943). *J. Biol. Chem.* **151,** 57.
Dedonder, R. (1952). *Bull. Soc. Chim. Biol.* **34,** 144, 157, and 171.
Drummond, G. S., Smith, E. E., and Whelan, W. J. (1970). *FEBS (Fed. Eur. Biochem. Soc.) Lett.* **9,** 136.
Drummond, G. S., Smith, E. E., and Whelan, W. J. (1972). *Eur. J. Biochem.* **26,** 168.
Edelman, J., and Bacon, J. S. D. (1951a). *Biochem. J.* **49,** 446.
Edelman, J., and Bacon, J. S. D. (1951b). *Biochem. J.* **49,** 529.
Edelman, J., and Dickerson, A. G. (1966). *Biochem. J.* **98,** 787.
Edelman, J., and Jefford, T. G. (1968). *New Phytol.* **67,** 517.
Elbein, A. D. (1969). *J. Biol. Chem.* **244,** 1608.
Fischer, E. H., and Stein, E. A. (1960). *In* "The Enzymes" (P. D. Boyer, H. Lardy, and K. Myrbäck, eds.), Vol. 4, pp. 313–343. Academic Press, New York.
Fischer, E. H., Heilmeyer, L. M. G., Jr., and Haschke, R. H. (1971). *Curr. Top. Cell. Regul.* **4,** 211.
French, D. (1960). *In* "The Enzymes" (P. D. Boyer, H. Lardy, and K. Myrbäck, eds.), Vol. 4, pp. 345–368. Academic Press, New York.
Frey-Wyssling, A. (1953). "Submicroscopic Morphology of Protoplasm." Elsevier, Amsterdam.
Frey-Wyssling, A., and Mühlethaler, K. (1965). "Ultrastructural Plant Cytology." Elsevier, Amsterdam.

Frydenberg, O., and Nielsen, G. (1966). *Hereditas* **54**, 123.

Frydman, R. B., and Cardini, C. E. (1964). *Biochem. Biophys. Res. Commun.* **14**, 353.

Gahan, L. C., and Conrad, H. E. (1968). *Biochemistry* **7**, 3929.

Goldemberg, S. H., and Maréchal, L. R. (1963). *Biochim. Biophys. Acta* **71**, 743.

Gonzales, N. S., and Pontis, H. G. (1963). *Biochim. Biophys. Acta* **69**, 179.

Greenwood, C. T. (1956). *Advan. Carbohyd. Chem.* **11**, 335.

Gunja-Smith, Z., Marshall, J. J., Smith, E. E., and Whelan, W. J. (1970). *FEBS (Fed. Eur. Biochem. Soc.) Lett.* **12**, 96.

Hanes, C. S. (1940a). *Proc. Roy. Soc., Ser. B* **128**, 421.

Hanes, C. S. (1940b). *Proc. Roy. Soc., Ser. B* **129**, 174.

Hanes, C. S., and Maskell, E. J. (1942). *Biochem. J.* **36**, 76.

Hawker, J. S., Ozbun, J. L., and Preiss, J. (1972). *Phytochemistry* **11**, 1287.

Hawker, J. S., Ozbun, J. L., Ozaki, H., Greenberg, E., and Preiss, J. (1973). *Arch. Biochem. Biophys.* **160**, 530.

Haworth, W. N., Peat, S., and Bourne, E. J. (1944). *Nature (London)* **154**, 236.

Hehre, E. J. (1951). *Advan. Enzymol.* **11**, 297.

Hirst, E. L. (1957). *Proc. Chem. Soc., London* p. 193.

Hobson, P. N., Whelan, W. J., and Peat, S. (1951). *J. Chem. Soc., London* p. 1451.

Igaue, I. (1962). *Agr. Biol. Chem.* **26**, 424.

Illingworth, B., and Brown, D. H. (1964). *Ciba Found. Symp.*, p. 336.

Iwata, S., and Fukui, T. (1973). *FEBS (Fed. Eur. Biochem. Soc.) Lett.* **36**, 222.

Jacobsen, J. V., Scandalios, J. G., and Varner, J. E. (1970). *Plant Physiol.* **45**, 367.

James, W. O. (1953). "Plant Respiration." Oxford Univ. Press, London and New York.

Jones, D. (1970). *Trans Brit. Mycol. Soc.* **54**, 351.

Kainuma, K., and French, D. (1970). *Biopolymers* **11**, 2241.

Kamogawa, A., Fukui, T., and Nikuni, Z. (1968). *J. Biochem. (Tokyo)* **63**, 631.

Katz, J. R., and Van Itallie, T. B. (1930). *Z. Phys. Chem., Abt. A* **150**, 90.

Kozak, L. P., and Bretthauer, R. K. (1970). *Biochemistry* **9**, 1115.

Krisman, C. R. (1972). *Biochem. Biophys. Res. Commun.* **46**, 1206.

Laetsch, W. M. (1974). *Annu. Rev. Plant Physiol.* **25**, 27.

Larner, J., and Villar-Palasi, C. (1959). *Proc. Natl. Acad. Sci. U.S.* **45**, 1234.

Lee, E. Y. C., and Braun, J. J. (1973). *Ann. N. Y. Acad. Sci.* **210**, 115.

Lee, E. Y. C., and Whelan, W. J. (1971). *In* "The Enzymes" (P. O. Boyer, ed.), 3rd ed., Vol. 5, pp. 191–234. Academic Press, New York.

Lee, E. Y. C., Mercier, C., and Whelan, W. J. (1968). *Arch. Biochem. Biophys.* **125**, 1028.

Lee, Y. P. (1960). *Biochim. Biophys. Acta* **43**, 18 and 25.

Leloir, L. F. (1964). *Proc. Plen. Sess. Int. Congr. Biochem., 6th, 1964*, Vol. 33, pp. 15–29.

Leloir, L. F., and Cardini, C. E. (1957). *J. Amer. Chem. Soc.* **79**, 6340.

Leloir, L. F., De Fekete, M. A. R., and Cardini, C. E. (1961). *J. Biol. Chem.* **236**, 636.

Linderstrøm-Lang, K., and Engel, C. (1938). *C. R. Trav. Lab. Carlsberg, Ser. Chem.* **21**, 243.

Marchessault, R. H., and Sarko, A. (1967). *Advan. Carbohyd. Chem.* **22**, 421.

Maruo, B., and Kobayashi, T. (1949). *J. Agr. Chem. Soc.* **23**, 115 and 120.

Maruo, B., and Kobayashi, T. (1951). *Nature (London)* **167**, 606.

Mencik, Z., Marchessault, R. H., and Sarko, A. (1971). *J. Mol. Biol.* **55**, 193.

Meyer, K. H., Brentano, W., and Bernfeld, P. (1940). *Helv. Chim. Acta* **23**, 845.

Mommaerts, W. F. H. M., Illingworth, B., Pearson, C. M., Guillory, R. J., and Seraydarian, K. (1959). *Proc. Nat. Acad. Sci. U.S.* **45**, 791.

Momotani, Y., and Kato, J. (1966). *Plant Physiol.* **41**, 1395.

Mordoh, J., Leloir, L. F., and Krisman, C. R. (1965). *Proc. Nat. Acad. Sci. U.S.* **53**, 86.

Mordoh, J., Krisman, C., and Leloir, L. F. (1966). *Arch. Biochem. Biophys.* **113**, 265.

Murata, T. (1970). *Nippon Nogei Kagaku Kaishi* **44**, 412.

Murata, T. (1972). *Nippon Nogei Kagaku Kaishi* **46**, 1.

Murata, T., and Akazawa, T. (1966). *Arch. Biochem. Biophys.* **114**, 76.

Murata, T., Minamikawa, T., and Akazawa, T. (1963). *Biochem. Biophys. Res. Commun.* **13**, 439.

Murata, T., Akazawa, T., and Fukuchi, S. (1968). *Plant Physiol.* **43**, 1899.

Neufeld, E. F., and Hassid, W. Z. (1963). *Advan. Carbohyd. Chem.* **18**, 309.

Nikaido, H., and Hassid, W. Z. (1971). *Advan. Carbohyd. Chem. Biochem.* **26**, 352.

Ozbun, J. L., Hawker, J. S., and Preiss, J. (1971a). *Biochem. Biophys. Res. Commun.* **43**, 631.

Ozbun, J. L., Hawker, J. S., and Preiss, J. (1971b). *Plant Physiol.* **48**, 765.

Ozbun, J. L., Hawker, J. S., and Preiss, J. (1972). *Biochem. J.* **126**, 953.

Paleg, L. G. (1960). *Plant Physiol.* **35**, 293.

Paleg, L. G. (1961). *Plant Physiol.* **36**, 829.

Parodi, A. J., Mordoh, J., Krisman, C. R., and Leloir, L. F. (1969). *Arch. Biochem. Biophys.* **132**, 111.

Pigman, W., ed. (1957). "The Carbohydrates." 1st ed. Academic Press, New York.

Pontis, H. G. (1966). *Arch. Biochem. Biophys.* **116**, 416.

Recondo, E., and Leloir, L. F. (1961). *Biochem. Biophys. Res. Commun.* **6**, 85.

Recondo, E., Dankert, M., and Leloir, L. F. (1963). *Biochem. Biophys. Res. Commun.* **12**, 204.

Rundle, R. E., Daasch, L., and French, D. (1944a). *J. Amer. Chem. Soc.* **66**, 130.

Rundle, R. E., Foster, J. F., and Baldwin, R. R. (1944b). *J. Amer. Chem. Soc.* **66**, 2116.

Sachs, J. (1862). *Bot. Ztg.* **20**, 365.

Schlubach, H. H. (1958). *Fortschr. Chem. Org. Naturst.* **15**, 1.

Schoch, T. J. (1942). *J. Amer. Chem. Soc.* **64**, 2957.

Schoch, T. J. (1945). *Advan. Carbohyd. Chem.* **1**, 247.

Shiefer, S., Lee, E. Y. C., and Whelan, W. J. (1973). *FEBS (Fed. Eur. Biochem. Soc.) Lett.* **30**, 129.

Slack, C. R., Hatch, M. D., and Goodchild, D. J. (1969). *Biochem. J.* **114**, 489.

Smith, W. L., Nakajima, T., and Ballou, C. E. (1975). *J. Biol. Chem.* **250**, 3426.

Sprague, G. F., Grimhall, B., and Hixon, R. M. (1943). *J. Amer. Soc. Agr.* **35**, 817.

Stepanenko, B. N. (1960). *Bull. Soc. Chim. Biol.* **42**, 1519.

Sutherland, E. W. (1952). *Phosphorus Metab.* **2**, 577–596.

Tanaka, Y., and Akazawa, T. (1970). *Plant Physiol.* **46**, 586.

Tanaka, Y., and Akazawa, T. (1971). *Plant Cell Physiol.* **12**, 493.

Tanaka, Y., Ito, T., and Akazawa, T. (1970). *Plant Physiol.* **46**, 650.

Thoma, J. A., Spradlin, J. E., and Dygert, S. (1971). *In* "The Enzymes" (P. O. Boyer ed.), 3rd ed., Vol. 5, pp. 115–189. Academic Press, New York.

Umemura, Y., Nakamura, M., and Funahashi, S. (1967). *Arch. Biochem. Biophys.* **119**, 240.

Varner, J. E., and Johri, M. M. (1968). *In* "Biochemistry and Physiology of Plant Growth Substances" (F. Wightman and G. Setterfield, eds.), p. 793 Runge Press, Ottawa.

Wessels, J. G. H. (1969). *Bacteriol. Rev.* **33**, 505.

Whelan, W. J. (1971). *Biochem. J.* **122**, 609.

Wolfrom, M. L., and El Khadem, H. (1965). *In* "Starch: Chemistry and Technology" (R. L. Whistler and E. F. Paschall, eds.), Vol. 1, p. 251. Academic Press, New York.

Yomo, H. (1960). *Hakko Kyokaishi* **18**, 603.

Yoshida, M., Fujii, M., Nikuni, Z., and Maruo, B. (1958). *Bull. Agr. Chem. Soc. Jap.* **21**, 127.

Zaslow, B. (1965). *In* "Starch: Chemistry and Technology" (R. L. Whistler and E. F. Paschall, eds.), p. 279. Academic Press, New York.

Zuber, M. W. (1965). *In* "Starch: Chemistry and Technology" (R. L. Whistler and E. F. Paschall, eds.), p. 43. Academic Press, New York.

13

Cell Wall Biogenesis

ARTHUR L. KARR

I. Introduction

What is meant by the term, cell wall? In general, it is the name that has been used to describe those portions of the plant cell lying outside the plasma membrane. Polysaccharides and proteins are the most abundant wall polymers in the young plant cell. Lignin and other polymers are added to the cell wall as the plant cell grows and differentiates.

A host of functions has been ascribed to the cell wall. The cell wall is the structural element of the plant and in woody plants forms the skeleton on which the rest of the plant is suspended. Cell wall materials are major constituents of the conducting vessels. The cell wall physically counteracts the osmotic pressure resulting from the cell contents and, thereby, restricts the size and shape of the plant cell. The cell wall must play an important role in cell growth and differentiation, since these events involve changes in cell size and shape. A portion of the cell wall acts as an intercellular cement that binds the cells together to form tissue. Finally, the cell wall may play a role in pathogenesis by specifically inhibiting critical pathogen-produced enzymes or by presenting a physical barrier to the invading pathogen. A number of other possible roles for the cell wall have been entertained. These range from whether the cell wall includes specific communication sites such as those found on the surface of bacterial cells to whether certain wall polymers can affect ice crystal formation and reduce cell damage caused by freezing.

It is a well-worn cliche that the cell wall is made up of a large number of complex polymers and, indeed, is mostly a mixture of ill-defined polymers. In contrast to this general impression of cell wall complexity, Talmadge et al. (1973) and Keegstra et al. (1973) have recently reported that the cell wall of suspension-cultured sycamore cells (*Acer pseudoplatanus* L.) is composed of a very limited number of major structural polymers. They suggest that these well-defined polymers are interconnected to form a single macromolecule. The present ideas about cell wall structure are discussed in detail in Chapter 9.

It is clear that the composition and structure of the cell wall are strictly controlled. The polysaccharide composition of the cell wall may vary from species to species but does not vary among varieties of a given species (Nevins et al., 1967). The amount of the glycoprotein component, extensin, present in the cell wall and the degree of glycosylation of this component are characteristic of the plant source from which the cell walls are isolated (Lamport, 1965; Lamport and Miller, 1971). Wilder and Albersheim (1973) have shown that only minor differences exist between the structurally important xyloglucan (Bauer et al., 1973) from the cell walls of sycamore and bean. The retention of the functionally important part of this molecule in these distant relatives again implies strict control of cell wall structure. Finally, Marx-Figini (1966, 1969) has shown that the chain length of the cellulose molecules in the secondary cell wall is precisely controlled.

There is abundant evidence that the composition of the cell wall changes during the growth and development of the plant cell. Examples of such changes include the specific loss of a glucan from the wall during growth in beans (Nevins et al., 1968), the deposition of glycoprotein in the wall during the elongation process in oat coleoptiles (Cleland, 1967), and the turnover of a xyloglucan during stem elongation in pea (Labavitch and Ray, 1973). Future examples will probably include more subtle changes such as the alteration of the conformation of preexisting polymers noted in seaweed (Lawson and Rees, 1970). These changes in chemical composition of the cell wall are synchronized with growth and developmental events. It seems clear that specific alterations in the cell wall are included in the program that controls cellular development.

This introduction has been devoted to a brief glimpse of the cell wall. The plant cell wall is a specific organelle whose composition and structure is strictly controlled by the cell. The cell wall plays an important role in physiological processes in plants. As new functional demands are placed on the cell wall, its composition is altered to meet these demands.

The process of cell wall biogenesis includes formation of the precursors of cell wall polymers, biosynthesis of those polymers, assembly of the polymers into the cell wall, and any subsequent alterations of the cell wall. The large number of events that must be coordinated during synthesis, assembly, and alteration of the plant cell wall implies a complex control system. Little is presently known about how such processes are controlled at the molecular level.

This chapter is devoted to what is currently the most productive approach to understanding cell wall biogenesis. This approach involves the isolation and study of the enzymes that catalyze the individual reactions important in cell wall biogenesis. The discussion in this chapter will be limited to the polysaccharide and protein components of the cell wall.

II. Cell Wall Language

The structures of the compounds discussed in this chapter are shown in Fig. 1. These include the three hexoses, D-glucose (I), D-galactose (II), and D-mannose (III). The pentoses are D-xylose (IV) and L-arabinose (V). Arabinose residues in the glycoprotein, extensin, and the arabinose residues of many cell wall polysaccharides are present in the furanose ring form. Xylose generally exists in the pyranose ring form. It is important to notice that the two pentoses, L-arabinose and D-xylose, are C_4-epimers. The two uronic acids are D-glucuronic acid (VI) and its C_4-epimer, D-galacturonic acid (VII).

In addition to the constituents of cell wall polysaccharides mentioned above, it will be helpful to know the structure of myo-inositol (VIII) and sucrose (IX). myo-Inositol is one of the nine possible stereoisomers of cyclohexanehexol. Sucrose is the major translocate in plants. Both of these compounds are thought to be intermediates in the formation of important sugar nucleotides [e.g., UDP-D-glucose (X)]. The sugar nucleotides are used as substrates in the synthesis of a large number of plant polysaccharides. A number of sugar nucleotides will be discussed in this chapter, and the common abbreviations for the nucleic acid bases will

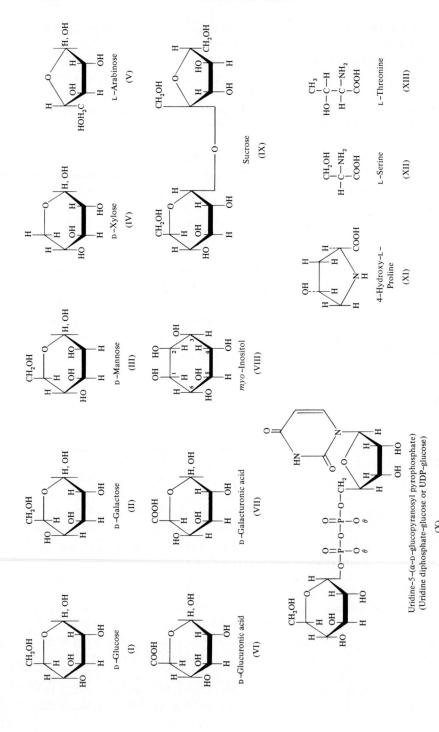

Fig. 1. Some compounds important to the structure and synthesis of the plant cell wall.

be used. These are U for uracil, G for guanine, A for adenine, T for thymine, and C for cytosine.

The peptide portion of extensin contains the hydroxyamino acids, 4-hydroxy-L-proline (XI), L-serine (XII), and L-threonine (XIII). The first two amino acids are known points of attachment between the carbohydrate and protein chains of the glycoprotein. No carbohydrate attachment to threonine has yet been demonstrated.

III. Synthesis of Compounds Used as Substrates for the Formation of Wall Polymers

A. General

Sugar nucleotides are the carbohydrate donors for the formation of cell wall polysaccharides. The sugar nucleotides can be synthesized directly from the monosaccharide, ATP, and the appropriate nucleoside triphosphate. In addition, the very important UDP-sugars can be synthesized from UDP-glucose by successive alterations of the glucose moiety. Five glucose nucleotides, including UDP-glucose, can be synthesized directly from sucrose by the enzyme sucrose synthetase. Finally, *myo*-inositol may be a critical intermediate in the formation of the sugar nucleotides containing glucuronic acid, galacturonic acid, arabinose, xylose, and the branched sugar, apiose. These are the pathways considered to be potentially important in the formation of the substrates for cell wall polysaccharide biosynthesis.

The arabinose side chain of the glycoprotein, extensin, is synthesized from UDP-L-arabinose. The enzymes responsible for the transfer of the other carbohydrates to extensin have not been isolated, but it is safe to assume that sugar nucleotides will be the glycosyl donors in these reactions. The protein backbone of extensin is thought to be synthesized on the ribosomes, but this is presently far from proved. Since synthesis of this protein by a cell-free system has not been demonstrated and the precursors of the protein are not known, extensin will not be discussed further in this section.

B. Sugar Nucleotide Synthesis and Sugar Interconversions

1. From Monosaccharides to Sugar Nucleotides

$$\text{D-Galactose} + \text{ATP} \rightarrow \text{D-galactose 1-P} + \text{ADP} \tag{1}$$

The first step in the formation of sugar nucleotides from monosaccharides is catalyzed by a class of enzymes commonly called kinases.

These enzymes catalyze reactions that result in the formation of the monosaccharide 1-P from the monosaccharide and ATP. An example of the reaction catalyzed by one such enzyme, galactokinase (Neufeld *et al.* 1960), is shown in Eq. (1).

The glucose phosphates are important intermediates in general metabolism, and glucose 1-P is available from a large number of sources. These sources include the reaction catalyzed by starch phosphorylase [Eq. (2)], a combination of the reactions catalyzed by hexokinase [Eq. (3a)] and phosphoglucomutase [Eq. (3b)], and a combination of the reactions catalyzed by sucrose synthetase [Eq. (4a)] and UDP-D-glucose pyrophosphorylase [Eq. (4b)].

$$\text{Starch} + \text{phosphate} \rightleftharpoons \text{D-glucose 1-P} \tag{2}$$

$$\text{D-Glucose} + \text{ATP} \rightleftharpoons \text{D-glucose 6-P} + \text{ADP} \tag{3a}$$

$$\text{D-Glucose 6-P} \rightleftharpoons \text{D-glucose 1-P} \tag{3b}$$

$$\text{Sucrose} + \text{UDP} \rightleftharpoons \text{UDP-D-glucose} + \text{fructose} \tag{4a}$$

$$\text{UDP-D-glucose} + \text{pyrophosphate} \rightleftharpoons \text{D-glucose 1-P} + \text{UTP} \tag{4b}$$

The second step in the formation of the sugar nucleotides by this pathway is catalyzed by a class of enzymes known as pyrophosphorylases. Examples of the reactions catalyzed by two such enzymes, UDP-D-glucose pyrophosphorylase (Ginsburg, 1958) and UDP-D-galactose pyrophosphorylase (Neufeld *et al.*, 1957), are shown in Eqs. (5) and (6).

$$\text{D-Glucose 1-P} + \text{UTP} \rightleftharpoons \text{UDP-D-glucose} + \text{pyrophosphate} \tag{5}$$

$$\text{D-Galactose 1-P} + \text{UTP} \rightleftharpoons \text{UDP-D-galactose} + \text{pyrophosphate} \tag{6}$$

The team of kinase and pyrophosphorylase (plus the glucose 1-P forming reactions) can account for the synthesis of most of the sugar nucleotides to be discussed in this chapter. Such a combination of enzymes could account for the formation of UDP-D-glucose, UDP-D-galactose, GDP-D-glucose, GDP-D-mannose, UDP-L-arabinose, UDP-D-galacturonic acid, and UDP-D-glucuronic acid.

2. INTERCONVERSION OF SUGARS

It can be seen from the structures in Fig. 1 that the monosaccharide constituents of cell wall polymers are closely related. Indeed, some differ only in the stereochemical configuration about a single carbon atom. Plant enzymes that catalyze reactions resulting in the interconversion of many of these monosaccharides are known. Such alterations generally (but not always) take place after the sugar nucleotide has been formed from the monosaccharide.

Enzymes that catalyze reactions resulting in alteration of the sugar portion of the UDP-sugars are widely distributed in plants. These enzymes include the C_4-epimerases, UDP-D-galactose-4-epimerase (Neufeld *et al.*, 1957) [Eq. (7)], UDP-D-galacturonic acid-4-epimerase (Feingold *et al.*, 1960) [Eq. (8)], and UDP-L-arabinose-4-epimerase (Neufeld *et al.*, 1957), [Eq. (9)].

$$\text{UDP-D-galactose} \rightleftharpoons \text{UDP-D-glucose} \tag{7}$$

$$\text{UDP-D-galacturonic acid} \rightleftharpoons \text{UDP-D-glucuronic acid} \tag{8}$$

$$\text{UDP-L-arabinose} \rightleftharpoons \text{UDP-D-xylose} \tag{9}$$

Other enzymes from plant sources capable of carrying out the alteration of the UDP-sugars include UDP-D-glucose dehydrogenase (Hassid, 1967) [Eq. (10)] and UDP-D-glucuronic acid decarboxylase (Neufeld *et al.*, 1958) [Eq. (11)].

$$\text{UDP-D-glucose} \rightarrow \text{UDP-D-glucuronic acid} \tag{10}$$

$$\text{UDP-D-glucuronic acid} \rightarrow \text{UDP-D-xylose} \tag{11}$$

The reactions above [Eqs. (7)–(11)] can be arranged (and often are) to give a pathway for the synthesis of the different UDP-sugars from UDP-D-glucose. The physiological significance of such a complete pathway is not known, and, therefore, the pathway is not emphasized here.

The work of Elbein and Hassid (1966) suggests the possible presence of another epimerase in plant systems. This enzyme would be a C_2-epimerase and catalyze the reaction shown below [Eq. (12)].

$$\text{GDP-D-mannose} \rightleftharpoons \text{GDP-D-glucose} \tag{12}$$

C. From Sucrose to Glucose Nucleotides

Sucrose falls into a class of compounds that have been termed high-energy compounds. The free energy of hydrolysis of the glycosidic bond in sucrose is approximately —6.6 kcal/mole (Barker and Hassid, 1951). This value is comparable to those of UDP-D-glucose (—7.6 kcal/mole) and ATP (—6.9 kcal/mole) and may be contrasted with the value of —3.0 kcal/mole for another disaccharide, maltose (Leloir *et al.*, 1960). Sucrose is a direct donor of fructose in fructan formation (Hestrin, 1949).

Two possible schemes of sucrose synthesis have been demonstrated in plants. These are the sucrose phosphate synthase–sucrose phosphatase system [Eq. (13)] and sucrose synthase [Eq. (14)].

UDP-D-glucose + fructose 6-P $\rightleftharpoons$ sucrose-P + UDP sucrose-P

$$\rightarrow \text{sucrose} + \text{phosphate} \qquad\qquad (13)$$

$$\text{NDP-D-glucose} + \text{fructose} \rightleftharpoons \text{sucrose} + \text{NDP} \qquad\qquad (14)$$

It should be noted in the sucrose synthase reaction (14) that, while N may be uridine, adenosine, thymidine, cytidine, or guanosine, UDP-D-glucose is the preferred substrate (Milner and Avigad, 1965; Grimes *et al.*, 1970).

The reaction catalyzed by sucrose phosphate synthetase has an equilibrium constant of over 3000:1 in favor of sucrose formation (Hassid, 1967). In the presence of sucrose phosphatase, this pathway for the formation of sucrose becomes irreversible. The two enzymes that catalyze the reactions shown in Eq. (13) are both present in the chloroplast (Delmer and Albersheim, 1970) and are most probably responsible for the accumulation of sucrose from photosynthetically fixed carbon (Hassid, 1967).

The sucrose synthase reaction [Eq. (14)] has an equilibrium constant between 1 and 8 and is readily reversible. In beans, sucrose synthetase is only present in nonphotosynthetic tissue (Delmer and Albersheim, 1970).

Grimes *et al.* (1970) have suggested that the *in vivo* function of sucrose synthetase is not sucrose formation but the formation of sugar nucleotides from sucrose. Their argument is based on the facts that sucrose synthase reaction does not greatly favor sucrose formation and that sucrose synthase is located in tissue normally associated with sucrose utilization. Since the reaction shown in Eq. (14) is reversible, the glucose moiety of sucrose could be transferred directly to nucleoside diphosphates to produce UDP-D-glucose and GDP-D-glucose for cell wall synthesis and ADP-D-glucose for starch biosynthesis. The function of TDP-D-glucose in higher plants has not been described, but it has been recently reported that CDP-D-glucose may be a glycosyl donor in the synthesis of a β-glucan (Tsai and Hassid, 1973). There is little question that sucrose synthetase could provide a source of glucose nucleotides (particularly UDP-D-glucose) for cell wall biosynthesis, but the physiological significance of such a pathway is still unclear.

An alternate pathway for the conversion of sucrose to polysaccharides has been considered. This pathway involves the coupling of the reactions catalyzed by sucrose synthase and UDP-D-glucose pyrophosphorylase, as shown in Eq. (4). Such a pathway for sucrose utilization has been suggested for beans (Delmer and Albersheim, 1970) and the starch granules of sweet corn (De Fekete and Cardini, 1964). The glucose 1-P formed in the reaction could be converted directly to glucose nucleotides by the action of pyrophosphorylases. In addition, the glucose 1-P could be converted to glucose 6-P by the action of phosphoglucomutase [Eq. (3b)] and then enter the cell wall through the *myo*-inositol pathway. There is one piece of correlative evidence that suggests that the second pathway for sucrose utilization [Eq. (4)] is physiologically significant. The two enzymes, sucrose synthase and UDP-D-glucose pyrophosphorylase, are found in the same parts of the plant and their catalytic activities rise and fall in unison in response to physiological changes in the plant tissue (Delmer and Albersheim, 1970). This is what would be expected if the two enzymes function together. The question that must be raised within the context of this chapter is whether the reactions in Eq. (4) represent a general pathway for sucrose utilization or one that is restricted to the conversion of sucrose to polysaccharides. There is presently no answer to this question.

D. From Inositol to Sugar Nucleotides

myo-Inositol is an excellent precusor for the sugar residues of cell wall polysaccharides. When plant cells are incubated with radiolabeled *myo*-inositol, a large portion of the label appears in the pentose and uronic acid residues of the cell wall. In a number of cases, over 50% of the label from *myo*-inositol has been recovered from the cell wall carbohydrates (Loewus *et al.*, 1962; Loewus, 1971).

How are the sugar nucleotide precursors for cell wall biosynthesis produced from *myo*-inositol? The first step involves the oxygenase-catalyzed cleavage of *myo*-inositol to D-glucuronic acid. This enzyme is known to occur in higher plants (Loewus *et al.*, 1962; Gruhner and Hoffman-Ostenhof, 1966). The D-glucuronic acid produced in this reaction is then converted to UDP-D-glucuronic acid by the action of D-glucuronic acid kinase and UDP-D-glucuronic acid pyrophosphorylase (Roberts and Rao, 1971). The resulting UDP-D-glucuronic acid can be converted to UDP-D-galacturonic acid, UDP-D-xylose, and UDP-L-arabinose by enzymes that have been described previously in this chapter.

Yet, it is clear that the label from radiolabeled carbohydates, such as glucose and sucrose, is also rapidly incorporated into the uronic acid and pentose residues of the cell wall. If *myo*-inositol is to be considered an important intermediate in the formation of cell wall precursors, then it is necessary to account for the conversion of glucose and sucrose into *myo*-inositol. The enzymes necessary for these conversions are known. A plant enzyme catalyzes a process of ring closure between $C_{(1)}$ and $C_{(6)}$

of D-glucose 6-P (Ruis *et al.*, 1967; Loewus and Loewus, 1971). The product of this reaction is L-*myo*-inositol 1-P. The L-*myo*-inositol 1-P is converted to *myo*-inositol by the action of a phosphatase (Loewus, 1969; Loewus and Loewus, 1971). The precursor of *myo*-inositol, D-glucose 6-P, can be synthesized from either sucrose or glucose.

There is a considerable amount of evidence in the literature that implicates *myo*-inositol in the formation of cell wall uronic acids and pentoses. More recently, Roberts and Loewus (1973) have reported that the presence of high concentrations of unlabeled *myo*-inositol significantly depresses the incorporation of label from [6-^{14}C]D-glucose into cell wall D-galacturonic acid in corn roots. These results suggest that the conversion of glucose to cell wall uronic acids involves an intermediate formation of *myo*-inositol. If *myo*-inositol is not an important intermediate in cell wall polymer biosynthesis, it will be necessary to explain why the plant cell possesses such an efficient system for converting *myo*-inositol into the precursors of cell wall polymers.

IV. Enzyme Systems Which Catalyze the Synthesis of Cell Wall Polysaccharides

A. General

The enzymes responsible for the formation of cell wall polysaccharides are membrane bound and are generally recovered in a particulate cell fraction. These enzymes have been given a number of common names in the literature. These names include polysaccharide synthetase, glycosyltransferase, and particulate enzyme system. Polysaccharide synthetases catalyze the transfer of sugars from sugar nucleotides to acceptors to produce polysaccharides [as in Eq. (15)]. The number of

$$\text{NDP-sugar} + \text{acceptor} \rightarrow \text{sugar-acceptor} + \text{NDP} \qquad (15)$$

enzymes necessary to carry out the transfer of a single sugar residue to a polysaccharide is unknown. Furthermore, the nature of the primer or acceptor is not known.

The polysaccharide synthetases possess a high degree of substrate specificity. The enzymes are specific for the base present in the sugar nucleotide. Enzymes that use UDP-D-glucose will not use GDP-D-glucose, etc. The enzymes are also specific for the sugar present in the sugar nucleotide. Enzymes which use UDP-D-glucose do not use UDP-D-galactose, etc. These enzymes also produce a chemically distinct product. A different enzyme is necessary for the formation of the different linkages possible with a single sugar (1,4-linked glucose versus 1,3-linked glucose), and a different enzyme is required for the formation of the different anomeric configurations of a single sugar (α-glucose versus β-glucose). These enzymes may also be specific for the anomeric configuration of the sugar in the sugar nucleotide (Clark and Villemez, 1972).

Polysaccharide synthetase activities are detected by measuring the transfer of radioactive labeled sugar from a sugar nucleotide to a polysaccharide. The labeled product is identified as a polysaccharide either by its immobility when subjected to electrophoresis in sodium tetraborate buffer or by its solubility properties. The product may be subjected to partial acid hydrolysis or acetolysis. These procedures result in the release of labeled disaccharides, trisaccharides, etc., from the polysaccharide. The linkages between sugars in the disaccharides and trisaccharides can be determined by using a combination of chromatographic and chemical methods. The anomeric configuration of the sugars in the polymers can be determined with stereospecific polysaccharide hydrolases. While this assay system has been valuable for the initial isolation and partial characterization of the polysaccharide synthetases, it has certain limitations. First, the analytical methods described above yield only structural information about labeled sugar residues and not about unlabeled sugar residues or possible noncarbohydrate portions of the product. Second, the methods do not allow differentiation between addition of a few sugar residues to the end of a preexisting polysaccharide (or addition of short side chains) and total synthesis of a polysaccharide. Third, the methods used generally do not give information about the degree of polymerization (DP) of the product. Finally, these methods do not provide proof of chemical identity between the *in vitro* product and the naturally occurring polymer.

B. Cellulose

Cellulose, a β-1,4-glucan, is synthesized by a large number of organisms ranging from bacteria to trees. It is present in the cell wall in the form of highly ordered structures called microfibrils. There are numerous reports in the literature that highly purified cellulose from plant cell walls contains small amounts of sugars other than glucose. It is not clear whether the presence of such sugars is due to their covalent attachment to the cellulose polymers or their occurrence in hemicellulose molecules, which have been shown to bind tightly to cellulose (Bauer *et al.*, 1973). Cellulose is present in both the primary and secondary cell wall of plant cells. Cellulose from the primary cell wall consists of shorter chains with a less precisely controlled degree of polymerization than cellulose polymers from the secondary wall (Marx-Figini, 1966, 1969; Spence and MacLachlan, 1972). Indeed, the differences in the degree of polymerization and the control of chain length between cellulose from the primary and secondary wall are sufficient to suggest that the two types of cellulose are formed by different mechanisms. Cellulose is an important structural polymer in the cell wall and could be important in determining the cell wall superstructure in the model (Keegstra *et al.*, 1973) of the primary cell wall (Villemez, 1974). While the mechanisms responsible for formation and orientation of cellulose microfibrils are not known, considerable information is available about enzymes capable of catalyzing the formation of celluloselike β-1,4-glucans.

The first reports of the synthesis of celluloselike molecule by a cell-free system from higher plants were by Elbein *et al.* (1964) and Barber *et al.* (1964). The enzyme is found tightly bound in a particulate cell fraction from mung beans (*Phaseolus aureus*). The glucosyl donor in the reaction is GDP-D-glucose, and the product was identified as a β-1,4-glucan using the types of procedures described above. Liu and Hassid (1970) have reported the solubilization and partial purification of this enzyme system from mung beans. It seems clear that mung beans and several other plants (Barber *et al.*, 1964) contain an enzyme system that can use GDP-D-glucose as a glycosyl donor for the formation of a β-1,4-glucan.

Barber *et al.* (1964) also reported that some substance present in a yeast cofactor concentrate stimulated the enzyme-catalyzed incorporation of glucose from GDP-D-glucose into polysaccharide. This substance was electrophoretically identical to and could be replaced by authentic GDP-D-mannose. When GDP-D-mannose, along with GDP-D-glucose, was included in the reaction mixture, a product with the solubility characteristic of cellulose was formed. A partial hydrolysate of the product synthesized when GDP-D-mannose was included was found to contain other saccharides in addition to the cellodextran series (glucose, cellobiose, cellotriose, etc.). It was later found that when this enzyme system was supplied with GDP-D-mannose alone, it could catalyze the synthesis of a glucomannan (Elbein and Hassid, 1966).

What is the *in vivo* function of this enzyme system? Barber *et al.* (1964) and Elbein and Hassid (1966) have interpreted these results to mean that the particulate system contains more than one enzyme that can use GDP-D-glucose. When GDP-D-glucose alone is present, an enzyme in the particulate system uses this compound as a substrate for cellulose synthesis. When both GDP-D-glucose and GDP-D-mannose are present, cellulose may be made, but an additional enzyme uses the GDP-D-glucose as one substrate for glucomannan synthesis. The *in vivo* functions of the enzymes that use GDP-D-glucose would be both the production of cellulose and a hemicellulose (glucomannan).

Not all workers have agreed with this interpretation. Villemez and Heller (1970) have posed the question, Is guanosine diphosphate-D-glucose a precursor of cellulose? They note that GDP-D-glucose may not be present in plant tissues that are synthesizing cellulose. They have also noted that while the level of catalytic activity of the enzyme system that uses GDP-D-glucose changes during growth and development of the plant, these changes do not coincide with periods of cellulose synthesis. Villemez (1970a, 1971) has also considered the kinetic consequences of having two enzymes that use the same substrate, GDP-D-glucose. He argues that if one enzyme is operative when GDP-D-glucose is present and both are operative when GDP-D-mannose is also present, then including both substrates in the reaction mixture should result in an increase in the initial rate of incorporation of glucose into polymers. He finds that, while the presence of GDP-D-mannose does increase the total incorporation of glucose into polysaccharides, it does not affect the initial rate of polymer formation from glucose. This suggests that only one enzyme in the enzyme mixture uses GDP-D-glucose as a substrate. Heller and Villemez (1972a,b) have solubilized (but not separated) both the mannosyltransferase and the glucosyltransferase activities from this particulate enzyme system. They have found that when only GDP-D-mannose is provided as a substrate to this soluble system that a β-1,4-mannan is produced, and they have confirmed that when only GDP-D-glucose is provided as a substrate a β-1,4-glucan is produced (Liu and Hassid, 1970). Heller and Villemez (1972b) have found with the soluble enzyme system that the enzyme-catalyzed incorporation of glucose from GDP-D-glucose into polysaccharide is a short-lived reaction that can be greatly extended if GDP-D-mannose is included in the reaction mixture. These results can be explained if it is assumed that continued action of the glucosyltransferase is dependent on the action of the mannosyltransferase and that a glucomannan is the normal product of this enzyme system. Furthermore, Heller and Villemez (1972b) argue that if cellulose is also produced when both GDP-D-glucose and GDP-D-mannose are provided as substrates, then cellobiose should be one major product of acetolysis of a polysaccharide mixture. Their results indicate that the amount of cellobiose recovered is much less than would be expected based on considerations of the glucosyltransferase activity present in the enzyme mixture. Finally, Villemez (1974) has determined the approximate molecular weight of the acetate derivative of the polysaccharide produced when only GDP-D-glucose is provided as a substrate. He finds this product to be a mixture of small polymers with an estimated peak degree of polymerization of about 40 (much smaller than naturally occurring cellulose). The addition of GDP-D-mannose to the reaction mixture results in the disappearance of the lower molecular weight polysaccharides and the appearance of higher molecular weight polymers. These results can be explained if it is assumed that the system functions in glucomannan synthesis and not glucan synthesis. C. L. Villemez (unpublished results) suggests that the glucans produced from GDP-D-glucose represent a series of molecular weight artifacts that result from premature termination of glucose transfer in the absence of sufficient mannose-containing acceptors. On the basis of these and other arguments, Villemez and co-workers have suggested that GDP-D-glucose is the *in vivo* precursor of a glucomannan and not cellulose.

Brummond and Gibbons (1965) have demonstrated the presence of enzymes that catalyze the transfer of glucose from UDP-D-glucose to produce a celluloselike polymer. Villemez *et al.* (1967) demonstrated a similar system from mung beans. They found that 80–90% of the glucose in the product synthesized by the mung bean enzyme was joined by β-1,4-linkages (the remainder of the glucose being joined by β-1,3-linkages). Clark and Villemez (1972) have described conditions under which all of the alkali-insoluble polysaccharide produced from UDP-D-glucose by the mung bean enzyme system is a β-1,4-glucan. Villemez (1974) has determined the approximate molecular weight of this glucan produced from UDP-D-glucose. He has found that the glucan is large with a portion, about 7.5%, having a molecular weight greater than 1.2×10^6. In contrast, Flowers *et al.* (1968) found that enzymes from mung beans could use UDP-D-glucose as a glucosyl donor for polysaccharide synthesis but that the product of this reaction did not contain β-1,4-linkages.

Ordin and Hall (1967, 1968) and Pinsky and Ordin (1969) have reported that an enzyme system from oat seedlings will catalyze the synthesis of celluloselike polymers from UDP-D-glucose. The product mixture contains some β-1,3-linked glucose but is mostly β-1,4-linked glucose. These workers found that the presence of high concentrations of UDP-D-glucose favored formation of β-1,3-linked glucose polymers, while low concentrations of the sugar nucleotide favored formation of β-1,4-linked glucose polymers. Tsai and Hassid (1973) have confirmed this observation. Tsai and Hassid (1971) have reported solubilization of this enzyme system from oat seedlings. They find that the particulate system contains two enzymes that use UDP-D-glucose. One catalyzes the transfer of glucose from UDP-D-glucose into a β-1,3-glucan and the other into a β-1,4-glucan.

Delmer *et al.* (1974) have provided evidence that an enzyme that uses GDP-D-glucose to produce a cellulose-like product is present in developing cotton fibers during the period when primary wall cellulose synthesis normally occurs. These workers have not excluded the possibility that this enzyme is involved in the formation of a hemicellulose. This enzyme is absent in older fibers undergoing active deposition of secondary wall cellulose.

C. Polymerization and Subsequent Alteration of the Uronic Acids

Pectin, a polymer rich in α-1,4-linked D-galacturonic acid residues, is an important constituent of the plant cell wall (Keegstra et al., 1973). The carboxyl groups on this polymer may be methylated.

A particulate enzyme system capable of producing polygalacturonic acid has been isolated from mung beans (Villemez et al., 1966). The D-galacturonosyl donor in this reaction is UDP-D-galacturonic acid. The 4-galacturonic acid nucleotide derivatives of adenine, cytidine, guanosine, or thymidine cannot be used as galacturonosyl donors by this enzyme system. The polymer produced by this enzyme system can be completely hydrolyzed with polygalacturonase to give D-galacturonic acid (Hassid, 1967). This probably represents only partial synthesis of a wall polymer, since the similar molecule from the cell wall of sycamore cells is known to contain rhamnose as part of the polymer backbone (Talmadge et al., 1973).

Methyl esterification takes place after the polygalacturonic acid molecule is formed. The particulate fraction that contains the D-galacturonosyltransferase also contains an enzyme responsible for the formation of the methyl ester derivative of polygalacturonic acid (Kauss and Hassid, 1967b; Kauss and Swanson, 1969). The methyl donor for this reaction is S-adenosyl-L-methionine.

An enzyme system that will catalyze the transfer of D-glucuronic acid from UDP-D-glucuronic acid to polysaccharides has been isolated from corn cobs (Kauss, 1967). The D-glucuronic acid-containing polymer produced in the in vitro reaction is similar to the polysaccharides in the hemicellulose B fraction obtained from corn cobs. The formation of the 4-O-methyl ether derivative of the polymerized D-glucuronic acid can be catalyzed by an enzyme present in the same particulate fraction (Kauss and Hassid, 1967a; Kauss, 1969a). The methyl donor in the reaction is S-adenosyl-L-methionine.

D. Synthesis of Other Polysaccharides

The particulate cell fraction from plants contains a number of polysaccharide synthetases in addition to those already mentioned. The reactions catalyed by these enzymes include the formation of a xylan from UDP-D-xylose, the formation of an arabinoxylan from UDP-D-xylose and UDP-L-arabinose (Bailey and Hassid, 1966), the formation of a galactan from UDP-D-galactose (McNab *et al.*, 1968; Panayotatos and Villemez, 1973), and the formation of a β-1,3-linked glucose polymer (Feingold *et al.*, 1958). The synthesis of a glucomannan was discussed along with cellulose. C. L. Villemez (unpublished results) has obtained evidence which indicates that this glucomannan (produced in the *in vitro* reaction) is one component of a larger molecule that may include noncarbohydrate components.

V. Intermediates in the Synthesis of Polysaccharides

Sugar nucleotides are used as substrates by the enzymes responsible for the formation of plant polysaccharides. Yet, it is not clear whether the sugar nucleotides are the direct glycosyl donors in the polysaccharide synthetase reactions. The synthesis of the bacterial cell wall polysaccharides (Anderson *et al.*, 1965) and a mannan from *Micrococcus lysodeikticus* (Scher and Lennarz, 1969) involve the formation of glycolipid intermediates. These intermediates serve as glycosyl donors in the formation of polymers.

The presence of similar intermediates as glycosyl donors in plant polysaccharide formation is an attractive possibility. Such intermediates could function as specific primers or acceptors for the formation of polysaccharides. Furthermore, subunit blocks of complex heteropolysaccharides could be assembled on such an intermediate and later transferred to produce polysaccharides or to be cross-linked into the cell wall. C. L. Villemez has provided evidence that such block polymerization mechanisms might be important in the formation of glucomannan (unpublished results).

Kauss (1969b) has demonstrated the presence of an enzyme in the particulate cell fraction which catalyzes the formation of a mannosyl lipid from GDP-D-mannose. The reaction is readily reversible. Villemez and Clark (1969) have isolated this mannosyl lipid and have found that the kinetics of its formation are consistent with the compound being an intermediate. Yet, attempts to use this mannosyl lipid as a direct glycosyl donor in polysaccharide synthesis have been less than successful. Clark and Villemez (1973) have reported that phytanol phosphate is an artificial acceptor for mannose in this reaction. Since phytanol is available in large amounts, use of this compound should facilitate isolation of the mannosyl transferase involved in mannolipid production.

Villemez (1970b) reported discovery of two types of compounds, a glycolipid and a glycoprotein, which could conceivably act as intermediates in the formation of a number of plant polysaccharides. The importance of these compounds in the biosynthesis of cell wall polysaccharides remains to be shown.

VI. Cytological Location of Polysaccharide Synthesis

The polysaccharide synthetases are isolated in a particulate cell fraction. Villemez *et al.* (1968) have shown that the enzymes responsible for the formation of polysaccharides from UDP-glucose (excluding callose synthesis), UDP-galactose, UDP-galacturonic acid, UDP-glucuronic acid, and GDP-glucose are present in a single particle. It had been previously reported by Whaley and Mollenhauer (1963) and others that the Golgi apparatus were the site of polysaccharide synthesis. Therefore, Villemez *et al.* (1968) attempted to demonstrate that the polysaccharide synthetase particle originated from Golgi membranes. These workers prepared a Golgi-rich fraction by the method of Morré *et al.* (1965). They found that the centrifugal fraction which should contain the Golgi membrane did not contain the enzymes responsible for polysaccharide synthesis. The polysaccharide synthetase activities were present in fractions containing larger particles. Some of the polysaccharide synthetase particles were found to be as large as whole cells. (No intact cells were present.) On the basis of these results, Villemez *et al.* (1968) suggested that the plasma membrane was the site of polysaccharide synthesis.

Ray *et al.* (1969) have attempted to determine the cellular location of a group of polysaccharide synthetases isolated from peas. The enzymes that they have studied are responsible for the formation of polysaccharides from GDP-glucose, UDP-glucose, and UDP-galactose and, in agreement with Villemez *et al.* (1968), these workers have found that the polysaccharide synthetase activities are present in a single particle. They tentatively identified these particles as originating from the Golgi apparatus. This tentative identification was later confirmed when Eisinger and Ray (1972) reported isolation of very pure dictyosome fractions that retained polysaccharide synthetase activity. These authors further reported that polysaccharide formed in the dictyosome fraction has a composition similar to the sum of the pectic and hemicellulose fractions of the cell wall but contains no true cellulose. Robinson and Ray (1973) later reported that synthesis of pectic wall polymers and hemicellulose takes place in the Golgi system but that cellulose synthesis takes place in a different cell compartment.

VII. Synthesis of the Cell Wall Glycoprotein, Extensin

Extensin is the name that has been given to a glycoprotein found in the cell wall of plants (Lamport, 1965). This polymer has been long thought to be of structural importance, but its specific home in the wall has only been recently suggested (Keegstra et al., 1973). The protein backbone of the glycoprotein is unique in that it contains a large number of hydroxyamino acids. Approximately 30% of the amino acid residues are 4-hydroxy-L-proline (Lamport, 1967). The hydroxyamino acids serve as points for the glycosidic attachment of carbohydrates to the protein backbone. Little is known about the synthesis of extensin. The protein may be assembled on the ribosomes by the normal mechanism of protein synthesis (Chrispeels, 1970). Proline is known to be the precursor of hydroxyproline in the protein (Holleman, 1967; Chrispeels, 1970) and hydroxylation of the peptide-bound proline is catalyzed by cytoplasmic enzymes (Chrispeels, 1970). The protein has been reported to be transported to the cell wall by a mechanism involving smooth membranes (Dashek, 1970), but more recently Gardiner and Chrispeels (1973) have reported that the Golgi system is involved in glycosylation and transport of the glycoprotein. Chrispeels (1969) has shown that the particulate cell fraction from carrots contains hydroxyproline-rich proteins and that this particulate extensin is rapidly transferred to the wall. The same particulate fraction from sycamore cells contains enzymes responsible for the synthesis of the arabinose oligosaccharide side chain of extensin (Lamport, 1971; Karr, 1972). These enzymes catalyze the transfer of L-arabinose from UDP-L-arabinose to the hydroxyproline-rich protein of the particulate fraction to produce glycoprotein. Both the ring form of the L-arabinose and the sequence of the linkages in the naturally occurring oligosaccharide are known. The product synthesized in the in vitro reaction appears to be identical to the naturally occurring oligosaccharide side chain of extensin. Because of the structural complexity of the tetrasaccharide, a number of enzymes must be necessary for the synthesis of the side chain (Karr, 1972). In addition, the oligosaccharide must be produced by the sequential transfer of monosaccharides of arabinose and not by the transfer of preformed oligomers.

VIII. Alterations of Cell Wall Polymers Which Occur Outside the Plasma Membrane

A large number of different enzyme activities are associated with the plant cell wall. These enzyme activities may be detected in the walls of whole plant cells, washed from the surface of cultured plant cells, or isolated with the cell walls after cells have been ruptured. A number of these enymes, particularly the glycosidases (Keegstra and Albersheim, 1970) and pectin methylesterase (Bryan and Newcomb, 1954), may be important for *in situ* alterations in cell wall polymers.

Certain physiological processes, such as growth, are accompanied by the specific removal of polymers from the cell wall. The cell wall-bound glycosidases are probably responsible for the removal of polysaccharides. Keegstra and Albersheim (1970) have shown that glycosidases isolated from the cell wall fraction of sycamore cells are able to catalyze the partial degradation of cell walls from the same source. Lee *et al.* (1967) have demonstrated *in vitro* autolysis of cell wall preparations. The enzyme responsible for the autolysis reaction is a glucanase (Kivilaan *et al.*, 1971). The released glucan has been partially characterized and found to be a lichenanlike polymer composed of 1,3- and 1,4-linked glucosyl units. The levels of such glycosidases in the cell wall change as the plant cell grows and develops (Nevins, 1970), but no overall relationship between the level of catalytic activity and functionally important changes in the cell wall can be presently drawn.

Nothing is known about how or where the component wall polymers are assembled to produce the final cell wall structure. Nothing is known about the forces responsible for the orientation of cellulose microfibrils. It is possible that the cell wall glycosidases are responsible for catalyzing transglycosylation reactions that result in the covalent cross-linking of wall polymers. While there is little evidence to support this role for the wall glycosidases, it is interesting that Murray (1971) has found that a cell wall glucanase can catalyze the transfer of glucose from *p*-nitrophenyl-β-glucose to produce oligosaccharides. Finally, it is possible that the glycoproteins, pectins, and hemicellulose are formed and cross-linked in blocks and transported to the cell wall where they react by a non-enzyme-catalyzed mechanism with cellulose microfibrils. This mechanism presupposes that the information for cell wall structure is inherent in the structure of the component polymers.

IX. Conclusion

The present knowledge of cell wall biogenesis consists of many pieces of information. While it is not possible to fit these pieces together and complete the puzzle of cell wall biogenesis, it is possible to sketch the hypothetical path followed by carbon on its trip from glucose (or sucrose) to the cell wall. Sugar nucleotides are synthesized starting with glucose (or sucrose) by some combination of the pathways described in this chapter. The sugar nucleotides serve as glycosyl donors for the synthesis of polysaccharides. The polysaccharides are transported to a point outside the plasma membrane by a mechanism that appears to involve the Golgi system. The polysaccharides are then incorporated into the cell wall.

It is difficult to draw more exact conclusions about cell wall biogenesis on the information presently available. Most work is involved with the individual biosynthetic steps that make up the cell wall biogenesis process. Until recently (Talmadge *et al.*, 1973; Bauer *et al.*, 1973; Keegstra *et al.*, 1973), the picture of the structure and organization of cell wall polymers was nebulous and workers found themselves studying the biosynthesis of natural products of unknown structure. For this and other reasons the physiological significance of most of the biosynthetic processes studied is not clear.

There is presently no precise information concerning either the control mechanisms that govern cell wall biogenesis or the interactions between cell wall biogenesis processes and general cellular metabolism. The number of steps involved in the formation of a polysaccharide from a sugar nucleotide is not known. It is not clear how cellular control is extended beyond the plasma membrane, or how the cell wall is formed from the component polymers. Indeed, the field of cell wall biogenesis provides more questions than answers, and one suspects that most of the major hypotheses about the operation of the biogenesis process have not yet been made.

REFERENCES

Anderson, J. S., Matsuhashi, M., Haskin, M. A., and Strominger, J. L. (1965). *Proc. Nat. Acad. Sci. U.S.* **53**, 881.

Bailey, R. W., and Hassid, W. Z. (1966). *Proc. Nat. Acad. Sci. U.S.* **56**, 1586.

Barber, G. A., Elbein, A. D., and Hassid, W. Z. (1964). *J. Biol. Chem.* **239**, 4056.

Barker, H. A., and Hassid, W. Z. (1951). *In* "Bacterial Physiology" (C. H. Werkman and P. W. Wilson, eds.), p. 548. Academic Press, New York.

Bauer, W. D., Talmadge, K. W., Keegstra, K., and Albersheim, P. (1973). *Plant Physiol.* **51**, 174.

Brummond, D. A., and Gibbons, A. P. (1965). *Biochem. Z.* **342**, 308.

Bryan, W. H., and Newcomb, E. H. (1954). *Physiol. Plant.* **7**, 290.

Chrispeels, M. J. (1969). *Plant Physiol.* **44**, 1187.

Chrispeels, M. J. (1970). *Plant Physiol.* **45**, 223.

Clark, A. F., and Villemez, C. L. (1972). *Plant Physiol.* **50**, 371.

Clark, A. F., and Villemez, C. L. (1973). *FEBS (Fed. Eur. Biochem. Soc.) Lett.* **32**, 84.

Cleland, R. (1967). *Plant Physiol.* **42**, 669.

Dashek, W. V. (1970). *Plant Physiol.* **46**, 831.

De Fekete, M. A. R., and Cardini, C. E. (1964). *Arch. Biochem. Biophys.* **104**, 173.

Delmer, D. P., and Albersheim, P. (1970). *Plant Physiol.* **45**, 782.

Delmer, D. P., Beasly, C. A., and Ordin, L. (1974). *Plant Physiol.* **53**, 149.

Eisinger, W., and Ray, P. M. (1972). *Plant Physiol.* **49**, 2.

Elbein, A. D., and Hassid, W. Z. (1966). *Biochem. Biophys. Res. Commun.* **23**, 311.

Elbein, A. D., Barber, G. A., and Hassid, W. Z. (1964). *J. Amer. Chem. Soc.* **86**, 309.

Feingold, D. S., Neufeld, E. F., and Hassid, W. Z. (1958). *J. Biol. Chem.* **233**, 783.

Feingold, D. S., Neufeld, E. F., and Hassid, W. Z. (1960). *J. Biol. Chem.* **235**, 910.

Flowers, H. M., Batra, K. K., Kemp, J., and Hassid, W. Z. (1968). *Plant Physiol.* **43**, 1703.

Gardiner, M. G., and Chrispeels, M. J. (1973). *Plant Physiol.* **51**, 60.

Ginsburg, J. (1958). *J. Biol. Chem.* **232**, 55.

Grimes, W. J., Jones, B. L., and Albersheim, P. (1970). *J. Biol. Chem.* **245**, 188.

Gruhner, K. M., and Hoffman Ostenhof, O. (1966). *Hoppe-Seyler's Z. Physiol. Chem.* **347**, 278.

Hassid, W. Z. (1967). *Annu. Rev. Plant Physiol.* **18**, 253.

Heller, J. S., and Villemez, C. L. (1972a). *Biochem. J.* **128**, 243.

Heller, J. S., and Villemez, C. L. (1972b). *Biochem. J.* **129**, 645.

Hestrin, S. (1949). *Wallerstein Lab. Commun.* **12**, 45.

Holleman, J. (1967). *Proc. Nat. Acad. Sci. U.S.* **57**, 50.

Karr, A. L. (1972). *Plant Physiol.* **50**, 275.

Kauss, H. (1967). *Biochim. Biophys. Acta* **148**, 572.

Kauss, H. (1969a). *Phytochemistry* **8**, 985.

Kauss, H. (1969b). *FEBS (Fed. Eur. Biochem. Soc.) Lett.* **5**, 81.

Kauss, H., and Hassid, W. Z. (1967a). *J. Biol. Chem.* **242**, 1680.

Kauss, H., and Hassid, W. Z. (1967b). *J. Biol. Chem.* **242**, 3449.

Kauss, H., and Swanson, A. L. (1969). *Z. Naturforsch. B* **24**, 28.

Keegstra, K., and Albersheim, P. (1970). *Plant Physiol.* **45**, 675.

Keegstra, K., Talmadge, K. W., Bauer, W. D., and Albersheim, P. (1973). *Plant Physiol.* **51**, 188.

Kivilaan, A., Bandurski, R. S., and Schulze, A. (1971). *Plant Physiol.* **48**, 389.

Labavitch, J., and Ray, P. M. (1973). *Plant Physiol.* **51**, 59.

Lamport. D. T. A. (1965). *Advan. Bot. Res.* **2**, 151.

Lamport, D. T. A. (1967). *Nature (London)* **216**, 1322.

Lamport, D. T. A. (1971). *Symp. Soc. Develop. Biol.* **30**.

Lamport, D. T. A., and Miller, D. (1971). *Plant Physiol.* **48**, 454.

Lawson, C. J., and Rees, D. A. (1970). *Nature (London)* **227**, 392.

Lee, S., Kivilaan, A., and Bandurski, R. S. (1967). *Plant Physiol.* **42**, 968.

Leloir, L. F., Cardini, C. E., and Cabib, E. (1960). *In* "Comparative Biochemistry" (M. Florkin and H. S. Mason, eds.), Vol. 2, p. 97. Academic Press, New York.

Liu, T., and Hassid, W. Z. (1970). *J. Biol. Chem.* **245**, 1922.

Loewus, F. (1969). *Ann. N.Y. Acad. Sci.* **165**, 577.

Loewus, F. (1971). *Annu. Rev. Plant Physiol.* **22**, 337.

Loewus, F., Kelly, S., and Neufeld, E. F. (1962). *Proc. Nat. Acad. Sci. U.S.* **48**, 421.

Loewus, M. W., and Loewus, F. (1971). *Plant Physiol.* **48**, 255.

McNab, J. M., Villemez, C. L., and Albersheim, P. (1968). *Biochem. J.* **106**, 355.

Marx-Figini, M. (1966). *Nature (London)* **210**, 754.

Marx-Figini, M. (1969). *J. Polym. Sci., Part C* **28**, 57.

Milner, Y., and Avigad, G. (1965). *Nature (London)* **206**, 825.

Morré, D. J., Mollenhauer, H. H., and Chambers, J. E. (1965). *Exp. Cell Res.* **38**, 672.

Murray, A. (1971). Ph.D. Thesis, Michigan State University, East Lansing.

Neufeld, E. F., Ginsburg, V., Putman, E. W., Fanshier, D., and Hassid, W. Z. (1957). *Arch. Biochem.* **69**, 602.

Neufeld, E. F., Feingold, D. S., and Hassid, W. Z. (1958). *J. Amer Chem. Soc.* **80**, 4430.

Neufeld, E. F., Feingold, D. S., and Hassid, W. Z. (1960). *J. Biol. Chem.* **235**, 906.

Nevins, D. J. (1970). *Plant Physiol.* **46**, 458.

Nevins, D. J., English, P. D., and Albersheim, P. (1967). *Plant Physiol.* **42**, 900.

Nevins, D. J., English, P. D., and Albersheim, P. (1968). *Plant Physiol.* **43**, 914.

Ordin, L., and Hall, M. A. (1968). *Plant Physiol.* **43**, 473.

Ordin, L., and Hall, M. A. (1967). *Plant Physiol.* **42**, 205.

Panayotatos, N., and Villemez, C. L. (1973). *Biochem. J.* **133**, 263.

Pinsky, A., and Ordin, L. (1969). *Plant Cell Physiol.* **10**, 771.

Ray, P. M., Shininger, T. L., and Ray, M. M. (1969). *Proc. Nat. Acad. Sci. U.S.* **64**, 605.

Roberts, R. M., and Loewus, F. (1973). *Plant Physiol.* **52**, 646.

Roberts, R. M., and Rao, K. M. K. (1971). *Fed. Proc., Fed. Amer. Soc. Exp. Biol.* **30**, 1117.

Robinson, D. G., and Ray, P. M. (1973). *Plant Physiol.* **51**, 59.

Ruis, H., Molinari, E., and Hoffmann-Ostenhof, O. (1967). *Hoppe-Seyler's Z. Physiol. Chem.* **348**, 1705.

Scher, M., and Lennarz, W. Z. (1969). *J. Biol. Chem.* **244**, 2777.

Spence, F. S., and MacLachlan, G. A. (1972). *Plant Physiol.* **49**, 58.

Talmadge, K. W., Keegstra, K., Bauer, W. D., and Albersheim, P. (1973). *Plant Physiol.* **51**, 158.

Tsai, C. M., and Hassid, W. Z. (1971). *Plant Physiol.* **47**, 740.

Tsai, C. M., and Hassid, W. Z. (1973). *Plant Physiol.* **51**, 998.

Villemez, C. L. (1970a). *Biochem. J.* **120**, 1.

Villemez, C. L. (1970b). *Biochem. Biophys. Res. Commun.* **40**, 636.

Villemez, C. L. (1971). *Biochem. J.* **121**, 151.

Villemez, C. L. (1974). *In* "Plant Carbohydrate Biochemistry" (J. B. Pridham, ed.). Academic Press, New York.

Villemez, C. L., and Clark, A. F. (1969). *Biochem. Biophys. Res. Commun.* **36**, 57.

Villemez, C. L., and Heller, J. S. (1970). *Nature (London)* **227**, 80.

Villemez, C. L., Swanson, A. L., and Hassid, W. Z. (1966). *Arch. Biochem. Biophys.* **116**, 446.

Villemez, C. L., Franz, G., and Hassid, W. Z. (1967). *Plant Physiol.* **42**, 1219.

Villemez, C. L., McNab, J. M., and Albersheim, P. (1968). *Nature (London)* **218**, 878.

Whaley, W. G., and Mollenhauer, H. H. (1963). *J. Cell Biol.* **17**, 216.

Wilder, B. M., and Albersheim, P. (1973). *Plant Physiol.* **51**, 889.

14

Lipid Metabolism

P. K. STUMPF

By tradition, lipids include the class of naturally occurring compounds that invariably partition into organic solvents, much to the disgust of the chemist and biochemist who are searching for more exotic substances.

However, in recent years, the techniques of gas-liquid chromatography and of thin layer chromatography coupled with new and quantitative degradation reactions have revolutionized the study of the complex mixtures of plant lipids. The reader should refer to Johnson and Davenport (1971), Kates (1972), and Christie (1973) for details concerning these methods.

In this chapter we shall outline some of the main aspects of lipid metabolism in higher plants. Substantial recent books of this area are by Mazliak (1968), Hitchcock and Nichols (1971), and Galliard and Mercer (1973).

I. Chemical Composition of Lipids

The important lipid classes in higher plants include the neutral lipids, the waxes, the glycerophosphatides and the phytosphingolipids, and the glycolipids.

A. Neutral Lipids

In higher plants the triacylglycerols are the major constituents of this class. They are found in exceptionally high concentrations in oil-containing seeds. Diacylglycerols and monoacylglycerols are relatively rare as normal members of the neutral lipids. Chemically, triacylglycerols are esters of the trihydric alcohol, glycerol, with one or more fatty acids. Plant triacylglycerols that are solid at room temperature are called fats and have as their major acid the saturated fatty acid, palmitic acid. Those that are liquid at room temperature are called oils and have as their major acids the unsaturated fatty acids, such as oleic, linoleic, and linolenic acids. Triacylglycerols are readily hydrolyzed under alkaline conditions to free glycerol and free fatty acids [Eq. (1)].

$$
\begin{array}{l}
\text{CH}_2\text{OCOR}_1 \\
| \\
\text{R}_2\text{COOCH} \\
| \\
\text{CH}_2\text{OCOR}_3
\end{array}
+ 3\text{H}_2\text{O} \xrightarrow{\text{OH}^-}
\begin{array}{l}
\text{CH}_2\text{OH} \\
| \\
\text{HOCH} \\
| \\
\text{CH}_2\text{OH}
\end{array}
+ \text{R}_1\text{COO}^- + \text{R}_2\text{COO}^- + \text{R}_3\text{COO}^- \quad (1)
$$

Since fatty acids are linked to the alcohol groups of glycerol by ester bonds, triacylglycerols are sensitive to reagents that can attack these bonds. Thus acid or alkaline conditions will hydrolyze triacylglycerols; hydroxylamine will split the ester linkage to free glycerol and hydroxamic acids, which can then be detected by a sensitive colorimetric method; sodium methoxide will rapidly cleave the ester linkages to yield methyl esters of fatty acids and free glycerol.

1. FATTY ACIDS

To a large extent the unusual physical and chemical properties of lipids are related to the long hydrocarbon chains of the fatty acids. These chains may be saturated, mono-, or polyunsaturated and may contain cyclic or polar functions (see Table I for typical examples of different types of fatty acids). Each of these functions greatly modifies the solubility and reactivity of the neutral lipid.

TABLE I

Important Fatty Acids in Higher Plants

Common name	Symbol	Structure
Common Fatty Acids		
Lauric	12:0	$CH_3(CH_2)_{10}COOH$
Myristic	14:0	$CH_3(CH_2)_{12}COOH$
Palmitic	16:0	$CH_3(CH_2)_{14}COOH$
Stearic	18:0	$CH_3(CH_2)_{16}COOH$
Arachidic	20:0	$CH_3(CH_2)_{18}COOH$
Palmitoleic	16:1(9)	$CH_3(CH_2)_5CH{=}CH(CH_2)_7COOH$
Oleic	18:1(9)	$CH_3(CH_2)_7CH{=}CH(CH_2)_7COOH$
Linoleic	18:2(9, 12)	$CH_3(CH_2)_4CH{=}CHCH_2CH{=}CH(CH_2)_7COOH$
Linolenic	18:3(9, 12, 15)	$CH_3CH_2CH{=}CHCH_2CH{=}CHCH_2CH{=}CH(CH_2)_7COOH$
Unusual Fatty Acids		
Tariric		$CH_3(CH_2)_{10}C{\equiv}(CH_2)_4COOH$
Sterculic		$CH_3(CH_2)_7C{=}C(CH_2)_7COOH$ $\diagdown\diagup$ CH_2
Chaulmoogric		$HC{=}CH$ $\diagdown$ $CH(CH_2)_{12}COOH$ $\diagup$ $H_2C{-}CH_2$
Ricinoleic		$CH_3(CH_2)_5CHOHCH_2CH{=}CH(CH_2)_7COOH$
Vernolic		$CH_3(CH_2)_4(CH{-}CHCH_2CH{=}CH(CH_2)_7COOH$ $\diagdown\diagup$ O
α-Eleostearic		$\overset{t}{}\quad\overset{t}{}\quad\overset{c}{}$ $CH_3(CH_2)_3CH{=}CHCH{=}CHCH{=}CH(CH_2)_7COH$

Structurally, the hydrocarbon chain has a zigzag configuration with spatial dimensions as indicated in Scheme 1.

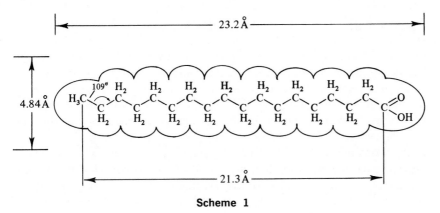

Scheme 1

When a double bond occurs in the chain, there is good evidence to suggest that the most stable configuration of the fatty acid is a slightly curving structure as represented below.

The cis isomer is the most commonly occurring geometric isomer, although in recent years several complex unsaturated fatty acids have been discovered which have double bond systems in the trans configuration (see the structure of α-eleostearic acid in Table I). When more than one double bond occurs in the hydrocarbon chain, it usually is separated from its neighboring double bond by a single methylene group, frequently called a nonconjugated system or a 1,4-diolefin system.

Again in higher plants there are exceptions. Thus, α-eleostearic acid, the chief acid in tung oil (70–80%), is isomeric with linolenic acid but differs by having a conjugated triene system with a trans,trans,cis-configuration. Unlike its nonconjugated isomer, linolenic acid, this conjugated acid elaidinizes very rapidly and is a solid at room temperature.

The interesting subject of how unique modifications of hydrocarbon chains are brought about in different plant species remains a challenging area of research for the plant biochemist. A study of the biosynthesis of functional groups, such as the acetylenic bond in tariric acid, the cyclopentene ring in chaulmoogric acid, the cyclopropene ring in sterculic acid, the epoxy ring in vernolic acid, the conjugated and nonconjugated double bond systems of α-eleostearic acid and linolenic acid, should reveal new and important reactions as well as suggest unique taxonomic aberrations in different species of the same family of plants.

As is well known, the vast majority of naturally occurring fatty acids have an even number of carbon atoms. However, when sensitive systems of analysis such as gas-liquid chromatography (GLC) are employed, significant amounts of straight-chain odd-carbon fatty acids from C_7 to C_{35} are detected. Branched-chain fatty acids are uncommon in higher plants. Oleic and palmitic acids are the most common fatty acids in plant lipids, although in specialized tissue, such as the chloroplast, linolenic acid is the dominant acid. Linoleic, palmitoleic, myristic, lauric, and stearic acids are present in smaller amount, although in some species these acids may again be the dominant acids.

The synthesis of triacylglycerols is located in the microsomal fraction of the cell. Essential for the synthesis is the formation of L-α-glycerol phosphate from free glycerol and ATP. The glycerokinase that catalyzes this reaction occurs as a soluble enzyme in the cell. The thiokinases and the acylating enzyme appear to be associated with microsomal particles. Time–course studies indicate that the route of synthesis of triacylglycerols in the avocado mesocarp is essentially the same as that found in animal tissues. Glycerol phosphate, synthesized by the phosphorylation of glycerol is acylated by acyl-CoA to yield phosphatidic acid, which in turn is dephosphorylated to give diacylglycerols. The diacylglycerols are then further acylated in the presence of acyl-CoA to give triacylglycerols (Barron and Stumpf, 1962).

2. SELECTED IMPORTANT CHEMICAL REACTIONS OF FATTY ACIDS

Three different chemical reactions will be described briefly. Two of them are of considerable importance in determining the structure of a straight-chain fatty acid, and the third is of great value in deciding whether or not a radioactively labeled fatty acid was synthesized *de novo* or simply elongated from a preexisting shorter-chain fatty acid. The reader should consult Gunstone (1967) for a good summary of reactions of the hydrocarbon chain of fatty acids.

a. Reduction. The double bond in a hydrocarbon chain is readily reduced by hydrogen gas under atmospheric pressure in the presence of either platinum or palladium as catalyst [Eq. (2)].

$$\diagdown C = C \diagup \xrightarrow[\text{H}_2]{\text{Pd}} \diagup \overset{\overset{\textstyle H}{|}}{C} - \overset{\overset{\textstyle H}{|}}{C} \diagdown \qquad (2)$$

An examination of the GLC pattern of the methyl esters of fatty acids before and after hydrogenation will give considerable information concerning the chain length of the fatty acids and the type of polyunsaturated fatty acid (18:1, 18:2, 18:3) in a mixture. Silver nitrate impregnated silica G thin-layer chromatography techniques are now used to separate classes of unsaturated fatty acids.

b. Oxidation. If a fatty acid appears to have one or more double bonds in its hydrocarbon chain, the biochemist must determine the precise position of the double bond in the chain. The permanganate–periodate oxidation method is one of several effective methods available to the investigator. In essence, the oxidation can be depicted as shown in reaction (3).

$$(3)$$

Periodate serves two functions: the first to cleave the vicinal hydroxyl groups to aldehydic groups, and the second to oxidize the manganate ion, formed during the first oxidation step, to permanganate. Table II summarizes the products of oxidation of a variety of fatty acids. These products are easily detected by GLC methods. Rapid ozonolysis techniques have been developed which have been scaled down to micro amounts of fatty acids, which are rapid, avoid tedious extraction procedures and allow determination of double bond positions. Partial reduction of a poly-

TABLE II

Permanganate–Periodate Oxidation Products of Some Fatty Acids

	Products of oxidation		
Acids	Monocarboxylic	Dicarboxylic	Position of double bond
Stearic	None	None	—
Oleic	Nonanoic	Azelaic	9
Linoleic	Caproic	Azelaic; malonic	9, 12
Linolenic	Propionic	Azelaic; malonic	9, 12, 15

unsaturated fatty acid with hydrazine, subsequent separation of mono-unsaturated fatty acid by $AgNO_3$, TLC techniques, and then ozonolysis allows the biochemist another dimension in determining with precision the double bond position in a fatty acid.

c. *Schmidt Degradation.* The third reaction involves the decarboxyla-tion of the carboxyl carbon of a fatty acid to CO_2 and a long-chain amine according to reaction (4).

$$RCH_2CH_2COOH + NaN_3 \xrightarrow[H_2SO_4]{heat} RCH_2CH_2NH_2 + CO_2 \qquad (4)$$

The long-chain amine and the carbon dioxide can be readily recovered and the radioactive content of each can be determined. The value of this type of degradation can be best appreciated if we postulate both a *de novo* synthesis of a fatty acid from [1-^{14}C]acetate and an elongation of a preexisting fatty acid with [1-^{14}C]acetate. In the first case we would obtain the following data by a Schmidt decarboxylation.

$$\overset{.}{C}\overset{.}{C}\overset{.}{C}\overset{.}{C}\overset{.}{C}\overset{.}{C}\overset{.}{C}\overset{.}{C}\overset{.}{C}\overset{.}{C}\overset{.}{C}\overset{.}{C}\overset{.}{C}\overset{.}{C}\overset{.}{C}\overset{.}{C}\overset{.}{C}COOH \rightarrow \overset{.}{C}\overset{.}{C}\overset{.}{C}\overset{.}{C}\overset{.}{C}\overset{.}{C}\overset{.}{C}\overset{.}{C}\overset{.}{C}\overset{.}{C}\overset{.}{C}\overset{.}{C}\overset{.}{C}\overset{.}{C}\overset{.}{C}\overset{.}{C}CNH_2 + \overset{.}{C}O_2$$

900 cpm   800 cpm   100 cpm

$$C = 100 \text{ cpm} \qquad \text{ratio} \frac{CO_2}{\text{total}} = \frac{1}{9}$$

$$\therefore \textit{ de novo } \text{synthesis}$$

In the second case

$$\overset{.}{C}CCCCCCCC\overset{.}{C}CCCCCCCCOOH \rightarrow CCCCCCCCC\overset{.}{C}CCCCCCCCNH_2 + \overset{.}{C}O_2$$

800 cpm   400 cpm   400 cpm

$$\overset{.}{C} = 400 \text{ cpm} \qquad \text{ratio} \frac{CO_2}{\text{total}} = \frac{1}{2}$$

$$\therefore \text{ primarily an elongation reaction}$$

B. The Waxes

Chemically a wax is defined as an ester of a higher fatty acid and a higher aliphatic alcohol. Recent work has shown, however, that in addition to these simple esters, waxes contain alkanes of odd carbon numbers ranging from C_{21} to C_{37}, long chain monoketones, β-hydroxyketones, β-diketones primary and secondary alcohols of even and odd number carbon atoms ranging from C_{22} to C_{32}, and very long-chain free fatty acids, ranging from C_{24} to C_{34}.

These compounds make up the thin waxy layers that coat the stems, leaves, flowers, and fruits of most plants. Wax apparently originates in the epidermal cells as oily droplets, passes through minute canals that penetrate through the cell walls, and crystallizes into rods or platelets when deposited on the surface of the tissue. Little is known about the physiological function, although presumably the waxy coating plays an important role in controlling the water balance of the plant. Sufficient information is now becoming available to make possible some logical postulations concerning the synthesis of these compounds. The student should consult the interesting account by Kolattukudy (1972) concerning further aspects of waxes in plants.

C. The Glycerophosphatides and the Phytosphingolipids

Several glycerophosphatides have now been isolated and identified from tissues. Their structures are given in Fig. 1. While phosphatidylcholine, phosphatidylethanolamine, and the phosphatidylinositides are found primarily in seed tissue, phosphatidylglycerol occurs to a large extent in leaf tissue. As much as 22% of the total phospholipid in leaf tissue is phosphatidylglycerol. Small amounts of diphosphatidylglycerol have been found in algae and in higher plants.

Although phosphatidic acid, the parent compound of the glycerophosphatides, has been isolated from cabbage leaves and other tissues, there is good evidence that its presence may merely be related to the action of a phospholipase on endogenous glycerophosphatides. However, Bradbeer and Stumpf (1960) and Barron and Stumpf (1962) have described two enzyme systems that synthesize phosphatidic acid. The first system [Eq. (5)] involves a diglyceride phosphokinase.

$$\alpha,\beta\text{-Diglyceride} + \text{ATP} \rightarrow \text{phosphatidic acid} + \text{ADP} \tag{5}$$

The second system [Eq. (6)] is similar to that described by Kennedy, namely, the acylation of α-glycerol phosphate.

$$\alpha\text{-Glycerol phosphate} + 2 \text{ RCO-CoA} \rightarrow \text{phosphatidic acid} + 2 \text{ CoA} \tag{6}$$

The occurrence of these two systems in higher plant tissues suggests that while phosphatidic acid may occur by the hydrolytic cleavage of preexisting glycerophosphatides, it may also be synthesized by well-defined systems; in intact tissue it probably participates in the synthesis of important members of the glycerophosphatide group.

Recently, Devor and Mudd (1971b) have studied the insertion of acyl-CoA's into lysophosphatidylcholine by spinach leaf microsomes and have shown that in general, saturated acyl-CoA derivatives were incorporated primarily into the 1-position of the glycerol moiety, while the unsaturated acyl-CoA's were inserted into the 2-position.

The biosynthesis of phospholipids in higher plants appears to be simi-

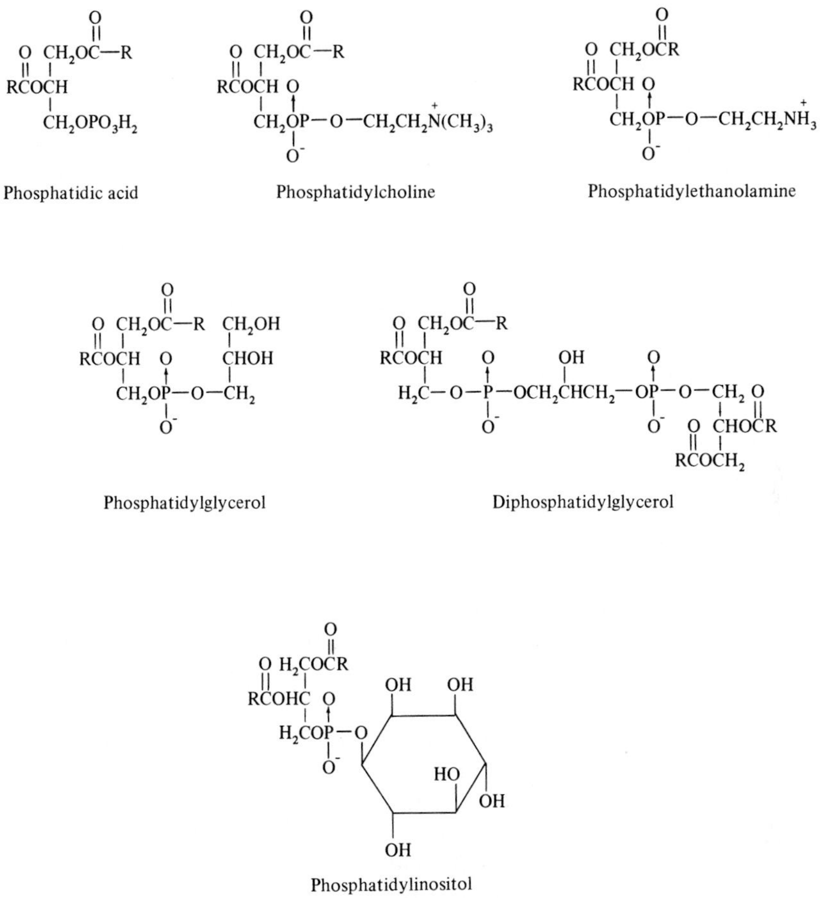

Phosphatidic acid Phosphatidylcholine Phosphatidylethanolamine

Phosphatidylglycerol Diphosphatidylglycerol

Phosphatidylinositol

Fig. 1. Structure for some common glycerophosphatids.

lar to animal systems. Devor and Mudd (1971a) have demonstrated, in spinach leaf microsomes, the presence of choline phosphotransferase that catalyzes reaction (7)

$$\text{Phosphorylcholine} + \text{CTP} \underset{}{\overset{\text{Mn}^{2+}}{\rightleftharpoons}} \text{CDP-choline} + \text{pyrophosphate} \qquad (7)$$

as well as the CDP-choline:diacylglycerol transferase [reaction (8)].

$$\text{CDP-choline} + \text{diacylglycerol} \underset{}{\overset{\text{Mn}^{2+}}{\rightleftharpoons}} \text{phosphatidylcholine} + \text{CMP} \qquad (8)$$

Recently, Macher and Mudd (1974) demonstrated in a variety of plant tissues reaction (9).

$$\text{CDP-ethanolamine} + \text{diacylglycerol} \underset{}{\overset{\text{Mn}^{2+}}{\rightleftharpoons}} \text{phosphatidylethanolamine} + \text{CMP} \qquad (9)$$

Similar studies by Moore *et al.* (1973) indicate that the synthesis of these phospholipids occurs exclusively in the membranes of the endoplasmic reticulum from the endosperm of young castor bean seedlings. In contrast, the synthesis of phosphatidylglycerol and phosphatidylinositol occurs in mitochondrial particles by the CDP-diacylglycerol pathway shown in Scheme 2 (Sumida and Mudd, 1970; Marshall and Kates, 1972).

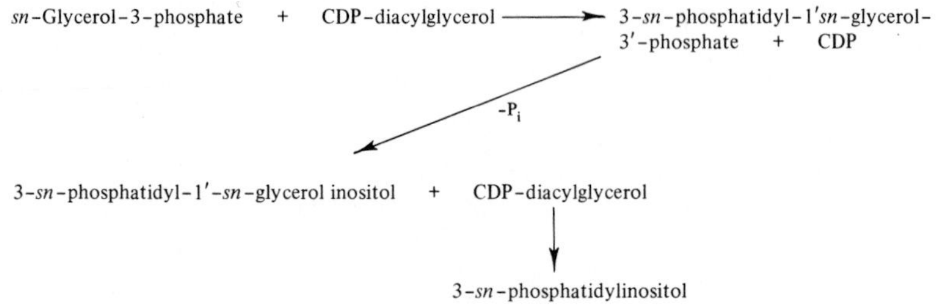

sn-Glycerol-3-phosphate + CDP-diacylglycerol ⟶ 3-*sn*-phosphatidyl-1′*sn*-glycerol-3′-phosphate + CDP

−P_i

3-*sn*-phosphatidyl-1′-*sn*-glycerol inositol + CDP-diacylglycerol

3-*sn*-phosphatidylinositol

Scheme 2

D. The Glycolipids

The most important members of the glycolipid class are monogalactosyl diacylglycerol and digalactosyldiacylglycerol, and these occur universally in higher plant tissues. Linolenic acid is the chief fatty acid associated with the diacylglycerol moiety. Glycolipids are the major lipid constituents in chloroplasts. No triacylglycerols occur in chloroplasts, and only a minor amount of phospholipids is present.

Neufeld and Hale (1964) were the first to show that spinach chloroplasts readily incorporate radioactivity from UDP-[^{14}C]galactose into chloroform-soluble material. After mild hydrolysis, one of the fractions identified was β-D-galactosyldiacyl glycerol.

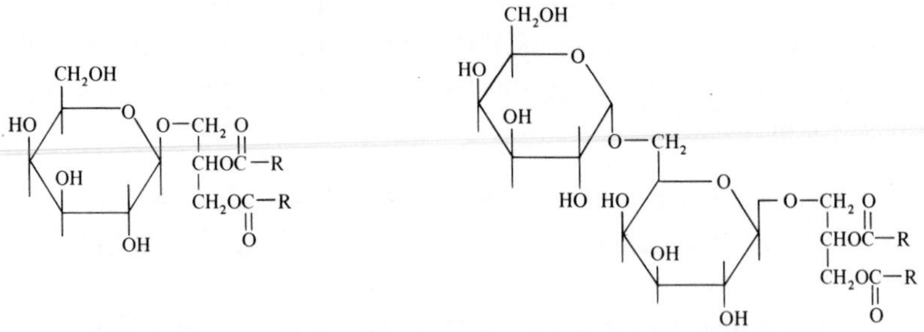

Monogalactosyl diacylglycerol

Digalactosyl diacylglycerol

Recently, Mudd and his co-workers have elucidated the biosynthesis of the galactosyl diacylglycerols in plants (Ongun and Mudd, 1968, 1970;

Mudd *et al.*, 1969). Evidence indicates that the enzyme responsible for the synthesis of monogalactosyl diacylglycerols is a transferase tightly bound to the chloroplast envelope membrane (Douce, 1974). The reaction involved presumably that shown in (10).

$$\text{UDP-galactose} + \text{diacylglycerol} \rightarrow \text{monogalactosyl diacylglycerol} + \text{UDP} \quad (10)$$

The enzyme responsible for the synthesis of digalactosyl diacylglycerol is, however, very loosely bound to the membranes and can be obtained in soluble form. It catalyzes the reaction shown in (11).

$$\text{UDP-galactose} + \text{monogalactosyl diacylglycerol} \rightarrow$$
$$\text{UDP} + \text{digalactosyl diacylglycerol} \quad (11)$$

An important anionic lipid in chloroplast lipids is the sulfolipid sulfoquinovosyl diglyceride.

Sulfoquinovosyl diglyceride

This compound is widespread in leaf tissue, although in small concentrations. Although it may play a structural role in the photosynthetic apparatus, little is, at present, known about the details of its biosynthesis and its function (Harwood, 1975).

II. Degradation of Fatty Acids

A. Hydrolysis of Triacylglycerol

When oil seeds with a high triacylglycerol content germinate, there is a rapid disappearance of lipid with a concomitant rise in sucrose. Lipase activity rises sharply in the early stages of germination and presumably participates in the stepwise hydrolysis of triacylglycerol to diacylglycerols, monoacylglycerol, and finally free glycerol and free fatty acids. Both castor bean and wheat germ lipases have been studied in some detail. An interesting general lipolytic acyl hydrolase has been recently described by Galliard (1971) from potato tubers which rapidly hydrolyzes acyl groups from both phospholipids and galactolipids.

Another group of lipases of considerable importance are those that hydrolyze the acyl groups of the galactosyl diacylglycerols. These enzymes occur in both chlorophyllous and nonchlorophyllous tissue and rapidly degrade these glycolipids to free acids and the galactosyl glycerols. Since lamellar membranes consist, in large part, of mono- and digalactosyl diacylglycerols, the activation of these hydrolases in isolated chloroplast systems can have a devastating effect on isolated chloroplast activities. Indeed, the simultaneous release of free fatty acids and the dissolution of the complex lipids of a membrane systems are commonly considered as major factors that make investigations on isolated plastid systems difficult. The control of these hydrolases *in vivo* is a major problem worthy of investigation.

B. α-Oxidation

Once free fatty acids are formed by lipase action on neutral lipids, several pathways are open in tissues for the complete oxidation of the hydrocarbon chain to CO_2 and water or for a partial oxidation and conversion to sucrose.

Free fatty acids from C_{13} to C_{18} may be readily attacked to yield a fatty acid with one less carbon atom and CO_2 and/or a D-α-OH fatty acid. The α-oxidation was first described in 1952 by Newcomb and Stumpf, later examined in some detail by Martin and Stumpf (1959) in germinated peanut cotyledons and by Hitchcock and James (1964) and by Hitchcock and Morris (1970) in pea leaves. A major difference emerged in these two systems, namely, an apparent H_2O_2 requirement in the seed system and molecular O_2 requirement in the leaf system. Also the supposition that an L-α-OH fatty acid served as an intermediate in the oxidative decarboxylation of a fatty acid was difficult to explain in terms of the usual stereospecific reactions catalyzed by enzymes.

Nevertheless, Hitchcock and James (1964) proposed and Markovetz *et al.*, (1972) presented evidence in support of a sequential oxidation and decarboxylation of a free fatty acid to a fatty acid containing one less carbon atom as illustrated in Scheme 3.

However, very recently Shine and Stumpf (1974) proposed a completely new sequence based on the following evidence.

1. Both the leaf and the seed systems required molecular oxygen for oxidative decarboxylation.

2. Moreover, both systems required a flavoprotein, such as glucose oxidase, and glucose which was not involved in the generation of H_2O_2

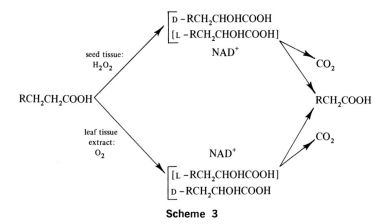

Scheme 3

but instead participated in the activation of molecular oxygen to a species capable of attacking the α-carbon of the free fatty acid.

3. Sensitive GLC techniques were not able to detect the formation of L-α-OH fatty acids from the parent substrate, a key requirement in the earlier pathway.

4. The substrate, [1-^{14}C]palmitic acid, was always more effective in forming $^{14}CO_2$ than was the proposed intermediate, [1-^{14}C]L-2-OH palmitic acid, which is in contradiction to the usual precursor–product relationship in L-2-OH palmitic were to be a precursor.

5. A key experiment resolving the problem was the use of glutathione peroxidase and GSH. This enzyme system has a wide specificity in reducing alkyl hydroperoxides to alkyl alcohols:

$$ROOH + 2\ GSH \rightarrow ROH + H_2O + G\text{–}SS\text{–}G$$

When this enzyme and GSH were added to a reaction mixture that normally converts [1-^{14}C]palmitic acid to the corresponding C_{15} fatty acid $+ CO_2$, and small amounts of D-2-OH palmitic acid, a dramatic decrease in $^{14}CO_2$ formation with a concomitant increase in D-2-OH palmitic acid occurred.

These results led Shine and Stumpf (1974), to the proposal outlined in Fig. 2, which differs markedly from the scheme suggested earlier. Two tightly coupled reaction sequences are proposed: (a) the molecular oxygen activation sequence and (b) the formation of a central intermediate, a D-2-hydroperoxyl fatty acid, which can then undergo one of two possible reactions, either a reduction to the corresponding D-2-hydroxy fatty acid, a deadend product, or a decarboxylation of the intermediate to CO_2 and a long chain aldehyde. The aldehyde, which can be detected during

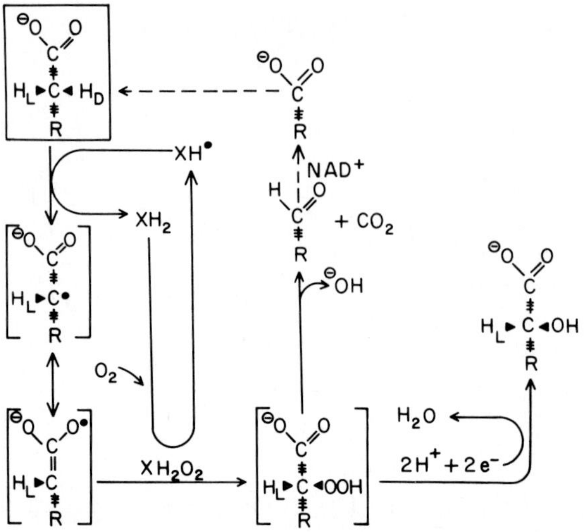

Fig. 2. The mechanism of α-oxidation in higher plants.

the reaction, is rapidly oxidized by a NAD⁺–long chain aldehyde dehydro-genase to the corresponding free fatty acid. This acid can then be recycled into the sequence for further α-oxidations.

Since lauric acid and the lower chain fatty acids are inactive sub-strates, only the longer chain fatty acids are susceptible to degradation.

What role does α-oxidation play in plant tissue? α-Oxidation may be a component of a system responsible for the synthesis of propionic acid from naturally occurring fatty acids, such as palmitic or stearic acid, by a combination of α- and β-oxidations as shown in reaction (12).

$$18:0 \xrightarrow[\alpha-\text{oxidation}]{} 17:0 + CO_2 \xrightarrow[\beta-\text{oxidation}]{} 3:0 + 7(2:0)(\text{acetate}) \quad (12)$$

This function may be of extreme importance, since propionic acid is a precursor of β-alanine, which in turn is a component of CoA and ACP (see Section II,D for propionic acid oxidation).

Another function could be that of by-passing blocking groups in sub-strates that would otherwise not be susceptible to β-oxidation [Eq. (13)].

$$R-CH_2-\overset{\overset{\displaystyle R'}{|}}{CH}-CH_2-COOH \longrightarrow \text{no } \beta-\text{oxidation}$$

$$\Big\downarrow \alpha-\text{oxidation} \qquad\qquad\qquad\qquad (13)$$

$$R-CH_2-\overset{\overset{\displaystyle R}{|}}{CH}-COOH + CO_2 \xrightarrow{\beta-\text{oxidation}} R-COOH + R'-CH_2-COOH$$

Indeed, Steinberg *et al.* (1967) have demonstrated in mammalian systems that α-oxidation is the essential step in bypassing a methyl group in the smooth oxidation of phytanic acid, a product in the metabolism of phytol. The small amounts of odd chain fatty acids found in nature may be formed by the α-oxidation pathway.

C. β-Oxidation

While α-oxidation probably plays a rather restricted 'role in the breakdown of fatty acids, β-oxidative reactions are presumably the principal mechanisms by which plant tissues degrade fatty acids. By means of β-oxidation, much of the energy inherent in the highly saturated hydrocarbon chain is trapped in the thioester bond of acetyl-CoA. In the higher plant cell, acetyl-CoA can then be further metabolized either via the tricarboxylic acid (TCA) cycle with a further extraction of available energy or by means of the glyoxylate cycle, which makes possible a net gain of carbon atoms for further synthetic purposes. Acetyl-CoA undoubtedly participates in many other reactions that we shall not list here.

There is much evidence to suggest that the β-oxidation system in plants is identical to that in animal tissues. It is now clear that high oil seeds, on germination, rapidly form glyoxysomes that contain the glyoxylate bypass enzymes as their principal components. In addition, all the β-oxidation enzymes are localized in this organelle (Hutton and Stumpf, 1969; Cooper and Beevers, 1969). Thus the fatty acids that are being released by lipase action from the triacylglycerol droplets in the germinating seed are rapidly and efficiently converted by the enzymes of the glyoxysome to acetyl-CoA and then to malic acid, which then feeds into the gluconeogenesis sequence of reactions responsible for the formation of phosphorylated hexoses. See Chapter 5 for a detailed discussion of these reactions.

We shall not discuss here the details of the β-oxidation enzymes. However, the student should consult standard textbooks for details concerning the four key enzymes, namely, the acyl-CoA dehydrogenase, enoyl-CoA hydrase, β-hydroxyacyl-CoA dehydrogenase, and β-ketoacyl-CoA thiolase. Figure 3 represents the sequence catalyzed by these enzymes and also indicates the possible fates for acetyl-CoA.

D. Odd-Chain Fatty Acid Oxidation

The methylmalonyl-CoA pathway is the major sequence for the metabolism of propionic acid in animal tissues. Kaziro and Ochoa (1964)

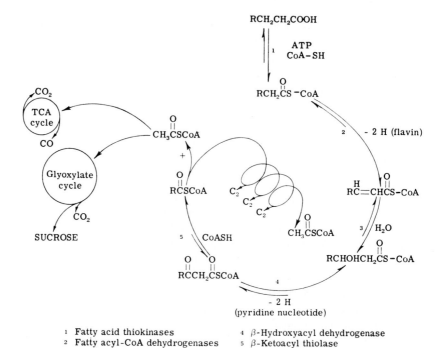

1 Fatty acid thiokinases   4 β-Hydroxyacyl dehydrogenase
2 Fatty acyl-CoA dehydrogenases   5 β-Ketoacyl thiolase
3 Enoyl hydrase

Fig. 3. The β-oxidation cycle.

have thoroughly explored this sequence and have demonstrated that the
first reaction required a carboxylation of propionyl-CoA to methylmalonyl-CoA with a subsequent isomerization through a coenzyme B_{12}
isomerase to succinyl-CoA [Eq. (14)].

$$CH_3CH_2CO-CoA \quad + \quad CO_2 \xrightarrow[\substack{carboxylase \\ (biotin)}]{propionyl-CoA} \quad \begin{array}{c} COOH \\ | \\ CH_3-CH \\ | \\ CO-CoA \end{array}$$

S-methylmalonyl–CoA

$$\begin{array}{c} COOH \\ | \\ CH_2 \\ | \\ CH_2 \\ | \\ CO-CoA \end{array} \xleftarrow[\substack{Co-B_{12}}]{isomerase} \begin{array}{c} COOH \\ | \\ H-C-CH_3 \\ | \\ CO-CoA \end{array} \Bigg\downarrow \text{racemase} \qquad (14)$$

R-methylmalonyl–CoA

Since coenzyme B_{12} has never been observed in plant tissues, it became of considerable interest to identify the system employed by higher plants to metabolize propionic acid. Employing mitochondrial particles from germinated peanut seedling, Giovanelli and Stumpf, in 1958, studied the oxidation of variously labeled propionic acid. On the basis of a kinetic analysis of the rate of $^{14}CO_2$ formation from [1-^{14}C]-, [2-^{14}C]-, and [3-^{14}C]propionic acid, the rate of entry of ^{14}C from these substrates into TCA cycle acids, the isolation and characterization of β-hydroxypropionic acid, and nonrequirement of CO_2 for the oxidation of propionic acid, these investigators proposed the sequence shown in Eqs. (15)–(20).

$$CH_3CH_2CO\text{-}CoA \xrightarrow{-2\,H} CH_2{=}CHCO\text{-}CoA \qquad (15)$$

$$CH_2{=}CHCO\text{-}CoA \xrightarrow{+H_2O} CH_2OHCH_2CO\text{-}CoA \qquad (16)$$

$$CH_2OHCH_2CO\text{-}CoA \xrightarrow{+H_2O} CH_2OHCH_2COOH + CoA \qquad (17)$$

$$CH_2OHCH_2COOH \xrightarrow{-2\,H} CHOCH_2COOH \qquad (18)$$

$$CHOCH_2COOH \xrightarrow{-2\,H,\,CoA} CoA\text{-}COCH_2COOH \qquad (19)$$

$$CoA\text{-}COCH_2COOH \longrightarrow CoA\text{-}COCH_3 + CO_2 \qquad (20)$$

Employing intact tissue (safflower, wheat seedlings, and peanut cotyledon slices), Hatch and Stumpf (1962b) confirmed that the modified β-oxidation reactions, based on work with isolated cell-free enzyme systems, indeed functioned in intact tissue. They were also able to demonstrate the synthesis of β-alanine from propionic acid and suggested that this amino acid could arise from propionic acid either by a β-amination of acrylyl-CoA or by transamination with malonic semialdehyde as the amino acceptor. Indeed, Meheriuk and Spencer (1967) have described a transaminase that carries out this reaction.

Since propionic acid is rarely found in plant tissues, the question arises as to its origin in plants. On the basis of what we know now concerning α- and β-oxidations we can postulate the mechanism shown in Eq. (21).

$$18{:}0 \xrightarrow{\alpha\text{-oxidation}} 17{:}0 \xrightarrow{\beta\text{-oxidation}} 7(2{:}0) \; + \; \text{propionic acid} \longrightarrow \beta\text{-alanine} \longrightarrow CoA \longrightarrow ACP$$

modified
β-oxidation

acetyl–CoA + CO_2

$$(21)$$

E. Lipoxygenase

The *cis,cis*-1,4-pentadiene system in linoleic and linolenic acids is readily susceptible to attack by lipoxygenase. The enzyme catalyzes the direct addition of oxygen to the *cis,cis*-1,4-pentadiene system with the formation of a *cis,trans*-1,3-butadiene hydroperoxide. The reaction sequence shown in Scheme 4 is suggested.

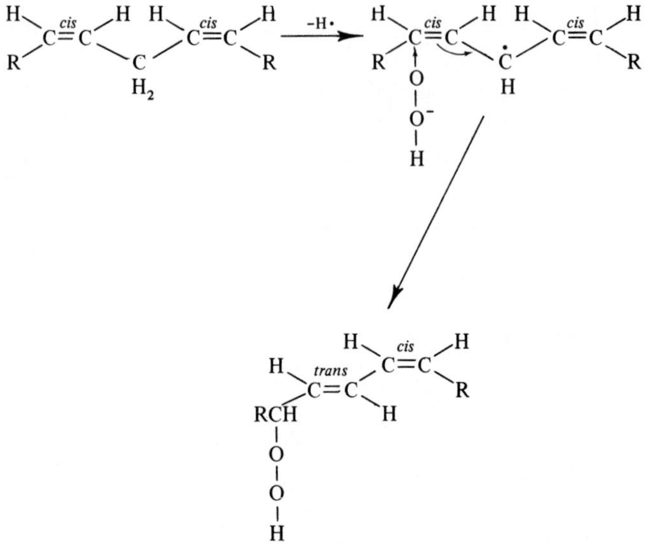

Hydroperoxide

Scheme 4

The enzyme has been found widely spread in plant seeds, roots, and leaf tissue. It does not occur in bacteria or in animal tissues. Soybean lipoxygenase has been crystallized to a homogeneous protein. Its molecular weight is 102,000, and its amino acid composition has been determined but shows no unusual amino acid composition nor any prosthetic group. Recently, it has been shown to be a ferriprotein with iron bound to the protein, presumably through a unique group of amino acids forming the binding site for the metal.

In the presence of molecular oxygen, all lipoxygenases catalyze the hydroperoxidation of long chain polyunsaturated fatty acids. One isoenzyme from soybean forms mainly 13-L-hydroperoxy-18:2 (10 *trans*, 12 *cis*), while corn and potato lipoxygenase form almost exclusively 9-D-hydroperoxy-18:2 (9 *cis*, 11 *trans*), and still other enzymes form mixtures of the two. The mechanism of action involves the stereospecific abstraction of a hydrogen atom from the ω-8-methylene group of the substrate. The removal of the D_R-hydrogen at this position leads to the formation of the ω-10-D-hydroperoxide isomer, whereas the formation of the ω-6-L-hydroperoxide isomer involves removal of the L_8-hydrogen of the ω-8-carbon.

Once the hydroperoxide is formed, another enzyme, linoleate hydroperoxide isomerase, first observed by Zimmerman (1966) catalyzes reaction (22)

$$RCHOOHCH\overset{t}{=}CHCH\overset{c}{=}CH(CH_2)_7COOH \rightarrow$$

$$RCHOHCOCH_2CH\overset{c}{=}CH(CH_2)_7COOH \quad (22)$$

which presumably is further metabolized by as yet undefined systems to products that can be utilized by cells.

Enzymes from potato tubers convert fatty acid hydroperoxide to a novel butadienylvinyl ether derivative of linoleic or linolenic acids, namely, colneleic and colnelenic acids, respectively (Galliard and Phillips, 1971), by the sequence shown in reaction (23).

$$CH_3(CH_2)_4CH\overset{c}{=}CH-CH_2-CH\overset{c}{=}CHCH_2(CH_2)_6COOH$$

$$\downarrow \text{lipoxygenase}$$

$$CH_3(CH_2)_4CH\overset{c}{=}CHCH\overset{t}{=}CHCH(OOH)\dot{C}H_2(CH_2)_6COOH \quad (23)$$

$$\downarrow \text{additional enzymes}$$

$$CH_3(CH_2)_4CH\overset{c}{=}CHCH\overset{t}{=}CHOCH\overset{t}{=}CH(CH_2)_6COOH$$

The ether derivative is further metabolized by enzymes in the potato to a series of carboxyl fragmentation products that are volatile flavor products observed when fresh plant tissues are cut.

It is obvious that lipoxygenase sets off a series of complex reactions with polyunsaturated fatty acid as substrates. How the products of these reactions fit with the overall scheme of cell metabolism must await further investigations.

F. Hydroxylation Reactions

In recent years, information concerning the formation of hydroxylated fatty acids has been rapidly accumulating. The important investigations of Coon, Sato, and Gunsalus have greatly clarified the roles of several components involved in bacterial and animal hydroxylation systems. Progress in the plant tissues has been less spectacular, since the reactions are far less active and the enzymes are quite unstable.

However, the hydrocarbon chain of fatty acids can now be hydroxylated by a number of specific enzymes that place the hydroxy group at specific carbons of the chain. Thus the α-oxidation sequence described above inserts a hydroxyl function on the second carbon, the oleyl-CoA hydroxylase of castor bean specifically adds a hydroxy group in carbon-12 of oleyl-CoA, and still another hydroxylase converts 1-palmityl-CoA to 10,16-dihydroxypalmityl-CoA. There is also some evidence that a conventional ω-hydroxylase exists in plant tissues.

The role of hydroxylated fatty acids will now be considered. The α-oxidation mechanism has already been discussed in Section II,A. Oleyl-CoA hydroxylase is responsible for the synthesis of ricinoleic acid, the principal fatty acid in the castor bean seed. A number of workers, including James, Canvin, and Yamada, have shown that the developing castor bean seed rapidly synthesizes ricinoleic acid in a critical short period in the maturing seed. The capacity for synthesis is completely missing in the germinating seed. Galliard and Stumpf, in 1966, carefully examined this system with cell-free preparations of the developing seed. They described a mixed function oxygenase localized in the microsomal fraction requiring NADH and molecular oxygen as well as oleyl-CoA as the only reactive substrate. These results are depicted in Scheme 5.

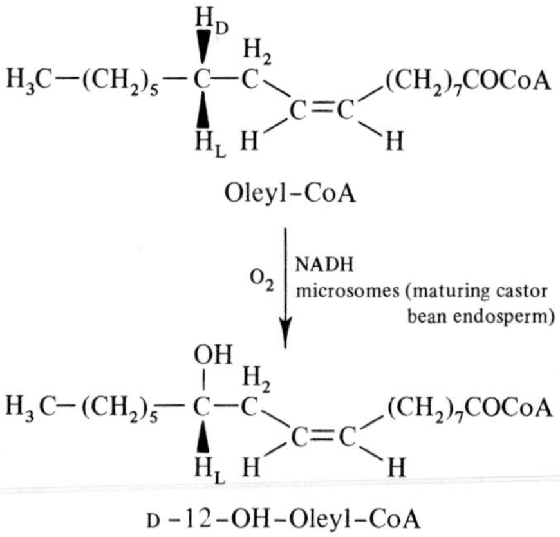

Scheme 5

Kolattukudy and his associates (1971) have described a 16-hydroxypalmityl-CoA hydroxylase, which converts 16-OH C_{16} to the 10,16-dihydroxypalmitic acid. This observation fits with the requirement of this compound as a component of cutin polymers, which consist of large polymers of 16-hydroxypalmitic and 10,16-dihydroxypalmitic acids in *V. faba* tissue. Presumably 16-hydroxypalmitic acid is formed by a ω-hydroxylase system, although the direct evidence for this reaction is still fragmentary.

III. Biosynthesis of Malonyl-CoA

Since malonyl-CoA plays such an important role in lipid synthesis, some remarks should be made concerning the several enzymes involved in controlling the level of the compound in the plant cell.

Malonyl-CoA is a highly reactive substrate in plant tissue. Hatch and Stumpf (1962a) have examined extracts in 12 plants and have observed that invariably at least six different enzyme activities involving malonyl-CoA can be observed. These activities are as follows:

A. Acetyl–CoA carboxylase

$$\text{Acetyl–CoA} + CO_2 \longrightarrow \text{malonyl–CoA}$$

B. Malonyl transcarboxylase

$$\text{Malonyl–CoA} + RCH_2CO\text{-CoA} \longrightarrow \text{acetyl–CoA} + R{-}\overset{\displaystyle COOH}{\underset{\displaystyle |}{C}}HCO\text{-CoA}$$

C. Malonic thiokinase

$$\text{Malonic} + ATP + CoA \longrightarrow \text{malonyl–CoA} + AMP + PP_i$$

D. Thiolesterase

$$\text{Malonyl–CoA} + H_2O \longrightarrow \text{malonic acid} + CoA$$

E. Decarboxylase

$$\text{Malonyl–CoA} \longrightarrow \text{acetyl–CoA} + CO_2$$

F. Fatty acid synthetase

$$\text{Acetyl–CoA} + n \text{ malonyl–CoA} \longrightarrow \text{long chain fatty acid} + n\,CO_2$$

It is obvious that in the push and pull of metabolic control, enzymes A, B, and C must be balanced against D and E. There is some evidence now available for plant systems that one of the rate-limiting steps in lipid synthesis is controlled by the activity of acetyl-CoA carboxylase. The plant can side-step this block by the efficient functioning of enzyme C. Shannon *et al.* (1963) have shown that oxalacetic acid can undergo an α-decarboxylation in root tissue [Eq. (24)].

$$\text{Oxalacetic} + \tfrac{1}{2} O_2 \xrightarrow[\text{Mn}^{2+}]{\text{peroxidase}} \text{malonic acid} + CO_2 \tag{24}$$

Since oxalacetic acid may be synthesized by a variety of systems in plants, it may be an important source of malonic acid.

The key enzyme in the synthesis of malonyl-CoA is acetyl-CoA carboxylase. In prokaryotic organisms, the enzyme consists of three proteins that can be readily separated by conventional protein fractionation techniques. The proteins are biotin carboxylase, biotin carboxyl carrier protein (BCCP) and a transcarboxylase and these are involved in the sequence shown in reactions (25) and (26).

$$\text{ATP} + CO_2 + \text{BCCP} \xrightarrow{\text{biotin carboxylase}} \text{ADP} + P_i + CO_2 \sim \text{BCCP} \quad (25)$$

$$CO_2 \sim \text{BCCP} + \text{acetyl-CoA} \xrightarrow{\text{transcarboxylase}} \text{BCCP} + \text{malonyl-CoA} \quad (26)$$

Each protein has been isolated from *E. coli* and has been well characterized (Alberts and Vagelos, 1968). Only BCCP contains biotin at 1 mole per mole of protein.

In animal systems, the enzyme exists as a tight, inactive, complex called a protomer (Moss and Lane, 1971). In the well-studied avian system, the complex has a molecular weight of 409,000, containing 1 mole of biotin per mole of protomer. This protomer is rapidly converted to a highly active polymer on exposure to polyanionic compounds such as citrate. This polymer has about 10–20 protomer units and is the fully active form of the enzyme. The protomer in turn can be further resolved irreversibly into four inactive subunits on exposure to urea, each subunit having a molecular weight of 100,000. In higher plants, Heinstein and Stumpf (1969) have purified and characterized the wheat germ acetyl-CoA carboxylase. This enzyme does not have the allosteric properties of the mammalian carboxylase and thus resembles the prokaryotic type, but like the mammalian system it readily polymerizes. The enzyme can be partially separated into three components, one of which is presumably BCCP, but all subunits are strongly self-aggregating and, therefore, difficult to purify. The specific activity of this system is identical to that of the avian system, while the *E. coli* enzyme has 1% of the specific activity of the eukaryotic systems. The acetyl-CoA carboxylase of the chloroplast will be described in a later section of this chapter.

IV. Biosynthesis of Long-Chain Saturated Fatty Acids

In 1952, Newcomb and Stumpf fed radioactive susbstrates to slices of cotyledons of maturing and germinating peanut seeds. In both systems, acetate was the most effective of several substrates tested for incorporation into long-chain fatty acids. Since then considerable progress has been made concerning the *de novo* synthesis of fatty acids in higher plants. Fatty acids are synthesized by a series of reactions that appear to be identical to the bacterial system as indicated in Scheme 6 in which ACP

$$\text{Acetyl-CoA} + \text{ACP-SH} \xrightarrow{\;①\;} \text{acetyl-S-ACP} + \text{CoA}$$
$$\text{Acetyl-S-ACP} + \text{Enz ③} \longrightarrow \text{acetyl-S-Enz ③} + \text{ACP}$$
$$\text{Malonyl-CoA} + \text{ACP-SH} \xrightarrow{\;②\;} \text{malonyl-S-ACP} + \text{CoA}$$
$$\text{Acetyl-S-Enz ③} + \text{malonyl-S-ACP} \xrightarrow{\;④\;} \text{acetoacetyl-S-ACP} + \text{Enz ③} + CO_2$$
$$\text{Acetoacetyl-S-ACP} + \text{NADPH} + H^+ \xrightarrow{\;④\;} \text{D}(-)-\beta\text{-hydroxybutyryl-S-ACP} + \text{NADP}^-$$
$$\text{D}(-)-\beta\text{-Hydroxybutyryl-S-ACP} \xrightarrow{\;⑤\;} \Delta^2\text{-}trans\text{-crotonyl-S-ACP} + H_2O$$
$$\Delta^2\text{-}trans\text{-Crotonyl-S-ACP} + \text{NADPH} + H^+ \xrightarrow{\;⑥\;} \text{butyryl-S-ACP} + \text{NADP}^+$$
$$\text{Butyryl-S-ACP} + \text{Enz ③} \longrightarrow \text{butyryl-S-Enz ③} + \text{ACP}$$
$$\text{Butyryl-S-Enz ③} + \text{malonyl-S-ACP} \longrightarrow \beta\text{-ketohexanoyl-S-ACP} + \text{Enz ③} + CO_2, \text{etc.}$$

① Acetyl transacylase   ④ β-Ketoacyl ACP-reductase
② Malonyl transacylase   ⑤ Enoyl ACP-hydrase
③ β-Ketoacyl ACP-synthetase   ⑥ Enoyl ACP-reductase

The final product of the *de novo* system is palmityl-ACP

Scheme 6

is acyl carrier protein. The final product of the *de novo* system is palmityl-ACP.

A. Distribution and Types of Synthetases

Unlike the mammalian systems that are localized in the cytoplasm as large soluble synthetase complexes and the cytoplasmic prokaryotic systems that are freely soluble and readily separable as discrete proteins, plant synthetases are associated with a large variety of organelles, such as chloroplasts, plastids, the endoplasmic reticulum, as well as the cytoplasm. Soluble synthetases can be readily isolated from disrupted organelles (Harwood and Stumpf, 1972).

Recent experiments by Harwood and Stumpf (1971) have clearly shown that there are at least three different types of synthetases, each being involved in the synthesis of a part of the complete hydrocarbon chain. These are depicted in reaction (27).

$$C_2 + 7C_3 \xrightarrow{\text{type I}} C_{16} + 7CO_2 \underset{\substack{\uparrow \\ C_3}}{\overset{\text{type II}}{\rightleftarrows}} \underset{\substack{\downarrow \\ CO_2}}{} C_{18} \underset{\substack{\uparrow \\ C_3}}{\overset{\text{type III}}{\rightleftarrows}} \underset{\substack{\downarrow \\ CO_2}}{} C_{20}, \text{etc.} \tag{27}$$

Type I is the *de novo* system that requires acyl carrier protein (ACP), NADPH, and NADH, if acetyl-CoA and malonyl-CoA are the two initial substrates. This system has been carefully studied using potato tuber tissue as a source of the enzyme by Huang and Stumpf (1971). The system can synthesize palmityl-ACP but it has lost all capacity for stearyl-ACP formation. Type II is the limited elongation system that is very sensitive to low levels of arsenite. Experiments by Harwood and Stumpf (1971) have clearly shown that while type II is sensitive to arsenite, at equivalent concentrations type I is not. Thus with a number of systems either under *in vitro* or *in vivo* conditions, the formation of the

C_{18} fatty acids can be sharply reduced with a concomitant accumulation of palmitic by the addition of arsenite.

The elongation system (type II) has now been examined in a number of tissues, including mature safflower seeds, avocado, and spinach chloroplasts (Jaworski et al., 1974). The system has been distinguished as separate from the de novo system (type I) by the following characteristics.

1. More heat labile than the de novo system
2. Requires only NADPH rather than both NADH and NADPH in the de novo system
3. Relatively insensitive to the antibiotic, cerulenin, whereas the de novo system is about 30-fold more sensitive.
4. The substrate is specifically palmityl-ACP

Stearyl-ACP is essentially inactive. Malonyl-ACP is the specific C_2 substrate. These results would suggest that the elongation system differs from the de novo in at least one if not two enzymes. Because of the difference in the cerulenin effect, which interacts specifically with the initial condensing enzyme responsible for the condensation of acetyl-ACP to malonyl-ACP to yield acetoacetyl-ACP, we can suggest that the initial condensing enzyme in the de novo system and the condensing enzyme in the elongation system that forms β-ketostearyl-ACP from palmityl-ACP and malonyl-ACP differ from each other. Also at least one of the two reductases in the elongation system must have an enzyme differing from the two reductases required for the de novo system.

Equally important is the observation that palmityl-CoA cannot serve as a substrate in the crude extract that readily elongates palmityl-ACP. Thus, the transacylase that could catalyze reaction (28)

$$\text{Palmityl-CoA} + \text{ACP} \rightleftharpoons \text{palmityl-ACP} + \text{CoA} \tag{28}$$

is absent in these extracts. This fact would explain the reasons for the inertness of the palmitic acid when added to in vivo systems. That is, palmitic acid can be readily activated to palmityl-CoA and then enter a limited number of reactions, such as β-oxidation or transfer to complex lipids, but it cannot flow into ACP-activated pathways. The same picture can be drawn for stearic acid. These results are summarized in Fig. 4. We shall have more to say about these relationships in Section VI.

B. ACP Requirement by Plant Synthetases

All plant extracts that synthesize fatty acids from acetyl-CoA and malonyl-CoA are completely dependent on the presence of ACP. Two

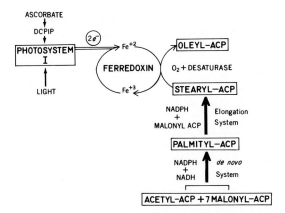

Fig. 4. Fatty acid synthesis, elongation and desaturation of stearyl-ACP to oleyl-ACP.

plant ACP's have been isolated and characterized, one from avocado mesocarp and one from spinach leaves. The characteristics of these ACP's as compared to prokaryotic ACP's are given in Table III.

Although the complete amino acid sequence has not been elucidated with spinach ACP, the core amino acid sequence adjacent to serine* the 4'-phosphopantetheine residue, namely, · · · Gly-Ala-Asp-Ser*-Leu-Asp · · · , is identical to that found in *E. coli* ACP (Matsumura and Stumpf, 1968).

Since *E. coli* ACP is readily available and is the most active of all ACP's tested, it has been employed in experiments with plant fatty acid synthetases. For *de novo* synthesis, *E. coli* ACP does not appear to modify the normal products of synthesis by plant synthetases. Moreover, its

TABLE III
A Comparison of Plant and Bacterial ACP's

Source[a]	MW	Amino acid residues	Relative[b] specific activities
E. coli	8,700	77	1
Arthrobacter sp.	9,500	81	1
Spinach leaf	9,500	88	0.3
Avocado mesocarp	11,500	117	0.3

[a] Simoni *et al.* (1967).
[b] With spinach chloroplast stroma as the source of the fatty acid synthetase assay system.

role in the initial monoenoic desaturation system is essential as is its role in the specific elongation for palmitic to stearic acid.

C. Specialized Systems

1. THE BIOGENESIS OF THE OIL DROPLET

High oil containing seeds have large amounts of triacylglycerols in the form of oil globules that appear under the electron microscope as uniform bodies with no apparent enclosing membranes. When immature cells are examined, nascent oil droplets can be observed distributed throughout the cytoplasm of developing cells and each contains vacuolelike inclusions. When these tissues are homogenized and fractionated by centrifugation, a low density pellicle is readily obtained, which contains over 80% of the total synthesizing capacity of the entire homogenate. When the pellicle is in turn examined by electron microscopy, the same inclusion bodies are observed as were observed in the intact cell. It can be shown that these inclusion bodies are the sites of the enzymes responsible for the total synthesis of fatty acid from acetyl-CoA. In addition, the enzymes required for the synthesis of triacylglycerols are also found associated with these granular bodies. It would appear, therefore, that the high oil plant cell has developed a pseudo-organelle for the purpose of forming oil droplets that contain all the enzymes necessary for the synthesis of triacylglycerols from acetate; as soon as a critical mass of triacylglycerols is achieved, these bodies with their associated enzymes are excluded into the ground substance of the cell with the simultaneous formation of the mature oil droplet (Harwood et al., 1971). However, other explanations have been proposed (see Gurr et al., 1974).

2. THE CHLOROPLAST SYSTEM

The present status of lipid biosynthesis in chloroplasts is summarized in Fig. 5. The acetate anion readily moves through the outer membrane into the stroma phase, (Jacobson and Stumpf, 1972), where it is rapidly converted by acetyl-CoA synthetase to acetyl-CoA. Acetyl-CoA then is transferred to ACP to form acetyl-ACP. In addition, malonyl-CoA is formed by the action of acetyl-CoA carboxylase. Both biotin carboxylase and the transcarboxylase occur exclusively in the stroma phase, with Mn^{2+} being the specific metal activator for the former enzyme while Mg^{2+} is the activator for the transcarboxylase. Biotin carboxylase is stable, and its activity can be readily followed in broken chloroplasts. However, the transcarboxylase is very unstable, since broken chloroplasts in Honda

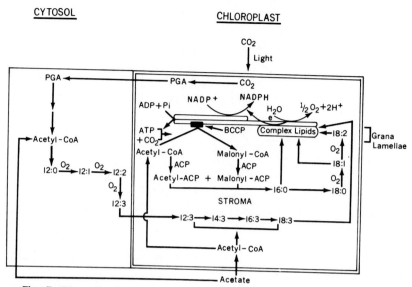

Fig. 5. Biosynthesis of fatty acid in spinach chloroplasts. PGA, phosphoglyceric acid.

medium show little if any activity. However, in the presence of 0.1 M sodium bicarbonate, chloroplasts may be disrupted, and the stroma phase shows high transcarboxylase activity. Some years ago an inhibitor was observed to be released when chloroplasts were disrupted, and this inhibition appears to bind with the transcarboxylase thereby inactivating it. In the presence of high bicarbonate concentrations, this inactivation effect is not observed. Indeed, after the initial steps of purification, the enzyme becomes stable, presumably because of the removal of the inhibitor.

The third protein, BCCP, is firmly associated with the lamellar membranes of chloroplasts from higher plants, although in the cytoplasm the BCCP is a soluble protein. The purpose of having BCCP closely associated with the lamellar membrane is not clear at present. Indeed, all the functional biotin in the chloroplast is lamellar membrane bound, although free biotin is present in significant amounts in the stroma. There is at present no evidence to suggest an additional function for the $CO_2 \sim$ BCCP besides that as a donor for the acetyl-CoA–malonyl-CoA system (Kannangara and Stumpf, 1972b). Figure 6 summarizes these observations.

Once malonyl-CoA is formed, it is rapidly transferred to ACP to form malonyl-ACP, which is in turn condensed with acetyl-ACP, reduced, dehydrated, and further reduced as indicated in Fig. 4 to yield

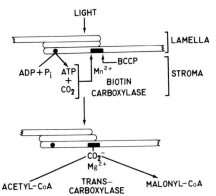

Fig. 6. The acetyl-CoA carboxylase system in spinach cloroplasts. BCCP, biotin carboxyl carrier protein.

the C_{16} acid. In chloroplasts, as in other plant tissues, the *de novo* system forms the C_{16} acid, which is then elongated to the C_{18} acid by the elongation system already discussed above, and then stearyl-ACP is desaturated to form oleyl-ACP. Further modifications of the C_{18} fatty acid presumably occur in the chloroplast to form linoleic and linolenic acids.

Light is an absolute requirement for fatty acid synthesis. Presumably this requirement is associated with the functioning of both photosystems I and II, which would generate ATP, NADPH, and oxygen, all required for fatty acid synthesis and desaturation by the intact chloroplast. With disrupted chloroplasts, on the other hand, no light requirement is observed, since the reducing potential is supplied by NADH and NADPH.

The chloroplast is a highly efficient organelle not only for the generation of ATP, NADPH, and oxygen and CO_2 fixation to yield sucrose and starch eventually but also for the *de novo* synthesis of fatty acids, which are presumably then inserted into the complex lipid to make up the lamellar membrane lipids. Chloroplasts have no triacylglycerol droplets in their structure, since the enzymes required for the formation of this lipid are missing in these organelles (Shine *et al.*, 1976).

V. Biosynthesis of Unsaturated Fatty Acids

In all eukaryotic systems, the aerobic mechanism for desaturation occurs, that is, a reductant and molecular O_2 are the components for the cis elimination of two hydrogens to form a cis double bond system. In yeasts, vertebrate tissues, fungi, and some bacteria, there is oxidative desaturation of a long-chain acyl-CoA to a monounsaturated acyl-CoA product [Eq. (29)].

$$\text{Stearyl-CoA} \xrightarrow[\text{O}_2]{\text{NADPH}} \text{oleyl-CoA} \qquad (29)$$

In all these tissues, the oxidative desaturase is associated with microsomal or membraneous particles. NADPH or NADH and O_2 are invariably required and cytochrome b_5 is the electron carrier coupling the reductant with the desaturase. In higher plants, it has been known for some time that avocado particles can form large amounts of [14C]oleic acid from [14C]acetate under aerobic conditions. However, under anaerobic conditions, mainly stearic and some palmitic acids are synthesized, thereby suggesting a possible precursor relationship between stearic and oleic acids (Mudd and Stumpf, 1962). Until recently, all attempts to prove this precursor relationship by a direct test, such as the conversion of [14C]stearic acid, [14C]stearyl-CoA, or chemically synthesized [14C]stearyl-ACP to [14C]oleic have failed. The reasons for previous failures are now apparent. In the first place, the stearyl desaturase is not microsomal, but rather a completely soluble enzyme. In the second place, the only active substrate is stearyl-ACP. A comment should be made here. Nagai and Bloch (1968) were the first to demonstrate that stearyl-ACP, when added to a chloroplast system containing NADPH-ferredoxin reductase, ferredoxin, and the desaturase + oxygen, was readily desaturated to oleic acid. However, the ACP was always chemically stearylated by a variety of methods, giving substrates of low reactivity. To circumvent this problem Jaworski and Stumpf (1974b) incubated [14C]malonic acid with extracts of maturing safflower seeds under anaerobic conditions in the presence of E. coli ACP and then proceeded to isolate [14C]stearyl-ACP. This enzymically formed product proved to be excellent as a substrate for desaturation, since only the single SH group of ACP was stearylated and this acyl group proved highly reactive. The stearyl-ACP desaturase occurs in the soluble fraction of extracts of maturing safflower seeds, maturing soybean seeds, in avocado mesocarp, in the stroma phase of spinach chloroplasts, and in cauliflower inflorescent tissue (Jaworski and Stumpf, 1974a). It requires ferredoxin and an electron donor such as NADPH. A more effective reducing system is an ascorbate–dichlorophenolindophenol (DCIP)–photosystem I (chloroplast)–ferredoxin system. In the presence of light, electrons derived from ascorbate are channeled through the photosystem I carrier system to ferredoxin. Reduced ferredoxin then in some manner interacts with molecular oxygen, the desaturase and stearyl-ACP leading to the desaturation reaction. These results are summarized in Fig. 4.

The desaturation of stearyl-ACP is probably the primary route for the synthesis of oleic acid in higher plants. However, the mechanisms for the synthesis of the two important C_{18} polyunsaturated fatty acids, namely linoleic [18:2(9,12)] and α-linolenic [18:3(9,12,15)] are controversial and somewhat unclear at the present time.

In 1963, James showed by feeding experiments with leaf tissue that there was a rapid conversion of oleic acid to linoleic and a much slower conversion of linoleic to α-linolenic acid. With the demonstration by Law that an α- or β-vaccenyl phosphatidylcholine served as a precursor for the formation of a β-cyclopropane acyl phosphatidylcholine in the presence of S-adenosylmethionine and a bacterial preparation, Gurr *et al.* (1969) as well as Talamo *et al.* (1973), Pugh and Kates (1973), and Baker and Lynen (1971) have provided evidence suggesting that an α- or β-oleyl phosphatidylcholine was indeed the substrate for a desaturation to form β-linoleyl phosphatidylcholine [Eq. (30)].

$$
\begin{array}{c}
\quad\quad\quad\quad\quad\quad\quad O \quad CH_2OCOR \\
\quad\quad\quad\quad\quad\quad\quad || \quad | \\
R-CH_2CH_2CH{=}CH(CH_2)_7-C-CH \quad O \\
\quad\quad\quad\quad\quad\quad\quad\quad\quad\quad | \quad || \\
\quad\quad\quad\quad\quad\quad\quad\quad\quad\quad CH_2OP-O-choline \\
\quad\quad\quad\quad\quad\quad\quad\quad\quad\quad\quad | \\
\quad\quad\quad\quad\quad\quad\quad\quad\quad\quad\quad O^- \\
\\
O_2 \;\big|\; NADPH \\
\\
\downarrow \quad O \quad CH_2COR \\
\quad\quad\quad\quad || \quad | \\
RCH{=}CHCH_2CH{=}CH(CH_2)_7COCH \quad O \\
\quad\quad\quad\quad\quad\quad\quad\quad\quad | \quad || \\
\quad\quad\quad\quad\quad\quad\quad\quad\quad CH_2OP-O-choline \\
\quad\quad\quad\quad\quad\quad\quad\quad\quad\quad | \\
\quad\quad\quad\quad\quad\quad\quad\quad\quad\quad O^-
\end{array}
\tag{30}
$$

The evidence is strong in support of this reaction. Unfortunately all the systems so far studied are membrane bound and thus difficult to purify. Moreover since oleyl-CoA is a very effective substrate for desaturation by these same systems, the question arises as to whether or not there are actually two desaturases responsible for the formation of linoleic acid, the first a membrane modifier that directly attacks the phospholipids of the membrane thereby altering the fluidity of the membrane while the second system is responsible for the synthesis of the bulk linoleic acid commonly associated with triacylglycerols. Thus we could expect chloroplasts, in which all the lipid is membrane associated, to have the membrane modifier enzyme that is directly converting β-oleyl phosphatidylcholine to β-linoleyl phosphatidylcholine, while in maturing seeds, such as the safflower, which has a high linoleic concentration associated with triacylglycerols, we would have a oleyl-CoA → linoleyl-CoA transformation with the product now serving as an acylating substrate for the synthesis of triacylglycerols. McMahon and Stumpf in 1966 and Vijay and Stumpf in 1971–1972 studied in some detail the conversion of oleyl-CoA to linolyl-CoA by microsomal preparations of maturing safflower seeds. The system was $(NADH + O_2)$-dependent although the photosystem I–ferredoxin–light reducing system was equally effective. The complete system is depicted in Fig. 7. The desaturase is highly specific for the oleyl-CoA; no other acyl thioester was desaturated.

The problem of the precise nature of the substrate required for the synthesis of 18:2(9,12) would be resolved once the desaturases have been solubilized and purified.

Once linoleic acid is synthesized, Harris and James (1965) and Trémolières and Mazliak (1974) have presented evidence that linoleic acid is then further desaturated to α-linolenic acid.

Recently, Kannangara and Stumpf (1972a) observed that with isolated spinach chloroplast small amounts of 18:3 were formed under anaerobic conditions, while oleic acid and linoleic synthesis were markedly diminished. Similar observations were made in the presence of cyanide. These results suggested a unique anaerobic, cyanide-insensitive pathway for α-linolenic acid synthesis. The pathway became clear when the newly synthesized [^{14}C]α-linolenic acid was degraded. The results revealed that only the carboxyl terminal end was labeled either in *in vitro* or *in vivo* conditions, with [^{14}C]acetate as the source of label. As was later shown, preexisting 16:3(7,10,13) was elongated by either acetyl-CoA or malonyl-CoA to α-linolenic acid. More data indicated that the synthesis of α-

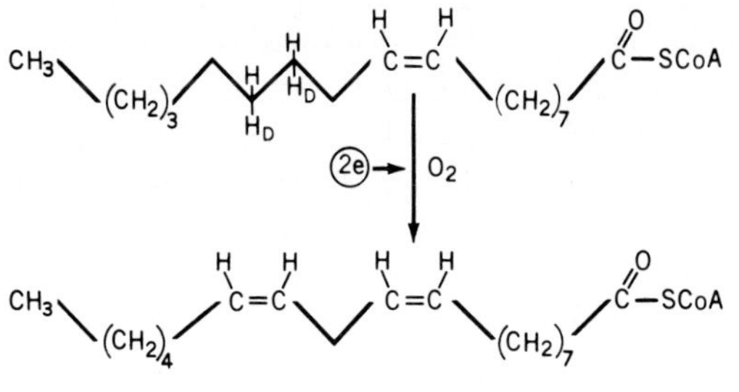

Ineffective substrates: Stearyl–CoA   18:0
                          Palmityl–CoA   16:0
                          Vaccenyl–CoA   18:1(11)
                          Elaidyl–CoA   18:1(9t)
                          Palmitoleyl-CoA 16:1(9)

Fig. 7. Oleyl-CoA desaturase of developing safflower seeds.

linolenic acid is completely separate from the $18:0 \rightarrow 18:1 \rightarrow 18:2$ pathway and that the desaturation occurs at the C_{12} level, where presumably a $C_{12} \rightarrow 12:1 \rightarrow 12:2 \rightarrow 12:3$ sequential desaturation occurs. The $12:3$ is then elongated to $14:3 \rightarrow 16:3 \rightarrow 18:3$. Thus there may be two pathways for α-linolenic synthesis, the first being the direct sequential desaturation of the C_{18} fatty acids and the second by the desaturation at C_{12} and then a subsequent elongation pathway (Jacobson et al., 1973a,b; Kannangara et al., 1973).

VI. Conclusion: A Discussion of the Interrelationship between Acyl-ACP's and Acyl-CoA's

Throughout this chapter, systems have been described that employ as substrates either derivatives of ACP or of CoA. Evidence is now clear that in the de novo and in the elongation pathways ACP derivatives are the only substrates involved, whereas for β-oxidation, the glyoxylate bypass, and for a number of acyl transferases, the CoA derivatives, are required.

The question now arises as to the mechanism that is employed by the plant cell to allow the flow of the products of the ACP system to the systems that employ the CoA derivatives. The simplest system would be a long chain acyl-ACP:CoA-acyl transferase which would catalyze the reaction:

$$C_{16-18}-\text{Acyl-ACP} + \text{CoA} \rightleftharpoons C_{16-18}\text{-acyl-CoA} + \text{ACP}$$

However, Jaworski et al. (1974) presented evidence that this system did not occur in a number of plant extracts. Furthermore, in vivo evidence that palmitic and stearic acids were not elongated or desaturated by a number of tissues, such as barley, safflower, spinach, and avocado, although these acids were readily activated to acyl-CoA's with subsequent insertion into complex lipids, etc., puzzled a number of investigators during the past decade.

Recently, Jaworski and Stumpf (1974a) noted that extracts of developing safflower seeds readily accumulated stearyl-ACP under anaerobic conditions where both the de novo and the elongation systems were functioning. Curiously, under conditions of desaturation, the product was always free oleic acid rather than the expected product, oleyl-ACP.

In reexamining this observation, Shine et al. (1976), established the presence of acyl-ACP thioesterases which had low activity for both palmityl-ACP and stearyl-ACP but a tenfold higher activity for oleyl-

ACP. It became apparent that here was an explanation for the earlier results of Jaworski et al. (1974a,b) and it also strongly suggested a key function for the acyl-ACP thioesterases, which normally would be considered as "nuisance" enzymes.

Figure 8 outlines a proposal in which the *de novo* and the elongation systems are defined as on the ACP track and all systems requiring acyl-CoA's as substrates are considered as being on the CoA track. These systems would include β-oxidation, glyoxylate bypass enzymes, and acyl transferases. Interconnecting these two tracks are two enzyme systems, the acyl-ACP thioesterases and acyl thiokinases, which together make-up the switching systems. This proposal logically (a) explains the observations of Jaworski et al. (1974a,b), (b) explains the *in vivo* data in which both palmitic and stearic acids were not desaturated or modified when fed to whole plant tissues [exceptions are the observations of Kolattukudy (1972) in epidermal cells where special elongation systems exist for the formation of C_{20-30} fatty acids as precursors of waxes], (c) describes a key function for both the acyl-ACP thioesterases and the acyl thiokinases, (d) suggests an explanation for the fact that stearic acid is always present in trace amounts in plant lipids, and (e) predicts that once the initial desaturation of stearyl-ACP to oleyl-ACP has occurred, all other desaturations would be on a CoA track or modifications thereof.

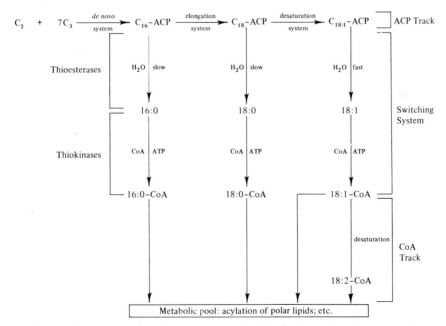

Fig. 8. The interrelationship between acyl-ACP's and acyl-CoA in plant cells.

Much progress has been made in elucidating the many systems involved in the biosynthesis and modifications of fatty acids during the past decade; it is expected that the next decade will markedly extend the body of information we now have and will provide the basic biochemical information that is essential for future investigations in genetic engineering of high and low lipid-containing seeds.

REFERENCES

Alberts, A. W., and Vagelos, P. R. (1968). *Proc. Nat. Acad. Sci. U.S.* **59**, 561.
Baker, N., and Lynen, F. (1971). *Eur. J. Biochem.* **19**, 200.
Barron, E. J., and Stumpf, P. K. (1962). *Biochim. Biophys. Acta* **60**, 329.
Bradbeer, C., and Stumpf, P. K. (1960). *J. Lipid Res.* **1**, 214.
Christie, W. W. (1973). "Lipid Analysis." Pergamon, Oxford.
Cooper, T. G., and Beevers, H. (1969). *J. Biol. Chem.* **244**, 3507 and 3514.
Devor, K. A., and Mudd, J. B. (1971a). *J. Lipid Res.* **12**, 403.
Devor, K. A., and Mudd, J. B. (1971b). *J. Lipid Res.* **12**, 412.
Douce, R. (1974). *Science* **183**, 852.
Galliard, T. (1971). *Biochem. J.* **121**, 379.
Galliard, T., and Mercer, E. I., eds. (1975). "Recent Advances in the Chemistry and Biochemistry of Plant Lipids." Academic Press, London.
Galliard, T., and Phillips, D. R. (1971). *Biochem. J.* **124**, 431.
Galliard, T., and Stumpf, P. K. (1966). *J. Biol. Chem.* **241**, 5806.
Giovanelli, J., and Stumpf, P. K. (1958). *J. Biol. Chem.* **231**, 411.
Gunstone, F. D. (1967). "An Introduction to the Chemistry and Biochemistry of Fatty Acids and Their Glycerides," 2nd ed. Chapman & Hall, London.
Gurr, M. I., Robinson, M. P., and James, A. T. (1969). *Eur. J. Biochem.* **9**, 70.
Gurr, M. I., Blades, J., Appleby, R. S., Robinson, M. P., and Nichols, B. W. (1974), *Eur. J. Biochem.* **43**, 281.
Harris, R. V., and James, A. T. (1965). *Biochim. Biophys. Acta* **106**, 456.
Harwood, J. L. (1975). *Biochim. Biophy. Acta* **398**, 224.
Harwood, J. L., and Stumpf, P. K. (1971). *Arch. Biochem. Biophys.* **142**, 281.
Harwood, J. L., and Stumpf, P. K. (1972). *Lipids* **7**, 8.
Harwood, J. L., Sodja, A., Stumpf, P. K., and Spurr, A. R. (1971). *Lipids* **6**, 851.
Hatch, M. D., and Stumpf, P. K. (1962a). *Plant Physiol.* **37**, 121.
Hatch, M. D., and Stumpf, P. K. (1962b). *Arch. Biochem. Biophys.* **96**, 193.
Heinstein, P. F., and Stumpf, P. K. (1969). *J. Biol. Chem.* **244**, 5374.
Hitchcock, C., and James, A. T. (1964). *J. Lipid Res.* **5**, 593.
Hitchcock, C., and Morris, L. J. (1970). *Eur. J. Biochem.* **17**, 39.
Hitchcock, C., and Nichols, B. W. (1971). "Plant Lipid Biochemistry." Academic Press, New York.
Huang, K. P., and Stumpf, P. K. (1971). *Arch. Biochem. Biophys.* **143**, 412.
Hutton, D., and Stumpf, P. K. (1969). *Plant Physiol.* **44**, 508.
Jacobson, B. S., and Stumpf, P. K. (1972). *Arch. Biochem. Biophys.* **153**, 656.
Jacobson, B. S., Kannangara, C. G., and Stumpf, P. K. (1973a). *Biochem. Biophys. Res. Commun.* **51**, 487.
Jacobson, B. S., Kannangara, C. G., and Stumpf, P. K. (1973b). *Biochem. Biophys. Res. Commun.* **52**, 1190.
James, A. T. (1963). *Biochim. Biophys. Acta* **70**, 9.

Jaworski, J. G., and Stumpf, P. K. (1974a). *Arch. Biochem. Biophys.* **162**, 158.
Jaworski, J. G., and Stumpf, P. K. (1974b). *Arch. Biochem. Biophys.* **162**, 166.
Jaworski, J. G., Goldschmidt, E. E., and Stumpf, P. K. (1974). *Arch. Biochem. Biophys.* **163**, 769.
Johnson, A. R., and Davenport, J. B., eds. (1971). "Biochemistry and Methodology of Lipids." Wiley (Interscience), New York.
Kannangara, C. G., and Stumpf, P. K. (1972a). *Arch. Biochem. Biophys.* **148**, 414.
Kannangara, C. G., and Stumpf, P. K. (1972b). *Arch. Biochem. Biophys.* **152**, 83.
Kannangara, C. G., Jacobson, B. S., and Stumpf, P. K. (1973). *Biochem. Biophys Res. Commun.* **52**, 648.
Kates, M. (1972). "Techniques of Lipidology." North-Holland Publ., Amsterdam.
Kaziro, Y., and Ochoa, S. (1964). *Advan. Enzymol.* **26**, 283.
Kolattukudy, P. E. (1972). *Progr. Chem. Fats Other Lipids* **13**, No. 2.
Kolattukudy, P. E., Walton, T. J., and Kushwaha, R. (1971). *Biochem. Biophys. Res. Commun.* **42**, 739.
Macher, B. A., and Mudd, J. B. (1974). *Plant Physiol.* **53**, 171.
McMahon, V., and Stumpf, P. K. (1966). *Plant Physiol.* **41**, 148.
Markovetz, A. J., Stumpf, P. K., and Hammarstrom, S. (1972). *Lipids* **7**, 159.
Marshall, M. O., and Kates, M. (1972). *Biochim. Biophys. Acta* **260**, 558.
Martin, R. O., and Stumpf, P. K. (1959). *J. Biol. Chem.* **234**, 2548.
Matsumura, S., and Stumpf, P. K. (1968). *Arch. Biochem. Biophys.* **125**, 932.
Mazliak, P. (1968). "Le métabolisme des lipides dans les plantes supérieures." Masson, Paris.
Meheriuk, M., and Spencer, M. (1967). Phytochem. **6**, 551.
Moore, T. S., Lord, J. M., Kagawa, T., and Beevers, H. (1973). *Plant Physiol.* **52**, 50.
Moss, J., and Lane, M. D. (1971). *Advan. Enzymol.* **35**, 321.
Mudd, J. B., and Stumpf, P. K. (1961). *J. Biol. Chem.* **236**, 2602.
Mudd, J. B., Van Vliet, H. H. D. M., and Van Deenen, L. L. M. (1969). *J. Lipid Res.* **10**, 623.
Nagai, J., and Bloch, K. (1968). *J. Biol. Chem.* **243**, 4626.
Neufeld, E. F., and Hall, C. W. (1964). *Biochem. Biophys. Res. Commun.* **14**, 503.
Newcomb, E. H., and Stumpf, P. K. (1952). *J. Biol. Chem.* **200**, 233.
Ongun, A., and Mudd, J. B. (1968). *J. Biol. Chem.* **243**, 1558.
Ongun, A., and Mudd, J. B. (1970). *Plant Physiol.* **45**, 255.
Pugh, E. L., and Kates, M. (1973). Biochim. Biophys. Acta **316**, 305.
Shannon, L. M., de Vellis, J., and Lew, J. Y. (1963). *Plant Physiol.* **38**, 691.
Shine, W. Mancha, M., and Stumpf, P. K. (1976). *Arch. Biochem. Biophys.* **172**, 110.
Shine, W. E., and Stumpf, P. K. (1974). *Arch. Biochem. Biophys.* **162**, 147.
Simoni, R. D., Criddle, R. S., and Stumpf, P. K. (1967). *J. Biol. Chem.* **242**, 573.
Steinberg, D., Herndon, J. H., Uhlendorf, B. W., Mize, C. E., Avigan, J., and Milne, G. W. A. (1967). *Science* **156**, 1740.
Sumida, S., and Mudd, J. B. (1970). *Plant Physiol.* **45**, 719.
Talamo, B., Chang, N., and Bloch, K. (1973). *J. Biol. Chem.* **248**, 2738.
Trémolières, A., and Mazliak, P. (1974). *Plant Sci. Lett.* **2**, 193.
Vijay, I. K., and Stumpf, P. K. (1971). *J. Biol. Chem.* **246**, 2910.
Vijay, I. K., and Stumpf, P. K. (1972). *J. Biol. Chem.* **247**, 360.
Zimmerman, D. C. (1966). *Biochem. Biophys. Res. Commun.* **23**, 398.

15

Nucleic Acid Metabolism

JOE L. KEY

I. Introduction

Although the flow of information from

$$\overset{\frown}{\text{DNA}} \quad \rightarrow \quad \text{RNA} \quad \rightarrow \quad \text{protein}$$
$$\text{(replication)} \quad \text{(transcription)} \quad \text{(translation)}$$

is a concept accepted as the central "dogma" of biology, the details of the transmission of information at each step are only beginning to be resolved. That RNA may serve as template for DNA synthesis is also becoming apparent. This chapter summarizes the enzymology of DNA replication and of RNA biosynthesis. (Details of the participation of

463

the various classes of RNA in protein biosynthesis are found in Chapter 16.) Additionally, the various classes of DNA and RNA are discussed in terms of complexity, properties, and cellular localization. The processing of large precursor molecules into the recognized functional RNA's (e.g., rRNA, tRNA, and mRNA) is discussed. While the processing schemes serve to point out the specificity that must reside in the hydrolysis of phosphodiester bonds of precursor RNA's to yield the functional RNA species, little is known about the enzymology of the process.

II. Enzymology of DNA Replication

A. DNA Polymerases of Bacteria

The DNA of bacteria is replicated by a semiconservative mechanism (Messelson and Stahl, 1958) as is the replication of DNA in nuclei, mitochondria, and chloroplasts in a wide range of eukaryotic organisms where this problem has been studied. At about the time semiconservative replication was being established, an enzyme, DNA polymerase I, was isolated and purified from *E. coli* (see Kornberg, 1969) that catalyzed the synthesis of DNA according to the following reaction:

$$\text{Primer DNA-3'-OH} + \begin{array}{l} \text{dATP} \\ \text{dCTP} \\ \text{dGTP} \\ \text{dTTP} \end{array} \xrightarrow[\substack{\text{Mg}^{2+} \text{ or Mn}^{2+} \\ \text{DNA template}}]{\text{DNA polymerase}} \text{DNA} + \text{pyrophosphate}$$

DNA polymerase I is a single polypeptide chain having a molecular weight of about 110,000, with single binding sites for the primer 3'-hydroxyl, the template DNA, and a deoxyribonucleoside triphosphate. The highly purified enzyme will utilize either native or single-stranded DNA as a template. The enzyme will also restore DNA which has had a strand partially degraded by a specific exonuclease to the complete double-stranded form. A segment of DNA terminating in a free 3'-hydroxyl is required as a "primer" in the reaction. The enzyme then carries out chain elongation, making a strand of DNA complementary to the template DNA. A double-stranded closed circular DNA molecule must be "nicked" by an endonuclease before DNA polymerase I can bind and catalyze the synthesis of DNA. The enzyme will bind single-stranded circular DNA and will initiate DNA synthesis if a short complementary primer is added. A single strand of DNA may fold back on itself providing both the primer 3'-hydroxyl and the template. Base composition and nearest neighbor analyses indicated that DNA polymerase I made correct complementary copies of the template DNA. Finally, this enzyme in conjunction with other requisite enzymes was shown to catalyze the *in vitro* synthesis of biologically active ΦX174 DNA (Goulian *et al.*, 1967).

Recent evidence suggests that the requirement for a free 3'-hydroxyl primer may be provided by RNA, since there is direct participation of RNA synthesis in DNA replication (e.g., Brutlag *et al.*, 1971). The coupling of DNA replication to RNA synthesis has further been demonstrated *in vitro* (e.g., Keller, 1972; Karkas, 1972). Additionally, the "Okazaki fragments," which appear to be involved in the discontinuous replication of DNA, have been shown to start with a piece of primer RNA (Sugino *et al.*, 1972; Sugino and Okazaki, 1973). Thus, while it seems clear that DNA replication is in some way connected to RNA synthesis, details of the association remain to be elucidated.

Although for many years DNA polymerase I was believed to be the enzyme responsible for DNA replication *in vivo*, there were several reasons to think that this might not be the case. The replication of double-stranded DNA *in vitro* by DNA polymerase I proceeded at a rate some two orders of magnitude below the *in vivo* rate. Additionally, this enzyme synthesized in a 5' to 3' direction, while *in vivo* both strands of DNA seemed to replicate from the same initiation point in the same direction. Also abnormal products with strange contortions (branches, etc.) of the DNA duplex were often synthesized. Finally, mutants of *E. coli* (Pol A⁻) were isolated (de Lucia and Cairns, 1969), which showed a low level of DNA polymerase I activity relative to the wild type; yet these mutants replicated their DNA and grew at normal rates. The DNA synthesizing activity was subsequently found in the cell membrane fraction of the DNA polymerase I-deficient mutant (Knippers and Strätling, 1970). This activity was solubilized and shown not to be sensitive to DNA polymerase I antiserum and to be sensitive to sulfhydryl reagents, while DNA polymerase I was completely insensitive. While the initial rate of *in vitro* DNA synthesis by DNA polymerase II approaches the *in vivo* rate, the duration of synthesis is short. Thus, this DNA polymerase II activity is markedly different from the Kornberg enzyme. As with DNA polymerase I, DNA polymerase II requires a primer containing a free 3'-hydroxyl group to initiate chain elongation, and the synthesis of DNA occurs in the 5' to 3' direction. Two DNA synthesizing activities have now been purified from the Pol A⁻ mutants of *E. coli* (Kornberg and Gefter, 1971). The major activity apparently corresponds to DNA polymerase II. The other DNA synthesizing activity (DNA polymerase III) differs from the major activity in Pol A⁻ mutants in that it is much more sensitive to sulfhydryl reagents and to thermal inactivation. A series of double mutants carrying one of the thermosensitive mutations for DNA synthesis and the Pol A⁻ mutation have been constructed (Gefter *et al.*, 1971). All of the mutant strains tested have normal DNA polymerase II activity. DNA polymerase III activity is, however, thermosensitive in those strains having thermosensitive mutations at the *dna E* locus. These results, along with those from other studies (e.g., Nüsslein *et al.*, 1971), indicate that DNA polymerase III is an enzyme required for chromosomal DNA replication in *E. coli*. The highest detectable activity of DNA polymerase III *in vitro* is much smaller than the polymerization rate *in vivo*. Thus while it is not clear if DNA polymerase III is the only enzyme involved in *in vivo* replication of the normal genomic DNA, it appears to be an essential component of the replication system. A mutant of *E. coli* deficient in DNA polymerase II has now been isolated from *E. coli* Pol A⁻ (Campbell *et al.*, 1972). The only polymerase activity detected in this mutant is DNA polymerase III. The mutant grows normally at 25° and 42°C and supports the growth of several bacteriophages, further supporting the view that DNA polymerase III is the enzyme involved in chromosomal DNA replication. However, it has recently been shown that the DNA of the colicinogenic factor E_1 of *E. coli* replicated at the restrictive temperature in the thermosensitive mutants for DNA polymerase III but not in Pol A⁻ mutants (Goebel, 1972). These results are suggestive that more than one mechanism of DNA replication may be operative in bacterial cells.

While DNA polymerase I may not function *in vivo* in chromosomal DNA replication, it does seem to have an important biological function in DNA repair. The most highly purified DNA polymerase I preparations contain nuclease functions.(The enzyme has recently been split into two parts by limited proteolysis. The large fragment has DNA polymerase I activity and the 3′ to 5′ exonuclease activity, while the smaller fragment has 5′ to 3′ exonuclease activity but no polymerase activity.) The multiple functions of DNA polymerase I thus allow it to excise thymine dimers and possibly mismatch regions of DNA and to repair lesions (Kelly *et al.*, 1969). Additionally, the DNA polymerase I-deficient mutants show increased sensitivity to UV irradiation, consistent with the idea of a repair function of DNA polymerase I. Other results indicate that DNA polymerase I functions in combination with the *rec* system of *E. coli* in the excision–repair process (Cooper and Hanawalt, 1972). DNA polymerase I-deficient mutants also show a decrease in the rate of joining of newly replicated DNA chains or "Okazaki fragments" (Okazaki *et al.*, 1971).

A detailed consideration of DNA replication can be found in recent reviews (Becker and Hurwitz, 1971; Klein and Bonhoeffer, 1972).

B. DNA Polymerase Activities of Higher Plants

In view of the above, it may not be surprising that the enzymology of DNA replication in eukaryotic organisms, and particularly in plants, is not detailed. There is a small number of reports describing the enzymatic synthesis of DNA for higher plant systems. A rather crude preparation from mungbean seedlings has been described which catalyzes the incorporation of a radioactive deoxyribonucleoside triphosphate into a DNA-like product dependent upon the presence of DNA, a divalent cation, and the simultaneous presence of the four deoxyribonucleoside triphosphates (Schwimmer, 1966). An enzyme that is active in DNA synthesis has been partially purified from maize (Stout and Arens, 1970). This enzyme is similar in activity to the DNA polymerases described above. Activity depends upon the simultaneous presence of DNA, Mg^{2+}, dATP, dCTP, dTTP, and dGTP. Heat-denatured DNA is a somewhat better template for this enzyme than native DNA. This activity is sensitive to sulfhydryl reagents. The enhancement of DNA synthesis by pancreatic deoxyribonuclease (an enzyme that cleaves phosphodiester bonds freeing a 3′-hydroxyl) over short treatment time indicates that a 3′-hydroxyl is required to prime the corn DNA polymerase, as is the case for bacterial and mammalian enzymes. Also, the enzyme is inhibited by the action of micrococcal nuclease, which produces 3′-phosphoryl groups during the hydrolysis of DNA. The base composition of the product is the same as that of the native primer template DNA, indicating that the bases of the product are specified by base complementarity with the template. DNA polymerase activity has also been studied in pollen grains of *Tradescantia* (e.g., Takats and Weaver, 1971) and in *Lilium* microspores (e.g., Howell and Hecht, 1971).

Several reports have appeared on the DNA synthetic activity of highly purified chloroplasts (e.g., Tewari and Wildman, 1967; Spencer and Whitfeld, 1969). The chloroplasts of spinach contain a firmly bound DNA-polymerizing activity as well as a soluble and readily leached DNA polymerase fraction. Whether these are, in fact, different activities is not established. The bound chloroplast enzyme is sensitive to sulfhydryl reagents. The "soluble" enzyme is more active on native template than on denatured DNA. The requirements for DNA synthesis are similar to those presented above for the bacterial DNA polymerases.

From these and many other studies, it appears that the chloroplast has the machinery for replication of its genome. Although apparently not studied with plant mitochondria, results from other systems indicate that mitochondria also contain the enzymatic machinery for DNA replication.

C. DNA Ligase

The enzyme DNA ligase catalyzes the synthesis of phosphodiester bonds within single-strand interruptions in DNA bounded by 3'-hydroxyl and 5'-phosphoryl groups. The ligases from various systems require an AMP donor (ATP or NAD) and transfer this moiety to the 5'-phosphoryl groups (see Becker and Hurwitz, 1971). The enzyme subsequently forms the 3',5'-phosphodiester bond with release of AMP. Although the exact biological roles of this enzyme are not known, it appears that ligase has a function in DNA replication, i.e., in joining newly replicated pieces of DNA. Current models of recombination and repair of DNA molecules require that polynucleotides be joined by 3',5'-phosphodiester bond formation, implying a ligase requirement.

A DNA ligase activity was recently reported for higher plants (Kessler, 1971). This enzyme shows an absolute requirement for ATP, with no activity in the presence of NAD.

D. RNA-Directed DNA Synthesis

A new excitement in DNA synthesis studies has come recently from experiments that suggest that RNA tumor viruses possess a DNA polymerase (or reverse transcriptase) activity which utilizes the viral RNA as template (see Temin, 1972). In addition to the fact that this finding may prove to have immense importance in the area of cancer research, this activity is not restricted to tissues infected with tumor viruses. Temin has held the view for many years that RNA-directed DNA synthesis might be involved in embryonic differentiation via gene amplification. RNA-dependent DNA polymerase activity has been detected in normal rat liver (Ward et al., 1972) and in chicken embryos (Kang and Temin, 1972). While some recent results have been interpreted to suggest that RNA-directed DNA synthesis may be involved in the amplification of ribosomal RNA (rRNA) genes of *Xenopus* oocytes (Crippa et al., 1971; Ficq and Brachet, 1971), there is some evidence that this may not be true (Bird et al., 1973). (Ribosomal DNA amplification will be discussed in Section IV.) To date there is no evidence of a specific RNA-dependent DNA polymerase activity in plants.

III. Enzymology of RNA Biosynthesis

In addition to functioning in its own replication, DNA serves as template for the production of RNA. In this process the nucleotide sequence of DNA is transcribed into the complementary sequence of bases in an RNA molecule of opposite chain polarity. In the case of DNA replication, complementary copies of both strands of the DNA duplex must be made, thereby forming upon completion of replication one identical copy of each chain of DNA leading to the formation of two molecules of the original DNA. When DNA participates in RNA synthesis *in vivo* (or *in vitro* under appropriate conditions discussed below), only one strand of the DNA duplex is transcribed into an RNA sequence at a given locus, and the DNA is totaly conserved, in contrast to the DNA polymerase reaction. The enzyme that accomplishes this DNA-dependent RNA synthesis according to the following reaction is RNA polymerase.

$$\begin{matrix} \text{GTP} \\ \text{ATP} \\ \text{CTP} \\ \text{UTP} \end{matrix} \xrightarrow[\substack{\text{DNA template} \\ \text{Mg}^{2+}\text{ or Mn}^{2+}}]{\text{RNA polymerase}} \text{RNA} + \text{pyrophosphate}$$

The sequence of nucleotides in the RNA is complementary, and of opposite polarity, to the nucleotide sequence of the strand of DNA which serves as its template.

Even though the basic reactions accomplished by RNA polymerase and DNA polymerase are similar, the complexity of RNA polymerases generally is much greater. RNA polymerases are generally large, multisubunit enzymes (Table I), and there are multiple forms of the enzyme, at least in eukaryotic organisms (e.g., Roeder and Rutter, 1970; Kedinger *et al.*, 1974; Weinmann and Roeder, 1974; Jacob, 1973). However, some phage-specific RNA polymerases and some mitochondrial RNA polymerases consist of a single, low molecular weight subunit (see Jacob, 1973; Chamberlin, 1974). Additional regulatory factors have been described for the bacterial enzyme which interact with the RNA polymerase to allow specific transcriptions (see Chamberlin, 1974). It is anticipated that a much more complex set of regulatory factors may exist in eukaryotic systems (see Chamberlin, 1970; Davidson and Britten, 1973).

A. The RNA Polymerase of Prokaryotes

The RNA polymerase of *E. coli* (see review by Burgess, 1971) can be purified in two states that differ significantly in their catalytic activity *in vitro*. The "core enzyme" consists of four major subunits (1β, $1\beta'$, 2α), as described in Table I. The purified enzyme usually contains variable amounts of a low molecular weight polypeptide (ω). If ω is a functional component of RNA polymerase, its role is unknown ("core

TABLE I

Subunit Structure of RNA Polymerases

Source	Subunits[a] (MW $\times 10^{-3}$)									
E. coli RNA polymerase[b]	165	155	85	40						
Thymus RNA polymerase I[c]	197	126	51	44	25	16.5				
Mouse myeloma RNA polymerase I[d]	195	117	60.5	50.5	27	16.5				
Soybean RNA polymerase I[e]	183	136	50	46	40	33	28			
Thymus RNA polymerase II[c]	190	150	35	25						
Mouse myeloma RNA polymerase IIa[d]	205	140	41	30	25	22	20	16		
Wheat germ RNA polymerase II[f]	220	140	45	40						
Maize RNA polymerase II[g]	220	160	35	25	20	17				
Soybean RNA polymerase IIb[h]	170	142	42	28	20	16	15.5	14		
Mouse myeloma RNA polymerase III[i]	155	138	89	70	52	43	41	34	29	19

[a] The subunit structures of the eukaryotic enzymes must be viewed as tentative.

[b] Data from Burgess, 1971. The 85×10^3 component is the σ regulatory subunit; the other subunits are present in a molar ratio of 1:1:2.

[c] Data from Kedinger et al., 1974. There are two forms of RNA polymerase I or A, the second having a 170×10^3 subunit instead of the 190×10^3 component.

[d] Data from Schwartz and Roeder, 1975. There are two forms of these enzymes differing in the MW of the large subunits.

[e] Data from Guilfoyle et al., 1976.

[f] Data from Jendrisak and Becker, 1974.

[g] Data from Mullinix et al., 1973.

[h] Data from T. J. Guilfoyle and J. L. Key, unpublished.

[i] Data from Sklar et al., 1975. There are also two forms of the type III RNA polymerase.

enzyme" devoid of ω is active in RNA synthesis). The complete or "holo-enzyme" contains an additional subunit, σ. While the core enzyme carries out DNA-dependent RNA synthesis, specific initiation of RNA chains requires the σ factor. Random initiation on both strands of the DNA duplex and synthesis of variable size RNA products are accomplished by the core enzyme alone. Sigma apparently binds to the core enzyme allowing for recognition of specific "initiation" sites on the DNA. After initiation of the new RNA chain is accomplished, σ is released and is then able to combine with a second core enzyme molecule and function in another round of chain initiation. An excellent example of the extent to which σ restricts initiation to specific sites on DNA comes from work on the *in vitro* transcription of fd phage DNA (Sugiura et al., 1970). They found that the holoenzyme initiated primarily only three different RNA chains of discrete size and initial base sequence from only one strand of the replicative form of the DNA. On the other hand, the products of the core enzyme were transcribed from both strands of the fd phage DNA, contained many initial sequences, and were very hetero-geneous in size.

Another factor that affects transcription of DNA by *E. coli* RNA polymerase functions in the regulation of bacterial genes subject to catabolite repression. During the transient repression following addition of glucose to the medium of growing bacterial cells, there is a decrease in the level of expression of many genes associated with sugar metabolism, and there is a marked decrease in the level of cyclic AMP in the cells. Furthermore, a single mutation has been shown to make the cells unable to express these same genes. A protein factor, catabolite gene-activating protein (CAP), has now been isolated which interacts with cyclic AMP. The CAP–cyclic AMP complex interacts with the complete RNA polymerase, or the DNA itself at or near the promoter locus, to allow for transcription of the catabolite-sensitive genes at a significant rate. The functioning of this complex has been studied *in vitro* using the synthesis of β-galactosidase as a model system (see, e.g., de Crombrugghe *et al.*, 1971).

Another factor, ρ, meets at least two criteria for a role in the normal transcription apparatus of *E. coli* (see Chamberlin, 1974). The ρ factor apparently functions in the chain termination process (see Burgess, 1971). In some cases normal physiological termination is achieved with ρ, while in other cases nonphysiological termination occurs (see Chamberlin, 1974).

B. The RNA Polymerases of Eukaryotes

1. MULTIPLE RNA POLYMERASES AND THEIR FUNCTIONS IN ANIMALS

Although several laboratories had earlier described RNA polymerase activities from many eukaryotic systems, the existence of multiple forms of RNA polymerase (Fig. 1A) in eukaryotic organisms was only recently established (Roeder and Rutter, 1969, 1970; see also Jacob, 1973; Chambon, 1975). At least three forms of RNA polymerase from sea urchin and calf thymus were resolved on DEAE–Sephadex columns. RNA polymerase I is localized in the nucleolus (Fig. 1B), while RNA polymerase II is found in the nucleoplasm (Fig. 1C). While the localization of RNA polymerase III was not established in these studies, its nuclear localization has now been shown (Weinmann and Roeder, 1974). The localization of RNA polymerase I within the nucleolus suggests that this enzyme may be involved in the transcription of rRNA, a process restricted to the nucleolus. The *in vitro* product of RNA polymerase I has the base compo-

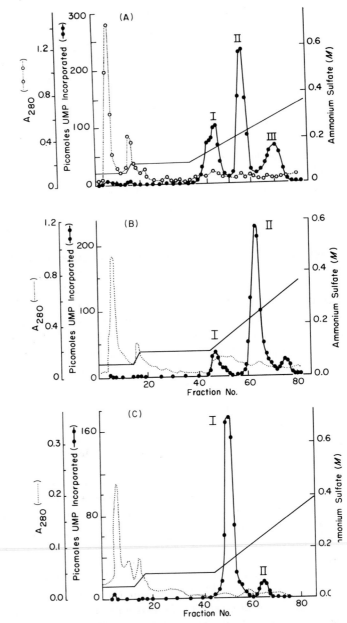

Fig. 1. Resolution of multiple forms of RNA polymerases by DEAE-Sephadex chromatography. (A) RNA polymerases from sea urchin. From Roeder and Rutter (1969). (B) Nucleoplasmic RNA polymerase from rat liver nuclei. (C) Nucleolar RNA polymerase from rat liver nuclei. (B) and (C) from Roeder and Rutter (1970).

sition and hybridization behavior expected of rRNA (Blatti et al., 1970), and studies with isolated nuclei also support this view (Reeder and Roeder, 1972; Zylber and Penman, 1971). These studies showed that the product of RNA polymerase II, on the other hand, has a DNA-like base composition and competes well with whole nuclear RNA in competition–hybridization experiments. This suggests that this enzyme synthesizes most if not all of the heterodisperse, rapidly labeled RNA's (HnRNA's) including mRNA. RNA polymerase III has been shown recently to transcribe the tRNA and 5 S RNA genes (Weinmann and Roeder, 1974).

The selective inhibition of the RNA polymerases by α-amanitin has contributed significantly to an understanding of the function of the different types (I, II, III). At low concentrations (below 1 μg/ml), RNA polymerase II is inhibited while I and III are unaffected (see, e.g., Lindell et al., 1970). At higher concentrations (20 to 50 μg/ml), polymerase III is inhibited (Weinmann and Roeder, 1974). Polymerase I is insensitive to much higher levels of the toxin (e.g., 200 μg/ml). The inhibition of RNA synthesis by α-amanitin results from the binding of the toxin to the enzyme and subsequent inhibition of phosphodiester bond formation (Cochet-Meilhac and Chambon, 1974). While the nuclear eukaryotic RNA polymerases are generaly insensitive to the rifamycins, which inhibit RNA chain initiation by the bacterial enzymes, some derivatives (e.g., AF/013) do inhibit chain initiation by the eukaryotic enzymes (see, e.g., Butterworth et al., 1971; Adman et al., 1972, Mandel and Chambon, 1974).

Recent work with antibodies prepared against purified calf thymus RNA polymerase I or A (Kedinger et al., 1974) shows that RNA polymerase I from widely different animals is structurally related. The antibodies against enzyme I have no effect on the activity of any of the purified RNA polymerase II (B) preparations thus far tested, and the inhibition detected with crude preparations gradually disappears as purification progresses. This may indicate that a component loosely bound to polymerase II may be lost during purification while the same or similar component remains bound to polymerase I during purification.

Each of the RNA polymerases (I, II, III, or A, B, C, respectively, as designated by different groups) occurs in at least two forms (Jacob, 1973; Kedinger et al., 1974; Weinmann and Roeder, 1974; Schwartz and Roeder, 1975; Chambon, 1975). Whether these different forms are of in vivo significance is still open to question. The two forms generally differ only in the molecular weight of one or two subunits (Weaver et al., 1971; Kedinger et al., 1974). The eukaryotic RNA polymerases that have been studied in sufficient detail are composed of multiple subunits (see, e.g., Table I); RNA polymerases I, and II, and III generally purify with more "subunits" than the *E. coli* enzyme. Unlike the *E. coli* enzyme which is known to be regulated by at least three factors with the function of other reported factors being of dubious significance (see Chamberlin, 1974), there is to date no definitive evidence on regulatory factors for the eukaryotic polymerases. Numerous "factors" that modulate the rate of synthesis of RNA *in vitro* have been reported (see Jacob, 1973; Chambon, 1975), but their physiological significance is not detailed. The similar subunit structure of the bacterial and eukaryotic enzymes might indicate that a similar, but probably more complex, set of "control elements" will be utilized in the regulation of the multiple eukaryote polymerases. The size and complexity of most eukaryote genomes relative to prokaryotes would seem to dictate a more elaborate regulatory system for transcriptional control (see Davidson and Britten, 1973).

There is a vast literature dealing with salt effects, template specificity and Mg^{2+} and Mn^{2+} preferences of RNA polymerases I and II (see Jacob, 1973). In general, however, polymerase II is more active at higher ionic strengths [e.g., 100–200 mM $(NH_4)_2SO_4$ optimum] than polymerase I (e.g., 25–50 mM optimum), and RNA polymerase I often prefers doubled-stranded DNA as template while polymerase II prefers single-stranded DNA. The ionic strength optima and Mg^{2+} or Mn^{2+} preferences vary with the relative concentration of DNA and whether native or denatured DNA is used as template (Gissinger *et al.*, 1974).

2. RNA POLYMERASE FROM PLANTS

A major problem in the study of RNA polymerases relates to the tight association of the RNA synthesizing activity with chromatin. A small amount of RNA polymerase activity has been solubilized from the crude chromatin fraction of pea by centrifugation in a high concentration (4 M) of cesium chloride (Huang and Bonner, 1962). This chromatin isolation procedure appears to yield a chromatin preparation containing primarily RNA polymerase I (Lin *et al.*, 1974). Most of the polymerase II activity is recovered from the chromatin supernatant.

A RNA polymerase from maize present in the 150,000 g (1 hour) supernatant of seedling homogenates was isolated and characterized (Stout and Mans, 1967, 1968). This enzyme was purified about a hundredfold by ammonium sulfate fractionation and DEAE–cellulose chromatography. The partially purified polymerase activity was essentially free of RNase, DNase, and polynucleotide phosphorylase activities. The four ribonucleoside triphosphates, DNA, and a divalent cation were required for RNA synthesis; the enzyme showed higher activity with Mn^{2+} than with Mg^{2+}. Actinomycin D, which specifically inhibits DNA-dependent RNA synthesis, inhibited the incorporation of [^{14}C]AMP by about 85% at 5 μg/ml. As expected, if pyrophosphate is a product of the reaction, inorganic pyrophosphate inhibited the reaction while orthophosphate did not. DNA from a number of species served as template, with maize DNA being intermediate in template effectiveness, and with heat denatured DNA being utilized more efficiently than native DNA. The product of the reaction was nondialyzable, acid insoluble, and hydrolyzed by ribonuclease and alkali. The product formed from either [α-^{32}P]ATP, [α-^{32}P]UTP, or [α-^{32}P]CTP yielded all four ^{32}P-labeled 2',3'-mononucleotides upon alkaline hydrolysis in proportions similar to that expected based on the nucleotide composition of the DNA template (i.e., A + U to G + C ratios in the RNA product were similar to the A + T to G + C ratios of the DNA template), indicating that both strands of the native DNA's were transcribed.

Three RNA polymerases have been fractionated from maize leaves (Bottomley *et al.*, 1971; Strain *et al.*, 1971). Two of these are of nuclear origin, while the other is localized in the chloroplast of maize leaves. RNA polymerase II is inhibited by α-amanitin, while RNA polymerase I is not; neither activity is affected by rifamycin. The inhibitor sensitivities are thus similar to those of RNA polymerase I and II of animals. These maize enzymes show greater activity with Mg^{2+} than with Mn^{2+}, in contrast to the RNA polymerases of animals which prefer Mn^{2+}. The soluble RNA polymerase of maize (Stout and Mans, 1967), which seems analogous to RNA polymerase II of other eukaryotes and wheat germ RNA polymerase II (Jendrisak and Becker, 1974), also exhibited maximum activity with Mn^{2+}. As noted above, these properties usually vary with specific assay conditions (see Gissinger *et al.*, 1974). The subunit structures of maize (Mullinix *et al.*, 1973) and wheat germ (Jendrisak and Becker, 1974) RNA polymerase II are shown in Table I; these enzymes appear similar to other eukaryotic RNA polymerases.

Multiple RNA polymerases have been fractionated on DEAE–cellulose columns from soybean (Horgen and Key, 1973); at least two of these are of nuclear origin and correspond to RNA polymerases I and II of other eukaryotic systems. RNA polymerases I and II have been purified to near homogeneity from soybean chromatin and nuclei (Guilfoyle *et al.*, 1976; T. J. Guilfoyle and J. L. Key, unpublished). These enzymes have different subunit structures with the possible exception that both possess a 28,000 MW subunit (Table I). RNA polymerase I is refractory to α-amanitin (Guilfoyle *et al.*, 1975), transcribes primarily rDNA (W. B. Gurley, C. Y. Lin, T. J. Guilfoyle, R. T. Nagao, and J. L. Key, unpublished), and is localized in the nucleolus (C. Y. Lin, T. J. Guilfoyle, Y. M. Chen, and J. L. Key, unpublished). The soybean RNA polymerase

II is found both in the nucleoplasm (Chen *et al.*, 1975) and in the soluble fraction of the cell (Lin *et al.*, 1974) and is totally inhibited by 0.1 μg/ml α-amanitin. A type III RNA polymerase has not been detected in the soybean system.

RNA polymerases have been fractionated and purified to varying degrees of homogeneity from a number of lower plants: the water mold, *Blastocladiella emersonii* (Horgen and Griffin, 1971); yeast (see, e.g., Adman *et al.*, 1972); and the slime molds, *Dictyostelium discoideum* (Pong and Loomis, 1973) and *Physarum polycephalum* (Burgess and Burgess, 1974).

Since the discovery of a polyadenylate polymerase activity in animal tissues (Edmonds and Abrams, 1960), several homopolymer polymerase activities have been reported (see, e.g., Twu and Bretthauer, 1971; Mans and Walter, 1971; Mans, 1971; Duda and Cherry, 1971; Haff and Keller, 1973; Burkard and Keller, 1974; Winters and Edmonds, 1973). The physiological significance of the poly(C) (Duda and Cherry, 1971) and poly(G) (Burkard and Keller, 1974) polymerases has not been established. On the other hand, the role of the poly(A) polymerases in the polyadenylation of pre-mRNA is assumed at this time (see Section V).

In general the poly(A) polymerases catalyze the addition of adenylate residues to the 3'-hydroxyl terminus of RNA molecules; in the case of maize, the enzyme will also adenylate single-stranded DNA and various deoxyhomopolymers (Mans, 1971). At least two poly(A) polymerase activities are present in yeast nuclei (Haff and Keller, 1973). Both enzymes have a strong preference for ATP, prefer Mn^{2+} to Mg^{2+}, and are strongly inhibited by GTP. One of these (I) adenylates RNA, poly(C), poly(A), poly(G), and poly(U). The other enzyme (II) uses only poly(A) sequences as primer. Neither of these enzymes is inhibited by α-amanitin (specific inhibitor of RNA polymerases II and III) nor by actinomycin D (inhibitor of DNA-directed RNA synthesis by RNA polymerases. The purified poly(A) polymerase from calf thymus is also highly specific for ATP, is inhibited by other ribonucleoside triphosphates, is more active with Mg^{2+} than Mn^{2+}, and is insensitive to α-amanitin (Winters and Edmonds, 1973).

HeLa cell nuclei have recently been shown to synthesize *in vitro* the normal poly(A) isolated from growing cells (Jelinek, 1974). This coupled with enzyme localization studies establishes that nuclei contain the requisite enzyme activities necessary for the posttranscriptional addition of poly(A); cytoplasmic synthesis of poly(A) could be accomplished by an enzyme such as that described in yeast (Twu and Bretthauer, 1971) [see Section V,B for details on poly(A) RNA].

IV. Characterization and Properties of DNA

The discussion here will be restricted to the complexity of plant and other eukaryotic DNA's as assessed by DNA reassociation studies, some distinguishing properties of nuclear, chloroplastic, and mitochondrial DNA, and the DNA that codes for rRNA (rDNA).

A. The Complexity of Eukaryotic DNA Based on Kinetics of Reassociation

One of the remarkable facts about DNA is that the separated complementary strands of DNA can specifically reassociate into the native duplex under appropriate conditions (see Britten and Kohne, 1968, for detailed discussion). Since reassociation of a pair of complementary strands of DNA results from a bimolecular collision, the expected half-period for their reassociation would be inversely proportional to the initial DNA concentration. Thus, the half-time for reassociation for a given total DNA concentration would be expected to be proportional to the number of different types of fragments and thus to the genome size of the particular organism. For a number of bacterial and bacteriophage DNA's, these expectations are borne out in that the half-time for reassociation is proportional to genome size (Britten and Kohne, 1968). While the DNA of these relatively small genomes (e.g., 2×10^5 nucleotide pairs for T2 phage DNA and 4.5×10^6 for *E. coli*) reassociated as expected for an ideal second-order reaction, such a reaction for DNA of plants and animals (genome sizes of about 10^9 to 10^{11} nucleotide pairs) was expected to require much longer times. Surprisingly varying amounts of rapidly reassociating DNA were observed with plant and animal DNA's. The reassociation of part of the DNA at a rate greater than expected on the basis of genome size shows that there are sequences present at concentrations greater than expected based on the genome size. The evidence to date indicates that for all eukaryotic organisms studied, except possibly some fungi, varying amounts of the genome (from 20 to 80% depending upon the organism and conditions of reassociation) are made up of many copies of similar or identical sequences of nucleotides, in some cases up to more than a million copies (e.g., Britten and Kohne, 1968; Britten and Davidson, 1969, 1971; Kohne, 1970).

A plot of the fraction of DNA present in the single-stranded form versus the initial DNA concentration–time parameter (C_0t in moles of nucleotides second per liter) is often used to present renaturation kinetics

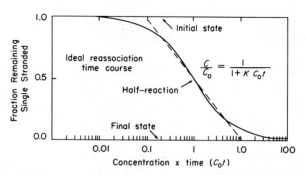

Fig. 2. Time course of an ideal, second-order reaction to illustrate the features of the log $C_o t$ plot. The equation represents the fraction of DNA that remains single-stranded at any time after the initiation of the reaction. For this example, K is taken to be 1.0, and the fraction remaining single-stranded is plotted against the product of total concentration and time on a logarithmic scale. From Britten and Kohne (1968).

for DNA. Such a plot for an ideal, second-order reaction is shown in Fig. 2, where reassociation occurs over a $C_o t$ range of about 10^2. Significant deviations from the theoretical are indicative of a heterogeneous reaction, i.e., renaturation of more than one family (or frequency distribution) of DNA base sequences is taking place.

An example of the kinetic complexity of plant DNA is shown in Fig. 3. The curve for reassociation of wheat DNA shows that there are at least two different populations of base sequences in addition to the

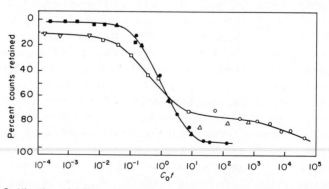

Fig. 3. Kinetics of DNA reassociation of wheat and *Bacillus subtilis* measured with hydroxyapatite. ^{3}H-labeled hexaploid wheat and ^{14}C-labeled *B. subtilis* (6700 cpm/μg) sheared DNA's were mixed and reassociated at 40°C in 62% formamide, 0.78 M NaCl, and 5×10^3 M sodium phosphate buffer pH 6.8. Filled circles, triangles, and squares represent *B. subtilis* DNA at 16, 4, and 0.3 μg/ml, respectively. Open circles, triangles, and squares represent wheat DNA at 6000, 300 and, 16 μg/ml, respectively, reassociated in the absence of *B. subtilis* DNA. Figure from Bendich and McCarthy (1970a).

10% that apparently reassociates very rapidly; the fraction reassociating over the range of C_ot's from 10^{-2} to 10^1 represents repeated sequence DNA, since it occurred at a C_ot value of 10^4 to 10^5 lower than expected for single copy DNA based on the genome size of wheat. That fraction reassociating from 10^3 to 10^5 C_ot is representative of unique or single copy DNA. (The internal standard *Bacillus subtilis* DNA shows reassociation over the range expected of an ideal second-order reaction and is typical of bacterial genomes that are composed essentially of unique sequences or single copy DNA under often used conditions of reassociation.) About 80% of the wheat DNA reassociated as repeated sequence DNA with varying degrees of homology (Bendich and McCarthy, 1970a).

DNA duplexes can be formed from DNA molecules that do not have complete base complementarity. The degree of base complementarity of reformed DNA duplexes can be assessed by their thermal stability. Native DNA melts or thermally denatures over a very narrow temperature range, the midpoint (T_m) being characteristic for each DNA and related to the base composition. Thermal denaturation profiles of reformed hybrids for the repeated sequence or fast renaturing DNA's usually show some to considerable deviation from that expected for a perfectly complementary duplex as evidenced by a broadening of the melt plot and a lowered T_m. However, in the case of some satellite DNA's (i.e., DNA that bands in cesium chloride gradients at different densities from the main band or major DNA component), there appear to be about 10^6 (mouse) and 1.5×10^6 (green monkey) copies of a base sequence representing 10 and 20% of the total genome of these organisms, while there is considerable mismatching of bases in the reassociated duplex of other satellite DNA's.

The organization of repeat and unique sequence DNA's within the genome of some organisms has been assessed (Davidson *et al.*, 1973; Graham *et al.*, 1974; Wilson and Thomas, 1974). In the case of *Xenopus* and sea urchin DNA's, repetitive and unique sequences are intimately interspersed in the majority of the DNA. About 50% of these genomes consist of a short-period pattern, with an average length of 300–400 nucleotide repetitive segments interspersed with about 1000 nucleotide non-repetitive segments; a similar organization of *Dictyostelium* DNA has been inferred (Firtel and Lodish, 1973). Long sequences of essentially unique or single-copy DNA make up another 20% of the genome of these organisms, while about 20% or so represents a longer period interspersed pattern. Up to about 10% of the genome is made up of relatively long regions of repetitive sequences. Another recently described feature of eukaryotic DNA relates to the "palindromes" (Wilson and Thomas, 1974), which consist of inverted repetitions that are very closely spaced

on the same DNA chain; thus given the antiparallel arrangement of the
duplex, these sequences read the same both backward and forward. These
palindrome structures certainly account for some of the very rapidly renaturing DNA observed in most DNA reassociation studies. Another possibly significant feature of DNA organization is the interspersion of poly-
deoxythymidylate sequences that have been described in the case of the
genome of *Dictyostelium* (Jacobson *et al.*, 1974). There are some
14,000–15,000 sequences of poly(dT) (25 nucleotides long) in this genome,
corresponding to the estimated number of structural genes (transcription
products) in *Dictyostelium* (Firtel and Lodish, 1973). Also, deoxyadenyl-
ate-rich regions (not more than 8000 per haploid genome in duck) have
been reported to be interspersed throughout animal DNA's (Shenkin and
Burdon, 1974; Bishop *et al.*, 1974).

The possible function of repeated sequence DNA is as intriguing as
the possible origin of this DNA. An elaborate theory of gene regulation
for higher organisms which ascribes regulatory functions to the repeated
sequence families of DNA has been proposed (Britten and Davidson,
1969, 1971) and elaborated upon recently (Davidson and Britten, 1973).

B. Nuclear, Chloroplast, and Mitochondrial DNA's

Much of the early work on unique organelle DNA's led to the conclusion that chloroplast DNA could be distinguished from nuclear DNA by
its bouyant density in cesium chloride equilibrium density gradient centrifugation. While other workers had suspected that the chloroplast DNA
might not separate from nuclear main band DNA in cesium chloride gradients, work with tobacco and spinach (Whitfeld and Spencer, 1968) and
lettuce (Wells and Birnstiel, 1969) clearly distinguished nuclear DNA
from chloroplast DNA by other criteria. DNA's prepared from highly
purified nuclei or chloroplasts had very similar buoyant densities in cesium chloride (Fig. 4). However, two other parameters confirmed that
the two DNA's shown in Fig. 4 were in fact different. First the nuclear
DNA of higher plants contains 5-methylcytosine while this base was not
detected in the chloroplast DNA preparations (Table II). A similar base
difference exists for DNA of *Euglena* and some lower plants. Secondly,
these DNA's could be distinguished by the kinetics of reassociation of
the single-stranded forms. The chloroplast component rapidly renatured
under reannealing conditions, while the nuclear DNA showed little or
no reassociation under identical conditions. These criteria, i.e., purity of
organelle preparation, the presence (or absence) of 5-methylcytosine, and
the rate of reassociation, permitted the identification of the major DNA
component of chloroplasts relative to nuclei.

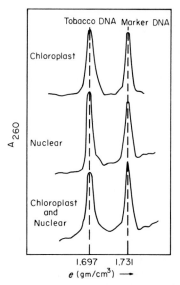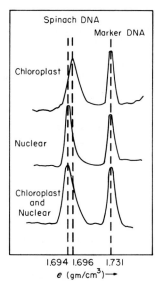

Fig. 4. Analytical density gradient centrifugation in CsCl of tobacco and spinach chloroplast and nuclear DNA's. Marker DNA is *M. lysodeikticus* DNA of density 1.731 gm/cm³. From Whitfeld and Spencer (1968).

Detailed studies on renaturation kinetics of chloroplast and mitochondrial DNA (Wells and Birnstiel, 1969) showed that chloroplast DNA is extensively reiterated. From kinetic measurements on reassociation

TABLE II

Base Composition of Tobacco and Spinach DNA[a]

	Tobacco DNA (mole %)		Spinach DNA (mole %)	
Base	Nuclear	Chloroplast[b]	Nuclear	Chloroplast[c]
Adenine	30.4	30.4	30.6	32.1
Thymine	28.3	29.0	28.9	27.6
Guanine	19.8	19.5	19.5	18.8
Cytosine	15.0	21.1	14.8	21.5
5-Methylcytosine	6.5	Not detected[d]	6.2	Not detected[d]

[a] Data from Whitfeld and Spencer (1968).

[b] Tobacco leaves were chopped, and chloroplasts purified on a discontinuous density gradient.

[c] Spinach leaves were chopped. The Triton-soluble fraction of the 2500 g pellet was used as the source of chloroplast DNA.

[d] Minimum level for detection of 5-methylcytosine is less than 0.5 mole % under the conditions used.

they concluded that the chloroplast DNA was made up of two populations, one representing about 10% of the DNA having a kinetic complexity of about 10^6 daltons and the major fraction showing a kinetic complexity of 1.2×10^8 daltons compared to a quantitative estimate of about 2×10^9 daltons of DNA in the lettuce chloroplast. These data are suggestive then that there are 20 or so copies of these sequences in the chloroplast. While the data are interpreted differently for the occurrence of more than one component of *Chlamydomonas* chloroplast DNA based on reassociation kinetics (Wells and Sager, 1971; Bastia *et al.*, 1971), the rate of renaturation shows a complement of unique nucleotide sequences of about 2×10^8 daltons in each case. This kinetic complexity is about 25-fold less than the DNA content of a single gamete chloroplast, indicating the presence of some 25 copies of a unique DNA sequence in the chloroplast of *C. reinhardtii*. Again renaturation studies on *Euglena* chloroplast DNA indicate a kinetic complexity of 2×10^8 daltons, and thus several copies per chloroplast (Stutz, 1970). These values for the kinetic complexity of chloroplast DNA have not been corrected for their relatively low $G + C$ contents (Wetmur and Davidson, 1968) and thus represent overestimates in size. More recent evidence shows that *Euglena* chloroplast DNA occurs in the circular form, having an average contour length of about 40 μm (Manning *et al.*, 1971). This length would correspond to a DNA molecule of about 9×10^7 daltons. At least a part of the chloroplast DNA of higher plants occurs as circular molecules of 44 μm (Manning *et al.*, 1972). Thus, all chloroplast DNA molecules may be about 9×10^7 dalton circles, present in 15 to 30 copies per chloroplast.

Another satellite DNA in plants having a buoyant density of about 1.706 has been assigned to the mitochondrion (Table III). As shown by the data in Table III, the density of mitochondrial or chloroplast DNA is much more constant in higher plants than is the density of nuclear main band DNA. The mitochondrial DNA of lettuce has a kinetic complexity of $>10^8$ daltons (Wells and Birnstiel, 1969), a complexity greater than that of the mitochondrial DNA from animal sources (approximately 10^7 daltons).

In addition to the nuclear main band DNA, chloroplast DNA, and mitochondrial DNA, the nuclear DNA of some plants contains a distinct satellite component having a density different from the main band DNA (see, e.g., Matsuda and Siegel, 1967; Ingle *et al.*, 1973; Bendich and Anderson, 1974). Of some 70 species of angiosperms investigated (Ingle *et al.*, 1973), 27 of 59 dicotyledonous plants contained satellite bands, while no such satellite bands were observed in 11 monocotyledonous species.

The melon satellite DNA has been rather extensively studied (Ingle

TABLE III

Buoyant Density of Nuclear, Chloroplast, and Mitochondrial DNA's from Higher Plants[a]

Source	Nuclear (main band) (gm cm^{-3})	Chloroplastal (gm cm^{-3})	Mitochondrial (gm cm^{-3})
Onion[b]	1.691	1.694	1.706
Swiss chard[b]	1.694	1.696	1.705
Spinach[b]	1.694	1.696	—
Lettuce[c]	1.694	1.697	1.706
Bean[c]	1.695	1.697	1.705
Tobacco[b]	1.695	1.697	—
Wheat[b]	1.702	1.698	—

[a] Buoyant density relates to the base composition of DNA by the following equation: $\rho = 1.660 + 0.98$ (mole fraction G + C).

[b] Data from Wells and Ingle (1970).

[c] Data from Wells and Birnstiel (1969).

et al., 1973; Bendich and Anderson, 1974). This high density satellite contains discrete components of widely differing T_m and complexity (see, e.g., Bendich and Anderson, 1974). About one-third of the satellite is high melting (T_m of 94°C) and is a simple repeating sequence of about 380 nucleotide pairs; the remaining two-thirds is low melting (T_m of 86.4°C), has a complexity 10^3 greater than the simple sequence, and is not covalently linked to the simple sequence. Based on the T_m of the reassociated components, the melon satellite DNA's are composed of essentially perfect copies, rather than less precisely matched repeat sequences as are found in most animal satellites. The melon rDNA could account for about 15% of this high-density satellite (the satellite is 30% of the total melon DNA). A similar fraction of pumpkin satellite consists of rDNA. Thus based on the nature of the melon satellite, it may be fortuitous that DNA's from plants containing high amounts of rDNA, e.g., pumpkin (Matsuda and Siegel, 1967) and cucumber (Bendich and McCarthy, 1970B) are generally characterized by having a distinct high-density satellite component even though the rRNA hybridizes to DNA in the position corresponding to the satellite component. However, rRNA hybridizes in the same DNA density region with DNA from plants lacking a distinct high density satellite.

C. Genes for Ribosomal RNA (rDNA)

The genes for cytoplasmic rRNA are present in multiple copies ranging from about 1 in mycoplasma, 4 to 30 in bacteria up to several thou-

sand in eukaryotes. In plants that have been studied, the number of
rRNA cistrons ranges from about 250 to 7000 for the haploid complement
(Table IV). The percentage of the total DNA which hybridizes with
rRNA varies from 0.02% for artichoke to about 1% for cucumber. Because of the large difference in genome size, there is no direct proportionality between percent DNA hybridizing with rRNA and the number of
ribosomal genes (e.g., only about one-tenth as much of the total onion
DNA hybridizes with rRNA as is the case for cucumber DNA, yet onion
contains about 50% more rRNA genes per haploid genome than cucumber). There is little species specificity among higher plants for hybridization of cytoplasmic rRNA to total DNA, indicative of very little if
any base sequence divergence in the rDNA (see, e.g., Matsuda and Siegel,
1967; Ingle et al., 1971).

The origin of chloroplast rRNA relative to nuclear and/or chloroplast DNA has been investigated by many laboratories. Under the hy-

TABLE IV

Number of rRNA Genes in Different Plants[a]

Plant	% DNA hybridized[b]	DNA (10^{-2} gm per telophase nucleus)	Genes per telophase nucleus	Ploidy of telo- phase nucleus	Genes per haploid comple- ment
Artichoke (Helianthus tuberosus)	0.022	24	1,580	6x	260
Swiss chard (Beta vulgaris var. cicla)	0.20	2.5	2,300	2x	1150
Maize (Zea mays)	0.18	7.5	6,200	2x	3100
Wheat (Triticum vulgare)	0.092	30	12,700	6x	2100
Cucumber (Cucumis sativum)	0.96	2	8,800	2x	4400
Onion (Allium cepa)	0.090	32	13,300	2x	6650
Pea (Pisum sativum)	0.17	10	7,800	2x	3900

[a] Data from Ingle and Sinclair (1972).

[b] A 2:1 mixture of 1.3×10^6 and 0.70×10^6 rRNA's was used. DNA prepared from the
various plant species was hybridized with 1.3×10^6 rRNA at 3 or 5 μg/ml for 2 hours in
6X SSC at 70°C. Wheat and cucumber DNA's were hybridized with pea rRNA, and the
other DNA's were hybridized with their homologous rRNA. (The percentage of DNA
hybridized is essentially independent of the plant rRNA.) The DNA of the nucleus was
determined by comparative Feulgen spectrophotometry, on the basis that pea contained
10×10^{-12} gm per telophase nucleus.

bridization conditions employed, cross-hybridization in both directions has often been observed. While chloroplast rRNA "hybridizes" to nuclear DNA to about the same extent as cytoplasmic rRNA (0.3% in the case of Swiss chard), it may anneal to the same sites as the cytoplasmic rRNA (Ingle et al., 1971). There is no additive effect when both chloroplast and cytoplasmic rRNA are added to the same hybridization reaction with nuclear DNA. More specificity is shown when chloroplast DNA is hybridized to rRNA. About 1.5% of Swiss chard chloroplast DNA hybridizes with chloroplast rRNA, while a value of 0.3% was obtained when cytoplasmic rRNA was tested. It is possible that the apparent cross-hybridization of the chloroplast and cytoplasmic rRNA with nuclear and chloroplast DNA's is the result of lack of specificity under the reaction conditions used, especially in the case of the low hybridization of cytoplasmic rRNA with the chloroplast DNA. That the nuclear DNA contains genes for chloroplast rRNA, however, must be considered an open question. The value of 1.5% hybridization of chloroplast DNA to chloroplast rRNA in Swiss chard corresponds to about 30 cistrons per chloroplast or some 6000 cistrons per cell (200 chloroplasts per cell). Similar values have been obtained for other plants.

With the demonstration that rDNA is amplified several thousand times in Xenopus oocytes (Brown and Dawid, 1968), considerable interest has been generated in possible gene amplification in other developing systems. While somatic cells of Xenopus laevis contain about 450 repeating genes for rRNA, the mature oocyte contains about 4000 times this amount of rDNA. While amplification of rDNA has been demonstrated for several animal systems, there is no definitive evidence for gross amplification in plants (Ingle and Sinclair, 1972). Some evidence had been presented which is consistent with deletion of about 30% of the rDNA during germination of wheat. Presumably, the rDNA had been amplified during embryogenesis. There is, however, no change in the percent of DNA hybridizing with rRNA in the maize embryo during germination or with wheat embryo DNA during embryo development and subsequent germination (Ingle and Sinclair, 1972). The large accumulation of ribosomes in response to auxin is likewise not associated with any significant change in the relative level of rDNA.

While the organization of the multiple copies of rDNA within the genome in plants has not been investigated, the organization of rDNA in Xenopus has been elegantly elaborated (Birnstiel et al., 1968; Brown and Weber, 1968; Brown et al., 1972 and references cited therein), and the rRNA cistrons have been shown to be associated with the nucleolus (or nucleolar organizer portion of the chromosome). The cistrons for the two large rRNA's (28 S and 18 S in Xenopus) alternate with one another in the rDNA and are interspersed with DNA of a higher $G + C$ content.

This portion of the rDNA is transcribed as one large RNA molecule which is the precursor to the rRNA. Additionally there are nontranscribed spacer regions between the repeat regions of the rDNA which are transcribed into the 28 S and 18 S rRNA's. The spacer DNA is remarkably uniform if not identical within a species as is the transcribed portion of the rDNA. However, between species (X. laevis and X. mulleri) some distinct differences in the spacer regions of rDNA exist. The 28 S and 18 S rRNA's appear to be identical based on renaturation kinetics and the fidelity of hybridization between the two species. There are detectable differences in nucleotide sequence in the transcribed portion of the spacer rDNA between these species based on hybridization of the precursor rRNA. Much larger differences in base sequence homology were observed for the nontranscribed portion of the spacer rDNA based on hybridization between species of complementary RNA transcribed from the total rDNA (the nontranscribed spacer DNA present in the repeating units of rDNA). While the 5 S rRNA is coordinately synthesized with 28 S and 18 S rRNA's, the 24,000 or so copies of the DNA coding for the 5 S rRNA in *Xenopus* are localized in a different region of the chromosome (Brown and Dawid, 1968; Brown and Weber, 1968) and are not amplified in oocytes. In the case of 5 S rRNA of *Xenopus*, hybridization occurs to DNA that bands to the low density side of the main band DNA. The 5 S rDNA has now been purified and shown to consist of repeating units of A–T- and G–C-rich regions using denaturation mapping by electron microscopy (Brown *et al.*, 1971). While the 5 S rRNA is 57% $G + C$, the 5 S rDNA is only about 35% $G + C$, thus accounting for its low density and denaturation map. The repeating unit of 5 S rDNA has a mass of about 5×10^5 daltons, whereas the 5 S rRNA has a MW of 84,000. Thus, one copy of the 5 S gene per repeating unit would correspond to about 16.8% of the 5 S rDNA, and this arrangement was confirmed by showing that 6.8% of the DNA (or 13.6% of its base pairs) hybridized with 5 S rRNA. It is assumed that the remaining DNA represents spacer. It is presently not known if any of this spacer DNA is transcribed during 5 S rRNA synthesis. Inheritance, mechanism of amplification, and evolutionary considerations of the rDNA have been discussed (Brown *et al.*, 1972). A detailed discussion of rRNA genes was also recently published (Birnstiel *et al.*, 1971).

V. RNA Metabolism

While progress is being made on the enzymology of RNA biosynthesis (Section III), much more effort has been devoted in recent years to studies on the characterization of *in vivo* short-time labeled RNA's. This approach was a result primarily of the early experiments that demonstrated the existence of a rapid turnover, DNA-like RNA in bacterial cells, and these experiments led ultimately to the conclusive demonstration of mRNA as the intermediate carrier of the genetic information of DNA (Astrachan and Volkin, 1958; Volkin, 1962). Similar studies in animal cells led to the demonstration of heterodisperse, rapidly-labeled RNA's (HnRNA), some of which had a nucleotide composition similar to DNA (DNA-like or D-RNA) and turned over with a relatively short half-life (see Darnell, 1968). In addition to the HnRNA, short-time labeled RNA's of animal cells also contained discrete high molecular weight RNA components that were later shown to be precursors to rRNA (see Burdon, 1971). The evidence from a wide variety of eukaryotic organisms shows that the mature rRNA's are transcribed from rDNA as a large precursor molecule that is processed to yield the rRNA characteristic of the particular organism.

A. Synthesis and Processing of rRNA

When plant tissues are labeled with [^{32}P]orthophosphate or some other appropriate RNA precursor, polyacrylamide gel analyses of total nucleic acid extracts yield patterns similar to that shown for carrot in Fig. 5 (Rogers *et al.*, 1970; Leaver and Key, 1970). Superimposed over the heterodisperse [^{32}P]RNA (discussed in Section V,B) are discrete [^{32}P]RNA species having molecular weights of about 2.9×10^6, 2.3×10^6, and 1.4×10^6. These components are labeled prior to the appearance of ^{32}P-labeled 25 S (1.3×10^6) and 18 S (7×10^5) rRNA's. The 1.4×10^6 component accumulates just prior to the appearance of the 7×10^5 rRNA, and the latter rRNA is labeled earlier than the 1.3×10^6 rRNA. Thus the kinetics of labeling of these discrete components are consistent with a precursor role in rRNA synthesis. In artichoke and pea, the largest detectable discrete component is 2.3×10^6 (Rogers *et al.*, 1970). In addition to the 2.3×10^6 and 1.4×10^6 components, a discrete 9×10^5 component is detected in artichoke and in pea (Fig. 6). These discrete components are enriched in crude "nuclear" preparations relative to total tissue nucleic acid extracts and have not been detected in polyribosome preparations or the postribosomal supernatant of carrot (Leaver and Key, 1970). Base composition analyses of these components support the view that they are precursors to rRNA (Table V). All have high G + C contents similar to rRNA, while the heterodisperse RNA's are AMP-rich. The preferential hybridization of these high molecular weight components to rDNA relative to main band DNA also supports

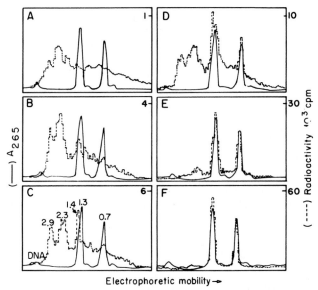

Fig. 5. Gel fractionation of total nucleic acids from carrot root discs. Labeling with [^{32}P]orthophosphate was for 10, 20, 40, 60, 120 and 240 minutes (A)–(F), respectively. Solid line, absorbance at 265 nm; dashed line, radioactivity. From Leaver and Key (1970).

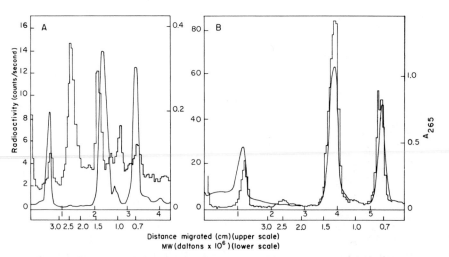

Fig. 6. Timing of rRNA synthesis in the cultured pea root tip. (A) Excised root tips. (B) Intact seedling. Solid line, absorbance at 265 nm; histogram, radioactivity. From Rogers *et al.* (1970).

TABLE V

Base Composition of Pea and Artichoke RNA Components[a,b]

MW of RNA (daltons × 10⁶)	Type of RNA	Source	Moles percent				
			C	A	G	U	G+C
Short label (45 to 60 minutes)							
>3	Heterogeneous	Pea	19.9	30.0	25.0	25.1	44.9
		Artichoke	19.5	32.0	26.9	21.8	46.4
<2	Heterogeneous	Pea	18.5	32.0	24.9	24.6	43.4
		Artichoke	18.9	33.3	26.8	21.2	45.7
2.3	Precursor (whole root)	Pea	20.5	27.7	28.8	22.9	49.3
2.3	Precursor (excised root)	Pea	20.5	26.3	29.4	23.8	49.9
		Artichoke	21.1	27.7	30.0	21.3	51.1
1.4	Precursor	Pea	20.4	27.5	31.5	20.6	51.9
		Artichoke	21.0	28.8	31.3	18.7	52.3
1.3	Ribosomal	Pea	20.1	28.0	31.8	20.1	51.9
		Artichoke	20.1	29.2	31.1	19.4	51.2
0.7	Ribosomal	Pea	19.8	27.9	28.3	23.3	48.1
		Artichoke	20.3	26.9	28.3	24.5	48.6
Long label (>2 hours)							
1.3	Ribosomal	Pea	21.8	24.9	32.2	21.1	54.0
		Artichoke	21.6	28.1	31.5	18.9	53.1
0.7	Ribosomal	Pea	21.1	25.9	28.5	24.4	49.6
		Artichoke	20.5	27.6	28.1	23.9	48.6
Weighted average of rRNA's (pea)			21.6	25.3	30.9	22.2	52.5

[a] Data from Rogers *et al.* (1970).
[b] Gel slices were taken from regions indicated by molecular weight. 1 to 3 peak slices were used for the ribosomal and precursor components and 2 to 5 for the heterogeneous components. Standard deviation were up to ±0.4 for peak fractions and ±0.8 for heterogeneous RNA.

the view that they are in fact precursors to rRNA (Grierson *et al.*, 1970).

These results indicate that the original transcription product of precursor rRNA has a different size in different plants. In pea, artichoke, wheat, and soybean, the largest detected component has a molecular weight of about 2.3×10^6; in carrot, tobacco and mung bean a larger component (2.7×10^6 to 2.9×10^6) is present. Of possibly more interest is the fact that the original transcription unit of different tissues of the mung bean have different molecular weights (Grierson and Loening,

1972). A 2.9×10^6 component was detected in the leaf, while the largest component in the root was 2.7×10^6. The other intermediates (2.5×10^6, 1.4×10^6, and 10^6) were identical between the two tissues. In the case of mung bean leaf, a 4.5×10^5 component was detected, which is considered to be a product of cleavage of the 2.9×10^6 precursor. Discrete low molecular weight products of processing of the precursor rRNA have not been detected in other plant tissues, nor has the expected 10^5 to 3×10^5 component of Fig. 7 been detected.

Recognizing that there may be some heterogeneity in molecular weights of the precursor rRNA between plants and between tissues of the same plant, Fig. 7 represents a proposed scheme for cytoplasmic rRNA synthesis in plants.

Precursors similar to those of higher plants have been studied in yeast (see, e.g., Udem and Warner, 1972), the green algae *Volvox* (Kochert, 1971), and in the slime mold *Dictyostelium* (Iwabuchi *et al.*, 1971) where the mature rRNA's have similar or identical molecular weights to those of higher plants. The precursor rRNA of *Euglena* may present a somewhat different situation (Brown and Haselkorn, 1971). The original transcription product of precursor rRNA has a molecular weight of 3.5×10^6. This large precursor is apparently cleaved to yield the 8.5×10^5 rRNA plus a 2.2×10^6 component which is processed to the 1.35×10^6 rRNA. The 8.5×10^5 rRNA appears to be processed from the 5′ end of the primary transcription product.

While there is some diversity of size among the rRNA precursors of plants, considerable diversity is encountered in animals. The precursors and processing in reptiles, amphibia, fish, and insects are similar to those discussed for plants, while birds and mammals have much larger original transcription products of rDNA (see, e.g., Perry *et al.*, 1970). The most detailed precursor work to date has been done with HeLa cells (see Burdon, 1971). Results from sequencing methods and fingerprint analysis of methyl-labeled dinucleotides of the HeLa cell rRNA's and their precursors (Maden *et al.*, 1972) support the basic scheme proposed for rDNA synthesis in HeLa cells (see Burdon, 1971). The 5.8 S rRNA of eukaryotic "80 S" ribosomes (see, e.g., Payne and Dyer, 1972), which is noncova-

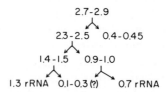

Fig. 7. Proposed scheme of processing of rRNA precursor in plants. Numbers are molecular weight in millions.

lently bound to the large (25 S or 1.3×10^6 in plants) rRNA, apparently is generated during the final cleavage of the immediate precursor to the 1.3×10^6 rRNA (see, e.g., Udem and Warner, 1972). While the original transcription product in HeLa cells has been generally characterized as a 45 S component (approximately 4.2×10^6), recent work has demonstrated the presence of both a 47 S and a 46 S component (Tiollais *et al.*, 1971). This observation, along with the heterogeneity seen in the mung bean rRNA (Grierson and Loening, 1972), is suggestive of the possibility that processing may occur so rapidly (possibly initiated during the transcription process) that the "original transcription unit" may often not be observed. Other possibilities to explain the observations are that the reiterated rRNA genes are different, or that the rRNA genes are identical but there are different termination or initiation sites in the rDNA (Grierson and Loening, 1972). These latter possibilities would seem not to apply, at least in the case of *Xenopus* where the rDNA has been studied in detail (see Section IV). The 5 S rRNA is synthesized as a separate component and thus is not a part of the large rRNA precursors.

The selective processing of precursor rRNA is not understood. The RNA apparently becomes associated with protein during transcription and remains as a ribonucleoprotein particle during processing in the nucleolus (see, e.g., Kumar and Warner, 1972). In the case of HeLa cells, the 45 S precursor is methylated, and all of the methyl groups of the 45 S component are conserved in the mature rRNA's (see, e.g., Weinberg and Penman, 1970). Either or both of these parameters could contribute to specificity of processing. No detailed methylation studies have been accomplished to date with plant systems.

Studies on rRNA synthesis in plant tissues, which show very different patterns of rRNA accumulation, may offer one approach to understanding the control of synthesis and processing of precursor rRNA in plants. When tissue slices of many, but not all, plants are excised and cultured in solution, the net synthesis of rRNA ceases. Loening (1965) has compared the state of excised tissue to a "step-down" culture of bacteria. In the excised pea root there is considerable accumulation of precursor rRNA (2.3×10^6, 1.4×10^6, and 9×10^5 components) with the formation of very little rRNA compared to the intact seedling root (Fig. 6). The processing of precursor rRNA is clearly more impaired in the excised root than is synthesis. Cells in the mature region of roots or stems also do not increase their rRNA content, but the processing of precursor rRNA seems not to be impaired in these cells relative to younger cells (Rogers *et al.*, 1970). Thus, the control of rRNA accumulation in old cells seems to relate more to synthesis than processing. In the case of the selective

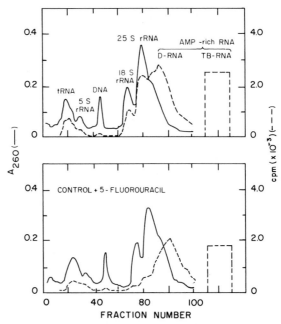

Fig. 8. The selective inhibition of rRNA synthesis by 5-fluorouracil. Total nucleic acids were purified from control (top) and 5-fluorouracil-treated (bottom) soybean hypocotyl after a 2-hour label in [^{32}P]orthophosphate and fractionated on MAK columns. From Key and Ingle (1968).

inhibition of rRNA accumulation by 5-fluorouracil (Fig. 8), there is no marked accumulation of precursor rRNA. In fact, the synthesis of precursor rRNA seems to be inhibited, but the greater inhibition of rRNA accumulation is also suggestive of an impaired processing mechanism.

Since different enzymes are responsible for rRNA and AMP-rich heterodisperse nuclear RNA (HnRNA) synthesis (Section III), a differential regulation of the enzymes may play an important role in the differential synthesis of the different types of RNA. It is well known that the availability of amino acids (in the form of aminoacyl-tRNA) is involved in the regulation of stable (rRNA and tRNA) RNA synthesis in bacteria (see, e.g., Edlin and Broda, 1968). While there is no evidence that the availability of amino acids plays a regulatory role in rRNA synthesis in plants, there is some evidence that this may be the case in other eukaryotic organisms (see, e.g., Franze-Fernandez and Pogo, 1971). It is known, however, that the pool of amino acids available for protein synthesis is rapidly depleted in excised plant tissues. Protein synthesis inhibitors, such as cychloheximide, also selectively inhibit rRNA accumulation in both plant (Key, 1966) and animal systems. All of the results taken together indicate a close coupling between protein synthesis and the synthesis and/or processing of precursor rRNA.

B. Synthesis and Processing of Heterogeneous Nuclear RNA (HnRNA)

After short exposures to radioactive RNA precursors, much of the newly synthesized RNA of animal (see Darnell, 1968) and plant (see Loening, 1968) tissues is heterodisperse with a molecular weight ranging from about 300,000 to several million and is commonly referred to as HnRNA (heterogeneous nuclear RNA). Most of this HnRNA of animal tissues turns over in the nucleus with a half-life of minutes (e.g., see Darnell, 1968; Brandhorst and McConkey, 1974). In the range of 2 to 10% of the HnRNA (the pre-mRNA) is processed and transported to the cytoplasm, where it becomes associated with polyribosomes (i.e., as mRNA). The exact relationship between total HnRNA and mRNA is not understood. While there is considerable support for the view that individual HnRNA molecules contain a structural gene sequence (see Darnell et al., 1973; Weinberg, 1973; Brawerman, 1974), there is no definitive evidence to choose between the alternatives that most or all HnRNA molecules contain an mRNA sequence plus other sequences that are degraded during processing (short half-life HnRNA) or whether a small fraction of the HnRNA molecules is composed primarily of mRNA sequences with the much larger fraction of HnRNA molecules undergoing rapid turnover (see Davidson and Britten, 1973). There is, however, considerable information accumulating on the nature of processing of the pre-mRNA fraction of HnRNA into cytoplasmic polyribosome-associated mRNA. The discovery of poly(A) sequences in HnRNA and the related methodologies (see Brawerman, 1974) have contributed significantly to the understanding of HnRNA metabolism. Some 20 to 40% of the HnRNA molecules contain poly(A) at the 3'-hydroxyl terminus. These poly(A) sequences range in size from 200 AMP residues in HeLa HnRNA down to 50 to 75 residues in some of the lower eukaryotes. In addition to these large poly(A) fragments, which are added posttranscriptionally, at least a part of the HnRNA molecules contain a short oligo(A) sequence (20 to 25 AMP residues) which is added during the transcription event (see, e.g., Firtel and Lodish, 1973; Nakazato et al., 1974). These findings on the origin of the large poly(A) and smaller oligo(A) fragments are consistent with the observations that there are no large tracts of poly(dT) in DNA but that there are many short tracts interspersed throughout the genome (Section IV,A). In addition to the nuclear synthesis of poly(A), there is some evidence for cytoplasmic polyadenylation of RNA (see, e.g., Wilt, 1973; Slater et al., 1973; Brawerman, 1974; Diez

and Brawerman, 1974). A part of the sea urchin egg mRNA's lack
poly(A), and upon fertilization these mRNA's are adenylated in the cytoplasm (Slater et al., 1973; Wilt, 1973). A similar situation may exist
in cotton where mRNA that is synthesized during embryogenesis, but
not translated until germination, is "stored" without poly(A); during
germination these mRNA's become adenylated (Walbot et al., 1974). Additionally, the poly(A) sequence of cytoplasmic mRNA is somewhat
shorter than nuclear poly(A) in some systems, and the terminal AMP
residues (on the average 7 or 8) of the poly(A) segment of mRNA turn
over rapidly in the cytoplasm (Diez and Brawerman, 1974). Most
eukaryotic mRNA's contain a poly(A) sequence at the 3'-terminus (see,
e.g., Adesnik et al., 1972). Of those mRNA's that have been studied in
any detail, only histone mRNA has been shown to lack poly(A) (see
Brawerman, 1974). More recent evidence indicates, however, that a larger
fraction of the mRNA's may not contain a poly(A) sequence (e.g., Milcarek et al., 1974; Nemer et al., 1974).

The RNA species in plants which correspond to HnRNA of animal
cells have variously been referred to as D-RNA (see e.g., Ingle et al.,
1965), mRNA (see, e.g., Chroboczek and Cherry, 1966), and AMP-rich
RNA's, including two species referred to operationally as D-RNA and
TB-RNA (Key et al., 1972). While a small percentage of these RNA's
undoubtedly functions as mRNA, much of it turns over with a relatively
short half-life. As shown in Fig. 5A, after short exposures to [^{32}P]ortho-
phosphate, much of the newly synthesized RNA of plants is heterogeneous corresponding to the HnRNA of animals (the AMP-rich RNA's
referred to above); with increasing times in labeled RNA precursor, the
relative proportion of label in rRNA increases while that in HnRNA decreases. These rapidly labeled, heterodisperse, AMP-rich RNA's partially
fractionate into two components on MAK columns as shown in Fig. 8
(see, e.g., Tester and Dure, 1967; Ewing and Cherry, 1967; Key and
Ingel, 1968; Johri and Varner, 1970). Further fractionation of these two
types of AMP-rich RNA (Key et al., 1972) results in the separation of
two distinct size classes of RNA which differ significantly in their base
composition, e.g., 30% versus 40% AMP (Table VI). The synthesis of
these AMP-rich RNA's is not affected by 5-fluorouracil, while the synthesis of rRNA and tRNA is very strongly inhibited (Fig. 8). More recent
results show that the major distinguishing feature between the two AMP-
rich RNA's of soybean relates to the presence of poly(A); the
D-RNA (30% AMP) does not contain a poly(A) sequence, while the
"TB-RNA" (40% AMP) has a poly(A) sequence at the 3'-terminus
(Key and Silflow, 1975). The soybean poly(A) RNA migrates on formamide gels as a broad band with a mean size of about 18 S. The poly(A)

TABLE VI

Base Composition of AMP-Rich RNA's from Total Tissue and Polyribosomes

	CMP (mole %)	AMP (mole %)	GMP (mole %)	UMP (mole %)
D-RNA[a]	20.3	30.6	23.1	26.0
"TB-RNA" [a]	17.2	40.7	19.0	23.1
Poly(A) RNA[b]	17.8	39.9	20.1	22.4
Poly(A)[b,c]	5.8	85.3	6.0	2.9
Poly(A)[b,d]	1.5[h]	98.5	—	—
Polyribosome-associated RNA[b,e]	16.8	41.6	19.4	22.2
Poly(A)[b,f]	5.7	85.4	5.1	2.8
Poly(A)[f,g]	3.7	87.4	5.1	3.8

[a] Data from Key et al. (1972).

[b] Key and Silflow (1975). These and other data identify the fraction of RNA referred to as "TB-RNA" as poly(A) RNA.

[c] Poly(A) RNA was digested with T_1 and pancreatic RNase followed by binding of the resistant material to poly(U) filters prior to base composition analyses.

[d] The poly(A) fragment was purified on polyacrylamide gels prior to base composition analyses; digestion was same as above.

[e] Polyribosome-associated RNA which bound to poly(U)–Sepharose.

[f] Polyribosome-associated RNA was digested and processed as in footnote c.

[g] Data from Higgins et al. (1973).

[h] The fraction reported as CMP is in fact "AMP" radioactivity resulting as a hydrolysis artifact.

sequences resulting from nuclease digestion of the poly(A) RNA migrates on these gels as a broad band overlapping the 4 S to 5.8 S regions with a mean size distribution somewhat greater than 5 S. Poly(A)-containing RNA's have now been reported for a few plant tissues (see, e.g., Higgins et al., 1973; Manahan et al., 1973; van de Walle, 1973; Sagher et al., 1974). While early work indicated that the polyribosome-associated mRNA was of the D-RNA type (i.e., about 30% AMP) in several plant tissues (see, e.g., Loening, 1965; Johri and Varner, 1970; Key et al., 1972), it appears likely that the poly(A) sequence was lost in those studies from some of the polyribosomal RNA during polyribosome preparation and/or subsequent RNA purification (Key and Silflow, 1975). Recent work (Higgins et al., 1973; Key and Silflow, 1975) shows that at least a part of the polyribosome-associated mRNA of plants contains a poly(A) sequence (see Table VI). While it is tempting to speculate that the D-RNA may serve as precursor to the poly(A) RNA, there is no detailed information on the relationship between D-RNA and poly(A) RNA.

C. Synthesis and Properties of tRNA

The availability of a specific assay for each species of tRNA and methods for purifying individual species of tRNA and for base sequencing the primary structure, and the implication of tRNA in "developmental regulation" at the level of translational control of protein synthesis have led to a vast amount of published literature relating to tRNA (see reviews by Novelli, 1969; Sueoka and Kano-Sueoka, 1970; Cramer, 1971; Gauss et al., 1971; Chambers, 1971).

Yeast alanine-tRNA was the first nucleic acid molecule to be sequenced (Holley et al., 1965). Following this major breakthrough in nucleic acid research, the primary structure (nucleotide sequence) of some 60 different tRNA's has been determined. These sequences are all compatible with the "cloverleaf" model for the secondary structure of tRNA. The tertiary structure of yeast phenylalanine-tRNA has recently been determined at 3 Å resolution by X-ray diffraction using isomorphous substitution (Robertus et al., 1974).

As with rRNA's and at least some mRNA's, tRNA's are transcribed as a larger precursor molecule (see, e.g., Altman and Smith, 1971; Burdon, 1971) and subsequently processed to the mature species. Sequence analysis of the precursor to an E. coli tyrosine-tRNA shows the presence of a large "extra" segment at the 5′ terminus and some additional nucleotides at the 3′ end. In addition to the processing of the precursor, many base modifications occur on the tRNA molecule after it is transcribed (see Gauss et al., 1971). The modifications include methylation to form mono- and dimethyl-substituted bases, thiolation, pseudouridine and dihydrouridine formation, and the attachment of various constituents to the base adjacent to the anticodon (3′ side) of several tRNA's. These modifications occur primarily after precursor processing based on the nucleotide sequence analysis of the tyrosine precursor tRNA of E. coli (Altman and Smith, 1971).

There has been some question as to whether the –C–C–A–OH (3′) terminus (the amino acid acceptor site) of tRNA's is transcribed or added after transcription. The tyrosine precursor tRNA described above contains this triplet linked to U in the expected position. However, an enzyme, –C–C–A–OH pyrophosphorylase, catalyzes the removal and addition of the –C–C–A–OH segment of tRNA molecules. The significance of the –C–C–A–OH "turnover" is not known, but the dynamics of the 3′ terminus is sufficient to account for significant precursor incorporation into RNA. It is not known whether the turnover of the acceptor terminus has a significant effect on the level of tRNA, possibly of specific tRNA's, which has amino acid acceptor activity. However, in preparations of plant tRNA's, where direct analysis of the 3' terminus has been made, some 80 to 90% of the molecules have an intact –C–C–A–OH (see, e.g., Vanderhoef et al., 1970). Also the data from studies where the level of amino acid acceptor activity for each of the 20 amino acids was measured for cotton tRNA show that about 90% of the tRNA molecules were acylated in vitro (Merrick and Dure, 1972).

The function(s) of the modified bases in tRNA, while receiving considerable attention, is not completely understood. tRNA's that have adenosine at the 3' end of the anticodon have, adjacent to this adenosine, a modified nucleotide, such as 2-thiomethyl-N^6-isopentenyladenosine, N^6-isopentenyladenosine, or base Y. Several of these modified adenosines are active cytokinins (see Skoog and Armstrong, 1970). Similarly tRNA's that have uridine as the 3' end base of the anticodon contain a carbamoylthreonine 6-amino-substituted purine in the corresponding position. The absence of one of these modifications on the base adjacent to the anticodon in these tRNA's leads to defective functioning (see, e.g., Hall, 1970; Ghosh and Ghosh, 1972). The tRNA's are acylated normally with the corresponding amino acid, but their function in protein synthesis is impaired. Binding of the aminoacyl-tRNA to ribosomes is less efficient in the case of seryl- and tyrosyl-tRNA's lacking the isopentenyl group (see Hall, 1970). Removal of base Y from phenylalanyl-tRNA changes the coding properties such that it recognizes UUC better than UUU (Ghosh and Ghosh, 1972) without affecting complex formation with GTP and the elongation factor. In the case of formylmethionine tRNA, a pseudouridylate, but not a 4-thiouridylate, appeared to be important for proper recognition by the aminoacyl-tRNA synthetase (Siddiqui and Ofengand, 1970).

It is known that the repression of the histidine operon is correlated directly with the in vivo level of a specific histidyl-tRNA (Lewis and Ames, 1972). Mutants that contained the same level of histidyl-tRNA as the wild-type organism but which were derepressed for histidine biosynthesis contained a histidine-tRNA that differed from the wild-type species of histidine-tRNA in having two uridine residues that were not modified to pseudouridine. This tRNA lacking pseudouridine must function reasonably well in protein synthesis, since it was acylated in vivo and the mutants grew very well. Thus base modifications in tRNA may be involved in many of the "regulatory" roles of aminoacyl-tRNA [e.g., control of amino acid biosynthesis (Singer et al., 1972), control of rRNA synthesis (Edlin and Broda, 1968), regulation of synthesis of aminoacyl-tRNA synthetases (Williams and Niedhardt, 1969), and allosteric regulation by a tryptophanyl-tRNA of a mutant form of tryptophan pyrrolase (Jacobson, 1971)] as well as in protein synthesis.

There are some 55 to 65 different species of tRNA (see, e.g., Gallo and Pestka, 1970) depending upon the organism and method used to fractionate the aminoacyl-tRNA. The number of different species which can be acylated with the same amino acid (isoaccepting tRNA's) varies from one up to about six. Some of these species read the same code word, while others have overlapping code word recognition. Excluding specific initiation and termination requirements, 31 tRNA's are sufficient to translate the genetic code, assuming a maximum use of "wobble" (Crick, 1966). Thus there are many more species of tRNA than appear necessary for protein synthesis. While it is not clear how much overlap there is between aminoacyl-tRNA's involved in protein synthesis per se and in the various "regulatory" functions, the "extra" species may well be involved primarily in regulatory roles.

The view that cell differentiation might be controlled in part by aminoacyl-tRNA is an outgrowth of the hypothesis that the translation of an mRNA is limited by a minor species of aminoacyl-tRNA (modulator tRNA) which is recognized by "modulating triplets" (see Sueoka and Kano-Sueoka, 1970). Changes in relative tRNA levels (i.e., a qualitative or quantitative change in the population of isoaccepting tRNA's for a given amino acid) associated with different conditions of growth or state of development of an organism or tissue have been taken as evidence that tRNA may be involved in developmental regulation. Examples where such changes have been noted are bacterial sporulation, virus infection of bacteria, tumor induction, response to hormones, tissue differentiation, and tissue maturation. Some differences in relative tRNA levels have been noted in plant tissues in going from more meristematic to more mature or fully developed tissues (see, e.g., Vold and Sypherd, 1968; Legocki and Wojciechowska, 1970; Bick *et al.*, 1970; Vanderhoef and Key, 1970). In a more detailed study of tRNA's of cotton using different tissues and different states of development (Merrick and Dure, 1972) only minor differences, which relate primarily to the contribution of chloroplast tRNA's, in total amino acid acceptor activity for the 20 amino acids were noted (Table VII). Additionally, no significant differences were observed in the relative levels of the isoaccepting tRNA's for a given amino acid. It should be pointed out that the methods used in the studies with cotton would not detect base modifications, e.g., alkylation and methylation, which often used methods might detect. It was thus concluded that developmental changes in cotton seedlings are not accompanied by changes in tRNA and thus in the capacity for translating individual code words.

A most interesting aspect of the work with cotton tRNA's relates to the development of the chloroplasts and their associated tRNA's

TABLE VII

Level of Acylation of Cotton tRNA with each Amino Acid[a,b]

Amino acid	Young embryo cotyle-dons	Dry seed cotyle-dons	Roots	Green cotyle-dons	Etiolated cotyle-dons	Chloro-plasts
			Source of tRNA			
Alanine	5.1	5.1	5.0	4.8	4.7	4.7
Arginine	9.0	8.7	8.7	9.3	9.3	9.5
Asparagine	1.4	2.5	2.5	1.3	1.4	2.1
Aspartic Acid	6.8	6.5	6.6	6.1	6.0	5.0
Cysteine	0.8	0.7	0.8	0.7	0.8	0.9
Glutamine	0.2	0.2	0.2	0.2	0.2	0.2
Glutamic Acid	2.0	2.1	2.3	2.2	2.3	3.0
Glycine	10.0	10.1	10.3	9.7	9.6	8.6
Histidine	3.4	3.3	3.3	3.7	3.7	3.6
Isoleucine	3.2	3.2	3.3	3.3	3.4	4.4
Leucine	10.0	9.8	10.2	11.2	11.0	11.1
Lysine	5.2	5.3	5.5	3.6	3.7	4.0
Methionine	3.5	3.3	3.5	4.7	4.7	5.8
Phenylalanine	4.6	4.6	4.7	5.0	5.1	6.7
Proline	4.6	4.6	4.4	4.1	4.0	3.2
Serine	3.4	3.2	1.3	3.8	3.7	4.1
Threonine	5.6	5.6	5.7	5.4	5.5	4.7
Tryptophan	2.1	1.9	1.9	2.3	2.3	1.9
Tyrosine	2.7	2.7	2.7	2.7	2.7	3.0
Valine	9.0	8.9	9.0	8.8	8.6	6.6
	92.6	92.3	91.9	92.9	92.7	93.1

[a] Data from Merrick and Dure (1972).

[b] Numbers are percent of total tRNA acylated by each amino acid.

during germination and greening. It is well known from early studies on *Neurospora* mitochondria (see, e.g., Barnett and Brown, 1967), *Euglena* chloroplasts (see e.g., Barnett *et al.*, 1969), and bean leaves (see e.g., Burkard *et al.*, 1970) that organelles contain a population of tRNA's that is different from the cytoplasm. The species of tRNA which were unique to cotton cotyledon chloroplasts were present, although in very small amounts, in roots and in embryo cotyledons which contain a small population of proplastids (Merrick and Dure, 1972). During germination of the cotton seed the proportion of the tRNA's unique to the chloroplast increased about sevenfold, while the level of cytoplasmic tRNA per cotyledon remained constant (there is no cell division in the cotyledons during germination). About 35% of the leucine-tRNA of cotton cotyledons was of chloroplast origin, in agreement with the level of bean leaf leucyl-tRNA, which hybridized specifically to chloroplast DNA (Williams and Williams, 1970). Additionally, the number of tRNA molecules per ribosome remained constant at 14 to 15 both in the cytoplasm and in chloroplasts during germination. The increase in chloroplast tRNA during germination occurred independent of light. This "preprogrammed" pattern of development of chloroplast tRNA's of cotton contrasts to the light-dependent development (environmentally induced development) of at least some of the chloroplast tRNA's of *Euglena* (Barnett *et al.*, 1969).

Another interesting feature of the work on plant tRNA's relates to methionine-tRNA's. Following the demonstration that formylmethionyl-tRNA is the initiator aminoacyl-tRNA for protein synthesis in bacterial systems, several studies have indicated that methionyl-tRNA is an initiator aminoacyl-tRNA in eukaryotic systems. In plants, one of two cytoplasmic methionyl-tRNA's of wheat (Leis and Keller, 1970) inserts methionine primarily into the N-terminal position (initiator tRNA), while the other inserts methionine primarily internally in the growing polypeptide chain (Marcus *et al.*, 1970). The methionine is apparently not formylated or otherwise modified on the initiator tRNA in this system (see Chapter 16 for details on initiator tRNA's and protein synthesis).

REFERENCES

Adesnik, M., Salditt, M., Thomas, W., and Darnell, J. E. (1972). *J. Mol. Biol.* **71**, 21.

Adman, R., Shultz, L. D., and Hall, B. D. (1972). *Proc. Nat. Acad. Sci. U.S.* **69**, 1702.

Altman, S., and Smith, J. D. (1971). *Nature (London), New Biol.* **233**, 35.

Astrachan, L., and Volkin, E. (1958). *Biochim. Biophys. Acta* **29**, 536.

Barnett, W. E., and Brown, D. H. (1967). *Proc. Nat. Acad. Sci. U.S.* **57**, 452.

Barnett, W. E., Pennington, C. J., Jr., and Fairfield, S. A. (1969). *Proc. Nat. Acad. Sci. U.S.* **63**, 1261.

Bastia, D., Chiang, K. S., Swift, H., and Siersma, P. (1971). *Proc. Nat. Acad. Sci. U.S.* **68**, 1157.

Becker, A., and Hurwitz, J. (1971). *Progr. Nucl. Acid Res. Mol. Biol.* **11**, 423.

Bendich, A. J., and Anderson, R. S. (1974). *Proc. Nat. Acad. Sci. U.S.* **71**, 1511.

Bendich, A. J., and McCarthy, B. J. (1970a). *Genetics* **65**, 567.

Bendich, A. J., and McCarthy, B. J. (1970b). *Proc. Nat. Acad. Sci. U.S.* **65**, 349.

Bick, M. D., Liebke, H., Cherry, J. H., and Strehler, B. (1970). *Biochim. Biophys. Acta* **204**, 175.

Bird, A., Rogers, E., and Birnstiel, M. L. (1973). *Nature (London), New Biol.* **242**, 226.

Birnstiel, M. L., Speirs, J., Purdom, I., Jones, K., and Loening, U. E. (1968). *Nature (London)* **219**, 454.

Birnstiel, M. L., Chipchase, M., and Speirs, J. (1971). *Progr. Nucl. Acid Res. Mol. Biol.* **11**, 351.

Bishop, J. O., Rosbash, M., and Evans, D. (1974). *J. Mol. Biol.* **85**, 75.

Blatti, S. P., Ingles, C. J., Lindell, T. J., Morris, P. W., Weaver, R. F., Weinberg, F., and Rutter, W. J. (1970). *Cold Spring Harbor Symp. Quant. Biol.* **35**, 649.

Bottomley, W., Smith, H. J., and Bogorad, L. (1971). *Proc. Nat. Acad. Sci. U.S.* **68**, 2412.

Brandhorst, B. P., and McConkey, E. H. (1974). *J. Mol. Biol.* **85**, 451.

Brawerman, G. (1974). *Annu. Rev. Biochem.* **43**, 621.

Britten, R. J., and Davidson, E. (1969). *Science* **165**, 349.

Britten, R. J., and Davidson, E. (1971). *Quart. Rev. Biol.* **46**, 111.

Britten, R. J., and Kohne, D. E. (1968). *Science* **161**, 529.

Brown, D. D., and Dawid, I. B. (1968). *Science* **160**, 272.

Brown, D. D., and Weber, C. S. (1968). *J. Mol. Biol.* **34**, 661 and 681.

Bottomley, W., Smith, H. J., and Boborad, L. (1971). *Proc. Nat. Acad. Sci. U.S.* **68**, 3175.

Brown, D. D., Wensink, P. C., and Jordan, E. (1972). *J. Mol. Biol.* **63**, 57.

Brown, R. D., and Haselkorn, R. (1971). *J. Mol. Biol.* **59**, 491.

Brutlag, D., Schekman, R., and Kornberg, A. (1971). *Proc. Nat. Acad. Sci. U.S.* **68**, 2826.

Burdon, R. H. (1971). *Progr. Nucl. Acid Res. Mol. Biol.* **11**, 33.

Burgess, B., and Burgess, R. R. (1974). *Proc. Nat. Acad. Sci. U.S.* **71**, 1174.

Burgess, R. R. (1971). *Annu. Rev. Biochem.* **40**, 711.

Burkard, G., and Keller, E. B. (1974). *Proc. Nat. Acad. Sci. U.S.* **71**, 389.

Burkard, G., Guillemaut, P., and Weil, J. H. (1970). *Biochim. Biophys. Acta* **224**, 184.

Butterworth, H. W., Cox, R. F., and Chesterton, C. J. (1971). *Eur. J. Biochem.* **23**, 229.

Campbell, J. L., Soll, L., and Richardson, C. C. (1972). *Proc. Nat. Acad. Sci. U.S.* **69**, 2090.

Chamberlin, M. (1970). *Cold Spring Harbor Symp. Quant. Biol.* **35**, 851.

Chamberlin, M. (1974). *Annu. Rev. Biochem.* **43**, 721.

Chambers, R. W. (1971). *Progr. Nucl. Acid Res. Mol. Biol.* **11**, 489.

Chambon, P. (1975). *Annu. Rev. Biochem.* **44**, 613.

Chen, Y. M., Lin, C. Y., Chang, H., Guilfoyle, T. J., and Key, J. L. (1975). *Plant Physiol.* **56**, 781.

Chroboczek, H., and Cherry, J. H. (1966). *J. Mol. Biol.* **19**, 28.

Cochet-Meilhac, M., and Chambon, P. (1974). *Biochim. Biophys. Acta* **353**, 160.

Cooper, P. K., and Hanawalt, P. C. (1972). *Proc. Nat. Acad. Sci. U.S.* **69**, 1156.

Cramer, F. (1971). *Progr. Nucl. Acid Res. Mol. Biol.* **11**, 391.

Crick, F. (1966). *J. Mol. Biol.* **19**, 548.

Crippa, M., and Tocchini-Valentini, G. P. (1971). *Proc. Nat. Acad. Sci. U.S.* **68**, 2769.

Darnell, J. E., Jelinek, W., and Molloy, G. (1973). *Science* **181**, 1215.

Darnell, J. E., (1968). *Bacteriol. Rev.* **32**, 262.

Davidson, E., and Britten, R. J. (1973). *Quart. Rev. Biol.* **48**, 565.

Davidson, E., Hough, B., Amenson, C., and Britten, R. J. (1973). *J. Mol. Biol.* **77**, 1.

de Crombrugghe, B., Chen, B., Anderson, W., Nissley, P., Gattesman, M., Pastan, I., and Perlman, R. (1971). *Nature (London), New Biol.* **231**, 139.

de Lucia, P., and Cairns, J. (1969). *Nature (London)* **224**, 1164.

Diez, J., and Brawerman, G. (1974). *Proc. Nat. Acad. Sci. U.S.* **71**, 4091.

Duda, C. L., and Cherry, J. H. (1971). *J. Biol. Chem.* **246**, 2487.

Edlin, G., and Broda, P. (1968). *Bacteriol. Rev.* **32**, 206.

Edmonds, M., and Abrams, R. (1960). *J. Biol. Chem.* **235**, 1142.

Ewing, E., and Cherry, J. H. (1967). *Phytochemistry* **6**, 1319.

Ficq, A., and Brachet, J. (1971). *Proc. Nat. Acad. Sci. U.S.* **68**, 2774.

Firtel, R., and Lodish, H. (1973). *J. Mol. Biol.* **79**, 295.

Franze-Fernandez, M. L., and Pogo, A. O. (1971). *Proc. Nat. Acad. Sci. U.S.* **68**, 3040.

Gallo, R. C., and Pestka, S. (1970). *J. Mol. Biol.* **52**, 195.

Gauss, D. H., von der Haar, F., Maelicke, A., and Cramer, F. (1971). *Annu. Rev. Biochem.* **40**, 1045.

Gefter, M. L., Hirota, Y., Kornberg, T., Wechsler, J. A., and Barnoux, C. (1971). *Proc. Nat. Acad. Sci. U.S.* **68**, 3150.

Ghosh, K., and Ghosh, H. P. (1972). *J. Biol. Chem.* **247**, 3369.

Gissinger, F., Kedinger, C., and Chambon, P. (1974). *Biochimie* **56**, 319.

Goebel, W. (1972). *Nature (London), New Biol.* **237**, 67.

Goulian, M. A., Kornberg, A., and Sinsheimer, R. L. (1967). *Proc. Nat. Acad. Sci. U.S.* **58**, 2321.

Graham, D., Neufeld, B., Davidson, E., and Britten, R. J. (1974). *Cell.* **1**, 127.

Grierson, D., and Loening, U. E. (1972). *Nature (London), New Biol.* **235**, 80.

Grierson, D., Rogers, M. E., Sartirana, M. L., and Loening, U. E. (1970). *Cold Spring Harbor Symp. Quant. Biol.* **35**, 589.

Guilfoyle, T. J., Lin, C. Y., Chen, Y. M., Nagao, R. T., and Key, J. L. (1975). *Proc. Nat. Acad. Sci. U.S.* **75**, 69.

Guilfoyle, T. J., Lin, C. Y., Chen, Y. M., and Key, J. L. (1976). *Biochim. Biophys. Acta* **418**, 344.

Haff, L. A., and Keller, E. B. (1973). *Biochem. Biophys. Res. Commun.* **51**, 704.

Hall, R. H. (1970). *Progr. Nucl. Acid Res. Mol. Biol.* **10**, 57.

Higgins, T., Mercer, J., and Goodman, P. (1973). *Nature (London), New Biol.* **246**, 68.

Holley, R., Apgar, J., Everett, G. A., Madison, J. T., Marquisee, M., Merrill, S. H., Penswick, M. R., and Zamir, A. (1965). *Science* **147**, 1462.

Horgen, P., and Griffin, D. P. (1971). *Proc. Nat. Acad. Sci. U.S.* **68**, 338.

Horgen, P. A., and Key, J. L. (1973). *Biochim. Biophys. Acta* **294**, 227.

Howell, S. H., and Hecht, N. B. (1971). *Biochim. Biophys. Acta* **240**, 343.

Huang, R. C., and Bonner, J. (1962). *Proc. Nat. Acad. Sci. U.S.* **48**, 1216.

Ingle, J., and Sinclair, J. (1972). *Nature (London)* **235**, 30.

Ingle, J., Key, J. L., and Holm, R. E. (1965). *J. Mol. Biol.* **11**, 730.

Ingle, J., Wells, R., Possingham, J. V., and Leaver, C. J. (1971). *In* "Autonomy and Biogenesis of Mitochondria and Chloroplasts" (N. K. Boardman, A. W. Linnane, and R. M. Smillie, eds.), p. 393. North-Holland Publ., Amsterdam.

Ingle, J., Pearson, G., and Sinclair, J. (1973). *Nature (London), New Biol.* **242**, 193.

Iwabuchi, M., Mizukami, Y., and Sameshima, M. (1971). *Biochim. Biophys. Acta* **228**, 693.

Jacob, S. (1973). *Progr. Nucl. Acid Res. Mol. Biol.* **13**, 93.

Jacobson, A., Firtel, R., and Lodish, H. (1974). *Proc. Nat. Acad. Sci. U.S.* **71**, 1607.

Jacobson, R. B. (1971). *Nature (London), New Biol.* **231**, 17.

Jelinek, W. R. (1974). *Cell* **2**, 197.

Jendrisak, J., and Becker, W. (1974). *Biochem. J.* **139**, 771.

Johri, M. M., and Varner, J. E. (1970). *Plant Physiol.* **45**, 348.
Kang, C. Y., and Temin, H. M. (1972). *Proc. Nat. Acad. Sci. U.S.* **69**, 1550.
Karkas, J. D. (1972). *Proc. Nat. Acad. Sci. U.S.* **69**, 2288.
Kedinger, C., Gissinger, F., and Chambon, P. (1974). *Eur. J. Biochem.* **44**, 421.
Keller, W. (1972). *Proc. Nat. Acad. Sci. U.S.* **69**, 1560.
Kelly, R. B., Atkinson, M. R., Huberman, J. A., and Kornberg, A. (1969). *Nature (London)* **224**, 495.
Kessler, B. (1971). *Biochim. Biophys. Acta* **240**, 330.
Key, J. L. (1966). *Plant Physiol.* **41**, 1257.
Key, J. L., and Ingle, J. (1968). *In* "Biochemistry and Physiology of Plant Growth Substances" (F. Wightman and G. Setterfield, eds.), p. 711. Runge Press, Ottawa.
Key, J. L., and Silflow, C. (1975). *Plant Physiol.* **56**, 364.
Key, J. L., Leaver, C. J., Cowles, J. R., and Anderson, J. M. (1972). *Plant Physiol.* **49**, 783.
Klein, A., and Bonhoeffer, F. (1972). *Annu. Rev. Biochem.* **41**, 301.
Knippers, R., and Strätling, W. (1970). *Nature (London)* **226**, 713.
Kochert, G. (1971). *Arch. Biochem. Biophys.* **147**, 318.
Kohne, D. E. (1970). *Quart. Rev. Biophys.* **3**, 327.
Kornberg, A. (1969). *Science* **163**, 1410.
Kornberg, T., and Gefter, M. L. (1971). *Proc. Nat. Acad. Sci. U.S.* **68**, 761.
Kumar, A., and Warner, J. R. (1972). *J. Mol. Biol.* **63**, 233.
Leaver, C. J., and Key, J. L. (1970). *J. Mol. Biol.* **49**, 671.
Legocki, A., and Wojciechowska, K. (1970). *Bull. Acad. Pol. Sci.* **18**, 7.
Leis, J. P., and Keller, E. B. (1970). *Biochem. Biophys. Res. Commun.* **40**, 416.
Lewis, J. A., and Ames, B. N. (1972). *J. Mol. Biol.* **66**, 131.
Lin, C. Y., Guilfoyle, T. J., Chen, Y. M., Nagao, R., and Key, J. L. (1974). *Biochem. Biophys. Res. Commun.* **60**, 498.
Lindell, T. J., Weinberg, F., Morris, P., Roeder, R. G., and Rutter, W. J. (1970). *Science* **170**, 447.
Loening, U. E. (1965). *Biochem. J.* **97**, 125.
Loening, U. E. (1968). *Annu. Rev. Plant Physiol.* **19**, 37.
Losick, R., Shorenstein, R. G., and Sonenshein, A. L. (1970). *Nature (London)* **227**, 910.
Maden, B. E. H., Salim, M., and Summers, D. F. (1972). *Nature (London), New Biol.* **237**, 5.
Manahan, C., App, A., and Still, C. (1973). *Biochem. Biophys. Res. Commun.* **53**, 588.
Mandel, J. L., and Chambon, P. (1974). *Eur. J. Biochem.* **41**, 379.
Manning, J. E., Wolstenholme, D. R., Ryan, R. S., Hunter, J. A., and Richards, O. C. (1971). *Proc. Nat. Acad. Sci. U.S.* **68**, 1169.
Manning, J. E., Wolstenholme, D. R., and Richards, O. C. (1972). *J. Cell Biol.* **53**, 594.
Mans, R. J. (1971). *Biochem. Biophys. Res. Commun.* **45**, 980.
Mans, R. J., and Walter, T. J. (1971). *Biochim. Biophys. Acta* **247**, 113.
Marcus, A., Weeks, D. P., Leis, J. P., and Keller, E. B. (1970). *Proc. Nat. Acad. Sci. U.S.* **67**, 168.
Matsuda, K., and Siegel, A. (1967). *Proc. Nat. Acad. Sci. U.S.* **58**, 673.
Merrick, W., and Dure, L. S., III. (1972). *J. Biol. Chem.* **247**, 7988.
Messelson, M., and Stahl, F. (1958). *Proc. Nat. Acad. Sci. U.S.* **44**, 461.
Milcarek, C., Price, R., and Penman, S. (1974). *Cell* **3**, 1.

Mullinix, K., Strain, G., and Bogorad, L. (1973). *Proc. Nat. Acad. Sci. U.S.* **70**, 2386.

Nakazato, H., Edmonds, M., and Kopp, D. (1974). *Proc. Nat. Acad. Sci. U.S.* **71**, 200.

Nemer, M., Graham, M., and Dubroff, L. M. (1974). *J. Mol. Bol.* **89**, 435.

Novelli, G. D. (1969). *J. Cell. Comp. Physiol.* **74**, Suppl. 1, 121.

Nüsslein, V., Otto, B., Bonhoeffer, F., and Schaller, H. (1971). *Nature (London), New Biol.* **234**, 285.

Okazaki, R., Arisawa, M., and Sugino, A. (1971). *Proc. Nat. Acad. Sci. U.S.* **68**, 2954.

Payne, P. F., and Dyer, T. (1972). *Nature (London), New Biol.* **235**, 145.

Perry, R. P., Cheng, T. Y., Freed, J. J., Greenberg, J. R., Kelley, D. E., and Tartof, K. D. (1970). *Proc. Nat. Acad. Sci. U.S.* **65**, 609.

Pong, S. S., and Loomis, W. R., Jr. (1973). *J. Biol. Chem.* **248**, 3933.

Reeder, R. H., and Roeder, R. G. (1972). *J. Mol. Biol.* **70**, 431.

Roebertus, J. D., Ladner, J. E., Finch, J. T., Rhodes, D., Brown, R. S., Clark, B. F. C., and Klug, A. (1974). *Nature (London)* **250**, 546.

Roeder, R. G., and Rutter, W. J. (1969). *Nature (London)* **224**, 234.

Roeder, R. G., and Rutter, W. J. (1970). *Proc. Nat. Acad. Sci. U.S.* **65**, 675.

Rogers, M. E., Loening, U. E., and Fraser, R. S. S. (1970). *J. Mol. Biol.* **49**, 681.

Sagher, D., Edelman, M., and Jakob, K. (1974). *Biochim. Biophys. Acta* **349**, 32.

Schwartz, L. B., and Roeder, R. G. (1975). *J. Biol. Chem.* **250**, 3221.

Schwimmer, S. (1966). *Phytochemistry* **5**, 791.

Shenkin, A., and Burdon, R. H. (1974). *J. Mol. Biol.* **85**, 19.

Siddiqui, M. A. Q., and Ofengand, J. (1970). *J. Biol. Chem.* **245**, 4409.

Singer, C. E., Smith, G. R., Cortese, R., and Ames, B. N. (1972). *Nature (London), New Biol.* **238**, 72.

Sklar, V. E. F., Schwartz, L. B., and Roeder, R. G. (1975). *Proc. Nat. Acad. Sci. U.S.* **72**, 348.

Skoog, F., and Armstrong, D. (1970). *Annu. Rev. Plant Physiol.* **21**, 359.

Slater, I., Gillespie, D., and Slater, D. W. (1973). *Proc. Nat. Acad. Sci. U.S.* **70**, 406.

Spencer, D., and Whitfeld, P. R. (1969). *Arch. Biochem. Biophys.* **132**, 477.

Stout, E. R., and Arens, M. Q. (1970). *Biochim. Biophys. Acta* **213**, 90.

Stout, E. R., and Mans, R. J. (1967). *Biochim. Biophys. Acta* **134**, 32.

Stout, E. R., and Mans, R. J. (1968). *Plant Physiol.* **43**, 405.

Strain, G. C., Mullinix, K. P., and Bogorad, L. (1971). *Proc. Nat. Acad. Sci. U.S.* **68**, 2647.

Stutz, E. (1970). *FEBS (Fed. Eur. Biochem. Soc.) Lett.* **8**, 25.

Sueoka, N., and Kano-Sueoka, T. (1970). *Progr. Nucl. Acid Res. Mol. Biol.* **10**, 231.

Sugina, A., and Okazaki, R. (1973). *Proc. Nat. Acad. Sci. U.S.* **70**, 88.

Sugino, A., Hirose, S., and Okazaki, R. (1972). *Proc. Nat. Acad. Sci. U.S.* **69**, 1863.

Sugiura, M., Okamoto, T., and Takanami, M. (1970). *Nature (London)* **225**, 598.

Takats, S. L., and Weaver, G. H. (1971). *Exp. Cell Res.* **69**, 25.

Temin, H. M. (1972). *Sci. Amer.* **226**, 25.

Tester, C. F., and Dure, L. S., III. (1967). *Biochemistry* **6**, 2532.

Tewari, K. K., and Wildman, S. G. (1967). *Proc. Nat. Acad. Sci. U.S.* **58**, 689.

Tiollais, P., Galibert, F., and Boiron, M. (1971). *Proc. Nat. Acad. Sci. U.S.* **68**, 1117.

Twu, J. S., and Bretthauer, R. K. (1971). *Biochemistry* **10**, 1576.

Udem, S. A., and Warner, J. R. (1972). *J. Mol. Biol.* **65**, 227.

Van de Walle, C. (1973). *FEBS (Fed. Eur. Biochem. Soc.) Lett.* **34**, 31.

Vanderhoef, L. N., and Key, J. L. (1970). *Plant Physiol.* **46**, 294.

Vanderhoef, L. N., Bohannon, R. F., and Key, J. L. (1970). *Phytochemistry* **9**, 2291.

Vold, B. S., and Sypherd, P. S. (1968). *Proc. Nat. Acad. Sci. U.S.* **59**, 453.

Volkin, E. (1962). *Fed. Proc., Fed. Amer. Soc. Exp. Biol.* **21**, 112.

Walbot, V., Harris, B., and Dure, L. S., III. (1975). *Symp. Soc. Develop. Biol.* **33**, 165.

Ward, D. C., Humphreys, K. C., and Weinstein, I. B. (1972). *Nature (London)* **237**, 499.

Weaver, R. F., Blatti, S. P., and Rutter, W. J. (1971). *Proc. Nat. Acad. Sci. U.S.* **68**, 2994.

Weinberg, R. A. (1973). *Annu. Rev. Biochem.* **42**, 329.

Weinberg, R. A., and Penman, S. (1970). *J. Mol. Biol.* **47**, 169.

Weinmann, R., and Roeder, R. G. (1974). *Proc. Nat. Acad. Sci. U.S.* **71**, 1790.

Wells, R., and Birnstiel, M. L. (1969). *Biochem. J.* **112**, 777.

Wells, R., and Ingle, J. (1970). *Plant Physiol.* **46**, 178.

Wells, R., and Sager, R. (1971). *J. Mol. Biol.* **58**, 611.

Wetmur, J. G., and Davidson, N. (1968). *J. Mol. Biol.* **31**, 349.

Whitfeld, P. R., and Spencer, D. (1968). *Biochim. Biophys. Acta* **157**, 333.

Williams, G. R., and Williams, A. S. (1970). *Biochem. Biophys. Res. Commun.* **39**, 858.

Williams, L. S., and Neidhardt, F. C. (1969). *J. Mol. Biol.* **43**, 529.

Wilson, D., and Thomas, C., Jr. (1974). *J. Mol. Biol.* **84**, 115.

Wilt, F. H. (1973). *Proc. Nat. Acad. Sci. U.S.* **70**, 2345.

Winters, M. A., and Edmonds, M. (1973). *J. Biol. Chem.* **248**, 4756.

Zylber, E., and Penman, S. (1971). *Proc. Nat. Acad. Sci. U.S.* **68**, 2861.

16

Protein Biosynthesis

ABRAHAM MARCUS

I. Introduction

The molecular constituents most directly defining the status of a cell are the proteins, a series of polypeptide chains made up of varying amounts of 20 different amino acids. In addition to their primary sequence, protein chains have secondary and tertiary structures that localize hydrophilic and hydrophobic areas and define sites for nucleophilic and electrophilic displacement. These latter features of proteins provide the basis of their unique catalytic activity and their molecular affinity. The functions served by proteins are numerous, ranging from catalysis of enzymatic reactions to participation in structural cellular elements. In some plant tissues, large amounts of stored proteins serve as reserves of energy and nitrogen to be utilized during periods of rapid growth. Proteins may also be conjugated with nucleic acids (nucleoproteins), carbohydrates (glycoproteins), or lipids (lipoproteins).

The processes whereby cells modify their biological development involve both changes in the cellular complement of individual proteins and modification of the activities of preexisting proteins. This chapter concerns itself primarily with the details of the protein synthetic process and

examines possible sites of the regulation of this synthesis. The controls
exerted at the level of protein function are considered in Chapter 10.

II. Transcription

The ultimate determinants of the amino acid sequence of a specific
protein are the nucleotide sequences of the DNA contained in the structural gene corresponding to that protein. The genetic information is transferred via complementary base pairing to RNA (see Chapter 15), which
then serves as the actual template for protein synthesis. This transfer
of genetic information, known as "transcription," is potentially a primary
site for the determination, both qualitative and quantitative, of the proteins to be synthesized. In bacterial cells, specific genetic loci, as well
as regulatory chemical entities (cyclic 3',5'-AMP, repressor proteins),
have been shown to participate in transcriptional control. In plant cells,
less fundamental information is available. In pea pod, the enzyme
phenylalanine ammonia-lyase has been shown to be induced by a variety
of compounds having one feature in common: the capacity to intercalate
with DNA (Hadwiger and Schwochau, 1971). This observation suggests
that induction may be controlled at the level of DNA.

III. Translation

The conversion of the genetic message into the polypeptide chain
is known as "translation." A general scheme describing the process is
shown in Fig. 1. Ribosomes attach to messenger RNA (mRNA) in such
manner as to allow an initiating nucleotide sequence (AUG of GUG)
on the mRNA to correctly direct the binding of an initiating aminoacyl-
tRNA to the ribosome. A second aminoacyl-tRNA directed by the nucleotide triplets at the 3' side of the initiating mRNA sequence is then attached to a "decoding" ribosome site. In all subsequent transitions, this
site serves as the acceptor for incoming aminoacyl-tRNA and is therefore
referred to as the "acceptor" or A site. Following the attachment of both
aminoacyl-tRNA's, a peptide bond is formed in a reaction in which the
carboxyl group of the initiating aminoacyl-tRNA attaches to the amino
group of the first aminoacyl-tRNA. The ribosome then moves a distance
equivalent to 3 nucleotides in the 5' to 3' direction relative to the mRNA.
The initiating tRNA is ejected, and the peptidyl-tRNA is translocated
from the A site to a second ribosomal site referred to as the "P" site.
Functionally, this ribosomal site (the P site) may be considered identical
to that occupied by the initiating aminoacyl-tRNA. The actual physical

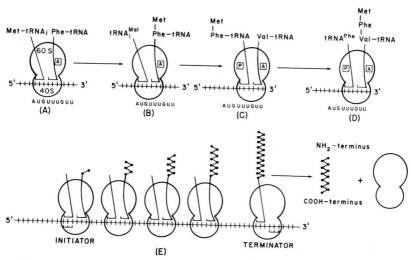

Fig. 1. Schematic model of protein chain biosynthesis. Binding of Met-tRNA$_i$ at the initiator site and Phe-tRNA at the ribosomal "A" site (A) is followed by peptide bond formation (B). Ejection of tRNA$_i^{Met}$, translocation of the peptidyl tRNA from the "A" to the "P" site, and the binding of Val-tRNA then occurs (C) and a new peptide bond is formed (D). This sequence is repeated until a termination codon is reached. The steady-state situation results in the formation of the polyribosome (E).

site occupied by the two species may, however, differ (Thach and Thach, 1971). A new aminoacyl-tRNA, coded for by the incoming mRNA triplet, now attaches to the vacant A site, and the process of peptide formation is repeated. In this manner the mRNA is translated in the 5′ to 3′ direction until a termination triplet is reached, whereupon both the ribosome and the completed polypeptide are released. The functioning complex in which several ribosomes are moving along the mRNA, each elongating a peptide chain, is referred to as a polyribosome (see Chapter 2). A further step in the translational process may be required if a polycistronic mRNA lacks termination signals between its cistrons. Such a situation occurs in HeLa cells translating poliovirus RNA (Jacobson *et al.*, 1970). In this case, a specific proteolytic cleavage of the translational product provides the discrete cellular proteins.

IV. The Genetic Code and Messenger RNA

As noted in Section III, the primary determinants of the amino acid sequence in a protein are the trinucleotide units of the particular mRNA. These trinucleotide codons form antiparallel Watson–Crick hydrogen bonds to an anticodon region in a tRNA that attaches a specific amino acid. A typical illustration is shown in Fig. 2 for wheat germ phenylalanyl-tRNA (Dudock and Katz, 1969). The salient feature of the system is that although the mRNA codon determines the particular amino acid, it does so by way of the tRNA "adapter." Further aspects of the "adapter" concept are elaborated in Section V.

The initial evidence for mRNA recognition by aminoacyl-tRNA was the classical demonstration that an *E. coli in vitro* system programmed with an mRNA consisting entirely of U residues [polyU] synthesized polyphenylalanine from phenylalanyl-tRNA (Nirenberg and Matthei, 1961). Subsequent experiments extended this observation, utilizing more sophisticated synthetic oligonucleotides and analyzing the amino acid

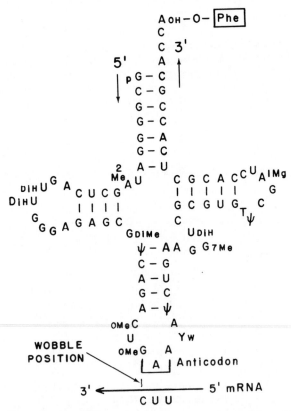

Fig. 2. Strucure of wheat germ tRNA^Phe in the cloverleaf model (Dudock and Katz, 1969). A theoretical mRNA codon is included to show the codon–anticodon interaction as well as the "wobble" position.

sequence of the peptides synthesized in response to these mRNA's (Khorana *et al.*, 1966). Typical data with repeating di- and trinucleotides are shown in Table I. The genetic code clearly utilizes nonoverlapping triplets and contains no punctuation between these triplets. Finally, an assay was developed which allowed a direct examination of the ribosomal binding of specific aminoacyl-tRNA's in response to nucleotide triplets of known sequences (Nirenberg and Leder, 1964). Cumulatively, these studies have resulted in the assignment of each of the triplet codons to particular amino acids (Table II).

A given amino acid may have several isoacceptor tRNA's that differ in their anticodon sequence and therefore respond to different mRNA codons. This observation explains the assignment of several different codons to a given amino acid (see Tables I and II), a phenomenon referred to as "degeneracy." That the existence of multiple isoacceptor tRNA species is not the complete explanation for degeneracy is apparent from the observation that even a single isoacceptor species of an aminoacyl-tRNA may bind to ribosomes in response to several different codons. The latitude of codon variation in this situation (i.e., for a single isoacceptor species) is, however, limited to the third nucleotide of the codon triplet. A theoretical model consistent with this observation has been developed by Crick (1966), and is called the "wobble" hypothesis. The basic concept of the hypothesis is that the nucleotide at the 5' end of the anticodon is not as spatially confined as the other two anticodon bases, and as such may form hydrogen bonds with one of several bases at the 3' end of the mRNA codon (see Fig. 2). Further details of both the genetic code and its degeneracy are described by Nirenberg (1970).

The universality of the genetic code has been extended to plants by

TABLE I

Peptide Synthesis with Synthetic Oligonucleotides[a]

	mRNA	Peptide product
Polymer	5' end → 3' end	NH$_2$ → COOH
(UC)$_n$	UCU CUC UCU CUC	Ser·Leu·Ser·Leu
(AG)$_n$	AGA GAG AGA GAG	Arg·Glu·Arg·Glu
(UG)$_n$	UGU GUG UGU GUG	Val·Cys·Val·Cys
(AC)$_n$	ACA CAC ACA CAC	Thr·His·Thr·His
(UAUC)$_n$	UAU CUA UCU AUC UAU CUA	Thr·Leu·Ser·Ile·Thr·Leu
(UUAC)$_n$	UUA CUU ACU UAC UUA CUU	Leu·Leu·Thr·Tyr·Leu·Leu

[a] After Khorana *et al.* (1966).

TABLE II

The Genetic Code[a]

1st	2nd				3rd
	U	C	A	G	
U	Phe	Ser	Tyr	Cys	U
	Phe	Ser	Tyr	Cys	C
	Leu	Ser	TERM	TERM	A
	Leu	Ser	TERM	Try	G
C	Leu	Pro	His	Arg	U
	Leu	Pro	His	Arg	C
	Leu	Pro	Gln	Arg	A
	Leu	Pro	Gln	Arg	G
A	Ile	Thr	Asn	Ser	U
	Ile	Thr	Asn	Ser	C
	Ile	Thr	Lys	Arg	A
	Met	Thr	Lys	Arg	G
G	Val	Ala	Asp	Gly	U
	Val	Ala	Asp	Gly	C
	Val	Ala	Glu	Gly	A
	Val	Ala	Glu	Gly	G

[a] The table shows the amino acid requirement of all the possible 64 trinucleotide colons. (TERM refers to terminator.) The triplets coding for initiation are AUG and GUG.

the analysis of amino acid replacement in the coat protein of tobacco mosaic virus (TMV) point mutations (Wittman and Wittman-Lebold, 1966), as well as by *in vitro* studies analogous to those carried out with the bacterial system (Basilio *et al.*, 1966). In the case of the TMV mutants, each of the amino acid replacements were completely consistent with a single nucleotide change in the standard codon assignment.

V. Synthesis of Aminoacyl-tRNA

The primary reactants in peptide chain formation are the aminoacyl-tRNA's. As noted in Section IV (see Fig. 2) the tRNA moiety serves as an "adaptor," transferring the nucleotide specificity of the mRNA to the amino acid to which the tRNA is attached. In this interaction, the amino acid itself is totally noninvolved. A direct demonstration of this point was shown by an experiment in which radioactive cysteinyl-tRNA[Cys] was desulfurated with Raney nickel, thereby converting it to Ala-tRNA^{Cys}. The converted product, when incubated in a cell-free system synthesizing hemoglobin, transferred radioactive alanine into positions normally occupied by cysteine. At the same time, no radioactivity was transferred into positions normally containing alanine (von Ehrenstein and Lipmann, 1963).

The reactions leading to the formation of aminoacyl-tRNA are shown in Fig. 3. A synthetase enzyme catalyzes the reaction between ATP and a specific amino acid, forming enzyme-bound aminoacyl adenylate. In the presence of a specific tRNA, the same synthetase enzyme transfers the amino acid to the terminal adenylic acid of the tRNA. The sequence of reactions leading to aminoacyl-tRNA synthesis provide an "activated" complex by utilizing the pyrophosphate bond energy of ATP, at the same time converting the amino acid into a form specifically directed to its placement within the peptide chain. Both the synthetase and the tRNA must have at least two different recognition sites: the synthetase for selecting both the amino acid and the tRNA, and the tRNA for recognizing synthetase and the appropriate messenger codon. In addition, tRNA presumably has a ribosome recognition site, sites for modification of specific bases within its sequence (such as those serving as substrates for the tRNA methylases), and a site for tRNA pyrophosphorylase (an enzyme that reversibly removes the 3'-CCA terminus). Considerable information is available on the primary and the secondary structure of tRNA (see Fig. 2). However, with the exception of the tRNA anti-

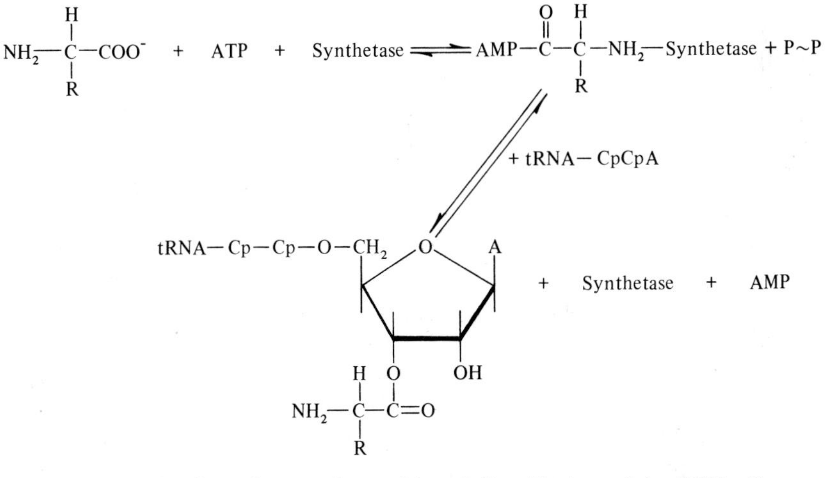

Fig. 3. Activation of an amino acid and its attachment to tRNA. Enzyme-bound aminoacyl adenylate formed by the reaction with ATP transfers the aminoacyl group to the adenylic acid at the 3' terminus of tRNA.

codon and the CCA amino acid attachment site at the 3′ end, little is known of the relation between tRNA structure and its various enzymatic specificities (Zachau, 1972).

In extracts of plant cells, two novel observations have been reported relevant to the formation of aminoacyl-tRNA. The first observation (Kanabus and Cherry, 1971) describes two leucyl-tRNA synthetases and six isoacceptor species of tRNA[Leu] in soybean cotyledons. Of the six tRNA[Leu] isoacceptor species, only two could be charged by one of the synthetases, while the other four isoacceptor species were aminoacylated solely by the second synthetase. Such discrimination by a synthetase within a species is as yet unreported in either mammalian or bacterial systems. The second novel observation (Peterson and Fowden, 1963) relates to a difference in the ability of proline synthetases from several plant species to "activate" the analog, azetidine-2-carboxylic acid. When this analog is supplied to a plant whose synthetase does not discriminate against it, it is incorporated into protein in place of proline and causes cell toxicity. Interestingly, in those species in which azetidinecarboxylic acid is a normal constituent, the synthetase does discriminate against the analog, thereby preventing its incorporation into protein.

VI. Amino Acid Polymerization

A. Initiation

As described in Section III, initiation of protein biosynthesis involves the attachment of ribosomes to mRNA together with the binding of an initiating aminoacyl-tRNA to the appropriate mRNA AUG sequence. Obviously, there are a large number of noninitiating mRNA AUG sequences (internal methionine codons in phase, as well as out of phase codons from an overlap of two sequential triplets). How does the ribosome specifically choose an initiating AUG sequence? There appear to be three possibilities: (1) the initiating sequence in mRNA encompasses a region greater than AUG, with the added nucleotide sequence being recognized either by an initiation factor (see below) or by the ribosome; (2) noninitiating AUG regions of mRNA are "masked" either by proteins or by the secondary structure of the RNA itself; or (3) the mRNA moves through the ribosome from its 5′ end until it reaches the first AUG sequence. This latter model requires that there be no AUG sequences in the mRNA region on the 5′ side of the first initiating codon, and that the ribosome be able to recognize the 5′ end of the mRNA.

With regard to the initiating aminoacyl-tRNA, prokaryotic and eukaryotic cells differ. In bacteria as well as in a number of organelles (chloroplasts and mitochondria) formylmethionyl-tRNA (fMet-tRNA) is the initiator (Marcker and Sanger, 1964), while in mammalian and plant cells a similar function is served by nonformylated Met-tRNA (Smith and Marcker, 1970; Marcus *et al.*, 1970b). Since most proteins do not have either fMet or Met at their N terminus, there must be an enzyme system for subsequent removal of the initiating amino acid. Such an enzyme system has thus far not been convincingly demonstrated in eukaryotes.

All cells, both prokaryotic and eukaryotic, contain at least two species of Met-tRNA. In wheat embryo, as in other eukaryotic cells, only one of these species, Met-tRNA$_i^{Met}$ (Leis and Keller, 1970), transfers Met to the N-terminal position of the peptide chain. The other species, Met-tRNA$_m^{Met}$, transfers Met solely to internal positions (Marcus *et al.*, 1970b). The specificity for Met-tRNA$_i^{Met}$ is determined by the initiation factors (see below), whereas with the internal species, Met-tRNA$_m^{Met}$, it appears that the elongation enzyme, EF1 [whose function is the alignment of incoming aminoacyl-tRNA on the ribosome (see section VI,B)], may provide specificity by discriminating against the initiating species of Met-tRNA, perhaps by possessing a low affinity for this species. Structurally, the only distinction possible between the two eukaryotic Met-tRNA species is in the tRNA moiety, and since both tRNA's recognize only the AUG triplet in mRNA, the distinction made by the initiation and elongation systems must be in an area other than the anticodon.

The specific ribosomal component that attaches to mRNA is the small subunit (30 S in bacterial cells and 40 S in mammalian and plant cells), and in *E. coli* at least, a subunit–mRNA complex can be formed as a *distinct* reaction, to be followed subsequently by the attachment of the initiating aminoacyl-tRNA (Herzberg *et al.*, 1969). Thereafter, there occurs the attachment of the large subunit (50 S in bacterial cells and 60 S in eukaryotes) forming the initiating ribosome–mRNA complex. This component, as well as some of the intermediate products, can be seen on sucrose gradients if either radioactive mRNA or radioactive initiating aminoacyl-tRNA is used in making the complex. The radioactivity, initially present at a position less dense than that of the 80 S ribosomes, migrates to a position in the gradient slightly heavier than that of the ribosomes when it is incorporated into an initiation complex (Iwasaki *et al.*, 1968; Greenshpan and Revel, 1969; Marcus, 1970).

The overall reaction, whereby the initiation complex is formed, requires, in bacterial cells, at least three protein factors as well as GTP

(Salas *et al.*, 1967; Revel *et al.*, 1968). Two of the factors (IF1, IF2) suffice for the ribosome binding of fMet-tRNA in response to AUG. The third factor (IF3) is required only for amino acid polymerization catalyzed by natural mRNA, and it has been suggested that this factor functions in the mRNA–ribosome attachment reaction, perhaps allowing discrimination between various mRNA's (Revel *et al.*, 1970; Heywood, 1970). Factor IF3 serves additionally to catalyze the dissociation of the 70 S bacterial ribosome into subunits (Sabol *et al.*, 1970). Such a reaction has not been found in eukaryotes and may indeed be unique to prokaryotes where a rapid growth rate would require formation of subunits at a rate faster than otherwise available by spontaneous dissociation.

In eukaryotic cells, the general scheme for protein chain initiation is similar to the bacterial process, although some major variations are apparent. As noted earlier, a more sophisticated recognition system must be available for distinguishing between the two Met-tRNA species. In addition, the first peptide bond formed in eukaryotic systems involves the interaction of aminoacyl-tRNA (free α-NH_2 on the amino acyl residue) as the donor moiety rather than peptidyl-tRNA (α-NH_2 is acylated) as in the bacterial system. Such a reaction may distinguish formation of the first peptide bond in the eukaryotic system from the ensuing chain elongation reactions. Experimentally, two eukaryotic systems have been described, one from reticulocytes (Shafritz and Anderson, 1970) and one from wheat embryo (Marcus *et al.*, 1968). In the wheat embryo system, analysis of initiation has been facilitated by the use of the specific inhibitor, aurintricarboxylic acid (ATA) (Marcus *et al.*, 1970a). As shown in Fig. 4, this reagent distinguishes between an amino acid incorporating reaction that requires ribosome–messenger attachment and one that does not. Tobacco mosaic virus RNA-dependent amino acid polymerization is completely inhibited by ATA, whereas polyribosome-catalyzed incorporation is essentially unaffected. With this specificity, an assay is available allowing the resolution of wheat embryo supernatant into two initiation factors and two elongation factors, all of which are required for amino acid incorporation catalyzed by natural plant mRNA (Seal *et al.*, 1972). The two initiation factors distinguish between the two species of Met-tRNA, catalyzing the ribosomal binding, exclusively, of Met-tRNA$_i$. The reactions require ATP as well as GTP. As in the bacterial systems, the detailed mechanisms are as yet unresolved.

B. Chain Elongation

As described in Section III (see also Fig. 1) the elongation reaction involves the codon-directed attachment of aminoacyl-tRNA to the ribo-

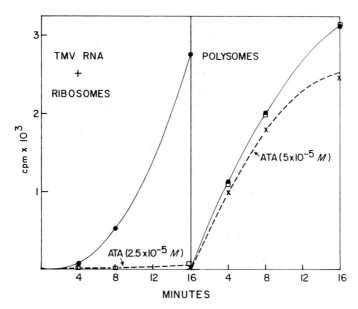

Fig. 4. Inhibition of ribosome–mRNA attachment by aurintricarboxylic acid (ATA) (Marcus *et al.*, 1970a). Polyribosome catalyzed amino acid incorporation where the ribosome–mRNA attachment has already occurred is unaffected by ATA, whereas the TMV RNA-catalyzed reaction is completely inhibited. The incubations contain either no ATA (filled circles), 2.5×10^{-5} *M* ATA (squares), or 5×10^{-5} *M* ATA (crosses).

somal acceptor, or "A" site, followed by peptide bond formation and translocation of the newly formed peptidyl-tRNA to the ribosomal "P" site. Simultaneously, the tRNA previously present at the "P" site is ejected and the ribosome moves, relative to the mRNA, a distance equivalent to three nucleotides. The study of this sequence of reactions has been made possible, by the use of defined synthetic oligonucleotides as mRNA's (Erbe *et al.*, 1969).

The enzymatic aspects of chain elongation have been established primarily by the study of poly(U)-dependent formation of polyphenylalanine. In the poly(U) system, chain elongation occurs just as in the translation of natural mRNA. However, the initial ribosome–mRNA attachment occurs without the need either of initiation factors or of nucleotides. The requirements, therefore, for phenylalanine polymerization in this system are exclusively for chain elongation, and in both mammalian and plant cells two soluble factors are required (Lin *et al.*, 1969; Siler and Moldave, 1969; Legocki and Marcus, 1970). One of these factors, EF1, catalyzes a GTP-dependent codon-directed binding of aminoacyl-tRNA to the ribosome at low concentrations of Mg^{2+}. Presumably an EF1–GTP–aminoacyl-tRNA complex is an intermediate in this reaction (Moon and Weissbach, 1972). In bacterial cells, the binding enzyme is composed of two entities referred to as Tu and Ts. The first reactant in the reaction, component Tu, forms a Tu–GTP–aminoacyl-tRNA complex. Transfer of the aminoacyl-tRNA to the ribosome results in cleavage of the GTP yielding Tu–GDP and P_i (Shorey et al., 1969). The GDP–Tu complex is of such apparent stability that GTP per se cannot replace the GDP so as to begin a new cycle of aminoacyl-tRNA attachment. Factor Ts catalyzes the GDP removal reaction forming a Tu–Ts intermediate in which the Ts component may then be replaced by GTP. Alternatively, the Tu-bound GDP may be phosphorylated in situ, bypassing the Ts requirement (Weissbach et al., 1971).

The second elongation factor is referred to as EF2 in mammalian and plant systems, or G factor in bacteria. The primary reaction catalyzed by the enzyme is a GTP-dependent translocation of peptidyl-tRNA from the "A" site, where it has been formed as a product of peptide bond formation, to the "P" site, where it can now act as a peptidyl donor, attaching its carboxyl onto the amino end of the incoming aminoacyl-tRNA (see Fig. 1). The two most distinctive reactions identifying the "translocase" enzyme are the puromycin reaction (Leder and Bursztyn, 1966) and the ribosome-dependent hydrolysis of GTP (Nishizuka and Lipmann, 1966). Puromycin (Fig. 5) structurally resembles aminoacyl–adenosine and thereby mimics an entering "A" site aminoacyl-tRNA. The basis of its reactivity in the "translocase" reaction is that only amino-

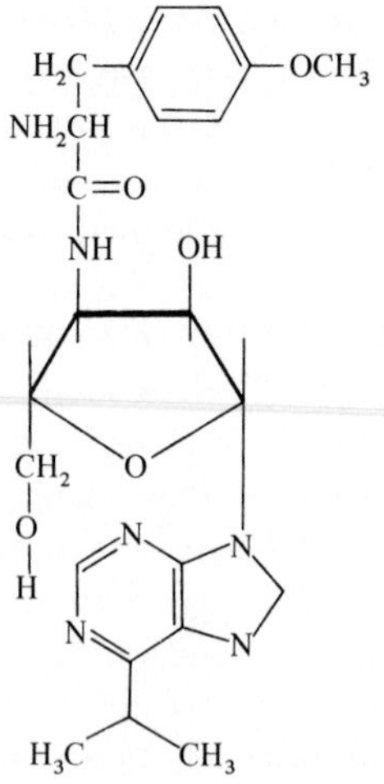

Fig. 5. Structure of puromycin.

acyl-tRNA or peptidyl-tRNA that is in the ribosomal "P" site can react with this reagent (Bretscher and Marcker, 1966). In any system, therefore, that contains peptidyl-tRNA in the "A" site, catalysis of translocation is a requisite for the puromycin reaction.

The "translocase" or EF2 reaction, is of additional interest because, concomitant with the translocation of peptidyl tRNA, two other reactions occur: ejection of tRNA and the movement of the mRNA. Two ribosomal proteins present in the 50 S subunit, L7 and L12, appear to be closely involved in these reactions (Lockwood et al., 1974). Several studies have attempted to dissect the sequence of the overall process. Experiments with the inhibitor bottromycin A2 led to the conclusion that the initial phase of this triphasic sequence is the release of tRNA catalyzed by GTP and the soluble factor and that the subsequent translocations are functions performed primarily by the larger ribosomal subunit (Tanaka et al., 1971). Alternatively, studies with model systems (Tanaka and Kaji, 1972) have suggested that the driving force of the translocation system is the transfer of peptidyl-tRNA from the "A" to the "P" site. Clearly the reaction is complex, and its mechanism is of considerable interest.

The formation of the peptide bond per se is essentially an exchange reaction between an ester and an amino group and probably requires no energy input. The reaction is catalyzed by the peptidyl transferase center on the large ribosomal subunit, and attempts to solubilize the activity have been unsuccessful. In practice, a "fragment" reaction, in which N-acetylated aminoacyl-CCA reacts with puromycin, is used as a model for peptidyl transferase (Monro et al., 1968; Gatica and Allende, 1971). The reaction utilizes only the 60 S subunit, is codon-independent, and requires the presence of 30% ethanol.

VII. Chain Termination

In a chemical sense, the primary reaction in the termination of the peptide chain is a modified transesterification with H_2O acting as the acceptor of the peptidyl chain. As such, at least one of the catalytic components is probably the ribosomal peptidyl transferase. The process has been studied in vitro by an assay in which fMet is released from a ribosome–AUG–fMet-tRNA complex. In bacteria, a terminator codon (UAA, UAG, or UGA) and one of two soluble factors, R1 or R2, each of which recognizes two of the three terminator codons, are required for the reaction. Presumably the appropriate codon directs the binding of the factor, which in turn causes the ribosomal peptidyl transferase to hydrolyze the peptidyl-tRNA. In reticulocytes, an analogous reaction has been obtained except that a tetranucleotide (UAAA, UAGA, or UGAA) is required as the termination codon, and no resolution of the soluble factors is apparent (Beaudet and Caskey, 1971). In plant systems, the reaction has as yet not been studied.

VIII. Formation of Completed Proteins

Upon completion and release from the ribosome, a protein chain assumes a specific three-dimensional structure, unique to the particular protein chain. The most prominent regular arrangement that has been observed by X-ray diffraction analysis is the α-helix, although most proteins contain both helical and nonhelical regions. The primary amino acid sequence of a given protein is the most important determinant of its folding pattern. Generally, a protein will assume a structure consistent with the energetically most favorable positionings of the side chains of its amino acids. This general concept, referred to as the "thermodynamic hypothesis" (Givol et al., 1965), is best supported by the observation that a protein may be denatured to give a randomly oriented polypeptide, often biologically inactive. Yet when allowed to renature, the original (native) conformation may be reformed. A classical illustration is the case of ribonuclease, where reductive denaturation disrupts 4 disulfide intrachain cross-links. During oxidative renaturation any one sulfhydryl group could potentially interact with any of the other 7 sulfhydryl groups. Yet the interaction is essentially specific with the original disulfide links reformed (White, 1961). Such a phenomenon is understandable only if the renatured structure is solely a consequence of specific polypeptide folding. The disulfide bridges might then be viewed primarily as stabilizing components.

An alternative concept would allow that a prosthetic group (e.g., a substrate or a coenzyme) ascertains the folding pattern of a given protein. Several instances have been described in which the presence of such prosthetic groups resulted in differences both in the rates of renaturation and in the actual protein structure achieved (Levi and Kaplan, 1971). Furthermore, even with the simpler disulfide renaturation systems, a microsomal enzyme has been found that catalyzes the rearrangement of disulfide bonds from improper to proper configurations (Givol et al., 1965).

Many proteins are regular aggregates of smaller protein chains, referred to as subunits. In most cases these subunits are held together by the same forces that order the three-dimensional structure of the polypeptide chain, i.e., hydrogen binding, ionic binding, and hydrophobic binding.

The same general principle would appear to be operative in the organization of the more complex structural elements, such as the cell membrane and cell wall. In some cases, however, covalent cross-links (disulfide and amide bonds) may provide the binding forces.

Protein biosynthesis is generally terminated when the polypeptide chain is released from the ribosome. In some instances, however, particularly that of the proteolytic enzymes, the primary biosynthetic product is an inactive proenzyme or zymogen that must be processed further to provide the active protein. This additional step allows further control of the appearance of the final protein. Two interesting illustrations are the formation of cocoonase in the emerging silkworm (Berger *et al.*, 1971) and the formation of collagenase in the metamorphosing tadpole (Harper *et al.*, 1971). Cocoonase (an enzyme that digests sericin, a serine-rich protein normally holding together the silk fibers of the cocoon) is synthesized in a special organ as an inactive "procacoonase," 9 days prior to the onset of moth emergence. Subsequently, this zymogen is transported to the surface of the moth and "activated" by proteolytic digestion. A similar situation occurs with collagenase, except that the precursor is made directly in the target organ, the tail of the tadpole. It remains in an inactive form until metamorphosis, when a proteolytic "activation" occurs and the liberated enzyme then aids in the resorbing of the tail.

In addition to affording another level of control, the requirement of a processing step in protein biosynthesis may also serve either to establish a specific three-dimensional configuration or to change the solubility characteristics so as to facilitate transport through cellular and extracellular spaces. Two such examples are those of insulin (Steiner and Clark, 1968) and collagen (Miller and Matukas, 1974), both of which are synthesized as larger precursors. In the case of insulin, processing of the proinsulin involves excision of an internal fragment resulting finally in a 2-chain product, cross-linked by disulfide bonds. Denaturation and renaturation experiments have shown that the final product, insulin, could not be renatured, whereas the proinsulin precursor when denatured regains full activity upon appropriate renaturation. It, therefore, appears that the function of the precursor is to attain a particular three-dimensional configuration utilizing the total amino acid sequence (particularly the section that is later excised). This configuration, once established, can now be maintained by the disulfide linkages, even in the absence of the full orientation provided by the primary amino acid sequence.

It is of interest that "processing" as a final biosynthetic step occurs also with messenger, ribosomal, and transfer RNA (see Chapter 15), and it may be that here too the advantage gained is the ability to establish a specific three-dimensional configuration.

IX. Regulation

With the advent of detailed information on the specific steps of protein biosynthesis, several possible sites of regulatory control become apparent. Briefly summarized, these include, modification of tRNA and enzymes forming aminoacyl-tRNA, specificity of factors determing attachment of mRNA to ribosomes, and perhaps factors controlling availability of mRNA to the ribosome translational system. One area that has been intensively examined relates to the finding of $N^6(\Delta^2\text{-isopentenyl})$adenine at a position adjacent to the 3′ end of the anticodon in a number of tRNA's (Hall, 1970). This compound is of particular interest for plant scientists because of its very potent cytokinin activity in tobacco callus culture assay. At the biochemical level, structural modification of the Δ^2-isopentenyl group of tRNA either by mutation or by chemical treatment strongly reduces the capacity of the tRNA to bind to ribosomes. Such results suggested that the Δ^2-isopentenyl group might somehow affect translation and as such provide a functional site for cytokinin activity. However, subsequent studies have shown that radioactive cytokinins are not significantly incorporated into tRNA. Moreover, tobacco callus cultures, dependent upon exogenous cytokinin for growth, synthesized cytokinins in tRNA, whether or not they were supplemented with exogenous cytokinins (Kende, 1971). Thus, although the presence of cytokinin in tRNA is an intriguing phenomenon, its significance in relation to growth promotion appears to be doubtful.

GENERAL REFERENCES

Baglioni, C., and Colombo, B. (1970). *In* "Metabolic Pathways" (D. M. Greenberg, ed.), 3rd ed., Vol. 4, pp. 277–351. Academic Press, New York.
Bosch, L. (1972). "Mechanism of Protein Synthesis and its Regulation." Amer. Elsevier, New York.
Boulter, D., Ellis, R. J., and Yarwood, A. (1972). *Biol. Rev. Cambridge Phil. Soc.* **47**, 113–175.
Haselkorn, R., and Rothman-Denes, L. B. (1973). *Annu. Rev. Biochem.* **42**, pp. 397–438.
McConkey, E. H. (1971). "Protein Synthesis," Vol. 1. Dekker, New York.
Watson, J. D. (1970). "Molecular Biology of the Gene." Benjamin, New York.

REFERENCES

Basilio, C., Bravo, M., and Allende, J. E. (1966). *J. Biol. Chem.* **241**, 1917.
Beaudet, A. L., and Caskey, C. T. (1971). *Proc. Nat. Acad. Sci. U.S.* **68**, 619.

Berger, E., Kafatos, F. C., Felsted, R. L., and Law, J. H. (1971). *J. Biol. Chem.* **246**, 4131.

Bretscher, M. S., and Marcker, K. A. (1966). *Nature (London)* **211**, 380.

Crick, F. H. C. (1966). *J. Mol. Biol.* **19**, 548.

Dudock, B. S., and Katz, G. (1969). *J. Biol. Chem.* **244**, 3069.

Erbe, R. W., Nau, M. M., and Leder, P. (1969). *J. Mol. Biol.* **39**, 441.

Gatica, M., and Allende, J. E. (1971). *Biochim. Biophys. Acta* **228**, 732.

Givol, D., DeLorenzo, F., Goldberger, R. F., and Anfinsen, C. B. (1965). *Proc. Nat. Acad. Sci. U.S.* **53**, 676.

Greenshpan, H., and Revel, M. (1969). *Nature (London)* **224**, 331.

Hadwiger, L. A., and Schwochau, M. E. (1971). *Plant Physiol.* **47**, 588.

Hall, R. H. (1970). *Progr. Nucl. Acid Res. Mol. Biol.* **10**, 57.

Harper, E., Bloch, K. J., and Gross, J. (1971). *Biochemistry* **10**, 3035.

Herzberg, M., Lelong, J. C., and Revel, M. (1969). *J. Mol. Biol.* **44**, 297.

Heywood, S. M. (1970). *Proc. Nat. Acad. Sci. U.S.* **67**, 1782.

Iwasaki, K., Sabol, S., Wahba, A. J., and Ochoa, S. (1968). *Arch. Biochem. Biophys.* **125**, 542.

Jacobson, M. F., Asso, J., and Baltimore, D. (1970). *J. Mol. Biol.* **49**, 657.

Kanabus, J., and Cherry, J. H. (1971). *Proc. Nat. Acad. Sci. U.S.* **68**, 873.

Kende, H. (1971). *Int. Rev. Cytol.* **31**, 301.

Khorana, H. G., Buchi, H., Ghosh, H., Gupta, N., Jacob, T. M., Kossel, H., Morgan, R., Narang, S. A., Ohtsuka, E., and Wells, R. D. (1966). *Cold Spring Harbor Symp. Quant. Biol.* **31**, 39.

Leder, P., and Bursztyn, H. (1966). *Biochem. Biophys. Res. Commun.* **25**, 233.

Legocki, A. B., and Marcus, A. (1970). *J. Biol. Chem.* **245**, 2814.

Leis, J. P., and Keller, E. B. (1970). *Biochem. Biophys. Res. Commun.* **41**, 765.

Levi, A. S., and Kaplan, N. O. (1971). *J. Biol. Chem.* **246**, 6409.

Lin, S. Y., McKeehan, W. L., Culp, W., and Hardestry, B. (1969). *J. Biol. Chem.* **244**, 4340.

Lockwood, A. H., Maitra, U., Brot, N., and Weissbach, H. (1974). *J. Biol. Chem.* **249**, 1213.

Marcker, K. A., and Sanger, F. (1964). *J. Mol. Biol.* **8**, 835.

Marcus, A. (1970). *J. Biol. Chem.* **245**, 962.

Marcus, A., Luginbill, B., and Feeley, J. (1968). *Proc. Nat. Acad. Sci. U.S.* **59**, 1243.

Marcus, A., Bewley, J. D., and Weeks, D. P. (1970a). *Science* **27**, 1735.

Marcus, A., Weeks, D. P., Leis, J. P., and Keller, E. B. (1970b). *Proc. Nat. Acad. Sci. U.S.* **67**, 1681.

Miller, E. J., and Matukas, V. J. (1974). *Fed. Proc., Fed. Amer. Soc. Exp. Biol.* **33**, 1197.

Monro, R. E., Cerna, J., and Marcker, K. A. (1968). *Proc. Nat. Acad. Sci. U.S.* **61**, 1042.

Moon, H. M., and Weissbach, H. (1972). *Biochem. Biophys. Res. Commun.* **46**, 254.

Nirenberg, M. (1970). *In* "Aspects of Protein Biosynthesis" (C. B. Anfinsen, Jr., ed.), Part A, pp. 215–241. Academic Press, New York.

Nirenberg, M. and Leder, P. (1964). *Science* **145**, 1399.

Nirenberg, M., and Matthei, J. (1961). *Proc. Nat. Acad. Sci. U.S.* **47**, 1588.

Nishizuka, Y., and Lipmann, F. (1966). *Proc. Nat. Acad. Sci. U.S.* **55**, 212.

Peterson, P. J., and Fowden, L. (1963). *Nature (London)* **200**, 148.

Revel, M., Herzberg, M., Becarevic, A., and Gros, F. (1968). *J. Mol. Biol.* **33**, 231.

Revel, M., Greenshpan, H., and Herzberg, M. (1970). *Eur. J. Biochem.* **16**, 117.

Sabol, S., Sillero, M. A. G., Iwasaki, K., and Ochoa, S. (1970). *Nature (London)* **228**, 1269.

Salas, M., Hille, M. B., Last, J. A., Wahba, A. J., and Ochoa, S. (1967). *Proc. Nat. Acad. Sci. U.S.* **57**, 387.

Seal, S. N., Bewley, J. D. and Marcus, A. (1972). *J. Biol. Chem.* **247**, 2592.

Shafritz, D. A., and Anderson, W. F. (1970). *J. Biol. Chem.* **245**, 5553.

Shorey, R. L., Ravel, J. M., Gardner, C. W., and Shive, W. (1969). *J. Biol. Chem.* **244**, 4555.

Siler, J., and Moldave, K. (1969). *Biochim. Biophys. Acta* **195**, 123.

Smith, A. E., and Marcker, K. A. (1970). *Nature (London)* **226**, 607.

Steiner, D. F., and Clark, J. L. (1968). *Proc. Nat. Acad. Sci. U.S.* **60**, 622.

Tanaka, N., Lin, Y. C., and Okuyama, A. (1971). *Biochem. Biophys. Res. Commun.* **44**, 477.

Tanaka, S., and Kaji, A. (1972). *Biochem. Biophys. Res. Commun.* **46**, 136.

Thach, S. S., and Thach, R. E. (1971). *Proc. Nat. Acad. Sci. U.S.* **68**, 1791.

von Ehrenstein, G., and Lipman, F. (1963). *Proc. Nat. Acad. Sci. U.S.* **49**, 669.

Weissbach, H., Redfield, B., and Brot, N. (1971). *Arch. Biochem. Biophys.* **144**, 244.

White, F. H., Jr. (1961). *J. Biol. Chem.* **236**, 1353.

Wittman, H. G., and Wittman-Liebold, B. (1966). *Cold Spring Harbor Symp. Quant. Biol.* **31**, 163.

Zachau, H. G. (1972). *In* "Mechanism of Protein Synthesis and its Regulation" (L. Bosch, ed.), pp. 173–217. Amer. Elsevier, New York.

17

Amino Acid Biosynthesis and Its Regulation

J. K. BRYAN

I. Introduction

Well over a hundred structurally diverse amino acids have been isolated from plants. Some of these compounds are found only in a few species, whereas others are widely distributed within the plant kingdom. The 18α-amino acids* and 2α-amino acid amides commonly present in proteins are among the amino acids that are universally synthesized by plants. The biosynthesis of these common or protein amino acids is the principal subject of this chapter. During the past few years it has become apparent that plants, as other organisms, possess the capability to regulate the biosynthesis of numerous cellular metabolites. Although our knowledge of the regulation of amino acid biosynthesis in multicellular plants is far less extensive than that which has been achieved in microorganisms, there is little doubt that a number of biosynthetic reaction sequences in plant cells are subject to regulation by their end product

---

* Unless otherwise specified the L-optical configuration of the amino acids discussed in this chapter should be assumed.

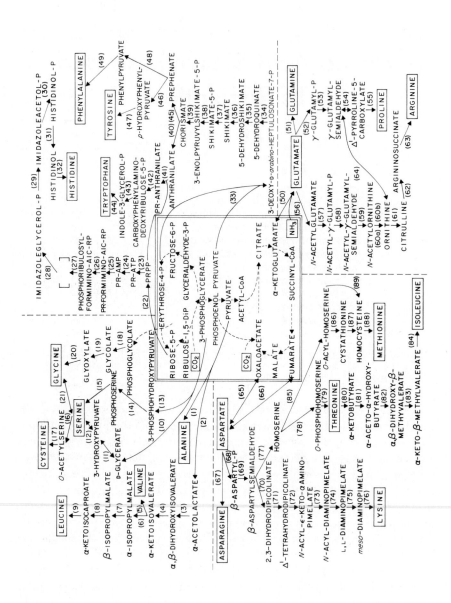

amino acids. Therefore, amino acid synthesis is most appropriately considered in context of those mechanisms that may regulate the flow of carbon and nitrogen through biosynthetic pathways.

The results of many studies with ^{14}C-labeled compounds indicate that the major metabolic pathways outlined in Fig. 1 are present in plant cells. It should, nevertheless, be recognized at the outset of this chapter that the information relating to many details of amino acid biosynthesis in plants is incomplete. Data derived from the study of other types of organisms have been incorporated in Fig. 1 and Table I in an attempt to provide a more comprehensive understanding of the probable pathways in plants. Conversely, details of most of the individual intermediate reactions have been omitted in order to emphasize overall metabolic relationships and regulatory phenomena. Specific limitations in our knowledge of plant metabolism and its regulation as well as alternative pathways for the synthesis of specific amino acids in plants are noted in the appropriate sections.

II. Origin and Mobilization of Amino Acid Precursors

A. Origin of the Amino Nitrogen

Plants may utilize several chemical forms of nitrogen for growth, but only nitrogen reduced to the level of ammonia is incorporated directly into organic substances. Nitrate or nitrite is reduced to ammonia during reactions catalyzed by a series of enzymes including nitrate and nitrite reductases (see Chapter 20). Cellular nitrogen balance is maintained in part by control of both uptake and reduction of nitrogen. An exogenous supply of nitrate generally results in the elevation of the levels of nitrate and nitrite reductases, while substitution of nitrate by a supply of re-

---

Fig. 1. Pathways and intermediates of amino acid biosynthesis. Information obtained directly from investigations of plants has been combined with data resulting from the study of microorganisms in the construction of this figure. The enzyme that catalyzes each numbered reaction is listed in Table I. Central metabolic pathways, such as glycolysis and the tricarboxylic acid cycle, are schematically indicated by the dashed arrows in the enclosed portion of the figure, and each quadrant contains those pathways associated with the synthesis of related amino acids. Not all physiologically reversible reactions are drawn with double-headed arrows, but those that are so indicated are discussed in the text. Abbreviations: PRPP, phosphoribosyl pyrophosphate; PR-ATP; phosphoribosyladenosine 5′-triphosphate; PR-AMP, phosphoribosyladenosine 5′-monophosphate; PR-FORMIMINO-AIC-RP, N-(5′-phospho-D-ribosylformimino)-5-amino-1-(5′-phosphoribosyl)-4-imidazolecarboxamide; CoA, coenzyme A; P, phosphate; [ ], undetermined structure.

TABLE I

Enzymes Associated with Amino Acid Biosynthesis[a,b]

1. Alanine aminotransferase
2. Acetohydroxy acid synthetase
3. α-Acetohydroxy acid isomeroreductase
4. α,β-Dihydroxy acid dehydratase
5. Branched chain aminotransferase
6. α-Isopropylmalate synthetase
7. Isopropylmalate isomerase
8. β-Isopropylmalate dehydrogenase
9. Leucine aminotransferase
10. 3-Phosphoglycerate phosphatase
11. D-Glycerate dehydrogenase
12. Serine aminotransferase
13. 3-Phosphoglycerate dehydrogenase
14. Phosphoserine aminotransferase
15. Phosphoserine phosphatase
16. Serine acetylase
17. O-Acetylserine sulfhydrylase
18. Phosphoglycolate phosphatase
19. Glycolate oxidase
20. Glycine aminotransferase
21. Serine transhydroxymethylase
22. Ribosephosphate pyrophosphokinase
23. Phosphoribosyl-ATP synthetase
24. Phosphoribosyl-ATP pyrophospho-hydrolase
25. Phosphoribosyl-AMP 1,6-cyclo-hydrolase
26. Phosphoribosylformiminoaminoimidazolecarboxamide ribonucleotide ketolisomerase
27. Amidotransferase
28. Cyclase
29. Imidazolylglycerol phosphate dehydratase
30. Imidazolylacetol phosphate aminotransferase
31. Histidinol phosphate phosphatase
32. Histidinol dehydrogenase
33. 3-Deoxy-D-arabino-heptulosate-7-phosphate synthetase
34. 5-Dehydroquinate synthetase
35. 5-Dehydroquinate dehydrogenase
36. Shikimate dehydrogenase
37. Shikimate kinase
38. 3-Enolpyruvylshikimate-5-phosphate synthetase
39. Chorismate synthetase
40. Anthranilate synthetase
41. Anthranilate-5'-phosphoribosyl-1-pyrophosphate phosphoribosyl transferase
42. Phosphoribosylanthranilate isomerase
43. Indoleglycerol phosphate synthetase
44. Tryptophan synthetase
45. Chorismate mutase
46. Prephenate dehydrogenase
47. Tyrosine aminotransferase
48. Prephenate dehydratase
49. Phenylalanine aminotransferase
50. Glutamate dehydrogenase
51. Glutamine synthetase
52. γ-Glutamokinase
53. γ-Glutamatesemialdehyde dehydrogenase
54. None—spontaneous reaction
55. Δ^1-Pyrroline-5-carboxylate reductase
56. N-Acetylglutamate synthetase
57. N-Acetyl-γ-glutamokinase
58. N-Acetylglutamate-γ-semialdehyde dehydrogenase
59. Acetylornithine aminotransferase
60A. Acetylornithinase
60B. Ornithine acetyltransferase
61. Ornithine transcarbamoylase
62. Argininosuccinate synthetase
63. Argininosuccinate lyase
64. Ornithine aminotransferase
65. Aspartate aminotransferase
66. β-Aspartase
67. Asparagine synthetase
68. β-Aspartokinase
69. β-Aspartate semialdehyde dehydrogenase
70. Dihydrodipicolinate synthetase
71. Dihydrodipicolinate reductase
72. Δ^1-Tetrahydrodipicolinate acylase
73. Acyl-diaminopimelate aminotransferase
74. Acyl-diaminopimelate deacylase
75. Diaminopimelate epimerase
76. Diaminopimelate decarboxylase
77. Homoserine dehydrogenase
78. Homoserine kinase
79. Threonine synthetase
80. Threonine dehydratase
81. See reaction (2)
82. See reaction (3)
83. See reaction (4)
84. See reaction (5)
85. Homoserine transacylase
86. Cystathionine-γ-synthetase
87. β-Cystathionase
88. Homocysteine methyltransferase
89. O-Acetylhomoserine sulfhydrylase

[a] Each number preceding enzyme name refers to the corresponding reaction presented in Fig. 1.

[b] Any of a number of comprehensive texts on enzymes should be consulted for formal enzyme nomenclature.

duced nitrogen, such as ammonia or certain amino acids, can result in a significant decrease in the levels of the reductases. These responses are superficially similar to induction and repression phenomena in microorganisms, but full details of the molecular mechanisms in plants remain to be established. For example, posttranscriptional control of the level of nitrate reductase in maize roots has been suggested by Wallace (1973).

Several reactions involving the direct assimilation of ammonia into organic compounds have been described in living systems. These include the reductive amination of keto acids, formation of the amides of glutamate and aspartate, direct amination of fumarate, and the synthesis of carbamoyl phosphate. Enzymes that could facilitate each of these reactions have been reported in various plants, but the reactions do not necessarily contribute equally to the assimilation of ammonia. The following factors must be considered in evaluating the relative contribution of different mechanisms of ammonia assimilation: (a) Demonstration of an enzyme in a cell establishes only the possibility that the reaction that it catalyzes may be metabolically functional. Conversely, the apparent absence or low activity of an enzyme must be viewed with caution due to the inherent difficulties of enzyme isolation and measurement. (b) Reactions that are readily reversible *in vivo*, such as the amination of fumarate, may or may not contribute to the net assimilation of ammonia. (c) Glutamine rather than ammonia may be the preferential substrate of enzymes catalyzing a number of reactions, including the synthesis of asparagine and carbamoyl phosphate. (d) Finally, the ubiquitous presence of aminotransferases in plant cells enables assimilated ammonia to move rapidly into many products. These considerations collectively dictate that the contribution of various mechanisms of ammonia assimilation can only be accurately evaluated by kinetic experiments with $^{15}NH_3$. It has been assumed, primarily on the basis of the careful kinetic experiments of Sims and Folkes with the yeast *Candida utilis*, that the synthesis of glutamate by reductive amination of α-ketoglutarate is the primary mechanism of ammonia assimilation in plant cells. Their revised results (Sims *et al.*, 1968) indicated that 75–80% of the ^{15}N assimilated followed this pathway, 10–12% was incorporated into glutamine, and the remainder utilized in the synthesis of carbamoyl phosphate. Nevertheless, the recent discovery of glutamate synthetase activity in extracts of several higher plants suggests an alternative mechanism of ammonia assimilation (Dougall, 1974; Fowler, *et al.*, 1974; Lea and Miflin, 1974). This involves the coupled activities of glutamine and glutamate synthetases as

$$\text{Glutamate} + NH_3 + ATP \rightleftharpoons \text{glutamine} + ADP + P_i + H_2O$$
$$\frac{\text{Glutamine} + \alpha\text{-ketoglutarate} + (2H) \rightleftharpoons 2 \text{ glutamate}}{NH_3 + ATP + \alpha\text{-ketoglutarate} + (2H) \rightleftharpoons \text{glutamate} + ADP + P_i + H_2O}$$

The net reaction would result in the formation of glutamate at the expense of ATP and reducing equivalents. The physiological reductant(s) may vary since NADH, NADPH, and reduced ferredoxin have proved to be substrates of different glutamate synthetases *in vitro*. In addition, the question of whether asparagine can serve as an alternate N donor in a reaction catalyzed by glutamate synthetase of any higher plant requires further clarification (Miflin and Lea, 1975).

B. Mobilization of the Amino Nitrogen

Once assimilated into an amino group, nitrogen may subsequently be distributed among a variety of metabolites in plant cells during transamination reactions. Transamination, catalyzed by an aminotransferase, is dependent upon the formation of a Schiff base between enzyme-bound pyridoxal phosphate and an amino donor; the derived enzyme–pyridoximine complex reacts in turn with a keto acid to form a new amino acid and to regenerate pyridoxal phosphate. The sum of these partial reactions is therefore

$$R_1 \cdot CH(^+NH_3)COO^- + R_2 \cdot CO \cdot COO^- \rightleftharpoons R_1 \cdot CO \cdot COO^- + R_2 \cdot CH(^+NH_3)COO^-$$

Since some plant aminotransferases are known to have a wide range of substrate specificity and many different keto acids have been found in plants, this general mechanism of nitrogen mobilization can participate in the synthesis of a large variety of amino acids. Considering only the major pathways for synthesis of the protein amino acids, 14 of the 89 reactions presented in Fig. 1 are catalyzed by aminotransferases (Table I). Both particulate and soluble aminotransferases have been isolated from plant cells, but the actual number of different proteins that are associated with the synthesis of the common amino acids has not been estimated.

C. Origin of the Carbon Skeletons

The carbon skeletons of the common amino acids are derived from a very few metabolic intermediates, each of which is associated with a central metabolic pathway. These pathways include reactions of carbon fixation, glycolysis, and the tricarboxylic acid cycle and are indicated, in much abbreviated form, by the enclosed central portion of Fig. 1. General groupings or families of amino acids have been delineated on the basis of their precursors. This is particularly convenient when considering branched pathways where two or more amino acids share a common precursor. Four general divisions of amino acids are indicated in Fig. 1 by the dashed lines which separate four quadrants. These include the pyruvate and serine families (upper left), histidine and the aromatic family (upper right), glutamate family (lower right), and the aspartate family (lower left). It should be emphasized that any such subdivision of metabolic phenomena is artificial and can oversimplify interactions that occur *in vivo*. For example, the nitrogen components of amino acids are not channeled into well-defined metabolic sequences. Furthermore, important interfamily relationships may not be easily recognized when the synthesis of the amino acids of a single family is considered out of the general context of cellular metabolism. Thus, serine is required for tryptophan biosynthesis (44),* aspartate condenses with citrulline to form argininosuccinate (62), and cystathionine formation is dependent upon cysteine (86). This interdependency among amino acids of different families provides one mechanism whereby a balanced synthesis of protein precursors might be approached. Certain regulatory mechanisms that act to integrate many facets of cellular metabolism are not associated with specific biosynthetic pathways. The adenylate energy charge, for example, will affect many cellular processes in all types of organisms (Chapman *et al.*, 1971). There is also evidence that photosynthetic carbon metabolism and nitrogen metabolism are integrated in plants. Ammonia has been observed to stimulate pyruvate synthesis and to curtail sharply the formation of sucrose in *Chlorella* cells, and it has been suggested that reduced pyridine nucleotide generated during photosynthesis is, in part, specifically channeled into reductive amination (Bassham, 1971). In addition, a number of enzymes related to amino acid biosynthesis are present in chloroplasts (Miflin, 1974).

III. Synthesis of Individual Amino Acids

A. Pyruvate and Serine Families

1. ALANINE

In plant cells the principal precursor of alanine, as well as leucine and valine, is pyruvate. This keto acid acts as the amino acceptor in the direct synthesis of alanine by transamination (1). Alanine aminotransferase activity has been identified in a number of different plants, and the enzyme has been partially purified from mung bean shoots (Gam-

---

* Numbers in parentheses used throughout the text refer to a reaction in Fig. 1 and the corresponding enzyme listed in Table I.

borg, 1965). This particular plant enzyme is of interest due to its virtual lack of substrate specificity; of the many potential amino donors tested, only serine, glycine, and threonine failed to serve as a substrate for alanine synthesis *in vitro*. With certain exceptions, enzymes with the potential to catalyze a number of different reactions are unlikely to be subject to highly specific regulatory controls. Many organic acids are strongly compartmentalized in plant cells (see Oaks and Bidwell, 1970). Therefore, alanine synthesis may be regulated simply by the availability of pyruvate.

Two alternate mechanisms of alanine biosynthesis have been established in some organisms. One of these is the reductive amination of pyruvate by a mechanism analogous to that catalyzed by glutamate dehydrogenase. Alanine dehydrogenase activity has been reported in plant cells, but the enzymes in question have not been purified. This is essential in order to provide unequivocal verification of this mechanism of alanine synthesis. Alanine can also be synthesized by β-decarboxylation of aspartate. β-Decarboxylases have been isolated and purified from a number of microorganisms but not from higher plants. The reaction catalyzed by these enzymes is

$$^-OOC \cdot CH_2CH(^+NH_3) \cdot COO^- \rightarrow CH_3CH(^+NH_3) \cdot COO^- + CO_2$$

[^{14}C]Alanine and $^{14}CO_2$ are produced in substantial amounts during the metabolism of [^{14}C]aspartate in some plants, and it would therefore be of interest to ascertain if this is due, at least in part, to β-decarboxylation.

2. VALINE, (ISOLEUCINE), AND LEUCINE

The biosynthesis of valine in both plants and microorganisms involves a series of reactions originating with the formation of α-acetolactate (2) in a thiamine pyrophosphate-dependent condensation of two molecules of pyruvate with the concomitant release of a molecule of CO_2. Subsequent transformations of the reaction product involve a combined isomerization and reduction (3), dehydration (4), and transamination (5) to yield valine. The biosynthesis of isoleucine, even though its precursor, α-ketobutyrate, is derived from aspartate, should be considered briefly in this discussion of the pyruvate family of amino acids. The same type of chemical transformations are involved in the synthesis of both isoleucine and valine [see (81), (82), (83), and (84)]. These parallel reaction sequences are catalyzed by the same enzymes in microorganisms and probably, plants. Although complete purification of these enzymes from plant sources has yet to be achieved, enzyme preparations from plants have been shown to utilize alternate substrates of reactions in the two pathways. For example, activity of the dehydratase (4, 83) has been detected in a variety of plants, and the ratio of the activity with α,β-dihydroxyisovalerate to that with α,β-dihydroxy-β-methylvalerate as alternate substrates does not change significantly during partial purification of the enzyme from *Phaseolus radiatus* (Satyanarayana and Radhakrishnan, 1964) or spinach (Kiritani and Wagner, 1970). Some indication that two forms of the isomeroreductase (3, 82) might be present in *P. radiatus* has been noted (Satyanarayana and Radhakrishnan, 1965). Although the terminal reactions in the synthesis of valine and isoleucine (5, 84) can each be catalyzed by an aminotransferase, an enzyme has been isolated from peas which facilitates the reductive amination of the respective keto acids (Kagan *et al.*, 1970). This could provide an alternative mechanism for the synthesis of these amino acids, but kinetic experiments with $^{15}NH_3$ are required to assess the relative importance of the two biosynthetic mechanisms.

α-Ketoisovalerate, when added to rose tissue culture cells, was a specific competitor of the *in vivo* synthesis of both valine and leucine from [^{14}C]glucose (Dougall and Fulton, 1967b), thus, providing evidence that this keto acid is a precursor of both of these amino acids. Data obtained from *in vitro* experiments with extracts of maize seedlings indicate that this plant contains all of the requisite enzymes for synthesis of leucine from α-ketoisovalerate (6, 7, 8, and 9) (Oaks, 1965c). With [^{14}C]valine as a precursor, addition of unlabeled DL-α-isopropylmalate to sorgum seedlings resulted in a reduction of the specific activity of leucine isolated from protein, but not soluble leucine (Butler and Shen, 1963). These observations support the concept that α-isopropylmalate is an intermediate in the synthesis of leucine and suggest that valine can be readily converted to its keto analog to provide carbon for leucine. The differences between soluble and protein leucine were undoubtedly due to a complex pool structure, a phenomenon that has been studied in some detail in maize (Oaks, 1965a).

Evidence that the biosynthesis of valine, leucine, and isoleucine is regulated by the end product amino acids in plants has been obtained in several laboratories. The results of different studies involving the use of isotopes are not, however, in complete agreement. For example, leucine, valine, and isoleucine regulate their own biosynthesis in barley embryos (Joy and Folkes, 1965) and maize seedlings (Oaks, 1965b), but the results of short-term experiments with rose tissue culture cells suggested that only the synthesis of valine and isoleucine was regulated (Fletcher and Beevers, 1971). When conflicting data are obtained in such *in vivo* experiments, it is difficult to ascertain whether different cells possess different regulatory mechanisms or whether the results are due to differences in the experimental procedures. At least two major problems have been encountered in interpreting the results of isotope competition experiments in plants. First, complex metabolic pool structures may mask regulatory phenomena (Oaks and Bidwell, 1970), and, second, isotopic evidence that may suggest that the synthesis of an amino acid is regulated does not normally distinguish between changes in enzyme activities and alterations in the rates of enzyme synthesis or degradation. There are few examples of critical evidence for the regulation of *de novo* enzyme synthesis in plant cells (in spite of the many reported changes in the apparent levels of specific enzymes) (Filner *et al.*, 1969). Consequently, an understanding of the regulatory mechanisms involved in the synthesis of amino acids in plants has thus far depended heavily on the demonstration of regulated enzymes *in vitro*. The role of metabolic compartmentation in cellular regulation and detailed analysis of apparent changes in enzyme levels are, nonetheless, problems that deserve considerable attention.

The most likely point of control of a metabolic pathway is the first unique reaction in a sequence leading to the synthesis of a specific product. In the case of leucine, this is the reaction catalyzed by α-isopropyl-malate synthetase (6), an enzyme that is specifically inhibited by leucine in extracts of maize roots (Oaks, 1965c). The evidence of *in vitro* inhibition and the results of *in vivo* isotope experiments in the same plant may be taken as support for the existence of an effective mechanism for conserving biosynthetic precursors in the presence of excess end product. The regulation of valine and isoleucine synthesis presents a more complex problem in that the same enzymes are involved in the synthesis of both products. In many microorganisms, this potential difficulty has been largely circumvented by multivalent repression, where each of the end products must be in excess before enzyme *synthesis* is reduced.* Such specific mechanisms affecting enzyme synthesis have not yet been established in higher plants.

Under the appropriate circumstances, synergistic and antagonistic effects of exogenous amino acids on plant growth can be interpreted in terms of specific pathways and their associated regulatory controls. The following specific effects on the growth of barley seedlings were observed: (a) The combination of valine plus leucine was inhibitory when each amino acid was employed at a concentration of 5×10^{-4} M, although

---

* The reader should refer to the general references of Umbarger (1969) and Stadtman (1970) for details and literature citations relating to regulatory phenomena in microorganisms and animals.

neither amino acid was inhibitory alone at this concentration. Each amino acid independently inhibited growth when used at 2×10^{-3} M. (b) Traces of isoleucine, but not leucine, relieved the inhibition due to valine. (c) Traces of both valine and isoleucine were required to relieve inhibition by leucine (Miflin, 1969). Among the possible explanations of these results was a complex set of regulatory controls governing the activity of α-acetohydroxy acid synthetase (2). This enzyme has now been isolated and partially purified from barley seedlings and shown to possess regulatory properties that are consistent with the observed pattern of growth inhibition (Miflin, 1971). The enzyme is maximally inhibited 50–70% by concentrations of either leucine or valine above 10^{-3} M. At low concentrations (5×10^{-5} to 5×10^{-4} M), the combination of leucine and valine is significantly more inhibitory than either amino acid alone. The observed relief of growth inhibition by specific amino acids can be explained by examination of Fig. 2, which diagramatically outlines the effects of the end products on the activities of the pathway enzymes. Excess leucine would result in a limitation of both valine and isoleucine, and growth inhibition could be relieved only by an exogenous supply of these amino acids. On the other hand, valine could be converted to its keto analog by deamination and therefore could supply carbon for the synthesis of leucine. If this occurred, only isoleucine would be required for relief of valine inhibition of growth. As noted above, end product inhibition of the leucine branch of the pathway has been reported in maize, and isoleucine inhibits the conversion of threonine to α-ketobutyrate in a number of plants (see Section III,D,4). These additional controls could also effectively enhance biosynthetic regulation (Fig. 2).

In different microorganisms, quite diverse patterns are associated with the regulation of the synthesis of specific families of amino acids. Whereas the regulatory properties of α-acetohydroxy acid synthetases

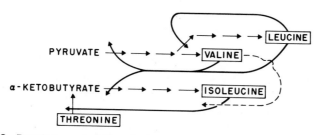

Fig. 2. Regulation of valine, isoleucine, and leucine biosynthesis. The sequential arrows represent enzyme-catalyzed reactions of the detailed pathways presented in Fig. 1, and each bold arrow depicts end product inhibition of the activity of a plant enzyme. The dashed arrow denotes the ability of valine to partially relieve inhibition of threonine dehydratase by isoleucine.

of several plant species appear to be similar to those of the barley enzyme (Miflin and Cave, 1972), diverse regulatory patterns have been demonstrated with other plant enzymes (Aarnes, 1974).

3. SERINE AND GLYCINE

Several pathways of serine synthesis have been identified in plants. This amino acid may be derived from 3-phosphoglycerate via two independent routes. One, termed the phosphorylated pathway, entails oxidation in a NAD⁺-linked reaction (13) and transamination of the product, 3-phosphohydroxypyruvate, to phosphoserine (14). The action of a specific phosphatase then yields serine (15). An alternate pathway involves loss of phosphate during the initial formation of D-glycerate from 3-phosphoglycerate (10), and is, therefore, referred to as the nonphosphorylated pathway. Glycerate is subsequently oxidized (11), and the keto acid product is transaminated (12) during the synthesis of serine according to this scheme. Glycine can be both a product of serine metabolism and a precursor of serine synthesis. With glycine as a precursor, serine transhydroxymethylase catalyzes the addition of a 1-carbon unit (21). In some plant tissues, an enzyme complex appears to facilitate the degradation of glycine to CO_2, NH_3, and N^5,N^{10}-methylenetetrahydrofolate. The folate derivative may then participate in the hydroxymethylation of a second molecule of glycine. These reactions are summarized by the following equation:

$$2 \text{ Glycine} \rightarrow NH_3 + CO_2 + \text{serine}$$

The conversion of glycine to serine with the concomitant evolution of CO_2 has received considerable interest as a major pathway in photorespiration (Kisaki and Tolbert, 1970). Formate and formaldehyde, which can be derived from the glycine precursor glyoxylate, are also potential indirect sources of the 1-carbon unit required for the synthesis of serine from glycine.

Evidence of these various pathways of serine biosynthesis in plants has been obtained in studies of individual enzymes and the distribution of ^{14}C after metabolism of several potential precursors. Glycine is almost certainly the major precursor of serine during photosynthesis of some plants (Rabson et al., 1962; Andrews et al., 1971). The ratio of 3-phosphoglycerate phosphatase (10) to phosphoglycolate phosphatase (18) activity has been shown to vary significantly between different plants (Randall et al., 1971). With the exception of corn leaves, those plants which utilize the C_4-type of photosynthetic metabolism (see Chapter 24) and exhibit negligible photorespiration are characterized by ratios ranging from 2:1 to 4:1. On the other hand, plants with typical C_3 metabolism exhibited ratios of 1:2 to 1:6. One interpretation of these data is that the C_3 plants have a greater capacity of utilizing glycine, derived from glycolate, as a serine precursor than C_4 plants.

In those cases where 3-phosphoglycerate is the precursor of serine, it is more difficult to assess the relative activity of the phosphorylated and nonphosphorylated pathways. It has been suggested, based on the relative levels of several enzyme activities in different tissues, that the phosphorylated pathway is important in rapidly proliferating plant cells, whereas the nonphosphorylated pathway may be more active in leaf tissue (Cheung et al., 1968). Although this suggestion is consistent with the levels of 3-phosphoglycerate phosphatase in C_3 plants noted above, both serine and glycine can be metabolized to sucrose in photosynthetic tissues—a process that most likely involves reversal of two of the reactions of the nonphosphorylated pathway (12), (11) and reentry into the mainstream of carbohydrate metabolism by phosphorylation mediated by glycerate kinase. Under these circumstances, this sequence of reactions would be gluconeogenic.

Glycine may be derived from serine with the simultaneous production of a 1-carbon unit (21). This process is a reversal of the serine transhydroxymethylase-catalyzed reaction noted above. Since 1-carbon units are utilized in a number of metabolic reactions, this pathway is likely to be of considerable importance. Substantial amounts of phosphoglycolate and glycolate can be generated during photosynthesis, and therefore these compounds provide an important alternate source of glycine precursors in plants. The mechanism responsible for the synthesis of phosphoglycolate may involve the reactions of O_2 with ribulose diphosphate. Its subsequent metabolism to glycolate is mediated by phosphoglycolate phosphatase (18), which is localized in chloroplasts. Glycolate can, in turn, be oxidized to glyoxylate (19) in peroxisomes isolated from photosynthetic tissues. This reaction, catalyzed by glycolate oxidase, requires the direct participation of oxygen and yields hydrogen peroxide as a byproduct. Glycine aminotransferase activity (20) is also associated with peroxisomes. At least as judged by the results of in vitro experiments, the transamination of glyoxylate appears to be largely unidirectional. The in vivo activity of this pathway during photosynthesis is supported by the observations that glycolate and glycine (and serine) can be uniformly labeled from $^{14}CO_2$ assimilated during periods when the carboxyl group of 3-phosphoglycerate remains labeled. Alternate mechanisms of glycine synthesis, including the demethylation of sarcosine and the conversion of threonine to glycine and acetaldehyde, have been reported in animal tissues. Such possibilities do not appear to have been examined in plant tissues.

Of the multiple pathways for the synthesis of serine and glycine only the phosphorylated one is known to be subject to end product regulation. Serine has been shown to inhibit a 3-phosphoglycerate dehydrogenase isolated from peas (Slaughter and Davies, 1968). Inhibition of this reaction has also been reported in microorganisms, whereas regulatory control of this pathway in animal cells is restricted to inhibition of phosphoserine phosphatase (15) by the reaction product, serine. Other factors that could influence the metabolism of glycine and serine in plant cells may be less specific. For example, the metabolism of these amino acids can be closely associated with photosynthesis and would therefore be influenced by those factors that affect photosynthetic activity. Since various biosynthetic enzymes are distributed among different subcellular organelles in plant cells (see Tolbert, 1971), it is possible that significant regulation of the metabolism of glycine and serine may be achieved by control of the rate and extent of the transport of intermediates and products of the pathways between various reaction sites.

4. CYSTEINE

[^{14}C]Serine is metabolized to [^{14}C]cysteine in cells of higher plants. The pathway in plants, like that of microorganisms, includes two reactions and the intermediate formation of O-acetylserine. Serine acetylase (16) activity has been demonstrated in extracts of a number of different plants (Smith and Thompson, 1969; Smith, 1972). Synthesis of cysteine from the acetylated intermediate involves the direct incorporation of sulfide and the release of acetate (17). O-Acetylserine was 60 times more effective as a precursor than serine when preparations of spinach were employed to catalyze the synthesis of cysteine from $Na_2{}^{35}S$ (Giovanelli and Mudd, 1967). Although it has been demonstrated that methylmercaptan or ethylmercaptan can replace sulfide in similar reactions catalyzed by enzymes from spinach (Giovanelli and Mudd, 1968) or turnip leaves (Thompson and Moore, 1968), the role of the products, S-methylcysteine or S-ethylcysteine, in plant metabolism is not clear.

Sulfide is synthesized from sulfate by a series of reactions involving activation and reduction in microorganisms and algae (see Chapter 19). Regulation of these pathways constitutes one of the major sites whereby the synthesis of sulfur amino acids may be controlled. No evidence of regulatory control of the first enzyme associated with sulfate reduction was obtained during experiments with several different plants (Ellis, 1969). Although such negative data are necessarily inconclusive, the uptake of sulfate, in higher plants, as contrasted to its reduction, may be regulated by specific amino acids. Results obtained with cultured tobacco cells suggested that sulfate assimilation was inhibited by sulfur amino acids (Hart and Filner, 1969).

B. Histidine and the Aromatic Family

1. HISTIDINE

The pathway of histidine biosynthesis has been elucidated in several microorganisms with the most extensive investigations being carried out with *Salmonella typhimurium* (Martin *et al.*, 1971). In these organisms, phosphoribosyl pyrophosphate, derived from ribose 5-phosphate and ATP (22), is condensed with ATP (23) and subsequently metabolized to imidazoleglycerol phosphate in reactions involving both cyclic and non-cyclic intermediates (24, 25, 26, 27, and 28). Imidazoleglycerol phosphate is dehydrated (29), and the product of this reaction is transaminated to form histidinol phosphate (30). The final two steps of the pathway involve the action of a specific phosphatase (31) and a complex NAD-linked oxidation reaction during which the α-carboxyl moiety is formed from $-CH_2OH$ (32). Until recently the evidence for this series of reactions in higher plant cells was minimal. Histidinol was initially shown to be an intermediate of histidine biosynthesis during isotope competition experiments with rose tissue culture cells (Dougall and Fulton, 1967b). However, activities of phosphoribosyl-ATP synthetase (23), imidazole-glycerol phosphate dehydratase (29), and histidinolphosphate phosphatase (31) have now been reported in extracts prepared from shoots of barley, oats, and peas (Wiater *et al.*, 1971). In addition, imidazoleglycerol was shown to accumulate in cells of rose tissue culture when they were treated with 3-amino-1,2,4-triazole (Davies, 1971). At least one effect of this herbicide is to inhibit imidazoleglycerolphosphate dehydratase (29). Although a considerable number of details remain to be established, it is reasonable to conclude that the pathway of histidine biosynthesis in higher plants is similar to that in microorganisms.

The results of isotopic competition studies with rose cells (Dougall, 1965) and barley seedlings (Joy and Folkes, 1965) indicate that histidine biosynthesis is subject to some form of product regulation. One mechanism that could account for the data obtained in the isotope experiments is inhibition of the activity of the first enzyme unique to the pathway of histidine biosynthesis. L-Histidine has been shown to inhibit this activity (23) in extracts of both oats and peas (Wiater *et al.*, 1971). These workers also attempted to determine whether the presence of triazole

during the germination of barley and oats would alter the level of imidazoleglycerolphosphate dehydratase in a fashion that could be interpreted
as derepression. Although plant growth was affected by the herbicide,
no evidence of repression or derepression was observed.

2. TRYPTOPHAN

Many aspects of aromatic amino acid synthesis and metabolism in
higher plants have been investigated, and a comprehensive review of the
early work has been published (Yoshida, 1969). The existence of the
multibranched pathway in plants (Fig. 1) is generally supported by the
results of isotope experiments, identification of specific intermediates, and
the isolation of a number of pathway enzymes. Nevertheless, some doubt
exists concerning specific portions of the pathway. For example, enzymes
catalyzing the four initial reactions of the linear prechorismate pathway,
(33–39), have been identified in plant extracts, but the nature of the reactions involved in the conversion of shikimate to chorismate in plants is
unknown. Unsuccessful attempts to isolate shikimate kinase (37) and
3-enolpyruvylshikimate-5-phosphate synthetase (38) from tobacco, pea,
and mung bean plants have been reported (Berlyn et al., 1970). Although
it is entirely possible that these enzymes are quite labile in plant extracts,
alternate pathways of chorismate synthesis should not be overlooked.
Chorismate is the substrate of two enzymes that act to define separate
branches of the metabolic pathway leading to tryptophan or to phenylalanine and tyrosine, respectively.

Shikimate, anthranilate, indoleglycerol phosphate, and indole are all
precursors of tryptophan in tobacco cells (Delmer and Mills, 1968a).
D-Tryptophan proved to be effective in reducing the endogenous biosynthesis of tryptophan from [^{14}C]sucrose. It has been suggested that this
unexpected result may be due to the presence of a racemase in tobacco
cells (Miura and Mills, 1971). The first reaction of the tryptophan branch
of the pathway involves addition of the amide nitrogen of glutamine to
carbon-2 of the chorismate ring structure and liberation of the pyruvyl
side chain from the adjacent carbon-3 (40). This complex reaction is
catalyzed by anthranilate synthetase. Each of the anthranilate synthetases, isolated from a wide variety of organisms, which has been examined
to date is extremely sensitive to inhibition by the pathway product tryptophan. The enzyme isolated from tobacco cells, for example, was reported to have an apparent K_i for tryptophan of 2.7×10^{-6} M (Belser
et al., 1971). Several investigations of tobacco anthranilate synthetase
are particularly relevant to the question of enzyme repression in plants.
No evidence of repression was observed when tryptophan was added to

tobacco cell cultures, even though it was established that the intracellular concentration of tryptophan was elevated (Widholm, 1971) and that the endogenous synthesis of tryptophan was reduced significantly (Belser *et al.*, 1971). Unfortunately, these data do not completely rule out the possibility that intracellular tryptophan concentrations specifically affect the synthesis or degradation of anthranilate synthetase in tobacco, since the enzyme may have been fully repressed or exogenous tryptophan may not be accessible to the site of repressor synthesis.

One or more of the enzymes that catalyze subsequent reactions in the pathway are commonly complexed with anthranilate synthetases in bacteria, although the nature and composition of these enzyme aggregates is highly variable from species to species. The activities of the enzymes (41, 42, 43) catalyzing the intermediate reactions have been measured in extracts of carrot cells (Widholm, 1973), but it is not known whether enzymes of the tryptophan pathway also exist as aggregates in plant cells. Several plant tryptophan synthetases, which catalyze the final reaction in the pathway (44), have, however, been studied. These include those derived from blue-green and green algae (Sakaguchi, 1970), tobacco (Delmer and Mills, 1968b), and peas (Chen and Boll, 1971, 1972). Although the plant enzymes are not inhibited by tryptophan, they are structurally similar to microbial tryptophan synthetases. In all cases, the enzymes can be dissociated into two distinct protein components, termed A and B, respectively. Both the A and B components are required for catalysis of the physiologically important reaction:

$$\text{Indoleglycerol phosphate} + \text{serine} \rightleftharpoons \text{tryptophan} + \text{glyceraldehyde 3-phosphate}$$

The B component alone can catalyze a direct condensation between indole and serine to form tryptophan, and the A component can catalyze the formation of indole and glyceraldehyde 3-phosphate from indoleglycerol phosphate. One of the interesting features of tryptophan synthetases is the apparent lack of significant evolutionary changes in structure. Thus, the results of *in vitro* complementation and antibody neutralization studies have demonstrated a significant degree of homology between the proteins isolated from plants and those isolated from microorganisms (Delmer and Mills, 1968b).

3. TYROSINE AND PHENYLALANINE

An alternate sequence of reactions leading from the branchpoint intermediate, chorismate, involves the formation of prephenate (45) and its subsequent metabolism *via* two independent pathways to form tyrosine (46, 47) or phenylalanine (48, 49) in microorganisms. Both prephenate dehydrogenase (46) and prephenate dehydratase (48) have been isolated from plants. Demonstration of these enzymes and aminotransferases, which are active with aromatic substrates, provides evidence for this bifurcated pathway in plants. Even though plant enzymes that can facilitate the hydroxylation of phenylalanine have been reported, there is considerable doubt concerning the synthesis of tyrosine from phenylalanine in plants (Davies, 1968). However, the presence of enzymes that catalize an alternate sequence of reactions leading from prephenate to tyrosine has been noted in tissues of higher plants (Jensen and Pierson, 1975). This pathway involves the transamination of prephenate to form pretyrosine, followed by a reductive decarboxylation that yields tyrosine.

Independent regulation of the synthesis of tyrosine and phenylalanine is achieved in a number of bacteria by the existence of two chorismate mutases (45)—one that is inhibited and repressed by tyrosine and complexed with prephenate dehydrogenase and one that is inhibited and repressed by phenylalanine and complexed with prephenate dehydratase. These enzyme aggregates provide a mechanism whereby prephenate can be channeled into separate reaction sequences. In contrast, fungi, several species of algae (Weber and Böck, 1969), and peas (Cotton and Gibson, 1968) appear to possess a single chorismate mutase that is inhibited by either tyrosine or phenylalanine. The synthesis of these amino acids is integrated with the synthesis of tryptophan in that the latter amino acid both activates the single chorismate mutase and prevents its inhibition by either of the other aromatic amino acids. Multiple chorismate mutases have been identified in several plants. One of the enzymes, resolved from mung bean seedlings, CM-1, resembles the previously described plant enzymes with respect to its regulatory properties (Gilchrist et al., 1972; Gilchrist and Kosuge, 1974). The activity of the other mutase, CM-2 was not affected by tryptophan, tyrosine, paraaminobenzoate, parahydroxybenzoate, serine, or indole-3-acetic acid. The collective properties of the three forms of chorismate mutase present in alfalfa (Woodin and Nishioka, 1973) suggest that the flow of carbon into amino acids and phenolic compounds could be effectively regulated by the concentration of various metabolites. The fact that no evidence of repression of chorismate mutase was detected in several plant tissue culture systems (Chu and Widholm, 1972) also tends to emphasize the importance of enzyme inhibition in the regulation of amino acid biosynthesis in higher plants. The end product-sensitive controls that could aid in the regulation of aromatic amino acid biosynthesis in plants are diagrammed in Fig. 3.

It should be noted that many of the diverse patterns of regulation of these pathways in microorganisms encompass regulation of the pre-

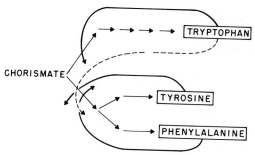

Fig. 3. Regulation of the synthesis of aromatic amino acids in plants. Each of the sequential arrows represents an enzyme-catalyzed reaction of the detailed pathways presented in Fig. 1. Bold arrows depict enzyme inhibition, and the dashed arrow symbolizes the ability of tryptophan to both activate chorismate mutase and prevent inhibition of this enzyme by either tyrosine or phenylalanine.

chorismate pathway. Potential regulatory properties of the first enzyme in this sequence were not reported when the enzyme was identified in plant extracts (Minamikawa, 1967), but highly purified enzyme from cauliflower was not inhibited by chorismate, phenylalanine, tyrosine, or tryptophan (Huisman and Kosuge, 1974). However, the enzyme that catalyzes the fourth reaction in the pathway, shikimate dehydrogenase, was obtained from pea seedlings and shown to be inhibited by a variety of organic compounds including anthranilate (Rothe, 1974). In view of the many aromatic compounds that are synthesized by higher plants, the pathways involved in the synthesis of the aromatic amino acids may be subject to a number of as yet undetected complex regulatory controls.

C. Glutamate Family

1. GLUTAMATE

The potential significance of glutamate biosynthesis with respect to ammonia assimilation was discussed in Section II,A. The reaction involved in this process is catalyzed by glutamate dehydrogenase (50) and can be written as

$$\alpha\text{-Ketoglutarate} + NH_3 + \begin{matrix} NADPH \\ \text{or} \\ NADH \end{matrix} \rightleftharpoons \text{glutamate} + H_2O + \begin{matrix} NADP^+ \\ \text{or} \\ NAD^+ \end{matrix}$$

Although the pH-independent equilibrium constant favors the synthesis of glutamate, the reaction is physiologically reversible and can constitute an important catabolic process. Many plants also contain glutamate decarboxylase, an enzyme that facilitates the synthesis of γ-aminobuty-rate. Under certain circumstances γ-aminobutyrate can also be metabolized to intermediates of the tricarboxylic acid cycle and can thus contribute to the catabolism of glutamate (Inatomi and Slaughter, 1971). The properties of glutamate dehydrogenase(s) isolated from different organisms differ markedly and have been studied extensively in an attempt to understand the regulation of the amphibolic reaction catalyzed by this enzyme (Goldin and Frieden, 1971). Microorganisms are generally characterized by the ability to synthesize NAD- and NADP-specific glutamate dehydrogenases, while mammals possess a single enzyme with dual coenzyme specificity. The mammalian enzymes are subject to regulation by a complex set of allosteric interactions that include those promoting an alteration in substrate specificity.

The intracellular concentrations of certain microbial glutamate dehydrogenases are regulated in such a manner as to suggest that the NAD-dependent enzymes are involved in catabolism (derepressed by glutamate) and that the NADP-dependent enzymes are important in assimilation of nitrogen (induced by NH_3). Although it is tempting to draw similar conclusions on the basis of reported changes in enzyme levels in higher plants, such conclusions are frequently difficult to document. The apparent induction of an NADH-linked glutamate dehydrogenase by nitrate or ammonia in *Lemna* is a relevant example of some of the difficulties that can be encountered. Even though the increase in enzyme activity was proportional to inducer concentration, careful analysis of the properties of the enzyme suggested that the apparent induction was not due to an increase in enzyme protein, but rather to a decrease in the sensitivity of the enzyme to inhibition by EDTA in the extraction media (Joy, 1971). Yet reexamination of this problem under different conditions led Shepard and Thurman (1973) to conclude that a glutamate dehydrogenase of *Lemna* was subject to induction.

Both photosynthetic and nonphotosynthetic plant cells have been reported to contain NAD- or NADP-specific glutamate dehydrogenases. Since very few of these enzymes have been purified and well characterized, the extent to which plants may possess enzymes with single or dual coenzyme specificity is not known. This problem is further complicated by reports of isozymes of glutamate dehydrogenase in plant tissues. The glutamate dehydrogenase purified to homogeneity from pea roots was reported to catalyze reductive amination with either NADPH or NADH as the coenzyme (Pahlich and Joy, 1971). This enzyme appears to be complex, as deaminating activity was only detectable with NAD^+ and the ratio of reaction rates with different coenzymes could be altered by various treatments of the enzyme. It has also been suggested in a preliminary report that the pea enzyme is activated by sulfate ions and subject to regulation by purine nucleotides (Pahlich, 1971). Since earlier work suggested that plant glutamate dehydrogenases do not exhibit the complex regulatory properties of mammalian enzymes, further studies are required before this question can be resolved. A unidirectional activation of an NADH-linked glutamate dehydrogenase, isolated and purified 250-fold from soy bean cotyledons, has been reported (King and Wu, 1971). Reductive amination catalyzed by this enzyme is stimulated up to 33% by glutamate, alanine, and/or aspartate. Glutamate was the least effective activator due to its action as a product inhibitor of the reaction at elevated concentrations. The reverse reaction was not stimulated by alanine or aspartate. Control of pea mitochondrial glutamate dehydrogenase by changes in the ratio of NAD to NADP has also been suggested (Davies and Teixiera, 1975). Data from a large number of different fungi suggest that both the specific regulatory properties and the coenzyme specificity of glutamate dehydrogenases are correlated with the taxonomic classification of these organisms (Lé John, 1971).

2. GLUTAMINE

The synthesis of glutamine is catalyzed by glutamine synthetase (51) and provides a route for ammonia assimilation. Glutamine is also required for the synthesis of a number of cellular metabolites, including several amino acids. For example, the amide nitrogen is incorporated into histidine (27), tryptophan (40), and arginine (via formation of carbamyl phosphate and its contribution to the synthesis of citrulline). Glutamine synthetases isolated from several organisms require Mg^{2+} or Mn^{2+} for catalysis of the following reaction:

$$\text{Glutamate} + NH_3 + ATP \rightarrow \text{glutamine} + ADP + P_i + H_2O$$

The active biosynthetic role of glutamine is reflected in the regulation of this enzyme activity. Most of the enzymes that have been examined are subject to partial inhibition by a variety of compounds that can be considered to be end products of glutamine metabolism. In those cases where the effects of each inhibitor are independent, the mechanism has been termed cumulative feedback inhibition. Detailed investigations of the glutamine synthetase of *E. coli* led to the discovery that the sensitivity of the enzyme to the allosteric effectors was subject to alteration by another regulatory mechanism involving adenylation of the enzyme (see Stadtman, 1970). A highly purified glutamine synthetase has been obtained from peas (cited in Tate and Meister, 1971). The properties of the different enzymes that were revealed during the comparative studies of Tate and Meister suggested that the pea enzyme possessed certain characteristics in common with microbial or animal glutamine synthetases but that it was not identical with the enzymes from either type of organism. The enzymes prepared from pea, rat liver and bovine brain were composed of fewer subunits than the microbial enzymes and were not subject to adenylation under the conditions tested. The enzymes from pea and microorganisms, but not the animal enzymes, were inhibited by AMP. All of the enzymes except pea glutamine synthetase were significantly inhibited by carbamyl phosphate. Subject to the qualification that the particular ion (Mg^{2+} or Mn^{2+}) employed in the reaction mixture could profoundly influence the interaction with allosteric effectors, the pea enzyme was also inhibited to some extent by glycine, alanine, tryptophan, histidine, and glucosamine 6-phosphate. The extent of cumulative effects of these inhibitors on the activity of the plant enzyme was not reported, but cumulative effects of metabolites on a partially purified enzyme from rice roots have been described (Kanamori and Matsumoto, 1972). The most probable form of regulation of the pea enzyme is control by energy charge (O'Neal and Joy, 1975). In green tissues, a significant proportion of the total glutamine synthetase activity appears to be associated with chloroplasts.

3. PROLINE AND HYDROXYPROLINE

Glutamate is a precursor of both proline and arginine in plants as evidenced by the results of a number of studies concerning the metabolism of [^{14}C]glutamate. Nevertheless, the initial reactions involved in the synthesis of these amino acids are not well established in plants. The major difficulty with respect to proline biosynthesis revolves around the mechanism(s) utilized for the synthesis of γ-glutamylsemialdehyde. The pathway outlined in Fig. 1 has been postulated for microorganisms (Baich, 1971). These reactions involve activation of the γ-carbonyl group of glutamate (52) and its subsequent reduction to the semialdehyde with release of the activating moiety, presumably phosphate (53). A similar pathway has been described for the synthesis of β-aspartylsemialdehyde. Ornithine, an arginine precursor and a product of arginine catabolism in those plants that possess an active arginase, can also be converted to γ-glutamylsemialdehyde in the presence of ornithine aminotransferase (64). This enzyme is known to be present in cells of several plants. Another route for the synthesis of γ-glutamylsemialdehyde is the deacetylation of N-acetyl-γ-glutamylsemialdehyde. The latter compound is considered to be an intermediate in the synthesis of arginine (see Section

III,C,4). If nonacetylated intermediates were specifically utilized for the synthesis of proline and acetylated intermediates utilized for the synthesis of arginine, the carbon derived from glutamate that is destined for proline could be isolated from that destined for arginine. Synthesis of both acetylated and nonacetylated semialdehyde compounds has been demonstrated in discs of swiss chard leaves (Morris *et al.*, 1969). However, neither compound could be established as the preferential precursor of proline biosynthesis. γ-Glutamylsemialdehyde can undergo spontaneous cyclization to form Δ^1-pyrroline-5-carboxylate (54), which is then reduced to proline in a pyridine nucleotide-linked reaction. A reductase (55) that catalyzes this reaction has been demonstrated in plants.

The synthesis of proline in cells of maize roots is subject to end product regulation (Oaks *et al.*, 1970). Even though specific regulatory mechanisms have not been established, the results of the experiments with maize are of interest. The synthesis of soluble proline from [^{14}C]acetate in root tips was inhibited nearly 50% in the presence of 10^{-4} M proline. When the identical experiment was performed with mature segments of roots, no inhibition was observed. In contrast, the synthesis of protein-bound proline was inhibited more than 90% in the root tips and 38% in the mature segments. One explanation of these results is that metabolic pools change during development, such that the accessibility of exogenous proline to the site of proline synthesis is not constant. Specific developmental changes in regulated enzymes, such as desensitization to end product inhibitors, may also occur. In either case, the possibility that regulation of amino acid synthesis may be altered as cells mature should be considered.

Hydroxyproline is found in a variety of plant proteins and is important in cell wall biosynthesis (see Chapter 13). Synthesis of this amino acid differs significantly from the synthesis of the other protein amino acids in that proline is hydroxylated after incorporation into peptide linkage. Oxygen is required for this reaction, and the product is 4-*trans*-L-hydroxyproline (Lamport, 1964). Evidence suggests that hydroxylation occurs after release of the polypeptides from the ribosomes and that the enzyme is analogous to a mixed function oxidase (Sadava and Chrispeels, 1971a,b).

4. ARGININE

It is possible to synthesize arginine from γ-glutamylsemialdehyde, the precursor of proline, but evidence suggests that the acetylated derivatives of glutamate play a predominant role in the synthesis of this basic amino acid. Initially it was found that N-acetylornithine was an effective competitor of arginine synthesis from [^{14}C]glucose in rose tissue culture cells and that extracts of these cells contained acetylornithine aminotransferase activity (59) (Dougall and Fulton, 1967a). Subsequently, evidence of the ability of plants to convert N-acetylglutamate to N-acetylglutamylsemialdehyde was obtained (Morris et al., 1969), and two reactions whereby N-acetylglutamate could be synthesized were demonstrated in extracts of radish leaves and *Chlorella* (Morris and Thompson, 1971). The acetyl moiety can be derived directly from acetyl coenzyme A and lost during the synthesis of ornithine according to the following two reactions:

$$S\text{-Acetyl coenzyme A} + \text{glutamate} \rightarrow N\text{-acetylglutamate} + \text{SH–coenzyme A} \quad (56)$$
$$N\text{-Acetylornithine} + \text{H}_2\text{O} \rightarrow \text{ornithine} + \text{acetate} \quad (60a)$$

Alternatively, a transacetylation may occur in which the acetyl group is recycled to glutamate:

$$N\text{-Acetylornithine} + \text{glutamate} \rightleftharpoons \text{ornithine} + N\text{-acetylglutamate} \quad (60b)$$

Either of these mechanisms may be characteristic of a given microorganism. The algae that have been studied possess enzymes that catalyze the more efficient cyclic process, and it was reported, without data, that the transacetylase system was also present in spinach leaves (Staub and Dénes, 1966). Although the available data provide strong support for the existence of the acetylated pathway in some, if not all, plants, the specific mechanisms utilized for the synthesis of N-acetylglutamylsemialdehyde (57, 58) and the distribution of alternate mechanisms of ornithine biosynthesis in different plants should be examined for further verification of the pathways.

The sequence of reactions involved in the synthesis of arginine from ornithine is well established in higher plants. Each of the appropriate enzymes, ornithine transcarbamoylase (61) (Kleczkowski and Cohen, 1964), argininosuccinate synthetase (62) (Shargool, 1971), and argininosuccinate lyase (63) (Rosenthal and Naylor, 1969) have been isolated and purified from higher plants. In addition, the results of both direct and isotope competition labeling experiments are consistent with the intermediate production of citrulline and argininosuccinate. It should be noted that an unusual pathway of arginine synthesis involving carbamoylaspartate and ornithine as intermediates has been identified but not examined in detail (Kleczkowski and Grabarek-Bralczyk, 1968).

In those microorganisms that utilize the transacetylase enzyme, regulatory control is exerted at the level of N-acetyl-γ-glutamyl phosphate synthesis (57); while in those organisms that possess an active acetylornithinase (60a), control is exerted at the level of N-acetylglutamate synthesis (56). Ornithine, citrulline, and arginine have been reported to inhibit the latter reaction in plant extracts (Morris and Thompson, 1971). An enzyme responsible for the synthesis of carbamoyl phosphate, which contributes to the synthesis of citrulline (61), has been isolated from pea seedlings (O'Neal and Naylor, 1968). This enzyme was inhibited by a number of metabolites including ornithine. Given the multiple sites for interaction of biosynthetic and degradative metabolism of arginine and proline in plants, it is likely that these pathways are regulated by several types of mechanisms including subcellular compartmentation (Oaks and Bidwell, 1970). For example, soybean argininosuccinate synthetase (62) responds to energy charge and the end product arginine in a manner that is characteristic of biosynthetic enzymes (Shargool, 1973a). The activity of this enzyme is reduced as energy charge is decreased and as the concentration of arginine is increased. There is also an indication that argininosuccinate lyase (63) is subject to inactivation *in vivo* in cultured soybean cells (Shargool, 1973b).

D. Aspartate Family

1. ASPARATE

Aspartate is normally considered to be synthesized by transamination of oxalacetate with glutamate acting as the primary amino donor. This reaction is catalyzed by aspartate aminotransferase (65), an enzyme that appears to be ubiquitous in plant cells. Oxalacetate could potentially be reductively aminated in a reaction similar to that catalyzed by glutamate dehydrogenase. Although aspartate dehydrogenase activity has been reported in plant cells, the enzyme has yet to be purified and characterized. A third potential mechanism of aspartate synthesis is the direct amination of fumarate with ammonia (66). With either oxalacetate or fumarate as the precursor, aspartate metabolism could be closely associated with the tricarboxylic acid cycle (mitochondria), dark CO_2 fixation (soluble phase), or photosynthetic CO_2 fixation in C_4 plants (chloroplasts). These obvious compartments plus the possibility of second level compartmentation, such as the existence of enzyme complexes, suggest that different biosynthetic mechanisms of aspartate formation could be utilized within the same cell. Since all of the reactions potentially involved in aspartate synthesis are reversible, it is also possible that some of these reactions may facilitate transport of amino and organic acids between cellular compartments rather than contribute to the net synthesis of aspartate.

2. ASPARAGINE

The amide asparagine is synthesized and accumulates to quite high concentrations in many plant cells. The results of experiments utilizing a variety of ^{14}C-labeled compounds indicate that aspartate is readily converted to asparagine, and it has been assumed that a direct conversion is catalyzed by asparagine synthetase (67). This enzyme catalyzes the following reaction in microorganisms:

$$\text{Aspartate} + NH_3 + ATP \rightarrow \text{asparagine} + AMP + PP_i$$

Early reports of asparagine synthetase activity in plant cells suggested that the plant enzyme catalyzed a different reaction in which the products were asparagine, ADP, and phosphate. However, the suggestion that the asparagine synthetase of certain plants such as yellow lupine seedlings (Rognes, 1970) and soybean cotyledons (Steeter, 1973) catalyzes a reaction analogous to that which occurs in animal cells has recently been confirmed (Rognes, 1975; Lea and Fowden, 1975). This reaction involves glutamine as the donor of the amide nitrogen:

$$\text{Glutamine} + \text{aspartate} + ATP \rightarrow \text{asparagine} + AMP + PP_i + \text{glutamate}$$

In certain plants, asparagine can be synthesized by a completely different pathway which involves cyanide as a precursor. Since [^{14}C,^{15}N]cyanide is incorporated directly into the amide group of asparagine, aspartate cannot be an intermediate of this pathway. The pathway involves the synthesis of β-cyanoalanine and its subsequent hydration as follows:

$$\text{Cysteine} + \text{cyanide} \rightarrow \beta\text{-cyanoalanine} + H_2S \qquad (\beta\text{-cyanoalanine synthetase})$$
$$H_2O + \beta\text{-cyanoalanine} \rightarrow \text{asparagine} \qquad (\beta\text{-cyanoalanine hydrolase})$$

Enzymes that catalyze both of these reactions have been isolated and partially purified from blue lupine seedlings (Hendrickson and Conn, 1969; Castric et al., 1972). The physiological significance of this pathway is not obvious, since a major source of cyanide would be required to account for the large amounts of asparagine synthesis in vivo. It may, however, serve as a mechanism of cyanide detoxification. This would be consistent with the observation that both cotton roots and lupine seedlings, which are capable of utilizing the cyanide pathway, also appear to synthesize asparagine from aspartate (Ting and Zschoche, 1970; Lever and Butler, 1971).

3. LYSINE

Two mutually exclusive pathways of lysine biosynthesis have been established. One pathway originates with aspartate and includes the intermediates shown in Fig. 1 (68–76). This sequence of reactions has been designated the diaminopimelate (DAP) pathway. The alternate pathway involves an initial condensation of α-ketoglutarate with acetyl-coenzyme A to form homocitrate. This intermediate is converted to α-aminoadipate in two reactions; α-aminoadipate is, in turn, metabolized to lysine. A total of seven enzymatically catalyzed reactions are involved in this pathway, which has been termed the α-aminoadipate (AAA) pathway. Vogel and his collaborators have utilized diagnostic labeling techniques in order to examine the distribution of these pathways among a large number of different organisms. The organism is allowed to incorporate specifically labeled ^{14}C metabolites and then the relative specific activity of aspartate and lysine isolated from protein is examined. For example, if [4-^{14}C]aspartate is employed, a functional DAP pathway would be indicated by approximately equal labeling of aspartate and lysine. In contrast, the labeled C-4 of aspartate would be lost as $^{14}CO_2$ in the conversion of oxalacetate to α-ketoglutarate prior to the synthesis of homocitrate in an organism possessing an active AAA pathway. Although this approach is subject to complications arising from metabolic compartmentation and regulatory phenomena, its use has led to the concept that the DAP pathway is functional in certain fungi, algae, and higher plants, whereas the AAA pathway is restricted to other fungi and euglenoids (Vogel et al., 1970). Verification of this distribution has been obtained in many cases, including higher plants, by isolation of specific enzymes (Shimura and Vogel, 1966). Studies among various fungi have revealed that the DAP pathway seems to be associated with those species that utilize cellulose as a cell wall component (Lé John, 1971).

With few exceptions very little is known about the nature of the intermediates or the enzymes that catalyze reactions of lysine biosynthesis in plants. For example, it is not known whether succinylated or acetylated intermediates of the DAP pathway are characteristic of plants. Examples of both types of intermediates occur among different species of microorganisms. Isotope competition results of experiments with plant cells indicate that lysine biosynthesis is regulated in vivo. In addition to influencing the activity of β-aspartokinase (68) (see Section III,D,5), lysine inhibits the activity of dihydrodipicolinate synthetase (70), the first enzyme unique to the synthesis of lysine (Cheshire and Miflin, 1975).

4. METHIONINE

Several different pathways may contribute to the synthesis of homocysteine, the immediate precursor of methionine (88) in higher plants. Direct sulfuration (89) of an acyl derivitive of homoserine (85) has been proposed (Giovanelli and Mudd, 1966) but the extent to which this pathway is functional *in vivo* remains to be established.

Although enzymes that catalyze the conversion of O-acyl homoserine to cystathionine (86) have been described, the recent results of Datko *et al.* (1974) suggest that O-phosphohomoserine is the most important precursor of cystathionine in green plants. Cystathionine can be hydrolyzed to homocysteine, pyruvate and ammonia in a β-elimination reaction (87). Catalysis of this reaction has been detected in extracts of several plants and a 400-fold purification of the spinach enzyme has been achieved (Giovanelli and Mudd, 1971). Evidence that cystathionine is an intermediate in methionine biosynthesis has also been obtained in rose tissue cultures and pea mitochondria (Clandinin and Cossins, 1974).

The final reaction in methionine biosynthesis is the methylation of homocysteine (88). Peas have been shown to contain enzymes that catalyze this reaction with three different methyl donors. These are S-adenosylmethionine, S-methylmethionine, and 5-methyltetrahydropteroyl-(N)-glutamate (Dodd and Cossins, 1969, 1970). Only the pteroyl compounds are considered to be involved in the net synthesis of methionine, and mono-, di-, and triglutamate derivatives are effective substrates of the vitamin B_{12}-independent enzyme in plants.

5. THREONINE AND ISOLEUCINE

Threonine synthesis is accomplished in a sequence of two reactions originating from the intermediate homoserine (78, 79). Since the enzymes catalyzing these reactions have not been isolated from plants, it can only be assumed that O-phosphohomoserine is the product of a reaction catalyzed by homoserine kinase and a substrate of threonine synthetase in plants as it is in microorganisms. Threonine is deaminated to yield α-ketobutyrate, which is a precursor is isoleucine (80). Many organisms possess two types of enzymes that can catalyze this deaminating dehydration. Degradative and biosynthetic threonine dehydratases are distinguished from one another on the basis of their unique patterns of regulation. The biosynthetic enzymes are generally subject to feedback inhibition by isoleucine, while the degradative enzymes are not inhibited by this amino acid. Multiple dehydratases have been reported in some plants, and isoleucine-sensitive enzymes have been isolated from several plant sources. A threonine dehydratase isolated from spinach has been extensively purified and shown to have properties that are similar to the biosynthetic dehydratases of several microorganisms (Sharma and Mazumder, 1970). In addition to being specifically inhibited by isoleucine, the enzyme is activated by monovalent cations including ammonia. Valine partially reverses inhibition by isoleucine and normalizes the sigmoid kinetics that are apparent in the presence of the end product inhibitor. A similar phenomenon involving other amino acids, including aspartate, has been reported for the enzyme isolated from pea seedlings (Blekman et al., 1971). The remaining reactions of isoleucine biosynthesis and their corresponding enzymes (81 to 84) have been considered in Section III,A,2 in relation to their role in valine biosynthesis.

A number of different experimental approaches have aided in validating the concept that aspartate is the general precursor of lysine, methionine, threonine, and isoleucine in plants. Several problems of aspartate metabolism do, however, remain to be examined in greater detail. For example, when [^{14}C]aspartate is fed to plants, the extent to which it is converted to the end product amino acids is frequently quite low. This appears to be due to active degradation of aspartate, a complex aspartate pool structure including the possibility of specific enzyme complexes, and the presence of end product sensitive enzymes that catalyze reactions prior to the origin of individual branches of the pathway. Two reactions are required for the synthesis of β-aspartylsemialdehyde, the first branch point metabolite of the pathway. Aspartate is initially phosphorylated in an ATP-dependent reaction, catalyzed by β-aspartokinase (68). In microorganisms, this reaction may be regulated in several different ways, such as by concerted feedback inhibition by lysine and threonine, or the existence of independently regulated isofunctional enzymes. The only aspartokinase isolated from maize is extremely sensitive to inhibition by lysine (apparent K_i of 2.9×10^{-5} M), and does not appear to be subject to concerted regulation by multiple pathway products (Bryan et al., 1970). Although several purification procedures have failed to separate completely the activity of this enzyme from that of homoserine dehydrogenase (77), the results of gel filtration experiments suggest that the two activities from maize seedlings are probably not catalyzed by the same protein as they are in two of the three aspartokinases of E. coli K$_{12}$ (J. K. Bryan, unpublished data). Independent attempts to isolate isofunctional aspartokinases from maize have also been unsuccessful (Cheshire and Miflin, 1973; J. K. Bryan, unpublished). The aspartokinase purified from wheat germ is also sensitive to inhibition by low concentrations of lysine, and the extent of inhibition appears to be enhanced in the presence of higher concentrations of threonine (Wong and Dennis, 1973a). A threonine sensitive aspartokinase has been isolated from peas (Aarnes and Rognes, 1974) while the aspartokinases of other plants appear to be inhibited by both threonine and lysine (Aarnes, 1974).

The second reaction involved in the synthesis of β-aspartylsemial-dehyde requires the activity of β-aspartate semialdehyde dehydrogenase (69), an enzyme that has been isolated from both peas (Sasaoka and Inagaki, 1960) and maize. The partially purified maize enzyme is not significantly affected by any of the end products of aspartate metabolism, but its activity is markedly dependent upon the presence of sulfhydryl compounds (C. E. Brunner and J. K. Bryan, unpublished data).

The second branch point intermediate in the aspartate pathway is homoserine. This compound is synthesized by reduction of the semialde-hyde in a reaction catalyzed by homoserine dehydrogenase (77). This enzyme has also been isolated from both peas (Sasaoka, 1961; Aarnes and Rognes, 1974) and maize (Bryan, 1969). The regulatory properties of the plant enzymes are somewhat similar to those of several microbial homoserine dehydrogenases. Threonine is a feedback inhibitor, although activity of the maize enzyme can also be inhibited significantly by cysteine and to a lesser extent by serine and aspartate. The extent to which any of these amino acids are inhibitory depends not only on the inhibitor and substrate concentrations but also on the specific coenzyme utilized in the reaction, NAD(H) or NADP(H), and the direction of the reaction being catalyzed. The extent to which threonine is an effective inhibitor also depends upon the stage of plant development from which the enzyme is isolated (Matthews et al., 1975). It should be noted that homoserine can also be produced from the hydrolysis of S-adenosylmethionine. This would not represent a net synthesis of homoserine from aspartate but may be of special significance in peas, which are characterized by high intracellular concentrations of homoserine (Grant and Voelkert, 1971).

The major regulatory interactions associated with enzymes of the aspartate pathway in plants are illustrated in Fig. 4. A lack of detailed information on certain aspects of the mechanisms of biosynthetic regula-tion in plants is apparent in this pathway. For example, independent regulation of the methionine branch of the pathway has not been eluci-dated in plants. A further complication arises from evidence that the characteristics of homoserine dehydrogenase are progressively altered during the growth of maize seedlings (Matthews et al., 1975; DiCamelli and Bryan, 1975). The high sensitivity of an apparently single asparto-

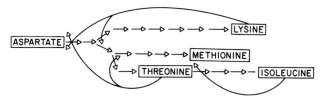

Fig. 4. Inhibition of enzymes of the aspartate pathway of amino acid biosynthesis in plants. Details of the pathways depicted by sequential arrows are presented in Fig. 1 and discussed in the text. Bold arrows indicate inhibition of enzyme activity by the end product amino acids.

kinase to inhibition by lysine, could conceivably result in a limitation of the alternate products of aspartate metabolism in maize. Such partial and potentially changing regulation, if truly characteristic of plants, would be in striking contrast to the highly integrated regulatory mechanisms that have been elucidated in microorganisms.

The highly specific effects of metabolites of aspartate on the growth of certain plants are consistent with the presence of concerted or multivalent inhibition of aspartokinase. The growth of liverwort gemmalings (Dunham and Bryan, 1969, 1971), rice tissue cultures (Furuhashi and Yatazawa, 1970), *Lemma* (Wong and Dennis, 1973b), and *Nimulus cardinalis* (Henke and Wilson, 1974) is synergistically inhibited by lysine *plus* threonine; while traces of methionine (or homoserine) relieve the inhibition. The available data obtained with isolated enzymes can be compared with the results of growth studies and provide at least circumstantial evidence that enzymes catalyzing identical reactions in different plants may exhibit substantially different regulatory properties.

IV. General Conclusions

Amino acid biosynthesis in plants follows the same general pattern that has been elucidated in microorganisms. Thus, ammonia is assimilated in a minimum number of reactions, and nitrogen is mobilized during transamination with a multitude of keto acids derived from a limited number of metabolically active precursors. This similarity among different organisms is extended by the apparent evolutionary conservation of many chemical reaction mechanisms and biosynthetic pathways. However, fewer data have frequently been employed in the characterization of specific pathways in multicellular plants than in microorganisms. In many cases the information derived from experiments with plants is sufficient to establish the gross nature of the pathway in question, but, when examined critically, it is obviously fragmentary. For example, for several of the intermediates of histidine synthesis—the pathway of chorismate biosynthesis from shikimate—the initial reactions involved in proline and arginine biosynthesis and the nature of the acyl intermediates in lysine synthesis have yet to be elucidated in any multicellular plant. Even though the concept of biochemical unity has been an invaluable aid in interpreting metabolic data, direct experiments are required for verification of many of the biosynthetic relationships among the common amino acids in multicellular plants.

A number of alternate pathways for the synthesis of specific amino acids have been identified in plants. An alternate pathway may be associated primarily with catabolism or may represent a truly amphibolic sequence of reactions. These possibilities were noted in the metabolism of serine, proline, arginine, and aspartate. Two or more biosynthetic pathways may coexist in a single cell, and each may be preferentially utilized depending upon the physiological state of the cell at a given time. It is conceivable, for example, that serine is derived from glycolate with the intermediate formation of glycine during photosynthesis in C_3 plants and that the synthesis of these amino acids originates with the glycolytic intermediate, 3-phosphoglycerate, in the dark. Cases of metabolic differentiation, in which the enzymes associated with a particular pathway are synthesized only in some of the cells of a plant, would represent another possibility for the existence of multiple pathways. This situation is difficult to evaluate due to the relative insensitivity of many enzyme assays and the low concentration of most biosynthetic enzymes in plant cells. Finally an "unusual" pathway for the synthesis of an amino acid may be restricted to certain plant species. An apparent example of this situation is the cyanide pathway of asparagine biosynthesis.

End product inhibition of enzymes in each of the most intensively studied pathways of amino acid biosynthesis in multicellular plants has been described. In some cases, the extent of this type of control appears to be adequate to regulate the flow of carbon through the pathways associated with a particular family of amino acids; while in others, there are unresolved questions concerning the possible regulation of key reactions. Contrast, for example, the controls that have been associated with the synthesis of valine, isoleucine, and leucine in plants (Fig. 2) with those that have been identified in the aspartate pathway (Fig. 4). It is also important to recall that the mechanisms that collectively regulate many biosynthetic pathways in microorganisms are specific for certain organisms. Therefore, even in those cases where the demonstrated controls appear to be complete, different regulatory patterns may be discovered in other plants. Currently our understanding of the mechanisms that regulate amino acid biosynthesis in plants is almost exclusively related to the demonstration of end product-sensitive enzymes. Attempts to demonstrate amino acid-dependent changes in the levels of specific biosynthetic enzymes in plants have been uniformly unsuccessful. Nevertheless, the apparent levels of biosynthetic enzymes do change in plants, particularly during development, and eventually the mechanisms responsible for such changes will be established. In addition, further insight into the potential role of intracellular compartmentation in regulation is likely to be obtained. Thus, it can be reasonably anticipated that multicellular plants will ultimately be characterized by a highly integrated series of regulatory controls—some of which may be unique to plants—which influence the synthesis of amino acids.

ACKNOWLEDGMENTS

Unpublished results from the author's laboratory were supported by Grant GB 29599 from the National Science Foundation.

GENERAL REFERENCES

Fowden, L. (1967). *Annu. Rev. Plant Physiol.* **18**, 85.

Greenberg, D. M., ed. (1969). "Metabolic Pathways," 3rd ed., Vol. 3. Academic Press, New York.

Meister, A. (1965). "Biochemistry of the Amino Acids," 2nd ed., Vols 1 and 2. Academic Press, New York.

Miflin, B. J. (1973). *In* "Biosynthesis and its Control in Plants" (B. V. Milborrow, ed.), pp. 49–68. Academic Press, New York.

Preiss, J., and Kosuge, T. (1970). *Annu. Rev. Plant Physiol.* **21**, 433.

Steward, F. C., ed. (1965). "Plant Physiology: A Treatise," Vol. 4A. Academic Press, New York.

REFERENCES

Aarnes, H. (1974). *Physiol. Plant.* **32**, 400.

Aarnes, H., and Rognes, S. E. (1974). *Phytochemistry* **13**, 2717.

Andrews, T. J., Lorimer, G. H., and Tolbert, N. E. (1971). *Biochemistry* **10**, 4777.

Baich, A. (1971). *Biochim. Biophys. Acta* **244**, 129.

Bassham, J. A. (1971). *Science* **172**, 526.

Belser, W. L., Murphy, J. B., Delmer, D. P., and Mills, S. E. (1971). *Biochim. Biophys. Acta* **237**, 1.

Berlyn, M. B., Ahmed, S. I., and Giles, N. H. (1970). *J. Bacteriol.* **104**, 768.

Blekman, G. I., Kagan, Z. S., and Kretovich, V. L. (1971). *Biokhimiya* **36**, 1050.

Bryan, J. K. (1969). *Biochim. Biophys. Acta* **171**, 205.

Bryan, P. A., Cawley, R. D., Brunner, C. E., and Bryan, J. K. (1970). *Biochem. Biophys. Res. Commun.* **41**, 1211.

Butler, G. W., and Shen, L. (1963). *Biochim. Biophys. Acta* **71**, 456.

Castric, P. A., Farnden, K. J. F., and Conn, E. E. (1972). *Arch. Biochem. Biophys.* **152**, 62.

Chapman, A. G., Fall, L., and Atkinson, D. E. (1971). *J. Bacteriol.* **108**, 1072.

Chen, J., and Boll, W. G. (1971). *Can. J. Bot.* **49**, 1155.

Chen, J., and Boll, W. G. (1972). *Can. J. Bot.* **50**, 587.

Cheshire, R. M., and Miflin, B. J. (1973). *Plant Physiol.* **51s**, 54.

Cheshire, R. M., and Miflin, B. J. (1975). *Phytochemistry* **14**, 695.

Cheung, G. P., Rosenblum, I. Y., and Sallach, H. J. (1968). *Plant Physiol.* **43**, 1813.

Chu, M., and Widholm, J. M. (1972). *Physiol. Plant.* **26**, 24.

Clandinin, M. T., and Cossins, E. A. (1974). *Phytochemistry* **13**, 585.

Cotton, R. G. H., and Gibson, F. (1968). *Biochim. Biophys. Acta* **156**, 187.

Datko, A. H., Giovanelli, J., and Mudd, S. H. (1974). *J. Biol. Chem.* **249**, 1139.

Davies, D. D. (1968). *In* "Recent Aspects of Nitrogen Metabolism in Plants" (E. J. Hewitt and C. V. Cutting, eds.), pp. 125–135. Academic Press, New York.

Davies, D. D., and Teixiera, A. N. (1975). *Phytochemistry* **14**, 647.

Davies, M. E. (1971). *Phytochemistry* **10**, 783.

Delmer, D. P., and Mills, S. E. (1968a). *Plant Physiol.* **43**, 81.

Delmer, D. P., and Mills, S. E. (1968b). *Biochim. Biophys. Acta* **167**, 431.

DiCamelli, C. A., and Bryan, J. K. (1975). *Plant Physiol.* **55**, 999.

Dodd, W. A., and Cossins, E. A. (1969). *Arch. Biochem. Biophys.* **133**, 216.

Dodd, W. A., and Cossins, E. A. (1970). *Biochim. Biophys. Acta* **201**, 461.

Dougall, D. K. (1965). *Plant Physiol.* **40**, 891.

Dougall, D. K. (1974). *Biochem. Biophys. Res. Commun.* **58**, 639.

Dougall, D. K., and Fulton, M. M. (1967a). *Plant Physiol.* **42**, 387.

Dougall, D. K., and Fulton, M. M. (1967b). *Plant Physiol.* **42**, 941.

Dunham, V. L., and Bryan, J. K. (1969). *Plant Physiol.* **44**, 1601.

Dunham, V. L., and Bryan, J. K. (1971). *Plant Physiol.* **47**, 91.

Ellis, R. J. (1969). *Planta* **88**, 34.

Filner, P., Wray, J. L., and Varner, J. E. (1969). *Science* **165**, 358.

Fletcher, J. S., and Beevers, H. (1971). *Plant Physiol.* **48**, 261.

Fowler, M. W., Jessup, W., and Sarkissian, G. S. (1974). *FEBS Lett.* **46**, 340.

Furuhashi, K., and Yatazawa, M. (1970). *Plant Cell Physiol.* **11**, 569.

Gamborg, O. L. (1965). *Can. J. Biochem.* **43**, 723.

Gilchrist, D. G., and Kosuge, T. (1974). *Arch Biochem. Biophys.* **164**, 95.

Gilchrist, D. G., Woodin, T. S., Johnson, M. L., and Kosuge, T. (1972). *Plant Physiol.* **49**, 52.

Giovanelli, J., and Mudd, S. H. (1966). *Biochem. Biophys. Res. Commun.* **25**, 366.

Giovanelli, J., and Mudd, S. H. (1967). *Biochem. Biophys. Res. Commun.* **27**, 150.

Giovanelli, J., and Mudd, S. H. (1968). *Biochem. Biophys. Res. Commun.* **31**, 275.

Giovanelli, J., and Mudd, S. H. (1971). *Biochim. Biophys. Acta* **227**, 654.

Goldin, B. R., and Frieden, C. (1971). *Curr. Top. Cell. Regul.* **4**, 77–117.

Grant, D. R., and Voelkert, E. (1971). *Can. J. Biochem.* **49**, 795.

Hart, J. W., and Filner, P. (1969). *Plant Physiol.* **44**, 1253.

Hendrickson, H. R., and Conn, E. E. (1969). *J. Biol. Chem.* **244**, 2632.

Henke, R. R., and Wilson, K. G. (1974). *Planta* **121**, 155.

Huisman, D. C., and Kosuge, T. (1974). *J. Biol. Chem.* **249**, 6842.

Inatomi, K., and Slaughter, J. C. (1971). *J. Exp. Bot.* **22**, 561.

Jensen, R. A., and Pierson, D. L. (1975). *Nature (London)* **254**, 667.

Joy, K. W. (1971). *Plant Physiol.* **47**, 445.

Joy, K. W., and Folkes, B. F. (1965). *J. Exp. Bot.* **16**, 646.

Kagan, Z. S., Polyakov, W. A., and Kretovich, W. L. (1970). *Enzymologia* **38**, 201.

Kanamori, T., and Matsumoto, H. (1972). *Arch. Biochem. Biophys.* **125**, 404.

King, J., and Wu, W. Y. F. (1971). *Phytochemistry* **10,** 915.

Kiritani, K., and Wagner, R. P. (1970). *In* "Methods in Enzymology" (H. Tabor and C. W. Tabor, eds.), Vol. 17A, pp. 745–750. Academic Press, New York.

Kisaki, T., and Tolbert, N. E. (1970). *Plant Cell Physiol.* **11,** 247.

Kleczkowski, K., and Cohen, P. P. (1964). *Arch. Biochem. Biophys.* **107,** 271.

Kleczkowski, K., and Grabarek-Bralczyk, J. (1968). *Acta Biochim. Pol.* **15,** 283.

Lamport, D. T. A. (1964). *Nature (London)* **202,** 293.

Lea, P. J., and Fowden, L. (1975). *Proc. Roy. Acad. Sci. B* in press.

Lea, P. J., and Miflin, B. J. (1974). *Nature (London)* **251,** 614.

Lé John, H. B. (1971). *Nature (London)* **231,** 164.

Lever, M., and Butler, G. W. (1971). *J. Exp. Bot.* **22,** 279.

Matthews, B. F., Gurman, A. W., and Bryan, J. K. (1975). *Plant Physiol.* **55,** 991.

Martin, R. G., Berberich, M. A., Ames, B. N., Davis, W. W., Goldberger, R. F., and Yourno, J. D. (1971). *In* "Methods in Enzymology" (H. Tabor and C. W. Tabor, eds.), Vol. 17B, pp. 3–44. Academic Press, New York.

Miflin, B. J. (1969). *J. Exp. Bot.* **20,** 810.

Miflin, B. J. (1971). *Arch. Biochem. Biophys.* **146,** 542.

Miflin, B. J. (1974). *Plant Physiol.* **54,** 550.

Miflin, B. J., and Cave, P. R. (1972). *J. Exp. Bot.* **23,** 511.

Miflin, B. J., and Lea, P. J. (1975). *Biochem. J.* **149,** 403.

Minamikawa, T. (1967). *Plant Cell Physiol.* **8,** 695.

Miura, G. A., and Mills, S. E. (1971). *Plant Physiol.* **47,** 483.

Morris, C. J., and Thompson, J. F. (1971). *Plant Physiol.* **47s,** 17.

Morris, C. J., Thompson, J. F., and Johnson, C. M. (1969). *Plant Physiol.* **44,** 1023.

Oaks, A. (1965a). *Plant Physiol.* **40,** 142.

Oaks, A. (1965b). *Plant Physiol.* **40,** 149.

Oaks, A. (1965c). *Biochim. Biophys. Acta* **111,** 79.

Oaks, A., and Bidwell, R. G. S. (1970). *Annu. Rev. Plant Physiol.* **21,** 43.

Oaks, A., Mitchell, D. J., Barnard, R. A., and Johnson, F. J. (1970). *Can. J. Bot.* **48,** 2249.

O'Neal, D., and Joy, K. W. (1975). *Plant Physiol.* **55,** 968.

O'Neal, D., and Naylor, A. W. (1968). *Biochem. Biophys. Res. Commun.* **31,** 322.

Pahlich, E. (1971). *Planta* **100,** 222.

Pahlich, E., and Joy, K. W. (1971). *Can. J. Biochem.* **49,** 127.

Rabson, R., Tolbert, N. E., and Kearney, P. C. (1962). *Arch. Biochem. Biophys.* **98,** 154.

Randall, D. D., Tolbert, N. E., and Gremel, D. (1971). *Plant Physiol.* **48,** 480.

Rognes, S. E. (1970). *FEBS (Fed. Eur. Biochem. Soc.) Lett.* **10,** 62.

Rognes, S. E. (1975). *Phytochemistry* **14,** 1975.

Rosenthal, G. A., and Naylor, A. W. (1969). *Biochem. J.* **112,** 415.

Rothe, G. M. (1974). *Z. Pflazenphysiol.* **74,** 152.

Sadava, D., and Chrispeels, M. J. (1971a). *Biochim. Biophys. Acta* **227,** 278.

Sadava, D., and Chrispeels, M. J. (1971b). *Biochemistry* **10,** 4290.

Sakaguchi, K. (1970). *Biochim. Biophys. Acta* **220,** 580.

Sasaoka, K. (1961). *Plant Cell Physiol.* **2,** 231.

Sasaoka, K., and Inagaki, H. (1960). *Mem. Res. Inst. Food Sci., Kyoto Univ.* **21,** 12.

Satyanarayana, T., and Radhakrishnan, A. N. (1964). *Biochim. Biophys. Acta* **92,** 367.

Satyanarayana, T., and Radhakrishnan, A. N. (1965). *Biochim. Biophys. Acta* **110,** 380.

Shargool, P. D. (1971). *Phytochemistry* **10**, 2029.

Shargool, P. D. (1973a). *FEBS (Fed. Eur. Biochem. Soc.) Lett.* **33**, 348.

Shargool, P. D. (1973b). *Plant Physiol.* **52**, 68.

Sharma, R. K., and Mazumder, R. (1970). *J. Biol. Chem.* **245**, 3008.

Shepard, D. V., and Thurman, D. A. (1973). *Phytochemistry* **12**, 1937.

Shimura, Y., and Vogel, H. J. (1966). *Biochim. Biophys. Acta* **118**, 396.

Sims, A. P., Folkes, B. F., and Bussey, A. H. (1968). *In* "Recent Aspects of Nitrogen Metabolism in Plants" (E. J. Hewitt and C. V. Cutting, eds.), pp. 91–114. Academic Press, New York.

Slaughter, J. C., and Davies, D. D. (1968). *Biochem. J.* **109**, 749.

Smith, I. K. (1972). *Plant Physiol.* **50**, 477.

Smith, I. K., and Thompson, J. F. (1969). *Biochem. Biophys. Res. Commun.* **35**, 939.

Stadtman, E. R. (1970). *In* "The Enzymes" (P. D. Boyer, ed.), 3rd ed, Vol. 1, pp. 397–459. Academic Press, New York.

Staub, M., and Dénes, G. (1966). *Biochim. Biophys. Acta* **128**, 82.

Streeter, J. G. (1973). *Arch. Biochem. Biophys.* **157**, 613.

Tate, S. S., and Meister, A. (1971). *Proc. Nat. Acad. Sci. U.S.* **68**, 781.

Thompson, J. F., and Moore, D. P. (1968). *Biochem. Biophys. Res. Commun.* **31**, 281.

Ting, I. P., and Zschoche, W. C. (1970). *Plant Physiol.* **45**, 429.

Tolbert, N. E. (1971). *Annu. Rev. Plant Physiol.* **22**, 45.

Umbarger, H. E. (1969). *Annu. Rev. Biochem.* **38**, 323.

Vogel, H. J., Thompson, J. S., and Shockman, G. D. (1970). *Symp. Soc. Gen. Microbiol.* **20**, 107–119.

Wallace, W. (1973). *Plant Physiol.* **52**, 197.

Weber, H. L., and Böck, A. (1969). *Arch. Mikrobiol.* **66**, 250.

Wiater, A., Krajewska-Grynkiewicz, K., and Klopotowski, T. (1971). *Acta Biochim. Pol.* **18**, 299.

Widholm, J. M. (1971). *Physiol. Plant.* **25**, 75.

Widholm, J. M. (1973). *Biochim. Biophys. Acta* **320**, 217.

Wong, K. F., and Dennis, D. T. (1973a). *Plant Physiol.* **51**, 322.

Wong, K. F., and Dennis, D. T. (1973b). *Plant Physiol.* **51**, 327.

Woodin, T. S., and Nishioka, L. (1973). *Biochim. Biophys. Acta* **309**, 211.

Yoshida, S. (1969). *Annu. Rev. Plant. Physiol.* **20**, 41.

18

Mineral Metabolism

D. W. RAINS

I. Introduction

Mineral metabolism might be defined as a process in which mineral nutrient elements are incorporated into plant metabolites. This very restrictive definition would exclude the large fraction of nutrients that function as catalytic agents and cofactors in enzymatic transformation of nonmineral substances in various anabolic and catabolic processes within plant cells. The essentiality of a mineral nutrient has been established in ways other than the demonstration of a metabolite consisting of a combined mineral element. No such metabolite has been found for one-fourth of the mineral elements now recognized as essential.

The functions of mineral nutrients can be grouped into some general categories. Mineral elements in ionic forms have been found to be important as osmotic regulators. Other functions include the maintenance of membrane integrity, and, in addition, many of the so-called micronutrients are required in minute quantities as cofactors in enzymatic reactions.

This chapter selects specific cations and anions known to be essential for plants and discusses their function in enzymatic processes and their role as structural components. The discussion will give primary importance to nutrients known to be essential for plants but will not be limited to those elements.

For extensive reading in the field of mineral metabolism the reader is referred to some very excellent reviews (Hewitt, 1963; Nason and McElroy, 1963; Epstein, 1965, 1972; Evans and Sorger, 1966; Wyn Jones and Lunt, 1967).

II. Essentiality

A. Criteria of Essentiality

Criteria of essentiality as proposed by Arnon and Stout (1939a) involve three main parts: (a) If the element is removed from the growth medium of a plant, the plant will be unable to complete its vegetative or life cycle. (b) The element is essential if its function is specific and cannot be replaced by other elements. (c) Essentiality is confirmed if the element is a necessary component of an essential metabolite. As with any set of rules, however, there are exceptions. Is an element essential if it removes the detrimental effects of another element present in excessive amounts? When an element is present in insufficient amounts, its function may be partially replaced by another element. This "sparing action" may be an essential part of the survival of certain organisms under adverse conditions (Nason and McElroy, 1963; Epstein, 1965; Bonds and O'Kelley, 1969).

B. Quantifying Essential Nutrients

Table I lists all of the elements known to be essential for higher forms of plant life in terms of relative numbers with respect to molybdenum. As can be seen, there is a break between chlorine and sulfur in the amounts considered to be essential, respectively, 100 and 1000 ppm. All the elements present at concentrations of 1000 ppm or higher are termed macronutrients.

It is possible to generalize as to the function of essential macronutrient elements. By making potassium the obvious exception, one can assign a role in either the structure of plants or as components of plant metabolites (Epstein, 1965; Evans and Sorger, 1966; Hewitt, 1963). This undoubtedly accounts for the relatively high requirements for these elements in plant metabolic processes.

TABLE I

Levels of Essential Elements Known to Be Critical for Growth of Multicellular Plants[a]

Element	Concentration in dry matter		Relative number of atoms with respect to molybdenum
	10^{-6} atoms/gm	μg/gm or ppm	
Micronutrients			
Molybdenum	0.001	0.1	1
Copper	0.1	6	100
Zinc	0.3	20	300
Manganese	1.0	50	1,000
Iron	2.0	100	2,000
Boron	2.0	20	2,000
Chlorine	3.0	100	3,000
Macronutrients			
Sulfur	30	1,000	30,000
Phosphorus	60	2,000	60,000
Magnesium	80	2,000	80,000
Calcium	125	5,000	125,000
Potassium	250	10,000	250,000
Nitrogen	1,000	15,000	1,000,000
Oxygen	30,000	450,000	30,000,000
Carbon	35,000	450,000	35,000,000
Hydrogen	60,000	60,000	60,000,000

[a] After Stout (1961) as presented by Price (1970).

The rather limited requirement for the micronutrients reflects their function in oxidation–reduction reactions and as enzymatic activators and cofactors.

III. Specific Function of Essential Nutrients

Since the roles of many of these nutrients have been established for many years, only the more recent and significant contributions to an understanding of these roles are cited.

A. Calcium

Calcium functions both as a structural component and as a cofactor for certain enzymes (Table II). The function of calcium in plants is treated more completely in a review by Wyn Jones and Lunt (1967).

TABLE II

Selected Roles of Calcium

---

1. Cell wall structure—calcium pectate (Tagawa and Bonner, 1957; Rasmussen, 1966; El Hinnawy, 1974)
2. IAA-stimulated cell wall elongation? (Bennet-Clarke, 1956; Rayle and Cleland, 1970)
3. Influences nonspecific growth responses (Hewitt, 1963; Purves, 1966; Helms, 1971)
4. Membrane structure and ion fluxes (Marinos, 1962; van Breeman, 1968; Epstein, 1972; Rains, 1972)
5. Protective role of Ca in ion transport and physiological processes (Wyn Jones and Lunt, 1967; Bonds and O'Kelley, 1969; Rains and Floyd, 1970; Nieman and Willis, 1971; Epstein, 1972)
6. Influences nitrate reductase (Paulsen and Harper, 1968)
7. Nodulation and nitrogen fixation (Banath et al., 1966; Lowther and Loneragan, 1968)
8. Some enzyme systems requiring Ca
   a. Amylase (Chrispeels and Varner, 1967)
   b. ATPase (Avron, 1962, 1967; Dodds and Ellis, 1966)
   c. Phospholipase D (Davidson and Long, 1958; Einset and Clark, 1958)

Epstein (1965, 1972) discussed calcium and ion transport, and Rains (1972) dealt specifically with the influence of calcium on ion regulation in saline environments.

1. Structural Functions of Calcium

Classically, calcium has been associated with the cell wall structure, and calcium pectate has been invoked as a material that binds together the cell walls of plants (Tagawa and Bonner, 1957; Rasmussen, 1966). The supposed removal of calcium with EDTA (ethylenediaminetetra-acetic acid) has been a common procedure in isolating individual cells of plants (Ginsburg, 1958; Jyung and Wittwer, 1964; El Hinnawy, 1974), and this treatment has been found to increase cell plasticity (Taylor and Wain, 1966). Recent work has suggested that protein as well as calcium may be involved in binding cell walls together (El Hinnawy, 1974).

The interaction of calcium with IAA (indoleacetic acid) stimulated cell elongation suggested to Bennet-Clarke (1956) the possibility that a calcium ionic bridge lends rigidity to cell walls. It was envisioned that IAA complexed calcium, thereby breaking the linkage and rendering the walls more elastic. Other investigators found no correlation between re-distribution of calcium and cell wall elongation when tissue was treated with IAA (Cleland, 1960). Low pH has been found to mimic IAA in the induced growth of plant cells. The mechanism for this effect has not been determined, but it has been suggested that auxin and hydrogen ions do not mediate their effect through the same mechanism (Rayle and Cleland, 1970).

Although somewhat speculative, there is a possible hypothesis that might explain many of the observations concerning inducible cell wall expansion. An increase in hydrogen ions within cell walls would decrease ionization of carboxyl groupings of the cellulosic fraction. This would decrease the amount of bonding of calcium between two associated carboxyls. It is possible that at the pH values (3.0) shown for optimally induced growth, only one carboxyl–calcium bond would be broken. One-half the carboxyl groups could be associated with hydrogen ions and the others could have calcium attached with residual positive charges. An increase in cation exchange capacity of the tissue should be observed. The observation that calcium is not redistributed when tissues grow in response to IAA might be explained by the above hypothesis. Cell wall extension induced by CO_2 and low pH is reduced in magnitude when calcium is included in the experimental solutions (Evans et al., 1971). This would also support the hypothesis that calcium is influencing cell wall rigidity through its effect on ionic bridging between associated cellulosic groupings. A clear understanding of the effect of calcium on cell wall metabolism remains to be established.

An appreciation of the involvement of calcium in growth responses has arisen from observations on symptoms of nutrient deficiency (Hewitt, 1963; Chapman, 1966). A plant found deficient in calcium shows stem collapse and subsequent termination of growth in the apical region. For example, until recently, damage to the stems of seedlings grown under certain conditions was attributed to damping-off disease. However, this damage was found to be due to calcium deficiency (Helms, 1971), and so-called stem rot of newly developing seedlings was virtually eliminated by adding appropriate amounts of calcium to the culture solutions.

Some growth responses show a negative correlation with calcium. It was found that calcium reduced elongation of cucumber hypocotyls, whereas most monovalent cations and some divalent cations stimulated elongation (Purves, 1966).

The relations between growth and calcium are very complex. Many of the processes in which calcium is known to function involve some aspect of growth. For example, calcium might influence growth directly or indirectly through cell division and middle lamellar deposition, ion regulation and related osmotic responses, and cell wall structures and membrane functions (Epstein, 1965, 1972; Rains, 1972).

2. MEMBRANES AND ION REGULATION

Calcium has been implicated in membrane functions for many years. Visually, this can be demonstrated by the occurrence of disarranged membrane structures of calcium-deficient systems (Marinos, 1962; Marschner and Gunther, 1964).

The ionic selectivity of artificially produced phospholipid membranes (cephalin and cholesterol) was altered by adding calcium to this model system (van Breemen, 1968). This alteration suggested that calcium was involved in spatially arranging the three-dimensional mode of the membrane. Calcium was found to decrease the exchange of sodium across nonpolar liquid membranes when various lipids, such as phosphatidylcholine, mediated ion transport. On the other hand, potassium had no significant effect on sodium exchange. It was concluded that calcium altered the affinity of membrane constituents for certain cations and thereby influenced the selectivity of the ion transport process (Kuiper, 1968).

The influence of calcium on membrane integrity has been measured indirectly by studying the fluxes of ions into and out of cells. Removal of calcium by complexing agents has greatly increased these fluxes (Foote and Hanson, 1964; van Steveninck, 1965).

Washing slices or cubes of certain tissues for various periods results in many physiological responses. One such response is alteration in ion fluxes (Laties, 1967). Calcium promotes the development of potassium absorption capacity in bean stem slices (Rains and Floyd, 1970). Calcium is also necessary to prevent leakage of potassium during the aging of beet slices, and calcium simultaneously enhances the development of boron absorption capacity (Wildes and Neales, 1971). The development of sodium absorption capacity by aging beet slices, however, does not require calcium (Poole, 1971).

These changes in ion fluxes suggest that membrane structure is altered during washing procedures, and changes have been observed in the endoplasmic reticulum (Jackman and van Steveninck, 1967). Exactly how calcium is involved when the physiological processes are altered during the washing of plant tissue (Laties, 1967; Floyd and Rains, 1971) is a moot question. Studies on leaf senescence indicate that the presence of calcium defers senescence by retaining the integrity of the membrane systems (Poovaiah and Leopold, 1973). Calcium apparently acts in a similar manner as some of the hormones that defer senescence.

Calcium ions can serve a protective function. Calcium protects plants from the injurious effects of hydrogen ions (Rains et al., 1964), high salt in the environment (Rains, 1972), and other potentially toxic ions present in the environment (Munns, 1965; Rains and Epstein, 1967a,b; Arnold, 1969). Nieman and Willis (1971) found that carrot discs lost protein when exposed to a medium with a high concentration of sodium. The addition of calcium substantially reduced protein loss and maintained active accumulation of the ions required by the plant. The action of calcium as a protective agent against inimical ions is not a highly specific function, but it can be replaced by the action of similar cations, such as strontium (Bonds and O'Kelley, 1969). Wyn Jones and Lunt (1967) have theorized that if the protective function of calcium was eliminated this element would be required in only small amounts and might be classified as a micronutrient.

3. PHYSIOLOGICAL AND BIOCHEMICAL FUNCTIONS

Nodulation and successful symbiotic nitrogen fixation requires relatively high concentrations of calcium. The impairment of nitrogen reduction in calcium-deficient plants is apparently not attributable to a reduced carbohydrate level or a decrease in translocation of reduced nitrogen compounds (Banath et al., 1966). The calcium requirements of leguminous species, such as subterranean clover, are higher for root infection and nodule initiation than for either nodule development or host plant growth (Lowther and Loneragan, 1968). The reduction of nitrogen in the nitrogen-fixing process would appear to be particularly sensitive to low levels of calcium as is the nodulation process.

Barley α-amylase has long been known to require calcium ions for activity (Chrispeels and Varner, 1967). Apparently, the different isozymes bind calcium ions with different affinities so that it is difficult to demonstrate a calcium requirement for some isozymes (Jacobsen et al., 1970).

Adenosine triphosphatases (ATPases) found in plant tissues have a varied response to calcium. Chloroplasts contain a calcium-activated, light-dependent ATPase (Avron, 1962). An enzyme isolated from plant cell walls required both magnesium and calcium (Dodds and Ellis, 1966). In contrast, Atkinson and Polya (1967) found that calcium decreased the activity of soluble and cell wall-bound ATPase, a finding substantiated by the observations of Fisher and Hodges (1969). This paradox may be resolved by determining the extent of interaction between calcium and magnesium. A high specificity for magnesium by ATPases may result in a negative effect by a chemically similar ion, such as calcium.

Calcium has been found to function in a number of other enzyme systems involved in phospholipid metabolism, for example, phospholipase D from cabbage (Davidson and Long, 1958) and from carrot (Einset and Clark, 1958).

B. Magnesium

Magnesium, like calcium, also serves as a structural component and is involved as a cofactor in many enzyme transfers.

1. Structural Functions

The function of magnesium in the chlorophyll structure is well known and documented (Bogorad, 1966; Nason and McElroy, 1963). The requirement for magnesium in chlorophyll was originally inferred from the observation of chlorosis when magnesium was limiting.

Magnesium has also been shown to be required for ribosome integrity (Goldberg, 1966; Tempest *et al.*, 1967). This might explain the observation that high amounts of magnesium appear to be associated with young growing tissue containing a high protein level.

2. Physiological and Biochemical Functions

Magnesium is commonly associated with transfer reactions involving phosphate-reactive groups (Nason and McElroy, 1963). Almost every phosphorylating enzyme in carbohydrate metabolism requires magnesium for maximal activity. Most reactions involving phosphate transfer from adenosine triphosphate (ATP) require this ion including ATP sulfurylase (Paynter and Anderson, 1974) and acyl-CoA synthetase (Young and Anderson, 1974). The conversion of 1,3-diphosphoglycerate to 3-phosphoglycerate does not require magnesium. In this reaction, catalyzed by phosphoglycerate kinase, a molecule of adenosine diphosphate (ADP) is converted to ATP without a demonstrated requirement for magnesium.

It has been suggested that magnesium forms a chelated structure with the phosphate groups, establishing the configuration that allows maximal activity in the transfer reactions (Shibko and Penchot, 1961). Magnesium has a marked influence on the energy of hydrolysis of high energy compounds such as adenosine phosphates. The magnesium complexes of ATP, ADP, and AMP (adenosine monophosphate) are formed with differing affinities, resulting in an ill-defined and complicated pattern of hydrolysis of these compounds (Alberty, 1968). It is highly probable that at the pH values found in most cells, a major portion of the adenosine phosphates are chelated with magesium (Lehninger, 1970). Magnesium, through the complexes formed with phosphate groups, has a controlling effect on the steady-state concentrations of high-energy phosphate groupings and undoubtedly influences the rate and extent of these all-important phosphate-transfer reactions.

Magnesium is also involved directly with potassium–sodium-stimulated ATPase activity (Fisher and Hodges, 1969; Hansson and Kylin, 1969; Sexton and Sutcliffe, 1969), which could imply a role for magnesium in ion transport mediated by ATPases.

C. Potassium

Studies concerned with the role of potassium in plant metabolism have produced a paradox. The relatively high amounts of potassium required for normal plant growth do not correlate with the observed functions of potassium. This element seems to function mostly as a catalytic agent in enzyme reactions (Table III). No potassium metabolite has been identified in plants (Epstein, 1972). Why, then, is potassium required in such large amounts by plants?

It has been proposed that potassium is not so much a catalytic agent but acts as an activator for the enzyme systems by maintaining a favorable ionic environment suitable for preserving the proper three-dimensional structure for optimal enzyme activity (Evans and Sorger, 1966).

1. PHYSIOLOGICAL FUNCTIONS OF POTASSIUM

Carbohydrate metabolism has been linked to potassium in the many studies on the effects of potassium deficiency in plants. Translocation of sugars was found to be closely linked to potassium (Hartt, 1969, 1970). Potassium-deficient sugarcane was demonstrated to have a slower sugar translocation even though the photosynthetic activity was not altered. Circulation of potassium around the sieve plate has been proposed as a mechanism for increasing translocation in sieve tubes (Spanner, 1958). A decrease in potassium content might therefore reduce translocation by depressing the potential across the sieve plates. Also, the light dependence of potassium-mediated translocation of sugars, independent of photosynthate buildup (Hartt, 1970), was related to the light en-

TABLE III

Functional Roles of Potassium

---

1. Translocation of sugars (Hartt, 1969, 1970)
2. Stomatal opening (Fujino, 1967; Fischer, 1968a,b; Humble and Hsiao, 1969, 1970)
3. Osmotic regulation (Bernstein, 1961, 1963; Rains, 1972)
4. Enzyme systems requiring potassium
   a. Acetic thiokinase (Hiatt and Evans, 1960)
   b. Aldolase (Rutter, 1964; Hsiao et al., 1970)
   c. Pyruvate kinase (Miller and Evans, 1957; Sorger et al., 1965)
   d. γ-Glutamylcysteine synthesis (Webster, 1953, 1956; Webster and Varner, 1954)
   e. Formyltetrahydrofolate synthetase (Hiatt, 1965a,b)
   f. Succinyl-CoA synthetase (Bush, 1969)
   g. Nitrate reductase induction (Nitsos and Evans, 1966; Oji and Izawa, 1969)
   h. Starch particulate synthetase (Murata and Akazawa, 1968; Nitsos and Evans, 1969)
   i. Activation of ATPase systems (Fisher and Hodges, 1969; Kylin and Gee, 1970)

---

hancement of potassium uptake observed in higher plants (Rains, 1968). Any factor that increases the transport of potassium could alter the electroosmotic potential between sieve tubes, thereby influencing sugar translocation.

Fujino (1967) observed an interaction between ATP, ATPases, and potassium in the opening of stomata. He proposed that energy in the form of ATP might be utilized to accumulate osmotically active ions such as potassium and regulate water flow.

At approximately the same time, Fischer (1968a) suggested that potassium might be required in the opening of *Vicia faba* stomata. He also showed a light response that was independent of CO_2, and he suggested ATP as a source of energy for accumulating potassium (Fischer, 1968b). Potassium uptake could be correlated with stomatal opening and the concentration for maximal opening was 10 mM potassium in CO_2-free air (Fischer and Hsiao, 1968).

Rubidium is as effective as potassium in stomatal opening in the light (Humble and Hsiao, 1969, 1970). In the light, the concentrations required for maximal stomatal opening are 100 times as great for sodium, ammonium, lithium, and cesium as for potassium and rubidium. In the dark, all ions are equally effective, although only at high concentrations. Light seems to mediate a high-affinity specific-transport system for potassium and rubidium, which increases the osmotic flow of water into the guard cells. Stomatal opening may be one of the first physiological processes in higher plants shown to have a specific requirement for potassium.

By using the X-ray electron probe microanalyzer, it is possible to locate spacially a relatively high concentration of potassium in the guard cells of opened stomata (424×10^{-14} gm equivalents K per stoma) as compared to closed stomata (20×10^{-14} gm equivalents K per stoma) (Humble and Raschke, 1971). Quantification of the potassium levels found in guard cells of opened stomata permits the conclusion that enough potassium accumulates to result in an increase of 16 bars osmotic pressure. The osmotic swelling produced was adequate to open the stomata. Inorganic anions have no significant effect, and it was concluded that organic anions are involved in balancing the potassium absorbed by the cells.

Potassium has a general function in the regulation of water in plant cells. This is not a specific function in most cases, although potassium has been shown to act in a rather selective way during osmotic adjustment of plants placed under stress (Bernstein, 1961, 1963; Rains, 1972). It was found that potassium was selectively absorbed by these plants and was one of the main ions involved in preventing the plant from losing water and becoming physiologically dry.

Early investigations pointed to a correlation between protein levels and potassium status of the plants. More recently, it has been found that protein actually increased in potassium-deficient plants relative to that in potassium-sufficient plants (Hsiao *et al.*, 1968, 1970). Only under conditions of severe potassium deficiency was a buildup of amino acids observed. However, experiments dealing with such a complex system as a nutrient-deficient plant are not rigorous enough to eliminate potassium completely from having a possible role in protein synthesis.

2. BIOCHEMICAL FUNCTIONS OF POTASSIUM

Evans and Sorger (1966) list some 46 enzymes in various plant, animal, and microbial species in which potassium is required for maximal activity.

This list was expanded by Suelter (1970), and a system of classifying various enzymic reactions activated by monovalent cations was presented. The classification system is consistent with the suggestions that monovalent cations (potassium, rubidium, ammonium, and sodium) interact with a substrate and enzyme to form a ternary complex that mediates the catalytic reaction.

In general, reactions involved in the phosphorylation of carboxyl groups or enolate anions and elimination through the formation of enol—keto tautomers or in reactions in which enol—keto intermediates are formed are activated by monovalent cations such as potassium (Suelter, 1970).

Acetic thiokinase from spinach leaves requires potassium for maximal enzymic activity (Hiatt and Evans, 1960). Rubidium and ammonium were similar to potassium in promoting activity. Sodium and lithium inhibited the enzyme system.

Aldolase from the green alga *Euglena gracilis* showed a requirement for potassium, but this is not generally the case for the enzyme from higher plants (Rutter, 1964). The aldolase in corn, however, was 30% less in specific activity when these plants were deficient in potassium (Hsiao *et al.*, 1970).

Pyruvate kinase is one of the few enzymes in which the requirement for potassium has been extensively studied and the mode of action characterized (Miller and Evans, 1957; Evans, 1963). In a series of investigations utilizing the immunoelectrophoretic and sedimentation properties of the enzyme, Sorger and co-workers (1965) concluded that cations known to enhance the activity of pyruvate kinase influence the conformation of the protein making up the enzyme. A paradoxical situation, however, has been shown with pyruvate kinase and potassium. It was found that when wheat plants were allowed to become deficient in potassium, pyruvate kinase activity increased (Sugiyama *et al.*, 1968). When the plants were furnished sufficient potassium the activity was reduced to "normal." They proposed that potassium might act as a regulator of this enzyme through repression of synthesis of pyruvate kinase.

γ-Glutamylcysteine synthesis requires potassium and is fairly specific for this cation (Webster, 1953, 1956). Rubidium could not readily replace potassium without reducing the activity of the enzyme. There is also a high specificity for potassium in the formation of glutathione from γ-glutamylcysteine and glycine (Webster and Varner, 1954).

Folic acid metabolism has been shown to require potassium. The system is not cationic specific, for rubidium and ammonium are equally effective. In spinach leaves, potassium promotes the formate-activating enzyme that converts formate and tetrahydrofolate to N^{10}-formyl tetrahydrofolate (Hiatt, 1965a,b).

Succinyl-CoA synthetase isolated from tobacco has been shown to have a potassium requirement of 0.06 M for maximal activity (Bush, 1969). Rubidium could replace potassium partially, as could sodium and ammonium, although they were less effective than rubidium.

The time required to induce nitrate reductase in *Neurospora crassa* was longer in the absence of potassium: 3 hours versus only 1 hour when potassium was sufficient. Again, rubidium could replace potassium partially, whereas sodium, lithium, and ammonium were without effect (Nitsos and Evans, 1966). The formation of nitrate reductase in rice seedlings also required potassium. The requirement was fairly specific, since substitution with rubidium gave only one-half as much activity as did potassium (Oji and Izawa, 1969).

Potassium was found to have little or no effect on nitrate reductase activity assayed from potassium-deficient corn plants (Hsiao *et al.*, 1970). It was concluded that the level of nitrate was the dominant factor in controlling the activity of this enzyme. Apparently potassium is more effective in the formation of nitrate reductase than in the maintenance of the enzyme system once it has been induced.

There is an absolute requirement for potassium by particulate starch synthetase isolated from sweet corn. Rubidium, ammonium, and cesium are 80% as effective as potassium, and sodium only 21% as effective (Nitsos and Evans, 1969). Tests of other species showed an absolute requirement for potassium. Potassium is essential for starch synthesis in sweet potato roots (Murata and Akazawa, 1968). The enzyme, ADP-glucose-starch transglucosylase has a K_m for potassium of 13.3 mM.

Potassium, through its role in ATPase activity, may be involved in ion transport (Skou, 1965). Generally, the activity of ATPase is enhanced by potassium and sodium, together or individually, and the hydrolysis of ATP might mediate the transfer of ions across biological membranes. Ion-stimulated ATPases were recently described in plant tissues (Gruener and Neumann, 1966; Fisher and Hodges, 1969): ATPases isolated from sugar beets and from mangroves (*Avicennia nitida*) require potassium plus sodium for maximal activity (Hansson and Kylin, 1969; Kylin and Gee, 1970). There is a close correlation between potassium absorption and ATPase activity in oat roots (Fisher et al., 1970). The implication that a potassium–sodium-stimulated ATPase is involved in ion transport is reinforced by the existence of relatively high concentrations of this enzyme in plant membrane fractions (Lai and Thompson, 1971).

The partial replacement of potassium by other ions in many physiological functions of plants may be attributed to its dual role. Potassium is involved in osmotic regulation. This osmotic role can be satisfied by ions with similar osmotic activity. The specificity of the function is manifested in delivery of the ion to the site of action. Penetration across cellular membranes at certain concentrations is a highly specific reaction (Epstein, 1965, 1966, 1972; Rains 1972). When whole cell physiology is studied—carbohydrate translocation and stomatal activity—specificity is demonstrable.

The isolation of enzymes and *in vitro* assays may require only a certain ionic strength and nonspecific osmotic regulation of the microenvironment of the protein. This could partially account for the relatively high levels of potassium required for many of the enzymes investigated. The specific role of an ion would then be directly dependent upon its chemical characteristics, such as energy of hydration and ionic crystal radii.

D. Iron

The essentiality of iron for plants and animals is unquestioned. Like other elements, it functions both as a structural component and as a cofactor for enzymic reactions. Oxidation–reduction reactions are most commonly associated with iron-containing systems. In animals the heme molecules responsible for oxygen transfer are made up of four pyrrole rings oriented around an atom of iron. The chemical properties of iron are responsible for its role in oxidation–reduction reactions. It is a transition metal, an element capable of existing in more than one oxidation state. Because of this property, it can accept or donate electrons according to the oxidation potential of the reactants.

In addition, iron combines with electron donors or ligands to form complexes. Chelation results when the ligand can donate more than one electron. Iron has been found to form stable chelates with molecules containing oxygen, sulfur, or nitrogen. Iron-containing organic ligands have a range of stabilities that are apparently dependent upon the oxidation state of this element (Price, 1968). The movement of electrons between the organic molecule and iron provides the potential for many of the enzymatic transformations in which iron is found to be essential (Table IV). This close association between structure and enzymatic function is such that a separation of the role of iron into these two categories would not be useful in the ensuing discussion.

1. PHYSIOLOGICAL ROLE OF IRON

In plants and animals a large portion of iron is found associated with porphyrins (Chance et al., 1968; Bendall and Hill, 1968). The iron porphyrins in animals are mainly heme molecules, whereas in plants cytochromes make up the majority of the iron porphyrins. Some of the iron-containing compounds, including enzyme systems, are as follows (Price, 1970): cytochromes b, b_6, c, c_2, and f; cytochrome oxidase complex; catalase and peroxidase; ferredoxin; ferrichrome; hematin; heme and leghemoglobin (Table IV).

Cytochromes are found in both respiratory and photosynthetic systems functioning in associated oxidation and reduction processes. These processes have been shown to be mediated by the interconversion of the ferrous and ferric states of iron. Characteristic absorption spectra identify the state of iron during the oxidation–reduction reactions (Chance et al., 1968).

TABLE IV

Functional Roles of Iron

---

1. Structural component of porphyrin molecules: cytochromes, hemes, hematin, ferrichrome, leghemoglobin (Price, 1968, 1970). Involved in oxidation–reduction reactions in respiration and photosynthesis
2. Structural component of nonheme molecules: ferredoxin (Price, 1970)
3. Enzyme systems
   a. Cytochrome oxidase (Chance et al., 1968; Price, 1968)
   b. Catalase (DeKock et al., 1960)
   c. Peroxidase (DeKock et al., 1960)
   d. Aconitase (Glusker, 1968)
   e. Chlorophyll synthesis, γ-aminolevulinate dehydratase (Carell and Price, 1965; Schneider, 1970); γ-aminolevulinate synthetase (Burnham and Lascelles, 1963); ferrochelatase (Jones, 1968)
   f. Peptidylproline hydrolase (Sadava and Chrispeels, 1971)
   g. Nitrogenase (Aleem, 1970)

If the distribution of iron is an indication of its role, then photosynthesis puts a great demand on iron. Various investigators (see Price, 1968) have estimated that as much as 75% of the total cell iron is associated with the chloroplasts, and in one case it was estimated that over 90% of the iron in leaves occurs with lipoprotein of the chloroplast and mitochondria membranes (Boichenko and Udel'nova, 1964). In contrast, iron accumulates in nuclei of root cells (Possingham and Brown, 1957). The localization of iron within the chloroplasts represents the relatively large contribution of cytochromes to the movement of electrons in the various photosynthetic reduction processes as well as the involvement of ferredoxin as an initial electron acceptor.

Involvement in the reduction of oxygen to water in the respiratory chain is the most commonly recognized function of iron-containing compounds. Other oxidation–reduction reactions take place in the cell, however, in which molecular oxygen is involved directly in the reaction. These include reactions catalyzed by catalase and peroxidase and the hydroxylation of proline in proline peptides.

The ferrous form of iron is required for the aconitase reaction in the tricarboxylic acid (TCA) cycle. Coordinate bonds form between the ferrous ion and the enzyme molecules. The citrate molecule is then attached in a highly specific manner that promotes the conversion of substrate to product (Glusker, 1968). The mechanism of citrate binding to the enzyme may be analogous to the formation of rather stable chelates formed between citrate and ferrous iron.

Iron was implicated in the synthesis of chlorophyll in plants when it was recognized that the yellowing of the leaves of iron-deficient plants was correlated with a reduction in chlorophyll. The question has been: where does iron have its effect on chlorophyll synthesis or maintenance?

In *Euglena*, for instance, iron is essential for conversion of coproporphyrinogen to protoporphyrin only when the growth of this alga is limited by low levels of iron in the substrate (Carell and Price, 1965). This was also true for enzymes involved in porphyrin synthesis [γ-aminolevulinate (AL) dehydratase, porphobilinogen deaminase, and synthesis of coproporphyrinogen and protoporphyrin from porphobilinogen]. As long as growth was independent of iron, so were these enzymes. It has been suggested that iron is involved directly in the conversion of coproporphyrinogen to protoporphyrin (Lascelles, 1961), but this was not supported by Carell and Price (1965). Iron was found to be necessary for γ-AL dehydratase isolated from spinach leaves (Schneider, 1970). Iron-deficient plants incorporated γ-ALA (γ-aminolevulinic acid) normally into chlorophyll but the incorporation of other metabolites into γ-AL was reduced (Marsh *et al.*, 1963). The synthesis of γ-AL was considered to be the control site for chlorophyll synthesis in iron-deficient systems. In *Rhodopseudomonas spheroides*, iron deficiency did not reduce the activity of γ-AL synthetase (Burnham and Lascelles, 1963). Chloroplast structure is dramatically disarranged in iron-deficient systems, indicating a role for iron in maintaining the physical shape necessary for chloroplast function.

In the dark, iron may be involved in regulating γ-ALA synthetase. An iron–protein complex apparently inhibits enzymatic synthesis of γ-ALA. If iron-specific chelators are added to the enzyme assay, synthesis proceeds in the dark (Duggan and Gassman, 1974). Chelators may mimic the effect of light in this instance. As in certain light-mediated reactions, chelators alter and retain a specific oxidation state of iron and in that way influence its interaction with protein and various substrates. More work is needed for a clearer understanding of the role of iron in chlorophyll synthesis.

Iron is also found as a component of various enzyme systems not related directly to the porphyrin structure. The enzyme ferrochelatase, which catalyzes the insertion of iron into protoporphyrin IX to give protoheme, obviously requires iron as a substrate (Jones, 1968).

Iron has been demonstrated to be essential for maximal activity of peptidylproline hydroxylase from carrot disks (Sadava and Chrispeels, 1971). In work with gibberellic acid-treated aleurone layers from barley, iron (in the ferric form) may be required for amylase synthesis to proceed at maximal activity (Goodwin and Carr, 1970).

Oxidation and reduction of inorganic nitrogen is related to iron metabolism. A decrease in nitrate reductase has been observed when iron is limiting, but that decrease is related to a corresponding reduction in the activity of cytochromes (Nicholas, 1961). The oxidation of inorganic nitrogen is directly linked with cytochromes in various bacteria, and has been found to be a part of the nitrate reductase system (Aleem, 1970).

The acquisition and translocation of iron in plants has been of considerable interest to plant physiologists and biochemists for many years (Price, 1968). The interest arises from the interaction observed between iron and other elements in the environment. In many situations, iron is not readily available to the organism but is nevertheless acquired in sufficient amounts so as not to be limiting for growth. Iron in the ferrous form is much more soluble when the pH of the medium is on the basic side. Plants mediate the reduction of $Fe^{3+} \rightarrow Fe^{2+}$. A reductant is excreted into the medium surrounding the roots, enabling the plant to solubilize sufficient iron for growth (Brown *et al.*, 1966; Ambler *et al.*, 1971). In solutions in which either iron-efficient or iron-inefficient plants had been cultured, a heat-labile factor was found which increased iron absorption by iron-stressed, iron-efficient soybeans (Elmstrom and Howard, 1970). Plants that were inefficient accumulators of iron, however, depressed iron absorption by efficient plants when grown in a mixed culture.

Translocation of absorbed iron also is influenced by organic metabolites in the plant. Citric acid may mediate the translocation of iron by forming a chelate with this element that is soluble and reduces interaction with other anions (Tiffin, 1966). There is an inverse relation between iron levels and citrate transport in the xylem of soybeans and tomatoes (Brown and Chaney, 1971). As iron is increased, citrate in the xylem decreases, implying, indirectly, that iron is combined with citrate during translocation and therefore that its concentration in the xylem is effectively reduced.

E. Manganese

Manganese is associated most commonly with its role in photosynthesis. It is also involved, however, in oxidation–reduction processes and decarboxylation and hydrolysis reactions. Manganese can replace magnesium in many of the phosphorylating and group-transfer reactions (e.g., glucose kinase, hexokinase, phosphoglucokinase, phosphoglucomutase, and adenosine kinase) (Nason and McElroy, 1963). Many of the enzyme reactions in the citric acid cycle require manganese for maximal activity. Most of the enzyme systems do not have a specific requirement for this metallic ion. In a majority of the enzyme systems listed above, magnesium is as effective as manganese in promoting enzyme transformations (Table V). One notable exception is the absolute requirement for manganese in the NAD-malic enzyme system found in leaves of aspartate-type C_4 plants (Hatch and Kagawa, 1974). Magnesium could not substitute for manganese in this enzymatic transformation. In the malic-type, C_4 plants, however, magnesium was readily substituted for manganese.

TABLE V

Functional Roles of Manganese

---

1. Electron transport in photosystem II (Cheniae, 1970)
2. Maintenance of chloroplast membrane structure (Teichler-Zallen, 1969; Constantopoulos, 1970)
3. Manganin (Dieckert and Rozacky, 1969)
4. Enzyme systems
   a. Chromatin-bound RNA polymerase (Duda and Cherry, 1971)
   b. Synthesis of tRNA-primed oligoadenylate (Walter and Mans, 1970)
   c. Synthesis of phosphatidylinositol (Sumida and Mudd, 1970a,b)
   d. Inactivation of IAA protectors (Stonier et al., 1968)
   e. NAD malic enzyme of aspartate-type C_4 plants (Hatch and Kagawa, 1974)

A protein containing manganese has been isolated from peanut seeds. The manganoprotein was assigned the trivial name of manganin (Dieckert and Rozacky, 1969). A molecule of manganin contained 1 atom of manganese and the protein had a molecular weight of 56,300. No functional role has been determined for this manganoprotein, but it appears to be analogous to the copper-containing blue proteins (Section IV,K).

The role of manganese that has been studied the most intensively is the one it plays in the evolution of oxygen in photosynthesis. Most aspects of photosynthesis seem to have been studied intensively, including manganese and its relation to oxygen evolution (Cheniae, 1970).

Teichler-Zallen (1969) found that manganese deficiency produced marked disorganization in chloroplast membrane structure, resulting in inhibition of photosystem II. The synthesis of fatty acids essential for formation of chloroplast lamellae was also disrupted in manganese-deficient *Euglena gracilis* Z (Constantopoulos, 1970). It was suggested that the Hill reaction was essential for synthesis of α-linolenic acid and that manganese deficiency reduced the activity of the Hill reaction.

There are other enzyme systems requiring manganese for maximal activity that do not involve phosphate group transfer or photochemical reactions.

Chromatin-bound RNA polymerase from root tissue of sugar beet has an absolute requirement for manganese or magnesium (Duda and Cherry, 1971). The data indicate that the system is more sensitive to manganese than to magnesium, although either divalent cation will meet the requirements of the enzyme system.

Another enzyme system in which manganese acts as a catalyst is in the synthesis of tRNA-primed oligoadenylate. In a crude enzyme extract from maize seedlings, manganese was essential for optimal activity (Walter and Mans, 1970).

Mitochondria isolated from cauliflower inflorescence were found to be capable of synthesizing phosphatidylinositol when myo-inositol and cytidine diphosphate diglyceride were added as substrates (Sumida and Mudd, 1970a). Manganese was required in a concentration $\frac{1}{20}$ that of magnesium for optimal activity, and the maximal rate of synthesis was 9 times as great with manganese as with magnesium. These investigators were also able to isolate the enzyme responsible for the formation of phosphatidylinositol, and manganese was found to be essential for optimal activity (Sumida and Mudd, 1970b).

There is considerable evidence that manganese influences the level of auxin in plant tissues. The original observations came from work on the effect of manganese toxicity on plant growth. It was suggested that the symptoms observed when cotton plants were exposed to excessive manganese were due to low levels of indoleacetic acid (IAA) (Morgan et al., 1966). The reduced levels were the result of an enhancement of IAA oxidase activity. This was brought about by a manganese-catalyzed destruction of IAA oxidase inhibitors. In morning glory, auxin protector I was inactivated by manganese (Stonier et al., 1968). The inactivation of auxin by manganese is thought to be the result of two mechanisms: first, the oxidation of the protectors, and then an acceleration of the oxidation of IAA by endogenous peroxidase. Manganous ion did not inactivate the protector; only manganic ion was effective in causing a decrease in IAA protector I. The interaction between manganese and IAA oxidase is not necessarily as straightforward as indicated in the above discussion. In dark-grown peas, IAA oxidase in the terminal buds did not require manganese. Addition of gibberellic acid (GA) to dwarf peas caused the IAA oxidase activity to reach the same level as in tall peas. If manganese was included in the assay, GA had no stimulatory effect (Ockerse and Naber, 1970). They suggested that the oxidase was peroxidative in nature.

F. Chlorine

Chlorine was the last element shown to be essential for most higher plants (Broyer et al., 1954; Johnson et al., 1957). No true metabolite containing chloride has been found in higher plants, although microorganisms do contain 5 and 6 carbon ring structures with chlorine attached (Petty, 1961).

Chlorine functions primarily in photosynthetic reactions. Even before it was shown to be essential for higher green plants, the evolution of photosynthetic oxygen by chloroplast fragments was found to require chloride (Warburg and Luttgens, 1946). Refinement of procedures for studying photosynthesis gave a better understanding of the mode of action of chlorine in light-dependent reactions. The above workers suggested that chloride ion is involved in the primary process of oxygen evolution. The effect of chloride on the Hill reaction placed the site of chloride activity in photosystem II. Chlorine-depleted systems showed normal rates of cyclic photophosphorylation and other photosystem I reactions but a disturbed photosystem II (Bové et al., 1963; Hind et al., 1969). The system was demonstrable only in an alkaline pH. Conclusive evidence was finally supplied by Izawa and co-workers (1969).

By using various inhibitors of photosynthetic electron-transfer reactions, it was concluded that chloride was acting on the oxidizing side of photosystem II near the water-splitting end. In chloride-deficient systems, hydroxylamine or ammonia substituted for chloride, and photolysis proceeded normally. These data together with previous data suggested that chloride was involved closely with oxygen evolution.

The actual function of chloride in photosystem II has not been established. It has been suggested that chloride functions on the oxidizing side of photosystem II (Heath, 1973) and that chloride might remove photochemically induced deleterious oxidants by stimulating electron transport. This would protect against photoinactivation (Cheniae, 1970).

Chloride is partially responsible for the opening of *Zea mays* stomata in the light. Within 2 minutes after the lights are turned on, both potassium and chloride move from the subsidiary cells of the stomatal apparatus to the guard cells. The subsequent osmotic uptake of water by the guard cells results in an increase in turgor and open stomata (Raschke and Fellows, 1971). The uptake of these two ions was not stoichiometric. Approximately one-half as much chloride as potassium was shuttled between the subsidary cells and the guard cells.

A general role for chloride in stomatal opening is not apparent at this time. As discussed previously (Section IV,C) organic anions, not chloride, appear to be responsible for electrically balancing the potassium taken up by the stomata of *Vicia faba* during opening (Humble and Raschke, 1971).

G. Boron

The essentiality of boron for all green plants is not absolute. It is required by higher green plants and diatoms but has not been shown to be essential for all species of green algae. The growth of *Chlorella* was found to be promoted by boron, but the requirement was not absolute (McIlrath and Skok, 1958). The growth-promoting effect of boron in *Chlorella* has been questioned (Bowen *et al.*, 1965). A boron requirement could not be demonstrated in *Scenedesmus* (Dear and Aronoff, 1968). After twelve successive subcultures, boron was reduced to values of less than 0.05 μM. Thus, if boron is essential at all, it is required in amounts of less than 0.05 μM. Until better purification procedures are available, one can assume that this element is not essential for most of the green algae species tested. In addition, boron has not been shown as an essential nutrient for the growth of fungi (Bowen and Gauch, 1966).

Early investigations of the function of boron centered on its effect on sugar transport (Gauch and Duggar, 1954). A complex, supposedly formed between sugar and boron, was thought to increase the translocation of sugar through the plant. Alternative suggestions have included an inhibition of starch synthesis, thereby maintaining sugars in easily translocated soluble forms, and an increase in synthesis of sucrose, which is the major compound translocated by plants (Duggar and Humphreys, 1960). Specifically, UDP-glucose (uridine diphosphate-glucose) pyrophosphorylase activity was increased by boron, and UDP-glucose levels were maintained. Sucrose was then synthesized by combining UDP-glucose and fructose through a transglycosylase reaction. The overall reaction is represented in Scheme 1.

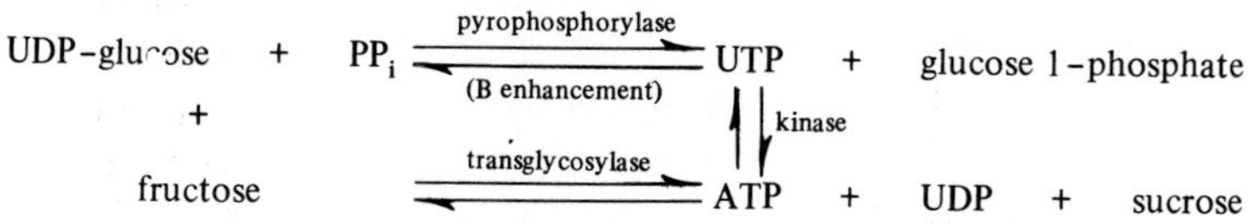

$$
\begin{array}{ccc}
\text{UDP-glucose} \;+\; \text{PP}_i & \xrightarrow[\text{(B enhancement)}]{\text{pyrophosphorylase}} & \text{UTP} \;+\; \text{glucose 1-phosphate} \\[4pt]
+ & \Big\updownarrow \text{kinase} & \\[4pt]
\text{fructose} & \xrightarrow{\text{transglycosylase}} & \text{ATP} \;+\; \text{UDP} \;+\; \text{sucrose}
\end{array}
$$

Scheme 1

In Scheme 1, boron maintains a level of substrate (UDP-glucose) adequate for the reaction to proceed and the sucrose to be formed.

This role of boron in carbohydrate translocation has not won universal acceptance, and the quest continues for other possible functions of this element.

Indications as to the functions of boron have arisen from studies on plants deficient in boron. Boron-deficient plants show a marked accumulation of phenolic acids. A buildup of these compounds causes necrosis and death. It has been suggested that boron might influence phenol synthesis (Lee and Aronoff, 1967). Since phenolic acids are synthesized from a compound formed by condensation of erythrose 4-phosphate (from pentose shunt) and pyruvate (glycosis), they investigated the effect of boron on these two pathways of carbon utilization. It was determined that boron combines with 6-phosphogluconate, the initial substrate for the pentose shunt. This complex inhibits the enzyme 6-phosphogluconate dehydrogenase, which ultimately results in a lowered level of erythrose 4-phosphate, which is required for phenolic acid synthesis. If boron is absent, the pentose shunt furnishes more substrate for phenol formation. Boron would thus regulate the pathways of carbon utilization, which in turn controls various intermediates that can function as substrates for the synthesis of phenolic acid. The situation in a boron-deficient plant is further complicated by an increase in the activity of synthesized phenolics in the cells. Adequate levels of boron in plant cells would normally complex the excess phenolics, rendering these potentially inimical compounds harmless. Boron deficiency, however, leads to an elevated level of phenols which is compounded by a reduction in the capacity of the detoxification mechanism.

Boron has other functions that are related to the observed reduction in the growth and death of meristematic regions of boron-deficient plants.

Root elongation ceased 12 hours after boron was removed from the culture solution of a tomato plant (Albert, 1965). The level of RNA in the root tips did not decrease until 24 hours after the boron was removed. Other symptoms of boron deficiency (tip browning and decreases in fluorescence of root tips) paralleled the reduction in RNA levels. The DNA content of these roots did not change over the period of the investigation. An alteration of RNA content could be expected to have some effect on the protein content of plants. Yih and Clark (1965) found that a reduction in root elongation was not reflected in changes in the protein or carbohydrate levels. Even after excised roots were exposed for 72 hours to zero boron, the protein and carbohydrate levels were not altered even though root elongation had ceased after 24 hours.

Since protein and carbohydrate levels are not altered over the experimental periods investigated, it can be reasoned that the decline in RNA levels must be one of the earliest biochemical changes observable in boron-deficient plants (Johnson and Albert, 1967). When boron-deficient plants were supplied with various nitrogen bases (thymine, guanine, and cytosine) root elongation was stimulated. Uracil, orotic acid, and adenine were not as effective as cytosine, guanine, and thymine. Barbituric acid and 6-azauracil inhibited root elongation in either the presence or absence of boron. Selected bases seemed to alleviate boron deficiency. Not only was root elongation stimulated but RNA levels decreased very slightly in the presence of the three effective bases. Barbituric acid and 6-azauracil induced symptoms similar to boron deficiency even though boron was present at normally adequate levels. Those workers suggested that boron must be involved in some aspect of nitrogen-base synthesis or utilization that would influence RNA metabolism. If this were so, one would expect some alteration in protein levels, but, as discussed previously, protein levels did not change in boron-deficient roots (Yih and Clark, 1965). It may be that individual proteins are affected, although this is not reflected in the total protein content of the tissue.

Boron and GA_3 have been found to influence the α-amylase activity in germinating seeds. Boron apparently has a regulatory role with synthesis of GA_3 (Cresswell and Nelson, 1973). Low levels of boron result in reduced amounts of GA_3 and these in turn alter the activity of α-amylase. Accompanying these changes were alterations in RNA metabolism. The data suggest there is an involvement of boron in RNA metabolism, which in turn manifests itself in a whole range of responses including protein synthesis and enzyme activities.

One has to be impressed with the multiplicity of functions that have been postulated for boron in plants, and also with the lack of agreement on whether any of these define the role of boron in plants. If it is ascertained that boron is not essential for all algae and if its function in higher plants is determined, a greater understanding could possibly emerge concerning the biochemical differences between algae and multicellular higher plants.

H. Molybdenum

Molybdenum was shown initially to be essential for fixation of gaseous nitrogen by *Azotobacter chroococcum*. It was later found also to be a necessary element in the fixation of nitrogen by legumes and some algae (Nason and McElroy, 1963). Molybdenum was demonstrated as an essential nutrient for nonleguminous higher plants by growing tomatoes in nutrient solutions containing no molybdate (Arnon and Stout, 1939b).

One of the most significant findings pertaining to the role of molybdenum in nitrogen utilization by plants was the interaction of molybdenum and the source of nitrogen. The amounts of molybdenum required were significantly less when the source of nitrogen was ammonia than when it was nitrate (Mulder, 1948; Arnon *et al.*, 1955; Sheat *et al.*, 1959). The role of molybdenum in the reduction of nitrate to ammonia was further clarified by the demonstration that when plants were grown with nitrite as a nitrogen source, molybdenum was not required although it stimulated growth when added (Sheat *et al.*, 1959). These findings indicated that molybdenum was necessary for the reduction of nitrate to nitrite, and that supplying either nitrite or ammonia effectively bypassed the initial reductive step and the molybdenum requirement (Table VI).

Nitrate reductase was initially isolated and characterized in higher plants from the leaves of *Glycine max* (Nason and Evans, 1953). This enzyme was found to be a sulfhydryl, metalloflavin adenine dinucleotide (metallo-FAD) protein containing molybdenum (Nicholas and Nason, 1955). With the demonstration of this element as a component of an

TABLE VI

Functional Roles of Boron and Molybdenum

Boron
1. Sugar translocation (Gauch and Duggar, 1954; Duggar and Humphreys, 1960)
2. Phenol metabolism (Lee and Aronoff, 1967)
3. RNA metabolism (Albert, 1965; Yih and Clark, 1965; Johnson and Albert, 1967)
4. GA and α-amylase activity (Cresswell and Nelson, 1973).

Molybdenum
1. Structural component of nitrate reductase (Nason and Evans, 1953; Notton and Hewitt, 1971)
2. Iron absorption and translocation (Berry and Reisenauer, 1967)

enzyme system, the essentiality of molybdenum has been firmly established, at least when nitrate is the nitrogen source.

Discussion in current literature on the role of molybdenum centers principally on the actual functional site of this element in the nitrate reductase enzyme complex. Two enzymatic activities can be identified when electrons are transferred from reduced nicotinamide adenine dinucleotide (NADH) to nitrate. The reductase complex consists of an FAD-dependent NADH diaphorase activity and a terminal nitrate reductase, requiring molybdenum. Molybdenum is required for the reduction of nitrate by the NADH-nitrate reductase complex isolated from *Chlorella* (Vega *et al.*, 1971). The complex is formed in the absence of molybdenum, but only the diaphorase is active. Transfer of electrons to nitrate from NADH requires molybdenum. *De novo* synthesis of the nitrate reductase apoenzyme does not require nitrate or molybdenum. Molybdenum is incorporated into the apoenzyme after it is released from the ribosomes (Notton and Hewitt, 1971). In normal (molybdenum-sufficient) plants, the ratio of diaphorase to nitrate reductase activity was approximately 50 to 1; molybdenum-deficient plants, in contrast, had a ratio of 750 to 1 (Notton and Hewitt, 1971). Diaphorase activity was influenced somewhat by molybdenum, but the mechanism of action in this enzyme system was thought to be separate from the mode of action of molybdenum in nitrate reductase. They completely refuted the contention of Paneque and Losada (1966) that molybdenum was not involved in the enzyme system responsible for the reduction of nitrate. This contention arose from the observation that cyanide was not effective in decreasing nitrate reductase activity. Notton and Hewitt were able to demonstrate a very tight binding of molybdenum by the enzyme, and this might explain the ineffectiveness of cyanide in removing molybdenum.

Another interesting aspect of nitrate reduction is the so-called "nitrate respiration." This is of considerable importance in microorganisms under anaerobic conditions where nitrate can replace oxygen as the terminal electron acceptor, and mediates respiration at an adequate rate (Nason and McElroy, 1963). Ulrich (1971) grew the green alga *Ankistrodesmus* in the absence of CO_2 and O_2 and found that noncyclic photophosphorylation was dependent upon nitrate for electron transfer. When the alga was maintained under N_2 gas at low light intensities, noncyclic photophosphorylation dominated over cyclic. High levels of DCMU [3-(3,4-dichlorophenyl)1,1-dimethylurea] were required to shift the system from the noncyclic to the cyclic mode. The role of molybdenum in this process is not clear at this time.

Investigations on higher plants have suggested a role for molybdenum separate from that of nitrate reduction. Molybdenum-deficiency symptoms can be demonstrated in plants that are supplied ammonia (Agarwala, 1952). Assuming that the pH of the culture solutions is maintained within reasonable limits and that there is no nitrification of ammonia, then molybdenum might have other functions.

There is an interesting interaction between molybdenum and iron in tomato plants (Berry and Reisenauer, 1967). The addition of adequate molybdenum enhances absorption and translocation of iron and also decreases the availability of iron compounds in the root media. With barley, the addition of adequate amounts of molybdenum increased the reductive capacity of the roots, thereby increasing the solubility of available iron and hence promoting absorption of this element by plant roots. On the other hand, molybdenum deficiency caused a decrease in the translocation of iron from veinal to interveinal tissue, the result of a decrease in the reduction of iron into more soluble forms. This also accounted for decreased absorption by roots. High levels of molybdenum similarly reduced iron uptake, although by a chemical mechanism external to the plant. The excess molybdenum became coated on iron oxide compounds, resulting in a decreased solubility (and hence availability) of iron for the plant.

I. Zinc

Zinc is associated most commonly with auxin and this appears to be one of its main functional roles. The classic work on this association, carried out by Skoog (1940), demonstrated that zinc-deficient plants were also deficient in auxin. He did not propose a direct relationship between zinc and auxin formation but suggested that zinc prevented oxidation of the hormone. The oxidizing capacity, in the form of peroxidase, was enhanced in zinc-deficient systems (Table VII).

Zinc appears to be necessary for the synthesis of tryptophan and through this compound auxin levels are affected (Tsui, 1948). This was

TABLE VII

Functional Roles of Zinc

1. Auxin metabolism (Skoog, 1940; Tsui, 1948)
   a. Tryptophan synthetase (Nason et al., 1951)
   b. Tryptamine metabolism (Takaki and Kushizaki, 1970)
2. Dehydrogenase enzymes; pyridine nucleotide, alcohol, glucose 6-phosphate and triose phosphate (for reference, see Price, 1970)
3. Phosphodiesterase enzyme from Avena (Udvardy et al., 1970)
4. Carbonic anhydrase (Gerebtzoff and Ramaut, 1970; Randall and Bouma, 1973)
5. Promotes synthesis of cytochrome c (Grimm and Allen, 1954; Brown et al., 1966)
6. Stabilizes ribosomal fractions (Prask and Plocke, 1971)

considered to be an indirect effect because the enzyme system responsible for oxidative deamination of tryptophan to IAA was not influenced by zinc deficiency; only the tryptophan levels were reduced.

The enzyme responsible for the synthesis of tryptophan from indole and serine was found to require zinc for maximal activity. The requirement for zinc was specific for the enzyme tryptophan synthetase, isolated from *Neurospora* (Nason et al., 1951). In a recent study the influence of zinc deficiency on the accumulation of tryptophan and tryptamine in maize seedlings has produced conflicting data (Takaki and Kushizaki, 1970). As the seedlings became more zinc-deficient, tryptophan levels increased. The highest levels were associated with the most deficient seedlings. This was true also for tryptamine. Since extremely zinc-deficient systems were utilized, this might have been a pathological response. The data suggest, although somewhat indirectly, a definite role for zinc in auxin metabolism, mediated through its influence on the synthesis of intermediates along the pathway of auxin formation.

Studies on zinc-deficient plants have led to suggestions of other possible functions of zinc in various enzymes. Some of the pyridine nucleotide dehydrogenases are zinc enzyme systems. Alcohol dehydrogenase requires zinc, as do glucose 6-phosphate and triosephosphate dehydrogenases (for references, see Price, 1970). It has been suggested that zinc is involved in binding NAD to the protein and stabilizing a tetramer made up of four units of the apoenzyme (Kagi and Vallee, 1960).

Udvardy and his co-workers (1970) have isolated from *Avena* leaf tissue a phosphodiesterase that has a requirement for zinc. Other divalent cations (magnesium, cobalt, and calcium) will also increase the activity of this enzyme system. This particular phosphodiesterase was effective in the hydrolysis of denatured DNA.

Another well-known enzyme system that requires zinc is carbonic anhydrase (Gerebtzoff and Ramaut, 1970; Price, 1970). Earlier investigators could not conclusively demonstrate zinc to be a component of the enzyme from plants. That has not yet been settled, and, in fact, zinc was recently considered to function as a coenzyme with carbonic anhydrase (Gerebtzoff and Ramaut 1970). Plants under severe zinc deficiency have carbonic anhydrase levels only 10% of normal (Randall and Bouma, 1973).

Zinc functions in processes other than as an enzyme activator or cofactor. This element has been found to promote the synthesis of cytochrome *c* in the fungus *Ustilago sphaerogena* (Grimm and Allen, 1954). It has been suggested that zinc was inducing *de novo* synthesis of cytochrome *c* and that new messenger RNA is released when the zinc effect is noted (Brown et al., 1966).

Zinc has also been shown to bind various cellular fractions. In studies on zinc-tolerant clones of *Agrostis tenuis* (Peterson, 1969; Turner, 1970), the amount of zinc associated with the cell wall fraction was much greater in the tolerant than the intolerant clone. Separating these clones into specific metal-tolerant groups demonstrated a high specificity of the metal for the binding sites (Turner and Marshall, 1971). The zinc could be bound by pectic materials in the cell wall and therefore fail to enter the cell and cause adverse effects such as zinc toxicity (Peterson, 1969).

A structural requirement for zinc has been demonstrated in *Euglena gracilis* (Prask and Plocke, 1971). When this alga was cultured under zinc-deficient conditions, cytoplasmic ribosomes disappeared from the cells. Zinc-sufficient cultures gave *Euglena* with a stable population of ribosomes and little turnover. Zinc content was four times as high in stable ribosomes as in ribosomes beginning to disappear in zinc-deficient cells. Intracellular zinc content was 14 times as high in the sufficient cells as in the deficient cells; thus zinc may be essential for maintenance of the tertiary and quaternary structure of ribosomes, thereby preventing breakdown by ribonucleases. The degree of specificity for zinc in this protective role has not been clearly established.

J. Copper

The essentiality of copper for higher plants was initially demonstrated in 1931 (Sommer, 1931; Lipman and Mackinney, 1931) and confirmed by Arnon and Stout (1939b).

Copper is commonly found in a group of enzymes in which oxygen is used directly in the oxidation of substrate. Much of the work on copper enzymes has been concerned with oxidases such as tyrosinase, laccase, and ascorbic acid oxidase. The ubiquitous feature is the addition of $\frac{1}{2} O_2$ to the substrate, with the formation of a more oxidized product (Table VIII). Copper is suggested to mediate these enzyme transformations by undergoing cyclic oxidation and reduction

$$Cu^{2+} \overset{e^-}{\rightleftharpoons} Cu^+$$

It has been proposed that copper atoms exist as a Cu^{2+}–Cu^{2+} pair in the oxidized protein, so that two electrons can be transferred to the terminal oxygen acceptor, water (Malkin *et al.*, 1969). This might also describe the mode of action of copper in the cytochrome oxidase system. The mechanisms of copper oxidase enzymes were thoroughly reviewed by Nason and McElroy (1963), and therefore are not dealt with in any greater detail in this chapter.

TABLE VIII

Functional Roles of Copper

---

1. Oxidase enzyme; tyrosinase, laccase and ascorbic acid (Price, 1970)

$$\text{General reaction: AH} + \tfrac{1}{2}\,O_2 \underset{\phantom{Cu^{2+}}}{\overset{Cu^{2+}}{\rightleftharpoons}} A + H_2O$$

2. Terminal oxidation by cytochrome oxidase (Nason and McElroy, 1963)
3. Photosynthetic electron transport mediated by plastocyanin (Katoh, 1960; Levine, 1969)
4. Indirect effect on nodule formation (Cartwright and Hallsworth, 1970)

Several blue proteins have been found in nonphotosynthetic systems: stellacyanin, azurin, mung bean blue protein (Shichi and Hackett, 1963), and umecyanin (Paul and Stigbrand, 1970). Umecyanin, isolated from horseradish root, was associated with peroxidase activity during extraction. The intensely blue protein contained one atom of copper per molecule of protein, which is very similar to that of other blue proteins. It was found to have no enzymatic activity toward substrates normally oxidized by copper-containing oxidase enzymes (Stigbrand, 1971). A blue protein has also been isolated from rice bran. It was found to be a glycoprotein, with the copper present in the cupric form (Morita *et al.*, 1971). The function of these copper-containing proteins is unknown, but they may be no more than a nondirected condensation of protein around this element during extraction.

Plastocyanin is a copper-containing protein that is involved in electron transfer in photosynthetic reactions (Katoh, 1960) and appears to have no function in the absorbance of light in the primary act of photosynthesis (Levine, 1969). A very sensitive enzyme assay procedure was utilized to demonstrate that plastocyanin was the only copper-containing compound in the photosynthetic electron transport system (Plesincar and Bendall, 1970).

There are many observations suggesting that copper has a specific inhibitory effect on photosynthetic reactions when it is added in toxic amounts. The Mehler reaction, a Hill reaction with oxygen as the electron acceptor, is inhibited by excessive copper (Habermann, 1969). Manganese reverses this inhibition, and glutathione enhances the manganese reversal. There is also a copper-specific toxicity in *Chlorella*, with the detrimental effect manifested only in the light (Nielsen *et al.*, 1969). It was proposed that copper was bound to the cytoplasmic membrane, preventing cell division and resulting in a buildup of photosynthate, effectively reducing further photosynthesis through a feedback inhibition mechanism. In diatoms there is a differential effect of excessive copper

(Nielsen and Wium-Andersen, 1971). Small amounts of copper are more detrimental to photosynthesis than to growth. Apparently the diatoms excrete organic matter that complexes copper, making the nutrient solution suitable for growth. The effect of copper on photosynthesis was much the same as that proposed for *Chlorella*.

Membrane damage occurs in *Chlorella* in the presence of excessive copper (Gross *et al.*, 1970). This damage could alter chlorophyll and carotenoid pigments, thereby reducing photosynthesis and respiration.

Another effect of copper on plant processes has been observed in the legume, subterranean clover. In this instance, the system was made copper-deficient, and nitrogen fixation, nodulation, and plant growth were studied (Cartwright and Hallsworth, 1970). Copper deficiency was found to have a marked effect on nodule development and nitrogen fixation, while leaf growth was reduced only slightly. The explanation of this phenomenon was that copper is critical in maintaining cytochrome oxidase. If the activity is reduced, then oxygen levels could increase in the nodule, which in turn would restrict nitrogen fixation. The more reduced is the total environment in the nitrogen-fixing system, the greater is the rate of reduction of atmospheric nitrogen.

IV. Other Nutrients

This section is concerned with selected nutrients that have not been demonstrated to be essential for many forms of plant life (Table IX).

Sodium

Demonstration of the essentiality of sodium for plant growth has been restricted to species normally found in high saline environments. Sodium is essential for *Atriplex vesicaria,* a plant common to the arid regions of the world (Brownell and Wood, 1957). Culture solutions containing less than 0.0016 ppm sodium produced deficiency symptoms in a relatively short time (20-day-old plants), and these plants died by day 35 (Brownell, 1965). Solutions containing 0.46 ppm sodium permitted adequate plant growth. There was a specific requirement for sodium during the recovery of deficient plants.

There is a sodium requirement for *Halogeton glomeratus* (Williams, 1960). Sodium is accumulated along with oxalic acid, and much of the sodium is present as sodium oxalate. The presence of large amounts of oxalate appears to be responsible for the poisonous nature of this plant.

When nonhalophytic plants are exposed to conditions of low potassium, sodium is found to have a beneficial effect on plant growth (Harmer

TABLE IX
Functional Roles of Other Nutrients

1. Sodium
   a. Oxalic acid accumulation (Williams, 1960)
   b. Potassium sparing action (Lehr, 1953; El-Sheikh et al., 1967)
   c. Stomatal opening (Willmer and Mansfield, 1969)
   d. Regulation of nitrate reductase activity (Brownell and Nicholas, 1967)
   e. Required for plants with C_4 photosynthetic pathway (Brownell and Crossland, 1972; Shomar-Ilan and Waisel, 1973)
   f. Induction of Crassulacean metabolism (Winter, 1973)
   g. Maintenance of water balance (Rains, 1972)

2. Selenium
   a. No functional role as yet established (Shrift, 1969)
   b. Sulfur metabolic analogs (Peterson and Butler, 1967)

3. Silicon
   a. Formation of silicon walls in diatoms (Lewin and Reimann, 1969)
   b. Reduce toxicity of other elements (Vlamis and Williams, 1967)

4. Cobalt
   a. Nitrogen fixation by symbiotic organisms (Reisenauer, 1960; Shaukat-Ahmed and Evans, 1959, 1961)
   b. Leghemoglobin metabolism (Wilson and Reisenauer, 1963; DeHertogh et al., 1964)
   c. Ribonucleotide reductase in Rhizobium (Cowles et al., 1969)

and Benne, 1945; Lehr, 1953). Sodium cannot completely replace potassium but appears to have a sparing action, with less potassium required for optimal growth. Sodium enhanced the growth of sugar beets whether potassium was adequate or deficient (El-Sheikh et al., 1967). It is of interest that sugar beets and Atriplex belong to the same family, Chenopodiaceae.

Sodium can partially replace potassium in many of the reactions known to require potassium (Epstein, 1965; Evans and Sorger, 1966). The role of potassium in stomatal opening can be partially replaced by sodium, although only at high ionic concentrations (Humble and Hsiao, 1969). There is a possible physiological role of sodium in the opening of Commelina stomata (Willmer and Mansfield, 1969), although the concentrations of sodium and potassium required in Commelina are 300 times those found for light-enhanced, potassium-specific opening in Vicia faba (Humble and Hsiao, 1969). The concentration of ions required for stomatal opening in Commelina are similar to the concentrations of sodium and potassium required for nonspecific, dark opening of Vicia stomata. The differences noted may be due to species differences.

When a previously sodium-deficient Atriplex is returned to a sodium-sufficient environment, a respiratory rise measured by O_2 consumption precedes the return to normal rates of growth (Brownell and Jackman, 1966). The effect was specific for sodium, and the CO_2 evolved was from fermentation processes. This is interpreted to mean that glycolysis was specifically affected by sodium deficiency in plants, thus showing a requirement for sodium for this process.

The alga *Anabaena cylindrica* was the first plant shown to have a sodium requirement (Allen and Arnon, 1955). The main visual symptom of the deficiency was the development of chlorosis. The chlorosis is due to nitrite toxicity from nitrate accumulation when sodium was deficient in the system (Brownell and Nicholas, 1967). Only small amounts of sodium were necessary when reduced nitrogen compounds were used as a nitrogen source. In the absence of sodium, nitrate reductase activity increases, resulting in a buildup of nitrite. Nitrogen assimilation is reduced in sodium-deficient cells. The general interpretation of these data is that sodium influences protein synthesis and in this way exerts a control on nitrogen metabolism.

Many plants that possess the C_4 dicarboxylic photosynthetic pathway require sodium as an essential nutrient (Brownell and Crossland, 1972). Plants from the same genus but characterized as C_3, Calvin–Benson type plants, do not respond to sodium and do not appear to have a general requirement for this element.

Sodium is apparently necessary for the expression of C_4 carbon fixation in certain plant species. *Aeluropus litoralis,* a halophytic plant species, was found to fix carbon via the Calvin–Benson pathway (C_3) when depleted of sodium. The inclusion of sodium at saline levels shifted the photosynthetic pathway from C_3 to C_4 (Shomar-Ilan and Waisel, 1973). It was also demonstrated that sodium influences the balance between phosphoenolpyruvate carboxylase (C_4) and ribulose-1,5-diphosphate carboxylase (C_3) in *Zea mays,* a nonhalophytic plant.

Sodium apparently has a similar regulatory function in other drought-tolerant species. Certain plant species show a requirement for Na for the expression of crassulacean acid metabolic (CAM) pathway. Two members of the Aizoacea family show an inducible CAM system only when exposed to sodium in salinized conditions (Winter, 1973; Winter and von Wilbert, 1972). If the plants are exposed to a NaCl medium, CO_2 uptake is increased in the dark and malate content increases in the leaves. The induction of CAM does not take place in the absence of sodium. Plants possessing the CAM system are characterized by high water use efficiency, an important characteristic in high salt environments.

No specific function for sodium has yet been satisfactorily demonstrated in higher plants, although the element is essential for the survival of many plants growing under conditions of high salt. Thus sodium is very important not only for the presumed specific functions but also for the nonspecific role it plays in maintaining a favorable water balance (Rains, 1972; Waisel, 1972).

Two other elements have been shown essential for some specialized functions in certain plants. These include silicon for cell wall rigidity of diatoms and cobalt for symbiotic nitrogen fixation by leguminous plants. Specific functions and references are listed in Table IX.

REFERENCES

Agarwala, S. C. (1952). *Nature (London)* **169**, 1099.
Albert, L. S. (1965). *Plant Physiol.* **40**, 649.
Alberty, R. A. (1968). *J. Biol. Chem.* **243**, 1337.
Aleem, M. I. H. (1970). *Annu. Rev. Plant Physiol.* **21**, 67.
Allen, M. B., and Arnon, D. I. (1955). *Physiol. Plant.* **8**, 653.
Ambler, J. E., Brown, J. C., and Gauch, H. G. (1971). *Agron. J.* **63**, 95.
Arnold, P. W. (1969). *In* "Ecological Aspects of Plant Nutrition" (I. H. Roison, ed.), pp. 115–25. Blackwell, Oxford.
Arnon, D. I., and Stout, P. R. (1939a). *Plant Physiol.* **14**, 371.
Arnon, D. I., and Stout, P. R. (1939b). *Plant Physiol.* **14**, 599.
Arnon, D. I., Ichioka, P. S., Wessel, G., Fujiwara, A., and Woolley, J. T. (1955). *Plant Physiol.* **8**, 538.
Atkinson, M. R., and Polya, G. M. (1967). *Aust. J. Biol. Sci.* **20**, 1069.
Avron, M. (1962). *J. Biol. Chem.* **237**, 2011.
Avron, M. (1967). *Curr. Top. Bioenerg.* **2**, 1-22.
Banath, C. L., Greenwood, E. A. N., and Loneragan, J. F. (1966). *Plant Physiol.* **41**, 760.
Bendall, D. S., and Hill, R. (1968). *Annu. Rev. Plant Physiol.* **19**, 167.
Bennet-Clarke, J. A. (1956). *In* "The Chemistry and Mode of Action of Plant Growth Substances" (R. L. Wain and F. Wightman, eds.), pp. 284–291. Butterworth, London.
Bernstein, L. (1963). *Amer. J. Bot.* **50**, 360 and 900.
Berry, J. A., and Reisenauer, M. M. (1967). *Plant Soil* **27**, 303.
Bogorad, L. (1966). *In* "The Chlorophylls" (L. P. Vernon and G. R. Seely, eds.), pp. 437–479. Academic Press, New York.
Boichenko, E. A., and Udel'nova, T. M. (1964). *Dokl. Akad. Nauk SSSR* **158**, 464.
Bonds, E., and O'Kelley, J. C. (1969). *Amer. J. Bot.* **56**, 271.
Bové, J. M., Bové, C., Whatley, F. R., and Arnon, D. I. (1963). *Z. Naturforsch. B* **18**, 683.
Bowen, J. E., and Gauch, H. G. (1966). *Plant Physiol.* **41**, 319.
Bowen, J. E., Gauch, H. G., Krauss, R. W., and Galloway, R. A. (1965). *J. Phycol.* **1**, 151.
Brown, D. H., Cappellini, R. A., and Price, C. A. (1966). *Plant Physiol.* **41**, 1543.
Brown, J. C., and Chaney, R. L. (1971). *Plant Physiol.* **47**, 836.
Brownell, P. F. (1965). *Plant Physiol.* **40**, 460.
Brownell, P. F., and Crossland, C. J. (1972). *Plant Physiol.* **49**, 794.

Brownell, P. F., and Jackman, M. E. (1966). *Plant Physiol.* **41**, 617.

Brownell, P. F., and Nicholas, D. J. D. (1967). *Plant Physiol.* **42**, 915.

Brownell, P. F., and Wood, J. G. (1957). *Nature (London)* **179**, 653.

Broyer, T. C., Carlton, A. B., Johnson, C. M., and Stout, P. R. (1954). *Plant Physiol.* **29**, 526.

Burnham, B. F., and Lascelles, J. (1963). *Biochem. J.* **87**, 462.

Bush, L. P. (1969). *Plant Physiol.* **44**, 347.

Carell, E. F., and Price, C. A. (1965). *Plant Physiol.* **40**, 1.

Cartwright, B., and Hallsworth, E. G. (1970). *Plant Soil* **33**, 685.

Chance, B., Bonner, W. D., and Storey, B. T. (1968). *Annu. Rev. Plant Physiol.* **19**, 295.

Chapman, H. D. (1966). *In* "Diagnostic Criteria for Plants and Soils" (H. D. Chapman, ed.), pp. 65–92. Univ. of California, Div. Agr. Sci.

Cheniae, G. M. (1970). *Annu. Rev. Plant Physiol.* **21**, 467.

Chrispeels, M. J., and Varner, J. E. (1967). *Plant Physiol.* **42**, 398.

Cleland, R. (1960). *Plant Physiol.* **35**, 581.

Constantopoulos, G. (1970). *Plant Physiol.* **45**, 76.

Cowles, J. R., Evans, H. J., and Russell, S. A. (1969). *J. Bacteriol.* **97**, 1960.

Cresswell, C. F., and Nelson, H. (1973). *Ann. Bot. (London)* [N.S.] **37**, 427.

Davidson, F. M., and Long, C. (1958). *Biochem. J.* **101**, 31.

Dear, J. M., and Aronoff, S. (1968). *Plant Physiol.* **43**, 997.

DeHertogh, A. A., Mayeux, P. A., and Evans, H. J. (1964). *J. Biol. Chem.* **239**, 2446.

DeKock, P. C., Commisiong, K., Farmer, V. C., and Inkson, R. H. E. (1960). *Plant Physiol.* **35**, 599.

Dieckert, J. W., and Rozacky, E. (1969). *Arch. Biochem. Biophys.* **134**, 473.

Dodds, J. A., and Ellis, R. J. (1966). *Biochem. J.* **101**, 31.

Duda, C. T., and Cherry, J. H. (1971). *Plant Physiol.* **47**, 262.

Duggan, J., and Gassman, M. (1974). *Plant Physiol.* **53**, 206.

Duggar, W. M., Jr., and Humphreys, T. E. (1960). *Plant Physiol.* **35**, 523.

Einset, E., and Clarke, W. L. (1958). *J. Biol. Chem.* **231**, 703.

El Hinnawy, E. (1974). *Z. Pflanzenphysiol.* **71**, 207.

Elmstrom, G. W., and Howard, F. D. (1970). *Plant Physiol.* **45**, 327.

El-Sheikh, A. M., Ulrich, A., and Broyer, T. C. (1967). *Plant Physiol.* **42**, 1202.

Epstein, E. (1965). *In* "Plant Biochemistry," 2nd ed. (J. Bonner and J. E. Varner, eds.), pp. 438–66. Academic Press, New York.

Epstein, E. (1966). *Nature (London)* **212**, 1324.

Epstein, E. 1972. "Mineral Nutrition of Plants: Principles and Perspectives." Wiley, New York.

Evans, H. J. (1963). *Plant Physiol.* **38**, 397.

Evans, H. J., and Sorger, G. J. (1966). *Annu. Rev. Plant Physiol.* **17**, 47.

Evans, M. L., Ray, P. M., and Reinhold, L. (1971). *Plant Physiol.* **47**, 335.

Fisher, J., and Hodges, T. K. (1969). *Plant Physiol.* **44**, 385.

Fisher, J. D., Hansen, D., and Hodges, T. K. (1970). *Plant Physiol.* **46**, 812.

Fischer, R. A. (1968a). *Science* **160**, 784.

Fischer, R. A. (1968b). *Plant Physiol.* **43**, 1947.

Fischer, R. A., and Hsiao, T. C. (1968). *Plant Physiol.* **43**, 1953.

Floyd, R. A., and Rains, D. W. (1971). *Plant Physiol.* **47**, 663.

Foote, B. E., and Hanson, J. B. (1964). *Plant Physiol.* **37**, 450.

Fujino, M. (1967). *Sci. Bull. Educ. Nagasaki Univ.* **18**, 1.

Gauch, H. G., and Duggar, W. M. (1954). *Md. Agr. Exp. Sta., Tech. Bull.* **A-80.**

Gerebtzoff, A., and Ramaut, J. L. (1970). *Physiol. Plant.* **23,** 574.

Ginsberg, B. Z. (1958). *Nature (London)* **181,** 398.

Glusker, J. P. (1968). *J. Mol. Biol.* **38,** 149.

Goldberg, A. (1966). *J. Mol. Biol.* **15,** 663.

Goodwin, P. B., and Carr, D. J. (1970). *Cytobios* **2,** 165.

Grimm, P. W., and Allen, P. J. (1954). *Plant Physiol.* **29,** 369.

Gross, R. E., Pugno, P., and Dugger, W. M. (1970). *Plant Physiol.* **46,** 183.

Gruener, N., and Neumann, J. (1966). *Physiol. Plant.* **19,** 678.

Habermann, H. M. (1969). *Plant Physiol.* **44,** 331.

Hansson, G., and Kylin, A. (1969). *Z. Pflanzenphysiol.* **60,** 270.

Harmer, P. M., and Benne, E. J. (1945). *Soil Sci.* **60,** 137.

Hartt, C. E. (1969). *Plant Physiol.* **44,** 1461.

Hartt, C. E. (1970). *Plant Physiol.* **45,** 183.

Hatch, M. D., and Kagawa, T. (1974). *Arch. Biochem. Biophys.* **160,** 346.

Heath, R. I. (1973). *Int. Rev. Cytol.* **34,** 49.

Helms, K. (1971). *Plant Physiol.* **47,** 799.

Hewitt, E. J. (1963). *In* "Plant Physiology" (F. C. Steward, ed.), Vol. 3, pp. 137–360 Academic Press, New York.

Hiatt, A. J. (1965a). *Plant Physiol.* **40,** 184.

Hiatt, A. J. (1965b). *Plant Physiol.* **40,** 189.

Hiatt, A. J., and Evans, H. J. (1960). *Plant Physiol.* **35,** 673.

Hind, G., Nakatani, H. Y., and Izawa, S. (1969). *Biochim. Biophys. Acta* **172,** 277.

Hsiao, T. C., Hageman, R. H., and Tyner, E. H. (1968). *Plant Physiol.* **43,** 1941.

Hsiao, T. C., Hageman, R. H., and Tyner, E. H. (1970). *Crop Sci.* **10,** 78.

Humble, G. D., and Hsiao, T. C. (1969). *Plant Physiol.* **44,** 230.

Humble, G. D., and Hsiao, T. C. (1970). *Plant Physiol.* **46,** 483.

Humble, G. D., and Raschke, K. (1971). *Plant Physiol.* **48,** 447.

Izawa, S., Heath, R. L., and Hind, G. (1969). *Biochim. Biophys. Acta* **180,** 388.

Jackman, M. E., and van Steveninck, R. F. M. (1967). *Aust. J. Biol. Sci.* **20,** 1063.

Jacobsen, J. V., Scandalios, J. G., and Varner, J. E. (1970). *Plant Physiol.* **45,** 367.

Johnson, C. M., Stout, P. R., Broyer, T. C., and Carlton, A. B. (1957). *Plant Soil* **8,** 337.

Johnson, D. L., and Albert, L. S. (1967). *Plant Physiol.* **42,** 1307.

Jones, O. T. G. (1968). *Biochem. J.* **107,** 113.

Jyung, W. H., and Wittwer, S. H. (1964). *Amer. J. Bot.* **51,** 437.

Kagi, J. H. R., and Vallee, B. L. (1960). *J. Biol. Chem.* **235,** 3188.

Katoh, S. (1960). *Nature (London)* **186,** 533.

Kuiper, P. J. C. (1968). *Plant Physiol.* **43,** 1372.

Kylin, A., and Gee, R. (1970). *Plant Physiol.* **45,** 169.

Lai, Y. F., and Thompson, J. E. (1971). *Biochim. Biophys. Acta* **233,** 84.

Lascelles, J. (1961). *Physiol. Rev.* **41,** 417.

Laties, G. G. (1967). *Aust. J. Sci.* **30,** 193.

Lee, S., and Aronoff, S. (1967). *Science* **158,** 798.

Lehninger, A. L. (1970). "Biochemistry, The Molecular Basis for Cell Structure and Function." Worth Publ., New York.

Lehr, J. J. (1953). *J. Sci. Food Agr.* **4,** 460.

Levine, R. P. (1969). *Annu. Rev. Plant Physiol.* **20,** 523.

Lewin, J., and Reimann, B. E. F. (1969). *Annu. Rev. Plant Physiol.* **20,** 289.

Lipman, C. B., and Mackinney, G. (1931). *Plant Physiol.* **5,** 593.

Lowther, W. L., and Loneragan, J. F. (1968). *Plant Physiol.* **43**, 1362.

McIlrath, W. J., and Skok, J. (1958). *Bot. Gaz. (Chicago)* **119**, 231.

Malkin, R., Malmstrom, B. G., and Vannguard, T. (1969). *Eur. J. Biochem.* **10**, 324.

Marinos, N. G. (1962). *Amer. J. Bot.* **49**, 834.

Marsh, H. V., Evans, H. J., and Matrone, G. (1963). *Plant Physiol.* **38**, 632.

Miller, G., and Evans, H. J. (1957). *Plant Physiol.* **32**, 346.

Morgan, P. W., Joham, H. E., and Amin, J. V. (1966). *Plant Physiol.* **41**, 718.

Morita, Y., Wadano, A., and Ida, S. (1971). *Agr. Biol. Chem.* **35**, 255.

Mulder, E. C. (1948). *Plant Soil* **1**, 94.

Munns, D. N. (1965). *Aust. J. Agr. Res.* **16**, 743.

Murata, T., and Akazawa, T. (1968). *Arch. Biochem. Biophys.* **126**, 873.

Nason, A., and Evans, H. J. (1953). *J. Biol. Chem.* **202**, 655.

Nason, A., and McElroy, W. D. (1963). "Plant Physiology" (F. C. Stewart, ed.), Vol. 3, pp. 451–536. Academic Press, New York.

Nason, A., Kaplan, N. O., and Coswick, S. P. (1951). *J. Biol. Chem.* **188**, 397.

Nicholas, D. J. D. (1961). *Annu. Rev. Plant Physiol.* **12**, 63.

Nicholas, D. J. D., and Nason, A. (1955). *Plant Physiol.* **30**, 135.

Nielsen, E. S., and Wium-Andersen, S. (1971). *Physiol. Plant.* **24**, 480.

Nielsen, E. S., Kamp-Nielsen, L., and Wium-Andersen, S. (1969). *Physiol. Plant.* **22**, 1121.

Nieman, R. H., and Willis, C. (1971). *Plant Physiol.* **48**, 287.

Nitsos, R. E., and Evans, H. J. (1966). *Plant Physiol.* **41**, 1499.

Nitsos, R. E., and Evans, H. J. (1969). *Plant Physiol.* **44**, 1260.

Notton, B. A., and Hewitt, E. J. (1971). *Plant Cell Physiol.* **12**, 465.

Ockerse, R., and Naber, J. (1970). *Plant Physiol.* **46**, 821.

Oji, Y., and Izawa, G. (1969). *Plant Cell Physiol.* **10**, 665.

Paneque, A., and Losada, M. (1966). *Biochim. Biophys. Acta* **128**, 202.

Paul, K. G., and Stigbrand, T. (1970). *Biochim. Biophys. Acta* **221**, 255.

Paulsen, G. M., and Harper, J. E. (1968). *Plant Physiol.* **43**, 775.

Paynter, D. I., and Anderson, J. W. (1974). *Plant Physiol.* **53**, 180.

Peterson, P. J. (1969). *J. Exp. Bot.* **20**, 863.

Peterson, P. J., and Butler, G. W. (1967). *Nature (London)* **213**, 599.

Petty, M. A. (1961). *Bacteriol. Rev.* **25**, 111.

Plesincar, M., and Bendall, D. S. (1970). *Biochim. Biophys. Acta* **216**, 192.

Poole, R. J. (1971). *Plant Physiol.* **47**, 735.

Poovaiah, B. W., and Leopold, A. C. (1973). *Plant Physiol.* **52**, 236.

Possingham, J. V., and Brown, R. (1957). *Nature (London)* **180**, 653.

Prask, J. A., and Plocke, D. J. (1971). *Plant Physiol.* **48**, 150.

Price, C. A. (1968). *Annu. Rev. Plant Physiol.* **19**, 239.

Price, C. A. (1970). "Molecular Approaches to Plant Physiology." McGraw-Hill, New York.

Purves, W. K. (1966). *Plant Physiol.* **41**, 230.

Rains, D. W. (1968). *Plant Physiol.* **43**, 394.

Rains, D. W. (1972). *Annu. Rev. Plant Physiol.* **23**, 367.

Rains, D. W., and Epstein, E. (1967a). *Plant Physiol.* **42**, 314.

Rains, D. W., and Epstein, E. (1967b). *Plant Physiol.* **42**, 319.

Rains, D. W., and Floyd, R. A. (1970). *Plant Physiol.* **46**, 93.

Rains, D. W., Schmid, W. E., and Epstein, E. (1964). *Plant Physiol.* **39**, 274.

Randall, P. J., and Bouma, D. (1973). *Plant Physiol.* **52**, 229.

Raschke, K., and Fellows, M. P. (1971). *Planta* **101**, 296.

Rasmussen, H. D. (1966). *Conn. Agr. Exp. Sta., Res. Rep.* **18**, 4.

Rayle, D. L., and Cleland, R. (1970). *Plant Physiol.* **46**, 250.

Reisenauer, H. M. (1960). *Nature (London)* **186**, 375.

Rutter, W. J. (1964). *Fed. Proc., Fed. Amer. Soc. Exp. Biol.* **23**, 1248.

Sadava, D., and Chrispeels, M. J. (1971). *Biochim. Biophys. Acta* **227**, 278.

Schneider, H. A. W. (1970). *Z. Pflanzenphysiol.* **62**, 328.

Sexton, R., and Sutcliffe, J. F. (1969). *Ann. Bot. (London)* [N.S.] **33**, 683.

Shaukat-Ahmed and Evans, H. J. (1959). *Biochem. Biophys. Res. Commun.* **1**, 271.

Shaukat-Ahmed and Evans, H. J. (1961). *Proc. Nat. Acad. Sci. U.S.* **47**, 24.

Sheat, D. C., Fletcher, B. H., and Street, H. E. (1959). *New Phytol.* **58**, 128.

Shibko, S., and Penchot, G. B. (1961). *Arch. Biochem. Biophys.* **93**, 140.

Shichi, H., and Hackett, D. P. (1963). *Arch. Biochem. Biophys.* **109**, 185.

Shomer-Ilan, A., and Waisel, Y. (1973). *Physiol. Plant.* **29**, 190.

Shrift, A. (1969). *Annu. Rev. Plant Physiol.* **20**, 475.

Skoog, F. (1940). *Amer. J. Bot.* **27**, 939.

Skou, J. C. (1965). *Physiol. Rev.* **45**, 596.

Sommer, A. L. (1931). *Plant Physiol.* **6**, 339.

Sorger, G. J., Ford, R. E., and Evans, H. J. (1965). *Proc. Nat. Acad. Sci. U.S.* **54**, 1614.

Spanner, D. C. (1958). *J. Exp. Bot.* **9**, 332.

Stigbrand, T. (1971). *Biochim. Biophys. Acta* **236**, 246.

Stonier, T., Rodriquez-Tormes, F., and Yoneda, Y. (1968). *Plant Physiol.* **43**, 69.

Stout, P. F. (1961). *Proc. 9th Annu. Calif. Fert. Conf.* pp. 21–23.

Suelter, C. H. (1970). *Science* **168**, 789.

Sugiyama, T., Goto, Y., and Akazawa, T. (1968). *Plant Physiol.* **43**, 730.

Sumida, S., and Mudd, J. B. (1970a). *Plant Physiol.* **45**, 712.

Sumida, S., and Mudd, J. B. (1970b). *Plant Physiol.* **45**, 719.

Tagawa, T., and Bonner, J. (1957). *Plant Physiol.* **32**, 207.

Takaki, H., Kushizaki, M. (1970). *Plant Cell Physiol.* **11**, 793.

Taylor, H. F., and Wain, R. L. (1966). *Ann. Appl. Biol.* **57**, 301.

Teichler-Zallen, D. (1969). *Plant Physiol.* **44**, 701.

Tempest, D. W., Dicks, J. W., and Meers, J. L. (1967). *Biochem. J.* **102**, 36.

Tiffin, L. O. (1966). *Plant Physiol.* **41**, 510.

Tsui, C. (1948). *Amer. J. Bot.* **35**, 172.

Turner, R. G. (1970). *New Phytol.* **69**, 725.

Turner, R. G., and Marshall, C. (1971). *New Phytol.* **70**, 539.

Udvardy, J., Marre, E., and Farkas, G. L. (1970). *Biochim. Biophys. Acta* **206**, 392.

Ullrich, W. R. (1971). *Planta* **100**, 18.

van Breeman, C. (1968). *Biochem. Biophys. Res. Commun.* **32**, 977.

van Steveninck, R. F. M. (1965). *Physiol. Plant.* **18**, 54.

Vega, J. A., Herrera, J., Aparicio, P. J., Paneque, A., and Losada, M. (1971). *Plant Physiol.* **48**, 294.

Vlamis, J., and Williams, D. E. (1967). *Plant Soil* **27**, 131.

von Marschner, H., and Gunther, I. (1964). *Z. Pflanzeneraehr., Dung., Bodenk.* **107**, 118.

Waisel, Y. (1972). "Biology of Halophytes." Academic Press, New York.

Walter, T. J., and Mans, R. J. (1970). *Biochim. Biophys. Acta* **217**, 72.

Warburg, O., and Luttgens, W. (1946). *Biokhimiya* **11**, 303.

Webster, G. C. (1953). *Plant Physiol.* **28**, 728.

Webster, G. C. (1956). *Biochim. Biophys. Acta* **20**, 565.

Webster, G. C., and Varner, J. E. (1954). *Arch. Biochem. Biophys.* **52**, 22.

Wildes, R. A., and Neales, T. F. (1971). *Aust. J. Biol. Sci.* **24**, 397.

Williams, M. C. (1960). *Plant Physiol.* **35**, 500.

Willmer, C. M., and Mansfield, T. A. (1969). *Z. Pflanzenphysiol.* **61**, 398.

Wilson, D. O., and Reisenauer, H. M. (1963). *Plant Soil* **19**, 364.

Winter, K. (1973). *Planta* **115**, 187.

Winter, K., and von Wilbert, D. J. (1972). *Z. Pflanzenphysiol.* **67**, 166.

Wyn Jones, R. G., and Lunt, O. R. (1967). *Bot. Rev.* **33**, 407.

Yih, R. Y., and Clark, H. E. (1965). *Plant Physiol.* **40**, 312.

Young, O. A., anl Anderson, J. W. (1974). *Biochem. J.* **137**, 435.

19

Sulfate Reduction

LLOYD G. WILSON AND ZIVA REUVENY

I. Introduction*

The reduction of inorganic sulfate to the level of sulfide and its in-corporation into amino acids is characteristic of plants and some micro-

* The following abbreviations are utilized throughout this chapter: ADP, adenosine 5′-diphosphate; AMP, adenosine 5′-phosphate; APS, adenosine 5′-phosphosulfate; ATP, adenosine 5′-triphosphate; BAL, 2,3-dimercapto-1-propanol; FAD and FADH₂,

organisms. The small scale reduction of sulfate required for the synthesis of cellular material is termed assimilatory sulfate reduction. This is in contrast to dissimilatory sulfate reduction in which large amounts of sulfate are reduced to sulfide. Dissimilatory sulfate reduction is limited to a small group of obligately anaerobic bacteria in which sulfate is the terminal electron acceptor for respiration. Because animals are dependent upon the reduced sulfur of amino acids, assimilatory sulfate reduction is similar in ecological importance to the reduction of carbon dioxide and nitrate.

This chapter is primarily concerned with assimilatory sulfate reduction and the incorporation of reduced sulfur into cysteine and methionine. The general pathway of reduction in fungi and bacteria has been fairly well established, but our knowledge is still fragmentary and very few studies of reduction in higher plants have appeared. Studies of sulfur metabolism are complicated by the ease with which nonenzymatic reactions of sulfur occur at various oxidation states. Consequently, some of the schemes and reactions based on *in vitro* studies may have little physiological importance.

II. Sulfate Activation

A. Sulfate-Activating System

Although inorganic sulfate is the principal form in which sulfur is taken up from the environment by plants and microorganisms, it must be enzymatically converted into an activated form before it can be incorporated into organic compounds. Sulfate activation, unlike sulfate reduction, is universal in the plant and animal kingdom and serves several biosynthetic functions. Activated sulfate is converted into (a) reduced sulfur compounds in plants and microorganisms, (b) a wide variety of sulfate esters in plants, animals, and microorganisms, and (c) sulfolipids in animal tissues and photosynthetic organisms.

There are two forms of active sulfate: adenosine 5'-phosphosulfate (APS) and 3'-phosphoadenosine 5'-phosphosulfate (PAPS) (Fig. 1).

---

oxidized and reduced flavin adenine dinucleotide, respectively; FMN and $FMNH_2$, oxidized and reduced flavin adenine mononucleotide, respectively; GSH, reduced glutathione; $GSSO_3H$, S-sulfoglutathione; MV and MVH, oxidized and reduced methyl viologen, respectively; $NADP^+$ and NADPH, oxidized and reduced nicotinamide adenine dinucleotide phosphate, respectively; PAP, 3'-phosphoadenosine 5'-phosphate; PAPS, 3'-phosphoadenosine 5'-phosphosulfate; PCMB, parachloromercuribenzoate; P_i, inorganic phosphate; PP_i, inorganic pyrophosphate; Tris, tris(hydroxymethyl)aminomethane.

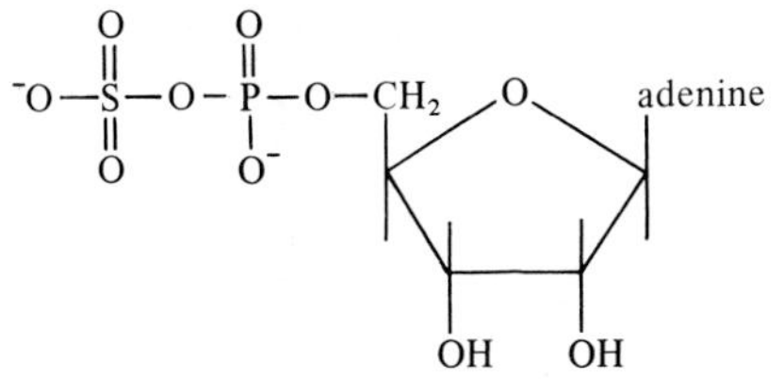

Adenosine 5'–phosphosulfate (APS)

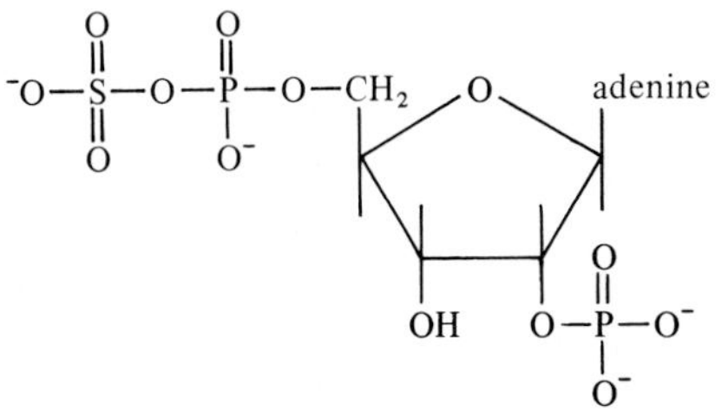

3'–Phosphoadenosine 5'–phosphosulfate (PAPS)

Fig. 1. Structure of APS and PAPS.

Both are mixed anhydrides of sulfuric and 5'-adenylic acid containing the sulfatophosphate bond. They are formed from sulfate and ATP in successive steps catalyzed by the enzymes ATP-sulfurylase and APS-kinase, respectively, as shown in Eqs. (1) and (2).

$$\text{ATP} + \text{SO}_4{}^{2-} \rightleftharpoons \text{APS} + \text{PP}_i \qquad \Delta F° = +11{,}000 \text{ cal} \qquad (1)$$
$$\text{APS} + \text{ATP} \rightleftharpoons \text{PAPS} + \text{ADP} \qquad \Delta F° = -6000 \text{ cal} \qquad (2)$$
$$\text{PP}_i \rightleftharpoons 2\text{P}_i \qquad \Delta F° = -5000 \text{ cal} \qquad (3)$$

The equilibrium is so strongly in favor of ATP that there is little net accumulation of APS in yeast extracts (Robbins and Lipmann, 1958) even in the presence of inorganic pyrophosphatase [Eq. (3)]. The reaction proceeds toward PAPS formation due to the combined free energy change of Eqs. (2) and (3). The three reactions together constitute the sulfate-activating system.

B. ATP-Sulfurylase

The enzyme ATP-sulfurylase (EC 2.7.7.4, ATP:sulfate adenyltransferase) catalyzes the reaction between ATP and inorganic sulfate in the presence of Mg^{2+}, resulting in the formation of APS and inorganic pyrophosphate as shown in Eq. (1).

ATP-sulfurylase is a soluble enzyme and has been detected in algae, fungi, bacteria, and liver. It was detected in higher plants by Asahi (1964), and its presence in extracts of roots, leaves, shoots and chloroplasts of higher plants is now firmly established (Adams and Johnson, 1968; Adams and Rinne, 1969; Ellis, 1969; Mercer and Thomas, 1969; Balharry and Nicholas, 1970; Shaw and Anderson, 1971; Onajobi et al., 1973).

Highly purified ATP-sulfurylases have now been obtained from yeast, *Penicillium chrysogenum* (Tweedie and Segel, 1971a,b), *Nitrobacter agilis* (Varma and Nicholas, 1971b), spinach (Shaw and Anderson, 1972) and rat liver.

These enzymes show the same broad pH optimum from 7.5 to 9.0, a specific nucleotide requirement for ATP, a requirement for divalent cations that can be satisfied by Mg^{2+}, Co^{2+}, or Mn^{2+}, a simple protein absorption with a maximum at 278 nm, and inhibition by the products of the forward reaction, APS and PP_i.

The molecular weights of the highly purified enzymes range from approximately 100,000 for yeast to 900,000 for rat liver sulfurylase. A partially purified sulfurylase from corn roots has a molecular weight of 42,000 based on gel chromatography (Onajobi et al., 1973). Amino acid analysis and extensive physical and kinetic studies have been carried out on the enzyme from *P. chrysogenum*. This enzyme is an octamer with a molecular weight of 425,000 to 440,000 and identical subunits of 56,000 MW with a single sulfhydryl per subunit. The wide variation in molecular weights suggest that the ATP-sulfurylases of different organisms consist of varying aggregates of subunits.

The sulfurylases with the lowest molecular weights (yeast, *P. chrysomenum*, and spinach) are relatively insensitive to freezing, and the yeast and spinach enzymes are relatively insensitive to sulfhydryl reagents. In contrast, the sulfurylases with higher molecular weights (*N. agilis*, MW 700,000; rat liver, MW 900,000) are sensitive to freezing, to Tris, and to sulfhydryl reagents. The sulfurylase activity of *Chlorella* (Schmidt, 1972b) as well as that of *N. agilis* and liver is increased by the presence of various thiols in the reaction medium, suggesting the requirement for a sulfhydryl group in the active enzyme.

ATP-sulfurylase specifically requires ATP, although deoxyadenosine triphosphate has some activity and competes with ATP in spinach sulfurylase. Sulfate is the only form of inorganic sulfur activated by the enzyme, but the enzyme can catalyze various reactions with group VI anions. Reaction with chromate, tungstate, or molybdate leads to a rapid pyrophosphorolysis of ATP resulting in the formation of AMP and PP_i as shown for molybdate in Eq. (4).

$$\text{ATP} + \text{H}_2\text{O} \xrightarrow{\text{MoO}_4{}^{2-}} \text{AMP} + \text{PP}_i \tag{4}$$

No anhydride linkage is detected, and large amounts of AMP and PP_i are formed. This reaction is the basis of a convenient assay technique for measurement of the forward reaction. In the case of selenate, the anhydride linkage can be demonstrated, and adenosine 5′-phosphoselenate (the analog of APS) can be isolated. The ratio of sulfate- to selenate-dependent pyrophosphate exchange is approximately constant during the purification of spinach sulfurylase, indicating that a single enzyme catalyzes both activities. Kinetic studies show that sulfate and selenate compete for the same site.

Kinetic studies with the sulfurylase of *P. chrysogenum* showed that a 1:1 complex of ATP and Mg^{2+} is the substrate for the forward reaction and that free ATP is a competitive inhibitor with respect to the complex (ATP-Mg^{2-}) and $\text{MoO}_4{}^{2-}$ $(K_i = 0.6$ to 1.25 mM). High Na^+ and K^+ concentrations apparently inhibit by increasing the free ATP concentration. Studies of initial velocity showed that there are no irreversible steps between the addition of the two substrates, ATP and $\text{MoO}_4{}^{2-}$. Furthermore, there is no detectable exchange of $[^{32}P]PP_i$ into ATP in the absence of $SO_4{}^{2-}$ or APS. Similarly there is no exchange of ^{35}S between APS and sulfate in the absence of the other substrates. This seems to rule out the double displacement (pingpong) mechanism proposed for liver sulfurylase (cf., Roy and Trudinger, 1970) in favor of a sequential mechanism in which both substrates combine with the enzyme before the products are released (Tweedie and Segel, 1971a).

C. APS-Kinase

The second step in sulfate activation is catalyzed by the enzyme APS-kinase (EC 2.7.1.25 ATP:adenylsulfate 3′-phosphotransferase). The reaction is a phosphorylation of the 3′ position of APS as shown in Eq. (2). This is essentially an irreversible reaction analogous to the hexokinase reaction.

The occurrence of this enzyme is largely a matter of conjecture. It is assumed to be present in all organisms which can form PAPS. PAPS formation has been demonstrated in *Chlorella* and a number of bacteria and fungi and in mammalian liver and kidney. It is not present in bacteria carrying out dissimilatory sulfate reduction (Michaels *et al.*, 1971), and its existence in higher plants has been a matter of dispute. Mercer and Thomas (1969) reported the occurrence of PAPS in bean and maize chloroplast fragments, but careful studies by Asahi (1964), Ellis (1969), and Balharry and Nicholas (1970) suggested that PAPS was not present in a wide variety of higher plant materials. Schmidt (1972b) has carefully compared sulfate activation in spinach and *Chlorella* and provided a possible solution for the controversy. He discovered that the APS-kinase of spinach requires activation by sulfhydryl and demonstrated the formation of PAPS from [^{35}S]sulfate by extracts of spinach chloroplasts upon the addition of cysteine to the reaction mixture. This requirement had not been determined earlier because the ATP-sulfurylase of higher plants, in contrast to that of *Chlorella*, does not require sulfhydryl activation.

Very little is known of the properties of APS-kinase. Only the enzyme from yeast has been purified to any extent. The reaction catalyzed by the enzyme has a broad pH optimum from 7.5 to 9.0 and requires Mg^{2+}, Co^{2+}, or Mn^{2+}. The affinity of the enzyme for APS is very high, and the reaction is inhibited by APS concentrations greater than 0.005 mM. The ability of the enzyme to function at low substrate concentrations and the free energy of the reaction result in very efficient trapping of the small amounts of APS formed by ATP-sulfurylase and allows PAPS to accumulate.

D. Intermediate Role of APS and PAPS

Once formed, APS and PAPS are substrates for a variety of reactions (Fig. 2). Depending upon the organism and nutritional conditions, the sulfuryl group may be reduced to sulfite, transferred, or replaced by phosphate, and PAPS may be reconverted to APS.

PAPS is the substrate reduced by most bacteria and fungi (Section

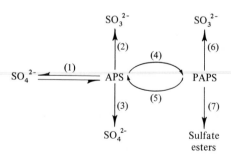

Fig. 2. Enzymatic reactions involving APS and PAPS. The enzymes are (1) ATP-sulfurylase, (2) APS-reductase, (3) ADP-sulfurylase, (4) APS-kinase, (5) 3'-phosphonucleotidase, (6) PAPS-reductase, and (7) sulfotransferase.

III,B), whereas APS is the substrate reduced by a few species of anaerobic bacteria (Michaels *et al.*, 1971) and apparently by *Chlorella* and higher plants (Section IV,B).

Sulfate is formed from APS by the reverse reaction of ATP-sulfurylase or by a reaction in which the sulfate group is displaced by inorganic phosphate as shown in Eq. (5).

$$\text{APS} + \text{P}_i \rightarrow \text{ADP} + \text{SO}_4{}^{2-} \tag{5}$$

This reaction, catalyzed by ADP-sulfurylase, has been reported in yeast, *Desulfovibrio desulfuricans*, *Thiobacillus thioparus*, and the Thiorhodaceae. The yeast enzyme has been purified 300-fold, does not require Mg^{2+}, has a strong affinity for APS, and is strongly inhibited by PCMB (Adams and Nicholas, 1972). Its occurrence and physiological significance are not known, but it has been suggested that it may play a role in the regulation of sulfation and reduction reactions by controlling the concentration of APS. Since it occurs in high activity in *T. thioparus*, it is believed to participate in the oxidation of thiosulfate by that organism and the generation of phosphate bond energy in ADP (Michaels *et al.*, 1971).

PAPS is reconverted to APS *in vitro* by means of a 3′-phosphonucleotidase found in *Salmonella pullorum* (Kline and Schoenhard, 1970), *Chlorella*, and spinach (Section IV,B).

Finally, a variety of sulfate esters may be formed by the transfer of the sulfuryl group of PAPS to suitable acceptors by means of a group of enzymes known as sulfotransferases (Roy and Trudinger, 1970). A general reaction for the formation of sulfate esters is shown in Eq. (6) where R may be either C or N.

$$\text{PAPS} + \text{R}-\text{OH} \rightarrow \text{R-OSO}_3{}^- + \text{PAP} \tag{6}$$

These transferases are widely distributed in animal tissues and are absolutely specific for PAPS. Sulfate esters that occur in algae and higher plants include choline sulfate, flavonoid sulfates, sulfated polysaccharides, and mustard oil glycosides. With the exception of choline sulfotransferase, the transferring enzymes have not yet been demonstrated in algae or higher plants.

Choline sulfate has been found in the higher fungi and the red algae and is widely distributed in higher plants. It is thought to function as a store of sulfur in higher fungi, since choline sulfatase, which releases sulfate and choline, is repressed by cysteine and derepressed during sulfur starvation in *Aspergillus nidulans* (Scott and Spencer, 1968). Derepression of choline sulfatase was also shown in *Neurospora* grown on limiting amounts of sulfur (Metzenberg, 1972). Choline sulfate may also function as a major sulfur reservoir in higher plants, since roots from sulfur-deficient plants form large amounts of labeled choline sulfate when transferred to solutions containing [^{35}S]sulfate (Nissen and Benson, 1961).

One other important sulfur metabolite is the plant sulfolipid, sulfoquinovosyl diglyceride, discovered by Benson and co-workers in the chloroplasts of all algae and higher plants examined. The deacylated sulfolipid accounts for up to 50% of all the soluble sulfur compounds in algae (Lee and Benson, 1972). Although it is a sulfonic acid rather than an ester, it is not derived from cysteic acid or from C_5 or C_6 sugars. Since its formation is inhibited by molybdate, it is suggested that it is formed directly during photosynthesis from phosphoenol pyruvate and PAPS or sulfite (Benson, 1971; Goodwin, 1971).

III. Assimilatory Sulfate Reduction in Fungi and Bacteria

A. Pathway of Reduction

Nutritional and enzymological studies with mutants and labeled substrates have established the general pathway by which sulfate is reduced to sulfide and incorporated into cysteine and methionine in assimilatory sulfate-reducing microorganisms. Sulfate reduction in algae (*Chlorella*) and higher plants is discussed separately (Section IV) since the general pathway for these photosynthetic organisms appears to differ slightly from that of fungi and bacteria. APS, PAPS, sulfite, and sulfide are intermediates in the pathway in fungi and bacteria, as shown in Fig. 3. The available evidence suggests that the sulfite level intermediate is a "bound sulfite" freely interchangeable with inorganic sulfite. Thiosulfate, when utilized as a source of sulfur, is generally converted into sulfite and sulfide

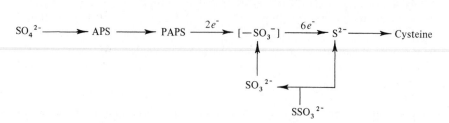

Fig. 3. Pathway of reduction from sulfate to cysteine in assimilatory sulfate-reducing fungi and bacteria.

before entering into the direct pathway of reduction. Sulfide is generally incorporated into cysteine (Sections V,A,B), but may be incorporated directly into homocysteine under certain conditions (Sections VI,A,B,C). Much of the evidence for this pathway is extensively reviewed by Thompson (1967) and Roy and Trudinger (1970). Some of the enzymological evidence, mainly obtained with yeast, is outlined below.

B. Activation

The reduction of sulfate to sulfite by cell-free extracts was first observed in bakers' yeast. An absolute requirement for ATP and inhibition by selenate and molybdate, specific inhibitors of sulfurylase, indicated that sulfate activation was involved in the reduction. In purified fractions PAPS could substitute for sulfate. APS could also serve as sulfur donor, but not as effectively as PAPS. Both were inhibitory at higher concentrations. The low equilibrium concentration of APS and the much higher affinity of the reductase for PAPS than for APS suggests that PAPS is the physiologically important substrate. The widespread occurrence of APS and PAPS in microorganisms suggested that sulfate activation was a general requirement for sulfate reduction. Mutants of yeast and *Salmonella*, which could grow on sulfite and which lacked ATP-sulfurylase or APS-kinase, provided convincing evidence that APS and PAPS were both necessary intermediates. Thus, the formation of APS and PAPS [Eqs. (1) and (2)] was established as the first step in sulfate reduction in assimilatory sulfate-reducing microorganisms.

PAPS does not occur in bacteria that carry out dissimilatory sulfate reduction, and APS is the activated intermediate in these anaerobic organisms. Since recent evidence suggests that APS is also the substrate for assimilatory sulfate reduction in higher plants (Section IV,B), the requirement for PAPS should be investigated in other microorganisms. A requirement for ATP and APS-kinase in the reduction of APS by a partially purified enzyme preparation of *Nitrobacter*, recently reported by Varma and Nicholas (1971a), supports the role of PAPS in sulfate reduction.

C. Sulfite Formation

The most detailed enzymatic studies of sulfite formation have been carried out with bakers yeast by Wilson *et al.* (1961). Two enzymes (A and B) and a heat-stable, low molecular weight protein disulfide (fraction C) are required for the reduction of PAPS to the sulfite level.

NADPH is the ultimate electron donor. The proposed reaction sequence is described in Eqs. (7) and (8).

$$\text{Fraction C-SS} + \text{NADPH} + \text{H}^+ \xrightleftharpoons{\text{enzyme A}} \text{fraction C(SH)}_2 + \text{NADP}^+ \tag{7}$$

$$\text{Fraction C(SH)}_2 + \text{PAPS} \xrightleftharpoons{\text{enzyme B}} \text{fraction C—SS} + \text{SO}_3{}^{2-} + \text{PAP} \tag{8}$$

Enzyme A has been purified 60-fold and fraction C to homogeneity. Enzyme B has not been purified or studied extensively. Enzyme A is a flavoprotein and couples the oxidation of NADPH with the reduction of the disulfide group of fraction C to a dithiol. Enzyme B (PAPS-reductase) catalyzes the reduction of PAPS to the sulfite level.

Earlier studies by Hilz *et al.* (1959) had demonstrated the participation of NADPH and PAPS in sulfite formation and suggested that the vicinal dithiol group of lipoic acid was involved [Eq. (9)].

$$\text{PAPS} + \text{Lip}\overset{\text{SH}}{\underset{\text{SH}}{\big\langle}} \longrightarrow \text{Lip}\overset{\text{S}-\text{SO}_3{}^-}{\underset{\text{SH}}{\big\langle}} + \text{PAP} \tag{9}$$

Attempts to isolate the postulated *S*-sulfolipoate intermediate were unsuccessful, but substrate amounts of reduced lipoic acid or lipoamide could replace NADPH as donor. Reduction by fraction C, however, is about 50 times as effective on a molar basis, suggesting that fraction C is the physiological donor.

There is some evidence that a disulfide bond of enzyme A, as well as a flavin component, is involved in the reduction of fraction C (Bandurski, 1965). Thus, enzyme A resembles a group of pyridine nucleotide disulfide oxidoreductases (thioredoxin reductase, glutathione reductase, and lipoamide dehydrogenase) in which the disulfide of a cystine as well a flavin participate in the electron transfer process (Ronchi and Williams, 1972).

The similarity of the fraction C–enzyme A and the thioredoxin–thioredoxin reductase systems is especially striking. Thioredoxin reductase couples the oxidation of NADPH with the reduction of thioredoxin. Thioredoxin is a heat-stable, low molecular weight protein disulfide that functions in yeast and *E. coli* as an intermediate electron carrier in the reduction of ribonucleotides to deoxyribonucleotides. Porque *et al.* (1970) have found that in yeast purified thioredoxin and thioredoxin reductase combine with partially purified enzyme B to carry out the reduction of PAPS to sulfite with NADPH. Thioredoxin acts catalytically in this system.

The thioredoxin–thioredoxin reductase system can also reduce methionine sulfoxide to methionine in the presence of purified methionine sulfoxide reductase. Because their system could participate in three different reductase systems, these workers have proposed a general scheme for the participation of thioredoxin in biological reductions with NADPH. Their studies do not establish that the two PAPS-reductase systems are identical, or that thioredoxin is the natural electron carrier, but they do show that reduced thioredoxin is another dithiol that can reduce PAPS *in vitro*.

There is good evidence that a bound form of sulfite is the normal intermediate in sulfate reduction in yeast. Torii and Bandurski (1967) have demonstrated the occurrence of a nondialyzable, low molecular weight protein (4000–8000) in yeast similar to fraction C which contains sulfite and is formed during incubation of partially purified enzymes A and B with [^{35}S]PAPS. The bound radioactive sulfite is exchangeable with free carrier sulfite as shown in Eq. (10) and can be released as radioactive SO_2 upon the addition of acid. This sulfonyl exchange reaction can be directly demonstrated by

$$X-^{35}SO_3^- + HSO_3^- \rightleftharpoons X-SO_3^- + H^{35}SO_3^- \qquad (10)$$

electrophoresis of reaction mixtures incubated in the presence and absence of carrier sulfite. When the purified $X-^{35}SO_3^-$ was incubated with a crude extract from yeast and then acidified, radioactive sulfide was obtained suggesting that the bound sulfite was a substrate for further reduction. NADPH was required for this reduction. These studies suggest that free sulfite is not a normal intermediate in sulfate reduction, but is freely exchangeable with a protein-bound intermediate.

A scheme for assimilatory sulfate reduction in yeast which is compatible with the available evidence and shows sulfite bound to fraction C is presented in Fig. 4. This scheme has not yet been confirmed by detailed enzymatic studies with other fungi or bacteria.

D. Sulfite Reduction in Yeast

Emzymes capable of reducing sulfite to sulfide have been isolated from a number of microorganisms. In yeast, *Escherichia coli*, and *Salmonella typhimurium* the sulfite-reducing activity has been purified to apparent homogeneity and the complete six electron reduction shown to be catalyzed by a single enzyme or enzyme complex.

The NADPH-sulfite reductase (hydrogen sulfide:NADP oxidoreductase, EC 1.8.1.2) of yeast is a large flavoprotein with a minimum

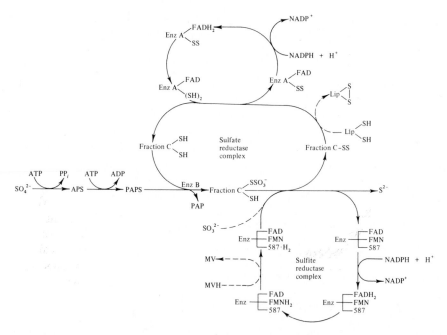

Fig. 4. Enzymatic pathway for the activation and reduction of sulfate in yeast according to present knowledge. The point at which alternate substrates [reduced lipoic acid, sulfite, reduced methyl viologen (MVH)] may be introduced are indicated by broken lines. The oxidized and reduced forms of chromatophore are represented by 587 and 587·H₂, respectively, in the sulfite reductase complex.

molecular weight of 350,000, a sedimentation coefficient of 14.8 S, and absorption peaks at 386, 455, and 587 nm. The enzyme contains 1 mole of FAD and FMN, 5 atoms of iron, and 2–3 moles of acid-labile sulfur. It catalyzes the reduction of sulfite to sulfide using NADPH as natural electron donor according to the following reaction [Eq. (11)].

$$3 \text{ NADPH} + 3 \text{ H}^+ + \text{SO}_3^{2-} \rightarrow 3 \text{ NADP}^+ + 3 \text{ H}_2\text{O} + \text{S}^{2-} \tag{11}$$

In accordance with its complex nature, the enzyme catalyzes a number of different reactions in addition to the reduction of sulfite. These include the NADPH-linked reduction of NO_2^-, NH_2OH, cytochrome c, and the dyes ferricyanide and 2,6-dichlorophenolindophenol. In addition, it catalyzes the reduced methyl viologen-linked reduction of sulfite, nitrite, hydroxylamine, and NADP$^+$. Our present knowledge of the interactions of this enzyme complex can be summarized in electron transport. This scheme, suggesting that intermediate electron carriers are involved in the flow of electrons from NADPH to sulfite, is included in Fig. 4 and is

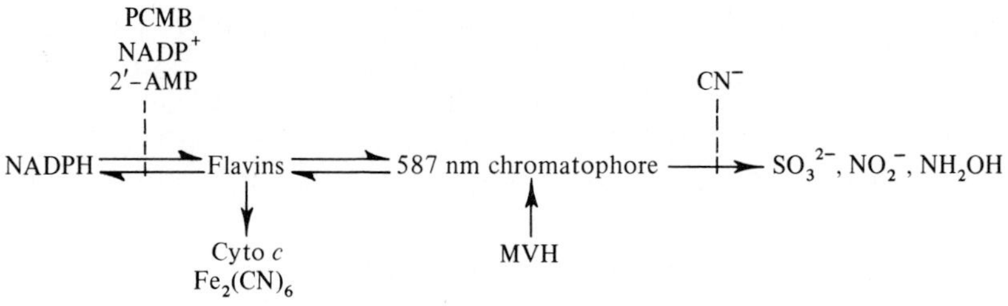

```
PCMB
NADP⁺
2'-AMP                                    CN⁻
  ┊                                        ┊
  ┊                                        ┊
NADPH ⇌┿═ Flavins ⇌═══ 587 nm chromatophore ─┿─► SO₃²⁻, NO₂⁻, NH₂OH
       │                              ▲
       ▼                              │
    Cyto c                          MVH
   Fe₂(CN)₆
```

Scheme 1

supported by extensive spectrophotometric, inhibitor, and mutant studies (Yoshimoto and Sato, 1970).

Cyanide, for example, combines with the NADPH- or MVH-reduced enzyme and completely inhibits sulfite reduction, but has little effect on cytochrome c reduction. Preincubation with sulfite prevents cyanide binding. Purified sulfite reductase of mutants unable to grow on sulfite lack NADPH-linked activities but are able to carry out sulfite reduction with MVH. These enzymes lack FMN or FAD and have smaller sedimentation coefficients than the wild-type enzymes. NADPH-linked activities are also much more sensitive to treatment with heat, ammonium sulfate, and low ionic strength than the MVH-linked sulfite reduction. These and other observations indicate that several intramolecular sites are involved in electron transfer.

Spectral properties of the 587 chromatophore and its reactivity with cyanide suggest an atypical heme-like component located at the sulfite-reducing terminus of the electron chain. This chromatophore also occurs in other purified sulfite reductases (Siegel *et al.*, 1971). It has been intensively studied in *E. coli* and is reported to be an iron-containing methylated urotetrahydroporphyrin with eight carboxylic side chains (Murphy *et al.*, 1973).

E. Thiosulfate Formation and Reduction

Inorganic thiosulfate is a relatively stable sulfur anion in which the two sulfur atoms (inner and outer) are not equivalent and occur at intermediate oxidation levels between sulfate and sulfide. Although it oc-

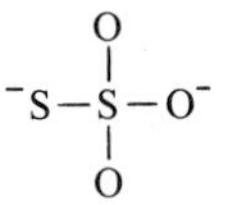

$$^-S-\overset{\displaystyle O}{\underset{\displaystyle O}{S}}-O^-$$

curs in the soluble extracts of many plants and microorganisms, its general role in metabolism is not clear.

An enzymatic system for the reduction of thiosulfate in yeast and *Salmonella* has been described, but its significance in sulfate reduction is questionable. Stoichiometric amounts of hydrogen sulfide and sulfite are produced in the presence of naturally occurring sulfhydryl compounds, such as cysteine, homocysteine and glutathione [Eq. (12)].

$$SSO_3{}^{2-} + 2\ RSH \rightleftharpoons RSSR + SO_3{}^{2-} + H_2S \qquad (12)$$

Although thiosulfate is an inhibitor of sulfate activation, Hilz *et al.* (1959) have shown that it does not compete with PAPS in the formation of sulfide. Furthermore, free thiosulfate is not formed during the reduction of sulfite to sulfide by purified yeast sulfite reductase. Mutants of *Salmonella typhimurium*, which cannot utilize sulfite, incorporate only the outer sulfur atom of thiosulfate and accumulate sulfite. These observations suggest that free thiosulfate is not a direct intermediate in sulfate reduction in these organisms.

In *Bacillus subtilis*, thiosulfate can be reduced to sulfite and sulfide by rhodanese (Roy and Trudinger, 1970). Evidence for the participation of thiosulfate in sulfate reduction in *Aspergillus nidulans* is reviewed by Trudinger (1969).

IV. Sulfate Reduction in Algae and Higher Plants

A. The Pathway of Reduction

Studies on spinach and *Chlorella* in several laboratories suggest that APS and protein-bound forms of sulfite and sulfide are the intermediates directly involved in sulfate reduction in photosynthetic tissues (Fig. 5). Sulfite and sulfide are freely exchangeable with these bound intermediates, but, as in bacteria, thiosulfate is probably cleaved to sulfite and sulfide before it can enter into the direct pathway of reduction.

This general pathway has been elucidated by studies of the incorporation of radioactive sulfate into sulfur-labeled intermediates by iso-

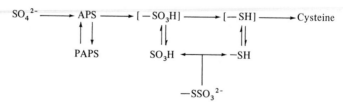

Fig. 5. Pathway of reduction from sulfate to cysteine in algae and higher plants.

lated chloroplasts and by fractionated chloroplast and leaf extracts of spinach (Schmidt and Trebst, 1969; Trebst and Schmidt, 1969). The presence of radioactive sulfur at the sulfite and sulfide level was demonstrated by adding carrier sulfite or sulfide, acidifying the reaction mixture, and passing the volatile radioactivity through $CdSO_4$ in lactic acid (absorbing H_2S) and KOH (absorbing SO_2). Isolated whole chloroplasts from spinach were able to reduce sulfate to the level of cysteine in the light at a rate of 3 μmoles per hour per mg of chlorophyll. Reconstituted systems of broken chloroplasts and chloroplast extracts reduced sulfate to the sulfite level and sulfite to the sulfide level in the light in the presence of ADP, P_i, $NADP^+$, ferredoxin, and glutathione. Extracts which could reduce sulfate in the dark required ATP and GSH for sulfite formation and ferredoxin for sulfide formation. These studies suggested that GSH functioned as reductant in sulfite formation and ferredoxin in sulfide formation (Fig. 6).

The properties of the labeled compounds formed *in vitro* by whole chloroplasts during photosynthetic sulfate reduction were determined by column chromatography and kinetic studies (Schmidt and Schwenn, 1971). Five ^{35}S-labeled compounds were eluted from a Dowex 1-nitrate column [APS, PAPS, and compounds designated as $RSSO_3H$, R'SSH, and P_2 (unknown)]. $RSSO_3H$ and R'SSH were proteins of about 4000 molecular weight, and P_2 was about 10,000. The radioactivity of $RSSO_3H$ was released as free sulfite or thiosulfate if preincubated with carrier sulfite or thiosulfate, respectively. The addition of mercaptoethanol also released free sulfite. These reactions suggest that $RSSO_3H$ is a "bound sulfite" with the sulfite attached to a thiol group of the protein. The radioactivity of R'SSH was released as H_2S upon the addition of sulfhydryl compounds, such as mercaptoethanol, and suggests that the radioactive sulfide is also bound to a thiol group on the same or a similar protein carrier.

The order of appearance of these compounds during short-time incubation in the light is as follows: SO_4^{2-}, APS, PAPS, P_2, $RSSO_3H$, and R'SSH. PAPS is not on the direct pathway (Section IV,B), and the na-

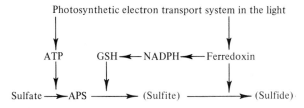

Fig. 6. Scheme for the reduction of sulfate to the sulfide level by broken chloroplasts and chloroplast extracts from spinach. Adapted from Schmidt and Trebst (1969).

ture of P_2 remains unclear since it is not formed by cell-free extracts that yield bound sulfide. It could be a product of PAPS metabolism. These kinetic studies support the general pathway of reduction outlined in Fig. 5.

B. Activation

For many years it was assumed that PAPS was the physiologically active substrate for reduction in all organisms carrying out assimilatory sulfate reduction. PAPS formation was readily demonstrated in *Chlorella* (Schiff and Hodson, 1970). The failure to detect APS-kinase activity in extracts of a wide variety of higher plants, however, led Ellis (1969) to postulate that APS might be the active substrate. Although PAPS formation has now been demonstrated in bean, maize, and spinach (Section II,C), recent studies strongly support the suggestion that APS is the activated intermediate directly involved in sulfate reduction in higher plants.

Hodson and Schiff (1971) showed that two enzymes were necessary to convert [^{35}S]PAPS to acid volatile radioactivity (sulfite) in *Chlorella*. Later they discovered that APS was an intermediate in this conversion and that one of the enzymes was a Mg^{2+}-dependent 3'-phosphonucleotidase that converts PAPS to APS (Tsang *et al.*, 1971). The other partially purified fraction was able to convert [^{35}S]APS directly into acid volatile, radioactive sulfite in the presence of certain thiols such as mercaptoethanol.

Similar results were obtained by Schmidt (1972a) using reduced glutathione. In this case the partially purified enzyme transferred radioactive sulfur from APS to glutathione and labeled S-sulfoglutathione was formed [Eq. (13)]

$$APS + GSH \rightarrow GSSO_3H + AMP \tag{13}$$

from which $^{35}SO_2$ could be obtained upon the addition of carrier sulfite. Extracts from spinach chloroplasts also formed S-sulfoglutathione (Schmidt, 1972b). GSH is assumed to be effectively replacing the unknown protein carrier (Section IV,A). The transferring enzyme has been termed APS-sulfotransferase; it is analogous to enzyme B of the yeast system (Section III,C).

C. Sulfite Formation

Early experiments by Fromageot and Perez-Milan and by Asahi and co-workers with excised leaves of tobacco and mung beans infiltrated with labeled sulfate and sulfite provided the first evidence that sulfite might be an intermediate in sulfate reduction in higher plants (Wilson, 1962).

Experiments with illuminated whole spinach chloroplasts have shown that sulfite formed from radioactive sulfate is bound to a low molecular weight protein, $RSSO_3H$ (Section IV,A), and can be released as free sulfite by exchange with carrier sulfite or thiosulfate. Free sulfite or thiosulfate was not detected in whole chloroplasts or cell-free extracts. The bound sulfite can be dialyzed or precipitated with ammonium sulfate and then released as SO_2. It is formed directly from APS in the presence of APS-sulfotransferase (Section IV,B). The bound sulfite intermediate of chloroplasts appears to be very similar to the low molecular weight bound sulfite of yeast (Section III,C).

The partially purified APS-sulfotransferase of *Chlorella* was found to have a molecular weight of about 330,000 by agarose column chromatography (Hodson and Schiff, 1971) and a pH optimum of about 8.5 (Schmidt, 1972a). Either GSH or dithiothreitol could be used as substrates in the reaction with the purified system, but other thiols, such as cysteine, homocysteine, BAL, or mercaptoethanol, were ineffective in the absence of glutathione. Since the natural acceptor is thought to be an –SH group on a low molecular weight protein, thiols apparently function as activators and substrate for this reaction [Eq. (14)].

$$APS + RSH \rightarrow RS{-}SO_3H + AMP \qquad (14)$$

Other characteristics of the natural acceptor are unknown, although Asahi (1964) has demonstrated the presence of a reducible disulfide protein in spinach chloroplasts with properties similar to those of fraction C of yeast.

Further study is necessary to establish the significance of the APS-sulfotransferase and the low molecular weight protein in view of the known reactivity of GSH and other thiols and disulfides (see Woodin and Segel, 1968; Winell and Mannervik, 1969). However, since free sulfite readily exchanges with the low molecular weight S-sulfoprotein, the compound appears to be the first reduced sulfur intermediate and the partially purified S-sulfotransferase can be termed an APS-reductase, analogous to the PAPS-reductase (enzyme B) of yeast (but see later review by Schiff and Hodson, 1973). APS-reductases have been partially purified and characterized in sulfate reducing bacteria (*Desulfovibrio* and *Desulfotomataculum*) and in the sulfur-oxidizing thiobacilli (see Michaels *et al.*, 1971).

D. Sulfite Reduction

Sulfite reductases of *Allium odorum* (Tamura, 1965) and spinach (Asada *et al.*, 1969) have been highly purified and partially characterized. These enzymes utilize the artificial dye, MVH (reduced methyl viologen), as electron donor and are able to carry out the complete six electron reduction of sulfite to sulfide [Eq. (15)].

$$6 \text{ MVH} + SO_3{}^{2-} \rightarrow 6 \text{ MV} + S^{2-} + 3 \text{ H}_2O \qquad (15)$$

These MVH-sulfite reductases can also reduce hydroxylamine, but, unlike the yeast enzyme, they cannot utilize NADPH and cannot reduce nitrite.

The MVH-sulfite reductase of spinach has been purified 492-fold and is homogeneous by ultracentrifugation. It has a molecular weight of 84,000 and absorption maxima at 278, 404, and 589 nm with a shoulder at 385 nm. The purified enzyme is extremely sensitive to reduced methyl viologen and is completely inactivated by the reduced dye in the absence of a protective protein, such as bovine serum albumin. Thiols (cysteine), disulfides (cystine), and sulfite can partially substitute for the protective protein if added prior to the reduced dye.

The purified sulfite reductase is also sensitive to treatment with PCMB, KCN, and CO. Even when protected by cystine, the MVH-reduced enzyme is completely inhibited by PCMB and KCN under anaerobic conditions. After dialysis to remove unreacted inhibitor, no sulfite- or hydroxylamine-reducing activity remains. The activity of oxidized enzyme is not affected by these inhibitors. PCMB inhibition is prevented by the presence of sulfite, but sulfite has no protective action against the inhibition by cyanide. The PCMB-sensitive site appears to be a thiol group. The addition of CO to MVH-reduced enzyme also inhibits sulfite and hydroxylamine reduction. In this case, the presence of sulfite prevents the inhibition of sulfite reduction but not hydroxylamine reduction. The inhibition is reversed by light.

The formation of complexes with cyanide and CO, the light reversal, the protective action of sulfite, the presence of iron, and the absorption spectra all suggest the presence of a chromatophore with the properties of a heme, although some of the properties are atypical.

The sulfite reductase of spinach is similar to that of yeast and other microorganisms in several respects. It reduces sulfite and hydroxylamine with MVH, contains iron, has an absorption maxima in the range 585–589 nm, and is inhibited by KCN. On the other hand, it has a lower molecular weight, lacks flavin, and cannot utilize NADPH as electron donor. The sulfite reductase from one of the yeast mutants similarly lacks flavin, is of low molecular weight, and utilizes MVH, but not NADPH, as reductant. Presumably, yeast and spinach have similar sulfite reducing sites, but differ in the way they couple with natural electron donors.

The natural electron donor for spinach and higher plants remains to be determined, although crude preparations can couple sulfite reduction with ferredoxin (Asada *et al.*, 1969; Schmidt and Trebst; 1969; Trebst and Schmidt, 1969). Recently, the nitrite reductase of *Chlorella*, a chloroplast enzyme, has been purified to homogeneity and shown to display a remarkable similarity to spinach sulfite reductase in molecular weight, iron and sulfur content, absorption spectra, KCN inhibition, and acceptor specificity (Zumft, 1972). Ferredoxin is a natural cofactor in nitrite reduction and couples directly with nitrite reductase when photosynthetically or chemically reduced. NADPH can also serve as reductant, but only when chloroplast NADP reductase and ferredoxin are present. A similar system may be involved in sulfite reduction. The demonstration that most of the sulfite reductase activity resides in chloroplasts (Mayer, 1967) supports this point of view. The opportunity for direct photosynthetic reduction as well as dark reduction via NADPH could explain why algae and higher plants might differ from fungi and bacteria in the manner in which they couple with electron donors.

The electron acceptor specificity of sulfite reductase requires further comment. The highly purified sulfite reductases of yeast, *Escherichia coli*, and *Salmonella typhimurium* reduce nitrite and hydroxylamine as well as sulfite. After partial purification, the nitrite and part of the hydroxylamine-reducing activity of spinach sulfite reductase is lost. A constant ratio of hydroxylamine to sulfite reducing activity remains during the final steps of purification. This suggests that sulfite and hydroxylamine are reduced by the same enzyme. Furthermore, both activities are inhibited by MVH, PCMB, KCN, and CO. There are some differences: sulfite protects the enzyme against MVH and CO, but hydroxylamine does not; the K_m is higher for hydroxylamine than for sulfite. If the same protein reduces both substrates, the properties of the catalytic site differs for the two substances or different sites are involved. A similar situation exists in the separation of nitrite and hydroxylamine reductases of spinach and *Chlorella* (Zumft, 1972).

Additional evidence that bound sulfite and bound sulfide may be intermediates in sulfate reduction has appeared. Using S-sulfoglutathione as a model substrate, Schmidt and Schwenn (1971) have investigated the reduction of the bound sulfite group. An ammonium sulfate fraction of spinach and *Chlorella* was able to carry out the reduction of S-sulfo-glutathione to the sulfide level [Eq. (16)].

$$GSSO_3H \xrightarrow[\text{ferredoxin}]{\text{NADPH}} GSSH \qquad (16)$$

NADPH served as the electron donor, but ferredoxin was required as previously shown for free sulfite. However, the S-sulfoglutathione reductase was not identical with sulfite reductase. It would not reduce free sulfite, and the sulfite reductase of Asada *et al.* (1969) could not reduce the S-sulfoglutathione.

E. Thiosulfate Formation and Utilization

The *in vivo* metabolism of thiosulfate has been intensively investigated in *Chlorella* by Schiff and co-workers (Schiff and Hodson, 1970). It is easily utilized as a source of sulfur, is formed from sulfate, and can be demonstrated in soluble extracts. Short-term uptake studies with differentially labeled thiosulfate have shown that the outer S moiety is rapidly taken up and incorporated into the cystine and methionine of proteins; reduction and incorporation of the inner SO_3 moiety occurs much more slowly. They concluded that exogenous thiosulfate undergoes dismutation to sulfite and sulfide before reduction. Most of the sulfite apparently returned to the sulfate level before incorporation into protein.

Cell-free extracts incubated with radioactive sulfate, ATP, Mg^{2+}, and NADPH or BAL as reductant resulted in the formation of acid volatile radioactivity. Radioactive products in the crude incubation mixture were shown to be PAPS and thiosulfate by column chromatography on Dowex-1-nitrate and by electrophoresis. Acidification of thiosulfate releases SO_2 as the acid volatile product. Similar enzyme extracts from several other strains and species of *Chlorella* and from a variety of other microorganisms including *E. coli, S. typhimurium,* and baker's yeast also yielded PAPS and thiosulfate, indicating the widespread occurrence of thiosulfate during sulfate reduction. However, degradation of the thiosulfate revealed that most of the radioactivity occurred in the sulfite moiety. This indicated that the sulfite moiety arose from sulfate reduction, but the S moiety originated from the enzyme extract. Since Schmidt and Schwenn (1971) have shown that free thiosulfate cochromatographs with their low molecular weight protein ($RSSO_3H$) on Dowex-1-nitrate and that free sulfite and thiosulfate are not formed in their crude extracts, it seems very likely that the S moiety is the thiol acceptor group of the low molecular weight bound sulfite.

An earlier proposal that thiosulfate was a direct intermediate in sulfate reduction has been revised as a result of recent studies with mutants and with purified enzymes. As noted above (Section IV,C), Schiff and co-workers and Schmidt have isolated an enzyme, APS-sulfotransferase, which transfers sulfate directly from APS to a bound form of sulfite. A number of mutants of *Chlorella* have been obtained which cannot grow upon sulfate and lack the APS-sulfotransferase. These mutants are able to grow upon thiosulfate, but canot form thiosulfate. Another mutant can grow on sulfate but cannot form thiosulfate; it also appears to lack the APS-sulfotransferase. These observations indicate that thiosulfate is not on the direct pathway of sulfate reduction, and throws some doubt upon the importance of the APS-sulfotransferase in sulfate reduction. Thus, the exact mode of utilization and formation of thiosulfate in *Chlorella* remains unclear.

V. Cysteine Biosynthesis

A. The Pathway

The formation of cysteine from sulfide and a 3-carbon acceptor is the point at which assimilatory sulfate reduction merges with amino acid metabolism. The carbon acceptor, serine, is first activated by acetyl-coenzyme A in a reaction catalyzed by serine transacetylase and then reacts with sulfide to give cysteine in the presence of O-acetylserine sulfhydrylase as shown in Eqs. (17) and (18).

$$\text{L-Serine} + \text{acetyl-CoA} \rightarrow O\text{-acetyl-L-serine} + \text{CoA} \tag{17}$$
$$O\text{-Acetyl-L-serine} + S^{2-} \rightarrow \text{cysteine} + \text{acetate} \tag{18}$$

These reactions were first demonstrated in *Escherichia coli* and *Salmonella typhimurium* by Kredich and Tomkins (1966) and have been reviewed by Smith (1971). Earlier studies had shown that serine could react directly with sulfide to form cysteine in yeast and a number of fungi and bacteria as well as spinach. Subsequent studies have shown that the rate of sulfhydrylation of activated serine is much higher (up to 20,000 times higher) than that of serine in many bacteria (Chambers and Trudinger, 1971) and is 60 to 100 times greater in spinach (Giovanelli and Mudd, 1967) and turnip leaves (Thompson and Moore, 1968). The extremely high activity of this system together with the direct demonstration of serine transacetylase activity in higher plants (Smith and Thompson, 1969) indicate that cysteine biosynthesis in bacteria and higher plants normally proceeds by way of activated serine. O-Acetylserine is also the preferred substrate for yeast (Wiebers *et al.*, 1967a; Thompson and Moore, 1968). The situation in *Neurospora* is not yet clear, since serine and O-acetylserine are utilized at the same rate and serine transacetylase has not been detected in crude extracts (see Kerr, 1971).

B. Cysteine Biosynthesis in Higher Plants

The evidence for the biosynthetic pathway for cysteine in higher plants is based exclusively on *in vitro* enzymatic studies. Since there is an apparent agreement in regard to the pathway and only limited information about eukaryotes, some comparisons with the well-studied bacterial systems are included in this discussion.

Both enzymes, serine transacetylase and O-acetylserine sulfhydrylase, have been demonstrated in extracts derived from seedlings, leaves, and roots of kidney beans (Smith and Thompson, 1971; Smith, 1972). Serine transacetylase activity was also found in a variety of plant extracts (Smith and Thompson, 1969), and O-acetylserine sulfhydrylase was located in spinach and turnip leaves (Giovanelli and Mudd, 1967; Thompson and Moore, 1968).

Both enzymes were separated and purified 50-fold from kidney beans (Smith and Thompson, 1971; Smith, 1972). Serine transacetylase activity was present in both the particulate (mitochondrial) and supernatant fractions, but O-acetylserine sulfhydrylase was only detected in the soluble fraction.

O-Acetylserine sulfhydrylase is characterized by a high degree of specificity toward the carbon acceptor and some lack of specificity toward the sulfhydryl donor. While O-acetylserine is significantly preferred over serine and O-acetylhomoserine, methylmercaptan and ethylmercaptan could substitute for sulfide to yield methylcysteine and ethylcysteine, respectively, although with much reduced rates compared with sulfide (Giovanelli and Mudd, 1967, 1968; Thompson and Moore, 1968; Smith and Thompson, 1971).

The observed lack of specificity of O-acetylserine sulfhydrylase from higher plants regarding the sulfhydryl donor, as observed *in vitro*, may imply that the actual sulfhydryl donor *in vivo* is a "bound sulfide" rather than free sulfide.

A preference for bound sulfide over free sulfide might be predicted from toxicity considerations. Sulfide is a well-known inhibitor of enzymes of aerobic respiration (James, 1953). Recently it has been shown that the inhibitory effects of sulfide in micromolar concentrations on activities of oxidative enzymes (such as cytochrome oxidase, catalase, and peroxidase) in rice root seedling extracts and on respiration of rice roots could account for the fact that H_2S is a yield-reducing factor in Louisiana rice fields, which contain up to 10^{-5} M H_2S (Allam and Hollis, 1972). The suggestion that H_2S is toxic in rice fields is supported by a unique biological detoxification mechanism postulated by Pitts et al. (1972). Beggiatoa, a sulfur-oxidizing bacterium found in these fields, may favor the rice plant by oxidizing the H_2S and, in turn, may be favored by a catalase-like activity surrounding the rice root tip.

Fractionation studies of spinach extracts suggest that there may be two species of O-acetylserine sulfhydrylase activity (Giovanelli and Mudd, 1968). This may be analogous to the situation in S. typhimurium, where crude extracts contain two separate fractions of O-acetylserine sulfhydrylase activity readily resolved by gel filtration. The first fraction, containing 5% of the total activity, is associated with serine transacetylase in a bifunctional protein complex called cysteine synthetase. The remaining 95% of the O-acetylserine sulfhydrylase activity is apparently in a free pool (Kredich and Tomkins, 1967). Both O-acetylserine sulfhydrylase activities were purified to near homogeneity, and they appear to be identical (Kredich et al., 1969; Becker et al., 1969). The importance of this in cysteine biosynthesis in Salmonella is not yet clear.

C. Regulation and Comparative Biochemistry

Serine transacetylase in Salmonella, is subject to feedback inhibition by very low concentrations (0.001 mM) of L-cysteine, but it is neither repressed by growth on L-cysteine, nor is it derepressed in response to sulfur starvation (Kredich and Tomkins, 1967; Kredich, 1971). In contrast, O-acetylserine sulfhydrylase is repressed during growth on L-cysteine (or sulfide) and is derepressed during growth on L-djenkolic acid, which is interpreted as derepression due to sulfur starvation (Kredich, 1971). The signal for induction (derepression) or repression is apparently mediated via O-acetylserine levels, which fluctuate according to the levels of the two enzymes involved in its biosynthesis and utilization (Kredich, 1971).

Attempts have been made to explore possible regulatory mechanisms involved in cysteine biosynthesis in higher plants based on two different approaches. One approach was to study the effects of sulfur amino acids on the in vitro activities of serine transacetylase and O-acetylserine sulfhydrylase. Results of these studies indicated that neither methionine nor cysteine can be considered as potential feedback inhibitors of serine transacetylase in kidney bean. O-acetylserine sulfhydrylase was also insensitive to end product inhibition (Smith and Thompson, 1971). The kinetic analysis of the cysteine inhibition, however, is complicated by nonenzymatic reactions of the cysteine added with O-acetylserine and acetyl coenzyme A.

The second approach was to study the effects of sulfur deficiency on the level of both enzyme activities as an indication of possible repression or derepression mechanisms. The results (Smith, 1972) indicated that the total extractable serine transacetylase activity was not affected by the growth of beans on sulfur-deficient medium. There was some increase in total O-acetylserine sulfhydrylase activity under the same conditions, but it was rather small (up to a 1.4-fold increase in 35-day-old roots) compared with the change observed in *Salmonella* (two- to threefold increase). A simple interpretation of the results with higher plants is difficult, however, in the absence of knowledge regarding endogenous storage pools, compartmentalization, and mobilization. Furthermore, the criteria used for determining sulfur starvation and the use of roots in this study could lead to ambiguous results.

The role of O-acetylserine in regulation of assimilatory sulfate reduction in higher plants has not yet been studied. However, the recent report of the participation of O-acetylserine in other pathways in higher plants (see note in Smith, 1972) may indicate that its level would not be determined solely by the sulfur status of the plant.

VI. Methionine Biosynthesis

A. Alternative Pathways

Methionine is an amino acid of special importance in view of its dual role as a component of proteins and a participant in many cellular reactions as methyl donor, initiator of protein synthesis, etc.

Two mechanisms have been proposed for the synthesis of methionine. One is by transsulfuration where reduced sulfur from assimilatory sulfate reduction is initially incorporated into cysteine (Section V) and then transferred to homocysteine and methionine via cystathionine. The other is direct sulfhydrylation where reduced sulfur is incorporated directly into a 4-carbon acceptor, activated homoserine, to yield homocysteine and methionine. These pathways are outlined in Fig. 7.

Cysteine and methionine are synthesized by a single route in transsulfuration and by different routes by direct sulfhydrylation. The significance of these alternative pathways is dependent upon the role of three enzymes, cystathionine γ-synthetase, β-cystathionase, and O-acetyl-

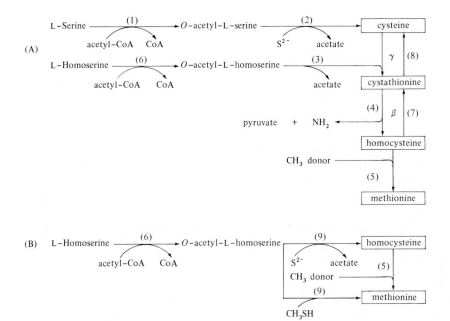

Fig. 7. Alternative pathways of methionine biosynthesis by (A) transsulfuration and (B) direct sulfhydrylation. The enzymes are (1) serine transacetylase, (2) O-acetylserine sulfhydrylase, (3) cystathionine γ-synthetase, (4) β-cystathionase, (5) transmethylase (6) homoserine transacetylase, (7) cystathionine β-synthetase, (8) γ-cystathionase, and (9) O-acetylhomoserine sulfhydrylase.

homoserine sulfhydrylase as discussed below. Transsulfuration refers to all of the reactions by which reduced sulfur is transferred from cysteine to homocysteine and from homocysteine back to cysteine, but we are primarily concerned with the reactions in the forward direction, toward homocysteine. Cystathionine, a thioether is the central intermediate in

$$\underset{a\quad\beta\quad\gamma\quad\quad\beta\quad a}{HOOC-\overset{\overset{\displaystyle NH_2}{|}}{CH}-CH_2-CH_2-S-CH_2-\overset{\overset{\displaystyle NH_2}{|}}{CH}-COOH}$$

transsulfuration. The cystathionine synthetase reactions form either β- or γ-thioether bonds and the cystathionase reactions cleave either β- or γ-thioether bonds.

B. Methionine Biosynthesis in Microorganisms

Enzymes catalyzing the reactions for both of the above pathways are present in bacteria (e.g., *E. coli* and *S. typhimurium*) and fungi (e.g., yeast and *Neurospora*).

1. BACTERIA

In bacteria the activation of homoserine in both reaction pathways is with succinate rather than acetate (see Smith, 1971). Cystathionine γ-synthetase, a pyridoxal phosphate-containing enzyme that catalyzes the formation of cystathionine via transsulfuration, has been obtained in pure form from *Salmonella*. This same enzyme also catalyzes direct sulfhydrylation of *O*-succinylhomoserine to homocysteine or methionine *in vitro*. However, studies with homocysteine-less mutants specifically lacking β-cystathionase activity confirm that transsulfuration is the functional pathway for *in vivo* methionine biosynthesis. The direct sulf-hydrylation reaction appears to be primarily an *in vitro* phenomenon reflecting lack of specificity of cystathionine γ-synthetase for cysteine (discussed by Kerr, 1971; de Robichon-Szulmajster and Surdin-Kerjan, 1971). Reverse transsulfuration, i.e., cystathionine β-synthetase and γ-cystathionase activities, has not been detected in any of the bacterial systems.

2. NEUROSPORA

The demonstration of *O*-acetylhomoserine sulfhydrylase activity in extracts of wild-type *Neurospora* which is inhibited *in vitro* by methionine (at 10 mM), reduced levels of the enzymatic activity in methionine-less mutants, and data from nutritional experiments (lack of transfer of label from cystathionine to methionine) led Wieber and Garner (1967a,b) to favor direct sulfhydrylation as the main pathway for methionine biosynthesis in *Neurospora*.

However, their experimental data do not exclude the possibility that they actually dealt with *O*-acetylserine sulfhydrylase rather than *O*-acetylhomoserine sulfhydrylase activity. This possibility is supported by the higher activity of *O*-acetylserine sulfhydrylase in this fraction, and similar behavior of *O*-acetylserine sulfhydrylase activity with respect to inhibition by methionine and mutant studies.

Later, rather extensive studies by Flavin's group (see Kerr and Flavin, 1970; Kerr, 1971) using 500-fold purified *O*-acetylhomoserine sulfhydrylase seem to exclude the involvement of this enzyme as the primary mechanism for *in vivo* methionine biosynthesis in favor of transsulfuration. This conclusion was based on mutant studies as well as studies of regulation of the enzymes involved. In *Neurospora*, unlike bacteria, *O*-acetylhomoserine sulfhydrylase and cystathionine γ-synthetase are two separate enzymes. Mutants defective specifically in the transsulfuration mechanism (either in cystathionine γ-synthetase or β-cystathionase activity) and containing normal levels of O-acetylhomoserine sulfhydrylase show almost complete dependence on an exogenous supply of homocysteine or methionine for growth. In addition, cystathionine γ-synthetase, but not O-acetylhomoserine sulfhydrylase, is very sensitive to control by S-adenosylmethionine ($K_i = 0.01$ mM) via feedback inhibition.

Studies of O-acetylhomoserine sulfhydrylase, especially the observations that the 500-fold purified enzyme can catalyze the biosynthesis of both homocysteine and methionine while the rate of the reaction with methylmercaptan as a substrate is 1.8 times the rate with sulfide, led Kerr (1971) to suggest that O-acetylhomoserine sulfhydrylase may function in the synthesis of methionine from methylmercaptan under conditions of sulfur deficiency according to Eqs. (19) and (20).

$$S\text{-Methylcysteine} \rightarrow \mathrm{CH_3SH} + \text{pyruvate} + \mathrm{NH_3} \qquad (19)$$
$$\mathrm{CH_3SH} + O\text{-acetylhomoserine} \rightarrow \text{methionine} + \text{acetate} \qquad (20)$$

S-Methylcysteine is presumably cleaved by γ-cystathionase [Eq. (19)], and then methioine is formed by the action of O-acetylhomoserine sulfhydrylase [Eq. (20)].

The existence of both activities in crude extracts of *Neurospora* (Moore and Thompson, 1967), the derepression (30-fold) of γ-cystathionase (Flavin and Slaughter, 1967), the persistence of O-acetylhomoserine sulfhydrylase activity under conditions of sulfur starvation, the accumulation of S-methylcysteine in *Neurospora,* as well as the capability of S-methylcysteine and methylmercaptan to serve as the sole source of sulfur for growth of methionineless mutants although never as well as methionine (Moore and Thompson, 1967) are all consistent with Kerr's suggestion. Additional support comes from a recent study of kidney beans. Doney and Thompson (1971) found that, although in normal plants methylcysteine is probably not a methyl or thiomethyl ($\mathrm{CH_3SH}$) donor, in sulfur-deficient beans the methyl group of methylcysteine can be a significant factor in the formation of methionine. However, since there is no known mechanism for the synthesis of S-methylcysteine or methylmercaptan except *via* methionine, the proposed route would not be an alternative pathway for direct synthesis of methionine from inorganic sulfur.

If direct sulfhydrylation is indeed limited to certain stress conditions, this would explain the discrepancy between the *in vitro* and *in vivo* activity of O-acetylhomoserine sulfhydrylase. It is also possible that the function of O-acetylhomoserine sulfhydrylase is to serve as a scavenger for potentially toxic amounts of sulfide or methylmercaptan (Kerr, 1971).

3. Yeast

Evidence accumulated for many years supports direct sulfhydrylation as the main pathway for methionine biosynthesis in yeast, in contrast to the transsulfuration pathway in *Neurospora*. This evidence (summarized in detail by de Robichon-Szulmajster and Surdin-Kerjan, 1971) consists of biochemical and genetic data. The genetic data are based on the observations that methionine-less mutants of yeast are defective in *O*-acetylhomoserine sulfhydrylase activity and that exogenously supplied cystathionine fails to support growth of methionine auxotrophs; cystathionine was shown to be actively transported. The biochemical evidence is primarily the inhibition and repression of *O*-acetylhomoserine sulfhydrylase by methionine. Failure to detect cystathionine γ-synthetase activity in yeast extracts further supported the direct sulfhydrylation pathway.

Savin and Flavin (1972), however, have definitely established the presence of cystathionine γ-synthetase activity in yeast extracts. Purification studies showed that cystathionine γ-synthetase and *O*-acetylhomoserine sulfhydrylase were two separate enzymatic activities, as in *Neurospora*. Although the ratio of rates of cystathionine γ-synthetase to *O*-acetylhomoserine sulfhydrylase activities is rather low in yeast compared with *Neurospora*, the rate of cystathionine γ-synthetase activity is higher than that of homoserine transacetylase and, thus, cystathionine γ-synthetase is probably not the rate-limiting factor for the transsulfuration pathway. Studies of both activities (cystathionine γ-synthetase and *O*-acetylhomoserine sulfhydrylase) showed that, unlike the situation in *Neurospora*, both enzymatic activities are absent from the same methionine auxotrophs, and both activities are reduced concomitantly in the presence of methionine in the growth media.

In addition, some *O*-acetylhomoserine sulfhydrylase-deficient mutants were able to grow on cysteine as the only sulfur source, and labeling experiments demonstrated the formation of cystathionine directly from cysteine rather than by reverse transsulfuration.

These results indicate that both pathways proposed for the biosynthesis of methionine are present in yeast. Which of these routes, transsulfuration or direct sulfhydrylation, is the major physiological path *in vivo* is yet to be determined.

C. Methionine Biosynthesis in Higher Plants

Studies of methionine biosynthesis suggest that both transsulfuration and direct sulfhydrylation mechanisms exist in higher plants, although transsulfuration appears to be the primary pathway involved in methionine biosynthesis from inorganic sulfur.

In vitro enzymatic studies indicate that the transsulfuration pathway in higher plants resembles that of bacteria in operating predominantly, perhaps exclusively, in the forward direction.

$$\text{Cysteine} \rightarrow \text{cystathionine} \rightarrow \text{homocysteine}$$

This is based on the following observations: (1) Crude extracts of spinach can synthesize cystathionine (with O-succinylhomoserine or O-acetylhomoserine as substrate) exclusively via cystathionine γ-synthetase, while no cystathionine β-synthetase activity could be detected (Giovanelli and Mudd, 1966). (2) Extracts of a range of higher plants can cleave cystathionine via β-cystathionase exclusively, while γ-cystathionase activity, if present, is always less than 0.5% of β-cystathionase activity (Giovanelli and Mudd, 1971). The failure to detect γ-cystathionase was not due to inhibitors in the plant extracts or failure to detect the assay reaction product.

β-Cystathionase activity is widespread among all plants and tissues tested including nonphotosynthesizing tissues. Purified β-cystathionase from spinach leaves has no detectable cystathionine β-synthetase activity.

First evidence for the existence of the direct sulfhydrylation pathway in higher plants came from work of Giovanelli and Mudd (1966) who stated, without showing data, that there was about a 20-fold increase in the rate of O-acetylhomoserine-dependent *in vitro* incorporation of labeled sulfur into a homocysteine derivative (S-adenosylhomocysteine) when [^{35}S]sulfide was used instead of [^{35}S]cysteine in crude spinach extracts.

However, later attempts to isolate the homocysteine-synthesizing system, i.e., O-acetylhomoserine sulfhydrylase activity, and separate it from O-acetylserine sulfhydrylase activity of spinach were unsuccessful. One ammonium sulfate fraction with both activities incorporated [^{35}S]sulfide into cysteine at two to three times the incorporation rate into homocysteine. A separate experiment indicated that the homocysteine was synthesized by sulfhydrylation of O-acetylhomoserine and not by β-cleavage of cystathionine (Giovanelli and Mudd, 1967). Another (second) ammonium sulfate fraction from spinach also had some O-acetylhomoserine sulfhydrylase activity (Giovanelli and Mudd, 1968). However the activity was only 1% of the activity of the O-acetylserine sulfhydrylase activity in the same fraction. When sulfide was replaced by methylmercaptan, the O-acetylhomoserine sulfhydrylase activity was about 5% of the O-acetylserine sulfhydrylase activity in this spinach fraction. It is possible that these negligible O-acetylhomoserine sulfhydrylase activities are only a reflection of a low level of nonspecificity of O-acetylserine sulfhydrylase toward the carbon acceptor. The isolation of O-acetylhomoserine sulfhydrylase as an independent enzyme in higher plants remains as a challenge for future work.

It may be worth noting that the O-acetylhomoserine sulfhydrylase of the second spinach fraction was about twice as active with methylmercaptan as with sulfide. It is possible that this enzyme could function under certain conditions as proposed for *Neurospora* (see above).

Further evidence for the role of the two alternative pathways in the biosynthesis of homocysteine and methionine was provided by *in vivo* studies.

Dougall and Fulton (1967) used the isotope competition technique to identify possible intermediates in the biosynthesis of protein amino acids in rose tissue culture. According to this technique, a compound that competes with glucose as carbon source for a particular amino acid is presumed to be an intermediate (or readily converted into an intermediate) in the biosynthetic route of this amino acid. Accordingly, some selected unlabeled compounds were tested for competition with [U–^{14}C]glucose (provided as a sole carbon source in the medium) as carbon contributors to methionine. Five unlabeled compounds were found to inhibit the incorporation of [^{14}C]glucose into protein methionine at millimolar concentrations as follows: homoserine (40% inhibition), O-succinylhomoserine (40%), O-acetylhomoserine (40%), cystathionine (50%), and homocysteine (70%). These results argue strongly for the participation of cystathionine, i.e., transsulfuration, in methionine biosynthesis in rose cells.

The participation of both O-succinylhomoserine and O-acetylhomoserine as presumed substrates for cystathionine formation in these competition studies with rose cells is in agreement with earlier findings that both compounds were substrates for cystathionine γ-synthetase in spinach extracts (Giovanelli and Mudd, 1966).

More recently, an attempt was made by Ngo and Shargool (1972) to distinguish between the two alternative pathways by using *in vivo* feeding experiments, where [^{35}S]sulfide or L-[^{14}C]homoserine were fed to germinating rape seeds, and the kinetics of subsequent incorporation of label into free sulfur amino acids were studied. The results (based on 80% recovery of the sulfur amino acids during extraction) show that when rape seeds germinate in the presence of [^{35}S]sulfide, there is an immediate uptake of label with the subsequent formation of all four sulfur amino acids: cysteine, homocysteine, cystathionine, and methionine. The kinetic data argue for transsulfuration as a major path between 5 and 21 hours of germination, i.e., cysteine is the first intermediate and acts as a donor of sulfur into all the other sulfur amino acids. However, during the first 5 hours, significant amounts of homocysteine are synthesized, presumably by direct sulfhydrylation. Similar results were obtained with the kinetic labeling experiments with L-[^{14}C]homoserine. Since seed germination studies of amino acid biosynthesis are complicated by hydrolysis of reserve proteins, pools, compartmentalization, etc., additional *in vivo* experiments with similar systems combined with *in vitro* studies are needed.

Further evidence that transsulfuration is involved in methionine biosynthesis in higher plants comes from independent studies on rhizobitoxine. Rhizobitoxine is a phytotoxin synthesized by certain strains of the soybean root nodule bacterium, *Rhizobium japonicum*. The toxin causes chlorosis in the host plant and in a variety of other plant species when added exogenously to culture solutions of seedlings (see Owens *et al.*, 1968). The mechanism by which rhizobitoxine inhibits the greening of leaves was suspected to involve methionine metabolism.

The discovery that rhizobitoxine inhibits the growth of *Salmonella typhimurium* led to a study of its possible mode of action in that organism (Owens *et al.*, 1968). The growth of both wild-type and a methionine auxotroph defective in cystathionine γ-synthetase supplemented with cystathionine is inhibited by millimolar concentrations of rhizobitoxine, while both are completely insensitive to the inhibition when either homocysteine or methionine is added simultaneously. L-Homoserine has no effect in overcoming the toxin effect. These results strongly suggested that, in *Salmonella*, the toxin inhibited the β-cleavage of cystathionine to form homocysteine, and therefore, the direct effect of the toxin on partially purified β-cystathionase of *Salmonella* was examined. The specific inhibition by very low toxin concentrations (10^{-8} to 10^{-7} M) strongly supports the idea that growth inhibition of *Salmonella* is caused by a methionine deficiency induced by rhizobitoxine inhibition of the β-cleavage of cystathionine to homocysteine by β-cystathionase.

The possibility that the toxicity of rhizobitoxine to higher plants was due to the same mechanism as in *Salmonella* was tested by Giovanelli *et al.* (1971), who studied the effect of rhizobitoxine on 400-fold purified β-cystathionase of spinach. Extensive kinetic studies showed that the rhizobitoxine, at slightly higher concentrations, also irreversibly inactivated the β-cystathionase of spinach. *In vitro* inhibition, however, does not prove that this is the *in vivo* mechanism by which the toxin causes pathology.

ACKNOWLEDGMENTS

Helpful discussions with Philip Filner and Gregory Dillworth and support by the U.S. Atomic Energy Commission under contract AT(11-1)-1338 are gratefully acknowledged.

REFERENCES

Adams, C. A., and Johnson, R. E. (1968). *Plant Physiol.* 43, 2041.
Adams, C. A., and Nicholas, D. J. D. (1972). *Biochem. J.* 128, 647.
Adams, C. A., and Rinne, R. W. (1969). *Plant Physiol.* 44, 1241.
Allam, A. I., and Hollis, J. P. (1972). *Phytopathology* 62, 634.
Asada, K., Tamura, G., and Bandurski, R. S. (1969). *J. Biol. Chem.* 25, 4904.
Asahi, T. (1964). *Biochim. Biophys. Acta* 82, 58.
Balharry, G. J. E., and Nicholas, D. J. D. (1970). *Biochim. Biophys. Acta* 220, 513.
Bandurski, R. S. (1965). *In* "Plant Biochemistry" (J. Bonner and J. E. Varner, eds.), 2nd ed., pp. 467–490. Academic Press, New York.
Becker, M. A., Kredich, N. M., and Tomkins, G. M. (1969). *J. Biol. Chem.* 224, 2418.
Benson, A. A. (1971). *In* "Structure and Function of Chloroplasts" (M. Gibbs, ed.), pp. 130–148. Springer-Verlag, Berlin and New York.
Chambers, L. A., and Trudinger, P. A. (1971). *Arch. Mikrobiol.* 77, 105.
de Robichon-Szulmajster, H., and Surdin-Kerjan, Y. (1971). *In* "The Yeasts" (A. H. Rose and J. S. Harrison, eds.), Vol. 2, pp. 335–418. Academic Press, New York,
Doney, R. C., and Thompson, J. F. (1971). *Phytochemistry* 10, 1745.
Dougall, D. K., and Fulton, M. M. (1967). *Plant Physiol.* 42, 941.
Ellis, R. J. (1969). *Planta* 88, 34.
Flavin, M., and Slaughter, C. (1967). *Biochim. Biophys. Acta* 132, 406.
Giovanelli, J., and Mudd, S. H. (1966). *Biochem. Biophys. Res. Commun.* 25, 366.
Giovanelli, J., and Mudd, S. H. (1967). *Biochem. Biophys. Res. Commun.* 27, 150.
Giovanelli, J., and Mudd, S. H. (1968). *Biochem. Biophys. Res. Commun.* 31, 275.
Giovanelli, J., and Mudd, S. H. (1971). *Biochim. Biophys. Acta* 227, 654.
Giovanelli, J., Owens, L. D., and Mudd, S. H. (1971). *Biochim. Biophys. Acta* 227, 671.
Goodwin, T. W. (1971). *In* "Structure and Function of Chloroplasts" (M. Gibbs, ed.), pp. 215–276. Springer-Verlag, Berlin and New York.
Hilz, H., Kittler, M., and Knape, G. (1959). *Biochem. Z.* 332, 151.
Hodson, R. C., and Schiff, J. A. (1971). *Plant Physiol.* 47, 300.
James, W. O. (1953). *Annu. Rev. Plant Physiol.* 4, 59.
Kerr, D. S. (1971). *J. Biol. Chem.* 246, 95.
Kerr, D. S., and Flavin, M. (1970). *J. Biol. Chem.* 245, 1842.
Kline, B. C., and Schoenhard, D. E. (1970). *J. Bacteriol.* 102, 142.
Kredich, N. M. (1971). *J. Biol. Chem.* 246, 3474.
Kredich, N. M., and Tomkins, G. M. (1966). *J. Biol. Chem.* 241, 4955.
Kredich, N. M., and Tomkins, G. M. (1967). *In* "Organizational Biosynthesis" (H. J. Vogel, J. O. Lapman, and V. Bryson, eds.), p. 189. Academic Press, New York.
Kredich, N. M., Becker, M. A., and Tomkins, G. M. (1969). *J. Biol. Chem.* 244, 2428.
Lee, R. F., and Benson, A. A. (1972). *Biochim. Biophys. Acta* 261, 35.

Mayer, A. M. (1967). *Plant Physiol.* **42**, 324.

Mercer, E. I., and Thomas, G. (1969). *Phytochemistry* **8**, 2281.

Metzenberg, R. L. (1972). *Annu. Rev. Genet.* **6**, 111.

Michaels, G. B., Davidson, J. T., and Peck, H. D. (1971). *In* "Flavins and Flavoproteins" (H. Kamin, ed.), Vol. 3, pp. 555–580. Univ. Park Press, Baltimore, Maryland.

Moore, D. P., and Thompson, J. F. (1967). *Biochem. Biophys. Res. Commun.* **28**, 474.

Murphy, M. J., Siegel, L. M., and Kamin, H. (1973). *J. Biol. Chem.* **248**, 2801.

Ngo, T. T., and Shargool, P. D. (1972). *Biochem. J.* **126**, 985.

Nissen, P., and Benson, A. A. (1961). *Science* **134**, 1759.

Onajobi, F. D., Cole, C. V., and Ross, C. (1973). *Plant Physiol.* **52**, 580.

Owens, J. D., Guggenheim, S., and Jilton, J. L. (1968). *Biochim. Biophys. Acta* **158**, 219.

Pitts, G., Allam, A. I., and Hollis, J. P. (1972). *Science* **178**, 990.

Porque, P. G., Baldesten, A., and Reichard, P. (1970). *J. Biol. Chem.* **245**, 2371.

Robbins, P. W., and Lipmann, F. (1958). *J. Biol. Chem.* **233**, 681.

Ronchi, S., and Williams, C. H. (1972). *J. Biol. Chem.* **247**, 2083.

Roy, A. B., and Trudinger, P. A. (1970). "The Biochemistry of Inorganic Compounds of Sulphur." Cambridge Univ. Press, London and New York.

Savin, M. A., and Flavin, M. (1972). *J. Bacteriol.* **112**, 299.

Schiff, J. A., and Hodson, R. C. (1970). *Ann. N.Y. Acad. Sci.* **175**, 556.

Schiff, J. A., and Hodson, R. C. (1973). *Annu. Rev. Plant Physiol.* **24**, 381.

Schmidt, A. (1972a). *Arch. Mikrobiol.* **84**, 77.

Schmidt, A. (1972b). *Z. Naturforsch. B* **27**, 183.

Schmidt, A., and Schwenn, J. D. (1972). *Proc. Int. Congr. Photosyn. Res., 2nd, 1971,* p. 507.

Schmidt, A., and Trebst, A. (1969). *Biochim. Biophys. Acta* **180**, 529.

Scott, J. M., and Spencer, B. (1968). *Biochem. J.* **106**, 471.

Shaw, W. H., and Anderson, J. W. (1971). *Plant Physiol.* **47**, 114.

Shaw, W. H., and Anderson, J. W. (1972). *Biochem. J.* **127**, 237.

Siegel, L. M., Kamin, H., Reuger, D. C., Presswood, R. P., and Gibson, Q. H. (1971). *In* "Flavins and Flavoproteins" (H. Kamin, ed.), Vol. 3, pp. 523–554. Univ. Park Press, Baltimore, Maryland.

Smith, D. A. (1971). *Advan. Genet.* **16**, 141.

Smith, I. K. (1972). *Plant Physiol.* **50**, 477.

Smith, I. K., and Thompson, J. F. (1969). *Biochem. Biophys. Res. Commun.* **35**, 939.

Smith, I. K., and Thompson, J. F. (1971). *Biochim. Biophys. Acta* **227**, 288.

Tamura, G. (1965). *J. Biochem. (Tokyo)* **57**, 207.

Thompson, J. F. (1967). *Annu. Rev. Plant Physiol.* **18**, 59.

Thompson, J. F., and Moore, D. P. (1968). *Biochem. Biophys. Res. Commun.* **31**, 281.

Torii, K. J., and Bandurski, R. S. (1967). *Biochim. Biophys. Acta* **136**, 286.

Trebst, A., and Schmidt, A. (1969). *Progr. Photosyn. Res.* **3**, 1510.

Trudinger, P. A. (1969). *Advan. Microbial Physiol.* **3**, 111.

Tsang, M. L., Goldschmidt, E. E., and Schiff, J. A. (1971). *Plant Physiol.* **47**, S20.

Tweedie, J. W., and Segel, I. H. (1971a). *J. Biol. Chem.* **246**, 2438.

Tweedie, J. W., and Segel, I. H. (1971b). *Prep. Biochem.* **1**, 91.

Varma, A. K., and Nicholas, D. J. D. (1971a). *Arch. Mikrobiol.* **78**, 99.

Varma, A. K., and Nicholas, D. J. D. (1971b). *Biochim. Biophys. Acta* **227**, 373.

Wiebers, J. L., and Garner, H. R. (1967a). *J. Biol. Chem.* **242**, 5644.

Wiebers, J. L., and Garner, H. R. (1967b). *J. Biol. Chem.* **242**, 12.

Wilson, L. G. (1962). *Annu. Rev. Plant Physiol.* **13**, 201.

Wilson, L. G., Asahi, T., and Bandurski, R. S. (1961). *J. Biol. Chem.* **236**, 1822.

Winell, M., and Mannervik, B. (1969). *Biochim. Biophys. Acta* **184**, 374.

Woodin, T. S., and Segel, I. H. (1968). *Biochim. Biophys. Acta* **167**, 78.

Yoshimoto, A., and Sato, R. (1970). *Biochim Biophys. Acta* **220**, 190.

Zumft, W. G. (1972). *Biochim. Biophys. Acta* **276**, 363.

20

Nitrate Metabolism

E. J. HEWITT, D. P. HUCKLESBY, AND B. A. NOTTON

I. Introduction

Nitrate is the principal source of nitrogen for most plants[*] growing under normal field conditions in fertile soils. Nitrification of ammonia is usually rapid when aeration, moisture content, and soil temperature are favorable and are compatible with good growth of plants (Russell, 1950). Nitrate is a more favorable form of nitrogen than ammonia for most plants (Nightingale *et al.*, 1931; Nightingale, 1937; Hewitt, 1966). Although nitrate assimilation is an important and characteristic physiological activity in practically all types of plants and microorganisms, there are exceptions in some mutant forms that are of great importance in understanding some structural and biosynthetic aspects of nitrate reductase. Environmental and nutrient conditions have a profound influence on the manifestation or the existence of the enzymes involved in reduction of nitrate to ammonia in plants and microorganisms.

The principal forms of inorganic nitrogen in plants are nitrate, nitrite, and ammonia. Concentrations of all three may vary considerably, but, in general, nitrite accumulation is rare and concentrations of ammonia are relatively low (between 0.004 and 0.01 M) (Nightingale, 1937; Hewitt *et al.*, 1957). Nitrate concentrations vary enormously and some-

[*] The word "plant" is used here to denote higher plants.

633

times unpredictably relative to the other two forms (Nightingale, 1937). Season, climate, plant age or development stage, nutritional status, management and species are all factors that affect nitrate content (Hewitt, 1966).

Although nitrate accumulation is not generally injurious, the large variations in cellular concentrations indicate that control of nitrate reduction depends on sensitive and complex mechanisms. Economically, the presence of nitrate is important, as it is potentially toxic either as in fodder or in some foodstuffs, e.g., spinach. Ingested nitrate causes methemoglobinemia in babies and even older humans due to the combination of nitrite produced by reduction in the liver with hemoglobin. Nitrate reduction is of interest, therefore, beyond its physiological importance to the plant.

II. Nitrate Reduction

A. Nitrate Reductase Enzymes

1. FUNCTIONAL AND KINETIC ASPECTS

Nitrate reductases (NR) are generally complex in component structure and mechanism. Most nitrate reductase systems show two distinct types of activity. These are nitrate reduction and dehydrogenase activity resulting in reduction of a variety of one or two electron acceptors including cytochrome c and ferricyanide (1 e^-) ; di- or trichlorophenolindophenol (DCPIP or TCPIP) and tetrazolium compounds (2 e^-). Pyridylium compounds, benzyl viologens (BV), or methyl viologen (MV) may possibly also be acceptors (1 e^-). Reduced flavin adenine dinucleotide (FADH) and flavin mononucleotide (FMNH) or reduced viologen (MV°, BV°) are also electron donors at the molybdenum site for NR activity (see below).

Nicholas and Nason (1954b) identified the dehydrogenase activity of NR by the reduction of TCPIP with reduced nicotinamide dinucleotide phosphate (NADPH) or reduced flavin mononucleotide (FMNH), even with cyanide-treated enzyme. Kinsky and McElroy (1958) found that *Neurospora crassa* NR preparations contained, in addition to a constitutive cytochrome c reductase, a substantial activity that was specifically induced by nitrate parallel with the induction of NR.

Dehydrogenase activities occur in the NR complex from *Chlorella* (Zumft *et al.*, 1969), spinach (Notton and Hewitt, 1971a). *Aspergillus nidulans* (Cove and Coddington, 1965), *Vicia faba* (Oji and Izawa, 1969a), and *Neurospora crassa* (Garrett and Nason, 1967).

The reaction for enzymic reduction of nitrate can be written as

$$AH_2 + NO_3^- \rightarrow A + NO_2^- + H_2O$$

for the couple

$$NO_3^- + 2\,H^+ + 2\,e^- \rightarrow NO_2^- + H_2O$$

E_0' at pH 7.0 equals $+0.425$ V.

When AH_2 represents $NAD(P)H$ ($E_0' = -0.322$), the reaction is

$$NAD(P)H + H^+ + NO_3^- \rightarrow NAD(P)^+ + NO_2^- + H_2O$$

and $\Delta E_0' = 0.75$ V, $F^o = -34.5$ kcal, and kEq $= 10^{25}$.

The reaction is virtually irreversible to the right. Nitrite is the stable product of enzymatic nitrate reduction at physiological pH values of plant, fungal, or bacterial NR's, animal aldehyde or xanthine oxidases, and also potato aldehyde oxidase. In the adenosine triphosphate- (ATP) dependent barley root particulate system in which pyruvate was the indirect electron donor, nitrate also yielded ammonia under conditions where separately added nitrite was only slowly reduced (Bourne and Miflin, 1970).

The reaction of the spinach enzyme (Eaglesham and Hewitt, 1971a) is an ordered sequential pingpong mechanism (bi–bi) (Cleland, 1963). NADH is attached and NAD^+ is released before nitrate combines.

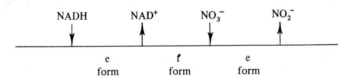

The enzyme is regarded as undergoing reversible (cyclic) transformation between two forms, e and f, at the stages shown.

This represents the kinetic behavior of the overall mechanism with NADH as donor. The mechanism of the other partial reactions is not known.

2. ELECTRON DONORS, INTERMEDIATE CARRIERS, OR COFACTORS

Electron donors include nicotinamide nucleotides NADH and NADPH for both types of reductive activity. Ferredoxin alone is apparently inactive for spinach NR (Ramirez *et al.*, 1964) but functions with the *Anabaena cylindrica* NR (Hattori and Myers, 1967). Reduced pyridylium compounds benzyl viologen (BV⁰) and methyl viologen (MV⁰) and flavin nucleotides are almost universally effective for NR action, but these probably donate electrons directly to the molybdenum, as shown below, and in more extensive schemes elsewhere (Hewitt, 1974).

The importance or specificity of different donors varies widely. NADPH is practically specific in the *Neurospora crassa* and other fungal systems compared with NADH (Nason and Evans, 1953). In soybean when plants are young or when cysteine is excluded from the extraction medium, NADPH and NADH appears equally effective as electron donors and flavin adenine dinucleotide (FAD) is required for maximum activity after purification (Evans and Nason, 1953; Beevers *et al.*, 1964). However phosphatase activity may convert inactive NADPH into NADH (Wells and Hageman, 1974). NR was induced in rice seedlings by nitrate and chloramphenicol (Shen, 1972a); however, while nitrate induced NADH–NR and NADH–cytochrome *c* activity, chloramphenicol induced NADPH–NR activity as well as NADH–NR activity. In *Lemna* plants the NADH- and NADPH-dependent activities show differential behavior toward amino acids and carbamyl phosphate (Young, 1967). In most higher plants NADH is the effective or near-specific nicotinamide nucleotide. As compared with soybean, many other higher plant NR's do not appear to require an added flavin, but Schrader *et al.* (1968) found both FMNH and NADH were equally effective as electron donors for NR obtained from leaves of maize, vegetable marrow, and spinach. However, there is disagreement regarding the apparent K_m for FMNH (Paneque *et al.*, 1965; Schrader *et al.*, 1968) and whether the flavin is the physiological or an incidental electron donor. In one view (Schrader *et al.*, 1968), which we favor, the low K_m for NADH (10^{-6} to 4×10^{-6} M) compared with that obtained for FMNH (2.9×10^{-4} M) as well as the possibilities for NR control inherent in the means of NAD⁺ reduction indicate NADH. However, the ready coupling of chloroplast reduction of FMN to NR was held to favor the flavin as the physiological donor in chlorophyllous cells (Paneque *et al.*, 1965). Another aspect of the flavin requirement appears to be that of stabilization of the *Chlorella fusca* protein. The NADH diaphorase activity of the fungal, algal, and plant enzymes but not the FMNH-nitrate reductase is selectively denatured by heating at 45°C for 5 minutes. In the presence of FAD, but not FMN, the denaturing effect of heating is partially prevented in *Chlorella* (Zumft *et al.*, 1970), in spinach (Palacian *et al.*, 1974), and in maize (Roustan *et al.*, 1974; see also Hewitt, 1974, 1975).

The participation of molybdenum as an essential factor in NR activity is universal, certain paradoxes being noted below. The association between molybdenum and NR metabolism was first recognized by Steinberg (1937), who prophesied the existence of a molybdenum-dependent NR in *Aspergillus niger* when the substitution of nitrate by ammonium compounds substantially eliminated the molybdenum requirement. The significance of molybdenum in nitrate reduction in higher plants was anticipated in the work of Hewitt and Jones (1947), Ducet and Hewitt (1954), and Mulder (1948). Nitrate was rapidly dissipated during a few hours by addition of molybdenum, and growth of plants was largely restored to normal compared with controls. The direct experimental identification of the functional role and constituent presence of the metal in the *Neurospora, Escherichia coli,* and soybean enzymes was first described by Nicholas and Nason (1954a,b,c, 1955a,b) and Nicholas *et al.* (1954).

Progressive purification of the *Neurospora* enzyme showed a correlation between increasing specific activity and molybdenum content in the protein. The partially purified enzyme was shown to possess two activities, namely, NR and reduction of redox dyes represented by DCPIP or flavins represented by FMN (dehydrogenase). The NR function was inactivated regardless of electron donor by dialysis of the enzyme against cyanide followed by removal of free cyanide by dialysis in molybdenum-free phosphate buffer. Activity was restored by subsequent incubation with 1–10 μM molybdate but not with other metal salts. The dehydrogenase was not inactivated. Nitrate inhibited the dehydrogenase activity except in the presence of cyanide or in the cyanide-treated enzyme. FAD (or FMN) was necessary for reduction of DCPIP by NADPH, and exogenous (substrate) concentrations of FMNH were able to reduce DCPIP in the presence of cyanide. Reduced DCPIP was able to reduce nitrate to nitrite with the normal enzyme but not when cyanide-treated. Mercurial reagents, which combine with free SH groups, severely inhibited all dehydrogenase activity and also NADH-NR, but had comparatively little effect on FMNH-NR activity. These experiments were interpreted as shown in Scheme 1 (redrawn from Nicholas and Nason, 1954a).

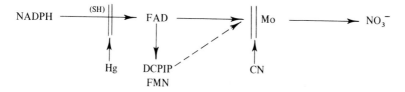

Scheme 1

Molybdate reduced with dithionite ($Na_2S_2O_4$) was reported to replace NADPH or FMNH as electron donors for NR, thereby indicating that the metal functioned through a valency change. Nicholas and Stevens (1955) showed that pentavalent molybdenum (Mo^V) was produced in this reaction. It was therefore proposed that the final stage in the reaction involved the reversible one electron change

$$Mo^{VI} + e^- \rightarrow Mo^V$$

Dialysis against cyanide removed molybdenum from the partially purified *Neurospora* preparation as shown by analysis. The NR of soybean root nodules (Cheniae and Evans, 1959) is irreversibly inhibited by cyanide. The spinach NR is reversibly inhibited by cyanide without removal of the molybdenum (Notton and Hewitt, 1971b). Molybdenum caused reactivation of fodder bean (*Vicia faba* L.) NR which had been dialyzed against 0.1–0.2 M NH_4OH (Pieve and Uvanova, 1969) but cyanide inhibition was freely reversible. Spencer (1959) found a similar reversible inhibition using wheat embryo NR.

The presence of molybdenum has been proved either by analysis, radioactive isotopes, physiological and genetic studies, or removal and restoration in *N. crassa* (Nicholas and Nason, 1954c; Garrett and Nason, 1969), soybean (Nicholas and Nason, 1955b), wheat (Anacker and Stoy, 1958), spinach (Notton and Hewitt, 1971a), *Chlorella* (Aparicio et al., 1971; Solomonson et al., 1975), *Escherichia coli* (Nicholas and Nason, 1955a; Forget, 1974), *Pseudomonas aeruginosa* (Fewson and Nicholas, 1961a), *Micrococcus halodenitrificans* (Rosso et al., 1973), and *Aspergillus nidulans* (Steinberg, 1939; Arst et al., 1970; Downey, 1973a). Evidence from inhibitor studies and specific restoration behavior strongly supports the contention that practically all NR complexes depend on the presence of molybdenum for the last electron transfer step.

Preincubation with nitrate decreases the sensitivity to cyanide, which initially behaves as a competitive inhibitor, $K_i = 0.2 \mu M$ (Relimpio et al., 1971), and azide is competitive with nitrate (Solomonson and Vennesland, 1972). Where the reaction is started under conditions in which enzyme, nitrate, and NADH are brought together before addition of cyanide, inhibition is initially weak but increases progressively during a few minutes. These results are consistent with the preferential chelation of the Mo^V state produced during each reaction cycle. Incubation with NADH before cyanide causes irreversible noncompetitive inhibition until a dehydrogenase acceptor, e.g., ferricyanide, is added (Vega et al., 1972). Activity with MV^o, BV^o, or FMNH is similarly inhibited or restored.

Iron is also a constituent metal of some NR's. At least two NR's appear to be intimately associated with cytochrome b, with a principal light absorption band about 557 nm. The *Neurospora crassa* enzyme crystallizes with cytochrome b_{557} (Garrett and Nason, 1967, 1969) in an overall complex of 228,000 daltons.

Evidence obtained with inhibitors, differential thermal inactivation, and light absorption changes indicated that the cytochrome links the NADPH–flavin-dependent moiety to the terminal molybdenum-dependent moiety of the complex shown in Scheme 2.

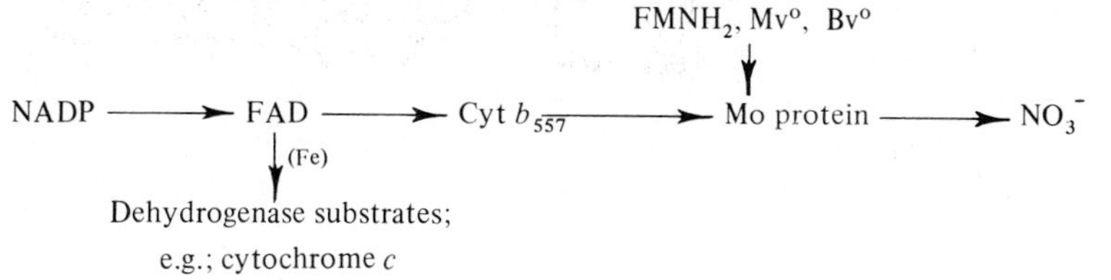

$$FMNH_2, Mv^\circ, Bv^\circ$$

NADP $\longrightarrow$ FAD $\longrightarrow$ Cyt $b_{\overline{557}}$ $\longrightarrow$ Mo protein $\longrightarrow$ NO_3^-

$\downarrow$ (Fe)

Dehydrogenase substrates;

e.g.; cytochrome c

Scheme 2

The effect of 8-hydroxyquinoline and o-phenanthroline suggest an additional role of iron in the dehydrogenase activity.

A cytochrome b_{557}-containing NR occurs in *Chlorella vulgaris* (Vennesland and Solomonson, 1972). There is now evidence in the system for the participation or presence of a flavin component (Solomonson *et al.,* 1975), and the enzyme has 2 cytochrome b_{557}, 2 FAD, and 2 molybdenum atoms in a protein of 356,000 daltons. In *Chlorella fusca,* Relimpio *et al.* (1971) presented evidence that FAD is the prosthetic group of the enzyme, but there is no indication of a heme protein. The absorption spectrum of the partially purified NR of *Aspergillus nidulans* (Cove and Coddington, 1965) had a peak at about 420 nm which is indicative more of a heme than of a flavin group; this was removed by further purification leaving a flavoprotein without heme (Downey, 1971). MacDonald and Coddington (1974) stated that cytochrome is probably present here. These fungi and green algae appear to have closely similar NR systems.

The NR system of *Escherichia coli* is very complex and apparently membrane-bound. However, it can be solubilized to a terminal homogeneous component, which contains several atoms of iron, without any heme component, one atom of molybdenum, and acid-labile sulfide equivalent to iron (Forget, 1974). In this system a separable b type cytochrome is the preceding electron donor in a sequence involving formate dehydrogenase, two cytochrome b_{557} of different redox potential, and nitrate reductase (Ruiz-Herrera and DeMoss, 1969) possibly containing a naphthoquinone or menadione-like carrier. *Pseudomonas aeruginosa* NR (Fewson and Nicholas, 1960) in partially purified state contains a moiety with about 60 atoms of iron and 1 atom of molybdenum. Cytochrome c is part of the electron donating system. The homogeneous NR (165,000 daltons) of *Micrococcus halodenitrificans* contains 1 molybdenum, 2 nonheme iron, and 4 labile sulphide atoms (Rosso *et al.*, 1973).

The NR of soybean nodule bacteroids was markedly stimulated by iron that accumulated during purification, independently of the response to and accumulation of molybdenum (Cheniae and Evans, 1960). Similar results were reported for *Ankistrodesmus braunii* (Zumft *et al.*, 1972).

The NR of *Achromobacter fischeri* (Sadana and McElroy, 1957) is unusual among all others in not apparently showing appreciable inhibition by cyanide when BV^0 was used as electron donor.

The formate dehydrogenase [Eq. (1)]–formate hydrogen lyase [Eq. (2)] complex of the coliform group of bacteria was shown to require selenium as

$$HCOOH + NAD^+ \rightarrow NADH + H^+ + CO_2 \qquad (1)$$
$$HCOOH \rightarrow CO_2 + 2\,H^+ + 2\,e^- \qquad (2)$$

selenite and also iron and molybdenum in the nutrient for production of the complex (Pinsent, 1954; Lester and DeMoss, 1971). Selenomethionine labeled with ^{75}Se was incorporated into the protein of the complex (Shum and Murphy, 1972). The importance of selenium in a system of this nature is apparently unique.

3. BIOSYNTHESIS, GENE CONTROL, COMPLEMENTATION, AND *in Vitro* ASSEMBLY OF NR

The effect of molybdenum on the biosynthesis of NR is not wholly clear. Omission of the metal from purified media used to grow *Neurospora crassa* (Nicholas *et al.*, 1954) resulted in lack of the enzyme in extracts that could not be reactivated by addition of the metal to *in vitro* cell-free extracts. Introduction of molybdenum by vacuum infiltration into excised leaves of molybdenum-deficient cauliflower, radish, and mustard plants resulted in steady appearance of enzyme activity over several hours (Hewitt and Afridi, 1959; Afridi and Hewitt, 1962, 1964; Hewitt *et al.*, 1967). In general, the overall rate of production of activity was linear with time. Addition of molybdenum to cell-free extracts resulted in no increase in activity. The presence of a wide range of antimetabolites, including cycloheximide (which interfere with protein synthesis), ribonuclease, and patulin inhibited the response to molybdenum as well as to nitrate, but chloramphenicol was comparatively inert. Inhibition of response to molybdenum occurred with L-azetidine-2-carboxylic acid and puromycin (Hewitt and Notton, 1967). The former is an analog of L-proline and may be incorporated in place of this residue during protein synthesis (Fowden, 1963). The latter causes premature termination of the extending polypeptide chain and release from the ribosome (Yarmolinsky and De La Habba, 1959). The results with puromycin were held to show that biosynthesis of the holoprotein depends on the presence *in vivo* of the constituent metal. The effect of L-azetidine-2-carboxylic acid indicated a loss of tertiary structure. Comparisons between the degree of inhibition obtained with several inhibitors for the response to nitrate or to molybdenum in terms of appearance of enzyme activity showed little differentiation (Afridi and Hewitt, 1965; Hewitt *et al.*, 1967). However, in the later work of Hewitt and Notton (1967), Notton (1972), Notton and Hewitt (1974), Hewitt *et al.* (1974), Notton *et al.* (1974), and Rucklidge *et al.* (1976), it now appears that a detectable level of apoenzyme, between 2 and 30% of normal holoprotein concentration, exists in molybdenum-deficient plants which can rapidly combine with molybdenum but only *in vivo*. The evidence is based on independent use of antibody to enzyme, restoration of activity in plant cell suspensions, or leaves, or radioactive metal incorporation in the presence of inhibitors of enzyme synthesis. Vega *et al.* (1971) reported that the restoration of NR activity in *Chlorella fusca* cells grown without molybdenum was not greatly inhibited by cycloheximide. There appeared to be a significant production of activity in a short period after adding molybdenum which was not influenced by the inhibitor. This reaction was followed by a more steady appearance of activity to much higher levels, which was inhibited by cycloheximide. It was concluded that substantial apoprotein was present in the molybdenum-deficient cells. Subramanian and Sorger (1972) found a normal concentration of nitrate inducible NADPH-cytochrome *c* reductase in *N. crassa* which sedimented like NR but was devoid of NR activity when obtained from molybdenum-deficient mycelia grown with ammonium tartrate and induced by nitrate. More work is needed to elucidate the discrepancies in these various experiments. Whereas the metal–protein complex is comparatively stable, the apoprotein may undergo breakdown *in vivo* more rapidly. It is perhaps relevant that whereas exchange of molybdenum and tungsten associated with the enzyme cannot be detected *in vitro* (Notton and Hewitt, 1971b), there is apparently such an exchange in the intact cells of tobacco callus cultures (Heimer *et al.*, 1969) or *N. crassa* (Subramanian and Sorger, 1972).

It is difficult to differentiate on the basis of existing evidence (Afridi and Hewitt, 1965; Vega *et al.*, 1971; Arst *et al.*, 1970) between the operation of a molybdenum-incorporating protein (enzyme), analogous to ferrochelatase, which is diluted out by cell extraction, and the possibility that the metal is required for continued synthesis or stability of NR protein subunits when held on a ribosomal complex that is initially saturated with apoprotein. The presence of a molybdenum-containing component being produced *in vivo* which can react *in vitro* with apoprotein or incomplete enzyme moieties is discussed by Rucklidge *et al.* (1976).

Tungsten is a competitive inhibitor of molybdate function in nitrate assimilation by *Aspergillus niger* (Higgins *et al.*, 1956) and of molybdate uptake and utilization in *Azotobacter vinelandii* (Keeler and Varner, 1958) as well as inhibiting growth of the same organism when nitrate is the sole nitrogen source (Takahashi and Nason, 1957). Heimer *et al.* (1969) examined the effect of tungstate on the formation of active NR in suspension cultures of tobacco XD cells and in intact barley shoots; in both cases the formation of the active enzyme was inhibited without apparently preventing nitrate uptake. The inhibition was reversed by addition of molybdate even in the presence of cycloheximide, possibly demonstrating some *in vivo* exchange of the two metals. By feeding radioactive tungsten to molybdenum-deficient spinach plants, Notton and Hewitt (1971c) showed incorporation of the tungsten into purified fractions normally rich in NR to produce a tungsten–protein lacking NR but having normal NADH dehydrogenase activity. No such accumulation occurred when an adequate supply of molybdenum was available. The specificity of the tungsten incorporation into apo-NR and the stability of the tungsten–protein binding were illustrated by several preparative and fractionation procedures (Notton *et al.*, 1972). Cyanide failed to remove the tungsten from the enzyme analog, and attempts to exchange protein-bound tungsten and free tungsten or molybdenum *in vitro* were unsuccessful (Notton and Hewitt, 1971b). By transferring ammonium-grown *Chlorella* to a nitrate regime containing radioactive tungsten, Vega *et al.* (1971) demonstrated the formation of a tungsten-labeled NR; unlike the spinach analog above, the association of the tungsten with the enzyme however did not survive electrophoresis on polyacrylamide, since practically all of the tungsten traveled to the electrophoretic front, but the complex was stable to gel filtration on agarose. Although L-azetidine-2-carboxylic acid inhibits active nitrate reductase formation (Hewitt and Notton, 1967), it does not prevent incorporation of tungsten into enzyme protein (Notton and Hewitt, 1974).

Sucrose density gradient analysis revealed the presence of three NADH-cytochrome *c* bands in barley (Wray and Filner, 1970), two of which (the 8 S and the 3.7 S bands) were inducible by nitrate irrespective of the presence or absence of tungstate in the culture medium. However, tungsten caused a superinduction of the dehydrogenase activity and prevented the formation of both the NADH-NR and the FMNH-NR activity in the 8 S fraction by nitrate induction, suggesting that in the presence of tungstate the 8 S enzyme complex is formed but has only dehydrogenase activity of apoprotein. Vega et al. (1971) demonstrated the same superinduction of dehydrogenase activity by tungstate using Chlorella.

When spinach plants were grown with nitrate and a low molybdenum supply and then transferred to solutions containing tungstate or molybdate for 24 hours before purifying with respect to NR, Notton and Hewitt (1971c) found no increase in dehydrogenase as a result of tungstate treatment as well as no increase in the NADH-NR level. When the plants were transferred to molybdate there was no increase in the dehydrogenase level, but there was a large increase in the NADH-NR level, supporting the idea of the preexistence of at least the dehydrogenase moiety and the superactivity of this dehydrogenase in the presence of tungsten as well as in the absence of molybdenum. Cauliflower plants grown in sterile culture with low molybdenum and with ammonium as a nitrogen source have negligible NADH-NR activity and a low dehydrogenase content. Infiltration of leaf tissue with nitrate alone results in an increase in the dehydrogenase content, whereas both molybdate and nitrate are required to allow simultaneous production of NADH-NR activity. Plants grown with a low molybdenum supply accumulated nitrate (Hewitt and Jones, 1947; Hewitt et al., 1949), and addition of tungstate to barley also caused nitrate accumulation (Wray and Filner, 1970). Under both these conditions NADH-NR is low, but the dehydrogenase content is high. This accumulated nitrate rapidly disappears on addition of molybdate (Ducet and Hewitt, 1954), and concurrently NADH-NR is induced. Nitrate, therefore, seems to be the inducer of the dehydrogenase moiety of the NR system in higher plants, algae, and fungi even in the absence of the prosthetic metal. When the extracts obtained from spinach plants grown on nitrate and a low molybdate supply before being transferred to tungstate or molybdate were examined by acrylamide gel electrophoresis and stained for NADH-dehydrogenase activity, up to nine separate bands were visible in the low molybdate treatment, which reduced to five or six on transfer to tungstate and to two or three on transfer to molybdate solution. The intensity of the dehydrogenase band in the region of NADH-NR increased on transfer to molybdate and especially on transfer to tungstate. As the total dehydrogenase content of the extracts as measured by reduction of DCPIP was approximately the same, it seems that addition of the metals may cause some aggregation of existing dehydrogenase in the tissue (Notton and Hewitt, 1971a,c). Although tungsten analogs of NR are inactive in this respect the formate dehydrogenase of Clostridium thermoaceticum utilizes tungsten in preference to molybdenum (Ljungdahl and Andreesen, 1975).

Molybdenum deficiency in cauliflower results in a condition described as whiptail (Hewitt and Jones, 1947), which is shown to result from the presence of nitrate at concentrations that would induce NR in normal plants (Hewitt and Gundry, 1970). When cauliflower plants were grown without molybdenum but with tungsten, the onset of the condition was considerably delayed without any increase in the NADH-NR activity of the tissue. An electron microscopic examination of the whiptail lesion area and corresponding areas in the tungsten-supplemented plants revealed stabilization by tungsten of the chloroplast structure, which was severely disrupted in the molybdenum-deficient tissue, especially with regard to the double membrane and vacuolization in the thylakoid stacks.

Gene control of NR in maize influences the production and stability of the enzyme (Warner et al., 1969). Crosses were made with two inbred lines of maize, and hybrids were backcrossed for two generations. A complex segregation including some F_4 progeny was fitted experimentally to a theoretical distribution 1:4:4:3:3:1 consistent with the operation of a double locus control. One controlled synthesis of the enzyme, and the other its stability. The proteins appeared to show in vitro differences related to their genetic origin. Thus the denaturation or other modes of deterioration of activity which occur during storage in vitro ranged from 15% per hour for the double dominant to 97% per hour for one parent at 29°C. The K_m values, pH, and temperature optima did not vary with genetic origin. It was considered possible that the differential in vitro stability resulted from differences in the effects of inactivating compounds (e.g., phenolics or proteases).

Mutant forms of Neurospora crassa, Aspergillus nidulans, and Escherichia coli have been isolated which are unable to grow on nitrate media. Sorger (1963, 1964, 1966) and Sorger and Giles (1965) discovered mutants of N. crassa nit-1 and nit-3 which lacked NR activity. Both possessed similar constitutive NADPH-cytochrome c reductase to the wild-type mycelia grown in the absence of nitrate.

In mycelia of nitrate-grown nit-1 and wild type there was a substantial increase in cytochrome c reductase induced by nitrate (Sorger, 1963). There was no increase in nit-2 or nit-3 mycelia. Only the nitrate-induced wild-type mycelia possessed NADPH-NR. In addition to NADPH-NR there was also a BV⁰-NR activity in the nitrate-induced wild-type which occurred in abnormally high amounts in both induced and noninduced nit-3 mycelia and was lacking from induced nit-1. In Aspergillus nidulans, by contrast, the BV⁰-NR activity and the NADPH-NR activities were coinducible and absent or present together in all wild-type and mutant strains tested (Pateman et al., 1967).

Sorger concluded that NR of N. crassa was comprised of two polypeptide moieties under separate gene control. The first of these mediated nitrate-inducible NADPH-cytochrome c reduction, while the second, which contains molybdenum, reduced nitrate with BV^0 as electron donor. Together the two moieties catalyzed the reduction of nitrate by NADPH.

Nason et al. (1970), showed that induced nit-3 mycelium contained also the FADH-NR activity which was lacking from nit-1. The excess of MV^0-NR activity, resembling BV^0-NR, in induced nit-3 preparations over that in the wild type was confirmed. The nit-3 product when purified had a MW of 160,000 (6.8 S) (Antoine, 1974).

These experiments show that the NADPH-NR, NADPH-cytochrome c reductase, FADH-NR, and $MV^0(BV^0)$-NR activities are equally inducible by nitrate in wild-type mycelia. They also sediment in the same manner in sucrose gradients and behave similarly during electrophoresis in acrylamide gels (Garrett and Nason, 1969). The cytochrome b_{557} component is present at all stages of purification. The gene control of this component is not known but it is produced by nit-1 and nit-3 mutants. The NADPH moiety is relatively sensitive to heat or mercurial reagents, and elimination of NADPH-NR activity by either of these means or mutation at nit-3 causes activation of the $MV^0(BV^0)$–Mo activity. The reciprocal changes may indicate either separation of subunits or exposure of the molybdenum site that promotes enhanced activity with $BV^0(MV^0)$.

Several mutant Aspergillus nidulans strains are unable to grow on nitrate (Cove and Pateman, 1963; Pateman and Cove, 1967; Pateman et al., 1967; Cove, 1970). At least six gene loci involved in eight or more heterokaryon groups comprising over 50 mutants are able to grow on nitrate when recombined. The A. nidulans NR resembles that in N. crassa in being NADPH-specific for NR and FAD-dependent for NR and dehydrogenase activities (Pateman et al., 1967) but the cytochrome component is disputed (Downey, 1971; MacDonald and Coddington, 1974). The wild-type mycelial extracts of A. nidulans also have xanthine dehydrogenase (XDH) activity (Pateman et al., 1964). In the wild type the NR and cytochrome c reductase activities are both inducible by nitrate. Several mutants produced by irradiation lacked both NR and XDH activities. Others lacked NR but possessed XDH activity. Another group showed an inducible response to nitrate for cytochrome c reductase but produced no NR activity. Of this group two lacked inducible XDH activity. The ability to produce serologically cross-reacting material to NR (NRCRM) was tested. This material was naturally induced by nitrate in the wild type. It was constitutive in 9 out of 11 of the group having a high level of constitutive cytochrome c reductase activity but lacking NR or XDH. In the other two of this group, NRCRM was inducible by nitrate, although NR was still not induced. NRCRM was also constitutive in two other mutants that were unable to produce additional (inducible) cytochrome c reductase activity but possessed high XDH. It was concluded (1) that NR and inducible cytochrome c reductase reside in the same protein, (2) that NR but not cytochrome c reductase activity requires additionally the production of a cofactor CNX, and (3) that XDH requires the same CNX cofactor that is produced under control of multiple gene locus. The CNX factor was concluded to be a molybdenum-containing component. Some mutants were found which possessed nitrate-inducible cytochrome c reductase but no NR in mycelia grown with a normal (3.3 μM) molybdenum level, but produced NR when given 33 mM molybdenum (Cove et al., 1964; Arst et al., 1970). Excess molybdenum allowed "repair" at the CNX lesion in some of the mutants. The "repaired" enzyme possessed a tenfold greater K_m (6×10^{-4} M) for nitrate than the normal enzyme (6×10^{-5} M). The K_m for NADPH was the same for both. Some mutants formed molybdenum–protein that was devoid of NR activity but reduced cytochrome c (Downey, 1973a).

Nitrate reduction, formate metabolism, and chlorate resistance are closely linked in $E.$ $coli$ in a complex system. The significance of the chlorate-resistant mutants may be explained by the fact that the formate NR system will reduce chlorate to chlorite, which is toxic. Mutants unable to reduce nitrate are therefore frequently chlorate-resistant (Glaser and DeMoss, 1971). A mutant (K16), which cannot grow on nitrate, was defective in formate dehydrogenase and formate NR, but MV⁰-NR was still produced abundantly by induction in the presence of nitrate (Showe and DeMoss, 1968). In another mutant (C98) unable to grow on nitrate, NADH-NR, formate NR, NAD-formate dehydrogenase, formate–cytochrome b reductase, and formate hydrogen lyase were all absent. The Chl D, chlorate resistant mutant of $E.$ $coli$ lacks formate hydrogen lyase and formate NR, but addition of high concentrations (10^{-4} M) molybdate to the medium resulted in "repair" of both the formate NR and hydrogen lyase systems (Glaser and DeMoss, 1971). The production of a molybdenum "processing" system under the control of the Chl D locus was postulated and regarded as analogous with the "repair" of the $A.$ $nidulans$ CNX mutants already described. Analogous "repair" occurs in a $Pseudomonas$ $aeruginosa$ mutant (Hartingsveldt and Stouthamer, 1972).

The demonstration that fungal or bacterial mutants are unable to grow because they lack one or other of apparently distinct protein moieties of the NR complex encouraged the idea that reconstitution of partial moieties produced by different mutants might be achieved by a process of in $vitro$ complementation as distinct from heterokaryon formation. This reaction was shown to occur for the $N.$ $crassa$ mutants described above: nitrate-inducible NADPH cytochrome c reductase and MV⁰-NR or FADH-NR could be combined to produce significant yield of wild type NR having all three activities (Nason et al., 1970). The complementation of induced nit-1 mycelial extracts with uninduced nit-3 resulted in production of significant NADPH-NR activity. This was actually decreased by inductive treatment of the nit-3 mutants. The excessive MV^0-NR in induced nit-3 mutants was noted above. No NADPH-NR was produced when uninduced nit-1 was used. Sucrose density sedimentation of the components of the complemented systems showed that the complemented activity had a sedimentation value ($s_{20,w}$) of 7.9 and 228,000 daltons, identical with the wild type for NADPH-NR, FADH-NR, and MV^0-NR and for the inducible NADPH-cytochrome c reductase. In induced nit-1 alone, the inducible NADPH-cytochrome c reductase was 4.5 S or 7.8 S, and in induced nit-3 alone the FADH-NR and MV^0-NR coincided at 6.8. The implication was that partial protein moieties of different molecular weights could be aggregated by complementation with a heavier complex displaying all the associated activities. Complementation was limited to pH 6.5–7.0 and was time- and temperature-dependent. The complemented enzyme had similar kinetic properties to the wild type. A. nidulans has not shown complementation in vitro, although heterokaryons yield NR activity (Downey, 1973b).

In vitro complementation of NR has been observed in E. coli (Azoulay et al., 1969) by reconstituting the enzyme from extracts of two chlorate-resistant mutants (Chl A_{15} and Chl B_{24}) lacking NADH-NR and formate hydrogen lyase. At pH 7–7.6 and 32°C under anaerobic conditions with appropriate protein concentration ratios, 10% of wild type formate-NR activity was reconstituted over a 2-hour period by mixture of the soluble proteins in separate extracts of nitrate-induced mutants. The reconstituted enzyme contained a cytochrome b component and was particulate like the wild type, but the K_m for nitrate was increased from 0.8 to 3 mM. In an E. coli C.98 mutant lacking formate hydrogen lyase, the activity was reconstituted by complementation of a practically inactive membrane fraction from either of the wild-type or mutant cells with a soluble fraction from the wild-type cells, whereas the soluble fraction from the C.98 mutant was inert in this reaction (Venables et al., 1968).

A reconstruction in vitro of the wild-type holoenzyme has been achieved in N. crassa nit-1 mutants (Ketchum et al., 1970; Nason et al., 1971; Lee et al., 1974) in a manner that is distinct from complementation of intercistronic mutants described above. In these reconstruction experiments, the induced nit-1 extract was incubated at pH 6.8 with an acid-treated and enzymically denatured fraction obtained from any of the following molybdoproteins: liver or milk xanthine oxidase; liver aldehyde oxidase; sulfite oxidase; Clostridium, Azotobacter, or Rhizobium nitrogenase molybdoferredoxins; *E. coli* formate dehydrogenase; and *Neurospora* wild type or plant NR's. The critical pH range was 2.5–3.0, and the fraction obtained was highly labile before combination with the nit-1 protein. It was suggested that all molybdoproteins have a detachable, common, small molybdenum peptide fragment that can be utilized *in vitro* by the NADPH-cytochrome *c* reductase component of NR in the mutants but which itself can only be synthesized *in vivo*. The *CNX* gene in *A. nidulans* (Pateman *et al.*, 1964) can then be regarded as coding for this component. The reconstituted NR system was similar in sedimentation, molecular weight, cytochrome b_{557} content, and K_m values to the wild-type enzyme. The NADPH-cytochrome *c* reductase of nit-1 ($s_{20,w} =$ 4.5 S) was converted after reconstruction to NR 7.9 S. A nitrate-inducible soluble form of the MV^0-NR moiety of nit-3 permitted reconstruction of the holoprotein when the acid treatment was done at pH 3.5 but was unable to complement nit-1 extracts without acid treatment, whereas a constitutive particulate MV^0-NR moiety, representing 10% of the inducible component in nit-3 extracts was able to complement the nit-1 cytochrome *c* reductase without acidifying to produce a soluble NR holoprotein (Nason *et al.*, 1970). The results here and elsewhere indicate that the cytochrome b_{557} component is synthesized by the nit-1 and nit-3 mutants and that FADH donation to the MV^0-NR is distinct from its role in NADPH-NR. The nit-1 product that reduces cytochrome *c* is devoid of molybdenum and appears to be an apoprotein sedimenting at 7.8 S (Ketchum and Downey, 1975) similar to that formed by molybdenum-deficient wild-type mycelia (Subramanian and Sorger, 1972). Inducible apoprotein produced by molybdenum-deficient spinach leaves, that sedimented at 3.7 S or 8.1 S, is also reconstituted *in vitro* by the acid-dissociated product of spinach NR (Rucklidge *et al.*, 1976).

B. Physiological Aspects

1. Respiratory and Assimilatory Systems

NR has been described either as *assimilatory* where the physiological activity is primarily the ultimate production of ammonia for metabolism or as *dissimilatory*, or by the preferred term *respiratory*, where the nitrate is primarily an electron acceptor in place of oxygen (cf. Hewitt and Nicholas, 1964) as found mainly in bacteria but reported in *Vigna* cotyledons (Kumada, 1953; Taniguchi, 1961). In both circumstances, the same product is nitrite, and the distinction between different proteins involved in the two activities in the same organism is doubtful often if any. There is no doubt however, that NR's may be obtained in a soluble state or aqueous phase when they are usually assimilatory but may also occur in a membrane-bound or particulate form in close association with cytochrome systems. Examples include the NR systems from *Micrococcus denitrificans* (Lam and Nicholas, 1969a), *Escherichia coli* (Iida and Taniguchi, 1959; Cole and Wimpenny, 1968; Showe and DeMoss, 1968), and *Pseudomonas aeruginosa* (Fewson and Nicholas, 1961b). In these situations, the NR systems are often associated with nitrate respiration and are sometimes subject to apparently different methods of control, e.g., suppression by oxygen but not ammonia (Van't Riet *et al.*, 1968), and the nitrite produced may be either unmetabolized or else reduced to nitric or nitrous oxides or nitrogen instead of to ammonia. In some organisms, nitrogen is assimilated either from organic sources, exogenous ammonia, or by soluble assimilatory NR activity that may be only 5–10% of the respiratory activity. Soluble respiratory pattern NR systems also occur as in *Spirillum itersonii* (Gauthier *et al.*, 1970), and in this example the activity is repressed by oxygen and derepressed by anaerobiosis in the absence of nitrate, which induces higher levels when present. The respiratory NR of *Aerobacter aerogenes* also appears to be readily extracted in a soluble form (van't Riet *et al.*, 1968) but is in fact also located in a membrane system (van't Riet and Planta, 1969). Only one type of NR having respiratory or assimilatory functions is probably present under different physiological conditions. This view has been derived from studies with *Aerobacter aerogenes* (van't Riet *et al.*, 1968), *Neurospora crassa* (Nicholas and Wilson, 1964), and *E. coli* (Murray and Sanwal, 1963) where serological tests failed to reveal any evidence of different proteins under different respiratory conditions. It is considered likely that the same protein is held in different complexes, possibly particulate and soluble, and complexed with different cytochromes and that menadione is a carrier in membrane-bound systems. This distinction in location probably accounts for the apparent differences in properties.

The contrasting view that distinct forms of NR activity occur in one organism has been presented (Pichinoty, 1966). The comparative aspects of two forms, NR A and NR B, were described from *Micrococcus denitrificans*. Form A was mainly particulate and produced only anaerobically in nitrate media, thus appearing to differ from the anaerobic form in other bacteria which is derepressed under these conditions without nitrate. Form B was mainly soluble and was produced constitutively under aerobic conditions, which did not inactivate or repress, and nitrate was not required for its production. Form A reduced chlorate as well as nitrate; form B could not, and chlorate was an inhibitor. Azide inhibited A competitively ($K_i = 1$ μM) and reversibly but inhibited B with mixed kinetics ($K_i = 63$ μM). Energies of activation for A were 12.8 kcal

(NO_3^-) and 7.5 kcal (ClO_3^-), and for B 24.3 kcal (NO_3^-). The K_m's (NO_3^-) did not differ drastically, neither utilized NADH as electron donor, and both reacted with BV^o, MV^o, and FMNH. Forms A and B were identified in other bacterial species. Several contain only one or the other, but either may occur in widely differing species. Where both occur, there is evidence for separate gene control of their biosynthesis in mutants. Pichinoty (1966) considered that the two forms were distinct proteins usually produced in different locations. Bacterial nitrate metabolism was classified into four groups.

1. Nitrate assimilation but no nitrate respiration, e.g., *Pseudomonas putida* strains
2. Nitrate respiration but no assimilation, e.g., *Providencia alcalifaciens* strains
3. Both assimilation and respiration of nitrate, e.g., *Aerobacter aerogenes* strain LIII-I, *Pseudomonas aeruginosa*, *P. fluorescens*, *Micrococcus denitrificans*
4. Neither assimilate nor respire with nitrate

2. Distribution in Organs and Cells of Plants

NR activity can usually be detected in all parts of a plant, and nitrate assimilation or reductase has been shown to occur in attached or excised root systems of many plants (White, 1933; Vaidyanathan and Street, 1959; Sanderson and Cocking, 1964; Wallace and Pate, 1965; Miflin, 1967, 1970a; Bourne and Miflin, 1970; Minotti and Jackson, 1970; Smith and Thompson, 1971; Li *et al.*, 1972).

Doubt has been expressed that certain plants do not in fact possess NR in roots. In *Xanthium pennsylvanicum*, the cocklebur, 95% of the total soluble nitrogen in the bleeding sap of decapitated plants is nitrate, but there is abundant NR in leaves (Wallace and Pate, 1967; Pate, 1973). Absence of root enzyme was inferred. It was once considered that nitrate assimilation occurred specifically or almost wholly in roots of some species, especially woody rosaceous plants (Eckerson, 1931). However, abundant NR is present in apple leaves as well as roots (Klepper and Hageman, 1969) and shows all the characteristics of NADH-NR systems found in other species.

Weissman (1972) found that leaves of *Helianthus annuus* (sunflower) plants had only 1.5% of the NR activity which could be extracted from roots. It was suggested that root-exported ammonia, or a product thereof, had a repressing effect on enzyme activity in leaves. Negligible NR activity could be detected in roots or leaves of Douglas fir (*Pseudotsuga menziesii*). It was suggested that mycorrhizal fungi were able to reduce nitrate before transfer to the Douglas fir (Li *et al.*, 1972), but this tree can be grown in sand culture with nitrate. NR activity may reach maximal and then decline to negligible levels as leaves age or may be very low in young leaves (Afridi and Hewitt, 1964). NR is present in embryos (Rijven, 1958), barley aleurone cells (Ferrari and Varner, 1969, 1970), germinated seedlings (Tang and Wu, 1957), maize scutellum (Hucklesby and Elsner, 1969), marrow cotyledons (Cresswell, 1961), and root nodules (Cheniae and Evans, 1960; Li *et al.*, 1972). Phenolic compounds severely inhibit NR when extracted from roots and sometimes also from leaves of some plants, and here the *in vivo* method of assay is especially useful (Jaworski, 1971), but prevention of enzymatic inactivation may be important (Wallace, 1975). Some calefuge species may have restricted NR activity even when given nitrate (Havill *et al.*, 1974).

There is considerable argument concerning the cellular distribution of NR in plants. In all simple extraction procedures, the bulk of NR activity of leaves appears in a "soluble" protein fraction. Losada and his associates (Del Campo *et al.*, 1963; Losada *et al.*, 1965; Paneque *et al.*, 1965; Ramirez *et al.*, 1964) have inferred a chloroplastic location based on the facility of flavins reduced by chloroplast systems to serve as electron donors for NR. However, Schrader *et al.* (1968) concluded that NADH is as good an electron donor or probably better than reduced flavins with a much lower K_m than flavins. As a result of nonaqueous chloroplast isolation (Ritenour *et al.*, 1967) or certain extraction-media (Dalling *et al.*, 1972a), use of chloroplast permeable metabolites in *in vivo* NR assays (Klepper *et al.*, 1971) and differential responses of nitrate and nitrite reductase to induction in the presence of chloramphenicol or cycloheximide (Beevers *et al.*, 1965; Schrader *et al.*, 1967; Sawhney and Naik, 1972; Stewart, 1972; Sawhney *et al.*, 1972), the conclusion is reached that NR is synthesized by cytoplasmic ribosomes and is outside the chloroplast, but is dependent also upon chloroplast function for its synthesis (see Hewitt, 1975). Ritenour *et al.* (1967) and Eaglesham and Hewitt (1971b) nevertheless considered that the NR activity could well be attached to the outer chloroplast membrane and that such a site was wholly consistent with their results and others (Coupé *et al.*, 1967; Grant and Canvin, 1970; Grant *et al.*, 1970; Miflin, 1974). Experiments by Lips (1975) suggest attachment of NR to particles resembling microbodies (see also Hewitt, 1975) where a phytochrome system appeared to influence the attachment of the enzyme and also its induction in other experiments (Jones and Sheard, 1972).

Miflin (1968, 1970a,b) presented evidence for location of NR and nitrite reductase in a particulate fraction in barley and pea roots which was distinct from mitochondria carrying cytochrome oxidase and fumarase and also distinct from particles (possibly peroxisomes) carrying catalase. The particulate NR utilized succinate as electron donor, which was not possible for the NR in the soluble fraction. The association of the two enzymes in the one particle was consistent with the integrated reduction of nitrate to ammonia observed for a particulate preparation from barley roots in the presence of ATP and pyruvate (Bourne and Miflin, 1970). ATP was considered to be required for the reduction of nitrite for which the physiological electron carrier from pyruvate was not identified but was replaced by MV^0; in the absence of pyruvate and ATP. BV^0 was an effective donor. The particles retained substrate concentrations of nitrate. Miflin (1970a) suggested that roots might contain both respiratory and assimilatory NR activity. The barley root particle would resemble the nitrosome postulated for yeast (Sims et al., 1968).

The intercellular distribution of NR and nitrite reductase in three species showing C_4 type photosynthesis, Zea mays, Gomphrena globosa, and Sorghum sudanense was studied (Mellor and Tregunna, 1971; Slack et al., 1969). The fraction enriched in the outer mesophyll cells of vascular bundles contained 2 to 9 times the total NR activity found in the bundle sheath cells. Nitrate accumulated in the mesophyll cells to between 3 and 30 times the concentration in bundle sheath cells. NR tended to be substantially more active in the mesophyll cells; absence of thylakoid grana in bundle sheath chloroplasts was related to failure to reduce nitrite, since in Gomphrena nitrite reductase was similar in bundle sheath and mesophyll cells and each had thylakoid grana. In mutant forms of maize, NR was absent from white albino leaves but present although decreased somewhat in pale green chlorino leaves (Sawhney et al., 1972). Nitrite reductase was present but greatly decreased in white albino leaves that contain proplastids but not differentiated chloroplasts (Walles, 1967).

3. Effects of Other Mineral Elements on Nitrate Assimilation

Calcium was reported to be required for a nitrite permease in chloroplasts (Paulsen and Harper, 1968). Magnesium deficiency causes loss of chlorophyll, and no doubt this results in impaired reduction ultimately by impaired reductant capacity, but perhaps also by interference in a light-dependent reaction for nitrate reduction or for stability of NR.

Potassium deficiency impaired the induction by nitrate of NR and nitrite reductase in rice seedlings or Neurospora crassa (Oji and Izawa, 1969b; Nitsos and Evans, 1966). Potassium could be partially replaced in decreasing order by rubidium or sodium, but induction of both enzymes was inhibited by ammonium ions. The requirement was attributed to the role of potassium in protein synthesis. Phosphate activates NR *in vitro* (see Hewitt and Nicholas, 1964).

Manganese deficiency often results in excessive nitrate accumulation in wheat, cauliflower, beans, tomato, maize, sunflower, and other plants (Burström, 1939; Leeper, 1941; Hewitt *et al.*, 1949; Subba Rao and Lal, 1955; Steward and Margolis, 1961–1962; Kretovich, 1965; Vielemeyer *et al.*, 1969). Nitrate concentrations may reach 10% of dry matter and are second only to those produced by molybdenum deficiency (Hewitt *et al.*, 1957). The suppression of NR activity may be related to the role of manganese in oxygen evolution (Cheniae and Martin, 1968, 1970; see also Hewitt, 1975). Manganese deficiency inhibited nitrite reduction of *Ankistrodesmus braunii* (Kessler, 1957a,b). There also appears to be a possible analogy between the effects of manganese described here and the effects of suppression of oxygen evolution by 3-(3,4-dichlorophenyl)-1,1-dimethyl urea (DCMU) in the absence of carbon dioxide in the carbon-starved cells of *Chlorella* (Ahmed and Morris, 1968) or in normal cells of *Chlamydomonas reinhardii* (Thacker and Syrett, 1972b).

C. Regulatory Aspects of Nitrate Reduction

Regulation of nitrate reduction is achieved in several ways. These include induction, repression and derepression of protein synthesis, reversible inactivation and activation, kinetic inhibiton, or allosteric control of the already formed enzyme, destruction or irreversible disappearance of enzyme activity, and permease or nitrate transport activity. These phenomena do not all occur in all organisms.

1. INDUCTION, REPRESSION, AND DEREPRESSION

The biosynthesis of NR is subject to repression, induction, and derepression in an independent manner. Induction by nitrate is almost universal. In a few instances, e.g., *N. crassa* (Evans and Nason, 1953), *Lemna minor* (Stewart, 1968) tobacco callus cells (Kelker and Filner, 1971), rice seedlings (Shen, 1972b), beans (Lips *et al.*, 1973), and barley (Kaplan *et al.*, 1974) nitrite appears to be equally effective, but not in many higher plants (Afridi and Hewitt, 1965; Beevers *et al.*, 1965). In *Escherichia coli* and *Aerobacter aerogenes* the particulate or membrane bound NR systems are repressed by oxygen (Showe and DeMoss, 1968; van't Riet *et al.*, 1968) but not by ammonia. However, the soluble form of NR considered to be involved in nitrate assimilation in *Aerobacter* is repressed by ammonia but not by oxygen (van't Riet *et al.*, 1968).

Both factors suppress activity in the fungus *Scopulariopsis brevicaulis* but in plants aerobic conditions are needed for continued production (Candela *et al.*, 1957). In most instances, ammonia does not repress the formation of higher plant NR in leaves (Hewitt and Afridi, 1959; Afridi and Hewitt, 1965; Beevers *et al.*, 1965; Bayley *et al.*, 1972). Repression by ammonia, or its products, appears to occur in barley and pea roots (Smith and Thompson, 1971; Pate, 1973), tobacco cultures (Filner, 1966; Kelker and Filner, 1971), and *Lemna minor* (Sims *et al.*, 1968; Ferguson and Bollard, 1969; Orebamjo and Stewart, 1975a). Repression by ammonia or amino acids is more usual in yeasts (Sims *et al.*, 1968), fungi (Pateman *et al.*, 1967; Cove, 1966), and green algae (Losada *et al.*, 1970). The effect of ammonia may sometimes be indirect, a product with acetate or CO_2 and light being the active factor in repression as deduced for *Chlamydomonas reinhardii* (Thacker and Syrett, 1972a) and *Chlorella vulgaris* (Morris and Syrett, 1963). Amino acid action is often specific, and some amino acids, e.g., glutamate, appear to derepress NR synthesis as in *Cyanidium caldarium* (Rigano, 1971). The induction of NR activity in *Lemna minor* is complex (Stewart, 1968). Two NR systems, one specific for NADPH and the other for NADH, are both induced by nitrate, but the NADPH-NR is especially derepressed in the presence of exogenous sucrose. The NADPH-NR is repressed by carbamyl phosphate and pyridoxamine phosphate as for the yeast nitrosome complex (Sims *et al.*, 1968), but the NADH-NR is unaffected, whereas ammonia may repress both, but at tenfold higher concentrations. In *Chlorella fusca* NR is derepressed by nitrogen deficiency after removal from repression by an amino acid medium (Vega *et al.*, 1971). In rice seedlings (Shen, 1972a,b), the specificity of inducible NADH-NR and NADPH-NR systems was different for induction by nitrate or chloramphenicol.

It is generally assumed that induction of NR activity is synonymous with *de novo* protein synthesis. The phenomena of activation by nitrate discussed below (Section II,C,3) nevertheless indicates that the criteria for protein synthesis must be carefully applied.

The first demonstrations of induction of NR by nitrate in higher plants were those of Hewitt *et al.* (1956), Candela *et al.* (1957), Tang and Wu (1957), and Rijven (1958). The induction of NR in several plant species has since been shown to fulfill most criteria for protein synthesis. The most extensive results have been obtained with cauliflower, radish, tobacco callus cells, spinach, and maize (Hewitt and Afridi, 1959; Afridi and Hewitt, 1962, 1964, 1965; Beevers *et al.*, 1965; Hewitt *et al.*, 1967; Notton and Hewitt, 1971a,c; Schrader *et al.*, 1968; Zielke and Filner, 1971). The triple labeling by ^{14}C, ^{3}H, and ^{15}N and buoyant density experiments of Zielke and Filner (1971) with tobacco callus cells supplied elegant confirmation of both *de novo* synthesis and simultaneous breakdown of preformed enzyme.

The role of nitrate in activation of nonfunctional enzyme as distinct from induction of synthesis is established. In *Chlorella fusca*, the greater part of the enzyme complex is actually present in nitrate-free (uninduced) cells (Vega *et al.*, 1971) and the principal response of intact cells of this alga to nitrate is that of activation, based on the incomplete inhibitory effect of cycloheximide on NR production following addition of nitrate and the time course of the appearance of activity.

The particular polypeptide or partial protein components actually induced by nitrate are not entirely clear. Not all NR activity is inducible, and a mutant Nir[c] of *A. nidulans* is constitutive for both NR and nitrite reductase, but these activities are nevertheless repressed by ammonia (Pateman and Cove, 1967). In *A. nidulans* (Cove, 1967, 1970, Cove and Pateman, 1969), the kinetics of appearance of NR activity during growth in the presence of nitrate were interpreted by supposing that nitrate-free NR protein is the repressor of NR synthesis. In the presence of nitrate repression by the protein is released, i.e., derepressed.

When tissues are deprived of nitrate, CO_2, or light the NR already present very often declines from a steady state (Hewitt *et al.*, 1956; Candela *et al.*, 1957; Hewitt and Afridi, 1959; Hageman and Flesher, 1960; Afridi and Hewitt, 1964; Huffaker *et al.*, 1966; Kannangara and Woolhouse, 1967; Notton, 1972). This decline, when examined, has been shown to follow first-order kinetics (Schrader *et al.*, 1968; Upcroft and Done, 1972; Shen, 1972a). Nitrite reductase loss also shows first-order kinetics, but usually with notably longer half-life (Heimer and Filner, 1970). It is also proved (Zielke and Filner, 1971) that synthesis of enzyme can continue for a period after removal of nitrate and during the process of simultaneous breadkown of existing enzyme. An NR inactivating enzyme has been found in maize roots, scutellum, and leaves and pea leaves and roots which is not, however, inhibited by nitrate (Wallace, 1973, 1974).

Cycloheximide or actinomycin D inhibited the decay when barley plants were transferred from light to darkness (Travis *et al.*, 1969), and simazine also stabilized NR (Ries *et al.*, 1967). Lowered temperatures similarly inhibited the decay of NR activity (Ritenour, 1964). Enzyme turnover, representing simultaneous synthesis and decline, is a common phenomenon (Schimke, 1966; Rechcigl, 1968). Rates of decline in terms of half-life ($t_{1/2}$) may range from 20 days for alanine aminotransferase (Segal *et al.*, 1969), to about 10–15 minutes for pyruvatephosphate dikinase in darkness where structural inactivation is probably occurring (Hatch and Slack, 1969a), or 60 minutes for δ-aminolevulinic acid synthetase in the absence of substrate (Marver *et al.*, 1966) and NR of *A. nidulans* (Cove, 1966). The stabilizing effect of cycloheximide has been observed for other enzymes [phenylalanine ammonia lyase in gherkin seedlings and potato tuber discs (Engelsma, 1967; Zucker, 1968) and UDP-galactose transferase in *Dictyostelium discoideum* (Sussman and Sussman, 1965)], and interpretation is obscure in terms of simple inhibition of general protein synthesis.

2. PERMEASE ACTIVITY

There is evidence for the idea that the entry into cells or movement of nitrate to the site of the enzyme-forming system is controlled by permease and transporter systems which may be sensitive to light or specific metabolites. The uptake of nitrate by *Spirodella oligorrhiza* is partially inhibited by ammonium ions (Ferguson and Bollard, 1969; Ferguson, 1969) independent of an inhibitory effect of ammonia or a product on NR formation. Similar inhibitory effects of ammonia on nitrate uptake have been found for potatoes (El-Shishiny, 1955), wheat (Weissman, 1951; Minotti *et al.*, 1969), and rye grass (Lyclama, 1963), but not for excised barley roots (Smith and Thompson, 1971).

The induction of nitrate permease activity in wheat was suggested by Minotti *et al.* (1968). A permease system was found in cultured tobacco XD cells (Heimer and Filner, 1970, 1971); this was induced by nitrate and inhibited by threonine, to which a mutant strain was tolerant. Tungstate did not inhibit permease formation, which presumably was not a nitrate reductase type of protein. Nevertheless, a Michaelis–Menten relationship between permease activity and nitrate concentration indicated a K_m of 0.4 mM.

When cells were transferred to a nitrate-free medium, NR activity declined immediately, whereas intracellular nitrate concentration remained relatively high. It appeared that the bulk of the previously absorbed nitrate was not available to maintain enzyme induction (Heimer and Filner, 1971). Threonine and L-methionine were found to inhibit nitrate uptake appreciably by excised pea roots (Sahulka, 1972). A nitrate-inducible permease system was shown to occur in maize (Jackson *et al.*, 1973). Separate inducing and inert storage pools were concluded to occur in pawpaw fruits (Menary and Jones, 1972). Nitrate present in the exocarp of ripe (senescing) fruit was thought to be unable to move from the vacuole to cytoplasm, even in the presence of light that was necessary for NR induction, whereas light and exogeneous nitrate supply resulted in enzyme induction. Separate metabolic (cytoplasmic) and storage (vacuolar) pools were inferred for tobacco callus cultures by Ferrari *et al.* (1973) from effects of washing and anaerobic conditions.

It is possible that nitrate-binding proteins are present in NR preparations of barley root (Bourne and Miflin, 1970), spinach leaf (Eaglesham, 1972), and *Chlorella* preparations (Vennesland and Jetschmann, 1971). Light appears to regulate the uptake of nitrate or the permeability of cells and the nitrate transport to the enzyme-forming system (Beevers *et al.*, 1965); this particular effect of light is fulfilled by fairly low intensities (Chen and Ries, 1968), and may follow a circadian rhythm in wheat (Upcroft and Done, 1972). Malate synthesis tends to balance nitrate reduction (Ben-Zioni *et al.*, 1970). The movement of malate from leaves to roots is suggested to control nitrate absorption by tobacco after decarboxylation to CO_2 and production of bicarbonate, which is excreted in exchange for nitrate uptake (Ben-Zioni *et al.*, 1971), while potassium ions alternately accompany nitrate and malate movements.

3. REVERSIBLE INACTIVATION AND ACTIVATION

In green algae represented by *Chlorella fusca, C. pyrenoidosa (C. vulgaris), Chlamydomonas reinhardii,* and *Cyanidium caldarium,* the NR system may be present in a latent form (Vennesland and Jetschmann, 1971; Solomonson and Vennesland, 1972; Jetschmann *et al.*, 1972) or is inactivated by ammonia in the medium (Losada *et al.*, 1970; Herrera *et al.*, 1972; Rigano, 1971; Rigano and Violante, 1972a,b) in a manner that is distinct from repression of synthesis. Inactivation occurs during incubation with NAD(P)H for *Chorella* and spinach NR (Moreno *et al.*, 1972; Vega *et al.*, 1972; Palacian *et al.*, 1974), whereas NADH is the specific donor for NR activity.

The latent or inactive enzyme in *Chlorella pyrenoidosa* and *Chlamydomonas reinhardii* can be activated, with respect to NAD(P)H as the donor, by incubation with nitrate for prolonged periods (Vennesland and Jetschmann, 1971; Herrera *et al.*, 1972; Moreno *et al.*, 1972) or very rapidly by ferricyanide (Herrera *et al.*, 1972; Jetschmann *et al.*, 1972) and possibly by cytochrome oxidase, as oxygen-dependent activation of crude preparations is reversibly inhibited by carbon monoxide in light. Whereas the NADH dehydrogenase function is not lost, the FMNH-NR activity was inactivated by ammonia *in vivo* and reactivated as described above. The latent BV^o-NR and the FMN(FAD)H-NR of *Cyanidium caldarium* obtained from cells given ammonia is activated by heating at 50°C for 20 minutes or instantly at 60°C, but the NAD(P)H-NR, and therefore also the dehydrogenase, is destroyed under these conditions.

Activation by nitrate also occurred *in vivo* when cells were incubated in the presence of cycloheximide, indicating a process that is not dependent on protein synthesis, whereas in ammonium-grown (and repressed) cells cycloheximide prevented induction by nitrate (Rigano, 1971). Activation of NR from *Dunaliella parva* is achieved by heating in the presence of FAD which is not required for the NAD(P)H-dependent activity thereby revealed (Heimer, 1975), but there is possibly some analogy with protective effects of FAD noted above. The inactivation *in vivo* by ammonia is attributed by Herrera *et al.* (1972) to its effect as an uncoupler of photosynthetic phosphorylation and, as a consequence, increasing the redox potential of the cell. Cyanide inhibits NR by binding very strongly to molybdenum in lower valence states (V or IV) produced by reduction, but this is reversed by ferricyanide and then is competitive with nitrate. Endogenous cyanide, which forms a 1:1 complex with NR, is regarded as physiologically active in this type of regulation in *Chlorella* (Solomonson, 1974), the physiological oxidant being unknown. Inactivation of latent enzyme *in vivo* in plants is less well established. However, cyanogenic glycosides may be involved in sorghum (Maranville, 1970), and ammonia appears to cause inactivation in *Lemna* (Orebamjo and Stewart, 1975b) as well as repression. The dehydrogenase site is probably the site of the physiological regulation (Hewitt, 1975).

a. Kinetic Regulation. Kinetic regulation of NR's as distinct from inactivation may also be important. Thus, most NR systems appear to be stimulated by phosphate (Nicholas and Scawen, 1956; Kinsky and McElroy, 1958; Spencer, 1959; Nelson and Ilan, 1969; Eaglesham and Hewitt, 1971a). The requirement is not absolute in the sense that all attempts to eliminate phosphate completely results only in decreasing activity by about 50%. The effect of phosphate is optimal at about 10–20 mM and is reversible.

The possibility that NR in green plants might be subject to allosteric regulation was examined by Nelson and Ilan (1969), and the inhibitory action of ADP on the tomato leaf enzyme was revealed. The kinetic mechanism of this inhibition was investigated for the spinach leaf enzyme (Eaglesham and Hewitt, 1971a,b). ADP was found to inhibit both competitively with NADH and noncompetitively with different K_i values. The noncompetitive action was completely and reversibly abolished by thiol compounds. The effect of the thiol was influenced in a negative manner by the presence of oxygen. Ribose 5-phosphate also inhibited noncompetitively, and its effect was abolished by glutathione.

The effective concentrations of ADP (1–10 mM) are difficult to reconcile with maximal physiological concentrations of about 150–250 μM, which do not change by more than a ratio of about 2:1 in relation to light and dark. Concentrations of NADH may be between 5 and 20 μM and therefore between K_m and five times K_m values (Eaglesham and Hewitt, 1971a,b), and maximal inhibition would not exceed 25%. In order for a physiologically effective control by ADP to be possible, some additional postulation of compartmentation seems necessary. Alternatively, the kinetics of a membrane-bound NR system *in vivo* may be different from those found in soluble systems *in vitro*. Concentrations of glutathione occurring in spinach chloroplasts increase in response to light (Hirose *et al.*, 1971); increases which reverse the noncompetitive aspect of the inhibition are similar. The effect of ADP appears to be quite general, having been reported also from barley leaves (Vunkova, 1971) and from wheat roots (Chang *et al.*, 1965). Other adenine nucleotides are less inhibitory, except cyclic 3',5'-AMP, which is similar in action to ADP. It is possible that the true inhibitor is another nucleotide that has not yet been identified. Evidence for constraint by kinetic regulation is provided by experiments showing that extracted NR *in vitro* activity may exceed several fold that formed by various *in vivo* assay methods (Ferrari and Varner, 1970; Klepper *et al.*, 1971; Wallace, 1975).

b. Hormonal Regulation. Light is reported to influence the synthesis of gibberellins in plants (Köhler, 1966). The effect of cytokinins on NR formation was discovered by Borriss (1967). The induction of 95% of NR activity was obtained in tobacco plants that had been deprived of nitrate and then placed in a nitrate-containing solution but kept in darkness if the plants were sprayed with an appropriate mixture of kinetin (10 ppm) and gibberellic acid GA_3 (200 ppm) (Lips and Roth Bejerano, 1969; Roth Bejerano and Lips, 1970). The ratio of kinetin to GA_3 was important, as excessive or relatively low concentrations of GA_3 decreased the induction to 1–4%. The hormone treatment, therefore, replaced the requirement for light in the induction [although not for rice (Ghandi and Naik, 1974)]. The higher the kinetin, the lower the GA_3 concentration required. When plants were decapitated, the NR activity of young leaves fell and was not restored by indoleacetic acid but was induced to over threefold the control value of normal plants when GA_3 was given to decapitated plants. *In vitro* activity was unaffected by the hormones. The need for light for induction of NR activity by nitrate in radish cotyledons was partially abolished by kinetin (Rijven and Parkash, 1971). Light and kinetin together produced greater induction than either alone, but the enzyme produced seemed less stable than when only one factor was operating. Abscisic acid inhibited the response to kinetin or light.

In embryos of *Agrostemma githago* (Kende *et al.*, 1971), effects of the cytokinin benzyladenine (BA) on NR induction were detected 30–60 minutes after application to freshly excised embryos without the addition of nitrate. The endogenous nitrate was 5 nmoles or less per embryo, which is possibly adequate as a potential inducer. Exogenous nitrate (50 mM) also caused induction of NR activity. When induced embryos were washed and transferred to water, the NR declined as expected. On transfer again to the benzyladenine induction was again observed, but transfer to nitrate (50 mM) produced no further response. Abscisic acid inhibited the response to nitrate or BA. Two possibilities considered were either that BA inhibits breakdown of NR already present, since proteolysis is inhibited by cytokinins (Tavares and Kende, 1970), or that preformed latent enzyme was activated. Kende and Shen (1972) found that cycloheximide inhibited induction in response to BA much more severely than induction by nitrate, whereas puromycin inhibited both equally. Buoyant density measurement of NR protein indicated *de novo* protein synthesis of NR in response to either nitrate or BA (Hirschberg *et al.*, 1972; Kende and Shen, 1972) or succinic acid-2,2-dimethylhydrazine (B9) (Knypl, 1974). The derepression by cytokinins of constitutive NR formation would explain these results. In other experiments with excised pea roots growing in nitrate-containing media, kinetin slightly depressed NR activity, and indoleacetic acid had no effect (Sahulka, 1972). Beevers and Hageman (1969) also reported that cytokinin treatment did not influence NR activity in radish cotyledons, and Afridi and Hewitt (1965) were unable to influence NR induction by kinetin in cauliflower leaf tissues.

4. ENVIRONMENTAL FACTORS

a. Effect of Light. The interrelationships between light and nitrate assimilation are very complex, and numerous observations date from the early work of Schimper (1888). The subject has been reviewed extensively in different contexts for higher plants (Burström, 1943, 1945; McKee, 1962; Beevers and Hageman, 1969, 1972; Hewitt, 1970, 1975) and for green algae (Kessler, 1953, 1957a, 1959; Bongers, 1956; Syrett, 1962).

Nitrate reduction by *Ankistrodesmus braunii* is dependent on carbon dioxide in the light or added glucose in the dark (Kessler, 1955a). In the light, the addition of nitrite results in oxygen evolution, and its reduction is stimulated by carbon dioxide but is not entirely dependent on this factor. In the dark, nitrite reductase requires the presence of oxygen and is also inhibited by 2,4-dinitrophenol (Kessler, 1955b; Ahmed and Morris, 1967, 1968).

The relations between light, carbon dioxide, and oxygen evolution and nitrate or nitrite are complex and varied. Thus, Van Niel *et al.* (1953), Kessler (1955a), and Morris and Ahmed (1969) found with *Chlorella pyrenoidosa* or *Ankistrodesmus* that nitrate and/or nitrite stimulated oxygen evolution in the presence of carbon dioxide in light, whereas this result was not obtained with other experiments on *Chlorella* species (Myers, 1949; Davis, 1953), *Scenedesmus* (Bongers, 1958), or *Dunaliella tertiolecta* (Grant, 1967). Nitrite may compete with $NADP^+$ as a Hill reagent for oxidation of ferredoxin. This explanation is supported by the fact that oxygen evolution, dependent on added nitrate in *Chlorella vulgaris*, is inhibited by carbon monoxide (Vennesland and Jetchmann, 1971) because carbon monoxide strongly inhibits nitrite reductase (Hucklesby *et al.*, 1970), whereas ferredoxin cannot function as an electron donor for NR of *Chlorella* (Zumft *et al.*, 1969) unlike that of *Anabaena cylindrica* (Hattori and Myers, 1967). Reduction of nitrate and nitrite was stimulated by carbon dioxide in *Ankistrodesmus* but not in *Chlorella* (Morris and Ahmed, 1969). DCMU, which prevents oxygen evolution in light, inhibited nitrate reduction in the light by *Chlorella* but had less effect on this process in *Ankistrodesmus*. Nitrite reduction in light was less affected by DCMU in both species. Lack of carbon dioxide or inhibition of photosynthesis by DCMU or darkness caused a disappearance of NR activity in cells of *Chlamydomonas reinhardii* (Thacker and Syrett, 1972b). It was concluded that either stability or synthesis of NR was dependent on a product of carbon dioxide assimilation or of acetate metabolism. Assimilation of ammonia was also prevented in the absence of carbon dioxide or DCMU; probably a product of ammonia inhibited nitrite or nitrate assimilation (Thacker and Syrett, 1972a). There is evidence in *Chlorella pyrenoidosa* for close regulation between production of ammonia from nitrate (or nitrite) and the direction of carbon dioxide from sucrose to amino acids (Kanazawa *et al.*, 1970).

Although flavin nucleotides or NAD^+ reduced by spinach chloroplasts ferredoxin and light can reduce nitrate by NR (Del Campo *et al.*, 1963; Losada *et al.*, 1965; Paneque and Losada, 1966; Paneque *et al.*, 1965), this result does not indicate that NR is a light-dependent mechanism or that photosynthesis is involved. The poor specificity of most NR's for NADPH (Beevers *et al.*, 1964) and the great disparity in K_m values of NADPH and NAD^+ in the chloroplast transhydrogenase system by which NAD^+ is also reduced argue against the likelihood of these systems being physiologically effective.

Light and dark appear to activate or inactivate certain enzymes (Zucker, 1965; Muller, 1970; Huffaker *et al.*, 1966; Hatch and Slack,

1969b; Bradbeer, 1969, 1970). The mechanism is often obscure, but is apparently quite rapid and possibly direct. The ability of previously illuminated cauliflower leaf tissues to form NR in response to nitrate was independent of light during the induction period (Afridi and Hewitt, 1965). Green radish and maize tissues similarly produced NR when given nitrate and incubated in darkness (Beevers et al., 1965), but induction was promoted by light and they suggested that this facilitated movement of nitrate to the enzyme-forming system. Induction of NR occurred during darkness of preilluminated leaves of barley plants given nitrate but was noticeably slower than that observed during continued illumination (Travis et al., 1970a,b). The effect of preillumination was exhausted after 24 hours, but nitrate continued to accumulate. No induction occurred if the nitrate was originally given in darkness to the etiolated plants or to germinating maize (Hageman and Flesher, 1960) unless only 3 days old (Travis and Key, 1971). The effect of light was shown to be at least twofold on induction of the enzyme in rye or oat seedlings (Chen and Ries, 1968) and on *Wolffia arrhiza* (Swader et al., 1975). Low light intensities promoted the rate of nitrate uptake of seedlings that were previously dark treated for several hours, and the response was saturated at about 3000 lux. Some uptake occurred in the dark, but no enzyme was induced. Enzyme induction continued to increase with increasing light intensities beyond 15000 lux. Induction continued for several hours after preilluminated plants were returned to dark, in agreement with original results of Afridi and Hewitt (1965) and the later results of Travis et al. (1970b). The effect of preillumination does not prevent the operation of a diurnal rhythm in the activity of NR. This rises to a maximum about noon in soybeans (Harper and Hageman, 1972). Diurnal rhythm also appears in maize (Hageman et al., 1961), peas (Wallace and Pate, 1965), and wheat (Upcroft and Done, 1972). In maize, light is not now considered necessary for induction if polyribosomes are abundant (Travis et al., 1970a; Travis and Key, 1971).

The questions of whether existing enzyme is activated by light (and inactivated by dark) as for pyruvatephosphate dikinase (Hatch and Slack, 1969b), whether the enzyme breaks down irreversibly in dark and is stabilized by light or a photosynthetic product, or whether normal loss is continually occurring and resynthesis is dependent on a product of illumination, or is inhibited by a product of dark metabolism are not answered by the experiments so far described. The phenylalanine ammonia lyase system that yields precursors of phenylpropanol derivatives involved in lignin and polyphenol synthesis is phytochrome-controlled as also is NR formation (Jones and Sheard, 1972, 1975). Among possible products coumarin, *trans*-cinnamic acid and *trans*-*o*-hydroxycinnamic acid were found to inhibit the induction of NR in maize (Schrader and Hageman, 1967). Rates of nitrate assimilation (not NR) in wheat were maximal in blue light, minimal in green light, and enhanced in red light, but carbon dioxide fixation showed the same action spectrum (Stoy, 1955). Wheat NR activity was lower in red than in blue light (Harper and Paulsen, 1968).

b. Effect of Temperature and Water Stress. Cauliflower lost NR during water shortage (Afridi and Hewitt, 1965); corn accumulated nitrate under drought conditions (Hanway and Englehorn, 1958). Accumulation of nitrate in corn seedlings or wheat under artificial drought conditions was found to be associated with reduced NR activity (Mattas and Pauli, 1965; Plaut, 1974). This response is associated with conversion of poly- to monoribosomes (Hsiao, 1970). High temperatures (25°–30°C) appear to impair NR activity in maize compared with 20°–25°C, and activity at 15°–20°C was six times that at 25°–30°C, whereas nitrate concentrations showed opposite trends (Younis *et al.*, 1965).

III. Nitrite Reduction

A. Nitrite as a Metabolite

Nitrite, the product of the nitrate reductase reaction, must be further reduced to ammonia before the constituent nitrogen can enter into organic combination. This task is accomplished by a specific enzyme. Nitrite, unlike nitrate, rarely accumulates in plant tissues, and cell concentrations are very low under normal conditions of plant growth. These low endogenous levels probably reflect the toxicity of nitrite—and more particularly of nitrous acid ($pK = 3.6$, 3% free acid at pH 5)—to plant tissues. Nitrite combines with heme proteins, and nitrous acid reacts with amino groups, and can interconvert nucleic acid bases (Mahler and Cordes, 1966). As may be expected in these circumstances, nitrite reductase functions efficiently at low substrate concentrations *in vitro* and shows high activity relative to nitrate reductase.

B. Nitrite Reductase Enzymes

1. ELECTRON DONORS AND ENZYME ASSAYS

Reduced nicotinamide nucleotides can function as electron donors to assimilatory nitrite reductase enzymes from several nonphotosynthetic organisms ($\Delta G^0 = -93.7$ kcal for ammonia formation). NADH-specific nitrite reductase has been extracted from *E. coli* (Kemp and Atkinson,

1966), and NADPH-specific enzymes have been extracted from *Aspergillus nidulans* (Pateman *et al.*, 1967), *Candida utilis* (Sims *et al.*, 1968), and *E. coli* (Lazzarini and Atkinson, 1961). Nitrite reductases from *Neurospora crassa* (Nicholas *et al.*, 1960) and *Azotobacter agile* (Spencer *et al.*, 1957; Lafferty and Garrett, 1974), can use both NADH and NADPH. It is possible that these enzymes may also accept electrons from reduced benzyl viologen as has already been shown for the *Neurospora crassa* nitrite reductase (Cook and Sorger, 1969; Garrett, 1972).

By contrast, the higher plant nitrite reductases show only very low activities with NADH and NADPH, although these can be increased to a certain extent by the addition of high concentrations of flavins (Hucklesby and Hewitt, 1970). The detection and assay of nitrite reductase from leaves remained baffling until the early 1960's when work in several laboratories showed that certain compounds having lower redox potentials than NAD(P)H can serve as effective electron donors for the enzyme. These are the synthetic dipyridilium dyes, benzyl viologen (BV) and methyl viologen (MV) (Hageman *et al.*, 1962; Cresswell *et al.*, 1962), which have E_0' —0.359 and —0.44 V, respectively, and the naturally occurring nonheme iron protein, ferredoxin (Huzisige and Satoh, 1961; Losada *et al.*, 1963; Hewitt and Betts, 1963; Joy and Hageman, 1966) with $E_0' = -0.432$ V. Another natural redox compound, phytoflavin, which has been prepared from *Anacystis* but not so far from higher plants, has also been shown capable of donating electrons to a higher plant nitrite reductase (Bothe, 1969). Ability to reduce nitrite with ferredoxin or MV^0 but not NAD(P)H as electron donor is also characteristic of nitrite reductase from the green alga *Chlorella* (Zumft *et al.*, 1969) and the blue-green alga *Anabaena* (Hattori and Uesugi, 1968).

Dithionite can be used for the reduction of all the above electron carriers except phytoflavin, which must be reduced with illuminated chloroplast preparations in order to serve as a donor for nitrite reduction (Bothe, 1969). Benzyl viologen can be reduced with palladium and hydrogen (Hageman *et al.*, 1962).

Diaphorase enzymes, hydrogenase systems, or illuminated grana may be used for the enzymatic reduction of viologens or ferredoxin. Partially purified NADPH-diaphorase (Shin *et al.*, 1963; Hewitt *et al.*, 1968) can be used to give 10–15% reduction of ferredoxin, providing that the NADPH is maintained in the reduced form. Ferredoxin reduction of 90–100% with consequent greatly increased rates of nitrite reduction can be attained by means of an illuminated grana system. In the method as usually used, photosystem 2 is inactivated by heat and electrons are supplied from an ascorbate–dichlorophenolindophenol couple (Hewitt *et al.*, 1968). This method is believed to be closely related to the natural nitrite reduction system (III 7a).

The reduction of nitrite to nitric oxide by nitrite reductase from various anaerobic organisms has been reported with a wide range of electron donors (Walker and Nicholas, 1961; Radcliffe and Nicholas, 1968) including MV^0, BV^0, NAD(P)H, flavins, various cytochromes, and such relatively weak reductants as pyocyanin ($E_0' = -0.034$ V) and methylene blue ($E_0' = +0.012$ V).

2. PHYSICAL AND CHEMICAL PROPERTIES

Dissimilatory nitrite reductases, which have generally been prepared from anaerobic bacteria, will not be considered at any length in this chapter (see Hewitt, 1974). The reaction product of these enzymes is commonly nitric or nitrous oxide which may contain c and/or a_2 (d) type hemes (Prakash and Sadana, 1972; Yamanaka and Okunuki, 1963; Newton, 1969; Lam and Nicholas, 1969b). Sometimes copper is present (Walker and Nicholas, 1961; Iwasaki and Matsubara, 1972; Iwasaki et al., 1963). All these characteristics are distinct from the known features of the assimilatory enzymes, and may for convenience be grouped as follows.

i. *Assimilatory enzymes from photosynthetic organisms.* These show a considerable unity of properties among diverse phyla. They contain iron but no flavin groups. Ferredoxin or viologen dyes function as electron donors.

ii. *Assimilatory enzymes from nonphotosynthetic organisms.* These can use NAD(P)H as electron donor. They have not yet been well described, but probably contain flavin and iron, which may be present as a siroheme group (Lafferty and Garrett, 1974).

a. *Enzymes from Photosynthetic Organisms.* Nitrite reductases from organisms belonging to diverse groups have been reported to have molecular weights of 60,000–70,000 (Hewitt et al., 1968; Hattori and Uesugi, 1968; Cardenas et al., 1972; Zumft, 1972). Methods used in purification characteristically yield a single peak of activity, although exceptions have been noted (Hucklesby et al., 1972; Zumft, 1972).

Cyanide and carbon monoxide inhibit nitrite reductase from *Cucurbita pepo*, while metal-chelating reagents have little effect (Cresswell et al., 1965). Nitrite reductases from spinach (Cardenas et al., 1972), vegetable marrow (Hucklesby et al., 1974), and *Chlorella* (Zumft, 1972) have recently been purified to homogeneity and shown to contain two atoms of iron per molecule. The spectrum of the enzyme in its oxidized form shows major absorption peaks at 380 and 570 nm (spinach) and at 384 and 573 nm (marrow and *Chlorella*). The spectrophotometric properties of the enzyme at first appeared to preclude the presence of a heme group. Murphy *et al.* (1974) have now presented evidence that the prosthetic group is an unusual heme ("siroheme") having a Soret band of low wavelength. Earlier reports of manganese participation in the nitrite reductase enzyme system have not been confirmed.

The *Chlorella* enzyme contains seven cysteine residues per molecule, some of which may participate in binding the iron atoms. Nitrite reductases from a great variety of organisms are inhibited by *p*-chloromercuribenzoate (with thiol reversibility), indicating the involvement of sulfhydryl groups. Some labile sulfur is also present (Zumft, 1972; Hucklesby *et al.*, 1974; Aparicio *et al.*, 1975) suggesting the presence of nonheme iron. The relationship of this to the siroheme is not understood at present.

The absence of absorbance by the enzymes from both spinach and *Chlorella* in the 450 nm region of the spectrum appears to confirm earlier conclusions based on indirect evidence (Ramirez *et al.*, 1966) that the enzyme structure does not include a flavin group.

The visible spectrum of nitrite reductase is to some extent similar to that of sulfite reductase from spinach; the siroheme group has been extracted from both spinach nitrite reductase and *E. coli* sulfite reductase (Murphy *et al.*, 1974). Other resemblances between sulfite and nitrite reductases have been known for some time. In particular, nitrite and sulfite reductases are the only enzymes known which mediate the transfer of six electrons. Both groups of enzymes accept electrons from ferredoxins and viologens, and it is possible that the leaf sulfite reductase is photosynthetically coupled (Tamura *et al.*, 1967). Sulfite reductase from *E. coli* (Kemp *et al.*, 1963) is able to reduce nitrite. However, the nitrite reductases of *Cucurbita pepo* (Hucklesby and Hewitt, 1970) and *Chlorella* (Zumft, 1972) do not reduce sulfite with BV° and MV°, respectively, as electron donors.

b. Assimilatory Enzymes from Nonphotosynthetic Organisms. These possibly exist as complexes of considerably larger size than the ferredoxin-nitrite reductases from photosynthetic organisms. There is some evidence for flavin stimulation of nitrite reductases from *Neurospora crassa* (Nicholas *et al.*, 1960; Lafferty and Garrett, 1974) and *Azotobacter agile* (Spencer *et al.*, 1957). These enzymes are sensitive to cyanide, and there is strong indication of a metal prosthetic group. Copurification of iron and copper with the *Neurospora* nitrite reductase was reported, and the spectrum of this enzyme (Lafferty and Garrett, 1974) suggests the presence of a heme constituent. The NADPH-specific nitrite-reducing enzyme from *E. coli*, which is physiologically a sulfite reductase, has a molecular weight of 670,000 and contains 4 FMN, 4 FAD, 20–21 atoms of iron, and 14–15 labile sulfides per molecule (Siegel et al., 1973). Three or four of the iron atoms are combined in a new type of heme structure that has been characterized as an iron tetrahydroporphyrin of isobacteriochlorin type (Murphy et al., 1973). This siroheme component is responsible for the absorbance peaks at 386 and 587 nm. The yeast sulfite reductase (Yoshimoto and Sato, 1968, 1970) appears to be a similar enzyme. Probably the siroheme group is characteristic of sulfite and nitrite reductases in general.

3. PRODUCTS AND INTERMEDIATES IN NITRITE REDUCTION

a. Ammonia as Product.   Ammonia is the product of many assimilatory nitrite reductase enzymes, and the reaction does not involve the formation of free intermediates.

$$NO_2^- + 6\,e^- + 6\,H^+ \rightarrow NH_3 + H_2O + OH^-$$

E_0' for $NO_2^-/NH_4^+ = +0.48$ and $\Delta F^\circ = -123$ kcal.

Reduction is visualized as a complex sequence of one or two electron transfers, generating a series of free radical intermediates none of which is released from the enzyme. Some speculations concerning the nature of these have been made by Kemp et al. (1963), Hewitt et al. (1968), Hewitt 1974, 1975), and Loussaert and Hageman (1974).

The transfer of a total of six electrons is an impressive feat for a relatively small enzyme. An earlier supposition, based on the hypothesis of Meyer and Schulze (1894), was that the reduction of nitrate to ammonia comprised a sequence of four steps, each removing two electrons.

$$NO_3^- \rightarrow NO_2^- \rightarrow N_2O_2^{2-} \rightarrow NH_2OH \rightarrow NH_3$$
$$\text{nitrate} \quad \text{nitrite} \quad \text{hyponitrite} \quad \text{hydroxylamine} \quad \text{ammonia}$$

Impetus was given to this theory by the discovery of nitrate reductase (product nitrite) and the realization that many organisms also show hydroxylamine reductase activity. A search for enzymes specifically reducing the intermediate compounds followed. Evidence against the free intermediate hypothesis has accumulated, however, and can be summarized as follows. Hyponitrite and hydroxylamine cannot usually be demonstrated in plant tissues or in systems reducing nitrite in vitro (Hucklesby and Hewitt, 1970); hyponitrite is not reduced by the enzymes; hydroxylamine reductases usually have K_m values that are often 10–100 times as great as K_m values for nitrite in nitrite reductases from the same source (Hewitt *et al.*, 1968), and nitrite severely inhibits hydroxylamine reduction; stoichiometric conversion of nitrite to ammonia is obtained using highly purified nitrite reductases from various sources (Hewitt *et al.*, 1968; Hucklesby and Hewitt, 1970; Ho and Tamura, 1973; Zumft, 1972; Prakash and Sadana, 1972); hydroxylamine equilibrated with *E. coli* nitrite reductase and $^{15}NO_2^-$ did not become labeled (Lazzarini and Atkinson, 1961) and no evidence for independent structural genes for nitrite, hyponitrite, or hydroxylamine reductases (Pateman *et al.*, 1967) could be found.

The conclusion of Fewson and Nicholas (1960) that nitrite assimilation involves nitric oxide as an intermediate is not confirmed by preliminary studies of *Cucurbita* ferredoxin-nitrite reductase (Hewitt, 1974) although bound nitroxides could be formed.

b. Function of Hydroxylamine Reductase. Hydroxylamine is reduced by purified nitrite reductases to some extent, and in the case of *Candida utilis* hydroxylamine reductase is induced by nitrite (Sims *et al.*, 1968). In maize scutellum the hydroxylamine reductase (approximately 1%) is associated with nitrite reductase after chromatography and is induced by nitrite (Hewitt, 1975; Hucklesby and Hageman, 1976). It is clear that nitrite reductases from some sources have a limited ability to reduce hydroxylamine (Hewitt *et al.*, 1968; Zumft, 1972; Prakash and Sadana, 1972), perhaps through some degree of chemical relationship of this compound to one of the free radical intermediates of nitrite reduction. It is probable that the widely observed enzymatic reduction of hydroxylamine is in fact catalyzed by nitrite reductase or by other enzymes which are not physiological hydroxylamine reductases. Higher plant sulfite reductases reduce hydroxylamine, presumably as nonphysiological attributes (Tamura, 1965; Asada *et al.*, 1968), and the same may be true of the heme-containing hydroxylamine reductase from leaves (Hucklesby *et al.*, 1970). Multiple hydroxylamine reductases, one of which appears to be a physiological sulfite reductase, have been described in *Neurospora* (Siegel *et al.*, 1965). The existence of enzymes whose function is to eliminate small quantities of hydroxylamine formed by hydrolysis of oximes and hydroxamates (both of which occur in tissues) cannot be dismissed.

4. KINETICS

The K_m values for nitrite reductase have been reported in the range 10^{-6} to 10^{-4} M. The cause of variation is obscure. This subject has been discussed by Hewitt *et al.* (1968).

The degree of reduction of electron donor (BV°) was found to be a major determinant of the activity of *Cucurbita* nitrite reductase. A point of unusual interest with this system is the direct proportionality between enzyme activity and the redox potential (*Eh*) of the benzyl viologen (Cresswell *et al.*, 1965). It is possible that this may eventually be found to be true of nitrite reductases from other sources using natural electron donors (see Hewitt, 1974).

5. REGULATION

a. Regulation through Enzyme Induction. Nitrite is commonly an inducer of the enzyme. Induction is also observed when tissues are infiltrated with nitrate; in this case it is difficult to identify the inducing ion which might be nitrate itself or nitrite generated as a consequence of the induction of nitrate reductase. Kinetic studies of induction by nitrate in radish cotyledons (Ingle *et al.*, 1966) appear to indicate the latter circumstance, i.e., sequential induction of the two enzymes. This is not necessarily true of all types of plant material. Evidence for nitrate as an inducing ion is provided by experiments with *Lemna*, where higher levels of nitrite reductase can be obtained with nitrate rather than nitrite—an observation that is difficult to explain if induction is sequential (Stewart, 1972). More persuasively, when nitrite reductases of cultured tobacco cells (Kelker and Filner, 1971) or *Lemna* (Stewart, 1972) are inactivated by the inclusion of tungstate in the culture medium, nitrite reductase is induced by nitrate. In mutants of *Aspergillus nidulans*, which lacked nitrate reductase activity, nitrate induced the nitrite reductase enzyme (Pateman *et al.*, 1967).

The degree of independence of induction of nitrate and nitrite reductases in response to either nitrate or nitrite varies with species. In the yeast, *Candida utilis* (Sims *et al.*, 1968), these enzymes, which are possibly associated with a nitrosome, show a high degree of coordinated response to either ion. Although some degree of coordination is seen in *Lemna* (Stewart, 1972), this is sufficiently flexible to permit the induction of nitrite reductase without nitrate reductase (Sims *et al.*, 1968); nitrite is also a more specific inducer for nitrite reductase in *Spirodela* (Ferguson, 1969). Nitrite is a poor inducer of nitrate reductase in radish cotyledons (Ingle *et al.*, 1966), cauliflower leaves (Afridi and Hewitt, 1964), and maize seedlings (Beevers *et al.*, 1965), but induces this enzyme very effectively in bean cotyledons (Lips *et al.*, 1973).

Repression of nitrite reductase synthesis by ammonia and/or amino acids has been reported for *Lemna* (Joy, 1969; Stewart, 1972), *Chlorella* (Losada *et al.*, 1970), tobacco, *Aspergillus nidulans*, and *Neurospora crassa*. Evidence that the *Neurospora crassa* enzyme is derepressible rather than substrate-inducible (Cook and Sorger, 1969) is to some extent contradictory to the findings of Garrett (1972).

Nitrite reductase synthesis *de novo* during induction has not been demonstrated using isotope incorporation techniques, but inhibitors of protein synthesis prevent induction of the enzyme. Experiments with cycloheximide and chloramphenicol (Schrader *et al.*, 1967; Stewart, 1968, 1972) suggest that both nitrate and nitrite reductases are formed on cytoplasmic rather than chloroplastic ribosomes; this is true of many other soluble chloroplastic enzymes (Ellis and Forrester, 1972).

b. Noninductive Regulation. The activity of nitrite reductase *in vitro* is characteristically 5–20 times as great as that of nitrate reductase, and the low levels of nitrite in tissues compared with nitrate suggest that nitrite reductase has an excess capacity *in vivo*. The reduction of nitrate is likely to be rate controlling in nitrate assimilation, and, logically, this step rather than nitrite reduction should be the site of regulation by feedback inhibitors, a phenomenon that is often associated with the first reaction in a metabolic sequence. Inhibition of the higher plant nitrite reductases by ATP, carbamyl phosphate or glutamine, asparagine, or other amino acids (singly or in concert) could not be demonstrated (Dalling, 1971; D. P. Hucklesby, unpublished).

Ammonia, amino acids, and carbamyl phosphate which inhibit nitrate reductases from *Lemna* do not alter the activity of nitrite reductase from this species (Sims *et al.*, 1968). The absence of feedback regulation of nitrite reductase activity cannot be generalized to all organisms. In the yeast, *Candida utilis*, nitrate and nitrite reductases respond, apparently in a highly coordinated manner to inhibition by ammonia, pyridoxamine phosphate, lysine, and a combination of amino acids chosen to simulate the natural amino acid pool (Sims *et al.*, 1968).

Nitrate metabolism may be relevant to regulatory mechanisms other than the feedback control of nitrate assimilation. There is some experimental evidence (Shin and Oda, 1966; Betts and Hewitt, 1966) for competition between nitrite and NADP for electrons provided by the photosynthetic light reactions. The rate of reduction of nitrite may also have an impact upon the Calvin cycle through depression of the rate of conversion of sugar diphosphates to monophosphates (Hiller and Bassham, 1965) or inhibition of carbonic anhydrase (Everson, 1970).

6. LOCALIZATION

a. Leaves. In leaves of *Zea mays* and *Sorghum* plants, which possess a C$_4$ type of photosynthesis, nitrite reductase activity is associated largely with the mesophyll (Mellor and Tregunna, 1971). The bundle sheath cells, which are wholly or partly lacking in chloroplast grana, have low activities of the enzyme.

There is general agreement that nitrite reductase is situated at least partly or principally in the chloroplasts (Ritenour et al., 1967; Grant et al., 1970; Betts and Hewitt, 1966; Ramirez et al., 1966; Dalling et al., 1972b). Grant et al. (1970) concludes that the larger part of the activity is located outside the chloroplast and argues that there are at least two sites for nitrite reduction in the cell. Illuminated chloroplasts (Swader and Stocking, 1971; Grant and Canvin, 1970; Magalhaes et al., 1974; Miflin, 1974a; Canvin and Atkins, 1974) reduce nitrite rapidly without the addition of cofactors or fractions containing other cell components, whereas the reduction of nitrate (at least in the case of Wolffia) requires the addition of a soluble extrachloroplastic factor.

Reported discrepancies in the location of nitrate and nitrite reductases might be explained if the enzyme were located in the peroxisomal microbodies, since these are fragile (and therefore difficult to prepare) and are closely associated with the chloroplasts. Evidence for the location of both enzymes in the microbodies has been presented by Lips and Avissar (1972), but nitrite reductase was not found in these particles by Miflin, (1974b).

b. Nonchlorophyllous Tissues. Nitrite reductase obtained by isotonic extraction of barley roots (Miflin, 1967, 1970a) and maize scutellum (Hucklesby et al., 1972) is found predominantly in supernatant fractions. A certain amount of activity is also associated with a particulate fraction (Bourne and Miflin, 1970, 1973; Miflin, 1970b), which may contain proplastids (Dalling et al., 1972b).

7. PHYSIOLOGICAL ELECTRON DONORS

a. Leaves. In leaves, the most widely held view at the present time regards nitrite reduction in vivo as directly coupled to electron transport at the level of ferredoxin, i.e., nitrite is a recipient of electrons from noncyclic phosphorylation (see Section II,B). Nitrite reduction should, therefore, be accompanied by oxygen evolution and ATP formation. The stoichiometry of the reaction, i.e., 1 mole nitrite reduced per $1\frac{1}{2}$ moles O_2 evolved per 3 moles ATP formed, has been demonstrated (Paneque et al., 1963). The leaf nitrite reductases closely resemble nitrite reductase from nonchlorophyllous tissues, which do not appear to contain ferredoxin (Dalling et al., 1973; Ida et al., 1974).

b. *Nonchlorophyllous Tissues.*   Nitrite reduction by intact nonchlorophyllous tissues is suppressed by oxygen deprivation and by uncouplers of phosphorylation (Elsner, 1969; Ferrari and Varner, 1971). There is some evidence that nitrite reduction might be coupled to the pentose phosphate pathway (Butt and Beevers, 1961), which would perhaps indicate NADPH-generating systems as the source of reducing power. Nitrite can be reduced by cell-free extracts of scutellum using glucose 6-phosphate as hydrogen donor and NADP and ferredoxin as intermediate electron carriers (Hucklesby *et al.*, 1972). The three enzymes necessary for this system are all present in the scutellum extracts, but the ferredoxin apparently is not and was supplied from a leaf source in these experiments. Attempts to replace ferredoxin with a factor obtained from the scutellum itself have not been successful. Probably the physiological nitrite system does not function in this manner at all, but is related to the cell-free system from barley roots described by Bourne and Miflin (1970). Particulate preparations reduced nitrate to ammonia in the presence of pyruvate, ATP, and various cofactors. Omission of ATP resulted in accumulation of nitrite and decrease in ammonia yield. The system has not been further characterized. It is not known whether ATP is involved directly or indirectly in this type of nitrite reduction.

REFERENCES

Afridi, M. M. R. K., and Hewitt, E. J. (1962). *Life Sci.* **1**, 287.
Afridi, M. M. R. K., and Hewitt, E. J. (1964). *J. Exp. Bot.* **15**, 251.
Afridi, M. M. R. K., and Hewitt, E. J. (1965). *J. Exp. Bot.* **16**, 628.
Ahmed, J., and Morris, I. (1967). *Arch. Mikrobiol.* **57**, 219.
Ahmed, J., and Morris, I. (1968). *Biochim. Biophys. Acta* **162**, 32.
Anacker, W. F., and Stoy, V. (1958). *Biochem. Z.* **330**, 141.
Antoine, A. D. (1974). *Biochemistry* **13**, 2289.
Aparicio, P. J., Cardenas, J., Zumft, W. G., Vega, J. Ma., Herrera, J., Paneque, A., and Losada, M. (1971). *Phytochemistry* **10**, 1487.
Aparicio, P. J., Knaff, D. B., and Malkin, R. (1975). *Arch. Biochem. Biophys.* **169**, 102.
Arst, H. N., MacDonald, D. W., and Cove, D. J. (1970). *Mol. Gen. Genet.* **108**, 129.
Asada, K., Tamura, G., and Bandurski, R. S. (1968). *Biochim. Biophys. Acta* **30**, 554.
Azoulay, E., Puig, J., and Couchoud-Beaumont, P. (1969). *Biochim. Biophys. Acta* **171**, 238.
Bayley, J. M., King, J., and Gamborg, O. L. (1972). *Planta* **105**, 15.
Beevers, L., and Hageman, R. H. (1969). *Annu. Rev. Plant Physiol.* **20**, 495.
Beevers, L., and Hageman, R. H. (1972). *Photophysiology* **7**, 85.
Beevers, L., and Flesher, D., and Hageman, R. H. (1964). *Biochim. Biophys. Acta* **89**, 453.

Beevers, L., Schrader, L. E., Flesher, D., and Hageman, R. H. (1965). *Plant Physiol.* **40**, 691.

Ben-Zioni, A., Vaadia, Y., and Lips, S. H. (1970). *Physiol. Plant.* **23**, 1039.

Ben-Zioni, A., Vaadia, Y., and Lips, S. H. (1971). *Physiol. Plant.* **24**, 288.

Betts, G. F., and Hewitt, E. J. (1966). *Nature (London)* **210**, 1327.

Bongers, L. H. J. (1956). *Meded. Landbouwhogesch. Wageningen* **56**, 1.

Bongers, L. H. J. (1958). *Neth. J. Agr. Sci.* **6**, 79.

Borriss, H. (1967). *Wiss. Z. Univ. Rostock, Math.-Naturwiss. Reihe* **16**, 629.

Bothe, H. (1969). *Progr. Photosyn. Res., Proc. Int. Congr., 1968* Vol. 3, p. 1483.

Bourne, W. F., and Miflin, B. J. (1970). *Biochem. Biophys. Res. Commun.* **40**, 1305.

Bourne, W. F., and Miflin, B. J. (1973). *Planta* **111**, 47.

Bradbeer, J. W. (1969). *New Phytol.* **68**, 233.

Bradbeer, J. W. (1970). *New Phytol.* **69**, 635.

Burström, H. (1939). *Planta* **29**, 292.

Burström, H. (1943). *Kgl. Lantbruks-Hoegsk. Ann.* **11**, 1.

Burström, H. (1945). *Kgl. Lantbruks-Hoegsk. Ann.* **13**, 1.

Butt, V., and Beevers, H. (1961). *Biochem. J.* **80**, 21.

Candela, M. I., Fisher, E. G., and Hewitt, E. J. (1957). *Plant Physiol.* **32**, 280.

Canvin, D. T., and Atkins, C. A. (1974). *Planta* **116**, 207.

Cardenas, J., Barea, J. L., Rivas, J., and Moreno, C. G. (1972). *FEBS (Fed. Eur. Biochem. Soc.) Lett.* **23**, 131.

Chang, T. J., Wang, J. C., and Tang, J. W. (1965). *Acta Phytophysiol. Sinica* **2**, 1, cited by Vunkova (1971).

Chen, T. M., and Ries, S. K. (1968). *Can. J. Bot.* **47**, 341.

Cheniae, G., and Evans, H. J. (1959). *Biochim. Biophys. Acta* **35**, 140.

Cheniae, G., and Evans, H. J. (1960). *Plant Physiol.* **35**, 454.

Cheniae, G. M., and Martin, I. F. (1968). *Biochim. Biophys. Acta* **153**, 819.

Cheniae, G. M., and Martin, I. F. (1970). *Biochim. Biophys. Acta* **197**, 219.

Cleland, W. W. (1963). *Biochim. Biophys. Acta* **67**, 104.

Cole, J. A, and Wimpenny, J. W. T. (1968). *Biochim. Biophys. Acta* **162**, 39.

Cook, K. A., and Sorger, G. J. (1969). *Biochim. Biophys. Acta* **177**, 412.

Coupé, M., Champigny, M. L., and Moyse, A. (1967). *Physiol. Veg.* **5**, 271.

Cove, D. J. (1966). *Biochim. Biophys. Acta* **113**, 51.

Cove, D. J. (1970). *Proc. Roy. Soc., Ser. B* **176**, 267.

Cove, D. J., and Coddington, A. (1965). *Biochim. Biophys. Acta* **110**, 312.

Cove, D. J., and Pateman, J. A. (1963). *Nature (London)* **198**, 262.

Cove, D. J., and Pateman, J. A. (1969). *J. Bact.* **97**, 1374.

Cove, D. J., Pateman, J. A, and Rever, B. M. (1964). *Heredity* **19**, 529.

Cresswell, C. F. (1961). Ph.D. Thesis, University of Bristol, Bristol, England.

Cresswell, C. F., Hageman, R. H., and Hewitt, E. J. (1962). *Biochem. J.* **83**, 38.

Cresswell, C. F., Hageman, R. H., Hewitt, E. J., and Hucklesby, D. P. (1965). *Biochem. J.* **94**, 40.

Dalling, M. J. (1971). Ph.D. Thesis, University of Illinois, Urbana.

Dalling, M. J., Tolbert, N. E., and Hageman, R. H. (1972a). *Biochim. Biophys. Acta* **283**, 505.

Dalling, M. J., Tolbert, N. E., and Hageman, R. H. (1972b). *Biochim. Biophys. Acta* **283**, 513.

Dalling, M. J., Hucklesby, D. P., and Hageman, R. H. (1973). *Plant Physiol.* **51**, 481.

Davis, E. A. (1953). *Plant Physiol.* **28**, 539.

Del Campo, F. F., Paneque, A., Ramirez, J. M., and Losada, M. (1963). *Biochim. Biophys. Acta* **66**, 450.

Downey, R. J. (1971). *J. Bacteriol.* **105**, 759.

Downey, R. J. (1973a). *Biochem. Biophys. Res. Commun.* **50**, 920.

Downey, R. J. (1973b). *Microbios* **7**, 53.

Ducet, G., and Hewitt, E. J. (1954). *Nature (London)*, **173**, 1141.

Eaglesham, A. R. J. (1972). Ph.D. Thesis, University of Bristol, Bristol. England.

Eaglesham, A. R. J., and Hewitt, E. J. (1971a). *Biochem. J.* **122**, 18.

Eaglesham, A. R. J., and Hewitt, E. J. (1971b). *FEBS (Fed. Eur. Biochem. Soc.) Lett.* **16**, 315.

Eckerson, S. H. (1931). *Contrib. Boyce Thompson Inst.* **3**, 405.

Ellis, R. J., and Forrester, E. E. (1972). *Biochem. J.* **130**, 28P.

El-Shishiny, E. D. H. (1955). *J. Exp. Bot.* **6**, 6.

Elsner, J. E. (1969). Ph.D. Thesis, University of Illinois, Urbana.

Engelsma, G. (1967). *Planta* **75**, 207.

Evans, H. J., and Nason, A. (1953). *Plant Physiol.* **28**, 233.

Everson, R. G. (1970). *Phytochemistry* **9**, 25.

Ferguson, A. R. (1969). *Planta* **88**, 353.

Ferguson, A. R., and Bollard, E. G. (1969). *Planta* **88**, 344.

Ferrari, T. E., and Varner, J. E. (1969). *Plant Physiol.* **44**, 85.

Ferrari, T. E., and Varner, J. E. (1970). *Proc. Nat. Acad. Sci. U.S.* **65**, 729.

Ferrari, T. E., and Varner, J. E. (1971). *Plant Physiol.* **47**, 790.

Ferrari, T. E., Yoder, O. C., and Filner, P. (1973). *Plant Physiol.* **51**, 423.

Fewson, C. A., and Nicholas, D. J. D. (1960a). *Nature (London)* **188**, 794.

Fewson, C. A., and Nicholas, D. J. D. (1960b). *Biochem. J.* **77**, 3P.

Fewson, C. A., and Nicholas, D. J. D. (1961a). *Nature (London)* **190**, 2.

Fewson, C. A., and Nicholas, D. J. D. (1961b). *Biochim. Biophys. Acta* **49**, 335.

Filner, P. (1966). *Biochim. Biophys. Acta* **118**, 299.

Forget, P. (1974). *Eur. J. Biochem.* **42**, 325.

Fowden, L. (1963). *J. Exp. Bot.* **14**, 387.

Garrett, R. H. (1972). *Biochim. Biophys. Acta* **264**, 481.

Garrett, R. H., and Nason, A. (1967). *Proc. Nat. Acad. Sci. U.S.* **58**, 1603.

Garrett, R. H., and Nason, A. (1969). *J. Biol. Chem.* **244**, 2870.

Gauthier, D. K., Clark-Walker, G. D., Garrard, W. T., Jr., and Lascelles, J. (1970). *J. Bacteriol.* **102**, 797.

Ghandi, A. P., and Naik, M. S. (1974). *FEBS (Fed. Eur. Biochem. Soc.) Lett.* **40**, 343.

Glaser, J. H., and DeMoss, J. A. (1971). *J. Bacteriol.* **108**, 854.

Grant, B. R. (1967). *J. Gen. Microbiol.* **48**, 379.

Grant, B. R., and Canvin, D. T. (1970). *Planta* **95**, 227.

Grant, B. R., Atkins, C. A., and Canvin, D. T. (1970). *Planta* **94**, 60.

Hageman, R. H., and Flesher, D. (1960). *Plant Physiol.* **35**, 700.

Hageman, R. H., Flesher, D., and Gitter, A. (1961). *Crop Sci.* **1**, 201.

Hageman, R. H., Cresswell, C. F., and Hewitt, E. J. (1962). *Nature (London)* **193**, 247.

Hanway, J. J., and Englehorn, A. J. (1958). *Agron. J.* **50**, 331.

Harper, J. E., and Hageman, R. H. (1972). *Plant Physiol.* **49**, 146.

Harper, J. E., and Paulsen, G. M. (1968). *Crop Sci.* **8**, 537.

Hartingsveldt, J., and Stouthamer, A. H. (1972). *Antoine van Leeuwenhoek J. Microbiol. Serol.* **38**, 447.

Hatch, M. D., and Slack, C. R. (1969a). *Biochem. J.* **112**, 549.

Hatch, M. D., and Slack, C. R. (1969b). *Biochem. Biophys. Res. Commun.* **34**, 589.

Hattori, A., and Myers, J. (1967). *Plant Cell Physiol.* **8**, 327.

Hattori, A., and Uesugi, I. (1968). *Plant Cell Physiol.* **9**, 689.

Havill, D. C., Lee, J. A., and Stewart, G. R. (1974). *New Phytol.* **73**, 1221.

Heimer, Y. M. (1975). *Arch. Microbiol.* **103**, 181.

Heimer, Y. M., and Filner, P. (1970). *Biochim. Biophys. Acta* **215**, 152.

Heimer, Y. M., and Filner, P. (1971). *Biochim. Biophys. Acta* **230**, 362.

Heimer, Y. M., Wray, J. L., and Filner, P. (1969). *Plant Physiol.* **44**, 1197.

Herrera, J., Paneque, A., Maldonado, J. Ma., Barea, J. L., and Losada, M. (1972). *Biochem. Biophys. Res. Commun.* **48**, 996.

Hewitt, E. J. (1966). "Sand and Water Culture Methods Used in the Study of Plant Nutrition," 2nd rev. ed., Tech. Commun. No. 22. Commonwealth Bureau of Horticulture and Plantation Crops, Farnham Royal, England.

Hewitt E. J. (1970). *In* "Nitrogen Nutrition of the Plant" (E. A. Kirby, ed.), Proc. Agr. Chem. Symp., p. 78. University of Leeds.

Hewitt, E. J. (1974). *M.T.P. Int. Rev. Sci. Biochem. Ser. 1* **11**, 199.

Hewitt, E. J. (1975). *Annu. Rev. Plant Physiol.* **26**, 74.

Hewitt, E. J., and Afridi, M. M. R. K. (1959). *Nature (London)* **183**, 57.

Hewitt, E. J., and Betts, G. F. (1963). *Biochem. J.* **89**, 20.

Hewitt, E. J., and Gundry, C. S. (1970). *J. Hort. Sci.* **45**, 351.

Hewitt, E. J., and Jones, E. W. (1947). *J. Pomol.* **23**, 254

Hewitt, E. J., and Nicholas, D. J. D. (1964). *In* "Moderne methoden der Pflanzenanalyse" (H. F. Linskens, B. B. Sanwal, and M. V. Tracey, eds.), Vol. VII, p. 67. Springer-Verlag, Berlin and New York.

Hewitt, E. J., and Notton, B. A. (1967). *Phytochemistry* **6**, 1329.

Hewitt, E. J., Jones, E. W., and Williams, A. H. (1949). *Nature (London)* **163**, 681.

Hewitt, E. J., Fisher, E. G., and Candela, M. (1956). *Long Ashton Agr. Hort. Res. Sta. [Univ. Bristol], Annu. Rep.* p. 202.

Hewitt, E. J., Agarwala, S. C., and Williams, A. H. (1957). *J. Hort. Sci.* **32**, 34.

Hewitt, E. J., Notton, B. A., and Afridi, M. M. R. K. (1967). *Plant Cell Physiol.* Tokyo, **8**, 385.

Hewitt, E. J., Hucklesby, D. P., and Betts, E. J. (1968). *In* "Some Recent Aspects of Nitrogen Metabolism in Plants" (E. J. Hewitt and C. V. Cutting, eds.), p. 47. Academic Press, New York.

Hewitt, E. J., Jones, R. W., Abbott, A. J., and Best, G. (1974). *Proc. Int. Congr. Plant Tissue & Cell Cult., 3rd,* Abst. 131.

Higgins, E. S., Richert, D. A., and Westerfield, W. W. (1956). *Proc. Soc. Exp. Biol. Med.* **92**, 509.

Hiller, R. G., and Bassham, J. A. (1965). *Biochim. Biophys. Acta* **109**, 607.

Hirose, S., Katsuji, Y., and Shibata, K. (1971). *Plant Cell Physiol.* **12**, 775.

Hirschberg, R., Hubner, G., and Borriss, H. (1972). *Planta* **108**, 333.

Ho, C. H., and Tamura, G. (1973). *Agr. Biol. Chem.* **37**, 37.

Hsaio, T. C. (1970). *Plant Physiol.* **46**, 281.

Hucklesby, D. P., and Hageman, R. H. (1976). *Plant Physiol.* **57**, 693.

Hucklesby, D. P., and Elsner, J. E. (1969). *Proc. Int. Bot. Congr., 11th, 1969* p. 96.

Hucklesby, D. P., and Hewitt, E. J. (1970). *Biochem. J.* **119**, 615.

Hucklesby, D. P., Hewitt, E. J., and James, D. M. (1970). *Biochem. J.* **117**, 30P.

Hucklesby, D. P., Dalling, M. J., and Hageman, R. H. (1972). *Planta* **104**, 220.

Hucklesby, D. P., James, D. M., and Hewitt, E. J. (1974). *Biochem. Soc. Trans.* **121**, 213.

Huffaker, R. C., Obendorf, R. L., Keller, C. J., and Kleinkopf, G. E. (1966). *Plant Physiol.* **41**, 913.

Huzisige, H., and Satoh, K. (1961). *Bot. Mag.* **74**, 178.

Ida, S., Mori, E., and Morita, Y. (1974). *Planta* **121**, 213.

Iida, K., and Taniguchi, S. (1959). *J. Biochem. (Tokyo)* **46**, 1041.

Ingle, J., Joy, K. W., and Hageman, R. H. (1966). *Biochem. J.* **100**, 577.

Iwasaki, H., and Matsubara, T. (1972). *J. Biochem. (Tokyo)* **70**, 645.

Iwasaki, H., Shidara, S., Suzuki, H., and Mori, T. (1963). *J. Biochem. (Tokyo)* **53**, 299.

Jackson, W. A., Flesher, D., and Hageman, R. H. (1973). *Plant Physiol.* **51**, 120.

Jaworski, E. G. (1971). *Biochem. Biophys. Res. Commun.* **43**, 1274.

Jetschmann, K., Solomonson, L. P., and Vennesland, B. (1972). *Biochim. Biophys. Acta* **275**, 276.

Jones, R. W., and Sheard, R. W. (1972). *Nature (London)* **238**, 221.

Jones, R. W., and Sheard, R. W. (1975). *Plant Physiol.* **55**, 954.

Joy, K. W. (1969). *Plant Physiol.* **44**, 849.

Joy, K. W., and Hageman, R. H. (1966). *Biochem. J.* **100**, 263.

Kanazawa, T., Kirk, M. R., and Bassham, J. A. (1970). *Biochim. Biophys. Acta* **205**, 401.

Kannangara, C. G., and Woolhouse, H. W. (1967). *New Phytol.* **66**, 553.

Kaplan, D., Roth-Bejerano, N., and Lips, S. H. (1974). *Eur. J. Biochem.* **49**, 393.

Keeler, R. F., and Varner, J. E. (1958). *In* "Trace Elements" (C. A. Lamb, O. G. Bentley, and J. M. Beattie, eds.), p. 297. Academic Press, New York.

Kelker, H. C., and Filner, P. (1971). *Biochim. Biophys. Acta* **252**, 69.

Kemp, J. D., and Atkinson, D. E. (1966). *J. Bacteriol.* **92**, 628.

Kemp, J. D., Atkinson, D. E., Ehret, A., and Lazzarini, R. A. (1963). *J. Biol. Chem.* **238**, 3466.

Kende, H., and Shen, T. C. (1972). *Biochim. Biophys. Acta* **286**, 118.

Kende, H., Hahn, H., and Kays, S. E. (1971). *Plant Physiol.* **48**, 702.

Kessler, E. (1953). *Arch. Mikrobiol.* **19**, 438.

Kessler, E. (1955a). *Nature (London)* **176**, 1069.

Kessler, E. (1955b). *Planta* **45**, 94.

Kessler, E. (1957a). *Arch. Mikrobiol.* **27**, 166.

Kessler, E. (1957b). *Planta* **49**, 435.

Kessler, E. (1959). *Symp. Soc. Exp. Biol.* **13**, 87.

Ketchum, P. A., and Downey, R. J. (1975). *Biochim. Biophys. Acta* **385**, 354.

Ketchum, P. A., Cambier, H. Y., Frazier, W. A., Madansky, C. H., and Nason, A. (1970). *Proc. Nat. Acad. Sci. U.S.* **66**, 1016.

Kinsky, S. C., and McElroy, W. D. (1958). *Arch. Biochem. Biophys.* **73**, 466.

Klepper, L., and Hageman, R. H. (1969). *Plant Physiol.* **44**, 110.

Klepper, L., Flesher, D., and Hageman, R. H. (1971). *Plant Physiol.* **48**, 580.

Knypl, J. S. (1974). *Z. Pflanzenphysiol.* **71**, 37.

Köhler, D. (1966). *Planta* **69**, 27.

Kretovich, W. L. (1965). *Plant Physiol.* **16**, 141.

Kumada, H. (1953). *J. Biochem. (Tokyo)* **40**, 439.

Lafferty, M. A., and Garrett, R. H. (1974). *J. Biol. Chem.* **249**, 7555.

Lam, Y., and Nicholas, D. J. D. (1969a). *Biochim. Biophys. Acta* **178**, 225.

Lam, Y., and Nicholas, D. J. D. (1969b). *Biochim. Biophys. Acta* **180**, 459.

Lazzarini, R. A., and Atkinson, D. E. (1961). *J. Biol. Chem.* **236**, 3330.

Lee, K. Y., Pan, S. S., Erickson, R. H., and Nason, A. (1974). *J. Biol. Chem.* **249**, 3941.

Leeper, G. W. (1941). *J. Aust. Inst. Agr. Sci.* **7**, 161.

Lester, R. L., and DeMoss, J. A. (1971). *J. Bacteriol.* **105**, 1006.

Li, C. Y., Lu, K. C., Trappe, J. M., and Bollen, W. B. (1972). *Plant Soil* **37**, 409.

Lips, S. H. (1975). *Plant Physiol.* **55**, 598.

Lips, S. H., and Avissar, Y. (1972). *Eur. J. Biochem.* **29**, 20.

Lips, S. H., and Roth-Bejerano, N. (1969). *Science* **166**, 109.

Lips, S. H,, Kaplan, D., and Roth-Bejerano, N. (1973). *Eur. J. Biochem.* **37**, 589.

Ljungdahl, L. G., and Andreesen, J. R. (1975). *FEBS (Fed. Eur. Biochem. Soc.) Lett.* **54**, 279.

Losada, M., Paneque, A., Ramirez, J. M., and Del Campo, F. F. (1963). *Biochem. Biophys. Res. Commun.* **10**, 298.

Losada, M., Paneque, A., Ramirez, J. M., and Del Campo, F. F. (1965). *Biochim. Biophys. Acta* **109**, 86.

Losada, M., Paneque, A., Aparicio, P. J., Vega, J. Ma., Cardenas, J., and Herrera, J. (1970). *Biochem. Biophys. Res. Commun.* **38**, 1009.

Loussaert, D., and Hageman, R. H. (1974). *Plant Physiol. Suppl. June 1974*, p. 65, Abstr. 367.

Lyclama, J. C. (1963). *Acta Bot. Neerl.* **12**, 361.

MacDonald, D. W., and Coddington, A. (1974). *Eur. J. Biochem.* **46**, 169.

McKee, H. S. (1962). "Nitrogen Metabolism in Plants." Oxford Univ. Press (Clarendon), London and New York.

Magalhaes, A. C., Neyra, C. A., and Hageman, R. H. (1974). *Plant Physiol.* **53**, 411.

Mahler, H. R., and Cordes, E. H. (1966). "Biological Chemistry," p. 774. Harper, New York.

Maranville, J. W. (1970). *Plant Physiol.* **45**, 591.

Marver, H. S., Collins, A., Tschudy, D. P., and Rechcigl, M. (1966). *J. Biol. Chem.* **241**, 4323.

Mattas, R. E., and Pauli, A. W. (1965). *Crop Sci.* **5**, 181.

Mellor, G. E., and Tregunna, E. B. (1971). *Can. J. Bot.* **49**, 137.

Menary, R. C., and Jones, R. H. (1972). *Aust. J. Biol. Sci.* **25**, 531.

Meyer, V., and Schulze, E. (1894). *Ber. Deut. Chem. Ges.* **17**, 1554.

Miflin, B. J. (1967). *Nature (London)* **214**, 1133.

Miflin, B. J. (1968). *Biochem. J.* **108**, 49P.

Miflin, B. J. (1970a). *Planta* **93**, 160.

Miflin, B. J. (1970b). *Rev. Roum. Biochim.* **7**, 53.

Miflin, B. J. (1974a). *Planta* **116**, 187.

Miflin, B. J. (1974b). *Plant Physiol.* **54**, 550

Minotti, P. L., and Jackson, W. A. (1970). *Planta* **95**, 36.

Minotti, P. L., Williams, D. C., and Jackson, W. A. (1968). *Soil Sci. Soc. Amer., Proc.* **32**, 692.

Minotti, P. L., Williams, D. C., and Jackson, W. A. (1969). *Planta* **86**, 344.

Moreno, C. G., Aparicio, P. J., Palacian, E., and Losada, M. (1972). *FEBS (Fed Eur. Biochem. Soc.) Lett.* **26**, 11.

Morris, I., and Ahmed, J. (1969). *Physiol. Plant.* **22**, 1166.

Morris, I., and Syrett, P. J. (1963). *Arch. Microbiol.* **17**, 32.

Mulder, E. G. (1948). *Plant Soil* **1**, 94.

Muller, B. (1970). *Biochim. Biophys. Acta* **205**, 102.

Murphy, M. J., Siegel, L. M., Tove, S. R., and Kamin, H. (1974). *Proc. Nat. Acad. Sci. U.S.* **71**, 612.

Murphy, M. J., Siegel, L. M., and Kamin, H. (1973). *J. Biol. Chem.* **248**, 2801.

Murray, E. D., and Sanwal, B. D. (1963). *Can. J. Microbiol.* **9**, 781.

Myers, J. (1949). *In* "Photosynthesis in Plants" (J. Franck and W. E. Loomis, eds.), p. 349. Iowa State Univ. Press, Ames.

Nason, A., and Evans, H. J. (1953). *J. Biol. Chem.* **202**, 655.

Nason, A., Antoine, A. D., Ketchum, P. A., Frazier, W. A., and Lee, D. K. (1970) *Proc. Nat. Acad. Sci. U.S.* **65**, 137.

Nason, A., Lee, K. Y., Pan, S. S., Ketchum, P. A., Lamberti, A., and De Vries, J. (1971). *Proc. Nat. Acad. Sci. U.S.* **68**, 3242.

Nelson, N., and Ilan, I. (1969). *Plant Cell Physiol.* **10**, 143.

Newton, N. (1969). *Biochim. Biophys. Acta* **185**, 316.

Nicholas, D. J. D., and Nason, A. (1954a). *J. Biol. Chem.* **207**, 353.

Nicholas, D. J. D., and Nason, A. (1954b). *J. Biol. Chem.* **211**, 183.

Nicholas, D. J. D., and Nason, A. (1954c). *Arch. Biochem. Biophys.* **51**, 311.

Nicholas, D. J. D., and Nason, A. (1955a). *J. Bacteriol.* **69**, 580.

Nicholas, D. J. D., and Nason, A. (1955b). *Plant Physiol.* **30**, 135.

Nicholas D. J. D., and Scawen, J. H. (1956). *Nature (London)* **178**, 1474.

Nicholas, D. J. D., and Stevens, M. (1955). *Nature (London)* **176**, 1066.

Nicholas, D. J. D., and Wilson, P. J. (1964). *Biochim. Biophys. Acta* **86**, 466.

Nicholas, D. J. D., Nason, A., and McElroy, W. D. (1954). *J. Biol. Chem.* **207**, 341.

Nicholas, D. J. D., Medina, A., and Jones, O. T. G. (1960). *Biochim. Biophys. Acta* **37**, 468.

Nightingale, G. T. (1937). *Bot. Gaz. (Chicago)* **98**, 725.

Nightingale, G. T., Addams, R. M., Robbins, W. R., and Schermerhorn, L. G. (1931). *Plant Physiol.* **6**, 605.

Nitsos, R. E., and Evans, H. J. (1966). *Plant Physiol.* **41**, 1499.

Notton, B. A. (1972). Ph.D. Thesis, University of Bristol, Bristol, England.

Notton, B. A., and Hewitt, E. J. (1971a). *Plant Cell Physiol.* **12**, 465.

Notton, B. A., and Hewitt, E. J. (1971b). *FEBS (Fed. Eur. Biochem. Soc.) Lett.* **18**, 19.

Notton, B. A., and Hewitt, E. J. (1971c). *Biochem. Biophys. Res. Commun.* **44**, 702.

Notton, B. A., and Hewitt, E. J. (1974). *J. Less Common Metals* **36**, 437.

Notton, B. A., Hewitt, E. J., and Fielding, A. H. (1972). *Phytochemistry* **11**, 2447.

Notton, B. A., Graf, L., Hewitt, E. J., and Povey, R. C. (1974). *Biochim. Biophys. Acta* **364**, 45.

Orebamjo, T. O., and Stewart, G. R. (1975a). *Planta* **122**, 27.

Orebamjo, T. O., and Stewart, G. R. (1975b). *Planta* **122**, 37.

Oji, Y., and Izawa, G. (1969a). *Plant Cell Physiol.* **10**, 665.

Oji, Y., and Izawa, G. (1969b). *Plant Cell Physiol.* **10**, 743.

Palacian, E., de la Rosa, F., Castillo, F., and Gomez-Moreno, C. (1974). *Arch. Biochem. Biophys.* **161**, 441.

Paneque, A., and Losada, M. (1966). *Biochim. Biophys. Acta* **128**, 202.

Paneque, A., Del Campo, F. F., Ramirez, J. M., and Losada, M. (1965). *Biochim. Biophys. Acta* **109**, 79.

Paneque, A., Del Campo, F. F., and Losada, M. L. (1963). *Nature (London)* **198**, 190.

Pate, J. S. (1973). *Soil Biol. Biochem.* **5**, 109.

Pateman, J. A., and Cove, D. J. (1967). *Nature (London)* **215**, 1234.

Pateman, J. A., Cove, D. J., Rever, B. M., and Roberts, D. B. (1964). *Nature (London)* **201**, 58.

Pateman, J. A., Rever, B. M., and Cove, D. J. (1967). *Biochem. J.* **104**, 103.

Paulsen, G. M., and Harper, J. E. (1968). *Plant Physiol.* **43**, 775.

Pichinoty, F. (1966). *Bull. Soc. Fr. Physiol. Veg.* **12**, 97.

Pieve, Ya. V., and Uvanova, N. N. (1969). *Dokl. Akad. Nauk SSSR* **184**, 1224.

Pinsent, J. (1954). *Biochem. J.* **57**, 10.

Plaut, Z. (1974). *Physiol. Plant* **30**, 212.

Prakash, O., and Sadana, J. C. (1972). *Arch. Biochem. Biophys.* **148**, 614.

Radcliffe, B. C., and Nicholas, D. J. D. (1968). *Biochim. Biophys. Acta* **153**, 545.

Ramirez, J. M., Del Campo, F. F., Paneque, A., and Losada, M. (1964). *Biochem. Biophys. Res. Commun.* **15**, 297.

Ramirez, J. M., Del Campo, F. F., Paneque, A., and Losada, M. (1966). *Biochim. Biophys. Acta* **118**, 58.

Rechcigl, M. (1968). *Enzymologia* **34**, 23.

Relimpio, A. Ma., Aparicio, P. J., Paneque, A., and Losada, M. (1971). *FEBS (Fed. Eur. Biochem. Soc.) Lett.* **17**, 226.

Ries, S. K., Chimiel, H., Dilley, D. R., and Filner, P. (1967). *Proc. Nat. Acad. Sci. U.S.* **58**, 527.

Rigano, C. (1971). *Arch. Mikrobiol.* **76**, 265.

Rigano, C., and Violante, U. (1972a). *Biochem. Biophys. Res. Commun.* **47**, 372.

Rigano, C., and Violante, U. (1972b). *Biochim. Biophys. Acta* **256**, 524.

Rijven, A. H. G. C. (1958). *Aust. J. Biol. Sci.* **11**, 142.

Rijven, A. H. G. C., and Prakash, V. (1971). *Plant Physiol.* **47**, 59.

Ritenour, G. L. (1964). Ph.D. Thesis. University of California, Davis, California.

Ritenour, G. L., Joy, K. W., Bünning, J., and Hageman, R. H. (1967). *Plant Physiol.* **42**, 233.

Roustan, J.-L., Neuberger, M., and Fourcy, A. (1974). *Physiol. Végétàle* **12**, 527.

Rosso, J. P., Forget, P., and Pichinoty, F. (1973). *Biochim. Biophys. Acta* **321**, 443.

Roth-Bejerano, N., and Lips, S. H. (1970). *New Phytol.* **69**, 165.

Rucklidge, G. J., Notton, B. A., and Hewitt, E. J. (1976). *Biochem. Soc. Trans.* **4**, 77.

Ruiz-Herrera, J., and DeMoss, J. A. (1969). *J. Bacteriol.* **99**, 720.

Russell, E. W. (1950). "Soil Conditions and Plant Growth," 8th ed. Longmans Green, New York.

Sadana, J. C., and McElroy, W. D. (1957). *Arch. Biochem. Biophys.* **67**, 16.

Sahulka, J. (1972). *Biol. Plant.* **14**, 308.

Sanderson, G. W., and Cocking, E. C. (1964). *Plant Physiol.* **39**, 416.

Sawhney, S. K., Naik, M. S. (1972). *Biochem. J.* **130**, 475.

Sawhney, S. K., Prakash, V., and Naik, M. S. (1972). *FEBS Lett.* **22**, 200.

Schimke, R. T. (1966). *Bull. Soc. Chim. Biol.* **48**, 1009.

Schimper, A. F. W. (1888). *Bot. Ztg.* **46**, 65.

Schrader, L. E., and Hageman, R. H. (1967). *Plant Physiol.* **42**, Suppl., 38.

Schrader, L. E., Beevers, L., and Hageman, R. H. (1967). *Biochem. Biophys. Res. Commun.* **26**, 14.

Schrader, L. E., Ritenour, G. L., Eilrich, G. L., and Hageman, R. H. (1968). *Plant Physiol.* **43**, 930.

Segal, H. L., Matsuzawa, T., Haider, M., and Abraham, G. J. (1969). *Biochem. Biophys. Res. Commun.* **36**, 764.

Shen, T. C. (1972a). *Plant Physiol.* **49**, 546.

Shen, T. C. (1972b). *Planta* **108**, 21.

Shin, M., and Oda, Y. (1966). *Plant Cell Physiol.* **7**, 643.

Shin, M., Tagawa, K., and Arnon, D. I. (1963). *Biochem. Z.* **338**, 84.

Showe, M. K., and DeMoss, J. A. (1968). *J. Bacteriol.* **95**, 1305.

Shum, A. C., and Murphy, J. C. (1972). *J. Bacteriol.* **110**, 447.

Siegel, L. M., Leinweber, F. J., and Monty, K. J. (1965). *J. Biol. Chem.* **240**, 2705.

Siegel, L. M., Murphy, M. J., and Kamin, H. (1973). *J. Biol. Chem.* **248**, 251.

Sims A. P., Folkes, B. F., and Bussey, A. H. (1968). *In* "Some Recent Aspects of Nitrogen Metabolism in Plants" (E. J. Hewitt and C. V. Cutting, eds.), p. 91. Academic Press, New York.

Slack, C. R., Hatch, M. D., and Goodchild, D. J. (1969). *Biochem. J.* **114**, 489.

Smith, F. W., and Thompson, J. F. (1971). *Plant Physiol.* **48**, 219.

Solomonson, L. P. (1974). *Biochim. Biophys. Acta* **334**, 297.

Solomonson, L. P., and Vennesland, B. (1972). *Biochim. Biophys. Acta* **267**, 544.

Solomonson, L. P., Lorimer, G. H., Hall, R. L., Borchers, R., and Bailey, J. L. (1975). *J. Biol. Chem.* **250**, 4120.

Sorger, G. J. (1963). *Biochem. Biophys. Res. Commun.* **12**, 395.

Sorger, G. J. (1964). *Nature (London)* **204**, 575.

Sorger, G. J. (1966). *Biochim. Biophys. Acta* **118**, 484.

Sorger, G. J., and Giles, N. H. (1965). *Genetics* **52**, 777.

Spencer, D. (1959). *Aust. J. Biol. Sci.* **12**, 181.

Spencer, D., Takahashi, H., and Nason, A. (1957). *J. Bacteriol.* **73**, 553.

Steinberg, R. A. (1937). *J. Agr. Res.* **55**, 891.

Steinberg, R. A. (1939). *Bot. Rev.* **5**, 327.

Steward, F. C., and Margolis, D. (1961–1962). *Contrib. Boyce Thompson Inst.* **21**, 393.

Stewart, G. R. (1968). *Phytochemistry* **7**, 1139.

Stewart, G. R. (1972). *J. Exp. Bot.* **23**, 171.

Stoy, V. (1955). *Physiol. Plant.* **8**, 963.

Subba Rao, M. S., and Lal, K. N. (1955). *Sci. Cult.* **21**, 319.

Subramanian, K. V., and Sorger, G. J. (1972). *Biochim. Biophys. Acta* **256**, 533.

Sussman, M., and Sussman, R. R. (1965). *Biochim. Biophys. Acta* **108**, 463.

Swader, J. A., and Stocking, C. R. (1971). *Plant Physiol.* **47**, 189.

Swader, J. A., Stocking, C. R., and Chin, H. L. (1975). *Physiol. Plant* **34**, 335.

Syrett, P. J. (1962). *In* "Physiology and Biochemistry of Algae" (R. A. Lewin, ed.), p. 171. Academic Press, New York.

Takahashi, H., and Nason, A. (1957). *Biochim. Biophys. Acta* **23**, 433.

Tamura, G. (1965). *J. Biochem. (Tokyo)* **57**, 207.

Tamura, G., Asada, K., and Bandurski, R. S. (1967). *Plant Physiol.* **42**, S36.

Tang, P. S., and Wu, H. Y. (1957). *Nature (London)* **179**, 1355.

Taniguchi, S. (1961). *Z. Allg. Mikrobiol.* **1**, 341.

Tavares, J., and Kende, H. (1970). *Phytochemistry* **9**, 1763.

Thacker, A., and Syrett, P. J. (1972a). *New Phytol.* **71**, 423.

Thacker, A., and Syrett, P. J. (1972b). *New Phytol.* **71**, 435

Travis, R. L., and Key, J. L. (1971). *Plant Physiol.* **48**, 617.

Travis, R. L., Jordan, W. R., and Huffaker, R. C. (1969). *Plant Physiol.* **44**, 1150.

Travis, R. L., Huffaker, R. C., and Key, J. L. (1970a). *Plant Physiol.* **46**, 800.

Travis, R. L., Jordan, W. R., and Huffaker, R. C. (1970b). *Physiol. Plant.* **23**, 678.

Upcroft, J. A., and Done, J. (1972). *FEBS (Fed. Eur. Biochem. Soc.) Lett.* **21**, 142.

Vaidyanathan, C. S., and Street, H. E. (1959). *Nature (London)* **184**, 531.

Van Niel, C. B., Allan, M. B., and Wright, B. E. (1953). *Biochim. Biophys. Acta* **12**, 67.

van't Riet, J., and Planta, R. J. (1969). *FEBS (Fed. Eur. Biochem. Soc.) Lett* **5**, 249.

van't Riet, J., Stouthamer, A. H., and Planta, R. J. (1968). *J. Bacteriol.* **96**, 1455.

Vega, J. Ma., Herrera, J., Aparicio, P. J., Paneque, A., and Losada, M. (1971). *Plant Physiol.* **48**, 294.

Vega, J. Ma., Herrera, J., Relimpio, A. Ma., and Aparicio, P. J. (1972). *Physiol. Veg.* **10**, 637.

Venables, W. A., Wimpenny, J. W. T., and Cole, J. A. (1968). *Arch. Mikrobiol.* **63**, 117.

Vennesland, B., and Jetschmann, K. (1971). *Biochim. Biophys. Acta* **329**, 554.

Vennesland, B., and Solomonson, L. P. (1972). *Plant Physiol.* **49**, 1029.

Vielmeyer, H. P., Fischer, F., and Bergmann, W. (1969). *Albrecht-Thaer-Arch.* **13**, 291.

Vunkova, R. (1971). *C.R. Acad. Bulg. Sci.* **24**, 263.

Walker, G. C., and Nicholas, D. J. D. (1961). *Biochim. Biophys. Acta* **49**, 350.

Wallace, W. (1973). *Plant Physiol.* **52**, 197.

Wallace, W. (1974). *Biochim. Biophys. Acta* **341**, 265.

Wallace, W. (1975). *Plant Physiol.* **55**, 774.

Wallace, W., and Pate, J. S. (1965). *Ann. Bot. (London)* [N.S.] **29**, 655.

Wallace, W., and Pate, J. S. (1967). *Ann. Bot. (London)* [N.S.] **31**, 213.

Walles, B. (1967). *In* "Biochemistry of Chloroplasts" (T. W. Goodwin, ed.), Vol. 2, p. 615. Academic Press, New York.

Warner, R. L., Hageman, R. H., Dudley, J. W., and Lambert, R. J. (1969). *Proc. Nat. Acad. Sci. U.S.* **62**, 785.

Weissman, G. S. (1951). *Amer. J. Bot.* **38**, 162.

Weissman, G. S. (1972). *Plant Physiol.* **49**, 138.

Wells, G. N., and Hageman, R. H. (1974). *Plant Physiol.* **54**, 136.

White, P. R. (1933). *Plant Physiol.* **8**, 489.

Wray, J. L., and Filner, P. (1970). *Biochem. J.* **119**, 715.

Yamanaka, T., and Okunuki, K. (1963). *Biochim. Biophys. Acta* **67**, 379.

Yarmolinsky, M. B., and De La Habba, G. L. (1959). *Proc. Nat. Acad. Sci. U.S.* **45**, 1721.

Yoshimoto, A., and Sato, R. (1968). *Biochim. Biophys. Acta* **153**, 555.

Yoshimoto, A., and Sato, R. (1970). *Biochim. Biophys. Acta* **222**, 190.

Young, M. (1967). B. Sc. Thesis. University of Bristol.

Younis, M. A., Pauli, A. W., Mitchell, H. L., and Stickler, F. C. (1965). *Crop Sci.* **5**, 321.

Zielke, H. R., and Filner, P. (1971). *J. Biol. Chem.* **246**, 1772.

Zucker, M. (1965). *Plant Physiol.* **40**, 779.

Zucker, M. (1968). *Plant Physiol.* **43**, 365.

Zumft, W. G. (1972). *Biochim. Biophys. Acta* **276**, 363.

Zumft, W. G., Paneque, A., Aparicio, P. J., and Losada, M. (1969). *Biochem. Biophys. Res. Commun.* **36**, 980.

Zumft, W. G., Aparicio, P. J., Paneque, A., and Losada, M. (1970). *FEBS (Fed. Eur. Biochem. Soc.) Lett.* **9**, 157.

Zumft, W. G., Spiller, H., and Yeboah-Smith, I. (1972). *Planta* **102**, 228.

21

Phytochrome

PETER H. QUAIL

I. Introduction

Phytochrome is a reversible biological switch, activated by light. A vast number of light-controlled plant responses are now believed to be phytochrome mediated. The actual mechanism by which the molecule transmits its sensory message to the cell, however, remains obscure.

A blue-green chromoprotein, phytochrome is present in small quantities in almost all potentially autotrophic plants. Two principal forms of the molecule—P_r (for red absorbing, $\lambda_{max} = 660$ nm) and P_{fr} (for far-red absorbing, $\lambda_{max} = 730$ nm)—are readily distinguishable on the basis of their absorption spectra (Fig. 1) and biological activity. P_r is considered biologically inactive, whereas P_{fr} is biologically active, i.e., capable of inducing a measurable biological response.

The two forms are reversibly interconvertible by red and far-red light ("phototransformation"). In addition, whereas the P_r form is stable in the dark, P_{fr} can revert thermally to P_r ("dark reversion") or undergo

683

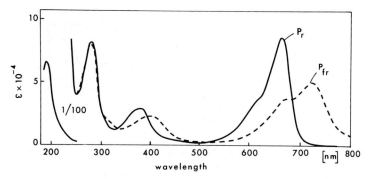

Fig. 1. Absorption spectra of purified oat phytochrome following saturating irradiations with red and far-red light. After Anderson *et al.* (1970).

an irreversible loss of photoactivity ("destruction"). These properties of the phytochrome system are traditionally summarized as in Scheme 1.

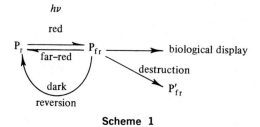

Scheme 1

Red light absorbed by P_r transforms the molecule to the P_{fr} form which in turn induces a biological response. Conversely, far-red light, dark reversion, and destruction provide alternative pathways for the removal of P_{fr} and thereby the potential for reversing induced responses. These properties enable the pigment to function as a reversible switch sensing the difference between light and dark.

In molecular terms, phytochrome might be considered as an effector capable of interacting with a cellular response system to produce a biological display. Phytochrome research can likewise be considered under two broad headings.

1. *Phytochrome, the molecule*—those studies concerned with the physical and chemical properties and behavior of the phytochrome molecule proper.

2. *Biological display*—investigations of the diverse and multiple phytochrome-mediated changes in cellular biochemistry and physiology. These changes are demonstrable at almost any level from the molecular to the gross morphogenetic.

II. A Chronology

The sequence of molecular events currently thought to occur from photon capture to biological display is summarized in Fig. 2. These events span the disciplines of radiation physics, photochemistry, biochemistry, and physiology.

Beginning with the capture of a quantum of light energy by P_r in about 10^{-17} seconds, the internal rearrangements of the molecule may be traced through several orders of magnitude in time, culminating in the formation of the metastable P_{fr} in about 4 seconds. This is the signal that initiates the chain of molecular events leading sooner or later to a measurable biological response. A set of displays have been selected with lag times from irradiation to response detection ranging from 15 seconds to several days (clear and hatched segments in Fig. 2). The duration of each response once initiated is indicated in the figure by the black segment, and ranges from about 10 minutes to 1 month. In addition to inducing a biological display, the P_{fr} molecule may itself undergo dark reversion ($t_{1/2} = 8$ minutes here, complete in 30 minutes), destruction ($t_{1/2} = 45$ minutes here, complete in 5 hours), or reconversion to P_r via the reverse phototransformation pathway (about 10^{-2} seconds from photon capture to completion). A scheme depicting the various intermediates encountered on the forward and reverse photoconversion pathways is included in Fig. 2. Some of these events will be examined in more detail in subsequent sections.

III. Phytochrome, the Molecule

A. Static Properties

The isolated phytochrome molecule consists of chromophore and protein with less than 4% carbohydrate having been reported (Roux and Lisansky, 1974). Data suggesting there is a single chromophore per monomer have been advanced (Tobin and Briggs, 1973). Phytochrome is stable and soluble between pH 6.2 and 8.0, but precipitates and is denatured above and below this range. The P_r form is, in general, considered to be more stable than the P_{fr} form.

The molecular weight of phytochrome has been controversial (Briggs and Rice, 1972), with estimates ranging from 18,000 to 359,000 appearing in the literature. Much of the published data on the properties of purified

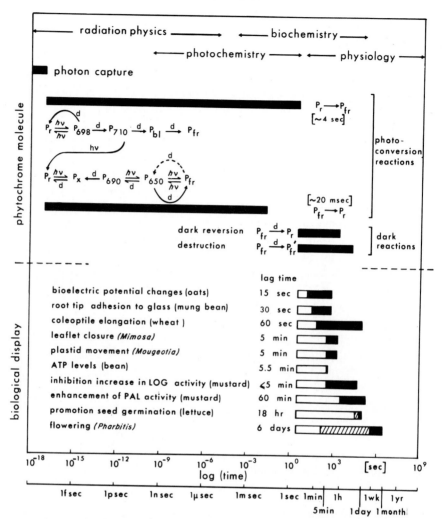

Fig. 2. Schematic representation of the kinetics of the light and dark reactions of the phytochrome molecule and of some phytochrome-mediated plant responses from light absorption to biological display. The duration of each process is indicated by the length of the horizontal bar. For the biological displays the onset of a measurable response is indicated by the beginning of the black segment. The clear areas represent the recorded lag times from presumed P_{fr} formation to response detection or escape from far-red reversibility. Hatched areas represent a period during which the response is not yet detectable but its ultimate appearance is no longer reversible by far-red. Lag time is the time from irradiation to response expression. A scheme of the intermediates encountered on the $P_r \rightarrow P_{fr}$ and $P_{fr} \rightarrow P_r$ photoconversion pathways is included (after Kendrick and Spruit, 1973). LOG, lipoxygenase; PAL, phenylalanine ammonia lyase.

phytochrome comes from studies on a 60,000 molecular weight species. Recent evidence suggests, however, that the native monomer is rather a polypeptide of 120,000 MW and that the 60,000 molecule is a discrete proteolytic product generated during the isolation procedure (Gardner *et al.*, 1971). The higher molecular weights observed are thought to be aggregates.

Published amino acid analyses of phytochrome indicate a high content of polar amino acids consistent with its water-solubility and suggestive of active surface properties. Current analyses show a high level of nonpolar residues as well, however, suggesting that regions of the molecule might also be capable of hydrophobic interactions (Roux and Yguerabide, 1973).

Available evidence indicates that the chromophore group is a linear tetrapyrrole. The proposed chromophore structures, their postulated linkages to the protein and a possible mechanism for their reversible photoisomerisation are presented in Fig. 3 (Rüdiger, 1972). An 11-amino acid peptide with chromophore attached has been isolated (Fry and Mumford, 1971). The sequence Leu-Arg-Ala-Pro-His-(Ser, Cys)-His-Leu-Glu-Tyr was reported (serine, cysteine order uncertain), and a possible thioether linkage of the chromophore to cysteine was suggested.

Differences between the P_r and P_{fr} species, apart from their visible absorption spectra (Fig. 1), have been reported from circular dichroism (CD) and optical rotatory dispersion (ORD) analyses (Kroes, 1970;

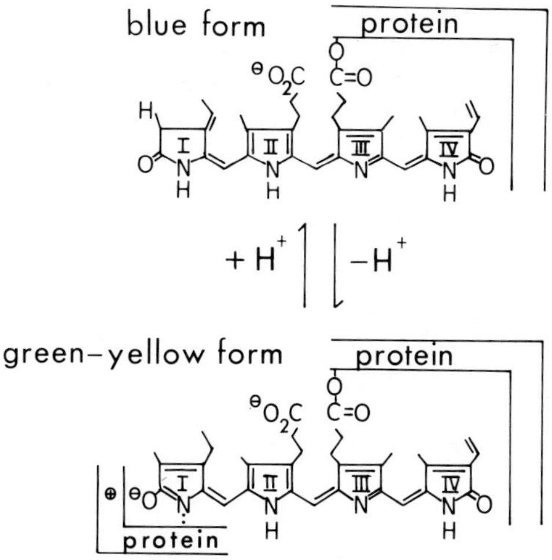

Fig. 3. Proposed structure for the phytochrome chromophore, its linkage to the protein, and possible phototransformation mechanism. The "blue form" is thought to correspond to P_r and the "green-yellow" form to P_{fr}. After Rüdiger (1972).

Hopkins and Butler, 1970). The data for the visible region have been interpreted as indicating a difference in the mode of chromophore–protein attachment. Several lines of evidence suggest that only relatively small changes in protein conformation occur upon phototransformation. Small differences in the CD and absorption spectra of P_r and P_{fr} in the UV region have been recorded (Hopkins and Butler, 1970), although the CD data have been disputed (Tobin and Briggs, 1973); P_{fr} fixed twice as much complement as P_r in a microcomplement fixation test (Hopkins and Butler, 1970); glutaraldehyde reacted with 13 lysine residues of P_r compared to 11 of P_{fr} out of a total of 27 (Roux, 1972); and an $s_{20,w}$ value of 5.1 S for P_r versus 5.0 S for P_{fr} has been claimed (Briggs and Rice, 1972). In addition, P_{fr} is more susceptible to denaturation by a variety of reagents, such as urea, sulfhydryl reagents, $(NH_4)_2SO_4$ precipitation, and proteolytic enzymes, than is P_r. In contrast, no differences between the two species were found as regards fluorescence spectra (Tobin and Briggs, 1973), electrophoretic mobility, binding to and elution from brushite, immunological properties on Ouchterlony plates, or gel filtration behavior (Briggs and Rice, 1972).

B. Photoconversion Reactions

Phytochrome is classified as a photochromic substance (Lhoste, 1972). On the basis of knowledge gained from other organic photochromic molecules, by analogy with the photosynthetic (Kamen, 1963) and visual (Kropf, 1972) pigments, and from flash photolysis and low temperature studies of the phytochrome molecule itself (Linschitz et al., 1966; Linschitz and Kasche, 1967), the time courses of the reversible phototransformation reactions of phytochrome presented in Fig. 2 can be rationalized.

Photon capture, the primary, physical act of the absorption of a quantum of light energy, is complete by 10^{-17} to 10^{-15} seconds. This will bring the molecule into its lowest excited singlet or multiplet state in 10^{-13} to 10^{-12} seconds. The lifetimes of the electronically excited states are generally sufficient to allow nuclear configurational rearrangements and the stabilization of the electronic excitation energy in a new chemical species before it is lost through reemission as fluorescence or phosphorescence (beginning 10^{-9} seconds). For phytochrome the available evidence indicates that the primary photoreaction results from direct electronic excitation of the chromophore leading to isomerization, although excitation energy transfer from aromatic residues of the protein is also known to occur with good efficiency (Pratt and Butler, 1970). The actual nature of the photochemical changes occurring during the lifetime of the elec-

tronically excited state is uncertain. Tautomerization of the pyrrole structure (Fig. 3) together with the possibility of cross-exchange of protons
between chromophore and protein is a currently favored hypothesis
(Lhoste, 1972). It is known for example that a large fragment of the
protein moiety is necessary to stabilize the isomerized chromophore.
Simultaneous cis–trans isomerization cannot be precluded, however. These
processes are considered to lie within the realm of radiation physics.

The photochemistry phase, which appears to lie between 10^{-9} and
5 seconds for phytochrome, completes the conversion of quantum energy
to free energy. Kinetic analysis following flash excitation (Linschitz et al.,
1966) indicates that the forward reaction ($P_r \rightarrow P_{fr}$) requires several
seconds to complete, whereas the reverse transformation ($P_{fr} \rightarrow P_r$) is
apparently complete by about 20 to 30 mseconds. The relatively long
times required for these reactions is indicative of their complexity. Indeed, several intermediates on separate pathways for the forward and
reverse reactions have been identified and characterized spectroscopically.
Various schemes have been advanced to explain the observed data. One
of the more recent and most detailed is presented in Fig. 2 (Kendrick
and Spruit, 1973). Although differing in detail, the different schemes
agree in broad terms that the primary photochemical intermediate (in
this case P_{698} on the $P_r \rightarrow P_{fr}$ pathway and P_{650} on the $P_{fr} \rightarrow P_r$ pathway)
results from isomerization of the chromophore only, without a change in
protein structure, and that the subsequent dark relaxations on both pathways involve conformational changes in the protein as well as rearrangements of the chromophore.

The kinetics of phototransformation have been shown to be first
order both in vivo (Schmidt et al., 1973) and in vitro (Butler, 1961).
The rate and extent of photoconversion is a function of the wavelength
(λ), intensity (I_λ) and duration (t) of irradiation, the extinction coeffi-
cients ($E_{r\lambda}$, $E_{fr\lambda}$) for P_r and P_{fr} at λ, and the quantum yields (ϕ_r, ϕ_{fr})
for P_r and P_{fr}, according to the expression (Butler, 1972)

$$\frac{d[P_{fr}]}{dt} = 2.3(I_\lambda E_{r\lambda}\phi_r[P_r] - I_\lambda E_{fr\lambda}\phi_{fr}[P_{fr}]) \tag{1}$$

From this it can be seen that at $t = \infty$ under continuous irradiation,
a photoequilibrium will be established with the ratio of P_{fr} to P_r remaining constant but with the pigment oscillating ("cycling") between the two
forms. The rate of cycling will be a function of the total absorption of the
two species. For the simple case of a phytochrome population of constant
size (such as for purified preparations or short term in vivo irradiations),

the ratio of P_{fr} to P_r will be irradiance independent but wavelength dependent

$$\frac{[P_{fr}]_\infty}{[P_r]_\infty} = \frac{E_{r\lambda}\phi_r}{E_{fr\lambda}\phi_{fr}} \tag{2}$$

The ratio becomes irradiance dependent, however, for continuous, long-term irradiations of intact tissue where the disparate metabolic turnover rates of P_r and P_{fr} (see Section III,C,2) have to be taken into account (Schäfer, 1975).

The wavelength dependence of both the rates of photoconversion (Fig. 4) and the short-term, photosteady state ratio of P_{fr} to P_r (Fig. 5) have been demonstrated experimentally (Butler *et al.*, 1964; Hanke *et al.*, 1969). The data are consistent with the measured absorption spectra.

It is important to note that total photoconversion of P_r to P_{fr} is not possible (Fig. 5). The maximum achievable conversion is 80% in the red region of the spectrum. This arises because there is no region of the absorption spectrum where P_r absorbs and P_{fr} does not (Fig. 1). On the other hand, for wavelengths longer than 725 nm, greater than 98% of the phytochrome is converted to the P_r form because the P_r absorbance is negligible compared to P_{fr} absorbance. Furthermore, it can be seen (Fig. 5) that the photosteady state ratio of P_{fr}/P_r can be conveniently

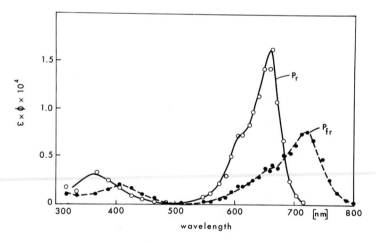

Fig. 4. Action spectra of photochemical transformations of P_r and P_{fr} in solution. The extinction coefficient ϵ is in liter mole^{-1} cm^{-1}, and the quantum yield ϕ is in mole einstein^{-1}. After Butler *et al.* (1964).

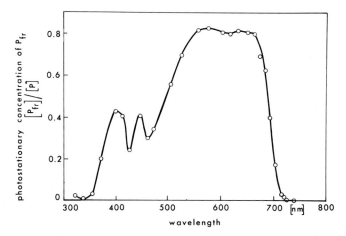

Fig. 5. Proportion of phytochrome in the P_{fr} form at photoequilibrium *in vivo* (*Sinapis* hooks) as a function of wavelength. After K. M. Hartmann and C. J. P. Spruit (in Hanke *et al.*, 1969).

manipulated by selecting monochromatic wavelengths between 660 and 730 nm. This procedure has been used to advantage in several physiological experiments.

For many years the only quantitative assay for phytochrome has been spectrophotometric. The photoreversible changes in absorbance (Fig. 1) are measured using the principle of differential spectrophotometry (Spruit, 1972). This procedure has the advantage that it can be used for the measurement of phytochrome *in vivo* as well as *in vitro*, but the disadvantages that it is unsuitable for use with green tissue and provides no index of the integrity of the molecule. For example, a photoactive core with a molecular weight as low as 15,000 has been reported (Walker and Bailey, 1970). Recently an immunocytochemical assay for phytochrome has been developed (Coleman and Pratt, 1974). This has the potential to overcome the above disadvantages.

C. Dark Reactions

The processes of "dark reversion" and "destruction" are the so-called dark reactions of phytochrome. Although poorly understood in molecular terms, the importance of these reactions appears to be the provision of a mechanism for the removal of the physiologically active form of phytochrome independently of light. Together with synthesis, the reversion and destruction processes constitute the metabolism of the pigment. Phytochrome is synthesized *de novo* in the P_r form (Quail *et al.*, 1973b), but it is the P_{fr} form that undergoes dark reversion and destruction (Frankland, 1972).

1. Dark Reversion

Whereas P_r is the thermodynamically stable form of phytochrome, P_{fr} is a metastable state (Lhoste, 1972). Thus, whereas P_r can only be transformed to P_{fr} by the absorption of light energy, P_{fr} may revert thermally to P_r in the dark.

Dark reversion has been observed both *in vivo* and *in vitro* (Frankland, 1972; Briggs and Rice, 1972). *In vivo* it occurs in most dicotyledons but not in monocotyledons, whereas *in vitro* phytochrome from both sources undergoes dark reversion. The reversion process appears to be first order and rapid at room temperature *in vivo*. For example a $t_{1/2} = 5$ to 10 minutes has been recorded for *Sinapis* and *Cucurbita* seedlings (Fig. 2) (Schäfer *et al.*, 1973; Schäfer and Schmidt, 1974). A second much slower phase has however also been observed in some cases both *in vivo* and *in vitro* (Briggs and Rice, 1972). The significance of these last observations is not understood.

Dark reversion *in vivo* in such plants as *Sinapis* and *Cucurbita* is virtually complete within 30 minutes at 25°C although only 15 to 20% of the spectrophotometrically detectable P_{fr} molecules are involved. The remainder continue to undergo "destruction" for a considerable period after reversion has ceased. Separate "reversion" and "destruction" pools of P_{fr} have been hypothesized to account for this apparent discrepancy (Schäfer and Schmidt, 1974).

Dark reversion is temperature dependent. On the basis of discontinuities in Arrhenius plots of the extent of reversion in *Cucurbita in vivo*, it has been suggested that the process is membrane associated (Schäfer and Schmidt, 1974). Dark reversion is insensitive to oxygen level and chelating agents, but is hastened *in vitro* by decreasing pH below 6.5 (consistent with the proton migration hypothesis of chromophore isomerization) and by reducing agents, such as NADH, dithionite, and reduced ferredoxin (800-fold increase in rate at $10^{-6} M$) (Briggs and Rice, 1972).

2. Destruction

The "destruction" (Frankland, 1972; Briggs and Rice, 1972) of phytochrome refers to the disappearance of spectrophotometrically detectable P_{fr} without the concomitant appearance of equimolar quantities of P_r. This process would appear to involve true degradation of the protein moiety of the pigment (Quail *et al.*, 1973b).

Destruction *in vivo* occurs in both monocotyledons and dicotyledons. In the former the reaction is zero order, being saturated at low levels of P_{fr}; in the latter it is first order. A short lag before the onset of destruction has been observed in some cases. The duration of the destruction process depicted in Fig. 2 is based on a $t_{1/2}$ of about 45 minutes measured for *Sinapis* (Marmé, 1969; Schäfer *et al.*, 1973). However, values of from 20 minutes for *Amaranthus* to 4 hours for *Daucus* have been recorded (Frankland, 1972).

Destruction of P_{fr} occurs both in the dark following brief irradiations and in continuous light. In the dark, the destruction process, unlike reversion, proceeds to completion, ultimately removing all unreverted P_{fr} molecules formed during irradiation. In the short term this leads to a rapid depletion of total phytochrome levels. The degree of depletion depends on the proportion of molecules converted to P_{fr} by the irradiation. Since this proportion never exceeds 80% P_{fr} even with saturating doses of red light, a residual P_r population is always retained irrespective of reversion or new synthesis. *De novo* synthesis of new P_r molecules does occur, however, replenishing the reduced pigment levels during the subsequent dark period (Quail *et al.*, 1973b). The evidence suggests that the synthesis of P_r (zero order) is a continuous process, itself unaffected by irradiation, and that the total pigment level is regulated against this background by the disparate degradation rate constants for P_r and P_{fr}.

This interpretation is also consistent with observations on phytochrome levels in continuous light. Phytochrome accumulates to high levels in dark-grown seedlings, ultimately reaching a plateau. Irradiation for extended periods induces a rate of decline in the total pigment level proportional to the steady state P_{fr} concentration (Frankland, 1972) and results ultimately in the establishment of a new plateau level (Fig. 6). The new plateau is the same for any given wavelength regardless of the starting level. The plateau in continuous light is considered to represent an equilibrium between synthesis and degradation (Schäfer *et al.*, 1972), although turnover under these conditions has not been directly demonstrated. The recovery in pigment level at the preirradiation rate, upon return to the dark after prolonged irradiation, supports this concept (Fig. 6).

Although the plateau level of P_{tot} with prolonged irradiations is a function of both wavelength and irradiance, the absolute level of P_{fr} is independent of both parameters (Schäfer and Mohr, 1974; Schäfer, 1975). This has extremely important implications for the manner in which phytochrome mediates the effects of long-term irradiations (Section IV,A,2).

Destruction is the predominant dark pathway for the removal of P_{fr}

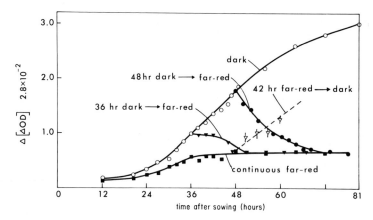

Fig. 6. Total phytochrome levels in *Sinapis* cotyledons as a function of time in the dark (◯), continuous far-red (■), 42 hours of far-red → dark (▽), 36 hours of dark → far-red (▼); 48 hours of dark → far-red (●). After E. Schäfer *et al.* (1972). *Photochem. & Photobiol.* **15**, 457. Reprinted with permission of Pergamon Press.

at least in seedlings at room temperature. On the other hand, the proportion of P_{fr} molecules undergoing dark reversion increases with increasing temperature at the expense of destruction (Schäfer and Schmidt, 1974). Furthermore, no discontinuities in Arrhenius plots of the destruction rates have been detected, suggesting that this process is not membrane associated. This emphasizes the separate nature of the reversion and destruction processes.

Destruction *in vivo* is inhibited by EDTA, sulfhydryl compounds, azide, cyanide, carbon monoxide, and the absence of oxygen but not by dinitrophenol at levels that uncouple respiration. It has been suggested on this basis (Hillman, 1967) that the process is oxidative and metal-dependent but not directly linked to respiration. Q_{10}'s of 2.7, 3.5, and 4.3 (Pratt and Briggs, 1966) for destruction *in vivo* have been considered suggestive of a noncatalyzed denaturation process. Immunocytochemical studies, however, have been taken as evidence of proteolysis of the protein moiety (Pratt *et al.*, 1974).

D. Localization

The distribution of phytochrome at the tissue and cellular level has been determined spectrophotometrically (Briggs and Siegelman, 1965) and immunocytochemically (Pratt and Coleman, 1971). In general, the highest pigment concentrations appear to be in meristematic or recently meristematic tissue, both in roots and shoots. Furthermore, phytochrome is most abundant in parenchyma cells immediately below the tip of dark-grown oat coleoptiles, while being absent from the tip cells themselves (Pratt and Coleman, 1971).

Approaches to the problem of intracellular localization fall into two broad categories: (a) location by inference from the measurement of phytochrome-induced responses having a spatial or vectorial component and (b) direct measurements of the pigment itself in attempts to determine the spatial or topographical location of the molecule.

The elegant use of polarized red and far-red microbeams has provided evidence that chloroplast movement in *Mougeotia* is controlled by phytochrome located and oriented in the outer cytoplasm, perhaps in the plasmalemma, of the cell (Haupt, 1972b). Similar conclusions have been reached on the basis of the directional growth of filamentous germ tubes of *Dryopteris* in polarized light (Etzold, 1965). The change in ion flux associated with phytochrome-mediated leaflet movement (Satter and Galston, 1973), root tip adhesion to glass surfaces (Tanada, 1968), and changes in bioelectric potentials (Newman and Briggs, 1972) are also indicative of phytochrome–cell surface interactions, but not necessarily that the pigment is a permanent membrane component. Phytochrome-controlled development of isolated etioplasts *in vitro* has been reported (Wellburn and Wellburn, 1973), and rapid, enzyme-mediated reduction of NADP to NADPH in response to red irradiation of a mitochondrial fraction *in vitro* has been claimed (Manabe and Furuya, 1974).

Attempts at direct measurements of subcellular phytochrome localization have employed both spectrophotometric and immunological techniques. Microspectrophotometric measurements led to a claim of phytochrome localization in the nucleus (Galston, 1968). Several aspects of these data have, however, been strongly questioned (Briggs and Rice, 1972). The cytochemical visualization of phytochrome antibody in fixed plant sections showed the photoreceptor to be distributed throughout the cytoplasm as well as in association with nuclei and plastids (Pratt and Coleman, 1971). Nonsaturating irradiations of maize coleoptile segments with red and far-red light polarized normal to the longitudinal axis were found to photoconvert about 20% more phytochrome than when polarized parallel to this axis (Marmé and Schäfer, 1972). This was interpreted as indicating that phytochrome is located and oriented in the plasmalemma. It is worth noting that the orientational rigidity of phytochrome implied from this, the *Mougeotia* (Haupt, 1972b) and the *Dryopteris* (Etzold, 1965) studies, are in direct contrast to the highly fluid rotational diffusion observed for rhodopsin in the visual receptor membrane (Cone, 1972).

Several attempts at subcellular localization have been made using standard extraction and fractionation procedures. Early claims of high levels of phytochrome being associated with mitochondria (Gordon, 1961) and plasmalemma (Marmé et al., 1971) can be attributed to precipitation of the protein itself at the pH's used (6.2 or less) (Siegelman and Butler, 1965; Hillman, 1967; Briggs and Rice, 1972). A low level (4%) of phytochrome reported to be pelletable at 40,000 g in extracts from dark-grown *Avena* at pH 7.4 (Rubinstein et al., 1969) has been corroborated by recent findings (Quail et al., 1973a). Another claim of 30–40% pelletability in homogenates of dark-grown, nonirradiated maize coleoptiles at pH 7.5 (Marmé et al., 1971) has, however, not been substantiated.

Recently, it has been observed that red irradiation prior to extraction substantially enhances the level of phytochrome subsequently associated with a pelletable fraction (Quail et al., 1973a; Quail and Schäfer, 1974). This association is dependent on pH and divalent cation concentration in the homogenate and can be induced by irradiation of extracts from dark-grown material as well as of the intact tissue (Marmé et al., 1973; Quail, 1974). Convincing evidence of interaction of the phytochrome with a specific, identifiable membrane fraction(s), is, however, lacking at the moment. The reported isolation of a phytochrome-containing membrane and "solubilized" receptor therefrom (Marmé et al., 1973, 1974; Marmé, 1975) has been attributed instead to the preferential electrostatic binding of P_{fr} to degraded ribosomal material (Quail, 1975). Thus, while the requirement for P_{fr} formation is indicative of some sort of specificity on the part of the phytochrome molecule itself, the existence of a correspondingly specific receptor(s) is yet to be demonstrated; as is the relationship of the binding response to a meaningful biological process. Of potential relevance to this question is the observation that phytochrome can mediate photoreversible conductance changes in artificial lipid membranes (Roux and Yguerabide, 1973). These kinds of observations have led to the suggestion that phytochrome might function as a stereospecific protein ligand capable of interaction with cellular membranes.

IV. Biological Display

A. Induction-Reversion and High Irradiance Responses

Two types of phenomena have been attributed to phytochrome—the so-called "induction-reversion" and "high irradiance" responses (Mohr, 1972). This terminology arises from the irradiation conditions under which the responses are observed. It is important to make a clear distinction between the two experimental approaches for the purpose of evaluating the various mechanistic interpretations of the data thus obtained.

1. INDUCTION-REVERSION RESPONSES

These are displays induced by a brief pulse of low intensity red light and reversed by a far-red pulse (Borthwick, 1972). This photoreversibility of the response is the classically accepted criterion for establishing phytochrome involvement in a process and is interpreted in qualitative terms as indicating that the P_{fr} form is biologically active, whereas P_r is inactive. Quantitatively, increases in the magnitude of the induced response with increasing light dose (where dose equals the total number of quanta) are interpreted as being a function of the degree of photoconversion of P_r to P_{fr} (Table I), i.e., the more quanta, the more P_r is converted to P_{fr} and therefore the greater the response.

Biological action spectra are interpreted similarly (Fig. 7). The magnitude of the response at different wavelengths is considered to be a function of the relative effectiveness of the quanta at those wavelengths in the photoconversion process. Basic to this interpretation is a demonstration that for any given wavelength the so-called law of reciprocity (intensity × time = constant) is valid for the light doses used (Table I). This establishes that the magnitude of the response at that wavelength is pro-

TABLE I

Anthocyanin Levels in *Sinapis* **24 hours after Irradiation with Various Doses of Red (658 nm) Light**[a,b]

Red light irradiance (*I*) (μW cm^{-2})	Anthocyanin (absorbance at 535 nm) after irradiation for		
	300 sec	30 sec	3 sec
675	0.158	0.141	0.107
67.5	0.140	0.110	0.069
6.75	0.107	0.069	—

[a] After Lange *et al.*, 1971.
[b] Values enclosed in dashed lines represent equal light doses, i.e., *I* × *t* = constant, where *I* is irradiance and *t* is time.

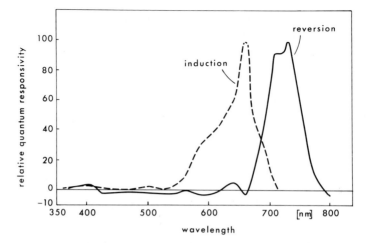

Fig. 7. Action spectra for induction and reversion of plumular hook opening in bean seedlings. After Withrow *et al.* (1957).

portional to the total number of incident quanta irrespective of the time (within limits) or intensity of the irradiation providing those quanta.

For phytochrome, the close agreement between the action spectra (Borthwick, 1972; Shropshire, 1972) of various biological responses, on the one hand (Fig. 7), and of the phototransformation reactions of the isolated pigment, on the other (Fig. 4), suggests a seemingly good correlation between P_{fr} level and response magnitude. Despite this, however, the majority of rigorous attempts to demonstrate a direct quantitative correlation between the spectrophotometrically detectable P_{fr} and the relevant biological response in the same system have been singularly unrewarding (Hillman, 1972). A recent possible exception may be lipoxygenase levels in *Sinapis* (Oelze-Karow and Mohr, 1973).

2. HIGH IRRADIANCE RESPONSES (HIR)

It is clear that the dose–response relationships outlined above would only be expected to hold for nonsaturating light doses, i.e., irradiations terminated prior to the establishment of photoequilibrium. Since the photosteady state P_{fr}/P_r ratios are expected to be irradiance independent [Schäfer, 1975; and Eq. (2)], no further increase in the response should result from further increases in dose once photoequilibrium is reached.

High irradiance responses, on the other hand (Mohr, 1969), do exhibit a strong irradiance dependence after photoequilibrium has been established. The higher the irradiance level, the greater the response.

These effects are observed where a photostationary state is rapidly established and maintained over relatively long periods by continuous irradiations. Reciprocity does not hold, and red/far red photoreversibility of
the response per se is not demonstrable in some cases (Hartman, 1966).
Furthermore, since continuous irradiation is necessary to sustain the response, photoreversibility of the high irradiance effect, in the sense previously used, is not applicable. The response rate simply reverts to the
control level when the irradiation ceases, without the necessity of a terminal antagonistic irradiation.

Several action spectra of HIR occur in the literature (Hendricks and
Borthwick, 1965; Mohr, 1969; Borthwick *et al.*, 1969). The most ex-
haustively investigated system is that of inhibition of lettuce hypocotyl
lengthening (Hartmann, 1967) (Fig. 8). These data were obtained following 18 hours continuous irradiation with monochromatic light of different
wavelengths and varying quantum flux densities. The single sharp, symmetrical peak at about 720 nm does not coincide with the absorption
maximum of either P_r or P_{fr}. The conclusion that phytochrome is responsible for mediating these light effects derives from another set of
observations using the same plant system (Hartmann, 1966). It was
demonstrated that irradiations with two wavelengths, which are relatively
ineffective when given separately (658 nm and 766 nm), are highly effective when given simultaneously at the appropriate quantum flux densities.
The maximum effect with these and other wavelength pairs was always
found to occur under conditions where the photoequilibrium ratio of
$[P_{fr}]/[P_{tot}]$ was about 0.03. This was in good agreement with the peak

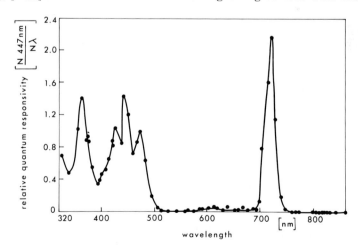

Fig. 8. Action spectrum for inhibition of lettuce hypocotyl lengthening under
continuous irradiation. After Hartmann (1967).

of activity obtained at 720 nm with single wavelength monochromatic light (Fig. 8)—a wavelength known to establish a photostationary state of about 3% P_{fr} (Fig. 5). Finally, the effectiveness of irradiation at 717 nm could be nullified by simultaneous irradiations with either 658 or 759 nm of sufficient intensity. These treatments would shift the photoequilibrium away from 3% P_{fr} toward higher or lower values, respectively.

Despite this apparently good quantitative correlation, the HIR is unlikely to be a function of the absolute P_{fr} level as this is independent of both irradiance and wavelength (Schäfer, 1975). The rate at which phytochrome oscillates between P_r and P_{fr} is, on the other hand, strongly controlled by both variables. This suggests that the HIR is some function of the cycling rate of phytochrome (Hartmann, 1966; Schäfer, 1975). The observation from dual wavelength experiments that the extent of the response is dependent on the total quanta absorbed by the two species supports this notion. Thus both the wavelength and irradiance dependence of the HIR have been ascribed in general terms to the photochromic nature of phytochrome. How phytochrome cycling can be translated into a biological display is a question yet to be answered however. An "excited form of P_{fr}" (P_{fr}^{*}) has been postulated to be the effector molecule (Schopfer and Mohr, 1972), but no direct evidence for such a species has been advanced. A recently proposed cyclic phytochrome–receptor model of phytochrome action can account in principle for both the irradiance and wavelength dependence of HIR as well as for induction–reversion phenomena (Schäfer, 1975).

3. A COMPARISON

Two fundamentally different concepts of the way in which phytochrome mediates the light effect in induction–reversion and high irradiance responses emerge from the interpretation of the effects of irradiance level in the two cases.

Irradiance effects in induction–reversion studies are interpreted as reflecting the effectiveness of the total dose of incident quanta in determining the degree of photoconversion. Implicit in this is some form of P_{fr}–response stoichiometry. Light is viewed simply as having thrown the switch and as having no direct role in the inductive function of P_{fr}, which can then proceed in the dark. The repeated photoreversibility of such responses as lettuce seed germination lend strong support to this argument (Borthwick et al., 1954).

Irradiance effects in the HIR, on the other hand, are considered to result from the sustained direct interaction of the photoreceptor with the incident excitation energy. This irradiance dependence and the requirement for continuous irradiation are together indicative that HIR are light-driven as distinct from being light-triggered, i.e., light appears to have a direct role in the inductive function of the pigment. These responses would appear therefore to have a requirement for a sustained light energy input. The possibility that photosynthesis or at least cyclic photophosphorylation might in some way be responsible has been raised on several occasions, but several lines of evidence mitigate against this (Mohr, 1972).

On the other hand, the apparent requirement for a sustained energy input is more consistent with a photocoupling than with a photosensing function of the pigment (Oesterhelt and Stoeckenius, 1973). This raises the prospect of a parallel between phytochrome and the bacteriorhodopsin of purple membrane. The latter pigment has been postulated to be a photocoupler. The light-induced, reversible deprotonation observed *in vitro* (cf. phytochrome chromophore, Fig. 3) is proposed to operate as a vectorial process *in vivo* generating a proton gradient across the membrane, i.e., the pigment acts as a light-driven proton pump. Since photosensing and photocoupling functions are not necessarily mutually exclusive properties of a pigment (Clayton, 1964), it is tempting to speculate that phytochrome may function in a dual capacity—as a photosensor in induction-reversion responses and as a photocoupler in high irradiance responses.

B. Kinetics of Phytochrome Action and Response Expression

1. RESPONSE CLASSIFICATION

Examination of the timing of phytochrome-mediated responses has yielded much valuable information as to the possible nature of the molecular mechanisms involved. The action of P_{fr} in inducing a biological response is generally formalized as

$$P_{fr} + X \rightleftharpoons P_{fr}X \rightarrow \rightarrow \rightarrow \text{biological response}$$

where the logical necessity of a reaction partner is symbolized by X regardless of its nature. The interaction of P_{fr} with X (the so-called "primary reaction" of phytochrome) is considered to trigger a sequence of molecular events culminating sooner or later in a biological response.

The initial triggering of those processes necessary for the development of a response can be referred to as phytochrome "action," and the appearance of a measurable change in the parameter being monitored can be called response "expression." The timing of phytochrome action and response expression is as variable as the number of biological displays. However, three major categories of phenomena are recognizable.

    i. Rapid action/rapid expression
    ii. Rapid action/delayed expression
    iii. Delayed action/delayed expression

Here "rapid" arbitrarily means 10 minutes or less and "delayed" 30 minutes or more after the initial photoconversion act.

The first category includes those responses where rapid phytochrome action can be implied from the kinetics of the response expression alone. (Obviously phytochrome action must either coincide with or precede the response monitored). Responses in the second category are those where rapid phytochrome action can be deduced from the rate at which the response escapes susceptibility to photoreversal by far-red light following a brief inductive red pulse. In this case, the actual response may not be expressed for hours or even days after irradiation, although the inevitability of its appearance has long since been irreversibly established. P_{fr} is described as having "potentiated" the response (Borthwick, 1972), and the escape from reversibility is viewed as the P_{fr}–triggered reaction chain having rapidly progressed beyond the step(s) directly under reversible phytochrome control. The responses in the third category also exhibit a distinctive lag between irradiation and response expression, but in addition they are readily reversed by far-red over relatively extended periods in the dark following the red pulse. Escape from reversibility does occur and can be a gradual, continuous process beginning in some cases more or less immediately after the inductive irradiation (Borthwick, 1972; Haupt, 1972a). However, the rate of escape is substantially slower than for responses in category ii. This is interpreted as indicating a requirement for the sustained presence of P_{fr} over a relatively long period in the dark to enable maximum response expression. A feature of these responses is that during the period when P_{fr} is required to perform its function, the competing reactions of dark reversion and destruction are effecting its often rapid removal.

It should be emphasized at this point that the timing of phytochrome action as deduced from response expression refers of necessity only to the participation of phytochrome in that particular response. It has been amply demonstrated (Mohr, 1972) that different parameters in the same system respond differently in time, direction, and magnitude to the same light treatments. A representative sample of responses belonging to the categories outlined above is included in the lower half of Fig. 2 and will be discussed briefly below.

2. RAPID ACTION/RAPID EXPRESSION RESPONSES

The most rapid phytochrome-mediated display so far reported is a change in electric potential in etiolated *Avena* coleoptiles (Newman and Briggs, 1972). A 10-second exposure to red light induces a far-red reversible change of 5 to 10 mV in the upper 1 cm of the coleoptile. Both the induction and reversion responses are detectable within 15 seconds of the start of irradiation.

A phytochrome-mediated change in surface potential had earlier been inferred from the red/far-red reversible adhesion of root tips to negatively charged glass surfaces [the so-called Tanada effect (Tanada, 1968)]. This was confirmed by direct measurement of the bioelectric potential across the root tip in parallel with adhesion measurements (Jaffe, 1968). The kinetics of the two responses show good agreement, and the potential changes although small (~ 1 mV) are in the right direction [red $\rightarrow$ positive; far-red $\rightarrow$ negative, the same as for *Avena* (Newman and Briggs, 1972)] to account for the adhesion data (Fig. 9). Both responses were detectable within 30 seconds of the start of irradiation. Additional evidence has been presented (Yunghans and Jaffe, 1972) that red light induces a H^+ efflux in this system. The magnitude of the charges involved has also been estimated using a platinum electrode (Racusen and Miller, 1972).

Exposure of grass coleoptiles to 15 seconds of red light induces an increase in growth rate in the tip region within 60 seconds of the start of irradiation (Weintraub and Lawson, 1972). This effect is substantially reduced but not completely reversed if followed immediately by far-red light. The response is insensitive to various inhibitors of transcription and translation.

Transfer of *Mimosa* plants from high intensity white light to darkness initiates leaflet closure within 5 minutes, the process being complete within 30 minutes. Far-red irradiation before transfer to darkness prevents closure, and the far-red effect is reversed by a subsequent red treatment. Photoreversibility is demonstrable through several cycles (Fondeville *et al.*, 1966). The interpretation is that P_{fr} induces closure in the dark, and P_r prevents it. These results have been confirmed with *Albizzia* and extended by electron microprobe analysis to show that a transfer of K^+ ions from the ventral to the dorsal motor cells of the pulvinus accompanies leaflet closure and vice versa during opening (Satter and Galston, 1973). Furthermore, closure is insensitive to actinomycin D and cycloheximide but is inhibited by anaerobiosis, NaN_3, and dinitrophenol.

In the alga *Mougeotia* following a brief red light pulse, a change

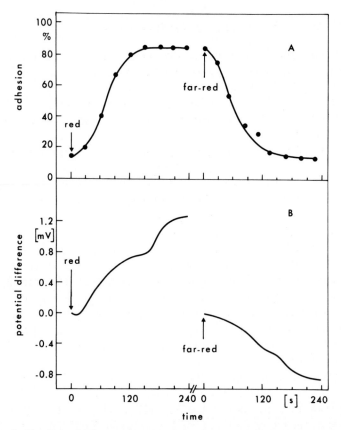

Fig. 9. Kinetics of (A) root tip adhesion to a negatively charged glass surface and (B) the development of a bioelectric potential across the root tip in response to irradiation with red or far-red light. After M. J. Jaffe (1968). *Science* **162,** 1016. Copyright 1968 by the American Association for the Advancement of Science.

in plastid orientation is evident in less than 10 minutes and is complete within 30 minutes (Haupt, 1972a). This effect can be fully reversed by far-red irradiation but only during the first minute after the red light. Thus potentiation of the response is considered to begin within 1 minute of photoconversion. Available data indicate that the effective phytochrome is located in or near the plasmalemma. Microscopic evidence coupled with the use of cytochalasin B and colchicine suggests that contractile fibrils but not microtubules might be involved in the chloroplast movement (Schönbohm, 1972).

Adenosine triphosphate (ATP) levels in bean buds irradiated with 5 minutes red light and returned to the dark rise sharply to a peak after 1 minute, followed by a slow decline to the control level again by 10 minutes (White and Pike, 1974). The effect is apparently reversible by 5 minutes of far-red, although interpretation is difficult because of the long far-red irradiation used (also 5 minutes).

Lipoxygenase accumulation in *Sinapis* is inhibited in apparently less than 5 minutes from the onset of irradiation with wavelengths between 660 and 724 nm (Oelze-Karow and Mohr, 1973). Similarly, enzyme accumulation appears to resume within 5 minutes of the return of 724 nm irradiated seedlings to the dark. Wavelengths of 727 nm or longer are without effect (at least in the short term). A comprehensive investigation of this system has led to a hypothesis that enzyme levels in the cotyledons are controlled by phytochrome in the hook through a highly cooperative threshold mechanism that responds to and is saturated by P_{fr} levels of the order of 1–2% of the total phytochrome. These interpretations have been questioned (Schmidt and Schäfer, 1974), however, since they are based to a large extent on calculations that ignore dark reversion. Spectrophotometric data from *Sinapis* hooks indicate that dark reversion is substantial and wavelength dependent, and for P_{fr} levels of 25% or less all the P_{fr} molecules formed appear to undergo reversion. On the other hand, rapid interorgan transfer of phytochrome signals are not without precedent in the literature (De Greef *et al.*, 1972).

3. RAPID ACTION/DELAYED EXPRESSION RESPONSES

The effect of red light in suppressing flowering in *Pharbitis nil* can only be partially reversed by far-red given 30 seconds after the beginning of red irradiation (Frédéricq, 1964). After 3 minutes the red light effect can no longer be reversed by far-red. Thus, although the flowering response itself is not expressed for many days, it is potentiated within seconds by P_{fr} formation. Similar results are obtained for flowering in *Chenopodium album* and *Kalanchoe*. Other responses in this category include the synergism between phytochrome and gibberellin in lettuce seed germination (Bewley *et al.*, 1968) and the so-called deetiolation of *Pisum* (Haupt, 1972a).

Phytochrome-mediated leaf unrolling is often quoted as one of the most rapid of the potentiated responses and as an example of intraorgan transmission of a phytochrome signal (Haupt, 1972a). When part of a leaf was irradiated with red light and the remainder kept darkened, the unirradiated portion exhibited the unrolling response (complete by 24 hours) when severed from the irradiated portion as early as 20 seconds from the beginning of irradiation (Wagné, 1965). More recent data do not support this interpretation, however (Kang and Zeevaart, 1968).

4. Delayed Action/Delayed Expression Responses

Included in this category are the classical red/far-red reversible responses, which led to the discovery of phytochrome and a vast catalog of other phenomena ranging from changes in cellular metabolism to patterns of growth and development (Mitrakos and Shropshire, 1972; Mohr, 1972). Irradiation of lettuce seed with 1 minute of red light, stimulates up to 100% germination recorded 24 hours later. This effect is reversible by far-red light up to 12 hours after the red irradiation but with an ever decreasing effectiveness. This suggests a continued requirement for P_{fr} during this period for full response expression.

Anthocyanin formation in *Sinapis* shows a similar requirement (Lange *et al.*, 1971). The lag from irradiation to response expression is 3 hours. During this time, the effect becomes decreasingly susceptible to reversal by far-red light. However, the rate of escape from reversibility is slow and is apparently never absolute, i.e., the continued presence of P_{fr} is apparently necessary to sustain continued anthocyanin formation even well past the lag period. The enzyme phenylalanine ammonia lyase shows an apparently similar pattern, except the lag to response expression is only of the order of 1 hour (Schopfer and Mohr, 1972).

C. Other Phenomena

The interaction of phytochrome with endogenous rhythms in the photoperiodic control of flowering (Vince, 1972), leaflet closure in *Albizzia* (Satter and Galston, 1973), and control of enzyme levels (Queiroz, 1972; Frosch and Wagner, 1973) is well documented. Plant hormones have been implicated in many responses controlled by phytochrome, but there is no convincing evidence that phytochrome exerts its effects via hormones (Black and Vlitos, 1972). Red/far-red reversible changes in acetylcholine levels in mung bean roots have been reported (Jaffe, 1970), but its role as a mediator of phytochrome effects is doubtful (Tanada, 1972; Kasemir and Mohr, 1972; White and Pike, 1974). Reversible changes in NAD kinase levels *in vitro* in response to red/far-red irradiations of extracts from *Pharbitis* and peas have been documented (Yamamoto and Tezuka, 1972).

D. Mechanism of Action

Attempts to rationalize the vast diversity of observed photoresponses in terms of a molecular mechanism of phytochrome action have resulted in recent years in two major hypotheses—the so-called differential gene activation hypothesis advanced by Mohr (1966) and the membrane permeability hypothesis of Hendricks and Borthwick (1967).

It could be argued that in its loosest sense the former hypothesis would be trivial. Few would assert that the gross morphogenetic changes induced by phytochrome would be likely to occur without some modification of gene expression. In its more stringent sense, on the other hand, the hypothesis would require a demonstration of direct P_{fr}–genome interaction. This problem has been approached mainly by measuring phytochrome-induced changes in enzyme activities (Mohr, 1972).

Criticisms of this approach are broadly twofold. First, the data so far obtained, while consistent in principle with the general framework of the hypothesis, provide no direct evidence that genetic activity is in fact altered, let alone that P_{fr} interacts directly with the genome. Second, there is a growing list of responses that seem unlikely to be accounted for by gene regulation. These are, in general, the rapid action/rapid expression phenomena. Not only are these responses more rapid than would be expected from genetic regulation, but in some cases at least, have proved insensitive to transcriptional and translational inhibitors (Weintraub and Lawson, 1972; Satter and Galston, 1973). Furthermore, the only enzyme showing a rapid enough response to be placed in this category is now postulated to be under the control of phytochrome in a separate organ (Oelze-Karow and Mohr, 1973).

Phytochrome-induced changes in membrane properties can, on the other hand, account for even the most rapid phytochrome-mediated phenomena. In fact, all of the most rapidly expressed responses are either direct surface phenomena or can be rationalized in terms of changes in membrane properties. On this basis, it has been suggested that the induced changes might in fact result from direct phytochrome–membrane interaction, i.e., the primary action of phytochrome might be the induction of such membrane changes (Hendricks and Borthwick, 1967). Of relevance to the requirement for a demonstration of a direct interaction of phytochrome with cellular membranes is the recently observed association of phytochrome with a particulate fraction following red irradiation of the tissue (Quail et al., 1973a; Quail and Schäfer, 1974).

While it has been pointed out that there is still a 15 second lag to account for (Briggs and Rice, 1972), no conceptual problems are apparent in the above proposal. Little imagination is required to envisage the potential for a multiplicity of secondary effects, perhaps in a variety of cellular membranes, emanating from a single, fundamental alteration in membrane properties. Such an alteration could afford the opportunity for changes in ion flux, activation of membrane-bound enzymes, altered compartmentalization, release of bound hormones, and so on. Furthermore, a phytochrome–membrane association affords the possibility of a photocoupling function for phytochrome in the HIR. Any or all of these might lead ultimately to altered gene expression.

The speculative nature of such proposals, however, does little more than emphasize our ignorance in this area. It has been pointed out, for example, that there is no evidence for the tacitly assumed premise in much of the phytochrome literature that there is a single, primary reaction (Mohr, 1972). Instead, a multiplicity of reactions in which phytochrome might directly participate is suggested. This would circumvent the second criticism of the gene regulation hypothesis outlined above and eliminate any potential conflict between the two major hypotheses. It is clear that the molecular mechanism(s) of phytochrome action is an entirely open question.

V. Conclusions

With the capture of a photon in 10^{-17} seconds, the P_r form of the phytochrome molecule begins a complex series of internal rearrangements culminating in the formation of the metastable but biologically active P_{fr} form within about 4 seconds. The living tissue responds to this photoconversion event in a measurable way as early as 10–15 seconds from the appearance of the first P_{fr} molecules. There follows over the next hours, days, or even weeks a vast cascade of monitorable responses, which range from changes in bioelectric potential and enzyme levels, through plastid development, cell expansion, germination, and flowering.

Two distinct classes of phenomena are attributed to phytochrome— the so-called induction-reversion and high irradiance responses. The former are light-triggered, whereas the latter are light-driven. This suggests that phytochrome may play a fundamentally different role in mediating the light effect in each case. Furthermore, three broad categories of responses are recognizable on the basis of their kinetic behavior. Such a categorization reflects the intrinsic properties of the responses themselves. The most rapid are membrane or surface phenomena; the majority are slow and could be rationalized in terms of gene regulation. The primary molecular mechanism of phytochrome action remains an unresolved question.

GENERAL REFERENCES

Briggs, W. R., and Rice, H. W. (1972). *Annu. Rev. Plant Physiol.* **23**, 293.
Butler, W. L., Hendricks, S. B., and Siegelman, H. W. (1965). *In* "Chemistry and Biochemistry of Plant Pigments" (T. W. Goodwin, ed.), pp. 197–210. Academic Press, New York.

Mitrakos, K., and Shropshire, W., Jr., eds. (1972). "Phytochrome." Academic Press, New York.

Mohr, H. (1972). "Lectures on Photomorphogenesis." Springer-Verlag, Berlin and New York.

Smith, H. (1970). *Nature (London)* **227,** 665.

REFERENCES

Anderson, G. R., Jenner, E. L., and Mumford, F. E. (1970). *Biochim. Biophys. Acta* **221,** 69.

Bewley, J. D., Black, M., and Negbi, M. (1968). *Planta* **78,** 351.

Black, M., and Vlitos, A. J. (1972). *In* "Phytochrome" (K. Mitrakos and W. Shropshire, Jr., eds.), pp. 517–550. Academic Press, New York.

Borthwick, H. A. (1972). *In* "Phytochrome" (K. Mitrakos and W. Shropshire, Jr., eds.), pp. 3–44. Academic Press, New York.

Borthwick, H. A., Hendricks, S. B., Toole, E. H., and Toole, V. K. (1954). *Bot. Gaz. (Chicago)* **115,** 205.

Borthwick, H. A., Hendricks, S. B., Schneider, M. J., Taylorson, R. B., and Toole, V. K. (1969). *Proc. Nat. Acad. Sci. U.S.* **64,** 479.

Briggs, W. R., and Rice, H. W. (1972). *Annu. Rev. Plant Physiol.* **23,** 293.

Briggs, W. R., and Siegelman, H. W. (1965). *Plant Physiol.* **40,** 934.

Butler, W. L. (1961). *In* "Progress in Photobiology" (B. C. Christensen and B. Buchmann, eds.), p. 569. Elsevier, Amsterdam.

Butler, W. L. (1973). *In* "Pytochrome" (K. Mitrakos and W. Shropshire, Jr., eds.), pp. 185–192. Academic Press, New York.

Butler, W. L., Hendricks, S. B., and Siegelman, H. W. (1964). *Photochem. & Photobiol.* **3,** 521.

Clayton, R. K. (1964). *In* "Photophysiology" (A. C. Giese, ed.), Vol. 2, pp. 51–77. Academic Press, New York.

Coleman, R. A., and Pratt, L. H. (1974). *Planta* **119,** 221.

Cone, R. A. (1972). *Nature (London), New Biol.* **236,** 39.

De Greef, J. A., Caubergs, R., and Verbelen, J. P. (1972). *Int. Congr. Photobiol., 6th, 1972 Book of Abstracts,* No. 172.

Etzold, H. (1965). *Planta* **64,** 254.

Fondeville, J. C., Borthwick, H. A., and Hendricks, S. B. (1966). *Planta* **69,** 357.

Frankland, B. (1972). *In* "Phytochrome" (K. Mitrakos and W. Shropshire, Jr., eds.), pp. 195–225. Academic Press, New York.

Frédéricq, H. (1964). *Plant Physiol.* **39,** 182.

Frosch, S., and Wagner, E. (1973). *Can. J. Bot.* **51,** 1529.

Fry, K. T., and Mumford, F. E. (1971). *Biochem. Biophys. Res. Commun.* **45,** 1466.

Galston, A. W. (1968). *Proc. Nat. Acad. Sci. U.S.* **61,** 454.

Gardner, G., Pike, C. S., Rice, H. V., and Briggs, W. R. (1971). *Plant Physiol.* **48,** 686.

Gordon, S. A. (1961). *Proc. Int. Congr. Photobiol., 3rd 1960* pp. 441–443.

Hanke, J., Hartmann, K. M., and Mohr, H. (1969). *Planta* **86,** 253.

Hartmann, K. M. (1966). *Photochem. & Photobiol.* **5,** 349.

Hartmann, K. M. (1967). *Z. Naturforsch. B* **22,** 1172.

Haupt, W. (1972a). *In* "Phytochrome" (K. Mitrakos and W. Shropshire, Jr., eds.), pp. 349–368. Academic Press, New York.

Haupt, W. (1972b). *In* "Phytochrome" (K. Mitrakos and W. Shropshire, Jr., eds.), pp. 553–569. Academic Press, New York.

Hendricks, S. B., and Borthwick, H. A. (1965). *In* "Chemistry and Biochemistry of Plant Pigments" (T. W. Goodwin, ed.), pp. 405–436. Academic Press, New York.

Hendricks, S. B., and Borthwick, H. A. (1967). *Proc. Nat. Acad. Sci. U.S.* **58**, 2125.

Hillman, W. S. (1967). *Annu. Rev. Plant Physiol.* **18**, 301.

Hillman, W. S. (1972). *In* "Phytochrome" (K. Mitrakos and W. Shropshire, Jr., eds.), pp. 573–584. Academic Press, New York.

Hopkins, D. W., and Butler, W. L. (1970). *Plant Physiol.* **45**, 567.

Jaffe, M. J. (1968). *Science* **162**, 1016.

Jaffe, M. J. (1970). *Plant Physiol.* **46**, 768.

Kamen, M. D. (1963). "Primary Processes in Photosynthesis." Academic Press, New York.

Kang, B. G., and Zeevaart, J. A. D. (1968). *Annu. Rep. MSU/AEC Plant Res. Lab., Mich. State Univ.* pp. 33–34.

Kasemir, H., and Mohr, H. (1972). *Plant Physiol.* **49**, 453.

Kendrick, R. E., and Spruit, C. J. P. (1973). *Photochem. & Photobiol.* **18**, 153.

Kroes, H. H. (1970). *Meded. Landbouwhogesch. Wageningen* **70-18**, 1.

Kropf, A. (1972). *Int. Congr. Photobiol., 6th, 1972 Book of Abstracts,* No. 022.

Lange, H., Shropshire, W., and Mohr, H. (1971). *Plant Physiol.* **47**, 649.

Lhoste, J.-M. (1972). *In* "Phytochrome" (K. Mitrakos and W. Shropshire, Jr., eds.), pp. 47–74. Academic Press, New York.

Linschitz, H., and Kasche, V. (1967). *Proc. Nat. Acad. Sci. U.S.* **58**, 1059.

Linschitz, H., Kasche, V., Butler, W. L., and Siegelman, H. W. (1966). *J. Biol. Chem.* **241**, 3395.

Manabe, K., and Furuya, M. (1974). *Plant Physiol.* **53**, 343.

Marmé, D. (1969). *Planta* **88**, 43.

Marmé, D. (1975). *J. Supramolec. Struc.* **2**, 751.

Marmé, D., and Schäfer, E. (1972). *Z. Pflanzenphysiol.* **67**, 192.

Marmé, D., Schäfer, E., Trillmich, F., and Hertel, R. (1971). *Eur. Annu. Symp. Plant Photomorphogenesis, 1971 Book of Abstracts,* p. 36.

Marmé, D., MacKenzie, J. M., Boisard, J., and Briggs, W. R. (1974). *Plant Physiol.* **54**, 263.

Mitrakos, K., and Shropshire, W., Jr., eds. (1972). "Phytochrome." Academic Press, New York.

Mohr, H. (1966). *Photochem. & Photobiol.* **5**, 469.

Mohr, H. (1969). *In* "Physiology of Plant Growth and Development" (M. B. Wilkins, ed.), pp. 509–516. McGraw-Hill, London.

Mohr, H. (1972). "Lectures on Photomorphogenesis." Springer-Verlag, Berlin and New York.

Newman, I. A., and Briggs, W. R. (1972). *Plant Physiol.* **50**, 687.

Oelze-Karow, H., and Mohr, H. (1973). *Photochem. & Photobiol.* **18**, 319.

Oesterhelt, D., and Stoeckenius, W. (1973). *Proc. Nat. Acad. Sci. U.S.* **70**, 2853.

Pratt, L. H., and Briggs, W. R. (1966). *Plant Physiol.* **41**, 467.

Pratt, L. H., and Butler, W. L. (1970). *Photochem. & Photobiol.* **11**, 503.

Pratt, L. H., and Coleman, R. A. (1971). *Proc. Nat. Acad. Sci. U.S.* **68**, 2431.

Pratt, L. H., Kidd, G. H., and Coleman, R. A. (1974). *Biochim. Biophys. Acta* **365**, 93.

Quail, P. H. (1974). *Planta* **118**, 357.

Quail, P. H. (1975). *Planta* **123**, 223.

Quail, P. H., and Schäfer, E. (1974). *J. Membrane Biol.* **15**, 393.

Quail, P. H., Marmé, D., and Schäfer, E. (1973a). *Nature (London) New Biol.* **245**, 189.

Quail, P. H., Schäfer, E., and Marmé, D. (1973b). *Plant Physiol.* **52**, 128.

Queiroz, O. (1972). *In* "Phytochrome" (K. Mitrakos and W. Shropshire, Jr., eds.), pp. 295–316. Academic Press, New York.

Racusen, R., and Miller, K. (1972). *Plant Physiol.* **49**, 654.

Roux, S. J. (1972). *Biochemistry* **11**, 1930.

Roux, S. J., and Lisansky, S. (1975). *Physiol. Plant* **35**, 85.

Roux, S. J., and Yguerabide, J. (1973). *Proc. Nat. Acad. Sci. U.S.* **70**, 762.

Rubinstein, B., Drury, K. S., and Park, R. B. (1969). *Plant Physiol.* **44**, 105.

Rüdiger, W. (1972). *In* "Phytochrome" (K. Mitrakos and W. Shropshire, Jr., eds.), pp. 129–141. Academic Press, New York.

Satter, R. L., and Galston, A. W. (1973). *BioScience* **23**, 407.

Schäfer, E. (1975). *J. Math. Biol.* **2**, 41.

Schäfer, E., and Mohr, H. (1974). *J. Math. Biol.* **1**, 9.

Schäfer, E., and Schmidt, W. (1974). *Planta* **116**, 257.

Schäfer, E., Marchal, B., and Marmé, D. (1972). *Photochem. & Photobiol.* **15**, 457.

Schäfer, E., Schmidt, W., and Mohr, H. (1973). *Photochem. & Photobiol.* **18**, 331.

Schmidt, W., and Schäfer, E. (1974). *Planta* **116**, 267.

Schmidt, W., Marmé, D., Quail, P., and Schäfer, E. (1973). *Planta* **111**, 329.

Schönbohm, E. (1972). *Ber. Deut. Bot. Ges.* **86**, 431.

Schopfer, P., and Mohr, H. (1972). *Plant Physiol.* **49**, 8.

Shropshire, W., Jr. (1972). *In* "Phytochrome" (K. Mitrakos and W. Shropshire, Jr., eds.), pp. 161–181. Academic Press, New York.

Siegelman, H. W., and Butler, W. L. (1965). *Annu. Rev. Plant Physiol.* **16**, 383.

Spruit, C. J. P. (1972). *In* "Phytochrome" (K. Mitrakos and W. Shropshire, Jr., eds.), pp. 77–104. Academic Press, New York.

Tanada, T. (1968). *Proc. Nat. Acad. Sci. U.S.* **59**, 376.

Tanada, T. (1972). *Plant Physiol.* **49**, 860.

Tobin, E., and Briggs, W. R. (1973). *Photochem. & Photobiol.* **18**, 487.

Vince, D. (1972). *In* "Phytochrome" (K. Mitrakos and W. Shropshire, Jr., eds.) pp. 257–291. Academic Press, New York.

Wagné, C. (1965). *Physiol. Plant.* **18**, 1001.

Walker, T. S., and Bailey, J. L. (1970). *Biochem. J.* **120**, 613.

Weintraub, R. L., and Lawson, V. R. (1972). *Int. Congr. Photobiol., 6th, 1972 Book of Abstracts,* No. 161.

Wellburn, F. A. M., and Wellburn, A. R. (1973). *New Phytol.* **72**, 55.

White, J. M., and Pike, C. S. (1974). *Plant Physiol.* **53**, 76.

Withrow, R. B., Klein, W. H., and Elstad, V. (1957). *Plant Physiol.* **32**, 453.

Yamamoto, Y., and Tezuka, T. (1972). *In* "Phytochrome" (K. M. Mitrakos and W. Shropshire, Jr., eds.), pp. 407–429. Academic Press, New York.

Yunghans, H., and Jaffe, M. J. (1972). *Plant Physiol.* **49**, 1.

22

Hormones[*]

J. E. VARNER and DAVID TUAN-HUA HO

[*] Sections I, II and III prepared by J. E. Varner, Sections IV and V prepared by David Tuan-hua Ho.

I. Ethylene

A. Appreciation of the Role of Ethylene as a Hormone

It seems likely that every plant tissue at some stage in its development produces, and responds to, ethylene. This appreciation of the importance of ethylene is of recent origin. Yet since 1901 (Neljubow, 1901) ethylene has been known to regulate plant growth and development. Its ability to promote abscission was discovered in 1913 (Doubt, 1917) and its ability to cause fruit ripening in 1924 (Denny, 1924), and since 1934 (Gane, 1934) it has been recognized as a product of plant tissue. Early clues to the possible role of ethylene as a hormone were the observations of ethylene-induced epinasty (Crocker and Zimmerman, 1932) and the accompanying ethylene-evoked inhibition of the lateral transport of auxin (Crocker, 1932).

B. Biosynthesis of Ethylene

L-Methionine is the only generally accepted precursor of ethylene in the tissues of higher plants. From the administration of methionine labeled with carbon-14 in various positions (Lieberman et al., 1966; Burg and Claggett, 1967) to apple slices, it was shown that C-1 is converted to CO_2, C-2 to several metabolites, C-3 and C-4 form ethylene (Fig. 1), the methyl carbon and the sulfur appear in S-methylcysteine, the methyl carbon in pectin, and the sulfur in the sulfur-containing amino acids (Fig. 1). The acceptance of methionine as an ethylene precursor followed the study of several model ethylene-forming systems. In the first of these, peroxidized linolenic acid in reactions catalyzed by Cu^+ (produced from Cu^{2+} and ascorbate) yielded ethylene and several other unsaturated and saturated hydrocarbon gases as well (Lieberman and Mapson, 1964). However, neither ^{14}C-labeled linolenic acid (Mapson et al., 1970) nor ^{14}C-labeled propanal (Baur and Yang, 1969) is converted by apple peel disks

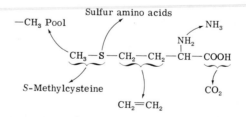

Fig. 1. Biosynthesis of ethylene from L-methionine precursor.

to ethylene, although the labeled compounds are taken up and converted to carbon dioxide.

In a second model system cupric ions and ascorbate catalyze the degradation of methionine to ethylene (Lieberman *et al.*, 1965). In this system, methional and 2-keto-4-methyl thiobutyrate are more effective as ethylene precursors than is methionine.

In a third model system (Yang *et al.*, 1967) FMN and light catalyze the conversion of methionine to ethylene (Fig. 2).

In a fourth model system purified horseradish peroxidases catalyze ethylene formation from methional or 2-keto-4-methyl thiobutyrate (Yang, 1967). This system requires either H_2O_2 or Mn^{2+}, a phenol, and sulfite as cofactors. Orthodiphenols are inhibitory, and methionine itself is not converted to ethylene.

Peroxidases that can work in this model system have been obtained from cauliflower, apple, and tomato tissue, and all of the components of this model system have been found in cauliflower florets (Mapson, 1970). These components include the peroxidase, glucose oxidase for the production of hydrogen peroxide, the methyl ester of *p*-coumaric acid (which meets the requirement for a monohydric phenol), and methane sulfinic acid (which provides the second required cofactor). These enzymes and

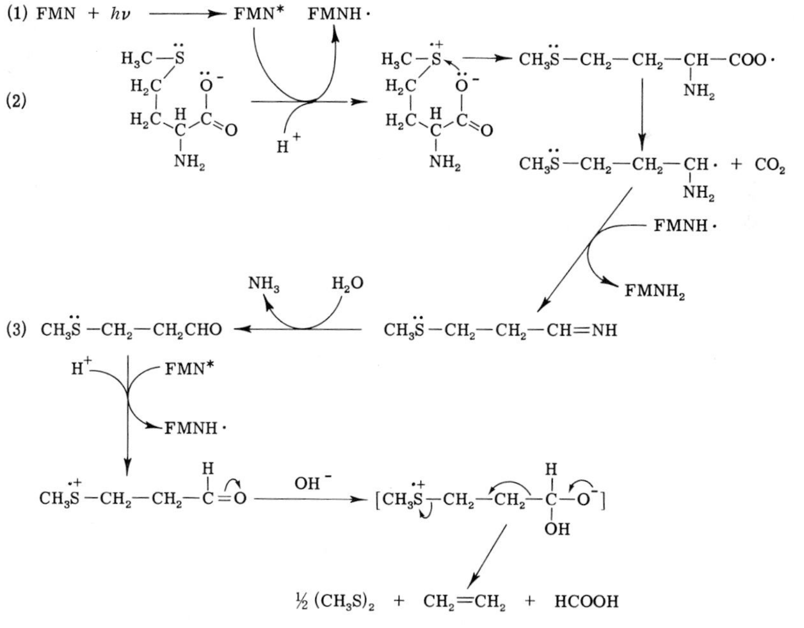

Fig. 2. A proposed scheme for the conversion of methionine to ethylene by flavin mononucleotide (FMN) and light. From Yang *et al.*, 1967.

cofactors might be expected to work *in vivo* for the production of ethylene (Fig. 3). Cauliflower floret tissue suspended in buffer solution does convert added 2-keto-4-methyl thiobutyrate to ethylene about four times more efficiently than it converts added methionine to ethylene. However, added methional is not converted to ethylene. Caution is advised in the interpretation of these experiments because of the possibility that the added substrates are converted to ethylene by an extracellular system formed by the leakage of peroxidases and the required cofactors from the tissue during the incubation (Lieberman and Kunishi, 1971).

Although these model systems are instructive about the possible chemistry of the biosynthesis of ethylene, the details of the *in vivo* conversion by higher plants of methionine to ethylene remain to be established. In addition to the possible ethylene precursors already mentioned ethanol, acetate, and acrylic acid have been given serious consideration (Abeles, 1972). None of these is generally accepted as an ethylene precursor in higher plants. However, fungi, notably *Penicillium digitatum*, also produce ethylene (Miller *et al.*, 1940; Biale, 1940) and probably not from methionine but more likely from glutamate (Yang, 1974).

Whatever the pathway for the biogenesis of ethylene, oxidative metabolism is required for ethylene formation (Hansen, 1942; Spencer, 1959). More specifically, oxygen (or, more likely, some metabolite or cofactor produced during electron transport through cytochrome oxidase to oxygen) is required at some point in the conversion of $[^{14}C]$methionine to $[^{14}C]$-ethylene (Baur *et al.*, 1971).

In view of the autocatalytic kinetics of some of the physiological

Enzymes

(1) *Transaminase*: Converts methionine to its α-keto analogue
$$CH_3SCH_2CH_2CH_2(NH_2)COOH \longrightarrow CH_3SCH_2CH_2CO-COOH$$

(2) *Glucose oxidase*: Produces H_2O_2 in oxidation of β-D-glucose
$$\beta\text{-D-glucose} + O_2 \longrightarrow \text{D-glucono-}\delta\text{-lactone} + H_2O_2$$

(3) *Peroxidase*: Uses H_2O_2 to produce C_2H_4 from KMBA in presence of 2 cofactors
$$CH_3SCH_2CH_2COCOOH \longrightarrow C_2H_4 + \text{products}$$

Cofactors

(1) *Phenolic*:   Esters of *p*-coumaric or other phenols having OH group in para position and conjugated system.

(2) *Sulfinic acid* :   Methanesulfinic acid
$$CH_3-S(OH)=O$$

Fig. 3. Proposed scheme for enzymatic formation of ethylene from methionine. KMBA, α-keto-γ-methyl thiobutyrate. From Mapson, 1970.

responses evoked by ethylene, it is interesting to consider the possibility that peptides having a C-terminal methionine (or methionine from such peptides) could be ethylene precursors (Ku and Leopold, 1970; Demorest and Stahman, 1971). Thus, during senescence and ripening, an increased concentration of ethylene precursors would be made available through protein degradation.

In many tissues treatment with auxins enhances ethylene production, apparently through the synthesis of more of the enzymes involved in the biosynthesis of ethylene (Abeles, 1966). The increased production of ethylene evoked by indoleacetic acid in subapical pea stem sections quickly (within 1–4 hours) ceases when the sections are transferred to indoleacetic acid-free medium. The rate of ethylene production by sub-apical pea stem tissue closely follows the levels of free auxin in the tissue (Kang et a., 1971). The level of free auxin is determined by the balance between the rate of uptake of auxin from the external medium and the rate of conjugation of the auxin in the tissue. Although both the auxin conjugation and the auxin decarboxylase systems increase in response to an increased external concentration of auxin, the conjugation system seems to be more important in determining the tissue concentration of free auxin (Kang et al., 1971). The rate of ethylene production may be a generally useful indicator of the tissue levels of the free auxin. Thus, more ethylene is produced by the dark side than by the light side of unilaterally illuminated plants (Abeles and Rubinstein, 1964) and more by the lower side than the upper side in horizontal stem tissue (Abeles, 1972).

C. Metabolism of Ethylene

The hormonal regulation of a physiological process is accomplished by a proportionate, continuous, and (up to a point) reversible mechanism. Thus, the tissue response to ethylene is proportional to the log of the concentration of ethylene present, and the removal of ethylene diminishes or stops the response except in the autocatalytic stage of responses, such as fruit ripening. Complete control of a tissue hormonal response, therefore, not only requires a mechanism for controlling the rate of biosynthesis and transport of the hormone but also requires a mechanism for controlling the rate of removal, detoxification, or degradation of the hormone. In the case of ethylene, removal is due to the diffusion of the gas from sites of high concentration—the sites of biosynthesis and the physiological receptor sites—into the surrounding tissue and into the atmosphere. In principle, the tissue level of the ethylene could be closely controlled by regulating only the rate of its biosynthesis.

Perhaps this is the point to deal with the proposition that ethylene is really not a hormone (Abeles, 1972). Basically, hormones mediate and integrate specific events that are of importance to the development of the whole organism. Thus, a tissue or an organ monitors osmotic pressure, metabolite concentration, temperature, day length, and developmental time and, when conditions are right, produces a hormone that is transported to the other tissues and organs to stir some or all of them into action appropriate for the moment. In order to fulfill the integrative function, the required intensity of the response may vary from day to day, hour to hour, and minute to minute. Therefore, the hormone and its receptor site must act not as a switch that is "on" or "off" but as a reversible regulator with an infinite number of positions between "on" and "off." However, through continued action of the hormone–receptor pair, the tissue may undergo an irreversible change in metabolism or developmental state and escape from the influence of that particular hormone–receptor pair.

Ethylene fails to fit the classical picture of a hormone because it apparently does not undergo directed transport. It does accomplish an integrative function by diffusing rapidly through the tissue in which it is produced, thereby exposing all cells in the tissue to ethylene and ensuring a reasonably uniform response of all cells in the tissue. Could such local coordination be accomplished with a less volatile agent? Probably not, because the transport mechanisms would quickly disperse the nonvolatile agent throughout the plant, and all tissues capable of showing a response would respond.

Consider the response of a plant to mechanical or chemical wounding. Ethylene formation by the damaged cells increases as a result of wounding and apparently evokes a wounding response in the cells surrounding the damaged cells. This response includes the increased activities of phenylalanine ammonia-lyase, polyphenol oxidase, and peroxidase activities (Abeles, 1972). These enzymes, and others, are thought to be involved in the wound healing that occurs. Assuming that this is an approximate description of what really happens, one can see the advantage of having a gaseous substance mediate the wound response. Every part of the plant has the capacity to respond to and heal a wound, but the response, by the simplest of devices, is limited to the damaged area.

The orderly abscission of senescent lower leaves may also represent a process that is effectively localized by the gaseous properties of the integrator.

D. Physiology and Mode of Action

Ethylene can regulate ripening, senescence, abscission, epinasty, swelling and elongation, hypertrophy, dormancy, hook closure, leaf expansion, flower induction, sex expression, and exudation (Abeles, 1972). As judged by dose–response relationships, the action of ethylene homologs, and the competitive action of carbon dioxide, each of these responses involves an identical ethylene receptor site.

The concentration of ethylene required to produce threshold effects in a variety of physiological responses is 0.01 ppm; half-maximal responses occur at 0.1 ppm. Saturation of the responses occurs at 10 ppm, and higher concentrations are generally not toxic. Also the relative effectiveness of ethylene homologs (Tables I and II) is approximately the same in the various responses. Double reciprocal plots of the physiological response as a function of ethylene concentration at different carbon dioxide concentrations indicate that carbon dioxide acts as a competitive inhibitor of ethylene in ethylene-mediated responses.

What are the further characteristics of the ethylene receptor? Formation of an active effector complex (as deduced from the data in Table II) requires that the effector have a terminal unsaturated bond. Activity of the effector is reduced by substitutions that might be expected to hinder approach to the unsaturated position, and activity is also reduced by substitutions that withdraw electrons from the unsaturated position. The order of physiological activity of olefins active in the ethylene bioassay is similar to their order in forming complexes with silver ions (Burg and Burg, 1967). This suggests that the receptor site includes a metal ion. This possibility is consistent with the observation that carbon monoxide,

TABLE I

Biological Activity of Ethylene and Other Unsaturated Compounds as Determined by the Pea Straight Growth Test[a]

Compound	K_A' relative to ethylene	Amount in gas phase (ppm) for half-maximum activity
Ethylene	1	0.1
Propylene	130	10
Vinyl chloride	2,370	140
Carbon monoxide	2,900	270
Vinyl fluoride	7,100	430
Acetylene	12,500	280
Allene	14,000	2,900
Methylacetylene	45,000	800
1-Butene	140,000	27,000
Vinyl bromide	220,000	1,600
Ethylacetylene	765,000	11,000

[a] From Burg and Burg (1967).

TABLE II

Biological Activity of Ethylene and Other Unsaturated Gases[a,b]

Compound	Relative concentrations for half-maximum activity			
	Inhibition of growth		Abscission	Epinasty
	Pea stem	Tobacco		
$CH_2{=}CH_2$	1	1	1	1
$CH_2{=}CH{-}CH_3$	100	100	60	500
$C{=}O$	2,700	1,600	1,250	5,000
$CH{\equiv}CH$	2,800	100	1,250	500
$CH_2{=}CH{-}CH_2{-}CH_3$	270,000	2,000	100,000	500,000

[a] From Yang (1974).
[b] CH_4, $CH_3{-}CH_3$, and $CH_3{-}CH{=}CH{-}CH_3$ are inactive.

at concentrations well below those required to inhibit cytochrome oxidase, mimics the physiological effects of ethylene. Carbon monoxide characteristically binds to (and usually inhibits) only those enzymes that incude a metal at their active site. The most likely possibilities for the metal at the ethylene receptor site would seem to be Cu^{2+}, Fe^{2+}, or Zn^{2+}.

Two facts suggest that the ethylene–receptor complex is readily dissociable: (1) most effects of ethylene cease soon after the removal of ethylene and (2) carbon dioxide acts as a competitive inhibitor. Alternate explanations for these facts can be devised; for example, ethylene might be covalently linked to its receptor with the protein moiety of the complex having only a short half-life. The kinetics of the carbon dioxide–ethylene "competition" might be fortuitous because there are, after all, many reactions involved between the formation of the ethylene–receptor complex and the measured physiological response. Nonetheless, the simplest explanation, the formation of a dissociable ethylene–receptor complex is most likely the best guide for further thinking about ways of identifying the receptor.

As already mentioned, oxygen is required for the biosynthesis of ethylene, and K_m, the Michaelis–Menten constant for O_2, is the same for ethylene biogenesis as it is for respiration. In addition, oxygen is required in the expression of the ethylene response. However, decreases in oxygen tension that do not lower the rate of respiration do lower the intensity of the ethylene response. The kinetics of the ethylene response at various oxygen tensions suggest that oxygen must bind with the ethylene receptor or bring about oxidation of the receptor before the receptor can form a physiologically effective complex with ethylene (Burg and Burg, 1967). The ideas derived from the kinetic models should be regarded as tentative and subject to further kinds of experimental verification.

There is as yet no clue to the intracellular localization of the ethylene receptor site(s). No one has yet reported a response to ethylene by a cell-free enzyme, organelle, or subcellular fraction. If there is a direct effect of ethylene on a membrane or membranes, it is most likely a highly specific effect that involves only a small fraction of the solutes that normally move through the membrane.

Further progress toward identification of the ethylene receptor and understanding of the mode of action of ethylene may well depend on the choice of system to be used for further study. Ripening fruit is an attractive material because a relatively large amount of tissue is available in each fruit and because all or nearly all cells present respond to the presence of ethylene. There is the disadvantage that the expression of the effects of ethylene treatment are visible only after hours or days, and it is therefore easy to confuse factors (e.g., inhibitors of protein synthesis) affecting the expression of the ethylene effect with factors required for the initial ethylene action.

The specialized cells involved in fruit and leaf abscission offer the advantage that the expression of the ethylene effect—abscission—apparently involves the synthesis and secretion of only a few cell wall dissolving enzymes, the principal one being cellulase. The effect of ethylene, at least in bean leaf explants, is at two stages (Rubinstein and Leopold, 1963; Abeles et al., 1971). In the first stage ethylene accelerates the aging of the abscission zone. This aging must include some development of the abscission zone cells, some preparation for the stage II that involves the synthesis (Lewis and Varner, 1970) and secretion of cellulase and, presumably, other cell wall softening enzymes. The effects of ethylene in the first stage can be delayed or prevented by the application of auxins to the abscission zone. The first stage is difficult to study because the changes occurring are largely unknown and not easily observable. The second stage, the synthesis and action of cell wall weakening enzymes, also requires ethylene. Progress through this stage is delayed by the application of auxin and is also delayed by the removal of ethylene. There is some evidence that ethylene is rather directly involved in the control of the secretion of cellulase during stage II in the bean explants (Abeles and Leather, 1971). This is of interest because it is possible to study secretion of enzymes separately from enzyme synthesis (Varner and Mense, 1972). Stage II can be quite short. The abscission of the flower pedicels of tobacco occurs after only 9 hours of ethylene treatment. During this treatment there is a proliferation of rough endoplasmic reticulum that is restricted to the abscission cells of the pedicel (Valdovinos *et al.*, 1971).

Just as fruit tissue responds to ethylene by producing the enzymes characteristic of ripening fruit and abscission, a great variety of tissues, in response to ethylene, produce characteristic enzymes. In many cases, perhaps in most, the increase in enzyme activity is most probably a result of enzyme synthesis rather than activation. In some tissues inhibitors of RNA synthesis will block the expression of ethylene action. It is reasonable to suppose that at least a part of the RNA synthesis is specifically required for the synthesis of the ethylene-evoked, tissue-specific enzymes. However, this has not yet been shown in any instance. Protein synthesis and RNA synthesis may be required to maintain cells in a healthy state competent to express a response to ethylene. Thus, there is necessarily some uncertainty in the interpretation of experiments in which an inhibitor prevents the usually observed response. Responses that may occur in too short a time after ethylene treatment to allow the involvement of RNA synthesis and protein synthesis are therefore of great interest (Warner, 1970; Burg *et al.*, 1971).

The etiolated pea seedling is an attractive experimental tissue because all parts of the seedling respond to ethylene: "stem growth slows, the hook tightens, the subapex swells and nutates horizontally, root growth slows and the zone of elongation swells, root hairs form, lateral root formation is inhibited, and the root tip bends plagiotropically" (Burg *et al.*, 1971). The subapical swelling occurs 3 to 4 hours after the application of ethylene, requires RNA synthesis and protein synthesis, and is accompanied by a marked decrease in the incorporation of hydroxyproline-containing peptides into the wall and an alteration in the birefringence pattern of the wall. Colchicine and vinblastin, agents known to disrupt microtubules, also cause swelling of the subapex and inhibit deposition of hydroxyproline peptides in the wall.

Subapical cells also swell in response to added benzimidazole, benzyladenine and kinetin. These cytokinins do not cause swelling by enhanced ethylene synthesis. Ethylene treatment decreases the rate of transport of indoleacetic acid (IAA) through the subapical tissue and the levels of diffusable and extractable IAA in the tissue. It is clear that the expression of the response of the subapex to ethylene is complex.

In many respects the effects of ethylene on tropistic and epinastic behavior (Burg *et al.*, 1971) are attractive phenomenological starting points for the search for the ethylene receptor—the primary site of action. These effects, hook tightening in etiolated seedlings, leaf epinasty, horizontal nutation in stems, plagiotropism in roots, apparently result from the inhibition by ethylene of the lateral transport of auxin and are

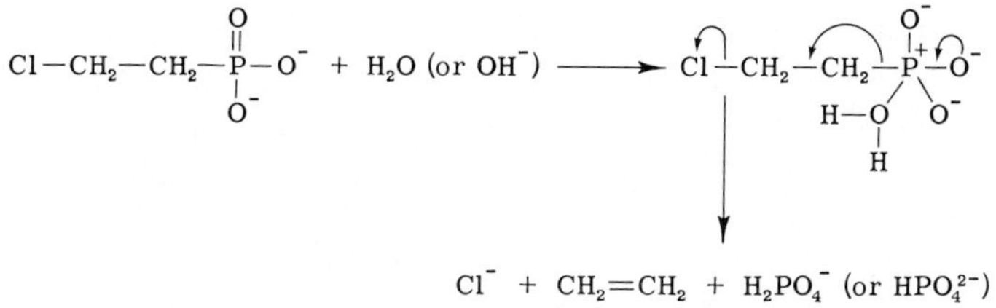

$$Cl-CH_2-CH_2-\overset{\overset{O}{\|}}{\underset{\underset{O^-}{|}}{P}}-O^- + H_2O \text{ (or } OH^-) \longrightarrow Cl-CH_2-CH_2-\overset{O^-}{\underset{\underset{\underset{H}{|}}{H-O}}{\underset{|}{P}}}\overset{}{\underset{O^-}{\overset{+}{P}}}-O^-$$

$$Cl^- + CH_2{=}CH_2 + H_2PO_4^- \text{ (or } HPO_4^{2-})$$

Fig. 4. Hydrolysis of 2-chloroethylphosphonic acid.

visible within minutes after the application of ethylene. Because the immediate effects on elongation of changed auxin concentrations are visible in 0–10 minutes and do not require RNA synthesis or protein synthesis, the effect of ethylene on lateral transport of auxin promises to be close to the primary site of action of ethylene.

As a part of the response of a tissue to ethylene, there is often a marked increase in the activity of one or more easily measurable enzymes (Abeles, 1972). In general, these increased enzyme activities seem to be at some distance in time from the initial site of ethylene action. Nonetheless these phenomena are of interest because they represent the expression of the tissue's capabilities. The increase in phenylalanine ammonia-lyase and peroxidase induced by ethylene in sweet potato root provides a system convenient for further study.

The regulation by ethylene of aging in the flowers of *Ipomoea tricolor* also provides a dramatic process, convenient for studies of the mechanism of action of ethylene. In this tissue ethylene promotes ethylene production, RNase synthesis and senescence of the corolla (Kende and Baumgardner, 1974).

The introduction of a compound, (2-chloroethyl)phosphonic acid (Maynard and Swan, 1963), which is stable below pH 4 and is slowly converted to ethylene (Fig. 4) after foliar application, has been of considerable commercial importance for the control of flowering, dormancy, abscission, ripening, disease resistance and latex production (de Wilde, 1971).

II. Cytokinins

A. Discovery

Cytokinin (Skoog *et al.*, 1965) is a generic name for substances that promote cytokinesis in cultured plant cells and also serve other regulatory functions similar to those of kinetin, the first chemically defined cytokinin.

The discovery of kinetin (Miller *et al.*, 1955a) derived from the use of excised tobacco pith tissue in culture and the observation that vigorous growth required, in addition to auxin, some factor present in coconut milk [van Overbeek *et al.* (1942) found that excised plant embryos required some coconut milk factor for centinued growth and cell division] and in extracts of yeast and malt (Jablonski and Skoog, 1954). This factor was deduced to have the properties of a purine, and testing of DNA hydrolysates led to the identification of N^6-furfurylaminopurine (kinetin Fig. 5) (Miller *et al.*, 1955b) as a factor that is effective at low concentrations (10^{-6} to 10^{-11} M). There was already a clue that one should expect such a factor to be a purine because it has been observed that the tendency for the formation of roots and buds on excised stem segments of tobacco was determined by the relative proportion of adenine and indoleacetic acid in the medium (Skoog and Tsui, 1948). The appreciation that a factor (or factors) in addition to indoleacetic acid could be involved in the promotion and suppression of the growth of differentiated cells (roots and buds) (Skoog *et al.*, 1942) stemmed from early work (Skoog and Thimann, 1934) on the inhibitory function of auxin in apical dominance—suppression of lateral buds—in beans and peas.

Zeatin, from immature corn kernels, was the first naturally occurring cytokinin to be isolated and identified (Letham *et al.*, 1964; Letham and Miller, 1965; Miller, 1961). Descriptions of the discovery of the known natural cytokinins (Table III) are given by Hall (1973), Kende (1971), and Skoog and Armstrong (1970).

B. Structure–Activity Relationships

Kinetin apparently does not occur naturally (Table III shows the known naturally occurring cytokinins), and its production in the laboratory as a degradative product of DNA has no known physiological parallel. In addition to kinetin there are hundreds of other biologically active synthetic cytokinins. The most effective of these are N^6-substituted adenine derivatives (Skoog *et al.*, 1967). Any substitutions of one atom for another in the adenine ring or for the N in the N^6-position results

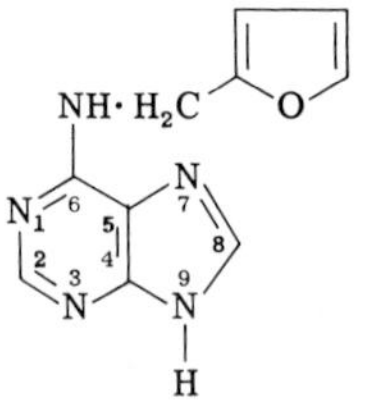

Fig. 5. Kinetin [6-(furfurylamino)purine].

TABLE III The Free Bases of Natural Cytokinins

	R_1	R_2	Chemical name	Common name or abbreviation	Concentration (M) for half-maximal response[a]
(I)	$-CH_2-CH=C\big(\begin{smallmatrix}CH_3\\CH_2OH\end{smallmatrix}\big)$	H	6-(4-Hydroxy-3-methyl-trans-2-butenyl)aminopurine	trans-Zeatin	5×10^{-9}
(II)	$-CH_2-CH=C\big(\begin{smallmatrix}CH_2OH\\CH_3\end{smallmatrix}\big)$	H	6-(4-Hydroxy-3-methyl-cis-2-butenyl)aminopurine	cis-Zeatin	10^{-7}
(III)	$-CH_2-CH_2-C\big(\begin{smallmatrix}H\ CH_2OH\\CH_3\end{smallmatrix}\big)$	H	6-(4-Hydroxy-3-methylbutyl)aminopurine	Dihydrozeatin	3×10^{-8}
(IV)	$-CH_2-CH=C\big(\begin{smallmatrix}CH_3\\CH_3\end{smallmatrix}\big)$	H	6-(3-Methyl-2-butenyl)aminopurine	IPA	10^{-8}
(V)	$-CH_2-CH=C\big(\begin{smallmatrix}CH_2OH\\CH_3\end{smallmatrix}\big)$	CH_3S-	6-(4-Hydroxy-3-methyl-2-butenyl)2-methylthioaminopurine	CH_3S-Zeatin (cis or trans)	10^{-8}
	$-CH_2-CH=C\big(\begin{smallmatrix}CH_3\\CH_2OH\end{smallmatrix}\big)$	CH_3S-	6-(4-Hydroxy-3-methyl-2-butenyl)2-methylthioaminopurine	CH_3S-Zeatin (cis or trans)	
(VI)	$-CH_2-CH=C\big(\begin{smallmatrix}CH_3\\CH_3\end{smallmatrix}\big)$	CH_3S-	6-(3-Methyl-2-butenyl)2-methylthioaminop urine	CH_3S-IPA	6×10^{-8}
(VII)	$-\overset{\displaystyle O}{\overset{\|}{C}}-NH-CH-CH-CH_3$ $\quad\quad HOOC\ \ OH$	H-	6-(Threonylcarbamoyl)purine	—	?

[a] Molar concentration for approximate half-maximal response in tobacco callus growth assay.

in a loss or complete elimination of activity in the tobacco tissue bioassay. When the N^6-substituent is an alkyl group, as in the N^6-alkylaminopurines, the optimum length of the side chain is five carbon atoms and N^6-pentylaminopurine has the same activity as kinetin. The activity of N^6-(3-methylbutyl)aminopurine is the same as that of N^6-pentylaminopurine. Introduction of a double bond to form N^6-(3-methyl-2-butenyl)aminopurine increases the activity tenfold. Introduction of the hydroxyl group to form the 4'-hydroxy derivative, N^6-(4-hydroxy-3-methyl-*trans*-2-butenyl)aminopurine, further increases the activity. Many N^6-(3-methylbutyl)aminopurine compounds have been synthesized and tested, and it appears that substituents tending to make the side chain more planar increase while those tending to make it less planar decrease biological activity (Hecht *et al.*, 1970). Modification of the purine ring with the 2-methylthio and/or the 9β-D-ribofuranosyl group leads to systematic decrements in biological activity.

Ring substituents in the N^6-position can also confer high cytokinin activity on adenine. The benzyl ring is most effective; the furfuryl, phenyl, and thienyl rings are less effective and the cyclohexyl ring much less effective. Other ring substituents produce cytokinin which have still less activity. (For a review of the extensive literature on the chemical structure/biological activity relationships of cytokinins see Leonard, 1974.)

Although zeatin has the highest activity of any of the natural cytokinins, some synthesized derivatives of zeatin have an even higher activity. For example, the formate, acetate, propionate and indole-3-acetate esters of 2-chlorozeatin have twice the activity (on a molar basis) of zeatin (Leonard, 1974). Also 8-methylbenzyladenine, 8-methylkinetin (Kulaeva *et al.*, 1968) and 6-(3-methyl-2-butenylamino)-8- methylpurine are more active than the unsubstituted parent compounds (Leonard, 1974).

Diphenylurea, originally isolated as one of the factors of coconut milk with cytokinin-like activity (Shantz and Steward, 1955), and other substituted ureas elicit a spectrum of responses quite similar to those elicited by the N^6-adenine derivatives (Bruce *et al.*, 1965). This, of course, suggests that the two classes of compounds act through the same or similar mechanisms (Kende, 1971), even though they appear to have few structural features in common.

C. Physiological Responses to Cytokinins

The response of excised tobacco pith tissue to exogenous cytokinins is only one of many responses of plant tissue to added cytokinins. The growth of cultured tissue from soybean cotyledons is about equally sensitive. Suspension cultures of some strains of tobacco cells and of *Acer pseudoplatanus* (Digby and Wareing, 1966) require added cytokinins for growth and may be advantageous for kinetic studies of growth. Soybean tissue in liquid culture synthesizes measurable and proportionate quantities of two deoxyisoflavones after only 24 hours of treatment with cytokinins (Miller, 1969). This response is not known to be related to the growth response: nonetheless it deserves further attention because it is observable in a relatively short time.

Cytokinins enhance DNA synthesis, and elongation in the hypocotyl of rootless soybean seedlings induces tuberization in excised potato stolons, overcomes the inhibitory effect of abscisic acid on the growth of *Lemna minor,* promotes the formation of tyramine methylpherase activity in roots of germinating barley and the formation of isocitrate lyase and protease activity in excised squash cotyledons, removes the thiamine requirement for growth of tobacco callus (Wis. #38) by inducing thiamine synthesis, induces auxin synthesis in tobacco tissue cultures, and enhances the activities of carboxydismutase and NADP-glyceraldehyde-phosphate dehydrogenase in etiolated rice seedlings, promote bud development, promote germination of some seeds, and promote the accumulation of nitrate reductase in some embryos (Skoog and Armstrong, 1970; Kende, 1971).

Since the observation that kinetin delays the senescence of detached cocklebur leaves (Richmond and Lang, 1957), it has been found that cytokinins delay senescence of detached leaves of many species.

It appears that one or more of the cytokinins is the hypothetical root hormone proposed by Chibnall (1939) to account for the fact that the aging of detached leaves was prevented or reversed by the formation of adventitious roots. Cytokinins have been found in the xylem exudate of many different plants. One of the natural causes of the aging and senescence that is associated with growth cessation and flowering may well be a decreased supply of cytokinins from the root to the shoot (see Chapter 23). The appearance of stress symptoms in the shoots of plants whose roots are exposed to low water potential, salt stress, or flooding may also result from the inability of such stressed roots to supply normal quantities of cytokinins to the shoots (Kende, 1971). In addition, cytokinin can, within limits, prevent the yellowing and senescence of heat-stressed leaves—both attached and detached.

Cytokinins promote the development of buds inhibited by adjacent or apical buds or by applied auxin. The growth patterns of tobacco callus is determined by the ratio of cytokinins, auxins, and gibberellins (Skoog and Miller, 1957). With no addition of hormones there is little growth.

Addition of indoleacetic acid allows growth with no differentiation. Addition of kinetin allows only growth. Addition of both kinetin and indoleacetic acid allows rapid growth and differentiation. Calli treated with relatively high ratios of indoleacetic acid to kinetin yield predominantly roots, while calli treated with relatively low ratios of indoleacetic acid to kinetin yield predominantly buds and leaves. If N^6-(3-methyl-2-butenyl)aminopurine is used alone, rather than kinetin, the callus grows rapidly and forms organs (leaves) apparently because it (in contrast to kinetin) induces indoleacetic acid syntheses in the callus.

Calli treated with high isopentenyladenine (IPA) to gibberellic acid (GA_3) ratios produce short green plants with rounded leaves, while those calli treated with high GA_3/IPA ratios produce slender etiolated plants with narrow leaves.

Bud formation in moss protonemata in culture is cytokinin-induced and is observable within 10 to 18 hours of addition of the cytokinin. The added cytokinin accumulates only in those cells that are capable of budding. Dedifferentiation of the buds occurs if the cytokinin is removed before the differentiated state has stabilized (Kende, 1971).

D. Molecular Basis for the Tissue Response

The finding of cytokinins in transfer RNA's (tRNA's) (Hall et al., 1966; Biemann et al., 1966; Madison et al., 1966) raised the exciting possibility that cytokinins might act by virtue of their incorporation into the structure of certain tRNA's. However, it now appears that exogenous cytokinins are not incorporated in significant amounts into tRNA's but rather that the substitution of the various side chains on the N^6 of adenine is accomplished after the tRNA molecule is formed. In addition, 9-methylbenzyladenine, which cannot be converted to the riboside triphosphate for incorporation into tRNA, is nonetheless active as a cytokinin (Kende and Tavares, 1968).

Tobacco callus cells synthesize N^6-(3-methyl-2-butenyl)adenosine by attachment of a side chain to an adenosine residue of preformed tRNA (Chen and Hall, 1969). A similar synthetic pathway occurs in other organisms. Thus, the turnover of the tRNA's is a possible source of cytokinins. However, the level of free cytokinins in pea root tips is 27 times as much as could be obtained by hydrolysis of the tRNA of the cells. One must therefore propose a high turnover rate of tRNA or else a second path for the biosynthesis of cytokinins-(presumably the addition of the appropriate side chain to adenine) (Short and Torrey, 1972).

Cytokinins apparently occur in certain tRNA's in all organisms. The first cytokinin identified as a constituent of tRNA was IPA (Zachau

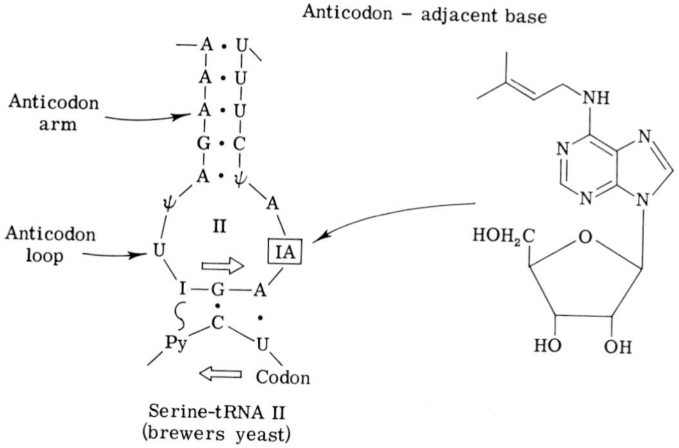

Fig. 6. The position of isopentenyladenosine in serine-tRNA. From Zachau et al., 1966.

et al., 1966; Hall et al., 1966). It was shown to be present only once in each of two tRNA^{Ser} species (Zachau et al., 1966) and was adjacent to the 3′ end of the anticodon in both species (Fig. 6). A short time later tRNA^{Tyr} was shown to contain CH_3S–IPA (Harada et al., 1968) adjacent to the 3′ end of the anticodon (Fig. 7).

The presence of the cytokinin base in this position helps to determine the conformation of the anticodon loop. Modification of the IPA of yeast tRNA^{Ser} by treatment with aqueous iodine impaired its binding by ribosomes but did not alter its ability to accept serine (Fittler et al.,

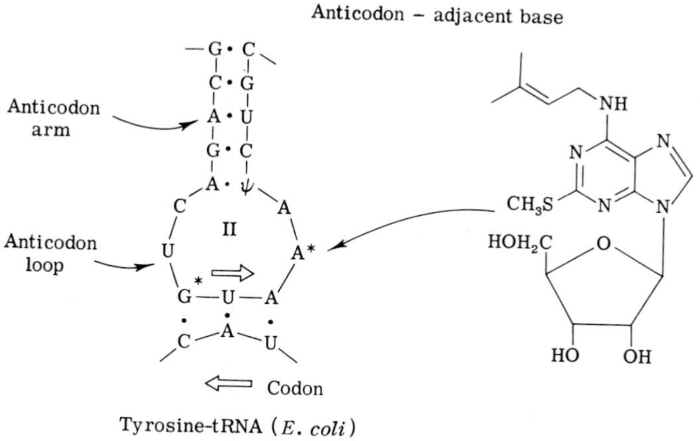

Fig. 7. The position of thiomethylisopentenyladenosine in tyrosine-tRNA. From Harada et al., 1968.

1968). Mutants of *E. coli* suppressor tRNA^{Tyr} that contain an adenosine in place of the CH$_3$S–IPA accept amino acids as well as the normal tRNA^{Tyr} but are ineffective in *in vitro* tests of suppression and are poorly bound by ribosomes (Gefter and Russell, 1969). Thus it is clear that cytokinin bases have an important role in certain tRNA's. However, this role does not seem to be associated with the role of cytokinins as hormones.

IPA and its derivatives occur only in those species of tRNA that respond to codons beginning with uridine (Armstrong *et al.*, 1969a,b; Peterkofsky and Jesensky, 1969).

Another modified adenosine, 6-(threonylcarbamoyl)purine, is found exclusively at the 3′ end of the anticodon of those species of tRNA that respond to codons beginning with adenosine (Skoog and Armstrong, 1970). Neither this base nor its riboside has growth-promoting activity in the soybean callus assay or in the tobacco callus assay. This lack of activity, however, may be due to lack of uptake into the cells because of the polarity of the side chain. Less polar ureidopurine analogs do have activity in the soybean callus assay, and those analogs most resembling 6-(threonylcarbamoyl)purine in three-dimensional and electronic configuration have the most activity (Dyson *et al.*, 1972a).

TABLE IV

Metabolic Reactions of Cytokinins in Plant Tissues

Reaction	Reference
Adenine in tRNA $\xrightarrow[\text{(EC 2.5)}]{\text{isopentenyl transferase}}$ IPA in tRNA	Chen and Hall, 1969
IPA riboside → zeatin riboside	Miura and Miller, 1969; Miura and Hall, 1973
IPA riboside → adenosine → adenine	Paces *et al.*, 1971; Whitty and Hall, 1972
IPA riboside → IPA	Whitty and Hall, 1974
IPA → N^6-(3-methyl-3-hydroxybutyl)aminopurine	Miura and Miller, 1969
Zeatin → zeatin riboside → zeatin ribotide	Sondheimer and Tzou, 1971
Dihydrozeatin → dihydrozeatin riboside → dihydrozeatin ribotide	
Zeatin riboside → adenosine ↓ zeatin	Paces *et al.*, 1971; Whitty and Hall, 1974
Zeatin → 7-glucosylzeatin	Parker *et al.*, 1972
Zeatin → 9-glucosylzeatin	Yoshida and Oritani, 1972; Letham, 1973

TABLE V

Metabolic Reactions of Some Synthetic Cytokinins in Plant Tissues

Reaction	Reference
Benzyladenine → benzyladenosine → benzyladenosine-5'-phosphate	McCalla et al., 1962; Dyson et al., 1972b
Benzyladenine → benzyladenine-7-glucoside	Deleuze et al., 1972
Kinetin → kinetin riboside → kinetin riboside monophosphate	Doree and Guern, 1973

E. Metabolism

The main features of cytokinin metabolism are shown in Tables IV and V.

F. Antagonists of the Cytokinins

The availability of a cytokinin antagonist could extend the study of cytokinins to tissues that produce their own cytokinins. A series of such antagonists have been synthesized (Skoog et al., 1973) by systematic modification of the side chain, by interchanging the C and N atoms of the 8- and 9-positions of the purine ring, and substitution with a methyl group of the equivalent of the 9-position of the purine nucleus (Fig. 8). One of the more potent of the antagonists is 3-methyl-7-(3-methylbutylamino)pyrazolo[4,3-d]pyrimidine (Fig. 8). This compound is not only devoid of cytokinin activity but also inhibits the growth of tobacco callus supplied optimal IPA or benzyl amino purine (BAP). It also inhibits the growth of a strain of tobacco callus that requires no exogenous cytokinin. It therefore seems likely that this class of antagonists will be useful in the study of those tissues that use endogenous cytokinins.

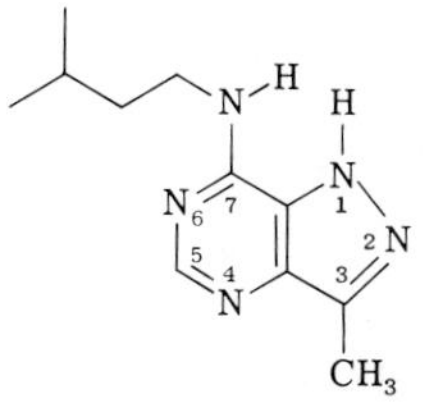

Fig. 8. 3-Methyl-7-(3-methylbutylamino)pyrazolo-[4,3-d]pyrimidine.

III. Auxins

A. Occurrence of Indoleacetic Acid and of Other Auxins

In spite of the widespread evidence for auxin (as determined by bioassay on chromatographically purified extracts) in plant tissues, rigorous chemical identification and quantitative estimation have seldom been made. However, the auxin in diffusates of maize coleoptile tips was identified by mass spectrometry as indoleacetic acid (IAA) (**I**) (Greenwood *et al.*, 1972). Methyl-4-chloroindolyl-3-acetate (**II**), 4-chloroindolyl-3-acetate (**II**) and methyl-4-chloroindolyl-3-acetyl-L-aspartate (**III**) have

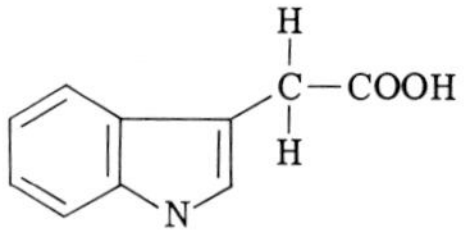

Indol-3-yl-acetic acid
(indole-3-acetic acid;
indoleacetic acid; IAA)

(I)

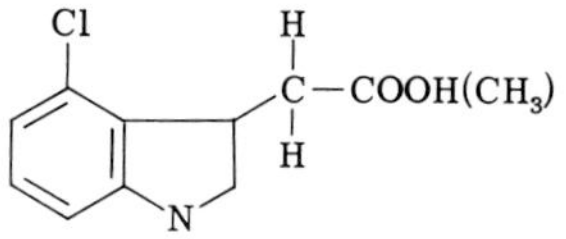

4-Chloroindolyl-3-acetic acid
(methyl-4-chloroindolyl-
3-acetate)

(II)

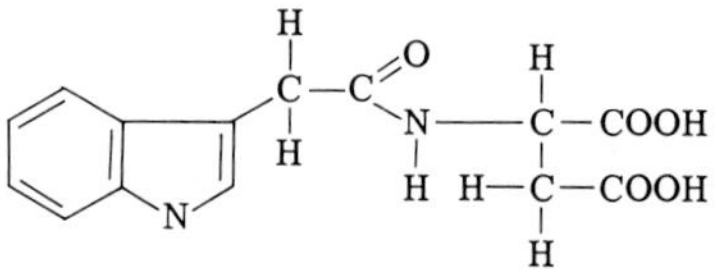

Indolyl-3-acetylaspartate

(III)

been reported to be present in immature pea seeds (Hattori and Marumo, 1972; Marumo *et al.*, 1968a,b). (No IAA or methyl-IAA was found.)

Also among the bound forms of IAA, IAA-glucose (Zenk, 1961), IAA-myoinsitols, IAA-myoinositol arabinosides and IAA-myoinositol galactosides (Ueda *et al.*, 1970), and IAA-glucans (Psornik and Bandurski, 1972) have been rigorously identified in maize seeds. Exogenous labeled IAA has been recovered as IAA-β-D-glucose (see Schneider and Wightman, 1974), IAA-myoinositol, IAA-glucans (Kopcewicz *et al.*, 1974), and IAA-aspartate (Andreae and Good, 1955; Schneider and Wightman, 1974).

Enzyme preparations from immature corn (Zea mays) kernels catalyze an ATP and CoA dependent formation from indole acetic acid and inositol or glucose of 2-*O*-(indole-3-aceytl)myoinositol, 1-DL-1-*O*-(indole-

3-acetyl)myoinositol, di-O-(indole-3-acetyl)myoinositol, tri-O(indole-3-acetyl)myoinositol, 2-O-(indole-3-acetyl)D-glucopyranose, 4-O-(indole-3-acetyl)-D-glucopyranose, and 6-O-(indole-3-acetyl)-D-glucopyranose (Kopcewicz *et al.*, 1974). The capacity of pea tissue to make IAA-aspartate from exogenous IAA is increased by pretreatment of pea stem sections with an active auxin, while inactive auxin analogs are ineffective in the pretreatment (Venis, 1972). There is evidence that phenylacetic acid is formed in shoot tissues of barley, maize, pea, tobacco, and tomato from fed [3-^{14}C]phenylalanine. Phenylalanine also occurs, according to bioassay and chromatographic evidence, in extracts of tomato and sunflower shoots not fed phenylalanine (Schneider and Wightman, 1974). Evidence for the possible occurrence of other nonindolic auxins is summarized by Schneider and Wightman (1974).

B. Biosynthesis of Indoleacetic Acid and of Phenylacetic Acid

Tryptophan is almost certainly the primary precursor of IAA (Schneider and Wightman, 1974). Although this has been thought likely since the observation that *Rhizopus suinus* cultures form IAA when tryptophan is supplied in the medium (Thimann, 1935), and although many *in vivo* experiments show the conversion of [^{14}C]tryptophan to [^{14}C]IAA, the accumulation of convincing evidence for this was slow because of the complications arising out of the presence of various microorganisms in most plant tissues (Libbert *et al.*, 1966). Work with sterile *Avena* coleoptile tissues (Black and Hamilton, 1971) (that indicates that exogenous tryptophan does not completely equilibrate with the endogenous pool of tryptophan important in IAA synthesis) and with sterile pea plants (that shows that in double-labeling experiments the ratio of ring label to side chain label in IAA and in tryptophan does not change irrespective of the addition of either labeled indole or labeled serine) demonstrates that in these tissues tryptophan is the precursor of IAA. In cucumber hypocotyls the conversion of [^{14}C]tryptophan to [^{14}C]IAA occurred under sterile conditions, and the growth response to tryptophan was as great under sterile conditions as under nonsterile conditions. Cucumber hypocotyls also convert [^{14}C]tryptamine to [^{14}C]tryptophol (indole ethanol) and the growth response to tryptophan and to tryptophol was not decreased by sterile conditions (Sherwin and Purves, 1969).

A comprehensive series of experiments in Wightman's laboratory (see Wightman, 1973) with labeled tryptophan with tomato shoots and with extracts of tomato shoot tissues leave little doubt that the enzymes are present in the extracts for the conversion of tryptophan to IAA by the indolepyruvate pathway and by the tryptamine pathway (Fig. 9) and that both pathways function *in vivo* with the indolepyruvate pathway

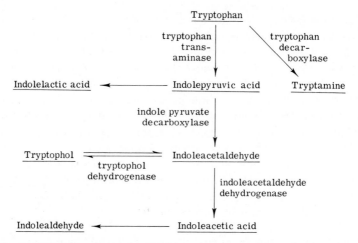

Fig. 9. Probable pathway for IAA biosynthesis in tomato shoots. From Wightman, 1973.

being quantitatively more important. Young expanding leaves, rather than the shoot tips, are the most active sites of IAA synthesis from tryptophan. Senescent leaves may also be a site of IAA synthesis (Sheldrake and Northcote, 1968).

Tomato shoots also convert [^{14}C]phenylalanine to [^{14}C]phenylacetic acid (Fig. 10) most probably by pathways similar to those for the conversion of tryptophan to IAA. And phenylacetic acid has been isolated from untreated tomato shoots (Wightman, 1973).

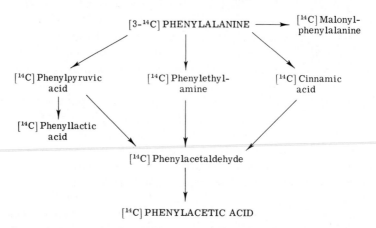

Fig. 10. Probable pathways for the metabolism of DL-phenylalanine to [^{14}C]-phenylacetic acid (and other phenyl compounds) in tomato shoots. From Wightman, 1973.

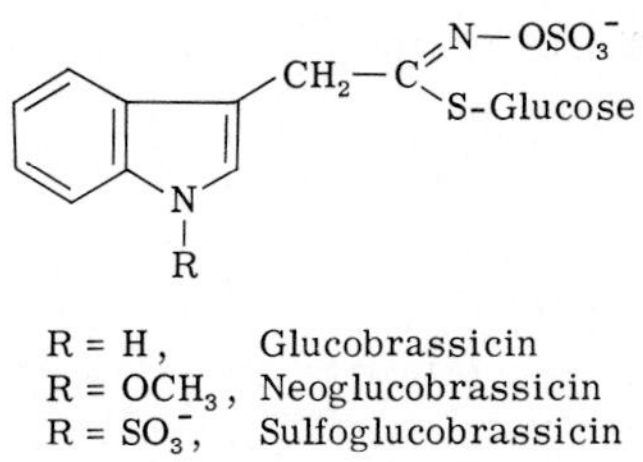

R = H,   Glucobrassicin
R = OCH_3, Neoglucobrassicin
R = SO_3^-, Sulfoglucobrassicin

Fig. 11. Structural formulas of the glucobrassicins. From Mahadevan and Stone, 1972.

Cruciferous plants have a high indole content (Kutacek and Kefeli, 1968) largely due to the presence of indole glucosinolates (Fig. 11). Woad (*Isatis tinctoria* L.) seeds contain as much as 0.23% glucobrassicin (Elliott and Stowe, 1971), and the roots of seedlings release glucobrassicin and neoglucobrassicin—possibly explaining the long-known deleterious effects of woad and of other crucifers on subsequent crops. Indoleacetaldoxime-derived from tryptophan is a precursor of the glucobrassicins (Fig. 12) (Kindl, 1968; Mahadevan and Stowe, 1972).

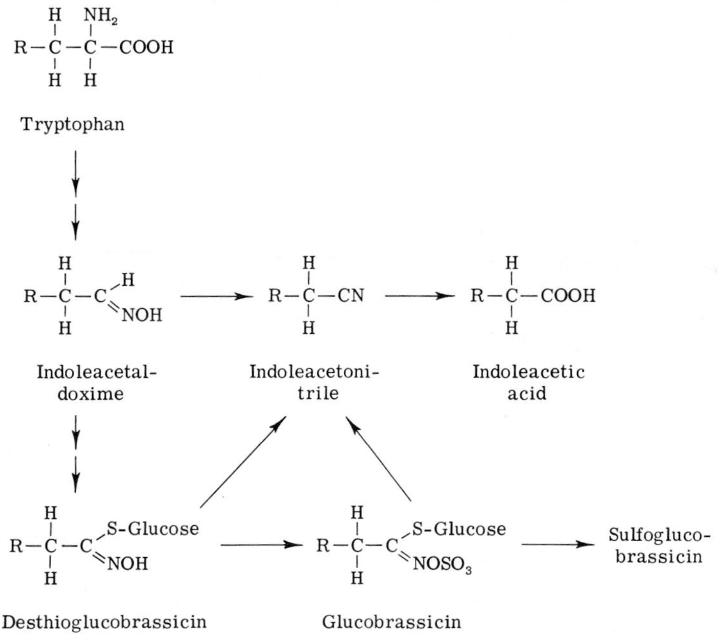

Fig. 12. Summary of the pathways from tryptophan to the indole glucosinolates and to IAA in crucifers. (Cystine is an effective source of the sulfide-sulfur, UDP-glucose is the apparent source of the —S-glucose, and presumably adenosine-5′-phosphosulfate (APS) is the source of the —OSO₃ groups.)

Apparently all crucifer tissues that can form indoleacetonitrile also can convert it to indoleacetic acid. Thus the tryptophan, indoleacetaldoxime, indoleacetonitrile, indoleacetic acid pathway must be one of the pathways for the biosynthesis of indoleacetic acid in the crucifers.

Indoleacetaldoxime promotes growth in wheat coleoptiles and in pea stem segments, and indoleacetonitrile promotes growth in a large number of different kinds of coleoptiles and hypocotyl tissues. Thus it is possible that the aldoxime–acetonitrile pathway has some physiological significance in noncruciferous tissues (see Schneider and Wightman, 1974).

C. Oxidation of Indoleacetic Acid

Indoleacetic acid is oxidized *in vivo* and in cell-free extracts to a series of compounds (Fig. 13) with methylene oxindole as the principal product. This oxidation is catalyzed by several of the peroxidase iso-

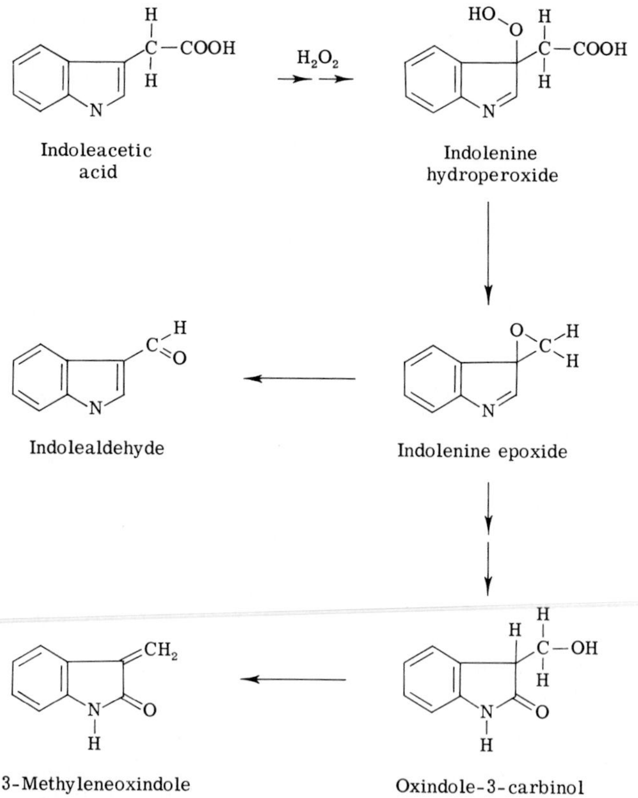

Fig. 13. Oxidative catabolism of indoleacetic acid. From Hinman and Lang, 1965.

zymes and apparently also by indoleactic acid oxidases that lack peroxidase activity (see Schneider and Wightman, 1974). These reactions are of interest because of their possible significance in the control of the physiological activity of indoleacetic acid either by destroying the active form (i.e., indoleacetic acid) or by producing the active form (methyleneoxindole) (Tuli and Moyed, 1969; Basu and Tuli, 1972). At the present methyleneoxindole is not generally accepted as having a function in growth regulation (Schneider and Wightman, 1974; Ray, 1974). Methyleneoxindole is not stable and reacts readily with sulfhydryl groups, and thus quantitative experiments with it are difficult.

Acceptance of the idea that indoleacetic acid oxidase is important *in vivo* in controlling the physiologically active concentration of indoleacetic acid allows for the possibility of a role of indoleacetic oxidase inhibitors (and activators), and such inhibitors or "auxin protectors" have been reported (Stonier *et al.*, 1970).

D. Physiological Responses of Tissues to Auxins

The classical effect of auxins is to promote cell enlargement. Whatever the primary site(s) of action in the production of this response, many of the observed effects of auxin are probably secondary to and supportive of cell enlargement. Sustained enlargement requires maintenance of all the machinery needed for cell enlargement and includes respiration, replacement of short-lived protein, RNA, etc.

Auxin is produced by and promotes growth in shoots and is transported basically by a polar transport system. Altered lateral transport accounts for geotopism and phototropism. Auxins also induce root formation, inhibit root elongation (through the mediation of ethylene produced in response to added auxin), induce vascular differentiation, control abscission, induce β-1,4-glucanases in pea roots, induce (through the mediation of ethylene) fading in flowers, prevent senescence in excised bean endocarp, induce tropic responses, enhance elongation of internodes, promote growth of the ovary into fruit, promote cell division and enlargement in callus tissue, bring about apical dominance, induce ethylene synthesis (ethylene inhibits elongation and promotes diametric expansion), promote cytoplasmic streaming, increase wall elasticity and plasticity, and cause hydrogen ion extrusion.

E. Possible Modes of Action of Auxin

Indoleacetic acid causes an increase in elongation rate in auxin-sensitive tissues within 10 to 15 minutes. Because an indoleacetic acid-dependent elongation is observable in the presence of cycloheximide, the primary action must not require new protein synthesis (Pope and Black, 1972).

Treatment of pea stem segments with indoleacetic acid doubles the rate at which they incorporate [^{14}C]glucose into cell wall polymers. A change in rate can be seen within 15 minutes of the hormone treatment (Abdul-Baki and Ray, 1971).

Treatment of pea stems with indoleacetic acid also causes a 2- to 4-fold increase within 1 hour in the particulate UDPglucose:1,4-β-glucan glucosyltransferase activity found in the extracts of the treated tissue (Ray, 1973). This increase is the result of enzyme activation rather than of enzyme synthesis.

In contrast to the effect of indoleacetic acid, ethylene treatment of etiolated peas decreases stem elongation and increases the incorporation of hydroxyproline-rich proteins into the cell walls (Sadava and Chrispeels, 1973).

Low pH (pH 3 to 4) was observed by Bonner in 1934 (cited in Rayle and Cleland, 1970) to induce elongation in auxin-sensitive tissue. This elongation begins within seconds of the change in pH and is not suppressed by metabolic inhibitors or lack of oxygen (Rayle and Cleland, 1970; Evans et al., 1971). It is now proposed that a primary effect of added auxin is to cause the cell to lower the pH of the medium that permeates the cell walls (Hager et al., 1971; Rayle and Cleland, 1972). This is presumed to result from a pumping out of H$^+$ ions, and auxin-dependent decrease in the pH of the medium surrounding oat coleoptile sections have been reported (Cleland, 1973; Rayle, 1973).

Whether the low pH activates enzymes already in the wall that mediate cell extension, e.g., glycosyltransferases (Johnson et al., 1974), or directly disrupts association among cell wall polymers is not known.

Low pH promotes (within 2 minutes) chloride uptake into Avena coleoptile cells. This uptake is not dependent on growth, i.e., it is not prevented by 0.3 M mannitol, but it is prevented by respiratory inhibitors. Auxin promotes an entirely similar chloride uptake (Rubinstein, 1974).

If an early effect of indoleacetic acid is to promote the secretion of wall-modifying enzymes and of wall precursors, it is obvious that a sustained response of a growing tissue to indoleacetic acid would require concomitant protein synthesis and RNA synthesis. However the immediate response (an increase in rate of cell extension) probably requires neither protein synthesis nor RNA synthesis. Thus although gene activation might be required for the sustained response, it would not be required for the primary response.

Cytokinins inhibit auxin-promoted elongation in auxin-sensitive tissues of dicotyl plants (Vanderhoef and Stahl, 1975, and references therein). In soybean hypocotyl segments, auxin evokes two overlapping growth responses. The first response begins after 12 minutes, reaches a maximum at 28 minutes, and decreases to a minimum at 50 minutes. The beginning of the second response is obscured by the first, but its maximum is at 71 minutes after auxin addition. Isopentenyladenine inhibits the second response but not the first. Low pH elicits the first response but not the second.

Many auxin analogs have been synthesized and tested for activity in various tissues. As a result it is known that the structural requirements for an auxin are (1) a ring with at least one double bond, (2) a side chain adjacent to the double bonds, and (3) a carboxyl group separated from the ring by one or two carbons. Although this tells us something about the shape of the receptor site, it has not been helpful in deducing possible mechanisms of action.

The auxin-induced elongation of pea stem sections is increased as much as 100% by micromolar concentrations of alkyl compounds that are 20 to 30 Å long. Longer and shorter molecules are less effective or ineffective (Stowe and Dotts, 1971). It is likely that this effect is the result of the interaction of the alkyl derivatives with one or more membranes, but it is not possible to conclude that auxin is in anyway involved in this interaction.

It is clear that we cannot yet explain the primary action of auxins in the promotion of elongation much less explain their effects in the many other responses to auxins.

IV. Abscisic Acid and Related Compounds

A. Discovery

The intensive studies about the physiology of abscission and bud dormancy during the 1950's led to the discovery of abscisic acid. Initially these studies were focused on the role of auxins. However, it later became apparent that another controlling factor formed when leaves reached senesence was capable to accelerates abscission. In 1955, Osborne published the first evidence for the existence in several plant species of a diffusable abscission-accelerating substances. Later works by various researchers confirmed the results, and endogenous abscission-accelerating substances became known to occur widely in higher plants. One of the most abundant sources of the abscission-accelerating substances is the fruit of cotton from which abscisic acid, then called abscisin II, was isolated and characterized by Addicott, Carns and their co-workers (Liu and Carns, 1961; Ohkuma *et al.*, 1963). In the early 1960's it was known that the bud growth of some woody perennials was partly regulated by some growth-inhibitory compounds (dormin). In sycamore and birch a change from long-day to short-day conditions produces a marked increase in dormin activity and the buds stop growing. Dormin-containing extracts of leaves from sycamore or birch plants grown under short-day condition inhibit the growth and induce the formation of resting buds in actively growing seedlings maintained under long-day condition. Wareing, Cornforth, and their associates (Cornforth *et al.*, 1965a) isolated dormin from sycamore leaves. From a comparison of the properties of dormin and of abscisin II isolated by Addicott's group, particularly the molecular weight, infrared spectra, and melting points, it was concluded that these two compounds were actually identical (Cornforth *et al.*, 1965b). Owing to the confusion generated from those different names for the same compound, such as abscisin, abscisin I, abscisin II, and dormin, by the original discoverers, it was recommended in 1967 (Addicott *et al.*, 1968) that abscisic acid be used as the trivial name for 3-methyl-5-(1'-hydroxy-4'oxo-2'-cyclyhexen-1'-yl)-*cis*-2,4-pentadienoic acid as shown in Fig. 14. Abscisic acid (ABA) has now been detected in several flowering plants, a fern, a horsetail, and a moss. So far ABA has not been found in liverworts or algae, and this may represent a taxonomic break of the occurrence of ABA (cited in Milborrow, 1974a).

B. Structure and Function Relationship

The chemical synthesis of ABA was accomplished not much later than the first isolation and characterization of ABA from natural sources (Cornforth *et al.*, 1965a; Roberts *et al.*, 1968). This chemical synthesis procedure has also been used to synthesize ABA analogs and ^{14}C-labeled ABA. As yet very few synthetic analogs have been found to possess significantly higher activity than ABA. The configuration of C-2 double bond must be cis. 2-*trans*-abscisic acid is almost inactive. The 2-trans isomers of a number of related compounds are also less active than the 2-cis isomers. 2-*trans*-Abscisic acid has been shown to be active in certain

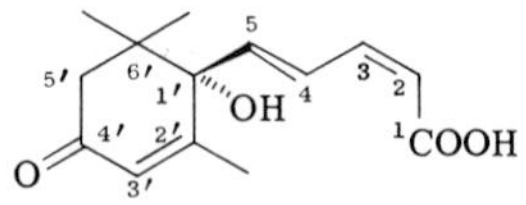

Fig. 14. (+)-Abscisic acid.

experiments; however, these experiments were carried out in light that causes isomerization of ABA. A free carboxylic group is necessary for activity because C-1 nitrile is inactive. Esters are always active; however, the hydrolysis of the ester bond is believed to be fast in most of the tissue tested. The cyclohexane ring must contain a double bond in an α- or β-position; however, the necessity of both 1'-hydroxyl group and 4'-oxo group is unclear. The only asymmetrical carbon in ABA molecule is C'-1. The (−)-enantiomer of ABA appears to be as active as the native (+)-ABA in almost all the tests except the closure of stomata, where (−)-enantiomer is inactive (Cummins and Sondheimer, 1973). Also, these two enantiomers may differ in uptake and metabolism rate in different tissues.

C. Biosynthesis

From the structure of ABA it is apparent that it is a terpenoid derived from mevalonate. This suggestion is further supported by the demonstration that mevalonate can be incorporated into ABA in several fruits (Noddle and Robinson, 1969). The origin of carbon atoms in ABA is very easily elucidated because each carbon atom in the isoprenoid units of ABA molecule corresponds to a carbon atom in mevalonate (Fig. 15). As for the origin and stereochemistry of hydrogen atoms, much more work was needed. ABA was isolated from plant tissue fed with mevalonate labeled with ^{14}C, or tritium, or both at special positions, such as [2-^{14}C, 2(R)-2-^3H], [2-^{14}C, 2(S)-2-^3H], [4(S)-4-^3H], [5(S)-5-^3H]. Either the $^3H/^{14}C$ ratio or the amount of ^3H label was followed when ABA was subject to certain chemical reactions that were supposed to remove hydrogen atoms at specific positions of the ABA molecule. For example, ABA isolated from avocado fruit labeled with [2-^{14}C, 4(R)-4-^3H]mevalonate has a $^3H/^{14}C$ ratio about 2:3 (Robinson and Ryback, 1969) (Table VI). Treatment with NaOH, which induces keto–enol tautomerism of C'-4 keto group and subsequently causes the exchange of C'-5 hydrogen with hydrogen in the medium, decreases the $^3H/^{14}C$ ratio to 3:1 (Robinson and Ryback, 1969). This result indicates that C'-5 hydrogen on ABA molecule is derived from 4(R) hydrogen of mevalonate. As expected, ABA isolated from [2-^{14}C, 4(S)-4-^3H]mevalonate fed tissue does not have any NaOH-releasable tritium label. Using this kind of approach, the origin and stereochemistry of essentially all the hydrogen atoms on ABA have been sorted out as shown in Fig. 15.

Two alternative pathways for the biosynthesis of ABA have been suggested (Fig. 16): (1) ABA is derived from a degradative product of carotenoids, in particular violaxanthin. (2) ABA is synthesized by a unique route through a C-15 precursor that is not a degradative product

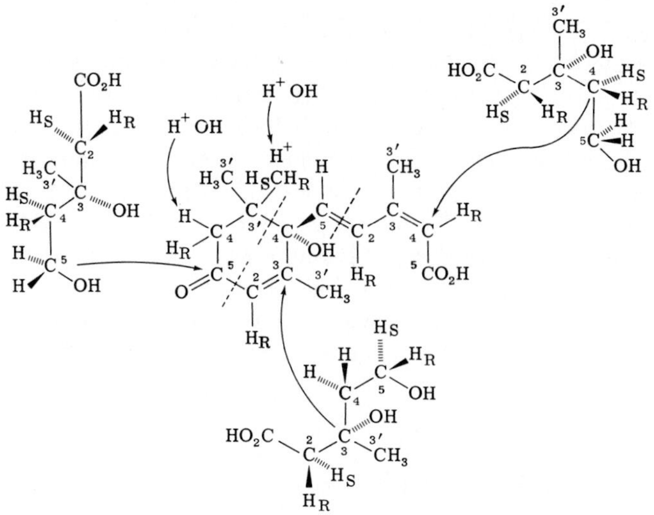

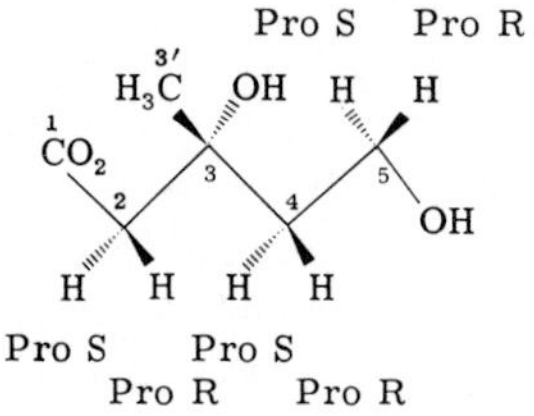

Fig. 15. The origins of H and C atoms in abscisic acid.

of carotenoids. The first possibility is readily apparent because ABA is structurally similar to the end portions of many carotenoid molecules. Further evidences supporting this pathway comes from the *in vitro* photoconversion of carotenoid (violaxanthin) to xanthoxin (Taylor and Smith, 1967). Conceivably, xanthoxin is then converted to ABA. Approximately equal amount of biologically active 2-*cis*-xanthoxin and inactive 2-trans isomer are generated this way. This ratio of cis to trans isomer is close to that found in the extract of dwarf bean and wheat. However, the *in vitro* photoconversion of violaxanthin to xanthoxin has very low yield (about 2%) and requires high light intensity. Therefore, some doubts have been cast on this postulated pathway. Recently it has been shown that soybean lipoxygenase is able to cleave violaxanthin to form xanthoxin with similar yield as photoconversion (Firn and Friend, 1972). However, the question of whether lipoxygenase is responsible for xanthoxin

TABLE VI

^{3}H/^{14}C Ratios in Abscisic Acid Biosynthesized by Avocado Fruit from [2-^{14}C, 4(R)4^{3}H]- and [2-^{14}C, 4(S)-4-^{3}H]Mevalonic Acid^a

	From [2-^{14}C,4(R)-4-^{3}H]mevalonate		From [2-^{14}C,4(S)-4-^{3}H]mevalonate	
	^{3}H/^{14}C	Presumed structure	^{3}H/^{14}C	Presumed structure
1. Isolated ABA	1.84:3 (2:3)		0.12:3 (0:3)	
2. Treatment with NaOH and chromatography	0.92:3 (1:3)		0.02:3 (0:3)	

^{a} Positions of ^{14}C are shown by asterisks, and those of ^{3}H are labeled by T.

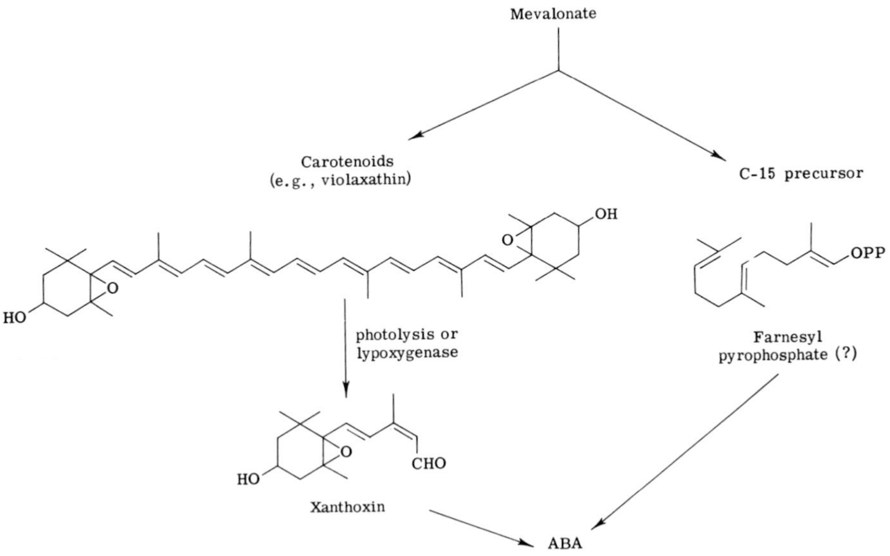

Fig. 16. Two alternative pathways for the biosynthesis of abscisic acid.

and ABA synthesis *in vivo* deserves further research. So far the strongest evidence against the idea that ABA is derived from carotenoid is provided by the work of D. R. Robinson (cited in Milborrow, 1974b). He incubated avocado fruit with [³H]mevalonate and [¹⁴C]phytoene, which is a precursor of caroteinoids. The tritium of mevalonate is incorporated into both ABA and caroteinoids. However, ¹⁴C was incorporated only into carotenoids, not into ABA. The interpretation of this result is that ABA is synthesized from mevalonate not involving a carotenoid intermediate. However, since this experiment was carried out with intact tissue, different cellular compartments can be involved. It is possible that phytoene cannot penetrate into the carotenoid pool, which may give rise to ABA. It was recently shown that ¹⁴C label of farnesyl phosphate, which is a potential C-15 precursor of ABA, is incorporated into ABA in avocado fruit (cited in Milborrow, 1974b). This result also favors the possibility that ABA is synthesized from mevalonate not via caroteinoids.

D. Catabolic Metabolism

When ABA solution is applied to tomato plant, both (+)- and (−)-enantiomer are conjugated with glucose to form polar, water-soluble abscisyl-β-D-glucopyranoside. The glucose ester of ABA has been found in two plant species, but free ABA is released by alkaline treatment of methanolic extracts of several other plants, presumably due to the hydrolysis of the glucose ester of ABA (cited in Milborrow, 1974a). Except in the peel of oranges matured on the tree (Goldschmidt *et al.*, 1973), the amount of the glucose ester of ABA never exceeds one-third of the amount of free ABA. The glucose ester of ABA is equally inhibitory as ABA on a molar basis when it is tested in a rice seedling bioassay, probably as a result of the hydrolysis of the ester to release free ABA. Like the glucose derivatives of gibberellins and auxins, the function of the glucose ester of ABA is still unknown. Because the glucose ester can be rapidly hydrolyzed to form free ABA, its role as a storage for extra ABA seems to be apparent. Although both (+)- and (—)-ABA are converted to glucose derivative, only the (+)-enantiomer is also hydrolyzed on one of the geminal methyl groups to form the unstable metabolite C (Milborrow, 1970) (Fig. 17). This compound rearranges easily to give phaseic acid by a nucleophilic attack of the 6'-hydroxymethyl group on C'-2 to form a saturated furan ring (Milborrow, 1969). The fast conversion from metabolite C to phaseic acid can thus account for the fact that previously isolated metabolite C has given phaseic acid. In bean embryo, phaseic acid can be further converted to "M-2" metabolite described by Walton and Sondheimer (1972a,b). M-2 has now been identified as 4'-dihydrophaseic acid (Tinelli *et al.*, 1973).

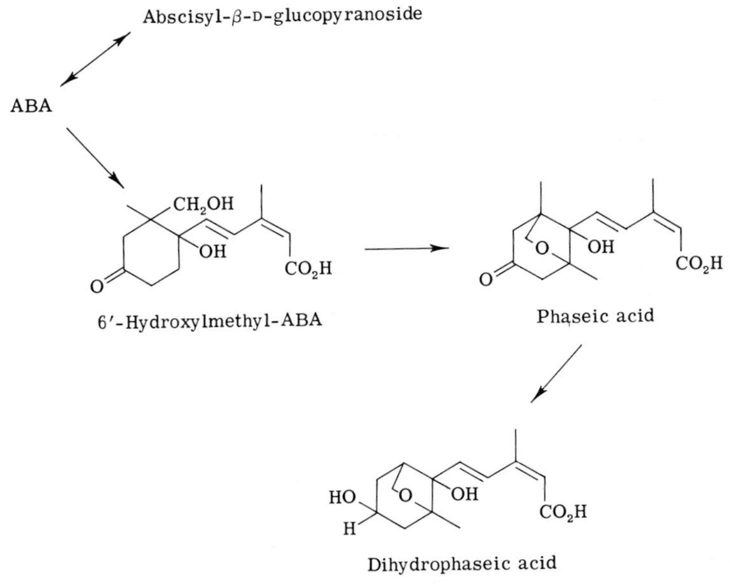

Fig. 17. Metabolism of abscisic acid.

E. Physiology and Mode of Action

One of the lines of research leading to the isolation of ABA was based on the discovery that a growth-inhibiting substance ("dormin") is able to cause bud dormancy in several woody plant species. [Exogenously supplied ABA has been shown to be able to prolong bud dormancy or induce dormancy of normally growing plants (El Antably et al., 1967).] When bioassay is used, there is a correlation of the content of inhibitory material in leaves grown in different day lengths and the ability of different leaf extracts to induce bud dormancy. However, chemical assay of ABA using gas-liquid chromatography does not always agree with the results of bioassays (Lenton et al., 1968), i.e., other factors besides ABA may be involved in the control of bud dormancy. The endogenous content of ABA decreases after stratification (Corgan and Martin, 1971), but whether this result from decreasing synthesis or increasing degradation is unknown.

Abscisic acid is an effective inhibitor of seed germination, and many dormant seeds contain ABA. This suggests that ABA is able to maintain seed dormancy, although it could not be the only factor controlling seed dormancy. As in dormant buds, the ABA content in some dormant seeds decreases after stratification (Sondheimer et al., 1968; Rudnicki, 1969). Apparently growth-promoting substances, such as gibberellins and cytokinins, also play a role in seed germination. Therefore, the balance between ABA and growth-promoting substances may be important in controlling the release of seed dormancy. For those seeds containing ABA, the content of ABA starts to increase during the development of seed and fruit. It has been shown that many fruits, cotyledon of avocado seeds, and, to a lesser extent, the endosperm of developing wheat seeds are capable of incorporating labeled mevalonate into ABA, indicating that the increase of ABA level in these tissues is not due to the transport of ABA from other parts of the plant (Milborrow and Robinson, 1973).

The effect of ABA on the germination of cotton and wheat seed has been studied in detail. The content of ABA in cotton fruit (mostly in the ovary wall) increases during seed development. Isolated developing embryos can be germinated precociously if ABA is washed away. The precocious germination can be prevented if the ABA or ovary extract is added back to the washed embryos. The inhibitory effect of ABA can be abolished by germinating the seeds in the presence of transcription inhibitors, such as actinomycin D, indicating the effect of ABA depends on continous synthesis of RNA (Ihle and Dure, 1970). Early stages of cotton seed germination are not sensitive to actinomycin D, indicating that the mRNA necessary for germination must be preexisting (Waters and Dure, 1966). Therefore, the effect of ABA on cotton seed germination appears to be via a posttranscriptional control point. Similar observations have been made in the germinating wheat seed (Chen and Osborne, 1970). Protein synthesis is initiated in the germinating wheat embryo immediately after inhibition, and it is suggested that this early protein synthesis is programmed by preexisting mRNA in the dry embryo because actinomycin D has no inhibitory effect on germination and masked mRNA in the form of ribonucleoprotein complex is present in dry wheat embryo. Gibberellic acid is a promoter, and ABA is an inhibitor of wheat seed germination. DNA–RNA hybridization experiment showed that no new mRNA or rRNA is associated with ribosomes, indicating that the hormones regulate the use of mRNA already stored in the embryo rather than controlling the *de novo* synthesis of mRNA.

ABA and structurally related compounds have been shown to reduce the growth of wheat coleoptiles. The incorporation of amino acids into proteins by polysomes isolated from coleoptile treated with ABA is strongly inhibited. However, total RNA synthesis is not altered by the hormone treatment of the coleoptile, indicating the effect of ABA in this system is also posttranscriptional (Bonnafous *et al.*, 1973).

Although ABA is also isolated from the extract of senesencing leaves and is able to induce abscission, the role of ABA in abscission is still unclear. Unless extremely high concentrations (500 to 1000 $\mu g/ml$) are used, exogenous ABA shows little ability to induce leaf abscission. Auxin and ethylene are also involved in the regulation of leaf abscission. ABA may have a more certain role in fruit abscission; application of ABA is able to accelerate abscission in peach, olive, citrus, apple, etc. Employing chemical assays for ABA, Davis and Addicott (1972) measured the level of ABA over the entire life span of the cotton fruit. They found high levels of ABA occurred in correlation with abortion and abscission of young fruit, with low germination, and with senesence and dehiscence of mature fruit.

The level of ABA in leaves increases dramatically when a plant is subject to water stress (Wright, 1969). When excised wheat leaves were kept in a wilted state, the content of ABA increased to as much as 40-fold in 4 hours (Wright and Hiron, 1969). Similar observations have been made with intact plants of several other species. It appears that only mild to moderate water stress is necessary to induce the ABA increase because the most rapid increase in ABA occurs when 5 to 10% of tissue fresh weight is lost. The increase of ABA level has been shown by Milborrow and Noddle (1970) to arise by *de novo* synthesis and not through release of a bound form. The high level of ABA is maintained in the wilted tissue no matter whether the wilt situation becomes a little less or more severe or remains constant. Recently it is reported that in the wilted leaves of *Phaeolus vulgaris*, an elevated level of ABA is maintained because the rate of synthesis and catabolic metabolism are both elevated and approximately equally (Harrison and Walton, 1975). The principal role of ABA in wilted plant is to close the stomata, thus preventing further loss of water (Cummins et al., 1971). Stomata open up when the guard cells are under turgor, which is caused by the accumulation of K^+ ion in the guard cells (Raschke, 1975). Abscisic acid is able to prevent the accumulation of K^+ within minutes of application.

Abscisic acid can reverse the effect of growth promoters, such as auxins, GA, and cytokinins, in certain plant tissues. The effect of GA on hydrolase synthesis and secretion in the aleurone layer of barley seed has been extensively studied. Therefore, the ability of ABA to prevent the tissue response to GA has been tested. Abscisic acid can reverse the GA-enhanced amylase and protease syntheses, membrane-bound polysome formation, incorporation of ^{32}P into membrane phospholipids, poly(A) RNA synthesis, and the activation of phosphorylcholine transferases (Yomo and Varner, 1971; Varner and Ho, 1976). However, ABA has no effect on general cellular metabolism as measured by oxygen consumption. Although ABA can prevent the response to GA, no direct effects of ABA in aleurone cells have been observed. However, the failure of the aleurone cells to respond to GA in the presence of ABA does not result from simple competition between these two hormones, because a high concentration of GA cannot completely overcome the ABA effects. Recently, it has been found that after 12 hours of GA treatment, α-amylase synthesis is no longer sensitive to transcription inhibitors, such as cordycepin (3'-deoxyadenosine) (Ho and Varner, 1974), while ABA at this stage can still effectively prevent the further synthesis of α-amylase. It appears that ABA selectively inhibits the synthesis of α-amylase while the synthesis of other protein remains normal. If cordycepin is added later than ABA, the effect of ABA is prevented (Fig. 18) indicating that the action of ABA depends on continuous synthesis of certain RNA (Ho and Varner, 1976). Abscisic acid has been shown to prevent the auxin enhanced cell elongation in Avena coleoptile within a lag period of a few minutes (Rehm and Cline, 1973).

Trewavas (1970, 1972) found in Lemna minor that benzyladenine increases both the synthesis and degradation rates of nucleic acids while ABA reduced only the rate of nucleic acid synthesis. However, ABA alters both synthesis and degradation of proteins, and benzyladenine influences only the rate of protein degradation.

F. Xanthoxin

Xanthoxin (Fig. 19) is not only a hypothetical precursor of ABA but a true endogenous inhibitor in many plant tissues (Firn et al., 1972).

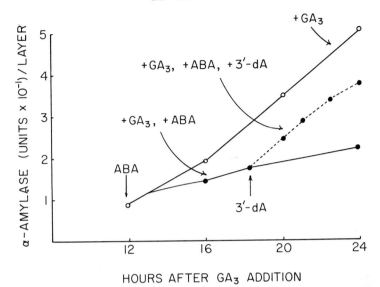

Fig. 18. Effect of abscisic acid (ABA) and transcription inhibitor [3'-deoxy-adenosine(3'-dA)] on the synthesis of α-amylase in barley aleurone layers. The arrows indicate the time when ABA or 3'-dA was added.

As described previously xanthoxin is probably derived from xanthophylls. However, the exact mechanism of the conversion from xanthophyll to xanthoxin is not yet known. The extent of the natural occurrence of xanthoxin throughout the plant kingdom appears to follow a pattern similar to that of ABA (Taylor and Burden, 1972). However, within the plant the distribution of xanthoxin appears to be limited, with young shoots as the best source. When present, the levels of xanthoxin appear to be no less than that of ABA (Taylor and Burden, 1972; Firn *et al.*, 1972).

The biological activities of xanthoxin have been tested in several systems including the elongation of wheat coleoptile and that of lettuce hypocotyl, cress seed germination, tobacco callus tissue growth, and bean petiole abscission (Taylor and Burden, 1972). Although 2-*trans*-xanthoxin is much less effective than the 2-cis isomer, the latter appears to be equally effective as ABA in most of these tests. When xanthoxin was supplied to cut tomato shoots, the level of ABA was drastically

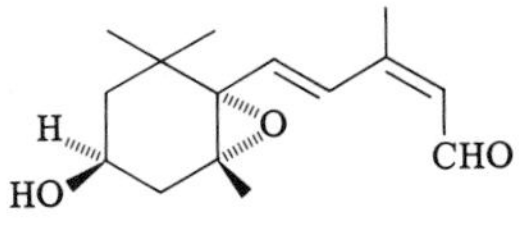

Fig. 19. Xanthoxin.

increased (Taylor and Burden, 1972). This suggests that the inhibitory effect of xanthoxin may result from its conversion into ABA. Xanthoxin supplied to the transpiration stream of detached leaves closes stomata rapidly (Raschke *et al.*, 1975). However, when tested with epidermal strips, xanthoxin has no effect on stomatal aperture. Therefore, this confirms the observations by Taylor and Burden that xanthoxin, which probably has no direct effect on guard cells, can be rapidly converted into ABA by plant tissues. [It was estimated that more than 50% of the xanthoxin can be converted into ABA in less than 5 minutes (Raschke *et al.*, 1975).]

V. Gibberellins

A. Discovery

The gibberellins are a large family of diterpenoid acids which can probably be found in all plant tissues. The discovery of this family of growth-regulating substances resulted from studies of the "bakanae" disease of rice whose symptom is the appearance of tall thin plants, easily distinguished from normal ones. This disease is caused by infection of a fungus [*Fusarium moniliforme (Gibberella fujikuroi)*]. In 1926, Kurosawa induced the bakanae disease on rice by treating the plants with culture medium of *F. moniliforme*. In 1938, Yabuta and Sumiki reported the isolation of crystalline, biologically active compounds, which they named gibberellins, from the culture medium of *F. moniliforme*. Most western researchers were not aware of the discovery of gibberellins until early 1950's. Since then the response of various plants to gibberellins has been tested, and a variety of bioassays and chemical assays have been developed. A few years later direct evidence for the existence of gibberellins in higher plants was obtained, and by now 45 different gibberellins had been isolated from various plant tissues and fungi. The abbreviation GA is used for gibberellins as a family. Individual gibberellins are designated as GA_1, GA_2, etc.

B. Biosynthesis

The gibberellins can be divided into two groups (Fig. 20): (1) the C_{20} gibberellins containing all the twenty carbon atoms of the parent *ent*-gibberellane, and (2) the C_{19} gibberellins that lack carbon-20. The C_{20} gibberellins are characterized by the presence of C-20 which may be present as CH_3 (e.g., GA_{20}), CH_2OH (e.g., GA_{15}), CHO (e.g., GA_{19}), or

CO_2H (e.g., GA_{28}). The C_{19} gibberellins are characterized by the presence of a C-19 to C-10 lactone bridge as in GA_3. Besides the known GA's themselves various derivatives, most of which are glucosylgibberellins, have been isolated from several plants.

The current knowledge of the biosynthesis of gibberellins is prinicipally derived from studies of cultures of *F. moniliforme* and partly from studies of higher plants such as wild cucumber (*Echinocystis macrocarpa*). Figure 21 summarizes the generally accepted biosynthetic pathway of gibberellins. Studies of the biosynthesis of gibberellins started when Birch and his associates (1958) showed that acetate and mevalonate are incorporated into GA_3, a fact that led to the suggestion that gibberellins are derived from a normal 3-cyclic diterpenoid that is a product of the mevalonate pathway. The route leading to the first cyclic product, kaurene, has been elucidated by West and co-workers, using the endosperm of immature seeds of *E. macrocarpa*. The acyclic precursor of kaurene was shown to be geranylgeranyl pyrophosphate, and copalyl pyrophosphate appears to be an intermediate in this conversion. Further metabolism of kaurene via a series of oxidation steps leads to the transformation of C-19 CH_3- into a $-COOH$ group. The resulting kaurenoic acid is then hydroxylated in the 7β–position.

The mechanism of contraction of the B ring from *ent*-kaurene to *ent*-gibberellane has been the subject of considerable interest and speculation. It is known from the tracer studies by Birch *et al.* (1959) that C-7 of the *ent*-kaurene is extruded in this ring contraction. In 1972, Hanson *et al.* presented evidence that ring contraction involves transfer of hydrogen from C-6 to C-7 as shown in Fig. 22. Two possible mechanisms have been suggested (Fig. 23): Mechanism A involves an initial oxidation step and leads directly to the GA_{12}-aldehyde, and mechanism B is nonoxidative and leads to the GA_{12}-monoalcohol as the primary product of ring contraction.

So far it has not been possible to demonstrate formation of a chemically identified gibberellins from kaurene or its derivatives in a higher plant. However, it is generally believed that kaurene can be converted into gibberellins in higher plants as in *F. moniliforme*.

The cyclic diterpene, steviol, which occurs as the glycoside, stevioside, in leaves of the South American composite *Stevia rebaudiana* is the 13-hydroxy derivatives of *ent*-kaurenoic acid. Steviol, which is active in several GA bioassays, has been suggested as a precursor of 13-hydroxy-GA's in higher plants. Steviol has been shown to be converted into a GA-like compound by wild type *F. moniliforme* (Ruddat *et al.*, 1965). It has recently been reported (Bearder *et al.*, 1975) that a mutant (B1-41a) of the wild type strain GF-1a of *F. moniliforme* is able to metabolize steviol

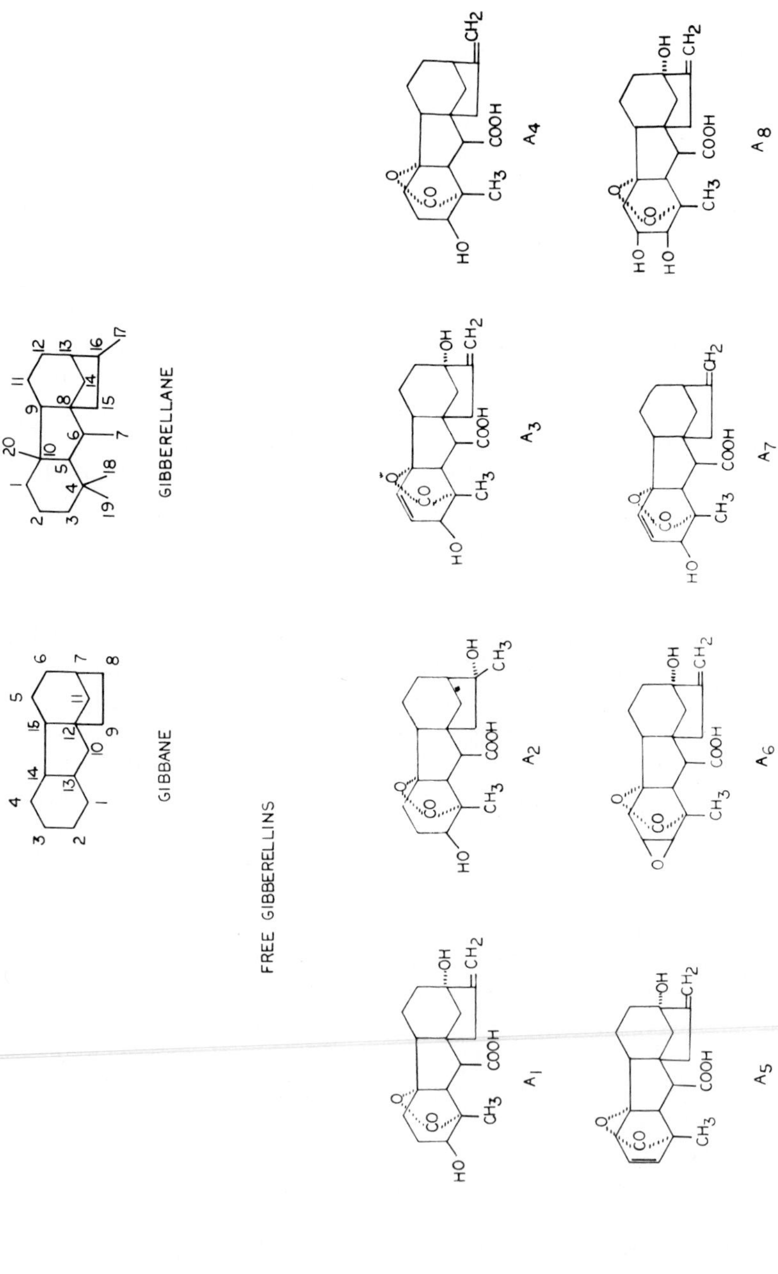

GIBBERELLANE

GIBBANE

FREE GIBBERELLINS

A_1

A_2

A_3

A_4

A_5

A_6

A_7

A_8

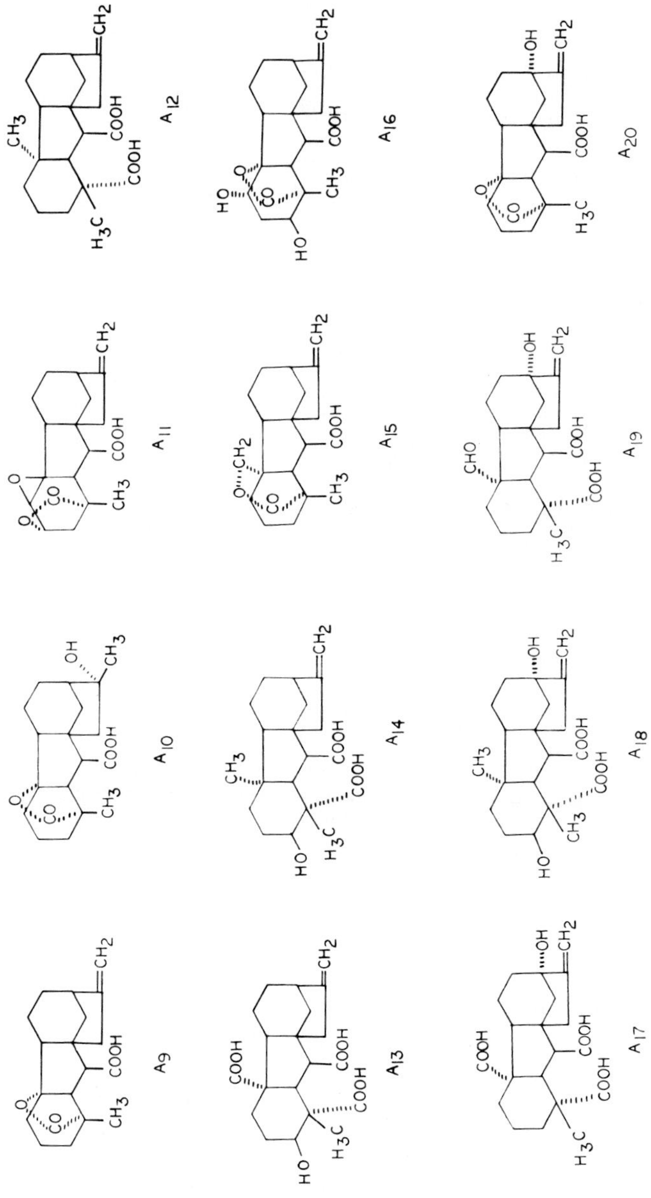

Fig. 20. The structural formulas of the gibberellins presently known.

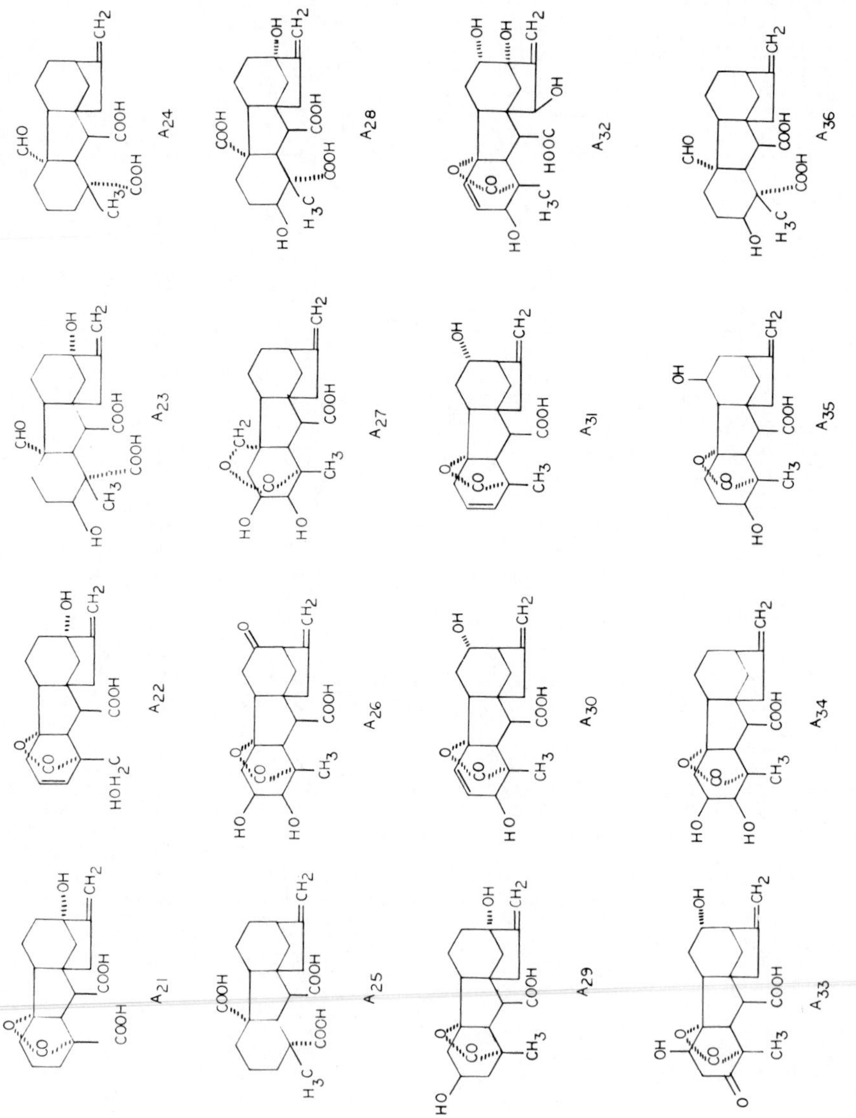
A21
A22
A23
A24
A25
A26
A27
A28
A29
A30
A31
A32
A33
A34
A35
A36

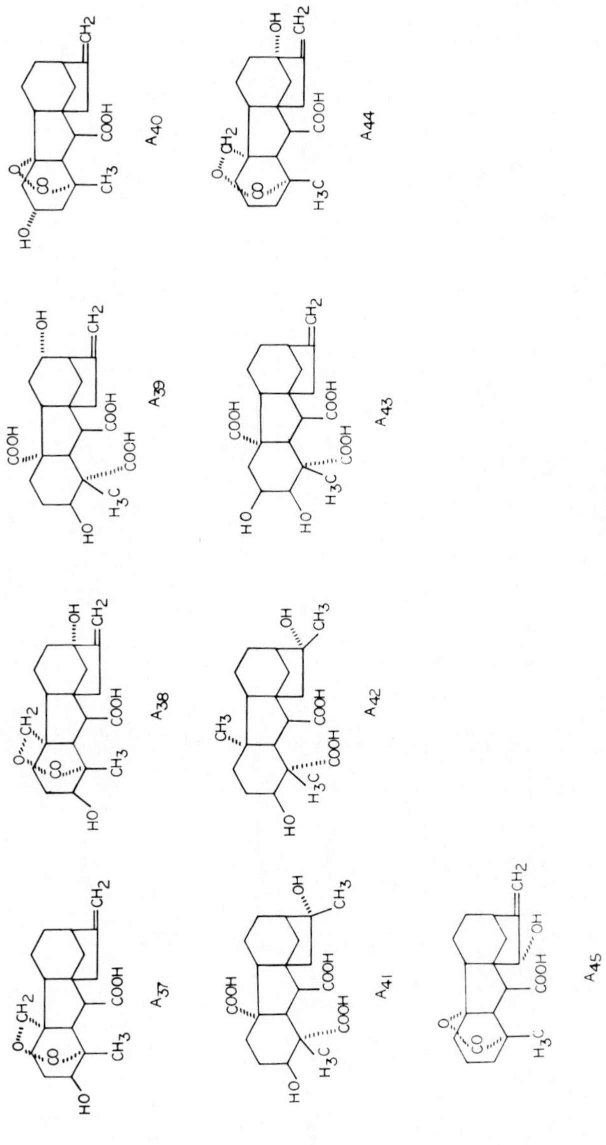

Fig. 20. (continued)

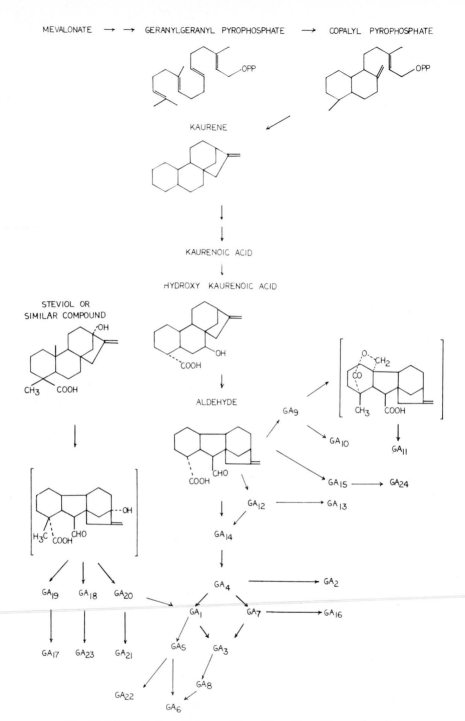

Fig. 21. Biosynthetic pathways of the gibberellins. After Lang, 1970.

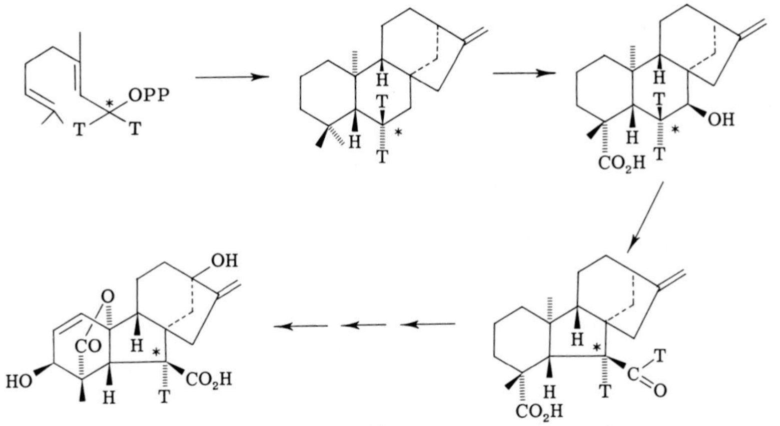

* = ^{14}C

Fig. 22. Incorporation of 1,1-ditritiogeranyl pyrophosphate into GA_{12}-aldehyde and GA_3. Asterisk denotes carbon-14. From Hanson *et al.*, 1972.

into several 13-hydroxylated higher plant GA's, such as GA_1, GA_{18}, GA_{19}, GA_{20}. However, whether steviol is a unique intermediate of GA biosynthesis in higher plants is still in doubt, because steviol has been found so far only in *S. rebaudiana*. Nonetheless, the capacity of the B1-41a mutant to metabolize steviol into higher plant GA's provides a useful model for the biosynthetic pathways to 13-hydroxylated GA's of higher plants.

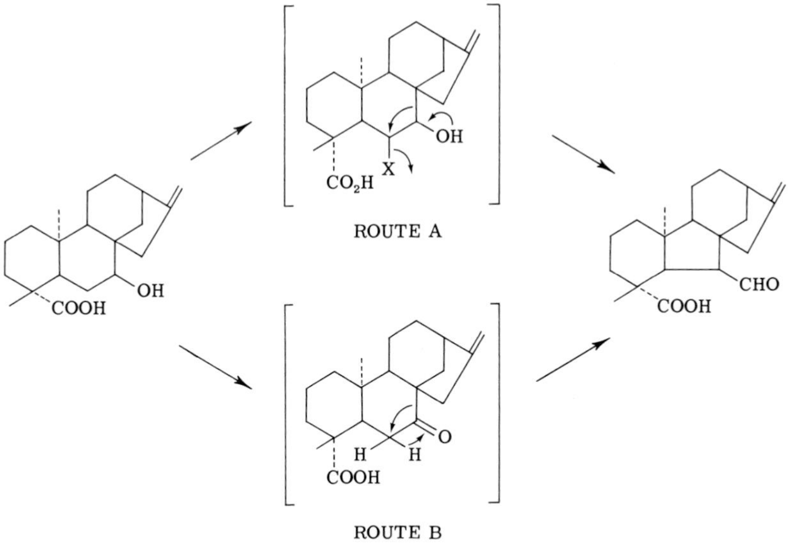

ROUTE A

ROUTE B

Fig. 23. Possible mechanisms for contraction of ring B.

Because gibberellins serve as growth-regulating substances it would be of potential agricultural importance if their metabolism could be regulated by synthetic compounds supplied exogenously.

A number of synthetic compounds have been found to reduce growth in many plants. Among these compounds, 2'-isopropyl-4'-(trimethylammonium chloride)-5'-methyl-piperidine carboxylate (AMO-1618), β-chloroethyltrimethylammonium chloride (CCC), and tributyl-2,4-dichlorobenzylphosphonium chloride (phosphon D) are the best known. Since the effect of these growth retardants can be counteracted by application of GA's, it was postulated and later shown that these compounds inhibit GA biosynthesis in both *F. moniliforme* and higher plants. Kaurene synthetase has been shown to contain two catalytic activities, i.e., the cyclization of geranylgeranyl pyrophosphate (activity A) and the cyclization of copalyl pyrophosphate (activity B) (Fall and West, 1971). Activity A of this enzyme appears to be the site of inhibition for the growth retardants (Shechter and West, 1969). However, whether there are additional sites of inhibition for the growth retardants is unknown.

C. Catabolic Metabolism

Usually plant tissues require very little GA for physiological response; thus, the lack of ready availability of GA with sufficient specific radioactivity tends to hamper studies on the catabolic metabolism of GA. In most of the plant tissues tested, a significant amount of GA is converted into water-soluble derivatives—the conjugated GA's or bound GA's. (As mentioned before most of the conjugated GA's are glycosyl derivatives.) Recently, a possible peptide-linked GA_1 has been reported in barley aleurone layers (Nadeau and Rappaport, 1974).

Two possible physiological functions of the conjugated GA's have been suggested, i.e., storage and transport (Lang, 1970). The evidence for the storage function rests on the observation of the interconversion between free (ethyl acetate soluble) and bound (water soluble) GA's during seed development and germination. During the development of pea seed, part of the GA is converted to the conjugated form, while during early germination the conjugated GA is reconverted to the free form (Barendse *et al.*, 1968). Because during early stages of germination the development of pea seedlings is not affected by AMO-1618, it appears that GA biosynthesis is not necessary at this stage and at least part of the GA needed is derived from the conjugated form. The transport function of conjugated GA is suggested by the presence of glycosyl-GA_8 in the bleeding sap from *Acer* (reviewed by Lang, 1970). The two possible functions of conjugated GA are not necessarily mutually exclusive because the conjugated GA found in the bleeding sap may represent storage GA's being transported to tissues in need of the hormone.

In two plant tisues, the germinating seeds of *Phaseolus vulgaris* (Nadeau and Rappaport, 1972) and barley aleurone layers (Nadeau *et. al.*, 1972), ABA substantially enhances the catabolic metabolism of GA_1. However, the change of overall rate of GA metabolism is insufficient to account for the rapid effect of ABA.

D. Physiology and Mode of Action

Among the many physiological responses of plant tissues to gibberellins, the GA_3 controlled formation and secretion of hydrolases in cereal grains, especially in barley, has been extensively characterized biochemically (Fig. 24).

Mobilization of endosperm reserves during the germination of cereal grains supplies nutrients for the growth of the embryo. This is accom-

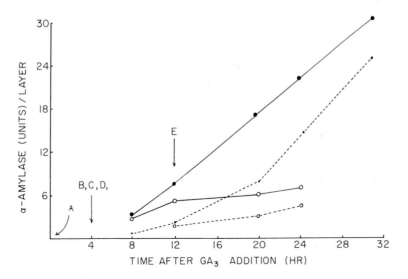

Fig. 24. Time course of GA_3-enhanced α-amylase synthesis, secretion, and related events in barley aleurone layers. ●——●, total α-amylase (+GA_3); ○——○, total α-amylase (−GA_3); ● - - - ●, α-amylase secreted into the medium (+GA_3); ○ - - - ○, α-amylase secreted into the medium (−GA_3). The arrows indicate the time when the following events occur. A, activation of phosphorylcholine-cytidyl transferase and phosphorylcholine-glyceride transferase; B, enhancement of membrane formation; C, enhancement of membrane-bound polysome formation; D, enhancement of poly(A) RNA synthesis; E, initiation of α-amylase secretion. The synthesis of α-amylase is no longer sensitive to transcription inhibitors.

plished by the action of several hydrolases, including α-amylase (EC 3.2.1.1) and protease (EC 3.4), which are synthesized in the aleurone tissue that surrounds the endosperm. The aleurone tissue of barley consists of three layers of homogeneous nondividing triploid cells. These cells respond to GA, which is formed in the embryo during the early stage of seed germination, by a series of morphological and biochemical changes (Yomo and Varner, 1971). The most prominent among these changes is the increase in α-amylase and protease activities after an 8- to 10-hour lag period. The GA_3 enhanced activities α-amylase, protease, and ribonuclease (EC 3.1.4) have been found to be due to the *de novo* synthesis (Filner and Varner, 1967; Jacobsen and Varner, 1967; Bennett and Chrispeels, 1972) of the enzyme proteins, and most of the α-amylase and protease are secreted into the endosperm after their synthesis. GA_3 also enhances the secretion and to a lesser extent the synthesis of β-1,3-glucanase (EC 3.2.1.6) and the release of acid phosphatase (EC 3.1.3.2) from the cell wall (Ashford and Jacobsen, 1974).

Because α-amylase becomes the most predominant protein synthesized in aleurone layer after several hours of GA_3 treatment, it is used as a marker to study the mechanism of the hormone-controlled enzyme formation in this system. RNA synthesis inhibitors, such as actinomycin D, 6-methylpurine (Chrispeels and Varner, 1967), and cordycepin (3'-deoxyadenosine) (Ho and Varner, 1974) can block GA_3-enhanced α-amylase synthesis. Zwar and Jacobsen (1972) demonstrated that the incorporation of radioactive nucleotides into polydispersed RNA is increased in the presence of GA_3, and it was reported recently that GA_3 enhanced the synthesis of RNA species containing polyadenylic acid in barley aleurone layer (Jacobsen and Zwar, 1974; Ho and Varner, 1974).

During the lag period before the production of hydrolase there is extensive proliferation of cellular membranes (principally endoplasmic reticulum) (Jones, 1969) and the formation of membrane bound polyribosomes (Evins, 1971). Concurrently the incorporation of choline and of ^{32}P into membrane phospholipids is enhanced about 4 hours after the addition of GA_3 (Koehler and Varner, 1973). Furthermore, the activity of two enzymes, phosphorylcholine transferase (EC 2.7.7.15) and phosphorylcholine-glyceride transferase (EC 2.7.8.2), which involved in the synthesis of lecithin (phosphatidylcholine), are enhanced within minutes of GA_3 addition (Johnson and Kende, 1971). All these events seem to indicate that membrane proliferation is a major GA_3 effect before the rapid increase of α-amylase activity. α-Amylase is a secretory protein and there is ample evidence for the participation of rough endoplasmic reticulum (RER) in the synthesis and exportation of secreted proteins in eukaryotic cells.

Thus, there are two alternative models that can account for the GA_3 enhanced synthesis of α-amylase (and other hydrolases):

1. The enhanced synthesis of α-amylase depends on the increased rate of formation (transcription and/or processing) of its specific mRNA. Membrane proliferation is a separate effect of GA_3.

2. The GA_3 enhanced membrane proliferation provides an increased number of sites available for the synthesis of α-amylase. Thus, the amount of α-amylase-specific mRNA need not be increased in the presence of GA_3.

The development of a cell-free assay for the α-amylase specific mRNA would make it possible to choose between these alternatives.

The increase in activity of phosphorycholine-glyceride transferase is apparently due to the activation because neither amino acids analogues nor cordycepin inhibit this increase (Ben-Tal and Varner, 1974). Phosphorylcholine-glyceride transferase activity is inhibited by Ca^{2+}. Therefore, the activity change of this enzyme could result from either changes in concentration of Ca^{2+} or the concentration of Ca^{2+}-complexing agents, such as citrate or phytate. Since aleurone tissue has a large storage of phytate, a redistribution of phytate or calcium or an early release of calcium to the medium due to a change in membrane permeability could account for the activation of phosphorylcholine-glyceride transferase. The activity of membrane-bound enzymes, such as phosphorylcholine-glyceride transferase, can be altered in the presence of a surfactant, such as lysolecithin, which is one of the products of the hydrolysis of storage phospholipids. The activation of phosphorylcholine-glyceride transferase begins within a short time after hormone treatment; therefore, the mechanisms of this activation process will provide useful information about the primary action of the hormone.

Gibberellins are able to induce normal growth in certain dwarf varieties of garden pea and some single gene dwarf mutants of maize (Phinney, 1956) and to enhance elongation growth in other plants, including rice, lettuce, wild oat, cucumber, wheat. Applications of GA to certain rosette long-day plants stimulate flower formation under noninductive daylength conditions (Lang, 1965). However, it has been argued that flower stimulation in these plants is probably due to the effect of GA on stem elongation (Lang, 1965). Apparently cell elongation rather than cell division is the principal cause of the GA enhanced growth because α-irradiated barley or wheat seedlings, in which cell division is completely inhibited, still show this response to GA. The GA enhanced cell elongation in lettuce hypocotyls coincides with the increase in the dry weight of cell wall (Srivastava *et al.*, 1975). Cell elongation is preceeded by extensive dictyosomal activity and proliferation of endoplasmic reticulum and polyribosomes. These changes may be prerequisites for the increase in the synthesis and secretion of extracellular macromolecules, which are the materials of cell walls. Therefore, it appears that the mode of action of gibberellin in lettuce hypocotyl elongation is similar to that of hydrolase synthesis and secretion in aleurone tissue of cereal grains.

In the elongating pea epicotyl, GA enhances RNA synthesis in isolated nuclei (Johri and Varner, 1968), and when applied to intact plants the hormone increases the level of RNA polymerase associated with chromatin without a detectable increase in the amount of DNA template available (McComb et al., 1970). This was later confirmed by DNA/RNA hybridization competition techniques. No distinct changes in hybridizable RNA were detected up to 50 hours after treatment of the seedlings with gibberellin (Thompson and Cleland, 1972). These observations seem to suggest that a massive gene activation is not required in the gibberellin enhanced cell elongation. However, it has been reported in cucumber hypocotyls that the template activity of isolated chromatin can be increased as a result of GA treatment of intact plants (Johnson and Purves, 1970).

Besides elongation growth GA is also able to regulate other morphological development in certain plants. Applications of GA to the mature form of *Hedera helix* induce morphological reversions to the juvenile form of growth. The juvenile form differs dramatically from the mature form phenotypically. DNA/RNA hybridization competition techniques were also used to detect possible gene derepression during these reversions. However, no difference in the RNA population isolated from the juvenile form and that from the mature form were detected (Rogler and Dahmus, 1974). This observation may imply a post transcriptional control mechanism during this GA mediated developmental process.

E. Hormone Receptors

In recent years much has been made of the role of hormone receptors in hormone action. At the present time, however, experimental evidence helpful in explaining the precise details of hormone action is lacking. It seems clear that hormone cannot act (that is the cell, tissue, organ, and plant cannot respond to the hormone) until the hormone is recognized by a receptor. Because of the observed narrow limits of specificity required for hormone activity, the receptor must be able to distinguish small differences in the hormone molecule. Proteins are the most likely candidates for such specific recognition tasks. Thus the search for hormone receptor proteins is underway, and such receptors have been reported. To date, no function has been established for any hormone receptor complex. Among the hormone receptor proteins one should expect to find are the hormone uptake system, enzymes involved in the final steps of biosynthesis of the hormone, enzymes involved in further metabolism (oxidation, glycosylation, conjugation, etc.) of the hormone, and, of course, the receptor responsible for the primary action of the hormone. Thus until the primary action is known, identification of the hormone receptor complex responsible can only be indirect and by elimination. We anticipate that photoaffinity labeled hormones will be most useful in the search for hormone receptors.

F. Cyclic AMP

Many animal hormones operate by regulating the intracellular level of cyclic AMP (cAMP), which then acts as a second messenger to elicit a variety of physiological effects. So far there has been no unambiguous detection of cAMP in higher plants (Lin, 1974). This may be due either to the insensitivity and nonspecificity of the assay procedures or to the absence of cAMP in higher plant tissues. The following indirect evidence also suggest that the presence of cAMP in plants is still in question.

The demonstration in higher plants of the presence of adenylate cyclase that catalyzes the formation of cAMP from ATP would be useful in answering the question of whether cAMP exists in higher plants. The formation of a small quantity of cAMP-like compound from a mixture of plant extract, ATP, Mg^{2+}, and NaF has been demonstrated. Since NaF is definitely required for the formation of this unknown compound and it has been shown that NaF in the presence of divalent ions is able to convert ATP into this unknown compound nonenzymatically, it is suggested that the unknown compound is either a fluoro derivative of cAMP or adenosine 5'-fluorophosphate; both have properties similar to authentic cAMP (Lin, 1974).

There is an enzyme capable of hydrolyzing cAMP in various higher plants (Lin and Varner, 1972; Amrhein, 1974). However, this phosphodiesterase can hydrolyze both 2',3'-cAMP and 3',5'-cAMP, and the activity toward the former is much higher than the latter. The activity of this enzyme is not affected by methylxanthines, such as caffeine and theophylline, which are inhibitors of animal 3',5'-cAMP phosphodiesterase. Since 2',3'-cyclic nucleotides are the primary products of RNA degradation in higher plants, it is suggested that the physiological role of phosphodiesterase in higher plants is to cleave the 2',3'-cyclic nucleotides into nucleoside monophosphates (Lin and Varner, 1972).

In animal tissues the regulatory properties of cAMP depend on the phosphorylation of crucial cellular proteins. Both phosphorylated proteins and protein kinase activity are present in higher plants (Keates, 1973; Trewavas, 1973). However, the extent of protein phosphorylation and the activity of protein kinase in the tissues of higher plants are not known to be affected by cAMP.

GENERAL REFERENCES

Abscisic Acid and related compounds
Addicott, F. T., and Lyon, J. L. (1969). *Annu. Rev. Plant Physiol.* **20,** 139.
Milborrow, B. V. (1974a). *Annu. Rev. Plant Physiol.* **25,** 259.
Milborrow, B. V. (1974b). *Recent Adv. Phytochem.* **7,** 57.

Auxin
Schneider, E. A., and Wightman, F. (1974). *Annu. Rev. Plant Physiol.* **25,** 487.
Wightman, F. (1973). *Biochem. Soc. Symp.* **38,** 247.
Ray, P. M. (1974). *Recent Adv. Phytochem.* **7,** 93.

Cytokinins
Chibnall, A. C. (1939). "Protein Metabolism in the Plant." Yale Univ. Press, New Haven, Connecticut.
Hall, R. H. (1970). *Progr. Nucl. Acid Res. Mol. Biol.* **10,** 57.
Hall, R. H. (1973). *Annu. Rev. Plant Physiol.* **24,** 415.
Helgeson, J. P. (1968). *Science* **161,** 974.
Kende, H. A. (1971). *Int. Rev. Cytol.* **31,** 301.
Leonard, N. J. (1974). *Recent Adv. Phytochem.* **7,** 21.
Skoog, F., and Armstrong, D. J. (1970). *Annu. Rev. Plant Physiol.* **21,** 359.

Ethylene
Abeles, F. G. (1972). *Annu. Rev. Plant Physiol.* **23,** 259.
Abeles, F. G. (1973). "Ethylene in Plant Biology." Academic Press, New York.
Burg, S. P., Apelbaum, A., Eisinger, W., and Kang, B. G. (1971). *Hort. Sci.* **6,** 359.
Lieberman, M., and Kunishi, A. T. (1971). *Hort. Sci.* **6,** 355.
Mapson, L. W. (1970). *Endeavour* **39,** 29.
de Wilde, R. C. (1971). *Hort. Sci.* **6,** 364.
Yang, S. F. (1974). *Recent Adv. Phytochem.* **7,** 131.

Gibberellins
Jones, R. L. (1973). *Annu. Rev. Plant Physiol.* **24,** 571.
Lin, P. P-C. (1974). *Adv. Cyclic Nucleotide Res.* **4,** 439.
Lang, A. (1970). *Annu. Rev. Plant Physiol.* **21,** 537.
MacMillan, J. (1974). *Recent Adv. Phytochem.* **7,** 1.
Stowe, B. B., and Yamaki, T. (1957). *Annu. Rev. Plant Physiol.* **8,** 181.
West, C. A. (1973). *In* "Biosynthesis and Its Control in Plants" (B. V. Milborrow, ed.), Chapter 7. Academic Press, New York.
Yomo, H., and Varner, J. E. (1971). *In* "Current Topics in Developmental Biology," (A. A. Moscona and A. Monray, eds.), Vol. 6, p. 111. Academic Press, New York.

REFERENCES

Abdul-Baki, A. A., and Ray, P. M. (1971). *Plant Physiol.* **47**, 537.

Abeles, F. B. (1966). *Plant Physiol.* **41**, 585.

Abeles, F. B. (1972). *Annu. Rev. Plant Physiol.* **23**, 259.

Abeles, F. B., and Leather, C. R. (1971). *Planta* **97**, 87.

Abeles, F. B., and Rubinstein, B. (1964). *Plant Physiol.* **39**, 963.

Abeles, F. B., Craker, L. E., and Leather, G. R. (1971). *Plant Physiol.* **47**, 7.

Addicott, F. T., Lyon, J. L., Ohkuma, K., Thiessen, W. E., Carns, H. R., Smith, O. E., Cornforth, J. W., Milborrow, B. V., Ryback, G., and Wareing, P. F. (1968). *Science* **159**, 1493.

Amrhein, N. (1974). *Z. Pflanzenphysiol.* **72**, 249.

Andreae, W. A., and Good, N. E. (1955). *Plant Physiol.* **30**, 380.

Armstrong, D. J., Skoog, F., Kirkegaard, L. H., Hampel, A. E., Bock, R. M., Gillam, I., and Gener, G. M. (1969a). *Proc. Natl. Acad. Sci. U.S.A.* **63**, 504.

Armstrong, D. J., Burrows, W. J., Skoog, F., Roy, K. L., and Soll, D. (1969b). *Proc. Natl. Acad. Sci. U.S.A.* **63**, 834.

Ashford, A. E., and Jacobsen, J. V. (1974). *Planta* **120**, 81.

Barendse, G. W. M., Kende, H., and Lang, A. (1968). *Plant Physiol.* **43**, 815.

Basu, P. S., and Tuli, V. (1972). *Plant Physiol.* **50**, 499.

Baur, A. H., and Yang, S. F. (1969). *Plant Physiol.* **44**, 189.

Baur, A. H., Yang, S. F., Pratt, H. K., and Biale, J. B. (1971). *Plant Physiol.* **47**, 696.

Bearder, J. R., MacMillan, J., Wels, C. M., and Phinney, B. O. (1975). *Phytochemistry* **14**, 1741.

Bennett, P. A., and Chrispeels, M. J. (1972). *Plant Physiol.* **49**, 445.

Ben-Tal, Y., and Varner, J. E. (1974). *Plant Physiol.* **54**, 813.

Biale, J. B. (1940). *Science* **91**, 458.

Biemann, K., Tsunakawa, S., Sonnenbichler, J., Feldmann, H., Dutting, D., and Zachau, H. G. (1966). *Angew. Chem.* **78**, 600.

Birch, A. J., Richards, R. W., and Smith, H. (1958). *Proc. Chem. Soc., London* p. 192.

Birch, A. J., Richards, R. W., Smith, H., Harris, A., and Whalley, W. B. (1959). *Tetrahedron* **7**, 241.

Black, R. C., and Hamilton, R. H. (1971). *Plant Physiol.* **48**, 603.

Bonnafous, J. C., Mousseron-Canet, M., and Olive, J. L. (1973). *Biochim. Biophys. Acta* **312**, 165.

Bruce, M. I., Zwar, J. A., and N. P. Kefford, (1965). *Life Sci.* **4**, 461.

Burg, S. P., and Burg, E. A. (1967). *Plant Physiol.* **42**, 144.

Burg, S. P., and Claggett, C. O., (1967). *Biochem. Biophys. Res. Commun.* **27**, 125.

Burg, S. P., Apelbaum, A., Eisinger, W., and Kang, B. G. (1971). *HortScience* **6**, 359.

Chen, C. M., and Hall, R. H. (1969). *Phytochemistry* **8**, 1687.

Chen, D., and Osborne, D. J. (1970). *Nature (London)* **226**, 1157.

Chibnall, A. C. (1939). "Protein Metabolism in the Plant." Yale Univ. Press, New Haven, Connecticut.

Chrispeels, M. J., and Varner, J. E. (1967). *Plant Physiol.* **42**, 398.

Cleland, R. E. (1973). *Proc. Natl. Acad. Sci. U.S.A.* **70**, 3092.

Corgan, J. N., and Martin, G. C. (1971). *HortScience* **6**, 405.

Cornforth, J. W., Milborrow, B. V., Ryback, G., and Wareing, P. F. (1965a). *Nature* (*London*) **205**, 1269.

Cornforth, J. W., Milborrow, B. V., and Ryback, G. (1965b). *Nature* (*London*) **206**, 715.

Crocker, W. (1932). *Proc. Am. Philos. Soc.* **71**, 295.

Crocker, W., and Zimmerman, P. W. (1932). *Contrib. Boyce Thompson Inst.* **4**, 177.

Cummins, W. R., and Sondheimer, E. (1973). *Planta* **111**, 365.

Cummins, W. R., Kende, H., and Raschke, K. (1971). *Planta* **99**, 347.

Davis, L. A., and Addicott, F. T. (1972). *Plant Physiol.* **49**, 644.

Deleuze, G. G., McChesney, J. D., and Fox, J. E. (1972). *Biochem. Biophys. Res. Commun.* **48**, 1426.

Demorest, D. M., and Staham, M. A., (1971). *Plant Physiol.* **47**, 450.

Denny, F. E. (1924). *J. Agric. Res.* (*Washington, D.C.*) **27**, 757.

de Wilde, R. C. (1971). *HortScience* **6**, 364.

Digby, J., and Wareing, P. F. (1966). *J. Exp. Bot.* **17**, 718.

Doree, M., and Guern, J. (1973). *Biochim. Biophys. Acta* **304**, 611.

Doubt, S. L. (1917). *Bot. Gaz.* (*Chicago*) **63**, 209.

Dyson, W. H., Hall, R. H., Hong, C. I., Dutta, S. P., and Chedda, G. B. (1972a). *Can. J. Biochem.* **50**, 237.

Dyson, W. D., Fox, J. E., and McChesney, J. D. (1972b). *Plant Physiol.* **49**, 506.

El-Antably, H. M. M., Wareing, P. F., and Hillman, J. (1967). *Planta* **73**, 74.

Elliott, M. C., and Stowe, B. B. (1971). *Plant Physiol.* **48**, 498.

Evans, M. L., Ray, P. N., and Reinhold, L. (1971). *Plant Physiol.* **47**, 335.

Evins, W. H. (1971). *Biochemistry* **10**, 4295.

Fall, R. R., and West, C. A. (1971). *J. Biol. Chem.* **246**, 6913.

Filner, P., and Varner, J. E. (1967). *Proc. Natl. Acad. Sci. U.S.A.* **58**, 1520.

Firn, R. D., and Friend, J. (1972). *Planta* **103**, 263.

Firn, R. D., Burden, R. S., and Taylor, H. F. (1972). *Planta* **102**, 115.

Fittler, F., Kline, L. K., and Hall, R. H. (1968). *Biochemistry* **7**, 940.

Gane, R. (1934). *Nature* (*London*) **134**, 1008.

Gefter, M. L., and Russell, R. L. (1969). *J. Mol. Biol.* **39**, 145.

Goldschmidt, E. E., Goren, R. R., Even-Chen, Z., and Bittner, S. (1973). *Plant Physiol.* **51**, 870.

Greenwood, M. S., Shaw, S., Hillman, J. R., Ritchie, A., and Wilkins, M. B. (1972). *Planta* **108**, 179.

Hager, A., Menzel, H., and Kraus, A. (1971). *Planta* **100**, 47.

Hall, R. H. (1973). *Annu. Rev. Plant Physiol.* **24**, 415.

Hall, R. H., Robins, M. J., Stasiuk, L., and Thedford, R. (1966). *J. Am. Chem. Soc.* **88**, 2614.

Hansen, E. (1942). *Bot. Gaz.* (*Chicago*) **103**, 543.

Hanson, J. R., Hawker, J., and White, A. F. (1972). *J. Chem. Soc., Perkin Trans. 1* p. 1892.

Harada, F., Gross, H. J., Kimura, F., Chang, S. H., Nishimura, S., and RajBhandary, U. L. (1968). *Biochem. Biophys. Res. Commun.* **33**, 299.

Harrison, M. A., and Walton, D. C. (1975). *Plant Physiol.* **56**, 250.

Hattori, A., and Marumo, S. (1972). *Planta* **102**, 85.

Hecht, S. M., Leonard, N. J., Schmitz, R. Y., and Skoog, F. (1970). *Phytochemistry* **9**, 1173.

Hinman, R. L., and Lang, J. (1965). *Biochemistry* **4**, 144.

Ho, D. T.-H., and Varner, J. E. (1974). *Proc. Natl. Acad. Sci. U.S.A.* **71**, 4783.

Ho, D. T.-H., and Varner, J. E. (1976). *Plant Physiol.* **57**, 175.

Ihle, J. N., and Dure, L., III. (1970). *Biochem. Biophys. Res. Commun.* **38**, 995.

Jablonski, J. R., and Skoog, F. (1954). *Physiol. Plant.* **1**, 16.

Jacobsen, J. V., and Varner, J. E. (1967). *Plant Physiol.* **42**, 1596.

Jacobsen, J. V., and Zwar, J. A. (1974). *Proc. Natl. Acad. Sci. U.S.A.* **71**, 3290.

Johnson, K. D., and Kende, H. (1971). *Proc. Natl. Acad. Sci. U.S.A.* **68**, 2674.

Johnson, K. D., and Purves, W. F. (1970). *Plant Physiol.* **46**, 581.

Johnson, K. D., Daniels, D., Dowler, M. J., and Rayle, D. L. (1974). *Plant Physiol.* **53**, 224.

Johri, M. M., and Varner, J. E. (1968). *Proc. Natl. Acad. Sci. U.S.A.* **59**, 260.

Jones, R. L. (1969). *Planta* **88**, 73.

Kang, B. G., Newcomb, W., and Burg, S. P. (1971). *Plant Physiol.* **47**, 504.

Keates, R. A. B. (1973). *Biochem. Biophys. Res. Commun.* **54**, 655.

Kende, H. (1971). *Int. Rev. Cytol.* **31**, 301.

Kende, H., and Baumgardner, B. (1974). *Planta* **116**, 279.

Kende, H., and Tavares, J. E. (1968). *Plant Physiol.* **43**, 1244.

Kindl, H. (1968). *Hoppe-Seyler's Z. Physiol. Chem.* **349**, 519.

Koehler, D. E., and Varner, J. E. (1973). *Plant Physiol.* **52**, 208.

Kopcewicz, J., Ehmann, A., and Bandurski, R. S. (1974). *Plant Physiol.* **54**, 846.

Ku, H. S., and Leopold, A. C. (1970). *Biochem. Biophys. Res. Commun.* **41**, 1155.

Kulaeva, O. N., Cherkasor, V. M., and Tret'yakova, G. S. (1968). *Dokl. Akad. Nauk USSR* **178**, 1204.

Kurosawa, E. (1926). *Trans. Natl. Hist. Soc. Formosa* **16**, 213.

Kutacek, M., and Kefeli, V. I. (1968). "Biochemistry and Physiology of Plant Growth Substances," p. 127. Runge Press, Ottawa.

Lang, A. (1965). *In* "Handbuch der Pflanzenphysiologie" (W. Ruhland, ed), Vol. 15, p. 1380. Springer-Verlag, Berlin and New York.

Lang, A. (1970). *Annu. Rev. Plant Physiol.* **21**, 537.

Lenton, J. R., Perry, V. M., and Sanders, P. F. (1968). *Nature (London)* **220**, 86.

Leonard, N. J. (1974). *Recent Adv. Phytochem.* **7**, 21.

Letham, D. S. (1973). *Phytochemistry* **12**, 2445.

Letham, D. S., and Miller, C. O. (1965). *Plant Cell Physiol.* **6**, 355.

Letham, D. S., Shannon, J. S., and McDonald, I. R. (1964). *Proc. Chem. Soc., London* p. 230.

Lewis, L. N., and Varner, J. E. (1970). *Plant Physiol.* **46**, 194.

Libbert, E., Wichner, S., Schiewer, U., Risch, H., and Kaiser, W. (1966). *Planta* **68**, 327.

Lieberman, M., and Kunishi, A. T. (1971). *HortScience* **6**, 355.

Lieberman, M., and Mapson, L. W. (1964). *Nature (London)* **204**, 343.

Lieberman, M., Kunishi, A. T., Mapson, L. W., and Wardale, D. A. (1965). *Biochem. J.* **97**, 449.

Lieberman, M., Kunishi, A. T., Mapson, L. W., and Wardale, D. A. (1966). *Plant Physiol.* **41**, 376.

Lin, P. P.-C. (1974). *Adv. Cyclic Nucleotide Res.* **4**, 439.

Lin, P. P.-C., and Varner, J. E. (1972). *Biochim. Biophys. Acta* **276**, 454.

Liu, W. C., and Carns, H. R. (1961). *Science* **134**, 384.

McCalla, D. R., Morre, D. J., and Osborne, D. J. (1962). *Biochim. Biophys. Acta* **55**, 522.

McComb, A. J., McComb, J. A., and Duda, C. T. (1970). *Plant Physiol.* **46**, 221.

Madison, J. T., Everett, G. A., and Kung, H. (1966). *Science* **153**, 531.

Mahadevan, S., and Stowe, B. B. (1972). *In* "Plant Growth Substances" (D. J. Carr, ed.), p. 117. Springer-Verlag, Berlin and New York.

Mapson, L. W. (1970). *Endeavour* **39**, 29.

Mapson, L. W., March, J. F., Rhodes, M. J. C., and Wooltorton, L. S. C. (1970). *Biochem. J.* **177**, 473.

Marumo, S., Hattori, H., Abe, H., and Munakata, K. (1968a). *Nature (London)* **219**, 959.

Marumo, S., Abe, H., Hattori, H., and Munakata, K. (1968b). *Agric. Biol. Chem.* **32**, 117.

Maynard, J. A., and Swan, J. M. (1963). *Aust. J. Chem.* **16**, 596.

Milborrow, B. V. (1969). *Chem. Commun.* p. 966.

Milborrow, B. V. (1970). *J. Exp. Bot.* **21**, 17.

Milborrow, B. V. (1974a). *Annu. Rev. Plant Physiol.* **25**, 259.

Milborrow, B. V. (1974b). *Recent Adv. Phytochem.* **7**, 57.

Milborrow, B. V., and Noddle, R. C. (1970). *Biochem. J.* **119**, 727.

Milborrow, B. V., and Robinson, D. R. (1973). *J. Exp. Bot.* **24**, 537.

Miller, C. O. (1961). *Proc. Natl. Acad. Sci. U.S.A.* **47**, 170.

Miller, C. O. (1969). *Planta* **87**, 26.

Miller, C. O., Skoog, F., von Saltza, M. H., and Strong, F. M. (1955a). *J. Am. Chem. Soc.* **77**, 1392.

Miller, C. O., Skoog, F., Okumura, F. S., von Saltza, M. H., and Strong, F. M. (1955b). *J. Am. Chem. Soc.* **77**, 2662.

Miller, E. V., Winston, J. R., and Fisher, D. F. (1940). *J. Agric. Res. (Washington, D.C.)*, **60**, 269.

Miura, G. A., and Hall, R. H. (1973). *Plant Physiol.* **51**, 563.

Miura, G. A., and Miller, C. O. (1969). *Plant Physiol.* **44**, 372.

Nadeau, R., and Rappaport, L. (1972). *Phytochemistry* **11**, 1611.

Nadeau, R., and Rappaport, L. (1974). *Plant Physiol.* **54**, 809.

Nadeau, R., Rappaport, L., and Stolp, C. F. (1972). *Planta* **107**, 315.

Neljubow, D. (1901). *Bot. Zentralbl., Suppl.* **10**, 128.

Noddle, R. C., and Robinson, D. R. (1969). *Biochem. J.* **112**, 547.

Ohkuma, K., Lyon, J. L., Addicott, F. T., and Smith, O. E. (1963). *Science* **142**, 1592.

Osborne, D. J. (1955). *Nature (London)* **176**, 1161.

Paces, V., Werstiuk, E., and Hall, R. H. (1971). *Plant Physiol.* **48**, 775.

Parker, C. W., Letham, D. S., Cowley, D. E., and MacLeod, J. K. (1972). *Biochem. Biophys. Res. Commun.* **49**, 460.

Peterkofsky, A., and Jesensky, C. (1969). *Biochemistry* **8**, 3798.

Phinney, B. O. (1956). *Proc. Nat. Acad. Sci. USA* **42**, 185.

Pope, D., and Black, M. (1972). *Planta* **102**, 26.

Psornik, Z., and Bandurski, R. S. (1972). *Plant Physiol.* **50**, 176.

Raschke, K. (1975). *Annu. Rev. Plant Physiol.* **26**, 309.

Raschke, K., Firm, R. D., and Pierce, M. (1975). *Planta* **125**, 149.

Ray, P. M. (1973). *Plant Physiol.* **51**, 609.

Ray, P. M. (1974). *Recent Adv. Phytochem.* **7**, 93.

Rayle, D. L. (1973). *Planta* **114**, 63.

Rayle, D. L., and Cleland, R. E. (1970). *Plant Physiol.* **46**, 250.

Rayle, D. L., and Cleland, R. E. (1972). *Planta* **104**, 282.

Rehm, M., and Cline, M. G. (1973). *Plant Physiol.* **51**, 93.

Richmond, A. E., and Lang, A. (1957). *Science* **125**, 650.

Roberts, D. L., Heckman, R. A., Hege, B. P., and Bellin, S. A. (1968). *J. Org. Chem.* **33**, 3566.

Robinson, D. R., and Ryback, G. (1969). *Biochem. J.* **113**, 895.

Rogler, C. E., and Dahmus, M. E. (1974). *Plant Physiol.* **54**, 88.

Rubinstein, B. (1974). *Plant Physiol.* **54**, 835.

Rubinstein, B., and Leopold, A. C. (1963). *Plant Physiol.* **38**, 262.

Ruddat, M., Heftman, E., and Lang, E. (1965). *Arch. Biochem. Biophys.* **111**, 107.

Rudnicki, R. (1969). *Planta* **86**, 63.

Sadava, D., and Chrispeels, M. J. (1973). *Dev. Biol.* **30**, 49.

Schneider, E. A., and Wightman, F. (1974). *Annu. Rev. Plant Physiol.* **25**, 487.

Shantz, E. M., and Steward, F. C. (1955). *J. Am. Chem. Soc.* **77**, 6351.

Shechter, I., and West, C. A. (1969). *J. Biol. Chem.* **244**, 3200.

Sheldrake, A. R., and Northcote, D. H. (1968). *Nature (London)* **217**, 195.

Sherwin, J. E., and Purves, W. K. (1969). *Plant Physiol.* **44**, 1303.

Short, K. C., and Torrey, J. G. (1972). *Plant Physiol.* **49**, 155.

Skoog, F., and Armstrong, D. J. (1970). *Annu. Rev. Plant Physiol.* **21**, 359.

Skoog, F., and Miller, C. O. (1957). *Symp. Soc. Exp. Biol.* **11**, 118.

Skoog, F., and Thimann, K. V. (1934). *Proc. R. Soc. London, Ser. B* **114**, 317.

Skoog, F., and Tsui, C. (1948). *Am. J. Bot.* **85**, 782.

Skoog, F., Schneider, C. L., and Malan, P. (1942). *Am. J. Bot.* **29**, 568.

Skoog, F., Strong, F. M., and Miller C. O. (1965). *Science* **148**, 532.

Skoog, F., Hamzi, H. W., Szweykowska, A. M., Leonard, N. J., Carraway, K. L., Fujii, T., Helgeson, J. P., and Loeppky, R. M. (1967). *Phytochemistry* **6**, 1169.

Skoog, F., Schmitz, R. Y., Bock, R. M., and Hecht, S. M. (1973). *Phytochemistry* **12**, 25.

Sondheimer, E., and Tzou, D. S. (1971). *Plant Physiol.* **47**, 516.

Sondheimer, E., Tzou, D. S., and Galson, E. C. (1968). *Plant Physiol.* **43**, 1443.

Spencer, M. (1959). *Can. J. Biochem. Physiol.* **37**, 53.

Srivastava, L. M., Sawhney, V. K., and Taylor, I. E. P. (1975). *Proc. Natl. Acad. Sci. U.S.A.* **72**, 1107.

Stonier, T., Hudek, J., Vaude-Stauve, R., and Yang, H.-M. (1970). *Physiol. Plant.* **23**, 775.

Stowe, B. B., and Dotts, M. A. (1971). *Plant Physiol.* **48**, 559.

Taylor, H. F., and Burden, R. S. (1972). *Proc. R. Soc. London, Ser. B* **180**, 317.

Taylor, H. F., and Smith, T. A. (1967). *Nature (London)* **215**, 1513.

Thimann, K. V. (1935). *J. Biol. Chem.* **109**, 279.

Thompson, W. F., and Cleland, R. (1972). *Plant Physiol.* **50**, 289.

Tinelli, E. T., Sondheimer, E., and Walton, D. C. (1973). *Tetrahedron Lett.* p. 139.

Trewavas, A. (1970). *Plant Physiol.* **45**, 742.

Trewavas, A. (1972). *Plant Physiol.* **49**, 47.

Trewavas, A. (1973). *Plant Physiol.* **51**, 760.

Tuli, V., and Moyed, H. S. (1969). *J. Biol. Chem.* **244**, 4916.

Ueda, M., Ehmann, A., and Bandurski, R. S. (1970). *Plant Physiol.* **46**, 715.

Valdovinos, J. G., Jensen, T. E., and Sicko, L. M. (1971). *Plant Physiol.* **47**, 162.

Vanderhoef, L. N., and Stahl, C. A. (1975). *Proc. Natl. Acad. Sci. U.S.A.* **72**, 1822.

van Overbeek, J., Conklin, M. E., and Blakeslee, A. F. (1942). *Am. J. Bot.* **29**, 472.

Varner, J. E., and Ho, D. T.-H. (1976). *In* "The Molecular Biology of Hormone Action" (D. Brown, ed.) (in press).

Varner, J. E., and Mense, R. M. (1972). *Plant Physiol.* **49**, 187.

Venis, M. A. (1972). *Plant Physiol.* **49**, 24.

Walton, D. C., and Sondheimer, E. (1972a). *Plant Physiol.* **49**, 285.
Walton, D. C., and Sondheimer, E. (1972b). *Plant Physiol.* **49**, 290.
Warner, H. L. (1970). Ph.D. Thesis, Purdue University, Lafayette, Indiana.
Waters, L. C., and Dure, L. S. (1966). *J. Mol. Biol.* **19**, 1.
West, C. A. (1973). *In* "Biosynthesis and Its Control in Plants" (B. V. Milborrow, ed.), Chapter 7. Academic Press, New York.
Whitty, C. D., and Hall, R. H. (1974). *Can. J. Biochem.* **52**, 789.
Wightman, F. (1973). *Biochem. Soc. Symp.* **38**, 247.
Wright, S. T. C. (1969). *Planta* **86**, 10.
Wright, S. T. C., and Hiron, R. W. P. (1969). *Nature (London)* **224**, 719.
Yabuta, T., and Sumiki, Y. (1938). *J. Agric. Chem. Soc. Jpn.* **14**, 1526.
Yang, S. F. (1967). *Arch. Biochem. Biophys.* **122**, 481.
Yang, S. F. (1974). *Recent Adv. Phytochem.* **7**, 131.
Yang, S. F., Ku, H. S., and Pratt, H. K. (1967). *J. Biol. Chem.* **242**, 5274.
Yomo, H., and Varner, J. E. (1971). *Curr. Top. Dev. Biol.* **6**, 111.
Yoshida, R., and Oritani, T. (1972). *Plant Cell Physiol.* **13**, 837.
Zachau, H. G., Dutting, D., and Feldman, H. (1966). *Angew. Chem.* **78**, 392.
Zenk, M. H. (1961). *Nature (London)* **191**, 493.
Zwar, J. H., and Jacobsen, J. V. (1972). *Plant Physiol.* **49**, 1000.

23

Senescence

LEONARD BEEVERS

I. Introduction

In discussing the topic of senescence, one is immediately confronted with the problem of terminology. In many instances, and particularly in studies relating to animals, the terms senescence and aging are used interchangeably and usually relate to the accumulation of somatic mutations and increase in metabolic failures which decrease the functional capacity and hasten the death of an organism (see relevant chapters in Woolhouse, 1967). In plant physiology and biochemistry, the term aging is used extensively, but not exclusively, to describe the changes that occur when slices of storage tissue, for example, are maintained for long periods in aerated solutions. Such a treatment usually results in an increased metabolic activity of the tissue as indicated by enhanced protein synthesis, increased respiration, and an enhanced capacity for ion uptake. These characteristics are more typical of younger tissue and, thus, this aging should be more accurately described as rejuvenation.

In other instances, aging is also used to describe the change from juvenile to adult form in those species showing different growth phases. In this case, aging is frequently accompanied by a change in vegetative morphology, and the adult form, in contrast to the juvenile form, is capable of flowering. In this respect, aging is synonymous with chronological development. In addition, the term aging has been used to describe the degradative running down of an organism (Wareing and Seth, 1967). However, in view of the ambiguity encountered in the use of the term aging to describe other nondeteriorative changes, it is felt that the deteriorative events which precede the death of a mature cell should be defined as senescence.

In this context, senescence has been most extensively used to characterize the changes that occur in leaves of deciduous trees in the fall. These changes occur at an accelerated rate in detached leaves, and this material has been used extensively in studies of plant senescence. Senescence also occurs in other organs and tissues of the plant. The cotyledons of seeds undergo senescence and death following a depletion of the storage reserves during germination.

In some cotyledons the reserve depletion and senescence may occur more or less simultaneously, and it is difficult to differentiate between metabolism associated with reserve utilization and the reactions associated with the senescence of the cotyledon. In seedlings showing epigeal germination, the cotyledons, after reserve depletion, may become photosynthetic and have an extended life span before the onset of senescence. Such cotyledons have provided a useful material for the study of senescence by Simon's group (Simon, 1967; Butler and Simon, 1970).

Ripening fruit demonstrate many of the characteristics typically classified as senescence, and metabolic changes during fruit ripening have been studied extensively. These have been reviewed by Rhodes (1970) and Sacher (1973). However, in many respects, the ripened fruit represents only a stage of senescence, and the fruit undergoes further deteriorative changes (senescence) following harvest.

In some instances, senescence occurs shortly after maturation of certain cells; such is the situation in the production of xylem elements or cells of the root cap. The senescence and death of these cells occurs while the bulk of the remaining tissues are viable and healthy. It thus appears that senescence and death at the cellular level are dictated by some internal control mechanism and cannot be accounted for by the deleterious effects of hazardous environment leading to increased metabolic failure and death. Further evidence for the internal regulation of senescence is provided by an examination of leaf senescence in annual plants, biennials or herbaceous perennials, and woody perennials. In the first group the whole plant senesces and dies during one growing season. Usually, the leaves of such annual plants senesce in a sequential fashion in which those leaves that are first formed by the growing apex commonly begin to senesce early in the life of the plant and before it has begun to flower.

Subsequently, following flowering of these annual types, the total plant senesces and dies. During the whole plant senescence, the leaves may senesce in a sequential manner from the base to the apex. This type of senescence may also be termed correlative senescence, since it is triggered or associated with developmental processes occurring elsewhere in the plant. In some biennials and herbaceous perennials, the leaves may show sequential senescence with an overall death of the aerial portions occurring at the end of the growing season. However, the underground perennating organs do not die out and serve as a source of new growth the following year. In the deciduous woody perennials the most prominent senescence is limited to the leaves; all the leaves usually senesce at the same time in a so-called synchronous senescence.

It becomes apparent in considering these various types of senescence that the event does not occur in a random fashion but is regulated. This regulation is even more refined when comparisons are made of the specific leaves of different species. In some species, senescence occurs uniformly over the leaf surface. In others, the distal regions of the leaf are the first to senesce, whereas in other species senescence commences at the basal area of the leaf.

The overall effect of the internal regulation of senescence is that cells and organs within the plant have different life spans, and different species demonstrate varying longevity. The orderly events of senescence within a plant provide a mechanism by which the internal cellular constituents can be depleted and transferred to other parts of the plant. In many instances processes are so regulated that following the senescence and transfer of materials from a particular plant part, the part is abscinded. The abscission is caused by a specialized senescence in which the death of defined cells is associated with the breakdown of adjoining cell walls and the formation of an abscission zone (Addicott, 1970).

II. Regulation of Senescence

In view of the regulated nature of senescence, it is logical to determine the mechanisms by which the process is controlled. For this purpose extensive studies have been made of detached leaves or leaf discs. Such material usually senesces at an accelerated rate in comparison to leaves attached to the plant.

In attempting to account for this accelerated senescence, it was originally proposed that the accelerated senescence of the detached leaf was associated with a depletion of the carbohydrates normally supplied by the plant from which the leaf was excised. However, Chibnall (1939) pointed out that Vickery's data showed that there was no direct relationship between soluble carbohydrate status of the leaf and the onset of senescence. It was nevertheless observed that illuminated leaves senesced at a slower rate than those kept in the dark. However, in view of the lack of relationship between senescence rate and carbohydrate status, it appeared unlikely that the retardation of senescence could be related to the production of photosynthate in the detached leaf. Goldthwaite and Laetsch (1967) have reported that the retardation of senescence of bean leaves by light was prevented by 3-(3,4-dichlorophenyl)-1,1-dimethyl urea (DCMU) and thus concluded that the light effect involved photosynthesis. In contrast, Haber *et al.* (1969) have concluded that in excised leaf tips of wheat the action of light in retarding senescence is insensitive to DCMU and thus was nonphotosynthetic. This finding agrees with the earlier report of Sugiura (1963) who demonstrated that the retardation of protein and chlorophyll loss in tobacco leaf discs by red light could be overcome by illumination with far-red light, thus indicating that the light retardation of senescence involved phytochrome. A more thorough demonstration of the involvement of phytochrome in the photocontrol of senescence in the liverwort *Marchantia* has been reported by De Greef *et al.* (1971).

In addition to this retardation of senescence by light, it has been demonstrated that senescence is delayed by the formation of roots on the petioles of detached leaves. Chibnall (1939) with considerable foresight suggested that the roots functioned to retard senescence by supplying some hormone that was essential for maintaining cellular integrity of the leaves. Preliminary experimental support for this concept was provided by the observations of Richmond and Lang (1957) that the senescence of detached *Xanthium* leaves could be delayed by the application of kinetin. Subsequently, it has been demonstrated that kinetin and related substances will delay the senescence of detached leaves or excised leaf discs in a wide variety of species (Osborne, 1962; Person *et al.*, 1957; Sugiura *et al.*, 1962; Wollgiehn, 1961). It has been shown that xylem exudate from *Helianthus* (Kende, 1964) or methanolic extracts of roots (Seth and Wareing, 1965) contain cytokinin-like materials that delay senescence of detached leaves. In addition to the kinetin retardation of senescence of leaves, it has been shown that auxins can function in this capacity (Osborne, 1965; Sacher, 1965). In other species, gibberellic acid (GA) is an effective retardant of senescence of detached leaves (Fletcher and Osborne, 1965; Beevers, 1966; Whyte and Luckwill, 1966). In contrast to the retardation of senescence by these compounds, it has been demonstrated that abscisic acid (ABA) accelerates the senescence of detached leaves (El-Antably *et al.*, 1967; Beevers, 1968).

These observations that senescence of detached leaves can be controlled by the exogenous application of growth regulators indicate that leaf senescence in the intact plant may be regulated by the balance of endogenous growth regulators. The leaves of many woody perennials show lower levels of endogenous auxins and gibberellins and a higher level of growth inhibitors under short days (Phillips and Wareing, 1959), which may contribute to the onset of senescence in the fall. Similarly, under drought conditions a depleted flow of materials from the roots would reduce the availability of gibberellins (Jones and Phillips, 1966) and cytokinins (Itai and Vaadia, 1965) normally provided by the root exudate and cause a lowering of the compounds in the aerial portions. Additionally, the senescence of water-stressed leaves may be accelerated by the accumulation of abscisic acid in such leaves (Wright and Hiron, 1969). Detached leaves showed a declining gibberellin content and an increase in abscisic acid-like compounds. Treatments that delayed senescence retarded the decline in gibberellin-like components and prevented the increase in ABA-like compounds (Chin and Beevers, 1970). Additional support for the role of endogenous growth regulators in controlling senescence is provided by the data of Mayak and Halevy (1970), which demonstrated a close correlation between endogenous cytokinin activity and longevity of cut roses.

III. Biochemical Changes during Senescence

Characteristically, during leaf senescence there is a decrease in chlorophyll, protein, and RNA level (Fig. 1). In detached leaves there is an accumulation of amino acids associated with protein depletion. In the attached leaf there is a net loss of nitrogen from the leaf as the products of protein depletion are exported to support growth and metabolism in other parts of the plant. Photosynthetic capacity declines during senescence, respiratory metabolism is retained at a fairly constant level during the degreening and yellowing phase. At the terminal period of senescence, the respiratory rate declines rapidly.

A. Protein

Experiments with mature leaves indicate that such leaves retain the capacity to incorporate label derived from the photosynthetic fixation of $^{14}CO_2$, radioactive amino acids, or ^{15}N-labeled nitrogenous compounds into protein at a time when the total protein content of the leaf tissue is remaining constant and even declining. Such observations indicate that

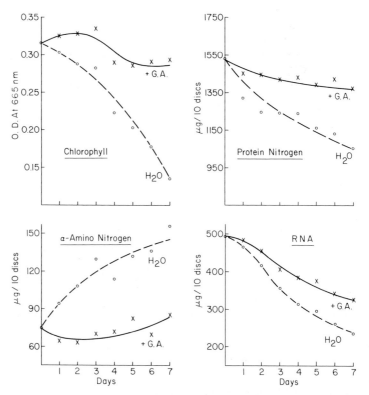

Fig. 1. The characteristic changes in metabolites occurring during the incubation of leaf discs of Nasturtium (*Tropaeolum majus*) on water or gibberellic acid (G.A.) 20 ppm.

the proteins of the mature leaf must be turning over and continued synthesis is counteracted by continued decay according to the reaction:

$$\text{Amino acids} \xrightarrow{\text{synthesis}} \text{protein} \xrightarrow{\text{degradation}} \text{amino acids}$$

The reaction is written in this manner in order to indicate that the amino acids utilized in protein synthesis may not be those derived from protein degradation (Bidwell *et al.*, 1964). At the same time, the possibility of recycling of some amino acids should not be completely excluded.

The phenomena of protein turnover and compartmentalization of amino acids into metabolic and inactive pools make the study of protein synthesis in plant tissue a frustrating task. In labeling studies, it is difficult to establish the specific radioactivity of the precursor pool, and it is equally difficult to assess the rate of protein breakdown. However, by the use of pulse labeling and chase techniques and mathematical manipulation, it has been assessed that the total protein is turned over at a rate of up to 2% per hour in leaf tissue. (Hellebust and Bidwell, 1963; Racusen and Foote, 1960).

The decline in protein content during senescence coupled with the observed accumulation of amino acids in detached leaves indicates that the normal turnover is arrested in some manner in the senescing leaf. Decreased protein synthesis or increased protein degradation would account for the decline in protein content. Simon (1967) indicates that in attached leaves the depletion of amino acids probably plays a key role in inducing a decline in protein content and senescence. This would appear to be particularly true in the case of sequential senescence where amino acids are translocated out of the older leaves to the regions of active growth and development. Other workers have attached greater importance to a decline in the capacity for protein synthesis as providing the initiation of senescence. It is frequently demonstrated that the capacity to incorporate exogenously supplied labeled amino acids declines in detached leaves or leaf discs but is retained or may be increased in leaf discs treated with senescing delaying growth regulators (Table I). On the basis of this type of evidence, it has been concluded (Osborne, 1962; Wollgiehn, 1967; Beevers, 1968) that senescence is caused by a failure of protein synthesis and that the growth regulators retard senescence by sustaining protein synthesis. However, in many of these investigations little consideration has been given to the fact that uptake of the radioactive precursor changes during senescence. In a detailed study of the change in protein synthesis in the ripening banana fruit, Sacher (1967)

TABLE I

Influence of Growth Regulator Treatment on the Capacity of Leaf Discs to Incorporate [^{14}C]Leucine into Proteins[a,b]

Treatment[c]	Total ethanol-soluble counts	Total counts in protein	% Counts incorporated
H_2O	26,116	11,899	31.2
Gibberellic acid	17,520	14,798	45.8
Kinetin	15,707	15,352	49.4
Abscisic acid	29,145	10,045	25.6

[a] Data from Beevers (1968).

[b] Nasturtium (*Tropaeolum majus*) leaf discs were incubated in the growth regulators for 24 hours, then transferred to radioactive [^{14}C]-leucine for 3 hours prior to analysis.

[c] Gibberellic acid, 20 ppm; kinetin, 20 ppm; abscisic acid, 2 ppm.

concluded that the decreased capacity to incorporate exogenously supplied leucine could be attributed to a changed internal permeability of the cell with the result that the applied exogenous labeled leucine was progressively diluted with leucine from internal cellular compartments as ripening progressed. In other instances, the incorporation of amino acids into protein appeared similar in both senescing and growth regulator-treated tissue; however, when analyses were made of the total uptake of radioactive precursor, it appeared that the senescing tissue was appreciably less efficient at incorporating available precursors into protein, suggesting a declining synthetic capacity in senescing tissue.

Of course an apparent decline in the capacity to incorporate exogenously supplied amino acids would also occur if the synthesized products were hydrolyzed more rapidly in the senescing tissue. If this was the case, then protein loss during senescence can be interpreted as an increased degradation rather than a decreased capacity for protein synthesis. Kuraishi (1968) indicated that the loss of radioactivity from [^{14}C]leucine labeled proteins in *Brassica rapa* L. leaf discs was retarded by kinetin treatment in comparison to water controls. The slow rate of release of radioactivity from the protein in the kinetin-treated leaf discs was not affected by supplying exogenous nonradioactive leucine or caseine hydrolysate, so the retention of radioactivity in the protein in the kinetin-treated discs could not be attributed to a recycling of the label. It was concluded that kinetin retarded the decomposition of protein. Tavares and Kende (1970), Shibaoka and Thimann (1970), Mizrahi *et al.* (1970), and Peterson and Huffaker (1975) have also stressed the important role of proteolysis in senescence and attribute the retardation of senescence by kinetin to its role in controlling protein degradation. Martin and Thimann (1972) suggest that the increase in protease in detached oat leaves may be induced by the accumulation of the amino acid serine and indicate that the proteolysis is brought about by serine-type proteases.

If the primary event of senescence is an increased proteolysis, then it would be useful to know the mechanisms by which proteins are degraded and to understand the reason for an accelerated protein degradation following leaf detachment. While the accumulation of amino acids during senescence is consistent with an extensive proteolysis, there is very little information available concerning enzymes capable of protein hydrolysis. Anderson and Rowan (1965) described a peptidase from tobacco leaves which hydrolyzed protein, and Beevers (1968) reported the occurrence of a similar enzyme in Nasturtium leaves. Anderson and Rowan (1965) indicated that there was no major increase in the proteolytic activity on a per unit fresh weight basis with physiological age. However, since total protein decreased in the older leaves, the specific activity of the peptidase increased with physiological age of the leaf. Peptidase activity extractable from detached leaves (leaf discs) increased as senescence proceeded (Anderson and Rowan, 1966; Beevers, 1968). Again the increase was more significant on a specific activity basis. Extractable proteolytic activity remained low in kinetin-treated leaf discs. This data can be interpreted as supporting the claim that protein loss and the accelerated senescence which occur in detached leaves are due to an increased protein degradation associated with increasing peptidase activity. However, Anderson and Rowan (1966, 1968) stress that the increase in proteolytic activity occurs after a measurable increase in free amino nitrogen content and a decrease in protein. Beevers (1968), in agreement with Anderson and Rowan, also demonstrated that kinetin prevented the increase in proteolytic activity and delayed the senescence of Nasturtium leaf discs; however, in other treatments that altered the senescence rate of the leaf discs, there was no relationship between protein loss and extractable proteolytic activity. The possibility exists that other proteolytic enzymes, not detected by the caseolytic assay, may increase during senescence. In fact, Balz (1966) has reported on the occurrence of an acid protease (hydrolyzing hemoglobin at pH 3.5) in tobacco seedlings. Immediately after detachment of leaves there was a production of a new particulate acid protease, which later appeared in the soluble fraction. Treatment of the leaves with kinetin delayed the synthesis of the acid proteases and other hydrolytic enzymes, and senescence was attributed to the degradation of the cytoplasm due to the action of hydrolytic enzymes released from the subcellular particle, the spherosome. In some instances (Peterson and Huffaker, 1975; Martin and Thimann, 1972) the senescence retarding effects of kinetin can be mimicked by cycloheximide an inhibitor of protein synthesis. It is postulated that cycloheximide prevents the synthesis of the proteolytic enzymes involved in senescence.

Of course, the onset of proteolysis and senescence could be achieved without any increase in protease activity. Such appears to be the case in *Perilla* leaves (Kannangara and Woolhouse, 1968) and attached tobacco leaves (Anderson and Rowan, 1966; De Jong, 1972). Also protease level does not increase during senescence of the corolla of ephemeral flowers of morning glory (*Ipomoea*), even though proteins are depleted rapidly in this organ (Matile and Winkenbach, 1971). Since the proteins are turning over, the plant cells are equipped with the necessary machinery for protein degradation. An increased proteolysis could occur if there was a changed accessibility of protease(s) to substrates. The increased substrate may represent enzymes whose substrates become depleted; some enzymes are more susceptible to proteolysis in the absence of substrates (Burchall, 1966). Alternatively, the increased proteolysis could arise from a changed compartmentalization in the senescing leaf such that previously isolated proteases are brought into contact with substrates. Such a situation would be analogous to the lysosomes of animal tissues. Balz (1966) equates the spherosomes of tobacco with the lysosomes, whereas Matile (1968) suggests that the functions of the lysosome are taken over by the vacuole in plant cells.

The question arises as to whether the release of latent hydrolases from the subcellular organelles would confer the selectivity for protein degradation that occurs during senescence. This question, of course, has to be satisfied by other proteases, also. The breakdown of proteins during senescence is a controlled event and not an uncontrolled lysis. Not all proteins are degraded at the same rate. Axelrod and Jagendorf (1951) initially demonstrated that the specific activity of various hydrolytic enzymes and peroxidase increased during senescence of detached tobacco leaves at a time when total protein was declining, and more recently Kannangara and Woolhouse (1968) have demonstrated a variable rate of loss of activity of several enzymes present in the soluble protein fractions in the course of senescence of leaves of *Perilla*. The observations that photosynthetic activity and respiratory rate decline at different rates during leaf senescence provide further evidence that not all proteins are degraded simultaneously. Ultrastructural studies to be discussed later further indicate that degradation follows a programmed sequence.

B. Ribonucleic Acid

The characteristic decline in protein content in senescing leaves is accompanied by a similar decline in RNA content (Fig. 1). The capacity to incorporate radioactive precursors into nucleic acids in senescing leaves is lower than that in detached leaves treated with growth regulators that retard senescence. On the basis of this type of evidence, it appears that the decline in RNA content during senescence is related to a decreased capacity for RNA synthesis (Osborne, 1962; Wollgiehn, 1967; Sacher, 1967; Beevers, 1968). However, as in the case of measurements of protein synthesis, it is difficult to interpret incorporation studies as providing an accurate assessment of RNA synthesis. The RNA in the plant cell is continuously being turned over.

$$\text{RNA precursors} \xrightarrow{\text{synthesis}} \text{RNA} \xrightarrow{\text{degradation}} \text{degradation products}$$

Thus the accumulation of radioactivity into RNA will be dependent upon the accessibility of the precursor to the site of synthesis, the precursor pool size, the rate of synthesis, the rate of degradation, and the extent of recycling. Trewavas (1970) has indicated that hormonal and nutritional status alter the turnover rate extensively by modifying both synthesis and degradation. The degradation products were not extensively recycled.

Although subject to the criticism that no measurements were made of the change in precursor pool size, the usually observed decline in the capacity to incorporate precursors into total RNA at the induction of senescence has been interpreted as indicating that senescence is caused by a failure of the DNA template to provide an effective template for RNA synthesis (Osborne, 1962). Hormonal treatments that delayed senescence would appear to maintain the template in a functional state (Fletcher and Osborne, 1965). In this regard Trewavas (1970) has indicated that benzyladenine increased the synthesis of RNA in *Lemna*, whereas abscisic acid markedly reduced the synthetic rate. These observations imply that senescence is regulated at the transcriptional level. Attempts have been made to demonstrate this both directly and indirectly. Srivastava (1968) prepared chromatin from senescing barley leaves and measured its capacity to catalyze RNA synthesis. With increasing age of the leaf there was an increase in the level of chromatin-associated ribonuclease, which made assessment of total RNA synthesis difficult. If consideration was given to the RNA loss due to chromatin-associated RNase, it appeared that chromatin from senescing leaves had a greater synthetic capacity than that from young leaves. Since the increase in chromatin-associated ribonuclease did not occur in nonsenescing kinetin treated leaves, it was postulated (Srivastava, 1968) that the onset of senescence was associated with the production of the chromatin-associated nucleases that degraded the newly synthesized RNA leading to a consequent decline in cellular functions. Various attempts have been made to demonstrate the decreased synthesis of a mRNA component in senescing tissue with a view of relating such a change to the onset of senescence. However, rather than a decreased synthesis of specific RNA component, it is usually found that the incorporation of radioactive precursors into all RNA components detected by sucrose density gradient centrifugation or methylated albumin Kieselguhr (MAK) column chromatography are reduced in control detached leaves in comparison to nonsenescing kinetin-treated leaves (Wollgiehn, 1967; Burdett and Wareing, 1968). These techniques, however, are insufficiently refined to detect the minor changes in mRNA species which might accelerate senescence. An alternate approach to detect changes in mRNA level is to determine the proportion of polysomes in ribosomal preparations. This approach has been used with limited success. In general, leaf tissue has a high ribonuclease content (see below), which makes isolation of ribosomal components difficult, and it is difficult to ascertain the extent of mRNA degradation that occurs during isolation. In spite of this hazard, ribosomal preparations have been made from senescing leaves. Srivastava and Arglebe (1967) indicated that the polysome content of ribosomal preparations declined during senescence. In contrast, the more impressive data of Eilam *et al.* (1971) and Callow *et al.* (1972) demonstrates that while there is a decline in polysome level during the growth of the leaf, the mature senescing leaves still contained polysomes. This presence of polysomes at later stages of senescence may indicate that mRNA is long lived or alternatively that mRNA synthesis may be sustained even when the leaves are yellowing and senescing rapidly. Foreseeably, the mRNA synthesis at this time could be coding for the production of the various hydrolytic enzymes that accumulate during senescence.

The observed decline in ribosome content during senescence could be attributed to a decreased rate of synthesis or an accelerated degradation. Studies of the incorporation of labeled precursors demonstrated a decline in the incorporation into all detectable RNA components (Wollgiehn, 1967; Burdett and Wareing, 1968). However, MAK column separation of RNA from senescent as opposed to mature leaves demonstrated an increase in the ratio of soluble RNA to ribosomal RNA, indicating an apparent greater lability of ribosomal RNA (Dyer and Osborne, 1971). Analysis of the changes in the levels of chloroplast and cytoplasmic ribosomal RNA's indicated that the two classes may be degraded at different rates during senescence of *Xanthium* leaves. The RNA extracted from senescing *Xanthium* leaves showed no detectable chloroplast, ribosomal, or transfer RNA; in contrast the RNA extracted from yellow leaves of *Vicia faba* and *Nicotiana tabacum* contained appreciable chloroplast RNA components. It should be pointed out, however, that the RNA preparations from mature *Xanthium* leaves contained much less chloroplast RNA than did similar extracts from *V. faba* and *N. tabacum.*

It is significant that Dyer and Osborne 1971) observed that tRNA was relatively stable during senescence. It has been proposed that the declining protein synthesis in senescing tissue might be due to a decreased translational ability associated with a changed tRNA complement (Strehler, 1967). The complement of isoaccepting leucyl-tRNA species changes during cotyledonary senescence (Bick et al., 1970); in addition, the capacity of the leucyl-tRNA synthetase to acylate certain isoaccepting species also changes during this period (Bick and Strehler, 1971; Kanabus and Cherry, 1971). This information indicates that senescence may be associated with the loss of the capacity to translate those mRNA species requiring the specific anticodon associated with a particular isoaccepting tRNA species. This would lead to a cessation in the synthesis of specific proteins. The possibility of other malfunctions at the translational level are provided by the demonstrations of Shugart and Barnett (1971). Transfer RNA extracted from senescing wheat leaves could be acylated with phenylalanine; however, the phenylalanyl-tRNA was unable to participate in polyuridylic acid-stimulated polyphenylalanine formation. This situation is reminiscent of that observed by Gefter and Russell (1969) demonstrating that while various isoaccepting tyrosyl-tRNA species could be charged with tyrosine, the capacity of the species to bind to ribosomes and hence function in polypeptide formation was dependent upon the degree of substitution of the base adjacent to the anticodon. This observation has greater relevance in view of the fact that the cytokinin, 6-(3-methyl-2-butenylamino)-9β-D-ribofuranosylpurine, has been shown to be a constituent of phenylalanyl-tRNA in *Escherichia coli* adjacent to the anticodon (Barrell and Sanger, 1969). Thus, the possibility exists that the phenylalanyl-tRNA from senescing leaves is deficient in this cytokinin, which would account for its restricted capacity to function in polypeptide formation.

In addition to regulating translation, it has been suggested that tRNA is involved in some manner in regulating the activity of endogenous protease activity in bacterial systems (Schlessinger and Ben-Hamida, 1966). Clearly, if a similar type of regulation occurred in higher plant cells, the observed changes in synthetase activity and isoaccepting tRNA species could lead to an alteration of protease activity.

Rather than invoking a reduced synthesis other authors attribute the decline in RNA content during senescence to an increase in ribonuclease. The observed accumulation of ribonuclease in detached leaves, which is prevented by kinetin treatment (Sodek and Wright, 1969) or increased by abscisic acid (Udvardy and Farkas, 1972), supports this proposition. However, the ribonuclease level does not invariably increase during senescence of attached leaves (Kessler and Engelberg, 1962) or detached leaves (Srivastava and Ware, 1965). In other instances, the

ribonuclease level declines in association with the overall decline in protein (Phillips and Fletcher, 1969). Additional difficulties in associating
ribonuclease level with the onset of senescence are encountered in the
observations of Sodek and Wright (1969) and Udvardy *et al.* (1967) that
the ribonuclease level in illuminated detached wheat and oat leaves was
higher than in detached leaves maintained in the dark. However, in spite
of the enhanced ribonuclease level, the illuminated leaves senesced (lost
chlorophyll) at a slower rate than those maintained in the dark. A further
complication in invoking the activity of ribonuclease with the onset of
senescence arises from the fact that there are at least two enzymes capable of hydrolyzing RNA in leaf tissue. One is particulate, sedimenting
at 20,000 g, whereas the other remains in the 20,000 g supernatant. The
increase in the particulate enzyme which occurred in the detached leaves
(but not in those treated with kinetin) was much less dramatic than that
observed in the soluble enzyme. Furthermore, the increase in activity of
the particulate enzyme was unaffected by illumination or chloramphenicol (Sodek and Wright, 1969; Udvardy *et al.*, 1967), suggesting that it
may arise from some inactive precursor. The increase in activity in the
soluble enzyme was prevented by cycloheximide (Udvardy *et al.*, 1969).
It should be noted that the most rapid increase in ribonuclease occurred
2–3 days after detachment (Sodek and Wright, 1969), that is, after the
initiation of protein, RNA, and chlorophyll breakdown. Again, however,
the possibility exists that the degradation of RNA could be achieved
without any major increase in tissue content of ribonuclease. The plant
cell at maturity contains sufficient RNase to degrade all of the cellular
RNA in only a few hours (Lewington *et al.*, 1967). Thus the declining
RNA content observed during senescence could be achieved by an increasing accessibility of the existing enzyme to its substrate. Although it has
been demonstrated that the soluble RNase from *Avena* leaf tissue
(Udvardy *et al.*, 1969) is relatively specific for purine residues and hydrolyzes purine nucleoside 2′,3′-cyclic phosphates, the mechanism of
RNA cleavage by the ribonuclease and the differential lability of the
RNA species are not known at present. There is no massive accumulation
of nucleotides in senescing leaves, and gel electrophoresis, MAK column
chromatography, or sucrose density gradient separation of RNA components from senescing leaves have failed to demonstrate the accumulation
of degradation products with intermediate molecular weight.

C. Lipids

The most characteristic visible feature of senescence in leaf tissue
is the decline in chlorophyll content as the leaf undergoes the sequence

of color changes, typical of senescence. (Fig. 1). As yet, no reliable mechanism has been proposed to account for the depletion of chlorophyll in senescing leaves. Holden (1961) has surveyed and characterized the chlorophyllase from various leaf tissues, but in view of the lack of relationship between enzyme level and chlorophyll bleaching, it appears that the enzyme may be involved in anabolic, rather than catabolic, processes.

In addition to the marked decline in chlorophyll, other lipids are degraded during senescence. The degradation appears to be regulated because not all lipids decline simultaneously. Draper (1969) demonstrated that galactolipids and sulfolipids (i.e., those located in the chloroplast) declined rapidly during the early phases of senescence, whereas other lipid components were depleted at later stages. The decrease in the galactolipid component in the senescing cucumber cotyledon was associated with an increase in free linolenic acid apparently arising from the hydrolysis of the chloroplast lipids. The enzymes involved in this hydrolysis have not been characterized. Sodek and Wright (1969) report that in detached wheat and barley leaves neutral esterase and lipase activity (measured at pH 7.5) declined during senescence. In nonsenescing kinetin-treated leaves, activity of these two enzymes was maintained during the course of the experimental period. In contrast Balz (1966) reported that the activity of a particulate acid lipase increased in detached tobacco leaves.

The observed accumulation of linolenic acid in the cucumber cotyledon and the buildup of α-tocopherylquinone observed by Barr and Arntzen (1969) indicate that although lipids were hydrolyzed, there is no extensive metabolism of the hydrolytic products. The α-tocopherylquinone appeared to accumulate in the osmiophilic globules that are characteristically formed in the chloroplast during senescence.

D. Photosynthesis

During leaf senescence there is a decline in photosynthetic capacity as measured by the ability to fix CO_2 (Woolhouse, 1967). This observation is not unexpected in view of the rapid decline in protein and chlorophyll content of the leaves. However, the mechanisms by which the capacity to incorporate CO_2 becomes limited has not been established. There have been very few studies aimed at discovering the sequence in which the partial reactions of photosynthesis become restricted. Woolhouse (1967) has indicated that there was a decline in content of fraction I protein which paralleled the decline in photosynthesis. Peterson and Huffaker (1975) have indicated that during the senescence of detached barley leaves the decline in ribulose-1,5-diphosphate carboxylase is accompanied by an increase in proteolytic activity. Ribulose-1,5-diphosphate carboxylase is the principal enzymatic component of fraction I. Brady *et al.* (1971) indicated that the decline in fraction I protein may also initially at least involve a simple degradation of fraction I into its component subunits, since the content of fraction I declined before any detectable decline in chlorophyll and protein. As senescence progressed, there was a decreased incorporation of supplied CO_2 into proteins with no incorporation in fraction I. This is in agreement with the data of Woolhouse (1967). However, label from $^{14}CO_2$ was still incorporated into other soluble proteins, suggesting that the cessation of protein synthesis during senescence may commence on chloroplast ribosomes. Kinetin treatment of leaves had a greater influence on incorporation into fraction I than any other proteins. Thus synthesis of fraction I was maintained by kinetin treatment.

The ultrastructural studies (Shaw and Manocha, 1965; Barton, 1966; Butler and Simon, 1971) demonstrate that there is a disruption of thylakoids and an accumulation of large osmiophilic globules in the chloroplast stroma. So it would be expected that photosynthetic electron transport and the accompanying photophosphorylation would be early casualties in the senescence process.

E. Respiration

Although the initial explanations for the onset of senescence were attributed to a failure of respiration associated with a depletion of respiratory substrates (see Chibnall, 1939, for a historical review of this aspect), it is usually found that respiratory activity is maintained at a fairly constant rate until the terminal phases of senescence when a rapid decline in respiration occurs (James, 1953). Throughout senescence, however, the respiratory quotient changes, indicating that as senescence progresses there is a change in respiratory substrates. The change in RQ is associated with the increased utilization, as respiratory substrate, of the amino acids that accumulate in the detached leaf (James, 1953; Chibnall, 1939). It would be interesting to determine if similar respiratory changes occur in attached leaves in which the products of protein hydrolysis are rapidly translocated to other parts of the plant. During the depletion of the carbon skeletons of the amino acids in respiratory metabolism, the amino nitrogen is initially conserved as amide nitrogen in the amides, aparagine and glutamine, which characteristically accumulate during senescence (Chibnall, 1939; Yemm, 1937).

While the respiratory metabolism as indicated by gas exchange measurements is sustained during senescence, it has been suggested that there may be a decreased production of adenosine triphosphate (ATP) due to a progressive uncoupling of electron transport and oxidative phosphorylation (see Varner, 1961). Such a situation would lead to an inadequate supply of ATP to support the synthetic events in the cell and senescence would ensue. Hanson *et al.* (1965) have indicated that ribonuclease effectively reduces the phosphorylative capacity of mitochondria from corn scutellum and that the *in vivo* accumulation of ribonuclease during senescence could bring about changes in both respiratory efficiency (ATP generation) and respiratory capacity (oxygen consumption). In addition, Hanson's group has demonstrated the inhibition of mitochondrial activity by linolenic acid (Baddeley and Hanson, 1967). It is possible that *in vivo* respiratory metabolism could be similarly inhibited by the fatty acids that accumulate during chloroplast breakdown. The observations of Baddeley and Simon (1969) that mitochondrial activity is inhibited by aqueous extracts from senescing leaves are consistent with this proposal. The inhibitory properties of the extracts, which were attributed to their free fatty acid content, increased with leaf age.

While changes in phosphorylative capacity of the mitochondria may be of significance at later stages of senescence, it appears that in the early stages of senescence of both intact and detached leaves there is adequate phosphorylation as indicated by the level of phosphorylated intermediates and the capacity to incorporate ^{32}P into phosphorylated intermediates and ATP (Berridge and Ralph, 1971; Adepipe and Fletcher, 1970). In studies of attached bean leaves, it was observed that kinetin treatment reduced the level of many phosphorylated intermediates; however, the incorporation of ^{32}P into phosphorylated intermediates increased in comparison to control leaves. Kinetin treatment apparently sustained or enhanced the turnover of metabolites (Adepipe and Fletcher, 1970). In chinese cabbage leaf discs, in contrast, Berridge and Ralph (1971) reported that kinetin treatment increased the ATP level but did not increase the incorporation of ^{32}P into this component. There was, however, increased incorporation of ^{32}P into chloroform-soluble material, indicating that kinetin caused an increase in the synthesis of lipids required for membranes and for structural cell components. In comparing the results of Berridge and Ralph (1971) with those of Adepipe and Fletcher (1970), recognition must be given to the fact that kinetin enhances the growth of chinese cabbage leaf discs, and, thus, some of the observed effects may relate to growth phenomena rather than to the metabolism of a mature cell.

While the above information indicates that an altered respiratory metabolism and a change in respiratory metabolites are associated with senescence, it also is evident that the onset of senescence is dependent upon a "normal aerobic respiratory metabolism" (James, 1953). The characteristic decline in metabolites in detached leaves is prevented by anaerobiosis. Various hypotheses have been advanced to account for this effect. One school of thought suggests that the requirement for respiration indicates that some product of oxidative metabolism triggers the onset of senescence, and such products fail to accumulate under anaerobic conditions (James, 1953).

An alternative explanation that has been advanced is that protein catabolism and senescence may be energy-requiring processes (Steinberg and Vaughan, 1956). However, in more recent studies with animal tissues (which show a similar retardation of protein breakdown by anaerobiosis) it was concluded that while there may be structural components involved in protein catabolism, the integrity of which requires metabolic energy, there was no direct involvement of energy in the actual degradation of proteins (Brostrom and Jeffay, 1970).

F. Membranes and Organelles

During senescence there are marked changes in the permeability characteristics. These changes have been studied extensively by Sacher and co-workers. However, there does not appear to be any uniformity in the changes which would allow for the conclusion that the onset of senescence is attributable to changed membrane properties. Senescence in bean endocarp or *Rheo* leaf was characterized by a loss of turgidity and an exchange of cellular materials into the intercellular spaces with an exudation of liquid from the external surfaces (Sacher, 1967). As dramatic as these changes were the cells, initially at any rate, still retained the capacity to be plasmolyzed and deplasmolyzed, and thus, have retained differential permeability. The change in cellular permeability did, however, influence the rate of uptake of exogenously supplied substrates, which as indicated previously, make it difficult to interpret incorporation studies. In the bean endocarp and *Rheo* leaves, Sacher (1967) reported that the changed cellular permeability was preceded by a decline in the capacity for RNA and protein synthesis (incorporation of radioactive precursors). In contrast, in banana, significant changes in permeability and protoplasmic compartmentalization precede the loss of nucleic acid and protein associated with fruit ripening. These permeability changes may contribute to the characteristic accelerated respiration, the so-called "respiratory climacteric" that precedes ripening and senescence in many fruits (see Rhodes, 1970).

In addition to changes in cellular permeability which are presumably attributable to changes in the plasmalemma and tonoplast allowing for an altered water influx and efflux, changes occur in other cellular membranes. These have been characterized by ultrastructural studies using the electron microscope. The remarkable pattern that has emerged from such studies is that regardless of the manner by which senescence is induced (drought, mineral deficiency, fungal or viral infection, attached or detached leaves), the ultrastructural changes occur in a fairly well-defined sequence (Butler and Simon, 1971). The first detectable changes are a decline in the ribosome population and an initiation of chloroplast breakdown. The chloroplast characteristically undergoes a breakdown of the stroma and a swelling and disintegration of the thylakoids accompanied by an increase in number and size of the osmiophilic globules. In contrast, the mitochondrial changes are much less dramatic. Early in senescence there may be a reduction in mitochondrial size with a swelling and reduction in the number of cristae. However, apparently functional mitochondria (as indicated by both respiratory and ultrastructural studies) are retained until the terminal phases of senescence. During senescence and accompanying the final breakdown of the chloroplasts, the endoplasmic reticulum swells, vesiculates, and disappears along with the Golgi apparatus. The tonoplast breaks down before the organelles have completely disintegrated, but the plasmalemma remains recognizable until cell death. The nucleus remains stable until the later stages of senescence and then undergoes degenerative changes marked by vesiculation of the nuclear membrane and a breakdown of the internal matrix. Spherosomes that were detected in the mesophyll cells of senescing leaves and cucumber cotyledons remain intact until late stages of senescence (Shaw and Manocha, 1965; Barton, 1966; Butler and Simon, 1971).

It is thus difficult to equate the onset of senescence with a release of the hydrolytic enzymes that have been associated with the spherosomes (Balz, 1966). Furthermore, if senescence was due to the release of hydrolases from a latent form in the spherosomes, it is difficult to envisage how the ordered sequence of changes so characteristic of senescence are achieved. It may be that the release of hydrolases from the spherosomes plays a greater role in the final stages of senescence following breakdown of the tonoplast.

It is significant, however, that acid phosphatase (an enzyme that has been used as a marker for lysosomes in animal tissues) accumulates extensively during senescence in many plants (Balz, 1966; De Leo and Sacher, 1970). The accumulation of acid phosphatase was prevented by treatment of the tobacco or *Rheo* leaves with kinetin or auxin, respectively. In contrast, abscisic acid treatment, which accelerated senescence of *Rheo* leaves, enhanced the production of the enzyme (Balz, 1966; De Leo and Sacher, 1970). A large proportion of the acid phosphatase extracted from *Rheo* leaf tissue was sedimentable by low-speed centrifugation. The acid phosphatase was extractable from this sedimented material by Triton X-100 or dilute salt solutions. The accumulation of the enzyme during senescence was prevented by inhibitors of protein and RNA synthesis, so it appeared that the increased enzyme level involved protein synthesis rather than an activation of a preexisting latent form (De Leo and Sacher, 1970). In addition to this association of acid phosphatase with a slow-speed sedimentable fraction which De Leo and Sacher (1970) indicated included cell debris and chloroplasts, Ragetti *et al.* (1966) have demonstrated the occurrence of an acid phosphatase in the chloroplasts. Such an enzyme according to Barton (1966) might be involved in the initiation of chloroplast degradation in leaf tissue.

IV. Reversal of Senescence

In drawing up any comprehensive explanation of senescence, recognition must be given to the fact that the event is in some instances reversible, and in other cases, the process can be temporarily suspended. Detached leaves that form roots regreen (Chibnall, 1954; Woolhouse, 1967) and removal of the aerial portions above a leaf on a plant undergoing sequential senescence results in a suspension of senescence and a regreening of the leaf (Woolhouse, 1967). This regreening is accompanied by a restoration of protein and RNA content (Wollgiehn, 1967; Callow and Woolhouse, 1973) and can be demonstrated in leaves that had previously undergone extensive senescence. There is a point at which the senescence cannot be reversed. This may be associated with the loss of tonoplast structure; however, there is currently no ultrastructural data on this point. Clearly, the breakdown of the tonoplast membrane and exposure of the cytoplasm to the vacuolar contents would represent a trauma from which recovery would be difficult.

Detached tobacco leaves treated with kinetin have been shown to regreen (Sveshnikova *et al.*, 1966), and gibberellic acid treatment of senescing leaf discs of Nasturtium (Beevers, 1968) and *Rumex* (Goldthwaite and Laetsch, 1968) prevents any further deteriorative changes (Fig. 2).

The above observations of senescence reversal or suspension indicate that if the decline in cellular constituents during senescence is caused by a sequential decrease in protein synthesis associated with translational or transcriptional failure, then such defects can be overcome. During the reversal of senescence, there must be a reinitiation of transcription and/or translation of those genes required for the synthesis of proteins depleted during the onset of senescence. Simultaneously or alternatively, there

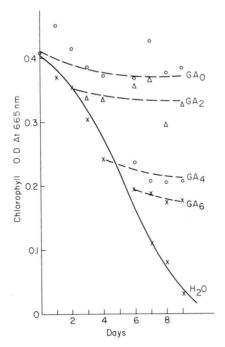

Fig. 2. The influence of delayed application of gibberellic acid (GA) 20 ppm on the chlorophyll loss from senescing leaf discs of Nasturtium (*Tropaeolum majus*). From Beevers (1968).

must be a repression of the synthesis or inhibition of the hydrolytic enzymes that accumulate during senescence.

V. Conclusions

Although it is accepted that senescence is characterized by a depletion of internal cellular constituents preceding the death of a mature cell, it is clear that at the present time there is little concensus on the manner by which this depletion is achieved. The data that have been described support either the concept of an increased degradation or a decreased synthetic capability. In the final analysis, both conditions probably occur simultaneously.

In spite of the lack of definitive conclusions concerning the mechanics of senescence, it is apparent that the process is regulated rather than being an uncontrolled destructive event. Significantly, the ultrastructural studies show that the nucleus is one of the last structures to undergo deterioration during senescence. The observations that senescence may be reversed or suspended indicate that the machinery for regulating protein synthesis is retained intact until later stages of senescence. It, therefore, appears that the ordered events of senescence are mediated by a programmed regulation of protein synthesis, originating at the nuclear level. In turn, the programming, at least in leaves, is regulated by the internal hormonal balance.

REFERENCES

Addicott, F. T. (1970). *Biol. Rev. Cambridge Phil. Soc.* **45**, 485.
Adepipe, N. O., and Fletcher, R. A. (1970). *Plant Physiol.* **46**, 614.
Anderson, J. W., and Rowan, K. S. (1965). *Biochem. J.* **97**, 741.
Anderson, J. W., and Rowan, K. S. (1966). *Biochem. J.* **98**, 401.
Anderson, J. W., and Rowan, K. S. (1968). *In* "Biochemistry and Physiology of Plant Growth Substances" (F. Wightman and G. Setterfield, eds.), pp. 1437–1446. Runge Press, Ottawa.
Axelrod, B., and Jagendorf, A. T. (1951). *Plant Physiol.* **26**, 406.
Baddeley, M. S., and Hanson, J. B. (1967). *Plant Physiol.* **44**, 1702.
Baddeley, M. S., and Simon, E. W. (1969). *J. Exp. Bot.* **20**, 94.
Balz, H. P. (1966). *Planta* **70**, 207.
Barr, R., and Arntzen, C. J. (1969). *Plant Physiol.* **44**, 591.
Barrell, R. G., and Sanger, F. (1969). *FEBS (Fed. Eur. Biochem. Soc.) Lett.* **3**, 275.
Barton, R . (1966). *Planta* **71**, 314.
Beevers, L. (1966). *Plant Physiol.* **41**, 1074.
Beevers, L. (1968). *In* "Biochemistry and Physiology of Plant Growth Substances" (F. Wightman and G. Setterfield, eds.), pp. 1417–1435. Runge Press, Ottawa.
Berridge, M. V., and Ralph, R. K. (1971). *Plant Physiol.* **47**, 562.
Bick, M. D., and Strehler, B. L. (1971). *Proc. Nat. Acad. Sci. U.S.* **68**, 224.
Bick, M. D., Liebke, H., Cherry, J. H., and Strehler, B. L. (1970). *Biochim. Biophys. Acta* **204**, 175.
Bidwell, R. G. S., Barr, R. A., and Steward, F. C. (1964). *Nature (London)* **203**, 367.
Brady, C. J., Patterson, B. D., Tung, H. F., and Smillie, R. M. (1971). *In* "Autonomy and Biogenesis of Mitochondria and Chloroplasts" (N. K. Boardman, A. W. Linnane, and R. M. Smillie, eds.), pp. 453–460. North-Holland Publ., Amsterdam.
Brostrom, C. O., and Jeffay, H. (1970). *J. Biol. Chem.* **245**, 4001.
Burchall, J. J. (1966). *Fed. Proc., Fed. Amer. Soc. Exp. Biol.* **25**, 277.
Burdett, A. N., and Wareing, P. F. (1968). *Planta,* **81**, 88.
Butler, R. D., and Simon, E. W. (1971). *Advan. Gerontol. Res.* **3**, 73.
Callow, J. A., Callow, M. E., and Woolhouse, H. W. (1972). *Cell Differentiation* **1**, 79.
Callow, M. E., and Woolhouse, H. W. (1973). *J. Exp. Bot.* **24**, 294.
Chibnall, A. C. (1939). "Protein Metabolism in the Plant." Yale Univ. Press, New Haven, Connecticut.
Chibnall, A. C. (1954). *New Phytol.* **53**, 31.
Chin, T. Y., and Beevers, L. (1970). *Planta* **92**, 178.

De Greef, J., Butler, W. L., Roth, T. F., and Frédéricq, H. (1971). *Plant Physiol.* **48**, 407.

De Jong, D. W. (1972). *Plant Physiol.* **50**, 733.

De Leo, P., and Sacher, J. A. (1970). *Plant Physiol.* **46**, 806.

Draper, S. R. (1969). *Phytochemistry* **8**, 1641.

Dyer, T. A., and Osborne, D. J. (1971). *J. Exp. Bot.* **22**, 552.

Eilam, Y., Butler, R. D., and Simon, E. W. (1971). *Plant Physiol.* **47**, 317.

El-Antably, H. M. M., Wareing, P. F., and Hillman, J. (1967). *Planta* **73**, 76.

Fletcher, R. A., and Osborne, D. J. (1965). *Nature (London)* **207**, 1176.

Gefter, M. L., and Russell, R. L. (1969). *J. Mol. Biol.* **39**, 145.

Goldthwaite, J. J., and Laetsch, W. M. (1967). *Plant Physiol.* **42**, 1757.

Goldthwaite, J. J., and Laetsch, W. M. (1968). *Plant Physiol.* **43**, 1855.

Haber, A. H., Thompson, P. J., Walne, P. L., and Triplett, L. L. (1969). *Plant Physiol.* **44**, 1619.

Hanson, J. B., Wilson, C. M., Chrispeels, M. J., Krueger, W. A., and Swanson, H. R. (1965). *J. Exp. Bot.* **16**, 282.

Hellebust, J. A., and Bidwell, R. G. S. (1963). *Can. J. Bot.* **41**, 969.

Holden, M. (1961). *Biochem. J.* **78**, 359.

Itai, C., and Vaadia, Y. (1965). *Physiol. Plant.* **18**, 941.

James, W. O. (1953). "Plant Respiration." Oxford Univ. Press (Clarendon), London and New York.

Jones, R. L., and Phillips, I. D. J. (1966). *Plant Physiol.* **41**, 1381.

Kanabus, J., and Cherry, J. H. (1971). *Proc. Nat. Acad. Sci. U.S.* **68**, 873.

Kannangara, C. G., and Woolhouse, H. W. (1968). *New Phytol.* **67**, 533.

Kende, H. (1964). *Science* **145**, 1066.

Kessler, B., and Engelberg, N. (1962). *Biochim. Biophys. Acta* **55**, 70.

Kuraishi S. (1968). *Physiol. Plant.* **21**, 78.

Lewington, R., Talbot, M., and Simon, E. W. (1967). *J. Exp. Bot.* **18**, 526.

Martin, C., and Thimann, K. V. (1972). *Plant Physiol.* **50**, 432.

Matile, P. H. (1968). *Planta* **79**, 181.

Matile, P. H., and Winkenbach, F. (1971). *J. Exp. Bot.* **22**, 759.

Mayak, S., and Halevy, A. H. (1970). *Plant Physiol.* **46**, 497.

Munro, H. N. (1970). *Mammalian Protein Metab.* **4**, 3–130.

Mizrahi, Y., Amir, J., and Richmond, A. E. (1970). *New Phytol.* **69**, 355.

Osborne, D. J. (1962). *Plant Physiol.* **37**, 595.

Osborne, D. J. (1965). *J. Sci. Food. Agr.* **16**, 1.

Person, C., Samborski, D. J., and Forsyth, F. R. (1957). *Nature (London)* **180**, 1294.

Peterson, L. W., and Huffaker, R. C. (1975). *Plant Physiol.* **55**, 1009.

Phillips, D. R., and Fletcher, R. A. (1969). *Physiol. Plant.* **22**, 764.

Phillips, I. D. J., and Wareing, P. F. (1959). *J. Exp. Bot.* **13**, 213.

Racusen, D., and Foote, M. (1960). *Arch. Biochem. Biophys.* **90**, 90.

Ragetti, H. W. J., Weintraub, M., and Rink, U. M. (1966). *Can. J. Bot.* **44**, 1723.

Rhodes, M. J. C. (1970). *In* "The Biochemistry of Fruits and their Products" (A. C. Hulme, ed.), Vol. 1, pp. 521–537. Academic Press, New York.

Richmond, A. E., and Lang, A. (1957). *Science* **125**, 650.

Sacher, J. A. (1965). *Amer. J. Bot.* **52**, 841.

Sacher, J. A. (1967). *Symp. Soc. Exp. Biol.* **21**, 269–303.

Sacher, J. A. (1973). *Annu. Rev. Plant Physiol.* **24**, 197.

Schlessinger, D., and Ben-Hamida, F. (1966). *Biochim. Biophys. Acta* **119**, 171.

Seth, A., and Wareing, P. F. (1965). *Life Sci.* **4**, 2275.

Shaw, M., and Manocha, M. S. (1965). *Can. J. Bot.* **43**, 747.

Shibaoka, H., and Thimann, K. V. (1970). *Plant Physiol.* **46**, 212.

Shugart, L. R., and Barnett, W. E. (1971). *Fed. Proc., Fed. Amer. Soc. Exp. Biol.* **30**, 1272 (abstr.).

Simon, E. W. (1967). *Symp. Soc. Exp. Biol.* **21**, 215–230.

Sodek, L., and Wright, S. T. C. (1969). *Phytochemistry* **8**, 1629.

Srivastava, B. I. S. (1968). *Biochem. J.* **110**, 683.

Srivastava, B. I. S., and Arglebe, C. (1967). *Plant Physiol.* **42**, 1497.

Srivastava, B. I. S., and Ware, G. (1965). *Plant Physiol.* **40**, 62.

Steinberg, D., and Vaughan, M. (1956). *Arch. Biochem. Biophys.* **65**, 93.

Strehler, B. L. (1967). *Symp. Soc. Exp. Biol.* **21**, 149–178.

Sugiura, M. (1963). *Bot. Mag.* **76**, 174.

Sugiura, M., Umemura, K., and Oota, Y. (1962). *Physiol. Plant.* **15**, 457.

Sveshnikova, I. N., Kulaeva, O. N., and Bolyakina, Y. P. (1966). *Sov. Plant Physiol.* **13**, 681.

Tavares, J., and Kende, H. (1970). *Phytochemistry* **9**, 1763.

Trewavas, A. (1970). *Plant Physiol.* **45**, 742.

Udvardy, J., and Farkas, G. L. (1972). *J. Exp. Bot.* **23**, 914.

Udvardy, J., Farkas, G. L., Marré, E., and Forti, G. (1967). *Physiol. Plant.* **20**, 781.

Udvardy, J., Farkas, F. L., and Marré, E. (1969). *Plant Cell Physiol.* **10**, 375.

Varner, J. (1961). *Annu. Rev. Plant Physiol.* **12**, 245.

Wareing, P. F., and Seth, A. K. (1967). *Symp. Soc. Exp. Biol.* **21**, 543–558.

Whyte, P., and Luckwill, L. C. (1966). *Nature (London)* **210**, 1360.

Wollgiehn, R. (1961). *Flora (Jena)* **151**, 411.

Wollgiehn, R. (1967). *Symp. Soc. Exp. Biol.* **21**, 231–246.

Woolhouse, H. W. (1967). *Symp. Soc. Exp. Biol.* **21**, 179–214.

Wright, S. T. C., and Hiron, R. W. (1969). *Nature (London)* **224**, 719.

Yemm, E. W. (1937). *Proc. Roy. Soc., Ser. B* **123**, 243.

III

Autotrophy

24

Photosynthesis: The Path of Carbon

M. D. HATCH

I. Introduction

Living organisms must expend energy both to attain and maintain thermodynamic status quo. For practically all organisms this energy is originally derived from sunlight via the process of photosynthesis. In obligate photoautrophs, including the higher plants, most algae, and some bacteria, light is the sole exogenous energy source, used primarily for the synthesis of organic compounds from CO₂, water, and other inorganic precursors. Heterotrophic organisms are absolutely dependent upon organic compounds formed in this way for their carbon and energy requirements. The interaction of these two major groups of organisms maintains the biospheric balance of CO₂, O₂, and water.

The processes by which photosynthetic cells absorb light energy and convert it to a utilizable chemical form are described in Chapter 25. Most commonly, this energy appears as the ubiquitous adenosine triphosphate (ATP), which contains phosphoric acid ester bonds with high free energies of hydrolysis, and reduced nicotinamide adenine dinucleotide phosphate (NADP) a pyridine nucleotide with a high reducing potential. In the simplest terms, the requirements of the reactions that produce ATP and NADPH are a source of electrons and light energy. The light energy is used to excite these electrons and hence to increase their reducing potential. In plants at least, water is the source of electrons and carbohydrate is the major primary product of CO_2 assimilation. Thus, the net reaction of photosynthesis can be represented simply by Eq. (1).

$$CO_2 + H_2O \xrightarrow{\text{light}} (CH_2O) + O_2 \qquad \Delta F = +112 \text{ kcal} \qquad (1)$$

Here one sees that an input of 112 kcal is required for each CO_2 ultimately appearing in carbohydrate. The primary source of this energy is light, but the immediate source is ATP and NADPH. The component reactions of the overall process represented in Eq. (1) will be described in Section II.

Both the photochemical and biochemical processes associated with the photosynthetic conversion of CO_2 to organic compounds occur in discrete subcellular organelles termed chloroplasts. A description of the structure of chloroplasts, and the relation of structure to function, is provided in Chapter 6. For the present purposes it is sufficient to note that the reactions leading to ATP and NADPH formation occur on the internal chloroplast membranes, termed lamellae or thylakoids. The reactions concerned with CO_2 assimilation occur in the interlamellae areas, termed the stroma. The products of photosynthesis move into the cytoplasm of photosynthetic cells and ultimately to the other parts of higher plants.

In the following sections the mechanism, regulation, and physiological significance of various pathways of photosynthetic CO_2 assimilation will be considered, together with the related process termed photorespiration. The relation between the variant pathways for CO_2 assimilation and other characteristics of higher plants will also be briefly discussed.

II. Pathways of CO_2 Fixation

A. Comparative Aspects

The following sections will describe three biochemical variants for photosynthetic assimilation of CO_2 in higher plants. The first of these to be recognized and elucidated was the process, termed here, the Calvin cycle (also known as the photosynthetic carbon reduction cycle or more recently as the C_3 pathway). Species with this pathway may be characterized experimentally by the pattern of labeling of metabolities formed from $^{14}CO_2$ and by the presence of several enzymes specific to the pathway (Calvin and Bassham, 1962; Bassham, 1964, 1965). In other species $^{14}CO_2$ is initially fixed into different intermediates and this process, termed here the C_4 pathway, requires the operation of several enzymes additional to those of the Calvin cycle. However, the terminal steps of photosynthesis in the latter species include the reactions of the Calvin cycle. In a third group of plants much of the net assimilation of CO_2 is due to a process termed crassulacean acid metabolism (CAM). In these plants, CO_2 is actually fixed in the dark and is initially stored as malic acid. However, it is ultimately converted to carbohydrate via the Calvin cycle during the following light period, at which time there may be little or no CO_2 assimilated directly from the air.

It was natural that earlier studies on the C_4 pathway, in particular, sought and emphasized differences rather than similarities between this process and the Calvin cycle. However, with current knowledge of these processes and CAM, it is now possible to make two important and unifying generalizations about photosynthetic metabolism in plants. These are, first, that the reactions of the Calvin cycle are apparently common to all plants and remain as the only known series of reactions capable of the net conversion of CO_2 to carbohydrate and, second, that where the various pathways differ is in the processes involved in moving externally derived CO_2 to the site of action of the Calvin cycle carboxylation reaction. In other words, the unique reactions of the C_4 pathway and CAM can be regarded as obligatory mechanisms for more effectively providing the Calvin cycle with CO_2. The following discussion will be considered within the framework of these concepts. Although the Calvin cycle is ultimately operative in all species it will be convenient to refer to those plants that fix CO_2 directly into this cycle, without the intervention of the C_4 pathway or CAM, as Calvin cycle species or C_3 species.

B. Calvin Cycle

1. REACTIONS AND ENZYMES

The present concept of the Calvin cycle has not changed significantly from the formulations of about ten years ago (Calvin and Bassham, 1962; Bassham, 1964). The reactions of the cycle, and the enzymes catalyzing those reactions, are depicted schematically in Fig. 1. Elucidation of the pathway was largely due to the efforts of Calvin, Benson, and co-workers. Although most of these studies were conducted with the alga *Chlorella*, there is now adequate evidence that the process is the same in higher plants. Various aspects relating to the operation of the Calvin cycle have been recently reviewed (Hatch and Slack, 1970a; Walker and Crofts, 1970; Bassham, 1971; Black, 1973).

It is convenient to consider the Calvin cycle commencing with in-

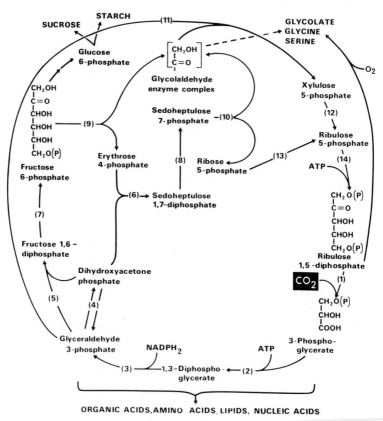

Fig. 1. Reactions and enzymes of the Calvin cycle. End products are indicated in capital letters. Glycolate, glycine and serine are also intermediates of the glycolate pathway (see Fig. 6) which cycles carbon back to 3-P-glycerate. The enzymes involved are (1) RuDP carboxylase; (2) 3-P-glycerate kinase; (3) NADP-glyceraldehyde-phosphate dehydrogenase; (4) triosephosphate isomerase; (5) and (6) sugardiphosphate aldolases; (7) and (8) sugardiphosphatases; (9), (10), and (11) transketolases; (12) and (13) pentosephosphate isomerases; (14) ribulose-5-phosphate kinase.

corporation of CO_2, a reaction involving the carboxylation of ribulose 1,5-diphosphate (RuDP) and catalyzed by RuDP carboxylase [Eq. (2)].

$$
\begin{array}{c}
\text{CH}_2\text{O(P)} \\
|\\
\text{C=O} \\
|\\
\text{CHOH} \\
|\\
\text{CHOH} \\
|\\
\text{CH}_2\text{O(P)}
\end{array}
\quad + \quad {}^*\text{CO}_2 \quad + \quad \text{H}_2\text{O} \quad \xrightarrow{\text{Mg}^{2+}} \quad
\begin{array}{c}
\text{CH}_2\text{O(P)} \\
|\\
\text{CHOH} \\
|\\
{}^*\text{COO}^-
\end{array}
\quad + \quad
\begin{array}{c}
\text{COO}^- \\
|\\
\text{CHOH} \\
|\\
\text{CH}_2\text{O(P)}
\end{array}
\quad + \quad 2\text{H}^+
\tag{2}
$$

RuDP $\qquad\qquad\qquad\qquad$ 3-P-glycerate

RuDP is carboxylated at the C-2, the keto carbon, and cleavage occurs between the C-2 and C-3 to give two molecules of 3-phosphoglycerate (3-P-glycerate) (Mülhofer and Rose, 1965), only one of which will be initially labeled from $^{14}CO_2$. There is now evidence for a bound C_6 intermediate (Seigel and Lane, 1973) but 3-P-glycerate is the first labeled product detectable during photosynthesis in $^{14}CO_2$.

Ribulose-1,5-diphosphate carboxylase has several unusual features. One of these is its relative inefficiency as a catalyst, indicated by the fact that it comprises about half of the soluble protein of leaves (Akazawa, 1970). This single protein entity was originally distinguished from the remainder of the leaf protein in the ultracentrifuge as a sharp and rapidly sedimenting peak, termed Fraction 1 protein (Singer et al., 1952). In its active form the protein is large (MW approximately 5×10^5) and highly oligomeric, probably consisting of 8 larger subunits and 8 smaller subunits (Kawashima and Wildman, 1970). RuDP carboxylase reacts with CO_2 rather than HCO_3^- (Cooper et al., 1969), and with normal isolation techniques has a high K_m for this substrate (200–500 μM). However, recent studies have demonstrated that the enzyme exists in a different form in vivo with a K_m CO_2 of about 15 μM (Bahr and Jensen, 1974a). RuDP carboxylase also catalyzes the oxygenation of RuDP to give phosphoglycolate and 3-P-glycerate (Bowes et al., 1971; Andrews et al., 1973; Lorimer et al., 1973), and the significance of this activity in relation to the process of photorespiration will be considered in Section IV.

The first two reactions involved in the further metabolism of 3-P-glycerate utilize ATP and NADPH provided by the light reactions of photosynthesis. 3-P-glycerate is phosphorylated by ATP to give 1,3-diphosphoglycerate, which is then reduced by NADPH to give glyceraldehyde 3-phosphate [Eq. (3)] via 3-P-glycerate kinase and NADP-

$$
\begin{array}{ccc}
\underset{\text{3-P-glycerate}}{\begin{array}{l}CH_2O(P)\\ \mid\\ CHOH\\ \mid\\ COO^-\end{array}} \;\underset{}{\overset{ATP\qquad ADP}{\rightleftarrows}}\;
\underset{\substack{\text{1,3-Diphospho-}\\\text{glycerate}}}{\begin{array}{l}CH_2O(P)\\ \mid\\ CHOH\\ \mid\\ C\!\!\underset{O(P)}{\overset{O}{\diagdown}}\end{array}} \;\underset{}{\overset{NADPH\qquad NADP^+}{\rightleftarrows}}\;
\underset{\substack{\text{Glyceraldehyde}\\\text{3-phosphate}}}{\begin{array}{l}CH_2O(P)\\ \mid\\ CHOH\\ \mid\\ CHO\end{array}} + P_i
\end{array}
\qquad (3)
$$

glyceraldehyde-3-phosphate dehydrogenase, respectively. The latter enzyme can be distinguished from its glycolytic counterpart by its specificity for $NADP^+$ rather than NAD^+.

Triosephosphate isomerase converts glyceraldehyde 3-phosphate to dihydroxyacetone phosphate, and these compounds are then condensed by fructosediphosphate aldolase to give fructose 1,6-diphosphate [Eq. (4)].

$$
\begin{array}{l}
\text{Dihydroxyacetone}\\ \text{phosphate}
\end{array}
\begin{array}{l}
CH_2O(P)\\ \mid\\ C\!=\!O\\ \mid\\ CH_2OH\\[4pt]
+\\[4pt]
CHO\\ \mid\\ CHOH\\ \mid\\ CH_2O(P)
\end{array}
\quad\rightleftarrows\quad
\begin{array}{l}
CH_2O(P)\\ \mid\\ C\!=\!O\\ \mid\\ CHOH\\ \mid\\ CHOH\\ \mid\\ CHOH\\ \mid\\ CH_2O(P)
\end{array}
\qquad (4)
$$

$$
\begin{array}{l}
\text{Glyceraldehyde}\\ \text{3-phosphate}
\end{array}
$$

$$
\begin{array}{c}
\text{Fructose 1,6-}\\ \text{diphosphate}
\end{array}
$$

An alkaline sugardiphosphate phosphatase, probably specific to photosynthesis, then catalyzes the formation of fructose 6-phosphate, which is in turn converted to glucose 1-phosphate by the combined action of phosphohexose isomerase and phosphoglucomutase. Although the enzymes responsible for glucose 1-phosphate formation from 3-P-glycerate have functional counterparts in glycolysis, they would be spatially separated and may prove to be different isoenzymes. The reactions involved in starch and sucrose formation from hexose phosphates are described in Chapters 11 and 12. Starch, and probably also sucrose (see Section II,B,3 and 4), are formed within chloroplasts.

It is inherent that photosynthetic assimilation of CO_2 in autotrophic organisms must involve a cyclic reaction sequence, since the compound originally carboxylated must be regenerated to sustain the process. Thus, as shown in Fig. 1, intermediates formed from 3-P-glycerate are utilized via a series of reactions to form RuDP. In fact, to maintain a carbon balance only one hexose molecule can be incorporated into sucrose or starch for each 6 CO_2 fixed. This would require six turns of the cycle and 6 RuDP. Thus, of the 12 molecules of 3-P-glycerate so formed, 10 molecules must be utilized to reform RuDP while the remaining 2 molecules can be removed from the cycle as a hexose molecule.

The cyclic phase of the process is initiated by reactions catalyzed by the enzymes transketolase and aldolase. In the transketolase reactions (reactions 9, 10, and 11 in Fig. 1) a glycolaldehyde radicle, derived from the C-1 and C-2 of the keto sugars fructose 6-phosphate or sedoheptulose 7-phosphate, is transferred to glyceraldehyde 3-phosphate to give xylulose-5-phosphate as shown in general form in Eq. (5).

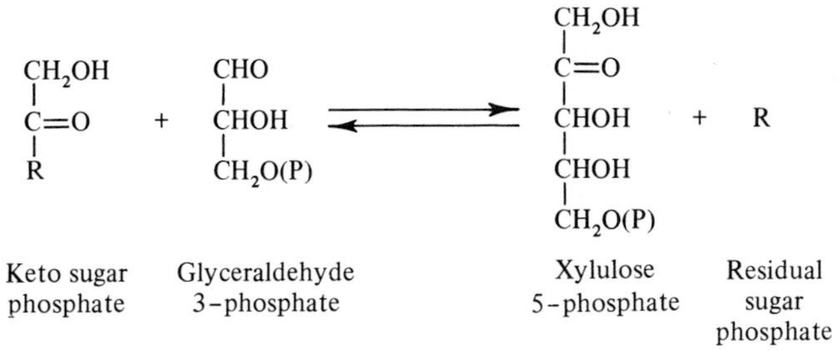

$$
\begin{array}{ccccccc}
& \text{CH}_2\text{OH} & & \text{CHO} & & \text{CH}_2\text{OH} & \\
& | & & | & & | & \\
& \text{C}=\text{O} & + & \text{CHOH} & \rightleftharpoons & \text{C}=\text{O} & + \quad \text{R} \\
& | & & | & & | & \\
& \text{R} & & \text{CH}_2\text{O(P)} & & \text{CHOH} & \\
& & & & & | & \\
& & & & & \text{CHOH} & \\
& & & & & | & \\
& & & & & \text{CH}_2\text{O(P)} & \\
\end{array}
\tag{5}
$$

| Keto sugar phosphate | Glyceraldehyde 3-phosphate | Xylulose 5-phosphate | Residual sugar phosphate |

Thiamine pyrophosphate is a bound cofactor of tranketolases and is the group to which glycoaldehyde binds. Erythrose 4-phosphate [R in Eq. (5)], the other product of the transkelolase reaction with fructose 6-phosphate, is condensed with dihydroxyacetone phosphate by an aldolase to give the C_7 keto sugar, sedoheptulose 1,7-diphosphate. From this compound, the other transketolase substrate sedoheptulose-7-phosphate is formed by the action of a specific sugardiphosphate phosphatase.

Ribose 5-phosphate and xylulose 5-phosphate, the products of the above reactions, are converted to ribulose 5-phosphate by pentosephosphate isomerases. Finally, RuDP is formed from ribulose 5-phosphate via the enzyme ribulose-5-phosphate kinase [Eq. (6)].

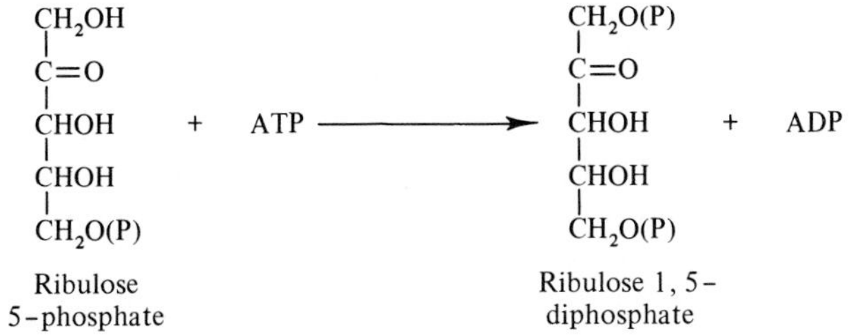

$$
\begin{array}{ccccc}
\text{CH}_2\text{OH} & & & \text{CH}_2\text{O(P)} & \\
| & & & | & \\
\text{C}=\text{O} & & & \text{C}=\text{O} & \\
| & & & | & \\
\text{CHOH} & + \quad \text{ATP} & \longrightarrow & \text{CHOH} & + \quad \text{ADP} \\
| & & & | & \\
\text{CHOH} & & & \text{CHOH} & \\
| & & & | & \\
\text{CH}_2\text{O(P)} & & & \text{CH}_2\text{O(P)} & \\
\end{array}
\tag{6}
$$

| Ribulose 5-phosphate | Ribulose 1, 5-diphosphate |

This is another of the enzymes specific to the photosynthetic process.

If the primary product of the Calvin cycle is regarded as hexose phosphate then Eq. (7) describes the stoichiometry of the overall process.

$$6\,CO_2 + 18\,ATP + 12\,NADPH \rightarrow \text{hexose-P} + 18\,ADP + 17\,P_i + 12\,NADP^+ \quad (7)$$

According to this equation, 3 ATP and 2 NADPH are required for each CO_2 fixed. Of course, additional ATP would be required for synthesis of starch and sucrose. As discussed in Section IV, the operation of the Calvin cycle in leaves is accompanied by a process termed photorespiration. This process markedly increases the real energy costs for net assimilation of CO_2.

Experimental evidence for the Calvin cycle has been documented on several occasions (Calvin and Bassham, 1962; Bassham, 1964, 1965) and will not be repeated in detail here. This evidence was provided by the results of radiotracer studies combined with information on the activity and location of the required enzymes. These interpretations have been supported by more recent radiotracer studies with *Chlorella* and isolated chloroplasts, conducted primarily to elucidate regulatory processes (Bassham and Kirk, 1968; Bassham, 1971). However, recent proposals for the operation of some alternative intermediates and reactions leading to the regeneration of ribulose 5-phosphate should be noted (Clark *et al.*, 1974). Later sections deal with more recent information about the regulation of the constituent enzymes (see Section III,B), the association of these enzymes with chloroplasts, and the capacity of isolated chloroplasts to conduct photosynthesis (see Section II,B,2).

2. Activity in Isolated Chloroplasts

Studies with isolated chloroplasts have contributed to understanding the Calvin cycle by providing ultimate proof for the location of enzymes, a system for studying details of the kinetics and regulation of the cycle, and the ways in which the process may interact metabolically with the cytoplasm *in situ*. Progress in these areas has been reviewed (Gibbs, 1967, 1971; Hatch and Slack, 1970a; Walker and Crofts, 1970; Heber, 1974).

Chloroplasts have proved to be difficult organelles to isolate in anything approximating a physically and metabolically intact state. The outer membrane, termed the envelope, is readily lost or damaged and this is invariably associated with the loss of soluble enzymes from the stroma. For many years the rates of CO_2 fixation or CO_2-dependent O_2 evolution per unit of chlorophyll observed with isolated chloroplasts remained at only a few percent of the rates for intact leaves. Various modifications of the isolation procedure have since provided preparations from a few species with activities approaching those observed *in vivo* (see Walker and Crofts, 1970). However, only rarely (Everson *et al.*, 1967; Bidwell *et al.*, 1970) has the pattern of labeling of products resulting from $^{14}CO_2$ fixation by isolated chloroplasts quantitatively approach the patterns observed with intact leaves, and retention of cytoplasm may account for these results (Winkerback *et al.*, 1972). With isolated chloroplasts most of the label generally remains in 3-P-glycerate and triose phosphates, while relatively little appears in starch and none in sucrose. Starch is certainly formed in chloroplasts in intact leaves, and possibly sucrose is as well (see Sections II,B,3 and 4). It seems likely that the disproportionate amount of fixed carbon appearing in the C_3 intermediates may be due to diffusion of these compounds to the surrounding medium (see Sections II,B,3 and 4).

Association of Calvin cycle enzymes with chloroplasts isolated in both aqueous (Latzko and Gibbs, 1968) and nonaqueous media, (Heber, 1970) has been amply demonstrated. As expected, those enzymes believed to be specific to the cycle are exclusive to chloroplasts, while chloroplasts contain variable proportions of those enzymes with counterparts in other metabolic processes. It is interesting to note that the content of Calvin cycle enzymes in isolated chloroplasts is not necessarily related to their capacity for photosynthesis (Latzko and Gibbs, 1968). This very likely indicates how important chloroplast integrity and the physical and chemical microenvironment within chloroplasts is to the proper and integrated operation of the cycle.

3. SECONDARY PATHWAYS AND END PRODUCTS

It is not easy to define just where photosynthetic metabolism ends and other cellular metabolism begins. The Calvin cycle is generally depicted as leading to carbohydrate synthesis because sucrose and starch are quantitatively the major primary products of photosynthesis in leaves. Of course, leaves must satisfy the carbon requirements of the whole plant so that much of the carbon assimilated by leaves is exported, mostly as sucrose. However, under some conditions algae and also developing leaves may incorporate substantial proportions of assimilated carbon into lipids, proteins, and other compounds (Bassham and Jensen, 1967). It is also common to see substantial amounts of fixed $^{14}CO_2$ appearing rapidly in some organic acids and amino acids with both algae and leaves.

One might reasonably define photosynthetic metabolism as that metabolism occuring within chloroplasts and depending upon light-generated energy. In addition to CO_2 assimilation, isolated chloroplasts have been shown to catalyze light-dependent synthesis of proteins from supplied amino acids, lipids from acetate, nucleic acids from nucleoside triphosphates, porphyrins from δ-aminolevulinate, and terpenes from mevalonate (see Kirk and Tilney-Basset, 1967; Kirk, 1970; also see relevant chapters). On the above criteria these processes may be considered as photosynthetic. However, the evidence would appear to be against these precursors being formed within chloroplasts, and hence for the chloroplasts being totally autonomous for the synthesis of the products formed from them. It appears more likely that these precursors are provided by extrachloroplastic reactions that perhaps in turn use compounds excreted from the chloroplasts. Of course, the possibility of a high degree of integration of processes involving a combination chloroplast and nonchloroplast reactions is not unreasonable. The glycolate pathway (see Section IV) and the C_4 pathway (see Section II,C) provide good examples of processes integrated in this way.

Figure 1 indicates some points at which carbon may leave the Calvin cycle other than as carbohydrate. Glycolate is rapidly labeled from $^{14}CO_2$ during photosynthesis by intact cells and is formed by isolated chloroplasts. It is at least mostly formed by an oxygenase reaction catalyzed by RuDP carboxylase and moves from the chloroplasts to peroxisomes where it is metabolized via the glycolate pathway (see Section IV). Some glycine and serine could arise as by-products of this pathway. Shah and Rogers (1969) have suggested that acetate for the synthesis of terpenes in the chloroplasts (carotenoids in particular), but not other sites in the cell, is derived from glycolate. Since the route proposed appears to involve some nonchloroplast reactions, a remarkably directed movement of acetate back to the chloroplasts may be required.

The other major form in which carbon can leave the chloroplasts is as the C_3 intermediates 3-P-glycerate and triose phosphates. However, as discussed further in Section II,B,4, the flux of these compounds is much more rapid than would be necessary for simply providing precursors of extrachloroplastic synthetic reactions. The compounds they could form include glycerol phosphate for lipid synthesis and phosphoenol pyruvate (PEP), pyruvate, and thence organic acids and amino acids by well-established nonchloroplast pathways (Bassham and Jensen, 1967). The enzymes responsible for the conversion of 3-P-glycerate to PEP and pyruvate are apparently not located in chloroplasts (Smillie, 1963; Heber, 1970). Whether PEP carboxylase, the enzyme responsible for C_4 acid formation, is in chloroplasts is uncertain, but oxalacetate formed via this reaction can be reduced in the chloroplasts. Leaves contain an NADP-specific malate dehydrogenase (Hatch and Slack, 1969b), and this enzyme is located in chloroplasts (Hatch and Slack, 1969b; Ting et al.,

1971). The special case of C_4 acid metabolism in relation to the C_4 pathway will be discussed in Section II,C.

Clearly, some doubts remain about the degree of metabolic autonomy of chloroplasts. There is evidence suggesting that chloroplasts may contain minor components of the total leaf complement of a number of enzymes normally assumed to operate in other cell compartments (Heber, 1970, 1974; Bidwell, et al., 1970). However, there are considerable technical difficulties in such studies and these conclusions cannot be accepted without reservation.

4. OUTSTANDING PROBLEMS

A longstanding problem, discussed in many occasions (Walker and Crofts, 1970; Hatch and Slack, 1970a; Black, 1973), concerns the apparent incapability of RuDP carboxylase to account for photosynthetic CO_2 fixation. Simply, the problem has been that with the K_m CO_2 observed for the isolated enzyme, its activity at CO_2 concentrations arising by diffusion of CO_2 from air would be much lower than observed photosynthesis rates. The likely resolution of this dilemma has been provided by recent studies showing that immediately after isolation, or after appropriate pretreatment, RuDP carboxylase exists in a form with a much lower K_m for CO_2 (Bahr and Jensen, 1974a, 1974b, 1974c; Badger et al., 1974) and close to that observed for photosynthesis by intact chloroplasts (Jensen and Bassham, 1966). This low K_m form of RuDP carboxylase also differs from the high K_m CO_2 form in vitro in respect to its maximum velocity and pH optimum. The satisfactory quantitative reconciliation of this problem will require careful consideration of all these factors (see Additional References for recent information on this topic).

Another contentious issue concerns the exact function of carbonic anhydrase in photosynthesis. Carbonic anhydrase catalyses the reversible hydration CO_2 to give bicarbonate, is particularly active in leaves of Calvin cycle species, and is at least largely associated with chloroplasts (Everson and Slack, 1968). Studies with the algae *Chlorella* and *Clamydomonas* show that, at normal CO_2 concentrations, the operation of carbonic anhydrase is apparently an absolute prerequisite for CO_2 fixation by RuDP carboxylase (Graham et al., 1971). Photosynthesis by isolated chloroplasts is also at least partially dependent upon carbonic anhydrase (Everson, 1970). The enzyme certainly appears to have some role in providing CO_2 to RuDP carboxylase, possibly by accelerating the movement of CO_2 across membranes (Enns, 1967). Whether this role is related to the proposed action of carbonic anhydrase in buffering the light-induced pH changes within chloroplasts (Everson and Graham, 1971) is not yet clear.

There are also some unresolved questions in relation to the degree of metabolic interaction between chloroplast and nonchloroplast cell compartments and the nature of the compounds responsible for the net movement of carbon from chloroplasts. In most leaves, sucrose is both the major end product of photosynthesis and the predominant form in which carbon is exported. Clearly, if sucrose is formed within chloroplasts then one need look no further for the compound responsible for the bulk of carbon efflux from these organelles. Doubts about the site of sucrose synthesis have been raised by studies suggesting that early labeled sucrose is not associated with chloroplasts (Heber and Willenbrink, 1964) and by the failure of isolated chloroplasts to form sucrose. Significantly, however, on the rare occasion when isolated chloroplasts have incorporated $^{14}CO_2$ into sucrose, this was associated with a pattern of labeling in other intermediates which much more closely resembled that observed with intact leaves (Everson et al., 1967; Bidwell et al., 1970). Perhaps there is some structural or functional factor missing from most chloroplasts preparations that affects sucrose synthesis either directly or via its effect on formation of other intermediates.

Other evidence supports the view that sucrose is synthesized within chloroplasts. Radiotracer studies have provided some evidence that sucrose is formed via sucrose phosphate (Bassham and Jensen, 1967), and sucrosephosphate synthase, but not sucrose synthase, has been found to be exclusively associated with chloroplasts isolated in nonaqueous media (Bird et al., 1965). Furthermore, Stocking et al. (1963) found the sucrose labeled in tobacco leaves after brief periods of photosynthesis in $^{14}CO_2$ was almost exclusively associated with chloroplasts isolated in nonaqueous media. The balance of evidence would appear to favor the view that sucrose is synthesized within chloroplasts, and hence is responsible for the major part of carbon transport from these organelles (see Additional References for recent references supporting the contrary view).

There is also evidence that 3-P-glycerate and dihydroxyacetone phosphate rapidly move both to and from chloroplasts (Bassham and Kirk, 1968; Heber, 1970, 1974). Opinions vary about the rate of movement of some other intermediates through the chloroplast envelope, but it seems that RuDP, sedoheptulose phosphates, and probably hexose phosphates are effectively retained. What might be the purpose of this rapid movement of some Calvin cycle intermediates between chloroplast and cytoplasm? Of course, if sucrose is synthesized in the cytoplasm, then the rapid excretion of a suitable precursor, such as dihydroxyacetone phosphate, would be essential. In any case, 3-P-glycerate and dihydroxyacetone phosphate could serve as the source of precursors for synthesis of proteins, lipids and many other compounds. However, the carbon demands for the latter reactions would be quantitatively small in mature leaves at least, and would not appear to justify the rapid rates of movement observed.

As pointed out elsewhere (Hatch and Slack, 1970a), the movement of labeled 3-P-glycerate and dihydroxyacetone phosphate from chloroplasts *in vivo* may simply reflect an equilibrium of cloroplast and cytoplasmic pools. Hence, there may be little or no net movement of these compounds from chloroplasts during steady-state photosynthesis. Such rapid exchange of metabolites could serve as a sensing device for metabolic regulation. Alternatively, cycling of these and other compounds between the chloroplasts and cytoplasm could provide an indirect means of transporting high-energy ester phosphate and reducing power to the cytoplasm. For instance, a cyclic shuttle of dihydroxyacetone phosphate and 3-P-glycerate, linked by chloroplast and cytoplasmic glyceraldehydephosphate dehydrogenase and 3-P-glycerate kinase, would operate effectively in this way (Heber and Santarius, 1970; Krause, 1971). It is generally agreed that pyridine nucleotides do not penetrate the chloroplast envelope (Bassham and Kirk, 1968; Heber, 1970), and it now appears that the movement of ATP and ADP is also very slow (Heber and Santarius, 1970).

C. C$_4$ Pathway

1. INTRODUCTORY COMMENTS

A retrospective search of the older literature reveals that certain tropical grasses and other plant species have some unusual features that might have indicated the operation of a modified metabolic process for photosynthesis (see Hatch and Slack, 1970b). These included high photosynthesis and growth rates, low photorespiration rates, an unusual leaf anatomy often associated with dimorphic chloroplasts, and a markedly reduced rate of water loss per unit of dry matter produced. In fact, these features are all intercorrelated, but this was not really appreciated until after it was shown that these species do utilize a unique process for photosynthetic CO$_2$ fixation.

The first clue that biochemical processes differing from the Calvin cycle might be operative in such species was provided by the observations that $^{14}CO_2$ initially labeled C$_4$ acids rather than 3-P-glycerate. These observations were made independently with sugarcane (Burr, 1962; Kortschak *et al.*, 1965) and maize (Karpilov, 1960; Tarchevskii and Karpilov, 1963). The first detailed account of the kinetics of labeling (Kortschak *et al.*, 1965) showed that radioactivity from $^{14}CO_2$ appeared initially in malate and aspartate and that there was a lag before the rapid labeling of 3-P-glycerate and other Calvin cycle intermediates. Hatch and Slack (1966) confirmed and extended these observations with sugarcane and, in a series of radiotracer and enzyme studies over the next three years, provided the basis for a detailed proposal for the pathway of CO_2 fixation (see Hatch and Slack, 1970a,b). These and other studies also showed that this process was operative in many other grasses and in species from several other families. Originally termed the C_4 dicarboxylic acid pathway (Hatch and Slack, 1968), the process is now referred to more simply as the C_4 pathway.

Not surprisingly, this field has moved rapidly in the past few years. The first reviews on the C_4 pathway (Hatch and Slack, 1970a,b; Karpilov, 1970; Walker and Crofts, 1970; Wolfe, 1970) are now outdated in many respects. More current interpretations have been provided in the proceedings of a recent symposium (see Hatch *et al.*, 1971) and reviews on photosysthesis (Hatch and Boardman, 1973; Black, 1973; see also Additional References).

2. REACTIONS AND ENZYMES

A symposium on the C_4 pathway held in December, 1970 (see Hatch *et al.*, 1971) revealed a large measure of agreement among those working in the field about the basic mechanism of the process. Typically, the chloroplasts of C_4 species are about equally distributed between two quite distinct cell types, and the fact that these cells have separate metabolic functions for photosynthesis was recognized at that time. These cells are generally arranged in two concentric layers around vascular bundles (see Fig. 2), the inner layer being termed bundle sheath cells and the outer layer mesophyll cells. The chloroplasts within each cell type differ morphologically, the degree of difference varying with species (see Chapter 7). The anatomy and chloroplast morphology of C_4 pathway species has been reviewed (Laetsch, 1971).

On the basis of more recent studies, C_4 species are now divisable into three subgroups according to the mechanisms they employ for decarboxylation of C_4 acids in bundle sheath cells. Therefore, it is no longer possible to describe the details of this process in a single comprehensive scheme. Instead, the general features of the pathway common to all C_4 species will be outlined (Fig. 2), followed by a detailed consideration of the steps involved (Figs. 3 and 4). At least the C_4 species examined to date all appear to be classifiable into one or other of the metabolic subgroups defined below.

The simplified scheme in Fig. 2 shows the basic reactions of C_4 photosynthesis. The primary assimilation of CO_2 occurs in mesophyll cells with PEP as the CO_2 acceptor and the C_4 acids, malate and aspartate, as the major products. Depending upon the species, one or other of these acids is then transported to the bundle sheath cells and decarboxylated, and the CO_2 so formed is refixed by the Calvin cycle. The C_3 compound remaining after C_4 acid decarboxylation is returned to the mesophyll cells and converted back to PEP—a step critical to the continued operation of the process. Later sections will consider evidence for this being the major or sole route for CO_2 assimilation in C_4 plants (Section II,C,3), and for the likely physiological advantages it offers (Section II,C,4).

Details of the reactions operative in mesophyll cells are shown in

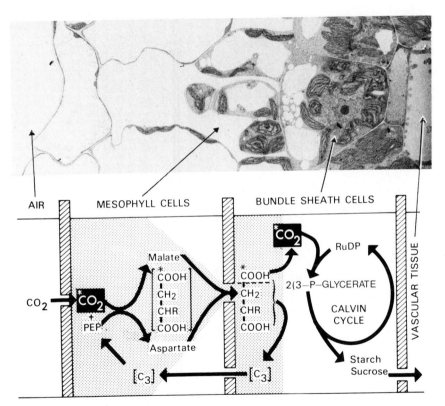

Fig. 2. Simplified scheme showing the key reactions of C_4 pathway photosynthesis and their intercellular location. Reactions in the shaded area are those unique to the C_4 pathway. The scheme is equated to an electron micrograph showing the cell arrangement of a C_4 leaf between the epidermis and a vascular bundle (see text).

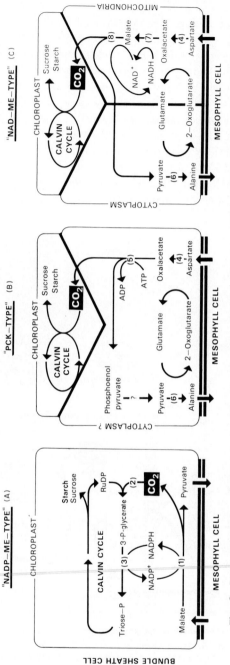

Fig. 3. Photosynthetic reactions in bundle sheath cells of different groups [(A) "NADP-ME-type," (B) "PCK-type," (C) "NAD-ME-type" (see text)] of C₄ pathway species and their intracellular location. The enzymes involved are (1) NADP-malic enzyme; (2) RuDP carboxylase; (3) 3-P-glycerate kinase and NADP-glyceraldehydephosphate dehydrogenase; (4) aspartate aminotransferase; (5) PEP carboxykinase; (6) alanine aminotransferase; (7) NAD-malate dehydrogenase; (8) NAD-malic enzyme. Heavy arrows indicate metabolites transported to or from mesophyll cells.

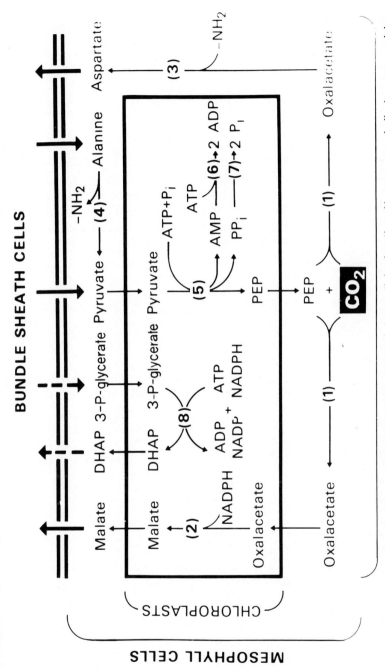

Fig. 4. C_4 pathway reactions in mesophyll cells and their intracellular location. Heavy arrows indicate movement to or from bundle sheath cells. "NADP-ME-type" species transfer malate and return pyruvate, while "PCK-type" and "NAD-ME-type" species transfer aspartate and return alanine. The enzymes involved are (1) PEP carboxylase, (2) NADP-malate dehydrogenase; (3) aspartate aminotransferase; (4) alanine aminotransferase; (5) pyruvate, P_i dikinase; (6) adenylate kinase; (7) pyrophosphatase; (8) 3-P-glycerate kinase and NADP-glyceraldehydephosphate dehydrogenase.

Fig. 4. CO_2 is initially assimilated in the cytoplasm via PEP carboxylase [Eq. (8)].

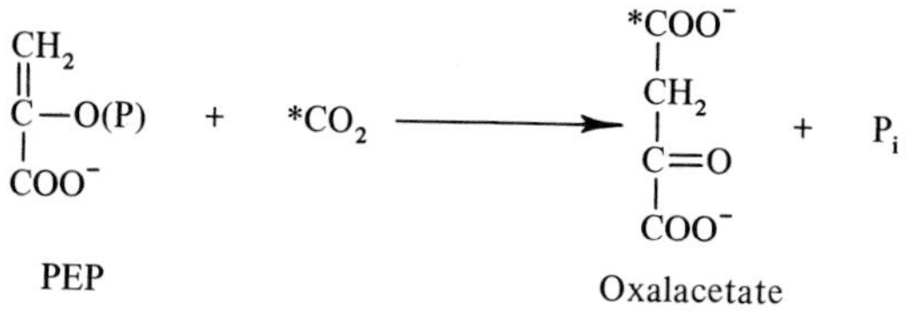

$$
\underset{\text{PEP}}{\begin{array}{c} CH_2 \\ \| \\ C-O(P) \\ | \\ COO^- \end{array}} + {}^*CO_2 \longrightarrow \underset{\text{Oxalacetate}}{\begin{array}{c} {}^*COO^- \\ | \\ CH_2 \\ | \\ C=O \\ | \\ COO^- \end{array}} + P_i \qquad (8)
$$

Labeling studies show that the product, oxalacetate, is rapidly interconverted with much larger pools of malate and aspartate in all C_4 species. However, for one group of C_4 species the net utilization of C_4 acids is via malate transported to bundle sheath cells. This malate is almost certainly formed from oxalacetate in chloroplasts via NADP-malate dehydrogenase [Eq. (9)], using photogenerated NADPH.

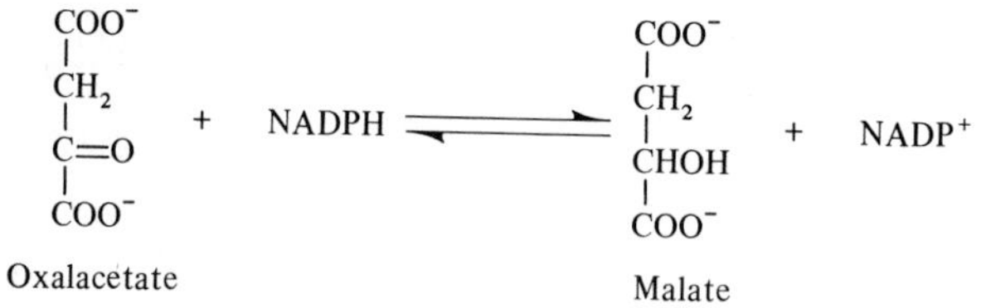

$$
\underset{\text{Oxalacetate}}{\begin{array}{c} COO^- \\ | \\ CH_2 \\ | \\ C=O \\ | \\ COO^- \end{array}} + NADPH \rightleftharpoons \underset{\text{Malate}}{\begin{array}{c} COO^- \\ | \\ CH_2 \\ | \\ CHOH \\ | \\ COO^- \end{array}} + NADP^+ \qquad (9)
$$

In other species aspartate is formed via aspartate aminotransferase [Eq. (10)] in the cytoplasm and transferred to the bundle sheath cells.

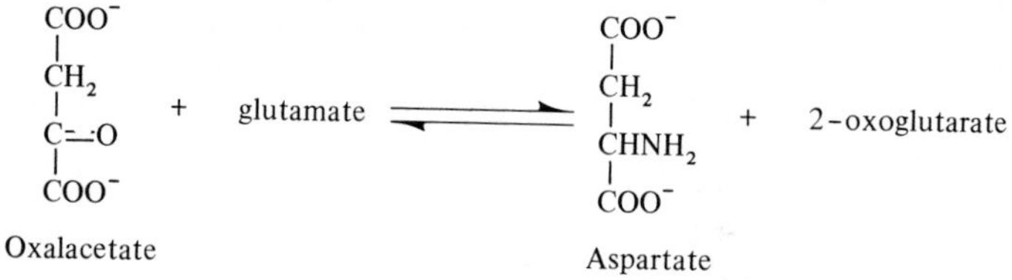

$$
\underset{\text{Oxalacetate}}{\begin{array}{c} COO^- \\ | \\ CH_2 \\ | \\ C=O \\ | \\ COO^- \end{array}} + \text{glutamate} \rightleftharpoons \underset{\text{Aspartate}}{\begin{array}{c} COO^- \\ | \\ CH_2 \\ | \\ CHNH_2 \\ | \\ COO^- \end{array}} + \text{2-oxoglutarate} \qquad (10)
$$

The other major function of mesophyll cells is to regenerate PEP from the C_3 compound returned from the bundle sheath cells. For those species moving malate to the bundle sheath cells pyruvate is the C_3 compound returned, while alanine is returned when aspartate is the acid transferred. The latter species convert alanine to pyruvate via an alanine aminotransferase [Eq. (11)] located in the cytoplasm.

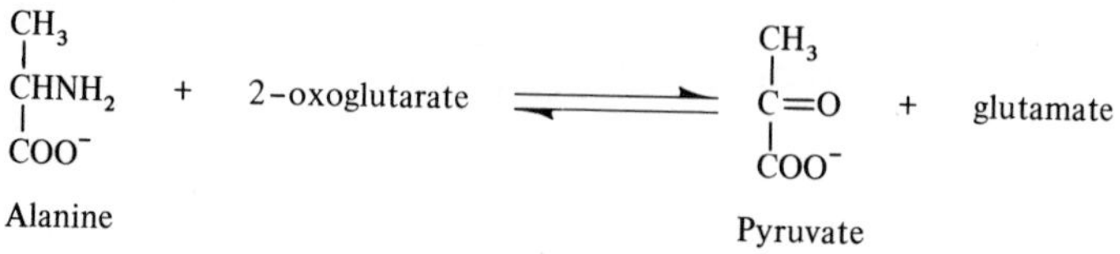

$$
\underset{\text{Alanine}}{\begin{array}{c} CH_3 \\ | \\ CHNH_2 \\ | \\ COO^- \end{array}} + \text{2-oxoglutarate} \rightleftharpoons \underset{\text{Pyruvate}}{\begin{array}{c} CH_3 \\ | \\ C=O \\ | \\ COO^- \end{array}} + \text{glutamate} \qquad (11)
$$

This reaction would couple with aspartate aminotransferase to provide the required amino group donors and acceptors. The pyruvate, derived either way, is converted to PEP in the chloroplasts by a unique enzyme named pyruvate, P_i dikinase [Eq. (12)].

$$
\begin{array}{c}
CH_3 \\
| \\
C=O \\
| \\
COO^-
\end{array}
\; + \; ATP \; + \; P_i \;\rightleftharpoons\;
\begin{array}{c}
CH_2 \\
\| \\
C-O(P) \\
| \\
COO^-
\end{array}
\; + \; AMP \; + \; PP_i \tag{12}
$$

Pyruvate   PEP

The mechanism of this reaction has been examined (Andrews and Hatch, 1969), and some aspects of its complex regulation are considered in Section III,C. Significantly, mesophyll chloroplasts also contain high levels of adenylate kinase and pyrophosphatase. As shown in Fig. 4, these enzymes would transform AMP and pyrophosphate, respectively, to compounds convertible to ATP via photosynthetic phosphorylation. Overall, the formation of PEP from pyruvate consumes the equivalent of two high-energy phosphate bonds.

The functionally unique nature of mesophyll chloroplasts should be emphasized. Besides the enzymes already mentioned, the mesophyll chloroplasts also contain a substantial part of the leaf complement of enzymes for the conversion of 3-P-glycerate to triose phosphates. They are probably responsible for reducing part of the 3-P-glycerate formed in bundle sheath cells (see Section II,C,5). However, they lack RuDP carboxylase and other enzymes of the Calvin cycle.

Mechanisms for C_4 acid decarboxylation in bundle sheath cells vary in different C_4 species (Fig. 3 and Table I). In some species, also typified by a deficiency of grana in bundle sheath chloroplasts, malate is decarboxylated in the bundle sheath chloroplasts via a NADP-specific malic enzyme [Eq. (13)].

$$
\begin{array}{c}
COO^- \\
| \\
CH_2 \\
| \\
CHOH \\
| \\
COO^-
\end{array}
\; + \; NADP^+ \;\rightleftharpoons\;
\begin{array}{c}
CH_3 \\
| \\
C=O \\
| \\
COO^-
\end{array}
\; + \; CO_2 \; + \; NADPH \tag{13}
$$

Malate   Pyruvate

While the CO_2, so formed, is fixed by the Calvin cycle and the pyruvate returned to the mesophyll cells, the NADPH is almost certainly recycled by coupling to the reducing step of the Calvin cycle. Such a coupling would provide half the total NADPH necessary for reducing the two molecules of 3-P-glycerate formed by fixation of CO_2. Species decarbox-

TABLE I

Activity and Location of C_4 Pathway Enzymes in Subgroups of C_4 Species[a]

Enzymes with similar activities in all groups of C_4 species

Enzymes	Activity (units/mg chlorophyll)	Location[b] Cell type	Within cell	Activity in C_3 species (units/mg chlorophyll)
PEP carboxylase	12-40	M	Cyto	0.4-1.5
Pyruvate,P_i dikinase	3-10	M	Chloro	0
Adenylate kinase	17-45	>M	Mostly chloro	0.5-1.0
Pyrophosphatase	20-60	>M	Mostly chloro	2-4
3-P-glycerate to triose-P enzymes	Similar to C_3	Both	Chloro	—
Other Calvin cycle enzymes	Similar to C_3	BS	Chloro	—

Enzymes with varying activity in subgroups of C_4 species

Enzymes	"NADP-ME-type" species Activity (units/mg chlorophyll)	Location[b] Cell type	Within cell	"PCK-type" species Activity (units/mg chlorophyll)	Location[b] Cell type	Within cell	"NAD-ME-type" species Activity (units/mg chlorophyll)	Location[b] Cell type	Within cell	Activity in C_3 species (units/mg chlorophyll)
NADP-malate dehydrogenase	8-17	M	Chloro	1-3	—	—	1-2	Both	Chloro	0.5-1.2
NADP-malic enzyme	9-14	BS	Chloro	0.3-0.4	—	—	0.2-0.8	—	—	0.1-0.8
Aspartate aminotransferase	5-7	>M	Chloro	45-60	Both	?	25-45	Both	Cyto,Mito	1-2.4
Alanine aminotransferase	2-4	—	—	38-45	Both	?	30-60	Both	Cyto	2-3
PEP carboxykinase	<0.2	—	—	10-14	BS	?	<0.2	—	—	<0.2
NAD-malic enzyme	0.2-0.4	—	—	0.2-0.5	—	—	5-9	BS	Mito	0.05-0.1

[a] Ranges of activity (a unit is 1 μmole/minute) are shown for "NADP-ME-type" C_4 species (*Zea mays, Sorghum sudanense, Pennisetum typhoides, Digitaria sanguinalis* and *Saccharum officinarum*), "PCK-type" type C_4 species (*Panicum maximum, Sporobolus fimbriatus, Chloris gayana*), "NAD-ME-type" C_4 species (*Atriplex spongiosa, Amaranthus edulis, Portulaca oleracea, Amaranthus edulis, Eragrostis curvula*), and C_3 plants (wheat, spinach, pea, soybean and *Atriplex patula*). Data was extracted from Hatch and Slack (1970a), Björkman and Gauhl (1969), Downton (1971), Andrews *et al.* (1971), Chen *et al.* (1971), Edwards and Black (1971a), Hatch and Mau (1973), Ting and Osmond (1973a), Hatch and Kagawa (1974a,b, and unpublished results). For comparison maximum photosynthesis rates for C_4 species range between 3 and 6 μmole min^{-1} mg chlorophyll^{-1}.

[b] The prefix > indicates the predominant location, "both" indicates approximately equal distribution between two cell types, and "vary" means intracellular location varies in the different cell types (see text). M, mesophyll; BS, bundle sheath; Chloro, chloroplast; Cyto, cytoplasm; Mito, mitochondria. Recent papers supporting data in this table are cited in Additional References.

ylating C_4 acids via NADP-malic enzyme will be referred to as "NADP-ME-type."

The remaining C_4 species contain little NADP-malic enzyme but very high aspartate aminotransferase and alanine aminotransferase activities, distributed about equally between mesophyll and bundle sheath cells (see Table I and Section III,C,3,b). For all these species, aspartate derived from mesophyll cells is apparently first converted back to oxalacetate in bundle sheath via aspartate aminotransferase [Eq. (10)] (Fig. 3). However, the subsequent fate of this oxalacetate varies in different species. In one group oxalacetate is directly decarboxylated by PEP carboxykinase [Eq. (14)],

$$
\begin{array}{c}
\text{COO}^- \\
| \\
\text{CH}_2 \\
| \\
\text{C=O} \\
| \\
\text{COO}^- \\
\text{Oxalacetate}
\end{array}
\;+\; \text{ATP} \rightleftharpoons
\begin{array}{c}
\text{CH}_2 \\
\| \\
\text{C-O(P)} \\
| \\
\text{COO}^- \\
\text{PEP}
\end{array}
\;+\; \text{ADP} \;+\; \text{CO}_2 \qquad (14)
$$

providing CO_2 for reassimilation via the Calvin cycle. At this time the exact location of this enzyme, the source of ATP for the reaction, and the immediate fate of PEP, are uncertain. There is reasonable grounds for proposing (Fig. 3) that PEP should give rise to pyruvate, that the pyruvate will be converted to alanine via alanine aminotransferase [Eq. (11)], and that alanine will be returned to the mesophyll cells. Not only would the latter reaction provide the necessary amino group coupling to sustain the operation of aspartate aminotransferase (see Fig. 3), but the return of alanine to the mesophyll cells would maintain a balance of amino groups between the two cell types. The plants utilizing this mechanism of C_4 acid decarboxylation will be designated "PCK-type" species.

The remaining C_4 species distinguished by their high aminotransferase activities lack significant levels of PEP carboxykinase (see Table I). The C_4 acid decarboxylation in these species is now accounted for by a NAD-malic enzyme [Eq. (15)] located in the bundle sheath mitochondria.

$$
\begin{array}{c}
\text{COO}^- \\
| \\
\text{CH}_2 \\
| \\
\text{CHOH} \\
| \\
\text{COO}^- \\
\text{Malate}
\end{array}
\;+\; \text{NAD}^+ \;
\underset{\substack{\text{CoA} \\ \text{or} \\ \text{Acetyl–CoA}}}{\overset{\text{Mn}^{2+}}{\rightleftharpoons}}
\begin{array}{c}
\text{CH}_3 \\
| \\
\text{C=O} \\
| \\
\text{COO}^- \\
\text{Pyruvate}
\end{array}
\;+\; \text{CO}_2 \;+\; \text{NADH} \qquad (15)
$$

As shown in Fig. 3, aspartate from mesophyll cells enters the bundle sheath mitochondria where it is converted to oxalacetate via aspartate aminotransferase [Eq. (10)]. The oxalacetate is then reduced to malate via NAD-malate dehydrogenase, and the malate is decarboxylated via NAD malic enzyme. The NAD+–NADH cycle that couples the latter reactions should be noted. Pyruvate moving from the mitochondria is then converted to alanine in the cytoplasm via alanine aminotransferase. In this instance, the coupling of the latter reaction with aspartate aminotransferase would require movement of 2-oxoglutarate and glutamate between mitochondria and the cytoplasm.

Theoretically, the Calvin cycle would utilize 3 ATP and 2 NADPH for each CO_2 fixed into hexose phosphate (Section II,B,1). The operation of the C_4 pathway would require two additional ATP, bringing the total requirement to 5 ATP and 2 NADPH for each CO_2 fixed. The additional ATP equivalents are used in the reaction catalyzed by pyruvate,P_i dikinase [Eq. (12)] in which both high-energy phosphate bonds of ATP are cleaved. However, as discussed in Section IV, when the influence of photorespiration is accounted for, the real energy costs for CO_2 assimilation by the Calvin cycle may be greater than for the C_4 pathway.

3. EXPERIMENTAL EVIDENCE

a. Radiotracer Studies. The present formulation of the C_4 pathway (Figs. 2–4) depends upon much the same kind of evidence as was used to formulate the Calvin cycle. Evidence for the sequence of metabolic events, disregarding for the moment their location, depends largely upon radiotracer studies combined with evidence on the activity of appropriate enzymes. Simple time–course and pulse–chase radiotracer studies of the incorporation of $^{14}CO_2$ into leaves under steady-state conditions have provided the following information.

1. $^{14}CO_2$ is incorporated at a steady rate from zero time. With light intensities near full sunlight, one generally finds a few percent of the fixed label in 3-P-glycerate after 1 or 2 seconds exposure to $^{14}CO_2$, but the remainder of the radioactivity in the C_4 acids, malate, aspartate, and oxalacetate (Kortschak *et al.*, 1965; Hatch and Slack, 1966; Johnson and Hatch, 1968; Hatch, 1971a,b). With increasing time in $^{14}CO_2$, 3-P-glycerate and then other Calvin cycle intermediates undergo a phase of rapid labeling, followed by the labeling of the photosynthetic end products, sucrose and starch. At lower light intensities label can remain undetectable in 3-P-glycerate for up to 5 seconds (Hatch *et al.*, 1967).

2. Initially, the radioactivity entering the C_4 acids from $^{14}CO_2$ is

almost exclusively in the C-4, while the first label entering 3-P-glycerate appears in the C-1 (Hatch and Slack, 1966; Hatch *et al.*, 1967; Johnson and Hatch, 1968, 1969; Hatch, 1971a). Label slowly enters the other carbons of the C_4 acids in a way consistent with it being derived by exchange of label between 3-P-glycerate and PEP, presumably via the enzymes enolase and 3-P-glycerate mutase (Hatch and Slack, 1966). Since C_4 pathway species are equipped with a special enzyme to convert pyruvate to PEP, there is no necessity for 3-P-glycerate to be a net source of PEP. In fact, if such a net conversion occurred it is difficult to visualize a scheme to accommodate the balance of carbon between the mesophyll and bundle sheath cells or within the Calvin cycle.

3. The C-4 of C_4 acids saturate with radioactivity after about 30 seconds in $^{14}CO_2$, and a relatively large internal CO_2 pool reaches saturation at about the same time (Hatch, 1971a). These carbons are saturated much more rapidly than the C-1 of 3-P-glycerate. The kinetics of labeling of the CO_2 pool was consistent with it being derived from the C-4 of C_4 acids and being the precursor of the carboxyl group of 3-P-glycerate.

In pulse–chase radiotracer studies leaves are exposed to $^{14}CO_2$ for a period and then transferred to unlabeled CO_2 while maintaining steady-state conditions. This type of experiment can provide more explicit information about reaction sequences. The following summarizes data provided by this procedure.

1. During the chase in CO_2 fixed radioactivity is retained in leaves but is rapidly lost from the C_4 acids, moves through 3-P-glycerate and other intermediates, and ultimately appears in sucrose and starch (Hatch and Slack, 1966; Johnson and Hatch, 1969; Chen *et al.*, 1971; Hatch, 1971a). Radioactivity lost from the C_4 acids is exclusively from the C-4 carboxyl (Johnson and Hatch, 1969; Hatch, 1971a). For "NADP-ME-type" species malate is the C_4 acid that loses label most rapidly, while label is lost more rapidly from aspartate with species that transport aspartate to bundle sheath cells for decarboxylation (Chen *et al.*, 1971; Hatch, 1971a). These results confirm the inference from time–course studies that the C_4 acids are rapidly turning over. With maize leaves this turnover entirely accounts for the radioactivity appearing in 3-P-glycerate from supplied $^{14}CO_2$ (Johnson and Hatch, 1969).

2. The rate of depletion of radioactivy from the internal CO_2 pool in leaves during a chase in CO_2 closely follows the loss from the C-4 of C_4 acids, but there is a lag before radioactivity is rapidly lost from the C-1 of 3-P-glycerate (Hatch, 1971a).

3. The rate of transfer of radioactivity from the C_4 acids to 3-P-glycerate and other intermediates is essentially the same during a chase

in normal air [0.03% (v/v) CO_2] and air containing 5% (v/v) CO_2 (Hatch and Slack, 1966). This supports the view that the intermediate CO_2 pool involved in this transfer is a special isolated pool that does not equilibrate readily with atmospheric CO_2.

Changes in the total leaf pools of certain compounds have also been examined during light–dark or CO_2 concentration transients. It should be noted, however, that for most intermediates the behavior of total leaf pools could be complicated by the inclusion of nonphotosynthetic pools. Farineau (1971) has shown that when CO_2 is removed from illuminated maize leaves an immediate decline in the aspartate pool is accompanied by an increase in the PEP pool. The 3-P-glycerate pool only begins to decline after the major drop in aspartate, and at the same time the RuDP pool begins to increase. This contrasted with the immediate responses of the 3-P-glycerate and RuDP pools in a Calvin cycle leaf under the same conditions. These observations are consistent with C_4 acids providing CO_2 to RuDP carboxylase.

b. *Enzymes and Their Location.* Interpretation of reaction sequences from radiotracer studies have been supported by the identification of appropriate enzymes. Data on the inter- and intracellular location of these enzymes has also provided clues about the partitioning of the various phases of the C_4 pathway. Information on the activity and location of C_4 pathway enzymes is summarized in Table I.

One group of enzymes apparently has a common and critical role in all C_4 species. These include PEP carboxylase, pyruvate,P_i dikinase and the enzymes ancilliary to the latter reaction, adenylate kinase and pyrophosphatase. Only pyruvate,P_i dikinase is unique to C_4 plants (Hatch and Slack, 1968). Although the remaining enzymes have functional counterparts in C_3 plants, their activities in C_4 plants are at least 10-fold and up to 100-fold higher (Table I). In all cases these activities in C_4 plants are adequate to account for the integral operation of the enzymes in photosynthesis. Activities of Calvin cycle enzymes are similar in C_3 and C_4 plants.

There is strong evidence for the total or almost total partitioning of PEP carboxylase and pyruvate,P_i dikinase in mesophyll cells. Conversely, Calvin cycle enzymes, except those responsible for converting 3-P-glycerate to triose phosphates, are confined to bundle sheath cells. There are considerable technical problems with the methods used for determining enzyme partitioning, and some data conflicting with the above conclusions (see Section II,C,5). Studies demonstrating at least about a 90% partitioning of these enzymes in the manner indicated (Slack *et al.*, 1969; Bjorkman and Gauhl, 1969; Edwards *et al.*, 1970; Berry *et al.*, 1970; Edwards and Black, 1971a; Huang and Beevers, 1972; Hatch and Kagawa, 1973) have been supported by more recent and definitive evidence (Chen *et al.*, 1973; Kanai and Edwards, 1973a,b; Hatch and Kagawa, 1973; Kagawa and Hatch, 1974a; see also Additional References). The approximately equal distribution of the photosynthetic enzymes for converting 3-P-glycerate to triose phosphates between the two cell types is commonly observed.

Phosphoenolpyruvate carboxylase is clearly not associated with mesophyll chloroplasts whereas pyruvate,P_i dikinase, and the major part of the mesophyll cell component of adenylate kinase and pyrophosphatase, are located in chloroplasts (Slack *et al.*, 1969; Hatch and Kagawa, 1973; Kagawa and Hatch, 1974a). The association of Calvin cycle enzymes with bundle sheath cell chloroplasts isolated in nonaqueous media has been demonstrated (Slack *et al.*, 1969).

The activity of the remaining enzymes listed in Table I vary in different C_4 species, but form distinct patterns in relation to the three subgroups of species defined above (see Fig. 3). Where a specific and integral role in C_4 photosynthesis has been assigned to these enzymes, their activities are very high compared with those in other groups of C_4 or C_3 plants. "NADP-ME-type" species are distinguished by their high NADP-malic enzyme activity also accompanied by much higher NADP-malate dehydrogenase activity. Most of the latter activity is located in mesophyll cells (Edwards and Black, 1971a), where it is confined to the chloroplasts (Slack *et al.*, 1969; Hatch and Kagawa, 1973; Kagawa and Hatch, 1974a). NADP-malic enzyme is located in the chlorplasts of bundle sheath cells (Slack *et al.*, 1969; Edwards and Black, 1971a; Chen *et al.*, 1973). Both "PCK-type" and "NAD-ME-type" species contain little NADP-malic enzyme activity, but very high activities of aspartate and alanine aminotransferases. These activities are partitioned about equally between mesophyll and bundle sheath cells, with the majority of each activity in each cell type being due to a separate and distinctive isoenzyme (Andrews *et al.*, 1973; Hatch and Mau, 1973; Hatch, 1973). These aminotransferases appear to be cytoplasmic enzymes, except for the bundle sheath cell aspartate aminotransferase isoenzyme of "NAD-ME-type" species, which is a mitochondrial enzyme (Hatch and Mau, 1973; Hatch and Kagawa, 1974b).

The feature distinguishing "PCK-type" and "NAD-ME-type" species is their high PEP carboxykinase (Edwards *et al.*, 1971) and NAD-malic enzyme (Hatch and Kagawa, 1974a,b) activities, respectively. PEP carboxykinase is at least very largely associated with bundle sheath cells, but its intracellular origin is uncertain. NAD-malic enzyme is located in the mitochondria of bundle sheath cells.

In support of these interpretations of the intercellular distribution of enzymes, isolated mesophyll cells have been shown to fix CO_2 into C_4 acids (see Black, 1973). They also evolve O_2 with stoichiometic production of malate when HCO_3^- and PEP are provided in the light (Salin et al., 1973). There are several reports of light-dependent fixation of HCO_3^- into Calvin cycle intermediates by bundle sheath cells, requiring the addition of ribose 5-phosphate and adenine nucleotides (see Black, 1973). Recently, high rates of HCO_3^- fixation without other additions, and also the incorporation of the C-4 carboxyl of C_4 acids into Calvin cycle intermediates, have been demonstrated with bundle sheath cells (Kagawa and Hatch, 1974b).

These enzymes, unique to the C_4 pathway, share a particular feature in common with Calvin cycle enzymes that provides further evidence for their involvement in photosynthesis. The fact that the activities of Calvin cycle enzymes and related photosynthetic enzymes are low in leaves of dark-grown plants, but increase severalfold along with chlorophyll when plants are illuminated, has been long recognized (Kirk and Tilney-Basset, 1967; Smillie and Scott, 1969). All the special enzymes of the C_4 pathway, including those not associated with chloroplasts, have been shown to share this characteristic. For the various enzymes, increases in activity of 10- to 15-fold have been observed within 48 hours after plants were illuminated (Hatch et al., 1969; Johnson and Hatch, 1970; Hatch and Mau, 1973; Hatch and Kagawa, 1974b).

c. Quantitative Contribution. The available evidence from a variety of experiments indicates that, where the C_4 pathway is operative, it is at least very largely responsible for the initial fixation of CO_2 assimilated from the atmosphere. Of course, minor contributions directly by the Calvin cycle will be difficult to prove or disprove, and may vary with species and conditions. Evidence for photosynthesis occurring largely or solely via the C_4 pathway is listed below.

1. As already detailed in Section II,C,2,a, simple time–course studies with leaves exposed to $^{14}CO_2$ show that the rate of labeling of C_4 acids is maximal from zero time, and is initially comparable to the total rate of fixation. Extrapolating curves for incorporation of label into 3-P-glycerate toward zero time show initial rates of labeling to be very low; under some conditions there is a lag of up to 5 seconds before label is detected in this compound. For eight species we examined (M. D. Hatch, unpublished) the proportion of fixed label in 3-P-glycerate after 1–2 seconds of steady-state photosynthesis ranged between 2 and 7%. Even without correcting for label derived via C_4 acids, this sets a low limit for the

maximum contribution of RuDP carboxylase to the direct assimilation of CO_2. However, more significant information can be obtained by plotting this time–course data as percent of total ^{14}C fixed. The curves for ^{14}C in the C-4 of C_4 acids extrapolate toward 100% at zero time, while those of ^{14}C in 3-P-glycerate plus products formed from 3-P-glycerate extrapolate to approximately zero, consistent with essentially all the $^{14}CO_2$ being assimilated via C_4 acids [see Hatch (1976) in Additional References].

2. Analysis of simultaneous time–course and pulse-chase radiotracer studies with maize leaves shows that essentially all the radioactivity entering 3-P-glycerate during exposure of leaves to $^{14}CO_2$ could be accounted for by radioactivity lost from the C-4 of C_4 acids (Johnson and Hatch, 1969).

3. Inhibitors of PEP carboxylase strongly inhibit CO_2 assimilation, by C_4 pathway leaves but not Calvin cycle leaves (Osmond and Avadhani, 1970).

4. Pulse-chase studies show that $^{12}CO_2$ concentrations as high as 5% (v/v) during the chase do not decrease the rate of transfer of label from C_4 acids to 3-P-glycerate (see Section II,C,3,a). This supports the view that there may be restricted access of external CO_2 to the bundle sheath cells of intact leaves, and therefore restricted access to RuDP carboxylase, *in vivo* (also see Section II,C,4).

d. Transport of Metabolites. Rapid fluxes of metabolites into and from chloroplasts and mitochondria and between mesophyll and bundle sheath cells are an essential feature of the C_4 pathway outlined in Figs. 2–4. For intercellular fluxes, Osmond (1971) has calculated that simple diffusive movement in the symplasm would be adequate. Significantly, the cell wall between mesophyll cells and bundle sheath cells contains an unusually large number of plasmodesmata that may provide the necessary channels for this rapid movement (Laetsch, 1971). Adequate diffusion fluxes into or from mesophyll chloroplasts (Hatch and Kagawa, 1973; Kagawa and Hatch, 1974a) and bundle sheath mitochondria of "NAD-ME-type" species (Kagawa and Hatch, 1974b and unpublished results) have been demonstrated for several key metabolites [for a detailed discussion of metabolite transport during C_4 photosynthesis see Hatch and Osmond (1976) in the Additional References].

e. Activities in Isolated Chloroplasts. As depicted in Fig. 4 the primary assimilation of CO_2 in all C_4 species occurs in the mesophyll cytoplasm via PEP carboxylase. For "PCK-type" and "NAD-ME-type" species other reactions of the pathway are also operative in nonchloroplast cell compartments. Therefore, simulation of total C_4 photosynthesis with isolated chloroplasts will be impossible, but several component phases of the process should be demonstrable. In accordance with the scheme in Fig. 4, preparations of mesophyll chloroplasts have been shown to mediate the rapid light-dependent transformation of pyruvate to PEP, 3-P-glycerate to dihydroxyacetone phosphate, and oxalacetate to malate (Hatch and Kagawa, 1973; Kagawa and Hatch, 1974a). These chloroplasts were devoid of RuDP carboxylase and did not fix CO_2. On the other hand, bundle sheath chloroplasts should assimilate CO_2 via the Calvin cycle. Attempts to prepare intact chloroplasts from isolated bundle sheath strands have been unsuccessful, presumably due to the severe procedures necessary to break these cells. However, goods rates of CO_2 assimilation into Calvin cycle intermediates have been observed with chloroplasts extracted from young primary leaves of maize (O'Neal *et al.*, 1972). Possibly, the bundle sheath cells are readily broken in these very young primary leaves so that these preparations could contain a mixture of intact chloroplasts from both cell types. Information on the photoactivities of broken chloroplast preparations from mesophyll and bundle sheath cells has recently been reviewed (Hatch and Boardman, 1973).

4. Physiological Significance

Superficially, the reactions unique to the C_4 pathway appear to perform a simple but somewhat pointless exercise, namely, to fix CO_2 in mesophyll cells, transport this CO_2 as C_4 acids to the bundle sheath cells, and there release it again where it is refixed by the Calvin cycle (see Fig. 2). What purpose might this process serve and what advantages might it offer to plants? For whatever reason, it appears these reactions serve to concentrate CO_2 in bundle sheath cells. Such a proposal (Bjorkman, 1971; Hatch, 1971b) has been supported by radiotracer studies demonstrating that a large CO_2 pool forms during photosynthesis, and has the kinetic characteristics of an intermediate between C_4 acids and 3-P-glycerate (Hatch, 1971a). The size of this pool was much larger than the CO_2 plus HCO_3^- calculated to form by diffusion equilibrium with $^{14}CO_2$, or the pool actually observed to form in the dark.

One rationale for the C_4 pathway depended upon the assumption that RuDP carboxylase has a high K_m CO_2 *in vivo* (Hatch, 1971b) but recent studies (Bahr and Jensen, 1974a, 1974b; Badger *et al.*, 1974) indicate that this is not so (see Section II,B,1). However, for C_4 species a mechanism for concentrating CO_2 in bundle sheath cells may still be essential for the adequate operation of RuDP carboxylase. The reason is that the higher stomatal diffusion resistance of C_4 leaves combined with their higher photosynthesis rates result in the development of much larger gradients of CO_2 between air and the leaf interior compared with C_3 plants. Calculations, based on diffusion resistance and photosynthesis measurements for several species (see Hatch and Osmond, 1976, in Additional References) indicate that the substomatal liquid phase CO_2 concentrations would be in the region of 6 μM for C_3 species compared with 1–2 μM for C_4 species. Presuming that there would be an additional diffusion gradient between mesophyll and bundle sheath cells of C_4 species, the CO_2 concentration at the site of RuDP carboxylase would be even lower. Projected activities for RuDP carboxylase at 6 μM CO_2 (based on recent data for the V_{max} and K_m CO_2 for this enzyme) would be adequate to account for photosynthesis of C_3 species. However, at about 1 μM CO_2, RuDP carboxylase activity would be deficient by several fold to account for maximum rates of photosynthesis of C_4 species but PEP carboxylase activity would be adequate to support the initial assimilation of CO_2. Significantly, if the intermediate CO_2 plus HCO_3^- pool observed to develop in C_4 species during steady-state photosynthesis is confined to bundle sheath cells (Hatch, 1971a), then the concentration of the CO_2 component (assuming a pH of 7.5) would be at least 80 μM. At such a concentration of CO_2 the potential activity of RuDP carboxylase would be more than adequate to account for C_4 photosynthesis [see Hatch and Osmond (1976) and Hatch (1976) listed in the Additional References to this chapter for a detailed development of these arguments].

Another suggested advantage of concentrating CO_2 in bundle sheath cells (Bowes et al., 1971; Bowes and Ogren, 1972) depends upon the consequences of RuDP carboxylase acting as an oxygenase as well as a carboxylase. This oxygenation reaction produces phosphoglycolate which is the substrate for the process known as photorespiration (see Section IV), and CO_2 produced by photorespiration reduces the net assimilation of CO_2 in C_3 plants by as much as 30 to 40%. It was reasoned that by concentrating CO_2 in bundle sheath cells, the C_4 pathway would serve to increase the CO_2 to O_2 ratio in these cells. Since CO_2 and O_2 act competitively as substrates for RuDP carboxylase (see Section IV), the effect of the higher CO_2 concentration in bundle sheath cells would be to reduce phosphoglycolate production, and hence the photorespiratory loss of CO_2. This could largely account for the higher net photosynthesis rates commonly found in C_4 plants. The fact that lowered atmospheric O_2 concentrations reduce photorespiration and increase net photosynthesis in C_3 plants, but have no effect on photosynthesis in C_4 plants, supports this conclusion (see Bjorkman, 1971).

Another special physiological feature of C_4 species, their economic use of water for growth (see Section V,B), can be explained by the operation of PEP carboxylase for the primary assimilation of CO_2. High stomatal resistances and consequent large CO_2 gradients between air and the leaf interior are features common to C_4 plants during photosynthesis (see Hatch et al., 1971). However, the high CO_2 affinity and activity of PEP carboxylase would permit rapid rates of CO_2 fixation in spite of low steady-state concentrations of CO_2 in mesophyll cells. As a consequence, these plants would lose less water per unit of CO_2 fixed compared with C_3 plants.

5. CONFLICTS AND UNRESOLVED PROBLEMS

In terms of the interpretations of the C_4 pathway outlined in Figs. 2–4 the major unresolved problem relates to the fate of PEP formed in "PCK-type" species. The status of this problem is considered in Section II,C,3,a. Another unresolved question is whether some variable or precise proportion of the 3-P-glycerate formed in bundle sheath cells must be returned to mesophyll cell chloroplasts for reduction to triose phosphates. In this regard, there is still uncertainty about the degree to which bundle sheath chloroplasts of "NADP-ME-type" species are deficient in the capacity to photoreduce NADP (see Hatch and Boardman, 1973). If this capacity is low or negligible in intact cells then about half the 3-P-glycerate would have to be reduced in mesophyll cells, with the remaining NADPH being provided by NADP-malic enzyme in the bundle sheath cells (see Fig. 3). For all groups of C_4 species a negative charged deficit would be generated in mesophyll cells by the continuing movement of dicarboxylic acids to bundle sheath cells with the return of a monocarboxylic acid. It is interesting to note that this imbalance would be restored by cycling half the 3-P-glycerate formed in bundle sheath cells to the mesophyll cells for reduction, since the product returned, dihydroxyacetone phosphate, would contain one less negative charge.

Problems concerning the transport of metabolites within and between cells have been considered in Section II,C,3,d. An unresolved dilemma is how the CO_2 concentrated in bundle sheath cells is effectively retained (see Section II,C,4). It costs at least 2 ATP (used by pyruvate,P_i dikinase) to move this CO_2 to bundle sheath cells, and it would be untenable to have a large part of this CO_2 lost by diffusion relative to that fixed by RuDP carboxylase [for more detailed discussion see Hatch and Osmond (1976) in Additional References].

A major conflict of evidence has centered around the quantitative partitioning of enzymes between cells, particularly the location of RuDP carboxylase and related Calvin cycle enzymes. Black (1973) has critically discussed claims that part or all of the Calvin cycle is located in mesophyll cells. These conflicting interpretations are all based on the use of an empirical procedure for distinguishing between mesophyll and bundle sheath cell constituents, involving a graded series of extractions of leaf tissue. The uses and limitations of this method, and the potential problems arising from its uncritical use, have recently been discussed (Black, 1973; Hatch and Kagawa, 1973). Critical evidence for the absence of Calvin cycle activity in mesophyll chloroplasts and for its exclusive location in bundle sheath cells is cited in Section II,C,3,b (also see papers listed in Additional References).

D. Crassulacean Acid Metabolism

The leaves of many species from the Crassulaceae and other families can rapidly assimilate CO_2 in the dark. The extent to which this process occurs depends upon leaf age and the prevailing environmental conditions. When operating in this mode, these species may fix little or no CO_2 from the atmosphere during the light period, primarily owing to the closure of stomata. The CO_2 assimilated in the dark appears mostly in malic acid, which accumulates in large quantities. Associated with this increase in malate is a decrease in starch, which is apparently the primary source of the acceptor for CO_2. In the following light period the carbon in malate, including that derived from CO_2, is metabolized and reappears predominantly as starch and other carbohydrates. Consideration of this process is included here because, while the primary fixation of CO_2 occurs in the dark, the total process is ultimately light-dependent and can provide the only means of maintaining a positive carbon balance.

Although the broad outline of a metabolic scheme to account for these transformations was formulated several years ago, the exact nature of some phases of the process remain to be resolved or confirmed. Figure 5 outlines the likely metabolic events involved in the dark and light phases of this process, generally called crassulacean acid metabolism (CAM). The reader is referred to accounts of the physiology and biochemistry of CAM (Ranson and Thomas, 1960; Beevers et al., 1966; Ting, 1971) and a recent review of the subject (Black, 1973).

Carbon dioxide fixation in the dark is mediated by PEP carboxylase [Eq. (8)], utilizing PEP originating from stored carbohydrate. PEP has generally been assumed to be derived by conversion of hexose phosphates to ribulose 5-phosphate via the pentose phosphate cycle, followed by RuDP formation [Eq. (6)] and then the carboxylation of this compound by RuDP carboxylase [Eq. (2)] to yield 3-P-glycerate. PEP could then be formed from 3-P-glycerate by the operation of 3-P-glycerate mutase

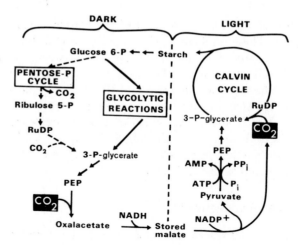

Fig. 5. Reactions of the light and dark phases of crassulacean acid metabolism (CAM). See text for details of the enzymes involved and for comments on the alternate route for 3-P-glycerate formation via ribulose 5-phosphate (dotted lines). For C_4 acid decarboxylation in the light some species probably utilize PEP carboxykinase instead of NADP-malic enzyme (see text).

and enolase. However, evidence for this proposal rests almost solely on the observation that malate formed from $^{14}CO_2$ contains about two-thirds of its label in the C-4 carboxyl and one-third in the C-1 (Bradbeer *et al.*, 1958). It was reasoned that the specific radioactivity of the C-1 carboxyl of 3-P-glycerate, and hence the C-1 of PEP, would reach only half that of the supplied $^{14}CO_2$, since only one of the two molecules of 3-P-glycerate formed via RuDP carboxylase derives its carboxyl carbon from CO_2. In contrast, the C-4 carboxyl would attain the same specific activity as the supplied $^{14}CO_2$. The observation that leaves of some CAM species contain glucose-6-phosphate dehydrogenase and 6-phosphogluconate dehydrogenase and that these enzymes are partially associated with chloroplasts (Garnier-Dardart, 1965; Mukerji and Ting, 1968) provided some support for this interpretation.

However, Sutton and Osmond (1972) have questioned the methods used for these earlier analyses, and have provided evidence that 90–95% of the label incorporated into malate is located in the C-4 carboxyl when several CAM species were exposed to $^{14}CO_2$ in the dark (Sutton and Osmond, 1972). This result is inconsistent with the double carboxylation mechanism and would suggest that glycolysis is the more likely route for PEP formation from starch (see Additional References). In Fig. 5, the former route is retained as an alternative, at least not completely eliminated for all species. The presence of adequate levels of PEP carboxylase and malate dehydrogenase for malate formation in the dark has been confirmed for a wide variety of CAM species (Dittrick *et al.*, 1973).

Malate decarboxylation in the light has generally been attributed to NADP-malic enzyme [Eq. (13)]. The pyruvate so formed is probably converted to PEP via pyruvate,P_i dikinase [Eq. (12)], now identified in CAM species (Kluge and Osmond, 1971). However, a recent survey has shown that some CAM species are deficient in NADP-malic enzyme but contain high levels of PEP carboxykinase [Eq. (14)] instead (Dittrick *et al.*, 1973). If the latter enzyme operates to decarboxylate C_4 acids then oxidation of stored malate to oxalacetate would be a prerequisite for malate utilization in the light, and the involvement of pyruvate,P_i dikinase would presumably be unnecessary.

The CO_2 derived from C_4 acids in the light is undoubtedly assimilated via the Calvin cycle, and most of the carbon in the remaining C_3 compound also ultimately appears as carbohydrate. This presumably occurs by conversion to 3-P-glycerate and the subsequent metabolism of this compound via the Calvin cycle (see Fig. 5).

Whether any of the reactions involved in malate formation in the dark occur in chloroplasts remains to be proved. However, at least part of the malic enzyme of cactus is associated with chloroplasts (Mukerji and Ting, 1968). Thus, if pyruvate,P_i dikinase is a chloroplast enzyme, as it is in C_4 pathway species (see Section II,C,3), then possibly all the reactions involved in the conversion of malate to carbohydrate may occur in chloroplasts.

As already indicated, the extent to which CAM operates depends upon the prevailing environmental conditions. Its operation in place of normal photosynthesis is favored by conditions where transpiration would be high. Since CAM species usually occur in arid areas, it would seem reasonable that the process has evolved to conserve water. Thus, by assimilating CO_2 in the dark when transpiration would be lower, stomata can be closed during the light period when increased temperatures combined with low humidity would result in rapid transpiration.

III. Regulation of Photosynthesis

A. General Aspects

Pathways of photosynthetic CO_2 fixation are complex cyclic or multicyclic processes that, in turn, are linked to energy-producing light reactions. Furthermore, these processes must be adaptable to the varying demands of cells or organisms for both the quantity and types of end product formed. It would be reasonable to anticipate the operation of control processes on many phases of photosynthesis. Accordingly, some form of feedback must operate both on the rate of CO_2 assimilation and on the partitioning of assimilated carbon into different products. The partitioning of carbon between the precursors of the CO_2 acceptor and end products must also be balanced. Other regulatory processes probably operate to integrate the various phases of the cyclic series of reactions and to accommodate the profound metabolic changes associated with fluctuations in light intensity and particularly light–dark transitions. Aspects of the regulation of photosynthesis have been recently reviewed (Hatch and Slack, 1970a; Preiss and Kosuge, 1970; Bassham, 1971, Walker, 1973; also see Chapter 10.

B. Calvin Cycle

Ribulose-1,5-diphosphate carboxylase would be a logical contender as a control site for CO_2 assimilation. There is evidence for light-mediated effects on RuDP carboxylase activity *in vivo*, probably partly due to the effect of changing Mg^{2+} and pH in the chloroplast stroma on the activity and substrate affinity of the enzyme (Preiss and Kosuge, 1970; Bassham, 1971; Walker, 1973). Activation of the enzyme by fructose 6-phosphate (Buchanan and Schurman, 1972) and inhibition by 6-phosphogluconate (Chu and Bassham, 1972) could also contribute to its light-mediated regulation *in vivo*. RuDP carboxylase activity could also be controlled indirectly by the supply of RuDP or CO_2. In this connection, the light-mediated activation of ribulose-5-phosphate kinase (Latzko *et al.*, 1970; Bassham, 1971) and carbonic anhydrase (Everson, 1971), and the inhibition of the kinase by 6-phosphogluconate (Anderson, 1973), could be significant. Two other enzymes, NADP-glyceraldhyde phosphate dehydrogenase (Ziegler *et al.*, 1969) and fructose diphosphate phosphatase (Buchanan *et al.*, 1967; Bassham and Kirk, 1968) are activated by light *in vivo*, possibly via NADPH and reduced ferredoxin, respectively. Bassham (1971) and Walker (1973) have discussed the probable significance of changes in activity of RuDP carboxylase and other Calvin cycle enzymes during light-dark transients (see references relating to RuDP carboxylase in Additional References).

Starch synthesis could be regulated by the supply of its precursor ADP-glucose through the activation of chloroplast ADP-glucose pyrophosphorylase by several photosynthetic intermediates and inhibition by AMP, ADP, and phosphate (Preiss and Kosuge, 1970). Sucrose synthesis may also be modulated by supply of precursors (Preiss and Kosuge,

1970), while its synthesis in *Chlorella* is inhibited by low concentrations of NH_4^+ (Bassham, 1971).

A slower-acting type of regulation operates to control the content of RuDP carboxylase in leaves in response to changes in the light intensity at which plants are growing (Björkman, 1970). An increase in light intensity has resulted in changes of up to 3-fold in 5 to 6 days and vice versa. These changes are accompanied by concomitant changes in maximum photosynthesis rates.

C. C₄ Pathway

Probably most of the regulatory processes cited above for Calvin cycle enzymes are also operative on these enzymes in C_4 pathway species. However, Steiger *et al.* (1971) have shown that the light–dark effects on NADP-glyceraldehyde phosphate dehydrogenase and ribulose-5-phosphate kinase occur in some C_4 pathway species but not others.

Enzymes specific to the C_4 pathway are also subject to regulation of various kinds. The enzymes concerned with the primary reactions of the pathway are inhibited by reaction products, PEP carboxylase by oxalacetase (Lowe and Slack, 1971) and pyruvate,P_i dikinase by PEP, AMP, and PP_i (Andrews and Hatch, 1969). These effects may be concerned with balancing the rates of reactions both within the first cycle and between the two cycles of the pathway (see Figs. 2–4). PEP carboxylase is activated by glucose 6-phosphate, which acts by increasing the affinity of the enzyme for PEP (Coombs *et al.*, 1973; Ting and Osmond, 1973a). Pyruvate,P_i dikinase (Hatch and Slack, 1969a) and NADP-malate dehydrogenase (Johnson and Hatch, 1970) are rapidly inactivated in darkened leaves and reactivated when leaves are illuminated. Since the activity of these enzymes also varies with changes in light intensity within the normal daily range, this kind of regulation is probably also important during photosynthesis. These effects are ultimately mediated by the reversible oxidation of enzyme thiol groups. However, for pyruvate,P_i dikinase at least the process is complex, being enzyme-catalyzed and controlled by P_i,pyruvate, AMP, and ADP.

There are also slower adaptive changes in the leaf content of pyruvate,P_i dikinase and PEP carboxylase in response to the prevailing light intensity at which plants are growing (Hatch *et al.*, 1969). Increases in activity in mature leaves may be up to 7- or 8-fold within a few days of increasing light intensity, and these changes are reversed when the light is reduced. As already mentioned, RuDP carboxylase levels change in Calvin cycle species under these conditions, but, significantly, its activity does not change in C_4 pathway species.

IV. Photorespiration and the Glycolate Pathway

A. Physiology of Photorespiration

The process termed photorespiration operates only in the light and is respiratory at least in the sense that it involves consumption of O_2 and evolution of CO_2. It is now clear that this process is intimately linked with photosynthesis both functionally and metabolically. The biochemistry and physiology of photorespiration has been considered in recent reviews (Jackson and Volk, 1970; Tolbert, 1971a; Black, 1973), a symposium (see Hatch et al., 1971), and a book (Zelitch, 1971).

The presence and approximate magnitude of photorespiration can be assessed in several ways. For instance, when leaves are allowed to photosynthesize in a closed system a steady CO_2 concentration is reached when CO_2 assimilation equals CO_2 evolution. This is known as the CO_2 compensation point and reflects the magnitude of photorespiration. The increase in net photosynthesis resulting from a decrease in the O_2 concentration to about 1% (v/v) has also been attributed, in part at least, to the elimination of photorespiration. Likewise, the burst of CO_2 that follows darkening of illuminated leaves has been attributed to a brief overshoot of photorespiration and may prove to be one of the better quantitative measures of the process (Bulley and Tregunna, 1971). Isotopic procedures have also been used to measure photorespiration (Jackson and Volk, 1970). The magnitude of photorespiration varies with species, leaf age, light intensity, CO_2 and O_2 concentrations, and temperature (see Jackson and Volk, 1970; Zelitch, 1971). Calvin cycle species growing under natural conditions lose between 20 and 40% of the CO_2 just fixed by photosynthesis via the process of photorespiration.

C_4 pathway plants have a CO_2 compensation point near zero and do not show enhancement of photosynthesis when the O_2 concentration is reduced. The extent to which these features are due to lack of photorespiration, rather than to effective refixation of photorespiratory CO_2 by PEP carboxylase, is still uncertain. However, there are reasons to propose that photorespiration rates will be lower in these species (see Section II,C, 4), and they contain lower activities of several of the key enzymes implicated in the process (Osmond and Harris, 1971; Rehfeld et al., 1970; Huang and Beevers, 1972).

B. Biochemistry

The fact that glycolate is the substrate utilized during photorespiration has been known for some time. However, there has been much uncertainty about the source of this glycolate and the function of the process responsible for its metabolism. Various possible routes of glycolate synthesis have been proposed (see Black, 1973; Zelitch, 1971) but currently there is strong evidence that the major or sole source of this compound is the oxygenation of RuDP, catalyzed by RuDP carboxylase (Ogren and Bowes, 1971; Bowes *et al.*, 1971; Bowes and Ogren, 1972; Andrews *et al.*, 1973; Lorimer *et al.*, 1973; also see Additional References). [Eq. (16)].

$$
\begin{array}{l}
CH_2O(P) \\
| \\
C=O \\
| \\
CHOH \\
| \\
CHOH \\
| \\
CH_2O(P)
\end{array}
\quad + \quad O_2 \quad + \quad H_2O \quad \longrightarrow \quad
\begin{array}{l}
CH_2O(P) \\
| \\
COO^-
\end{array}
\quad + \quad
\begin{array}{l}
CH_2O(P) \\
| \\
CHOH \\
| \\
COO^-
\end{array}
\quad + \quad 2H^+ \quad (16)
$$

RuDP   P-glycolate   3-P-glycerate

Some doubts about the capacity of this reaction to account for phosphoglycolate production *in vivo* have been resolved with the demonstration that the physiological form of RuDP carboxylase (see Section II,B,4) has similar pH optima for carboxylation and oxygenation, and an oxygenase activity about 30% of the carboxylase activity at physiological CO_2 and O_2 concentrations (Bahr and Jensen, 1974c). Badger *et al.* (1974) have confirmed these observations and clearly demonstrated that O_2 and CO_2 compete for RuDP so that the relative rates of oxygenation and carboxylation depend upon the ratio of CO_2 to O_2. As shown in Fig. 6, phosphoglycolate formed in chloroplasts is converted to gly-

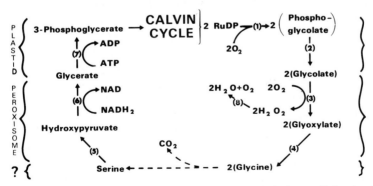

Fig. 6. Reactions of the glycolate pathway and their intracellular location. The enzymes involved are (1) RuDP carboxylase (oxygenase activity), (2) phosphoglycolate phosphatase, (3) glycolate oxidase, (4) and (5) aminotransferases, (6) glycerate dehydrogenase, (7) glycerate kinase, and (8) catalase. The location and nature of the reactions converting glycine to serine are uncertain (see text).

colate by a specific phosphoglycolate phosphatase located in chloroplasts (Tolbert, 1971a).

Reactions involved in the subsequent metabolism of glycolate to glycerate, with the probable exception of the glycine to serine steps, occur in subcellular particles called peroxisomes (Tolbert, 1971a). Similar organelles occur in most tissues and are alternatively called microbodies or glyoxysomes depending upon their orgin, enzyme content, and particular function (Beevers, 1971; Tolbert, 1971a). The process metabolizing glycolate, known as the glycolate pathway, is outlined in Fig. 6. Glycolate entering the peroxisome is oxidized to glyoxylate via the enzyme glycolate oxidase [Eq. (17)]. This reaction consumes a molecule of oxygen and produces H_2O_2, which is cleaved by catalase [Eq. (18)].

$$\begin{array}{l} CH_2OH \\ | \\ COO^- \end{array} + O_2 \longrightarrow \begin{array}{l} CHO \\ | \\ COO^- \end{array} + H_2O_2 \qquad (17)$$

Glycolate   Glyoxylate

$$H_2O_2 \longrightarrow H_2O + \tfrac{1}{2}O_2 \qquad (18)$$

As a result of these reactions one atom of oxygen is consumed for each molecule of glycolate oxidized, or one molecule of oxygen for two molecules of glycolate. It is more convenient to consider the scheme with the latter stoichiometry, since two molecules of glycolate must be metabolized to give one molecule of glycerate (Fig. 6). Glyoxylate is converted to glycine via an aminotransferase utilizing glutamate or serine as the amino donor. Isolated peroxisomes contain all three of these enzymes and readily convert glycolate to glycine (Tolbert, 1971a).

Reactions involving 2 molecules of glycine, or a molecule each of glycine and glyoxylate, have been proposed to account for the formation of serine with accompanying evolution of CO_2 (Tolbert, 1971b). The process favored involves a series of reactions in which an intermediate C_1 tetrahydrofolate derivative is formed by transfer of the methyl group of glycine leaving the carboxyl as CO_2. After transformation to 5,10-methylenetetrahydrofolate, the C_1 can be transferred to a second molecule of glycine to give serine, with the C-3 (hydroxymethyl) originating from the C_1. The latter reaction is catalyzed by serine hydroxymethyltransferase and most of this enzyme in leaf extracts appears to be associated with mitochondria (Tolbert, 1971b). The evidence that peroxisomes are not responsible for this phase of the process rests primarily on their inability to metabolize added glycolate beyond glycine.

Peroxisomes, however, do contain the enzymes necessary for the conversion of serine to glycerate (Tolbert, 1971a). Serine is converted to hydroxypyruvate via an aminotransferase and hydroxypyruvate is reduced to glycerate via an NAD-specific glycerate dehydrogenase [Eq. (19)].

$$
\begin{array}{c}
\text{CH}_2\text{OH} \\
| \\
\text{C}{=}\text{O} \\
| \\
\text{COO}^-
\end{array}
\;+\; \text{NADH} \;\rightleftarrows\;
\begin{array}{c}
\text{CH}_2\text{OH} \\
| \\
\text{CHOH} \\
| \\
\text{COO}^-
\end{array}
\;+\; \text{NAD}^+
\tag{19}
$$

Hydroxypyruvate   Glycerate

Finally, glycerate is phosphorylated via a specific glycerate kinase in chloroplasts to yield 3-P-glycerate, which can then be metabolized by the Calvin cycle.

The first evidence that a significant amount of photosynthetically fixed carbon passes through this pathway was provided by radiotracer studies (see Tolbert, 1963; Zelitch, 1964; Jackson and Volk, 1970). These studies showed that $^{14}\text{CO}_2$ fixed during photosynthesis rapidly appears in glycolate pathway intermediates, and that this carbon ultimately reappears in carbohydrate. Provision of specifically labeled intermediates provided further information about the sequence of reactions and showed that evolved CO_2 was derived originally from the C-1 (carboxyl) of glycolate and the C-3 of serine from the C-2 of glycolate. As already indicated, studies on the identification and location of enzymes have provided information about the nature and intracellular site of the individual reactions involved (see Tolbert, 1971a,b).

C. Function

The identification of the oxygenation of RuDP as the source of glycolate provides the basis for the first plausible and consistent rationale for photorespiration. As suggested by Lorimer and Andrews (1973), phosphoglycolate production via RuDP carboxylase appears to be an unavoidable consequence of this enzyme operating in air. With all its ingenuity, evolution has apparently been unable to modify this enzyme to reduce the effectiveness of O_2 as an alternative substrate to CO_2. Instead, the glycolate pathway operates as a kind of metabolic salvage process to recover at least a major part of the carbon diverted into glycolate by oxygenation of RuDP (Tolbert, 1971b; Lorimer and Andrews, 1973).

Irrespective of its function, photorespiration is costly. In terms of carbon metabolism some 20 to 40% of the total carbon fixed by photosynthesis is almost immediately lost again as CO_2 via photorespiration (see Tolbert, 1971a; Black, 1973). Therefore, the real energy cost in terms of ATP and NADPH required for CO_2 assimilation could increase by up to twice the theoretical requirement for the Calvin cycle of 3 ATP and 2 NADPH for each CO_2 fixed (Hatch, 1970).

V. Photosynthetic Pathways and Other Characteristics

A. Carbon Isotope Discrimination

One of the more recently discovered characteristics distinguishing C_4 pathway and Calvin cycle species is the differing ^{12}C to ^{13}C ratios in their organic carbon. This varying discrimination against ^{13}C is relevant in the present context because it presumably reflects the operation of different physical or chemical events leading to CO_2 assimilation. Species utilizing the C_4 pathway discriminate less against ^{13}C than Calvin cycle species (Bender, 1971; Smith and Epstein, 1971; also see Black, 1973). The δ $^{13}C\%_o$, a measure of this discrimination, ranges between -10 and $-18\%_o$ for C_4 species compared with -24 to $-30\%_o$ for Calvin cycle species. Notably, the δ $^{13}C\%_o$ values for plants capable of dark CO_2 fixation via CAM are similar to those for C_4 plants when assimilation is largely via the CAM mode, but similar to C_3 plants when daytime photosynthesis is the main source of assimilated CO_2 (Bender *et al.*, 1973; Osmond *et al.*, 1973).

Recent investigations explain the differences in isotope content of these plants in terms of differences in discrimination shown by PEP carboxylase and RuDP carboxylase toward $^{13}CO_2$ (Whelan *et al.*, 1973). Varying growth conditions, including varying initial ratios of isotope in the source of CO_2, and the differing isotope content of different classes of compounds, would contribute to the spread of δ $^{13}C\%_o$ values within C_3 and C_4 groups.

B. Ecology and Physiology

There are few ecological or geographic situations that are unique for species with a particular photosynthetic pathway, but some broad generalizations can be made (see Hatch *et al.*, 1971). Calvin cycle species can be found at practically all latitudes but predominate in the temperate and subtemperate region. They occur in both shaded and open habitats, but are less common in more arid situations. The C_4 pathway species are largely confined to tropical and subtropical regions. They occur in moist and arid locations, but often predominate in more arid areas. They are rarely found in shaded habitats. Species with CAM are largely restricted to arid tropical and subtropial areas.

Not surprisingly, there is an obvious correlation between the geographic and ecological occurrence of these different groups of plants and their special physiological features. Thus, in average conditions, C_4 pathway species are about twice as economical in the use of water per unit of dry matter fixed as Calvin cycle species, and this difference becomes even more pronounced as temperature and light intensity increase (see Black *et al.*, 1969; Downes, 1969, 1970). The C_4 pathway species also have much higher light and temperature optima for photosynthesis (Björkman, 1971). Under respective optimal conditions, the maximum photosynthesis rates per unit of chlorophyll for C_4 pathway species is generally about twice or more the rates for Calvin cycle species (see Hatch and Slack, 1970b; Hatch *et al.*, 1971). Although this difference is not manifested in terms of growth under all conditions (Bull, 1971; Slatyer, 1970; Gifford, 1974), its influence is apparent in the field performance of a wide range of crop and pasture species (Cooper, 1970; Stewart, 1970). A high potential for photosynthesis, and hence growth, could provide an advantage in two different situations. One would be for immediate survival in a highly competitive situation associated with ideal growing conditions. The other would be in generally arid situations where suitable conditions for growth may occur only intermittently, and rapid establishment may be vital for completion of a life cycle. Of course, under other conditions survival may be better served by economizing on water use at the expense of achieving maximum photosynthesis rates.

In Section II,C,4 arguments were advanced to explain some of the special physiological features of C_4 plants in terms of their unique biochemical processes. Their higher capacity for photosynthesis was related to reduced photorespiration, due in turn to their ability to concentrate CO_2 in bundle sheath cells. The higher water use efficiency of these species was related to the operation of PEP carboxylase, and their consequent ability to rapidly fix CO_2 inspite of the high resistance to diffusion of gases through the leaf stomata.

C. Taxonomy

Probably all plant families contain species that photosynthesize directly via the Calvin cycle. C_4 pathway species are more restricted in distribution but are now known to occur in at least fifteen families. These include the monocotyledonous families Gramineae and Cyperaceae and the dicotyledonous families, Acanthaceae, Aizoaceae, Amaranthaceae, Boraginaceae, Capparidaceae, Caryophyllaceae, Chenopodiaceae, Compositae, Euphorbiaceae, Nyctaginaceae, Portulaceae, Scrophulariaceae and Zygophyllaceae. Known C_4 pathway species have recently been listed [see Downton (1975) in Additional References). Species with CAM are also distributed among a number of plant families. These families, and those with C_4 pathway representatives occur in several advanced but divergent orders (Evans, 1971). Despite this divergence, there is a distinct taxonomic relationship between the C_4 pathway and CAM at this level. In fact, species with the C_4 pathway and CAM can occur within the same family and even the same genus (Downton, 1971).

Currently, it would seem that no families are exclusive for C_4 pathway species, although such exclusivity probably prevails at the generic level and possibly within tribes or subfamilies. For instance, the Pooid subfamily of Gramineae is apparently exclusive to Calvin cycle species, whereas the species of the Panicoid and Eragrostoid subfamilies are overwhelmingly of the C_4 pathway type. On the other hand, C_4 pathway and Calvin cycle species can occur within the same genus (e.g., *Panicum, Cyperus, Atriplex, Kochia, Euphorbia, Alternanthera, Bassia, Evolvulus, Mullugo,* and *Suaeda*).

D. Evolution and Genetics

Most evolutionary arrangements place families with the C_4 pathway and CAM among the most highly evolved. Furthermore, there is no evidence for the operation of either process in primitive plants or algae. Therefore, it would seem reasonable to propose that species with the C_4 pathway or CAM evolved from Calvin cycle species, probably in response to selection pressures associated with situations where higher temperatures, higher light, and restricted water prevail (see Section V,B).

Earlier evidence suggested that evolution of these processes did not involve major genetic changes. For instance, species with the C_4 pathway or CAM can be found in the same genus as Calvin cycle species, and both processes apparently evolved separately in several families. For the C_4 pathway, at least, this view was confirmed when a C_4 pathway *Atriplex* species was successfully hybridized with a Calvin cycle *Atriplex* (Björkman *et al.*, 1971). Using the C_4 pathway species as the female parent, the F_1 was intermediate morphologically, anatomically, and in respect to the levels of some key photosynthetic enzymes. However, photosynthesis rates were much lower than for either parent, and while some CO_2 was fixed into C_4 acids, this carbon was not transferred to 3-P-glycerate and sugars as it is in normal C_4 pathway species. There was segregation for most characters among the F_2 and F_3 hybrids, but maximum photosynthesis rates in normal air were low in all individuals examined. These included some with high PEP carboxylase, a leaf anatomy typical of the C_4 parent, and the capacity to incorporate $^{14}CO_2$ into

C_4 acids. However, other studies indicated that an integrated C_4 pathway was not operative in those species and that most of the net assimilation of CO_2 was due directly to the Calvin cycle.

What qualitative genetic differences exist between Calvin cycle and C_4 pathway species is not known. Of the enzymes involved in the unique reactions of the C_4 pathway only pyruvate,P_i dikinase appears likely to be absent from Calvin cycle species (see Section II,C,2). Although the other enzymes have functional counterparts in leaves of Calvin cycle species, their activity in C_4 pathway species are at least 10 and up to 100 times higher (see Table I); it is assumed that this additional activity represents that involved specifically in C_4 pathway photosynthesis. This component of activity can have a different intracellular location (see Section II,C,3b) and, being involved in a different metabolic process, may also require different kinetic and regulatory characteristics. These photosynthetic enzymes could therefore be comprised of genetically unique isoenzymes, evolved specifically to operate in the C_4 pathway. Some support for this view has been provided by recent studies on the aminotransferases (Hatch and Mau, 1973), PEP carboxylase (Ting and Osmond, 1973b), and NAD-malic enzyme (Hatch and Kagawa, 1974b) from C_4 species. Of course, mutations may also be necessary for other special features associated with this pathway, such as the transport processes or the anatomical and chloroplast modifications.

REFERENCES

Akazawa, T. (1970). *Progr. Phytochem.* **2**, 107.

Anderson, L. E. (1973). *Biochim. Biophys. Acta* **321**, 484.

Andrews, T. J., and Hatch, M. D. (1969). *Biochem. J.* **114**, 117.

Andrews, T. J., Johnson, H. S., Slack, C. R., and Hatch, M. D. (1971). *Phytochemistry* **10**, 2005.

Andrews, T. J., Lorimer, G. H., and Tolbert, N. E. (1973). *Biochemistry* **12**, 11.

Badger, M. R., Andrews, T. J., and Osmond, C. B. (1974). *Proc. Int. Congr. Photosyn., 3rd 1974*, p. 1421. Elsevier Amsterdam.

Bahr, J. T., and Jensen, R. G. (1974a). *Plant Physiol.* **53**, 39.

Bahr, J .T., and Jensen, R. G. (1974b). *Biochem. Biophys. Res. Comun.* **57**, 1180.

Bahr, J. T., and Jensen, R. G. (1974c). *Arch. Biochem. Biophys.* **164**, 408.

Bassham, J A (1964). *Annu. Rev. Plant Physiol.* **15**, 101.

Bassham, J. A. (1965). *In* "Plant Biochemistry" (J. Bonner and J. E. Varner, eds.), 2nd ed., p. 875. Academic Press, New York.

Bassham, J. A. (1971). *Science* **172**, 526.

Bassham, J. A., and Jensen, R. G. (1967). *In* "Harvesting the Sun" (A. San Pietro, F. A. Greer, and A. T. Army, eds.), p. 79. Academic Press, New York.

Bassham, J. A., and Kirk, M. (1968). *In* "Comparative Biochemistry and Biophysics of Photosynthesis" (K. Shibata *et al.*, eds.), p. 365. Univ. of Tokyo Press, Tokyo.

Beevers, H. (1971). *In* "Photosynthesis and Photorespiration" (M. D. Hatch, C. B.

Osmond, and R. O. Slatyer, eds.), p. 483. Wiley (Interscience), New York.

Beevers, H., Stiller, M. L., and Butt, V. S. (1966). *In* "Plant Physiology" (F. C. Steward, ed.), Vol. 4B, p. 119. Academic Press, New York.

Bender, M. (1971). *Phytochemistry* **10**, 1239.

Bender, M., Rouhani, I., Vines, H. M., and Black, C. C. (1973). *Plant Physiol.* **52**, 427.

Berry, J. A., Downton, W. J. S., and Tregunna, E. B. (1970). *Can. J. Bot.* **48**, 777.

Bidwell, R. G., Levin, W. B., and Shepard, D. C. (1970). *Plant Physiol.* **45**, 70.

Bird, I. F., Porter, H. K., and Stocking, C. (1965). *Biochim. Biophys. Acta* **100**, 366.

Björkman, O. (1970). *In* "Prediction and Measurement of Photosynthetic Productivity" (I. Setlik, ed.), p. 267. Pudoc, Wageningen.

Björkman, O. (1971). *In* "Photosynthesis and Photorespiration" (M. D. Hatch, C. B. Osmond, and R. O. Slatyer, eds.), p. 18. Wiley (Interscience), New York.

Björkman, O., and Gauhl, E. (1969). *Planta* **88**, 197.

Björkman, O., Nobs, M., Pearcy, R., Boynton, J., and Berry, J. (1971). *In* "Photosynthesis and Photorespiration" (M. D. Hatch, C. B. Osmond, and R. O. Slatyer, eds.), p. 105. Wiley (Interscience), New York.

Black, C. C. (1973). *Annu. Rev. Plant Physiol.* **24**, 253.

Black, C. C., Chen, T. M., and Brown, R. H. (1969). *Weed Sci.* **17**, 338.

Bowes, C., and Ogren, W. L. (1972). *J. Biol. Chem.* **247**, 2171.

Bowes, C., Ogren, W. L., and Hageman, R. H. (1971). *Biochem. Biophys. Res. Commun.* **45**, 716.

Bradbeer, J. W., Ranson, S. L., and Stiller, M. L. (1958). *Plant Physiol.* **3**, 66.

Buchanan, B. B., and Schurman, P. (1972). *FEBS (Fed. Eur. Biochem. Soc.) Lett.* **23**, 157.

Buchanan, B. B., Kalberer, P. P., and Arnon, D. I. (1967). *Biochem. Biophys. Res. Commun.* **29**, 74.

Bull, T. A. (1971). *In* "Photosynthesis and Photorespiration" (M. D. Hatch, C. B. Osmond, and R. O. Slatyer, eds.), p. 68. Wiley (Interscience), New York.

Bulley, N. R., and Tregunna, E. B. (1971). *Can. J. Bot.* **49**, 1277.

Burr, G. O. (1962). *Int. J. Appl. Radiat. Isotop.* **13**, 365.

Calvin, M., and Bassham, J. A. (1962). "The Photosynthesis of Carbon Compounds." Benjamin, New York.

Chen, T. M., Brown, R. H., and Black, C. C. (1971). *Plant Physiol.* **47**, 199.

Chen, T. M., Campbell, W., Dittrich, P., and Black, C. C. (1973). *Biochem. Biophys. Res. Commun.* **51**, 461.

Chu, D. K., and Bassham, J. A. (1972). *Plant Physiol.* **50**, 224.

Clark, M. G., Williams, J. F., and Blackmore, P. F. (1974). *Catal. Rev.* **9**, 35.

Coombs, J., Baldry, C. W., and Bucke, C. (1973). *Planta* **110**, 95.

Cooper, J. P. (1970). *Herb. Abstr.* **40**, 1.

Cooper, T. G., Filmer, D., Wishnick, M., and Lane M. D. (1969). *J. Biol. Chem.* **244**, 1081.

Dittrich, P., Campbell, W. H., and Black. C. C. (1973). *Plant Physiol.* **52**, 357.

Downes, R. W. (1969). *Planta* **88**, 216.

Downes, R. W. (1970). *Aust. J. Biol. Sci.* **23**, 775.

Downton, W. J. S. (1971). *In* "Photosynthesis and Photorespiration" (M. D. Hatch, C. B. Osmond, and R. O. Slatyer, eds.), p. 3. Wiley (Interscience), New York.

Edwards G. E., and Black, C. C. (1971a). *In* "Photosynthesis and Photorespiration"

(M. D. Hatch, C. B. Osmond, and R. O. Slatyer, eds.), p. 153. Wiley (Interscience), New York.

Edwards, G. E., and Black, C. C. (1971b). *Plant Physiol.* **47**, 149.

Edwards, G. E., Lee, S. S., Chen, T. M., and Black, C. C. (1970). *Biochem. Biophys. Res. Commun.* **39**, 389.

Edwards, G. E., Kanai, R., and Black, C. C. (1971). *Biochem. Biophys. Res. Commun.* **45**, 278.

Enns, T. (1967). *Science* **155**, 44.

Evans, L. T. (1971). *In* "Photosynthesis and Photorespiration" (M. D. Hatch, C. B. Osmond, and R. O. Slatyer, eds.), p. 130. Wiley (Interscience), New York.

Everson, R. G. (1970). *Phytochemistry* **9**, 25.

Everson, R. G. (1971). *In* "Photosynthesis and Photorespiration" (M. D. Hatch, C. B. Osmond, and R. O. Slatyer, eds.), p. 275. Wiley (Interscience), New York.

Everson, R. G., and Graham, D. (1971). *In* "Photosynthesis and Photorespiration" (M. D. Hatch, C. B. Osmond, and R. O. Slatyer, eds.), p. 281. Wiley (Interscience), New York.

Everson, R. G., and Slack, C. R. (1968). *Phytochemistry* **7**, 581.

Everson, R. G., Cockburn, W., and Gibbs, M. (1967). *Plant Physiol.* **42**, 840.

Fairneau, J. (1971). *In* "Photosynthesis and Photorespiration" (M. D. Hatch, C. B. Osmond, and R. O. Slatyer, eds.), p. 202. Wiley (Interscience), New York.

Garnier-Dardart, J. (1965). *Physiol. Veg.* **3**, 215.

Gibbs, M. (1967). *Annu. Rev. Biochem.* **36**, 757.

Gibbs, M., ed. (1971). "Structure and Function of Chloroplasts." Springer-Verlag, Berlin and New York.

Gifford, R. M. (1974). *Aust. J. Plant Physiol.* **1**, 107.

Graham, D., Atkins, C. A., Reed, M. L., Patterson, B. D., and Smillie, R. M. (1971). *In* "Photosynthesis and Photorespiration" (M. D. Hatch, C. B. Osmond, and R. O. Slatyer, eds.), p. 267. Wiley (Interscience), New York.

Hatch, M. D. (1970). *In* "Prediction and Measurement of Photosynthetic Productivity" (I. Setlik, ed.), p. 215. Pudoc, Wageningen.

Hatch, M. D. (1971a). *Biochem. J.* **125**, 425.

Hatch, M. D. (1971b). *In* "Photosynthesis and Photorespiration" (M. D. Hatch, C. B. Osmond, and R. O. Slatyer, eds.), p. 139. Wiley (Interscience), New York.

Hatch, M. D. (1973). *Arch. Biochem. Biophys.* **156**, 207.

Hatch, M. D., and Boardman, N. K. (1973). *In* "Chemistry and Biochemistry of Herbage" (G. W. Butler and R. W. Bailey, eds.), Vol. 2, p. 25. Academic Press, New York.

Hatch, M. D., and Kagawa, T. (1973). *Arch. Biochem. Biophys.* **159**, 842.

Hatch, M. D., and Kagawa, T. (1974a). *Arch. Biochem. Biophys.* **160**, 346.

Hatch, M. D., and Kagawa, T. (1974b). *Aust. J. Plant Physiol.* **1**, 357.

Hatch, M. D., and Mau, S. (1973). *Arch. Biochem. Biophys.* **156**, 195.

Hatch, M. D., and Slack, C. R. (1966). *Biochem. J.* **101**, 103.

Hatch, M. D., and Slack, C. R. (1968). *Biochem. J.* **106**, 141.

Hatch, M. D., and Slack, C. R. (1969a). *Biochem. J.* **112**, 549.

Hatch, M. D., and Slack, C. R. (1969b). *Biochem. Biophys. Res. Commun.* **34**, 589.

Hatch, M. D., and Slack, C. R. (1970a). *Annu. Rev. Plant Physiol.* **21**, 141.

Hatch, M. D., and Slack, C. R. (1970b). *Progr. Phytochem.* **2**, 35.

Hatch, M. D., Slack, C. R., and Johnson, H. S. (1967). *Biochem. J.* **102**, 417.

Hatch, M. D., Slack, C. R., and Bull, T. A. (1969). *Phytochemistry* **8**, 697.

Hatch, M. D., Osmond, C. B., and Slatyer, R. O., eds. (1971). "Photosynthesis and Photorespiration." Wiley (Interscience), New York.

Heber, U. (1970). *In* "Transport and Distribution of Matter in Cells of Higher Plants" (K. Mothes *et al.*, eds.), p. 151. Abh. Deut. Akad. Wiss. Berlin.

Heber, U. (1974). *Annu. Rev. Plant Physiol.* **25**, 393.

Heber, U., and Santarius, K. A. (1970). *Z. Naturforsch. B* **25**, 718.

Heber, U., and Willenbrink, J. (1964). *Biochim. Biophys. Acta* **82**, 393.

Huang, A. H., and Beevers, H. (1972). *Plant. Physiol.* **50**, 242.

Jackson, W. A., and Volk, R. J. (1970). *Annu. Rev. Plant. Physiol.* **21**, 385.

Jensen, R. G., and Bassham, J. A. (1966). *Proc. Nat. Acad. Sci. U.S.* **56**, 1095.

Johnson, H. S., and Hatch, M. D. (1968). *Phytochemistry* **7**, 375.

Johnson, H. S., and Hatch, M. D. (1969). *Biochem. J.* **114**, 127.

Johnson, H. S., and Hatch, M. D. (1970). *Biochem. J.* **119**, 273.

Kagawa, T., and Hatch, M. D. (1974a). *Aust. J. Plant Physiol.* **1**, 51.

Kagawa, T., and Hatch, M. D. (1974b). *Biochem. Biophys. Res. Commun.* **59**, 1326.

Kanai, R., and Edwards, G. E. (1973a). *Naturwissenschaften* **60**, 157.

Kanai, R., and Edwards, G. E. (1973b). *Plant Physiol.* **51**, 1133.

Karpilov, Y. S. (1960). *Tr. Kazan. Agricult. Inst.* **41**, 1.

Karpilov, Y. S. (1970). *Proc. Moldavian Inst. Irrigation Veg. Res.* **11**, (Pt. 3), 1.

Kawashima, N., and Wildman, S. G. (1970). *Biochem. Biophys. Res. Commun.* **41**, 1463.

Kirk, J. T. O. (1970). *Annu. Rev. Plant Physiol.* **21**, 11.

Kirk, J. T. O., and Tilney-Bassett, R. A. E. (1967). "The Plastids." Freeman, San Francisco, California.

Kluge, M., and Osmond, C. B. (1971). *Naturwissenshaften* **58**, 414.

Kortschak, H. P., Hartt, C. E., and Burr, G. O. (1965). *Plant Physiol.* **40**, 209.

Krause, G. H. (1971). *Z. Pflanzenphysiol.* **65**, 13.

Laetsch, W. M. (1971). *In* "Photosynthesis and Photorespiration" (M. D. Hatch, C. B. Osmond, and R. O. Slatyer, eds.), p. 309. Wiley (Interscience), New York.

Latzko, E., and Gibbs, M. (1968). *Z. Pflanzenphysiol.* **59**, 184.

Latzko, E., Garnier, R., and Gibbs, M. (1970). *Biochem. Biophys. Res. Commun.* **38**, 1140.

Lorimer, G. H., and Andrews, T. J. (1973). *Nature (London)* **243**, 359.

Lorimer, G. H., Andrews, T. J., and Tolbert, N. E. (1973). *Biochemistry* **12**, 18.

Lowe, J., and Slack, C. R. (1971). *Biochim. Biophys. Acta* **235**, 207.

Mukerji, S. K., and Ting, I. P. (1968). *Phytochemistry* **7**, 903.

Müllhofer, G., and Rose, I. A. (1965). *J. Biol. Chem.* **240**, 1341.

Ogren, W. L., and Bowes, G. (1971). *Nature (London)* **230**, 159.

O'Neal, D., Latzko, E., and Gibbs, M. (1972). *Plant Physiol.* **49**, 607.

Osmond, C. B. (1971). *Aust. J. Biol. Sci.* **24**, 159.

Osmond, C. B., and Avadhani, P. N. (1970). *Plant Physiol.* **45**, 228.

Osmond, C. B., and Harris, B. (1971). *Biochim. Biophys. Acta* **234**, 270.

Osmond, C. B., Allaway, W. G., Sutton, B. G., Troughton, J. H., Queiroz, O., Luttge, U., and Winter, K. (1973). *Nature (London)* **246**, 41.

Preiss, J., and Kosuge, T. (1970). *Annu. Rev. Plant. Physiol.* **21**, 133.

Ranson, S. L., and Thomas, M. (1960). *Annu. Rev. Plant Physiol.* **11**, 81.

Rehfeld, D. W., Randall, D. D., and Tolbert, N. E. (1970). *Can. J. Bot.* **48**, 1219.

Salin, M. L., Campbell, W. H., and Black, C. C. (1973). *Proc. Nat. Acad. Sci. U.S.* **70**, 3730.

Seigel, M. I., and Lane, M. D. (1973). *J. Biol. Chem.* **248**, 5486.

Shah, S. P. J., and Rogers, L. J. (1969). *Biochem. J.* **114**, 395.

Singer, S. J., Eggman, L., Campbell, J. M., and Wildman, S. G. (1952). *J. Biol. Chem.* **197**, 233.

Slack, C. R., Hatch, M. D., and Goodchild, D. J. (1969). *Biochem. J.* **114**, 489.

Slatyer, R. O. (1970). *Planta* **93**, 175.

Smillie, R. M. (1963). *Can. J. Bot.* **41**, 123.

Smillie, R. M., and Scott, N. S. (1969). *In* "Progress in Molecular and Subcellular Biology" (F. E. Hahn, ed.), p. 136. Springer-Verlag, Berlin and New York.

Smith, B. N., and Epstein, S. (1971). *Plant. Physiol.* **47**, 380.

Steiger, E., Ziegler, I., and Ziegler, H. (1971). *Planta* **96**, 109.

Stewart, G. A. (1970). *J. Aust. Inst. Agr. Sci.* **26**, 85.

Stocking, C. R., Williams, G. R., and Ongun, A. (1963). *Biochem. Biophys. Res. Commun.* **10**, 416.

Sutton, B. G., and Osmond, C. B. (1972). *Plant Physiol.* **50**, 360.

Tarchevskii, I. A., and Karpilov, Y. S. (1963). *Fiziol. Rast.* **10**, 229.

Ting, I. P. (1971). *In* "Photosynthesis and Photorespiration" (M. D. Hatch, C. B. Osmond, and R. O. Slatyer, eds.), p. 169. Wiley (Interscience), New York.

Ting, I. P., and Osmond, C. B. (1973a). *Plant Physiol.* **51**, 439.

Ting, I. P., and Osmond C. B. (1973b). *Plant Physiol.* **51**, 448.

Ting, I. P., Rocha, V., Mukerji, S. K., and Curry, R. (1971). *In* "Photosynthesis and Photorespiration" (M. D. Hatch, C. B. Osmond, and R. O. Slatyer, eds.), p. 534. Wiley (Interscience), New York.

Tolbert, N. E. (1963). *Nat. Acad. Sci.—Nat. Res. Counc., Publ.* **1145**, 648.

Tolbert, N. E. (1971a). *Annu. Rev. Plant Physiol.* **22**, 45.

Tolbert, N. E. (1971b). *In* "Photosynthesis and Photorespiration" (M. D. Hatch, C. B. Osmond, and R. O. Slatyer, eds.), p. 458. Wiley (Interscience), New York.

Walker, D. A. (1973). *New Phytol.* **72**, 209.

Walker, D. A., and Crofts, A. R. (1970). *Annu. Rev. Biochem.* **39**, 389.

Whelan, T., Sackett, W. M., and Bendict, C. R. (1973). *Plant Physiol.* **51**, 1051.

Winkenback, F., Parthasarathy, M. V. and Bidwell, R. G. S. (1972). *Can. J. Bot.* **50**, 1367.

Wolfe, F. T. (1970). *Advan. Front. Plant Sci.* **26**, 161.

Zelitch, I. (1964). *Ann. Rev. Plant Physiol.* **15**, 121.

Zelitch, I. (1971). "Photosynthesis, Photorespiration and Plant Productivity." Academic Press, New York.

Ziegler, H., Ziegler, I., Muller, B., and Dorr, I. (1969). *Prog. Photosynthesis Res.* **3**, 1636.

ADDITIONAL REFERENCES

The following are some recent references that relate to important issues discussed in this chapter and which appeared after this chapter was written and revised.

Ribulose-1,5-diphosphate carboxylase, affinity for CO_2, oxygenase activity, and glycolate production:

Badger, M. R., and Andrews, T. J. (1974). *Biochem. Biophys. Res. Commun.* **60**, 204.

Laing, W. A., Ogren, W. L., and Hageman R. H. (1974). *Plant Physiol.* **54**, 687.

Laing W. A., Ogren, W. L., and Hageman, R. H. (1975). *Biochemistry* **14**, 2269.

Andrews, T. J., Badger, M. R., and Lorimer, G. H. (1975). *Arch. Biochem. Biophys.* **171**, 103.

Site of sucrose synthesis:
Bird, I. F., Cornelius, M. J., Keys, A. J., and Whittingham, C. P. (1974). *Phytochemistry* **13**, 59.
Heber, U. (1974). *Annu. Rev. Plant Physiol.* **25**, 393.

C_4 pathway photosynthesis reviews:
Laetsch, W. M. (1974). *Annu. Rev. Plant Physiol.* **25**, 27.
Hatch, M. D. and Osmond, C. B. (1976). *In* "Encyclopedia of Plant Physiology," (New Series) (U. Heber and C. R. Stocking, eds.). Springer-Verlag, Heidelberg. In press.
Hatch, M. D. (1976). In "CO_2 Metabolism and Productivity," (C. C. Black and R. H. Burris, eds.). University Park Press, Baltimore. In press.

Inter- and intracellular location of C_4 pathway enzymes and processes:
Ku, S. B., Gutierrez, M., and Edwards, G. E. (1974). *Planta* **119**, 267.
Chen, T. M., Dittrick, P., Campbell, N. H., and Black, C. C. (1974). *Arch. Biochem. Biophys.* **163**, 246.
Gutierrez, M., Kanai, R., Huber, S. C., Ku, S. B., and Edwards, G. E. (1974). *Z. Pflanzenphysiol.* **72**, 305.
Gutierrez, M., Huber, S. C., Ku, S. B., Kanai, R., and Edwards, G. E. (1975). *In* "Proceedings Third International Congress on Photosynthesis," (M. Avron ed.), pp. 1219–1230. Elsevier, Amsterdam.
Kagawa, T., and Hatch, M. D. (1975). *Arch. Biochem. Biophys.* **167**, 687.
Huber, S. C., and Edwards, G. E. (1975). *Plant Physiol.* **55**, 835.

Differing C_4-pathway metabolism and subdivision of C_4 plants:
Gutierrez, M., Gracen, V. E., and Edwards, G. E. (1974). *Planta* **119**, 279.
Hatch, M. D., Kagawa, T., and Craig, S. (1975). *Aust. J. Plant Physiol.* **2**, 111.

Crassulacean acid metabolism:
Kluge, M., Kriebitzsch, C., and Willert, D. J. (1974). *Z. Pflanzenphysiol.* **72**, 460.
Sutton, B. G. (1975). *Aust. J. Plant Physiol.* **2**, 377, 389.

Photorespiration, see references above to RuDP oxygenase:
Chollet, R., and Ogren, W. L. (1975). *Bot. Rev.* **41**, 137.
Mahon, J. D., Fock, H., and Canvin, D. (1974). *Planta* **120**, 125, 245.
Cockburn, W., and McAulay, A. (1975). *Plant Physiol.* **55**, 87.

List of C_4 pathway species:
Downton, W. J. S. (1975). *Photosynthetica* **9**, 96.

25

Photosynthesis: The Path of Energy[*]

BESSEL KOK

---

[*] Abbreviations used in this chapter: ADP, adenosine diphosphate; ATP, adenosine triphosphate; CCP, carboxyl cyanide-m-chlorophenyl hydrazone; Cyt, cytochrome; DAD, diaminodurol; DCMU, 3-(3,4-dichlorophenyl)-1,1-dimethylurea; DPIP, DPIPH$_2$ 2,6-dichlorophenolindophenol and its reduced form; EDTA, ethylenediaminetetraacetic acid; ESR (EPR), electron spin resonance; FMN, FMNH$_2$, flavine mononucleotide and its reduced form; NAD$^+$, NADH, nicotinamide adenine dinucleotide and its reduced form; NADP$^+$, NADPH nicotinamide adenine dinucleotide phosphate and its reduced form; PCMB, p-chloromercuribenzoate; P$_i$, orthophosphate; PMS, phenazine methosulfate, PPNR, photosynthetic pyridine nucleotide reductase; PQ, plastoquinone; TMPD, N,N,N',N'-tetramethyl-p-phenylenediamine.

I. Introduction

The bulk of energy conversion on earth is carried out by oxygen-evolving plants (higher plants and algae). The main functional pigments, chlorophyll a and the accessory pigments, absorb all wavelengths shorter than 700 nm—or about half of the solar spectrum. Under optimal conditions, photosynthesis can convert up to 30% of absorbed radiant energy into chemical energy. For various reasons, however, under natural conditions $\leq 1\%$ of the solar energy reaching the earth's surface is routed through the plant kingdom and supports life, the remainder being wasted as heat. Still, the total energy conversion by photosynthesis exceeds by many times the total industrial output of man.

In plants CO_2 and H_2O are converted to organic material and O_2. This is accomplished through two complex series of reactions. The first involves the production by light of O_2 and stable high-energy compounds—reduced pyridine nucleotide (and perhaps other reductants) and adenosine triphosphate. In subsequent dark processes these high-energy compounds are used to reduce atmospheric CO_2 to carbohydrate, proteins, lipids, etc. (see Chapter 24). In this chapter, we shall discuss the transformations which convert light energy into chemical reducing power.

Light-driven electron transport is a structure-dependent process. The pigments, grouped in functional "units" with photochemical conversion centers and other catalysts, are located in membrane-bounded flattened sacs (thylakoids). In higher plants, these structures are concentrated in special organelles—the chloroplasts (see Chapter 6).

Much of our present knowledge of photosynthetic electron transport rests on observations of isolated chloroplasts. Such preparations can perform a host of photoconversions, which are often simpler and more amenable to study than the whole cell process.

Plants are able to produce molecular O_2 and at the same time a reductant as strong as or stronger than molecular H_2. As the potential* of the hydrogen electrode (E_m) is -0.42 V and that of the oxygen electrode $+0.81$ V, a total chemical potential > 1.2 V must be created through photosynthesis. Red quanta of about 700 nm (the wavelength that initiates the conversion) represent about 1.8 eV or, since 1 eV corresponds to 23 kcal/mole, about 40 kcal per quantum mole. If each quantum were to move one electron equivalent against the energy gradient over the full span of 1.2 V, the efficiency would be almost 70%. As we shall see, however, plants are not so efficient.

Photosynthetic bacteria, which will be given only cursory attention in this chapter, do not evolve O_2, generate only weak oxidizing power,

* All potential values quoted in this chapter are E_0 at pH 7.

and presumably have only a single photosystem (similar to photosystem I of plants, see Section IV,A).

The initial step in photosynthesis is the absorption of a quantum of light by a molecule of chlorophyll a or one of the accessory pigments leading to an excited state. Because of the dense packing of the light-harvesting pigments in the lamellae, the excited state can travel from one pigment molecule to the other until it hits a special chlorophyll a molecule, which carries out the conversion of light into chemical energy. Groups ("units") of roughly 200 harvesting pigment molecules serve each of these "trapping centers." The latter are specially bound in close prox-imity to appropriate electron donor and acceptor molecules. Figure 1 shows a simple but plausible scheme for the initial photoconversion. An electron of the trapping center chlorophyll a molecule, raised to an excited state by a quantum of red light, reduces electron acceptor A. Chloro-phyll a is left oxidized and is returned to its original state by electron donor D_{red}. As a result a photon has moved an electron against the en-ergy gradient; the backreaction $AHD^+ \rightarrow A^+DH$ could do work, if prop-erly coupled to an energy-requiring process, or would yield heat if allowed to run by itself.

Another aspect is important; the excited state Chl^* might contain all the energy of the photon which produced it, but with a lifetime of only 1 nanosecond, its stability is extremely limited. The chemical poten-tial of the products of the photochemical reaction is considerably less than that of the original excited state, but their stability is much greater.

The electron transfer chain is actually much more complex than Fig. 1 indicates. To cover the entire 1.2 V span and generate in addi-tion $\simeq 0.65$ eV ATP energy, two light reactions operate "in series" (using 2 photons per equivalent moved through the chain) as is visualized in

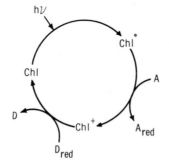

Fig. 1. Schematic illustration of a photoprocess in which chlorophyll (Chl) excited by a photon ($h\nu$) donates an electron to acceptor molecule A, and is subsequently reduced by donor molecule D_{red}. The net result is that A is photo-chemically reduced by D_{red}.

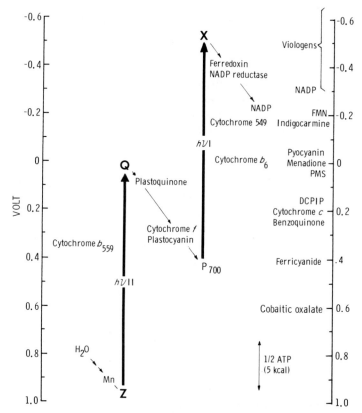

Fig. 2. Left: Potential diagram of the two photoacts. Chloroplast electron carriers are arranged according to their normal potential. Right: Examples of electron donors and acceptors in chloroplast reactions arranged according to their normal potential (see text).

Fig. 2. The first light reaction, photoact I, receives mainly photons that are absorbed by chlorophyll a. It produces a very strong reductant X⁻ and a weak oxidant P⁺. Photoact II is sensitized by both chlorophyll a and accessory pigments. It produces a weak reductant Q⁻ and a very strong oxidant Z⁺. To generate an O_2 molecule, four oxidizing equivalents (Z⁺) must cooperate, which requires a complex sequence of enzymatic dark reactions. At the reducing end, X⁻ does not reduce NADP directly, but via two enzymatic steps.

It thus is no surprise that in addition to the four primary photoproducts, only one of which has been identified so far, numerous other electron carriers are involved: Mn enzyme(s), plastoquinone(s), cytochromes, plastocyanine, ferredoxin, NADP reductase, and others. In Fig.

2, these are arranged according to their respective redox potentials, indicative of their sites of operation. In addition to and in parallel with the oxidation and reduction of specific intermediates, light energy is conserved in yet another way. In the double membrane, the individual reaction chains are arranged in an ordered oriented way. Both photochemical reactions move electrons from the inside to the outside. Oxygen is evolved and protons accumulate inside the thylakoid. The concerted action of many chains thus creates electric and concentration gradients across the membrane, which are utilized to drive the formation of ATP (one per equivalent moved through the chain).

Phosphorylation not only occurs "coupled" to electron transport from water to ferredoxin but it can also accompany abbreviated electron transport paths, which involve only one of the photochemical reactions or cyclic paths that yield no net change of electron donors or acceptors.

Photosynthesis is a very complicated process, the ramifications of which cover many areas of science. The following sections discuss a number of selected topics. Our aim is to provide the reader with sufficient background to fruitfully consult the original literature and its generally extensive review coverage.

II. Early Events

A. Absorption, Fluorescence

A light quantum or photon is a discrete package of light energy

$$E = h\nu$$

in which h is Planck's constant and ν is the frequency ($\nu = c/\lambda$, where c is the speed of light and λ the wavelength). For wavelengths of light utilized with high efficiency by green plant photosynthesis ($\sim$650 to 700 nm) this energy is about 1.8 eV/quantum, and a gram-mole of these quanta (1 einstein or 6.02×10^{23} quanta) has an energy of about 40 kcal.

Absorption of a photon by a pigment results in the promotion of an electron into a new orbital, with the absorbing molecule gaining an amount of energy equal to that which was present in the quantum. This excited state is unstable, however, and returns to the low-energy ground state with a definite decay or lifetime. The chance of a molecule absorbing a passing quantum is characterized by its extinction coefficient, i.e., its effective capture area or cross section. Figure 3 shows the absorption spectrum (solid line) of chlorophyll a in solution. The strong red and

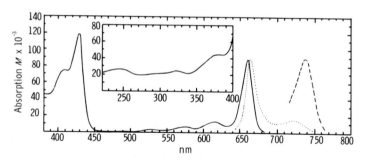

Fig. 3. Absorption spectra of chlorophyll a dissolved in ether (solid line). (Data from Holt and Jacobs, 1954.) Fluorescence emission spectrum (dotted line). Red absorption band of microcrystals (dashed line).

blue absorption bands at 660 and 430 nm (the first and second singlet excited state) have a high molar extinction coefficient (about 10^5 liters mole^{-1} cm^{-1} or 1.7×10^{-16} cm^2 per molecule).

An excited molecule can lose its energy in several ways. In the case of fluorescence, the electron returns to the ground state with emission of a quantum of about the same energy (wavelength) as the one that was absorbed. In a free atom, the electronic transition can appear unperturbed, the absorption and the emission occurring at one and the same wavelength, ν_0. In a complex molecule such as chlorophyll, excitation can lead to one of several slightly different energy states because other parts of the molecule are influenced besides the specific electron transition. These additional energy levels, which broaden the transition (ν_0) to a band of frequencies ($\nu_0 + \Delta\nu$), are rapidly dissipated, leaving the system with an amount of energy ν_0. When the electron falls back, some of the energy of the excited state is again diverted to the "nucleus" and wasted. Thus, the frequency of the emitted quanta (ν_e) is generally smaller than that of the exciting quantum ($\nu_e < \nu_0$).

Figure 3 shows that the red absorption band is some 15–20 nm wide and the fluorescence emission (dotted line) somewhat displaced to the long wave side of ν_0. Note in Fig. 3 that the fluorescence of chlorophyll a occurs exclusively in the red. The absorption of blue light results in exactly the same fluorescence as the absorption of red quanta. The second excited singlet state (430 nm) is extremely unstable (lifetime $\leq 10^{-12}$ second), and before fluorescence emission can occur, it converts into the first and lowest singlet. In this radiationless transition, the energy difference is wasted as thermal energy of the molecule.

The transition from the lowest singlet to the ground state is not always accompanied by the emission of a quantum. In dilute solution in organic solvents the fluorescence yield (quanta emitted/quanta absorbed) is only 0.3 for chlorophyll a and 0.1 for chlorophyll b (Livingston, 1960). The remaining excitations are dissipated as thermal energy.

The fluorescence yield can be greatly affected ("quenched") by the interaction of the pigment with other molecules at the moment the absorption act occurs. In the chloroplast, chlorophyll fluoresces much less ($\sim \frac{1}{10}$) than in solution. Since the fluorescence yield is an indicator of the lifetime of the excited state, this implies trapping mechanisms that quickly deprive the pigments of their energy—this time not to be dissipated as heat but put to work in photochemistry. Since one expects the fluorescence yield to be lower, the more efficient the photochemical trapping, this yield should be a useful indicator of energy flow. This expectation appears to be fulfilled in bacterial photosynthesis and in photosystem II of green plants (see Section IV,B). For unexplained reasons, the fluorescence from the pigment which sensitizes photosystem I is weak and unaffected by the state of the traps.

B. Primary Events

To do chemical work the energy of the excited state must be "caught" and converted into a stable and manageable form. In gases or solutions, an excited molecule can only meet a reaction partner by collision. Collisions, however, are rather infrequent during the brief lifetime of a singlet excited state. Photochemical conversions in solution, therefore, would be unlikely if it were not for the fact that in most pigment molecules longer-lived, metastable (triplet) excited states occurred.

The possibility has been considered that photosynthesis is initiated by the triplet excited state of chlorophyll. However, no evidence for this has been found in intact chloroplasts. To the contrary, it appears that the reactive triplet states are deliberately annihilated by the carotenoids that are ubiquitously dispersed among the chloroplast pigments. Chlorophyll triplets are efficiently transferred to carotenoids in which a rapid degradation to heat occurs (Chessin et al., 1966). The chlorophyll of carotenoidless plants is quite susceptible to photobleaching (Griffith et al., 1965).

It has become evident that in the photosynthetic apparatus the need for long-lived excitations is alleviated by special structural arrangements. A fraction of the pigment shows spatial orientation (Olson, 1963) and several phenomena have been observed which are reminiscent of "solid state" events—photoconductivity, trapping of photoproducts at low temperature, and delayed light emission.

Photosynthetically active materials reemit light long after cessation of an illumination (Strehler and Arnold, 1951). The spectral composition of this delayed light is that of the fluorescence, and both emissions originate exclusively from photosystem II (see Section IV,B). This delayed light has a very low intensity, which decreases continuously with time. Its decay consists of several components with different time constants ranging from microseconds to minutes. The emission is influenced by the state of the trapping and O_2 evolving centers of system II and by high-energy states of the thylakoid.

Evidently, in the lamellar structure the moieties that participate in the primary events are rigidly localized. Those that should react are held in proximity, and those that should not react are kept apart. Reactions that do not involve diffusion tend to be largely unaffected by temperature. Thus, observations at low temperature, if interpreted cautiously, can be useful in the study of "early" events.

III. Quantum Capture and Distribution

A. Binding States of Chlorophyll a, Accessory Pigments

The absorption properties of a pigment vary with its environment. The location of the red band of chlorophyll a dissolved in organic solvents ranges between 660 and 672 nm, depending upon the refractive index and polarity of the solvent. In the condensed crystalline state, the red maximum is shifted as far out as 740 nm (cf. Fig. 3). Intermediate locations are found depending upon the degree of aggregation and the environment. Aggregated chlorophyll tends to be nonfluorescent at room temperature but shows a long wave emission (720 nm) upon cooling to low temperature (e.g., to 77°K using liquid nitrogen). It is not certain in which physical state(s) the chlorophyll occurs in the dense layers of the lamellae.

During the greening process in higher plants, the red maximum of chlorophyll a (made from protochlorophyll) appears first at 684 nm and then shifts to 673 nm. In mature lamellae the red band comprises several partly overlapping components. Analyses of absorption and action spectra, of the fluorescence emissions at low temperature, and differential extraction experiments all show that chlorophyll a occurs in differently bound forms. French (see Brown, 1972) could fit a large number of plant spectra by assuming that they consisted of mixtures of four specific chl a bands.

Besides chlorophyll a, a number of other, so-called "accessory," pigments contribute to the light harvesting process. Chlorophyll b, present in about one-third the concentration of chlorophyll a in green plants, and chlorophyll c, characteristic for brown algae, differ slightly from chlorophyll a. Their red absorption bands are at the short wave side of the chlorophyll a maxima *in vitro* as well as *in vivo* ($\alpha_{max} \simeq 650$ and $\simeq 640$ nm, respectively).

Nonchlorophyllous, short wave-absorbing accessory pigments also occur abundantly in the plant kingdom. All photosynthetic tissues contain several carotenoids such as β- and α-carotene, lutein (reviewed by Goodwin, 1960). The various classes of "brown" algae contain special xanthophylls, such as fucoxanthin and peridinin, as accessory pigments.

Red and blue-green algae contain high concentrations of the water-soluble protein–bilin complexes phycoerythrin and phycocyanin. Although only two chromophores are involved, slightly different protein carriers or binding modes cause this pigment fraction to show an array of band locations (Fig. 4).

It has been amply proved that all the enumerated pigments contribute to the light harvesting process in photosynthesis. The ubiquitous carotenoids such as carotene and lutein, are relatively inefficient in this aspect and appear to serve another purpose as well—the protection against damage by light.

The bilin pigments and fucoxanthin are found only in aquatic organisms. They absorb in the middle of the spectrum (500–600), where chlorophyll absorption is low and transmission by (sea) water is highest. Engelmann's (1884) original interpretation that these accessory pigments serve to fill this absorption gap still seems largely correct.

In all plants, photosystems I and II have a somewhat different pigment complement. Chlorophyll a is a major constituent in both, but at least one long wave-absorbing state is typical for system I. System II generally contains more, if not most, of the accessory pigments.

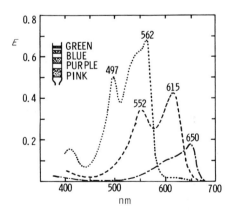

Fig. 4. Absorption spectra of chromoproteins of the red alga *Porphyra perforata* separated on $Ca_3(PO_4)_2$. Dotted line: phycoerythrin. Dashed line: phycocyanin. Dot-dashed line: allophycocyanin. (From Haxo *et al.*, 1955.)

B. Quantum Transfer between Pigments, The Photosynthetic Unit

One important aspect of the light-collecting system of photosynthesis which has been elucidated largely by fluorescence measurements is the transfer of quanta from pigment to pigment (Duysens, 1952). An excited molecule can pass its excitation on to a neighboring one, provided the partners are sufficiently close to each other, as is the case in the chloroplast lamellae. Such transfer by inductive resonance allows a red quantum to "visit" several hundred chlorophyll a molecules in the chloroplast during its lifetime of 5×10^{-9} second.

This transfer occurs not only between like molecules but also between unlike ones as long as the receiver absorbs at the same or slightly longer wavelengths than the emitter. For example, in solutions containing both chlorophyll b (red absorption band at 640 nm) and chlorophyll a (red band at 660 nm), one can irradiate with a wavelength that excites mainly the b component; however, the fluorescence of chlorophyll a, not that of b, is observed. The a component, with its slightly lower excitation level, drains the energy from the b component. Transfer in opposite direction is less efficient because it requires thermal energy to raise the frequency of the quantum. Thus, in the lamellar pigment arrays quanta tend to flow toward the longest wave absorption band. The system I array contains a minor long wave chlorophyll a fraction (5%) which absorbs at 700 nm (Butler, 1962). This is the same location as the red band of P700, which initiates a photoconversion and thus irreversibly traps incoming photons. Figure 5 illustrates this funneling of quanta received by many molecules into

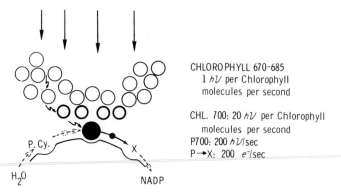

CHLOROPHYLL 670-685
1 $h\nu$ per Chlorophyll
molecules per second

CHL. 700: 20 $h\nu$ per Chlorophyll
molecules per second
P700: 200 $h\nu$/sec
P→X: 200 e^-/sec

Fig. 5. Hypothesis of the concentration of quanta into the trapping center of photosystem I in two steps, each yielding approximately a tenfold increase of excitations per unit time. The thin open circles symbolize the bulk of the light harvesting chlorophyll; the heavy circles, the minor long wave "chlorophyll 700" fraction. The filled circle symbolizes the trapping center P 700 which donates an electron to acceptor X and subsequently receives one from plastocyanin (P. Cy.).

a single molecule, like a lens system concentrating a light beam into its focal point. As a result, the trapping center receives ~200 times more photons per unit time than if it were exposed by itself, i.e., the photochemical rate is that much faster. Each group of pigment molecules collaborating to enhance the optical cross section of the trapping center is called a "photosynthetic unit."

The concept of the unit arose from measurements of Emerson and Arnold (1932), who determined the amounts of O_2 evolved in series of flashes. Each flash was bright enough to elicit maximum response and so brief (10^{-5} seconds) that the excitations could not be processed during the flash itself. If the flashes were spaced far enough apart to allow all dark processes to run to completion and restore the trapping centers to the "open" state, the oxygen yield per flash became maximal and constant regardless of the temperature. One O_2 was evolved per 2000 chlorophyll molecules present in the sample of algae used. Two experiments of this type, made at different temperatures, are shown in Fig. 6.

Since it takes about 10 quanta to produce an O_2 (≥ 4 in each photosystem) this implies that in a single flash only 1 photon can be fixed per 200 chlorophyll molecules present, i.e., there is only one trap per 200 chlorophyll molecules. If we further assume that the traps of photosystems I and II are sensitized by an equal number of chlorophyll molecules the abundance of the conversion centers of each system should be 1 per 400 chlorophyll molecules. This is indeed the approximate concentration of P700, cytochrome f, plastocyanin, Q, and other mediators of the electron transport chain (see Fig. 2).

C. Interactions between Units and Reaction Chains

A chloroplast thylakoid contains a large number of electron transport chains, each equipped with two sets of harvesting pigment, two trapping

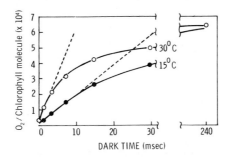

Fig. 6. Yield of O_2 per flash as a function of dark time between flashes at two temperatures. Dashed slopes indicate the respective rates of O_2 evolution in strong continuous light and show a zero-order time course during the initial phase. (From Kok, 1956.)

centers (I and II), and the necesary enzymes to produce O_2, NADPH$^+$, and ATP. The question arises as to what extent these chains or parts thereof interact. The pigment units of system II are not rigidly separated; photons absorbed in one unit which find the trapping center closed can move on to a neighboring unit (Joliot and Joliot, 1964) and, as has been speculated, also can move to system I. No interunit transfer seems to occur in system I. The O_2 generating enzymes, one per system II center, operate strictly independently of each other, i.e., there is no migration of plus charges.

On the reducing side of system II, the electron transport to system I is mediated by a relatively abundant pool of plastoquinone. The system II centers can communicate with each other and with system I centers via this pool (Siggel *et al.*, 1972).

Still another mode of collaboration between units rests on the fact that the thylakoid is a closed membrane. Several of the reactions, notably the two photoacts and the interconnecting chain, occur "across" the membrane. Thus, the concerted action of many electron transport chains in the vesicle membrane leads to a high-energy state based on charge and concentration differences between the inner and the outer phase of the thylakoid.

D. Action Spectra, Enhancement

Figure 7 shows the effectiveness spectrum of photosynthesis in green algae measured by Emerson and Lewis (1943). The quantum yield of O_2 evolution varies with wavelength, which indicates that not all pigments contribute equally to the process. A small ($\sim$25%) dip at 480 nm seems to indicate that carotenoids are not fully effective, a still

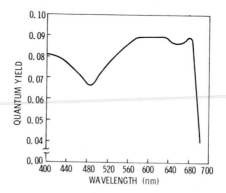

Fig. 7. Dependence of the quantum yield of *Chlorella* photosynthesis on wavelength. (From Emerson and Lewis, 1943.)

smaller dip at 650 nm might imply the same for chlorophyll b. Beyond 690 nm, the yield drops severely, as if an inactive long wave pigment were present.

Figure 8 shows the absorption spectrum and the action spectrum of O_2 evolution (the rate as a function of wavelengh) observed with a red alga. The ratio between the two curves would yield the relative effectiveness spectrum. In green light ($\sim$550 nm), where the accessory pigment phycoerythrin is the main absorber, this ratio is much higher than in red and blue light, where absorption by chlorophyll a and carotenoid is dominant, which suggests that the latter pigments are quite ineffective. Emerson (1958) later observed "enhancement": two wavelengths when given together, can be more effective than the sum of their individual effects ($V_{1+2} > V_1 + V_2$). These observations have found a ready explanation in the hypothesis of two cooperating photosystems equipped with different pigment assemblies; the quantum yield of the overall process will be optimal when the two systems absorb an equal fraction of the incident light and will be low in wavelengths that are preferentially absorbed by one of the two systems. Two wavelengths, one mostly absorbed by system II and the other mostly absorbed by system I, each by itself used rather inefficiently, will complement each other resulting in "enhancement."

Figure 9 illustrates that long wave quanta are not really lost, but are used quite efficiently by one of the photosystems. Using isolated chloroplasts, the wavelength dependence of the quantum yield of NADP reduction was observed under two conditions: (a) simultaneously with O_2 evolution and (b) in the presence of DCMU to abolish O_2 evolution and reduced indophenol to reduce oxidized P700 (Fig. 2). We note in case (a) that

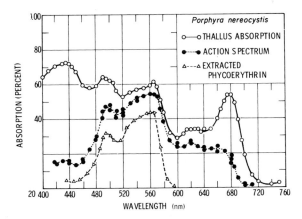

Fig. 8. Absorption and action spectra measured with a red alga, *Porphyra nereocystis* (after Haxo and Blinks, 1950). At each wavelength the effectiveness is the ratio between the two upper curves.

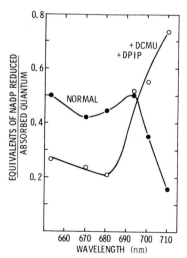

Fig. 9. Quantum yield of NADP⁺ reduction by isolated chloroplasts as a function of wavelength. Filled circles, normal oxygen-evolving system; open circles, in the presence of DCMU, DPIP, and ascorbate. (From Hoch and Martin, 1963).

the efficiency decreases precipitously at wavelengths >690 nm. On the other hand, in case (b) where only system I is operative, the efficiency rises at wavelengths >690 nm and approaches 1 Eq/hv. Also the converse observation has been reported: A quantum requirement close to one for the electron transport from water to DCIP, presumably a system II reaction (Sun and Sauer, 1971).

E. Chromatic Absorption Changes and Rate Transients

More direct evidence for two series connected photoacts, as shown in Fig. 2, came from spectroscopic observations of intermediates of the electron transport chain. Figure 10 shows how in a blue-green alga the accessory pigment phycocyanin sensitizes the reduction of P700, while light absorbed by chlorophyll a and carotenoids causes its oxidation. Similar "push and pull" effects have been reported for cytochrome f, plastoquinone and the yield of fluorescence, which reflects the redox state of Q, the trapping center of system II (Duysens, 1964). Also explained by the series scheme are the so-called "chromatic rate transients" first reported by Blinks (1957). In these experiments two monochromatic light beams were used, one mostly absorbed by accessory pigment and the other by chlorophyll a. Their intensities were adjusted to yield identical

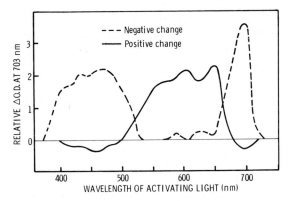

Fig. 10. Effects of wavelength upon the redox state of P700 in whole cells of the blue-green algae *Anacystis*. Negative change (photooxidation of the pigment) was measured without and positive change (photoreduction) with a background of 700 nm light. (From Kok and Gott, 1960.)

steady-state rates of O_2 evolution. When such beams were alternated, transitory changes of the rate (undershoots or overshoots) were observed. The explanation is that a pool of electron carriers located between the two photoacts (Fig. 2) can store as many as 10 electron equivalents per trapping center. In excess system I light, this pool (and primary reductant Q) becomes oxidized, since its photoreduction by system II is too slow. If now a system II light is given, it initially operates with maximum efficiency until, due to a shortage of system I photons, the intermediates in the chain, including Q, become more reduced. A "gush" of O_2 accompanies the photoreduction of this pool, and a rise of the fluorescence yield occurs which reflects the conversion of Q to Q⁻. In isolated chloroplasts the converse experiment is also possible; after a preillumination with system II light, which reduces the pool, the rate in system I light is temporarily high and a gush of reduction of NADP or viologen can be observed.

F. Distribution of Quanta

Figures 8 and 10 indicate that in blue-green and red algae, the pigment assemblies that sensitize the two photosystems are rather different. System I is sensitized mainly by chlorophyll a, and system II mainly by bilin pigment. Note, however, that in white (sun) light the two photosystems will receive very nearly equal amounts of light, i.e., an optimal distribution.

In green plants the relative distribution of the pigments appears more equal. Figure 11 shows the action spectra of the two photosystems. The maxima are displaced; some long wave component(s) are devoted to system I, while system II comprises more short wave components and most of the chlorophyll b.

As yet, the mechanism of quantum distribution is not well understood. One concept assumes that the two photosystems have their own independent pigment groups ("separate packages"). This concept is supported by much evidence, but it does not explain without further assumptions the flatness of the effectiveness spectrum in green cells below 690

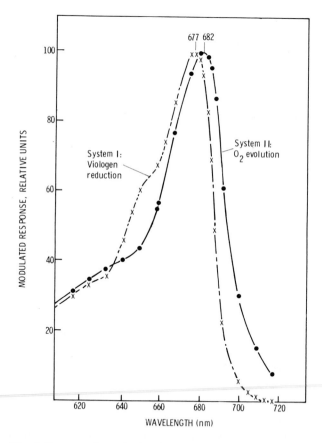

Fig. 11. Action spectra of the two photosystems observed in isolated chloroplasts with the modulated polarograph. Broken line (crosses), O₂ evolution with NADP⁺ as the electron acceptor; solid line, viologen reduction in the presence of DCMU and reduced diaminodural as the electron donor. (From Joliot *et al.*, 1968.)

nm. One explanation (Joliot *et al.*, 1968) assumes that the reaction chain in which the traps are restored

$$P^+ + Q^- \underset{k_{-1}}{\overset{k_1}{\rightleftharpoons}} P + Q$$

has a low equilibrium constant ($K \leq 10$). This assumption predicts the observed quantum yield spectrum but does not agree with the midpoint potentials, which are generally assigned to P and Q.

The alternate, "spill-over" concept of quantum distribution (Myers, 1963) assumes that, except for a long wave system I component, most pigment is shared by the two systems; a quantum preferably travels to a system II trap, but finding it closed, can move on to a system I trap. This concept predicts a flat quantum yield spectrum of O_2 evolution ($\lambda < 690$ nm), but fails to explain the poor quantum yield of NADP reduction at wavelengths <690 nm observed in the presence of DCMU and reduced DPIP in Fig. 9.

In red and blue-green algae, the relative abundance of the various pigments can vary considerably, depending upon the wavelengths in which the organisms are grown (Jones and Myers, 1964). In addition, it appears also in green cells that the quantum flow from some of the pigment can shift from one system to the other (Bonaventura and Myers, 1969; Murata, 1969). Such shifts, induced by the wavelength of illumination, are always in the direction of optimal quantum distribution. In isolated chloroplasts the ion content of the medium can elicit similar effects. The shifts thus might be related to conformational changes of the thylakoid and underly some of the uncertainties discussed above (see Myers, 1971).

IV. The Photosystems

A. Photosystem I

Photosystem I generates a strong reductant and a weak oxidant. It is a complex containing ~200 harvesting chlorophylls, ~50 carotenoids, a specially bound chlorophyll molecule P700, one cytochrome *f*, one plastocyanin, two cytochrome b_{563}, and one or two membrane-bound ferredoxin molecules.

Upon photochemical excitation or chemical oxidation an electron is lost from P700 resulting in the bleaching of its absorption band at about 700 nm and partial bleaching of its blue absorption band (430 nm) (Fig. 12). The high quantum yield ($1/\phi \cong 1$) of the photooxidation and its occurrence at very low temperatures classify it as a direct photochemical event.

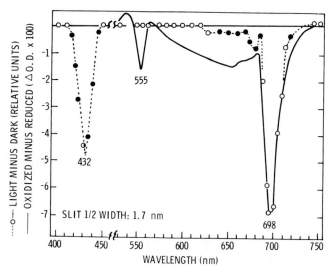

Fig. 12. Difference spectra measured with chloroplasts from which 80% of the chlorophyll was extracted with aqueous acetone. Solid line, oxidized minus reduced spectrum which shows the α band of cytochrome *f* (555 nm) and the red band of P700 (698 nm). For clarity the short wave part of this spectrum was omitted because changes due to the Soret (blue) bands of the two pigments overlap. Dotted line, light minus dark spectrum showing both bands of P700. (From Kok, 1961.)

P700 behaves as a one-electron redox agent with a pH-independent normal potential of +0.43 V. In the oxidized state, it shows a typical electron spin resonance ("signal I," Commoner, 1961; Beinert and Kok, 1963). While at low temperatures the photooxidation is irreversible, at room temperature P⁺700 is rapidly (<2 mseconds) reduced by plastocyanin and/or cytochrome *f*. Cytochrome *f*, discovered by Davenport and Hill (1952), is a high potential (~340 mV) *c*-type cytochrome (MW ~50,000 per heme) with an absorption band at 554 nm in chloroplasts.

Plastocyanin, first found by Katoh (1960) is a low molecular weight (21,000) copper–protein containing 2 gm-atoms Cu/mole protein. The oxidation reduction potential of this single electron transferring protein is constant ($E_m = +0.37$ V) between pH 5.4 to 9.9; at a more acidic pH, the potential increases 0.06 V per pH unit. The copper of the protein is EPR detectable; spectroscopically the oxidized form is blue with a broad absorption band at 597 nm. We generally assume that both cytochrome *f* and plastocyanin mediate electron transport to P700. The reason why there are two electron donors to P700 is still unresolved. Spectroscopic evidence at room temperature (Hiyama and Ke, 1971) and ESR observations at low temperature (Malkin and Bearden, 1971) suggest that the primary electron acceptor(s) of system I is ferredoxin. Presumably a fraction (1 per trapping center) of the ferredoxin is tightly bound to the chloroplast lamella; another fraction is loosely bound and tends to leave the chloroplasts during preparation (San Pietro and Lang, 1958). Chloroplast ferredoxin is a small nonheme iron protein (molecular weight $\sim$12,000) containing 2 gm-atoms of Fe per mole of protein and 1 labile sulfur per iron and has a midpoint potential of -0.43 V (Tagawa and Arnon, 1962).

The photoreduction of NADP requires both the soluble ferredoxin and a flavoprotein (ferredoxin-NADP-reductase). In addition to its mediation of NADP reduction this reductase can function as a transhydrogenase, diaphorase, and NADPH-cyt f reductase. It could therefore mediate cyclic operation of system I (Avron and Jagendorf, 1956). The cytochrome b_6, was described by Hill (1954) to have its α band at 563 nm and its midpoint potential at -0.06 V. Its role, like that of the other chloroplast cytochromes, is still rather obscure. There is some evidence that it mediates ATP formation in a ferredoxin–flavoprotein-dependent cyclic phosphorylation which is antimycin A sensitive (Tagawa $et\ al.$, 1963).

ELECTRON ACCEPTORS AND DONORS OF SYSTEM I, CYCLIC OPERATION

While photoreduction of NADP requires both the soluble ferredoxin and ferredoxin-NADP reductase, the photoreduction of other agents, such as methemoglobin and cytochrome c, requires only soluble ferredoxin. Also, under anaerobic conditions ferredoxin can transfer electrons to hydrogenase, resulting in the evolution of molecular H_2 (Arnon, 1963).

Numerous artificial electron acceptors can be reduced directly by the primary reductan of system I, i.e., without mediating enzymes. These include high potential agents, such as indophenol dye, PMS, ferricyanide, and quinones. The photoreduction of low potential oxidants, such as FMN and viologen dyes, can be observed (a) directly in chloroplast suspensions from which oxygen is carefully removed or (b) indirectly, as a light-induced net consumption of O_2 which, in the complete O_2 evolving system, proceeds as follows so that 1 O_2 is consumed per 4 equivalents of acceptor V which are reduced and reoxidized.

$$2\,H_2O + 2\,V \xrightarrow{h\nu} 2\,VH_2 + O_2$$
$$2\,VH_2 + 2\,O_2 \rightarrow 2\,V + 2\,H_2O_2$$
$$\overline{2\,H_2O + O_2 \rightarrow 2\,H_2O_2}$$

The primary photoreductant reacts with oxygen relatively slowly in the same way as the artificial low potential acceptors, so that chloroplasts without added acceptor show a small O_2 uptake upon illumination (Mehler, 1951).

In the complete system, P700$^+$ is reduced indirectly by Q$^-$ the photoreductant of system II, via the electron carriers linking the trapping centers. If system II is not sensitized (in long wave light) or inhibited by a specific poison such as DCMU, P700 can be reduced effectively by artificial electron donors (Vernon and Zaugg, 1960), such as reduced indophenol dye, PMS, and DAD. The sites of entry of such electron donors are not known and probably vary with the preparation and the concentration of the donor. In low concentration, DPIPH$_2$ and PMSH might reduce plastocyanin; at higher concentration the latter agent reduces P700$^+$ directly. Such agents, being able to function both as donor and acceptor, can "short circuit" the photoact in a "cyclic" operation of system I. This cycle can be shown by illuminating the system with series of light flashes and spectroscopically observing the turnover of P700; it is instantaneously oxidized by each flash and reduced in the subsequent dark period.

Alternatively, low potential acceptors such as NADP (with ferredoxin and reductase as mediators) can be photoreduced, while DPIPH$_2$ (and ultimately ascorbate, which is usually added to keep the dye reduced) becomes photooxidized. This "open ended" operation of system I thus results in a net gain of chemical energy. With an autooxidizable electron acceptor such as viologen (V), oxygen is consumed in the light. If DH$_2$ (e.g., DPIPH$_2$) represents the donor and V the acceptor, the following reactions take place so that 1 O_2 is consumed per 2 equivalents of V, which are reduced and reoxidized.

$$
\begin{aligned}
DH_2 + V &\xrightarrow{h\nu} VH_2 + D \\
VH_2 + O_2 &\rightarrow V + H_2O_2 \\
\hline
DH_2 + O_2 &\rightarrow D + H_2O_2
\end{aligned}
$$

In addition, with some (but not all) electron donors, either open or cyclic operation of system I can be coupled to the generation of ATP from ADP and/or the formation of a proton gradient across the thylakoid membrane. With some mediators, notably PMS, the rate of ATP formation can be extremely high (>2500 ATP per chlorophyll molecule per hour (Avron, 1962). For donor systems where the stoichiometry can be determined, a ratio $P/2\,e \simeq -0.5$ has been observed (Schwartz, 1966; Izawa and Good, 1968). As discussed in Section V, there might be more than one cyclic, ATP generating path.

System I is relatively stable, even in the presence of some detergents. Such detergents disrupt membrane structures, cause a loss of soluble proteins such as plastocyanin and NADP reductase, and permit a physical isolation of system I from system II.

Using differential centrifugation of digitonin-treated chloroplast particles, Boardman and Anderson (1964) obtained fractions that were enriched in system I (the smaller particles) or system II (the heavier ones). Similar procedures using other detergents or physical means of disruption (such as the French press) have yielded essentially the same results but with varying extents of separation and purity of the two photosystems. Preparations enriched in P700 have been obtained, but little progress has been made toward purification of the reaction center.

B. Photosystem II

Photosystem II is intimately connected with the O_2 evolving system and generates a weak reductant. It is a structure-bound complex, composed of ~200 chlorophylls, ~50 carotenoids, a trapping chlorophyll, an unidentified primary electron donor (Z), a primary electron acceptor, (Q) that might be plastoquinone, ~4 plastoquinone equivalents, 6 Mn atoms, and 2 cytochromes $b559$.

Since the liberation of an O_2 molecule from water requires four oxidizing equivalents of an average potential of 0.8 V we assign the primary photooxidant Z at least this high a midpoint potential. The potential of primary reductant Q, is generally thought to be 0 V.

A small flash induced absorption decrease at 680 nm which rapidly returns in subsequent darkness $t_{1/2} \sim 0.2$ mseconds) has been interpreted as a bleaching of "chlorophyll a II" (Döring et al., 1967). Probably the role of this chlorophyll ("P680") in system II is analogous to that of P700 in system I.

Two light-induced absorption changes have been associated with the photoreductant of system II: A short lived (0.6 mseconds) increase of absorption at 325 nm (Stiehl and Witt, 1967) might reflect the photochemical formation of plastosemiquinone and its rapid disappearance in a dismutation reaction. A decrease of absorption at ~550 nm (Knaff and Arnon, 1969) is probably due to a carotenoid band shift and only indirectly associated with the photoreduction of Q (Erixon and Butler, 1971).

In contrast to photosystem I, system II can be inactivated by a number of specific inhibitors. The herbicide DCMU is frequently used to block the reoxidation of Q^- and thus chemically isolate system II from system I.

The redox state of Q determines to a large extent the yield of chlorophyll a fluorescence (F). With all traps open, or receptive, one finds a residual fluorescence yield (F_0), which implies that not all excitations can reach the traps. With all traps closed (either Z or Q in the wrong redox state and thus unreceptive), the yield reaches a maximum (F_{max}). Thus the variable part of the fluorescence yield (F_0 to F_{max}) reflects the degree of opening of the traps. In first approximation the magnitude of ($F_{max} - F$) is proportional to the efficiency of O_2 evolution. Since under most conditions strong oxidant Z is in the reduced state, the fluorescence yield ($F - F_0$) reflects the redox state of primary reductant Q [for "quencher" according to Duysens and Sweers (1963)]. F becomes maximal (Q reduced) under reducing conditions or in strong rate saturating light. It becomes minimal (F_0) upon oxidation in the dark, or in far red light that excites primarily system I.

For instance, the following observations can be made with chloroplasts that have been in dark for a few minutes.

(a) In the presence of DCMU, a specific poison for system II, a brief flash (10^{-5} seconds) raises the fluorescence from F_0 to F_{max}, indicating that a single charge separation can still take place. The return to F_0 in subsequent dark is slow, however.

(b) In the absence of DCMU and in the presence of an electron acceptor, the fluorescence is also raised by a flash, but now it returns rapidly in the dark (~ 0.2 mseconds); Z^+ and Q^- are processed very quickly and in a series of flashes O_2 is evolved.

(c) In the absence of DCMU and of an electron acceptor, the fluorescence return in a series of flashes is normal initially but becomes incomplete after some 10–15 flashes, and the O_2 evolved per flash becomes small.

In the last experiment, a "gush" of O_2 evolution is observed at the onset of illumination (Blinks and Skow, 1938). One O_2 is evolved per ~ 40 chlorophylls, about ten times more than in a single flash, which induces only one turnover of the system (Joliot, 1961). Like the chromatic rate transients, this "gush" reveals the pool of ~ 10 equivalents of plastoquinone located between the two photoacts (Amesz, 1965; Witt et al., 1965).

Although plastoquinone is the most abundant carrier in the photosynthetic electron transport chain, it comprises only a fraction of the total quinones of the chloroplast (Bishop, 1959; Ogren et al., 1965). The absorption spectrum of plastoquinone A is shown as a dashed line in Fig. 13. The solid line illustrates the absorption change induced by a 0.1 second illumination with strong red light (~ 660 nm) which causes reduction of the quinone. A brief flash, which converts each system II trap once, reduces only $\sim \frac{1}{10}$ of the pool. The quinone pool becomes slowly oxidized in darkness and rapidly in long wave light (720 nm).

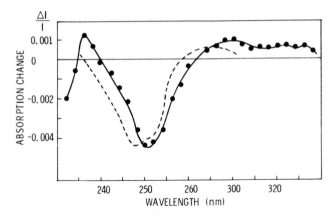

Fig. 13. Absorption changes in the UV-induced in isolated chloroplasts by alternating strong red light (660 nm ~0.1 seconds) with 720 nm light (~0.4 seconds) (solid line). Difference of the absorption spectra of reduced and oxidized plastoquinone A dissolved in ethanol (dashed line). (From Stiehl and Witt, 1967.)

In very bright light, the quinone pool is reduced with a half-time of ~10 mseconds, which implies a reaction time of about 1 msecond for an equivalent to move from water to quinone. Since its oxidation by system I is slower [~10 mseconds, per equivalent (Section VI,B)] the quinone pool becomes reduced in strong, rate saturating light.

The emerging picture seems to be that two quinone molecules are intimately associated with Q, while the others are closer to system I. The two photoacts liberate and consume single electrons. Passage through the two quinone pools is presumably the rate limiting reaction. It occurs across the thylakoid membrane and involves pairing and protonation, at the outside of the membrane, followed by unpairing and deprotonation at the inside, where protons accumulate (see Stiehl and Witt 1969, Radmer and Kok, 1973, Bouges-Bocquet, 1973, Velthuys and Amesz, 1974).

The role of cytochrome $b559$ first observed by Lundegarth (1952) is as yet unclear. It appears to be closely associated with system II (Boardman and Anderson, 1967) which, dependent on conditions, can cause either its reduction or oxidation (see Hind and Olson, 1968). No rapid turnover of this cytochrome has as yet been observed, and thus it seems to be located in a side path. Its normally high midpoint potential of ~0.37 V is readily and irreversibly lowered to +0.06 V in chloroplasts by certain "uncouplers" (CCCP, antimycin A), NH_2OH, temperature shock, Tris extraction, and detergents (Bendall and Sofrova, 1971; Cramer et al., 1971). We, therefore, suspect that membrane phenomena determine its redox state as well as its light-induced absorption changes.

1. Artificial Donors and Acceptors for System II

Several high potential electron acceptors, such as benzoquinone and ferricyanide, can be reduced by system II. However, since reduction by system I is more rapid, it is the predominant path. On the other hand, the strong oxidizing power of system II appears to be rather shielded from external reductants. Few if any external reductants are known to act readily without damage to the system. Hydroxylamine, for instance, can serve as an electron donor, but in the process it removes the Mn of the O_2 enzyme.

The O_2 evolution system discussed below is quite fragile in the presence of detergent and is lost upon heating to 50°C for a few minutes. Therefore, the isolation of O_2 evolving system II particles has proved to be quite difficult. However, the photosystem itself is relatively stable and even survives treatment with detergents. Such system II particles still show the 680 nm absorption changes and carry out the photooxidation of cytochrome $b559$ and of artificial donors, such as ascorbate, hydroquinone, phenylcarbazid, and hydroxylamine (Trebst et al., 1963; Yamashita and Butler, 1968).

2. Oxygen Evolution

Each system II reaction center appears to be equipped with its own enzyme system, which accumulates and processes four presumably identical plus charges (Z^+) to evolve an O_2 molecule from water:

$$4 Z^+ + 2 H_2O \rightarrow 4 Z + O_2 + 4 H^+$$

Without catalysis the four successive oxidation steps would involve widely different free energies and unstable intermediates of very different potentials.

As mentioned above, the O_2 evolving system is fragile and dependent upon structural (membrane) integrity. The oxidation of water, but not of artificial donors, requires the presence of chloride or other anions of strong acids (Izawa et al., 1969). Manganese appears to be a specific constituent of the O_2 evolving system; it occurs in an abundance of about 6 protein-bound atoms per trapping center. Mild heating or extraction procedures can remove four of these Mn atoms with a parallel loss of O_2 evolving capacity. However, the loss of this Mn fraction does not impair the photooxidation of electron donors other than water (Cheniae, 1970).

The formation of the O_2 evolving Mn catalyst requires light absorbed by system II (Cheniae and Martin, 1967). Once photoactivated in a multiquantum process, it remains permanently active in subsequent darkness.

In very weak light the efficiency of O_2 evolution decreases (the light curve is S shaped), and, after a dark period of a few minutes, O_2, evolution does not start immediately in the light. These intensity and time lags are due to the limited stability of O_2 precursors that are reduced in dark processes.

The cooperation of oxidizing equivalents in the light and their loss in the dark has been analyzed by measuring the O_2 yields of flashes given in sequences with a spacing of 0.1–1 seconds (Joliot, 1968). Each flash produces one "plus charge" in each trapping center so that the events proceed in discrete steps. Figure 14 shows such a sequence of O_2 flash yields observed with isolated chloroplasts after a 40 minute dark period. The yield oscillates with a period of four, indicating a 4-step process. The first two yields are negligible, which indicates that the final and penultimate O_2 precursors have disappeared in the dark. Unexpectedly, however, the third (and not the fourth) flash yield is maximal, indicating that in darkness a precursor accumulates which needs three (not four) further oxidation steps. The evidence is that each trapping center, plus

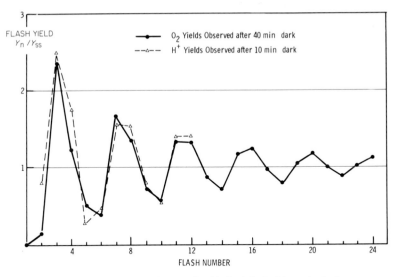

Fig. 14. Flash yields of O_2 observed with isolated chloroplasts in a sequence of flashes, spaced 1 second apart, given after a 40 minute dark period (solid line). (From Forbush *et al.*, 1971.) Individual yield values (Y_n) are normalized to the steady-state yield Y_{ss}. Flash yields of proton liberation measured in a similar experiment (dashed line). An uncoupling concentration of methylamine was present to restrict the proton events to those associated with the decomposition of water (From Fowler and Kok, 1974.)

its O_2 evolving enzyme, acts as a self-sufficient unit. This suggests a linear process in which the O_2 center (S) goes through four oxidation states (Kok *et al.*, 1970).

$$S_0 \xrightarrow{Z^+, k_0} S_1{}^+ \underset{k_{-2}}{\overset{Z^+, k_1}{\rightleftarrows}} S_2{}^{2+} \underset{k_{-3}}{\overset{Z^+, k_2}{\rightleftarrows}} S_3{}^{3+} \xrightarrow{Z^+, k_3} S_3{}^{4+} \xrightarrow{k_4} S_0 + O_2$$

S_3 and S_2 are unstable (k_{-3}, k_{-2}) and return to the S_1 state which, like the S_0 state is stable in darkness.

As also shown in Fig. 14, the protons liberated in the process are released in synchromy with the oxygen, i.e., the actual decomposition of water appears to occur in a terminal, concerted mechanism (step k_4), after the system has accumulated four plus charges (Fowler and Kok, 1974). The protons (and therefore also O_2) appear to be released inside the thylakoid so that they contribute to the pH gradient and the subsequent formation ATP.

The various reaction steps are rapid; the delay between photochemical excitation of the S_3 state and the appearance of molecular oxygen ($k_3 + k_4$) is ~0.8 mseconds (Joliot, 1966). The relaxation times of the other steps is still faster: 0.2–0.4 mseconds. Obviously, the O_2 evolution system does not limit the rate of photosynthesis in strong light.

In very weak light, however, where photons arrive infrequently at the trapping centers, the "deactivation" reactions

$$S_3 \xrightarrow{k_{-3}} S_2 \xrightarrow{k_{-2}} S_1$$

interfere. The rate of deactivation varies considerably with the material and its prehistory. In isolated chloroplasts, the half-life of the S_3 state can be more than 1 minute (Forbush *et al.*, 1971).

In whole cells in weak light, this half-life is a few (~3) seconds. Thus, in this instance the efficiency of light conversion would be half-maximal at an intensity in which each center is hit every ~3 seconds (or, for a unit of 200 chlorophylls per center, each chlorophyll is hit only once per 10 minutes). This corresponds to an intensity as low as 10^{-4} of full sunlight. On the other hand, we recall that in very strong light the process begins to saturate when the centers are hit every 5–10 mseconds. Thus, photosynthetic light conversion proceeds efficiently over a 1000-fold intensity range.

V. Photophosphorylation

A. Energy Conservation Sites

The photoreactions generate a higher chemical potential than that required for the oxidation of water and the reduction of ferredoxin. This additional energy can be conserved in ATP (Arnon *et al.*, 1954; Frenkel, 1954). The standard free energy of the conversion of ADP and P_i to ATP is about 10 kcal/mole. Illuminated chloroplasts can establish high ATP/ADP ratios, corresponding to a "phosphate potential" of ~15 kcal/mole (Kraayenhof, 1969). If this potential were generated in a one equivalent reaction, this reaction would conserve up to 0.66 eV. If per ATP formed two (or three) equivalents were to move through a conservation site, this value would be 0.33 (or 0.22) eV.

Though many early measurements of phosphorylation stoichiometry yielded values of 1 ATP per 2 e^- transferred to NADP, more recent reports indicate ratios that are consistently greater than 1 (~1.3, Izawa and Good, 1968). Moreover when NADP is replaced by electron acceptors (benzoquinone, oxidized *p*-phenylenediamine) which are presumably reduced directly by system II, this ratio is decreased by one-half (Saha *et al.*, 1971). Such data might locate two sites of ATP formation between O_2 evolution and NADP reduction: one within the O_2 evolution process and the other within the electron transport chain between the photoacts. It should be kept in mind that the overall ATP/2 e^- ratio depends not only upon the number of sites but also upon the mechanism (i.e., the amount of energy conserved at the sites and the efficiency with which this energy is converted to ATP).

As we will discuss later (see Section VI), photosynthetic growth requires 5–6 ATP's for each carbon incorporated into new cell stuff (≥ 3 ATP/2 e^-). Thus, unless its phosphorylation efficiency is higher *in vivo* than *in vitro*, noncyclic electron transport seems unable to meet the overall ATP demand. This apparent ATP deficit might be overcome with phosphorylations via "cyclic" pathways 3 and 4 mentioned below. These would occur at the expense of "extra" photons (which yield no net O_2 and reducing power) and thus decrease the overall efficiency of autotrophic growth.

The precise mechanism(s) for the conservation of energy into ATP are still being debated. There is considerable evidence that energy storage in a proton gradient (Mitchell, 1966) precedes ATP formation. Several plausible energy conserving loci in the electron transport can be mentioned.

1. The reaction chain between the photoacts, $Q^- \rightarrow P700^+$, which is known to pump protons into the thylakoids, thus forming a proton gradient (See Fig. 2)

2. Within the O_2 evolution process, which occurs inside the thylakoid, so that protons accumulate and contribute to the gradient

3. The reaction chain between Fd^- and $P700^+$ via cytochrome b_{563},

which encompasses a potential change of ~0.8 V. This cyclic electron transport thus contains two energetically feasible sites

4. Within an oxidative path, i.e., within reactions associated with a light enhanced uptake of O_2 observed in whole cells

B. Uncouplers

Photophosphorylation, a structure-dependent process, is rather labile and influenced by the conditions and procedures of isolation and storage of chloroplasts. Numerous agents (loosely termed uncouplers) are known which in one way or another prevent ATP formation while still allowing electron transport to proceed. Equations (1)–(3) show a two-step sequence of ATP formation.

$$P700^+ + Q^- + I \rightleftharpoons P700 + Q + \sim I \tag{1}$$
$$\sim I + ADP + P_i \underset{Mg^{2+}}{\rightleftharpoons} ATP + I \tag{2}$$
$$\sim I \rightarrow I \tag{3}$$

Energy in an electron transport reaction is conserved in $\sim I$, an intermediate or a state of high energy [Eq. (1)]. $\sim I$ is either consumed in the formation of ATP [Eq. (2)] or degraded to I [Eq. (3)]. If neither of the forward reactions of Eq. (2) or (3) can proceed, $\sim I$ accumulates and the electron transport in Eq. (1) is inhibited.

Even in chloroplasts exhibiting high phosphorylation efficiency, a low (basal) rate of transfer of electrons through Eq. (1) is observed in the absence of ADP and P_i. This basal rate reflects the decay of $\sim I$ in reaction (3), which now limits the overall rate of the system. The addition of $ADP + P_i$ removes the rate limitation imposed by $\sim I$ until, in the absence of an ATP consuming process, most of the ADP is phosphorylated and a high "phosphate potential" and a high ratio of $\sim I/I$ are reestablished. As a consequence, the rate of electron transfer through Eq. (1) again returns to the low basal rate.

Figure 15 illustrates the effect of methylamine on the rate of ferricyanide reduction in saturating light. The observations were made in the absence and presence of ADP and P_i, and the rates of ATP formation observed in the latter case are also plotted. Without uncoupler, the basal rate of ferricyanide reduction is about doubled by the addition of ADP and P_i, and one ATP is formed per 2 equivalents of ferricyanide reduced. Increasing concentrations of the amine inhibit ATP formation but stimulate the rate of electron flow. This probably reflects an acceleration of reaction (3), the loss of a high-energy precursor. Concentrations of ammonia and amines, which inhibit ATP formation, also dissipate the pH

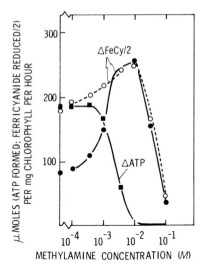

Fig. 15. Effect of the uncoupler methylamine upon the rate of ferricyanide (FeCy) reduction in the absence (filled circles) or presence (open circles) of phosphorylation reagents. Rate of ATP formation accompanying reduction in the latter experiment (filled squares). (From Avron and Shavit, 1963.)

gradient, which presumably mediates in the energy conversion process. A number of agents that allow the free exchange of ions through membranes ("ionophores," such as gramicidin) also uncouple phosphorylation.

The stimulation of the rate of electron flow by either uncouplers or phosphorylating agents (ADP, Mg^{2+}, P_i) (Fig. 15) is observed only when the rate of photon absorption exceeds the rate limitation imposed by ~I. (Note in Fig. 15 that concentrations of uncoupler higher than needed to abolish phosphorylation severely inhibit the rate of electron transport.)

Figure 16 illustrates some effects of pH. (These effects can vary considerably with the type of buffer used.) The basal electron transport shows a maximal rate at pH 8.5–9.0. The rate coupled to ATP formation responds similarly but drops less rapidly with decreasing pH. In the presence of methylamine, the rate drops at high pH, its optimum can extend to lower pH values than observed in Fig. 15. Consequently the effect of uncoupling on the rate can vary between strong enhancement ($\leq$ 50 times) at low pH and little effect at high pH.

Figure 17 shows the action of phlorizin. Agents of this type inhibit the final ADP phosphorylating step [Eq. (2)] and thus annihilate the rate stimulation by ADP, etc. The rate of either basal or amine uncoupled electron transport is not affected. Note that in this experiment the ratio ATP/2 $e^- = 2$ and is constant if calculated on the basis of the extra, ADP-stimulated, electron transport.

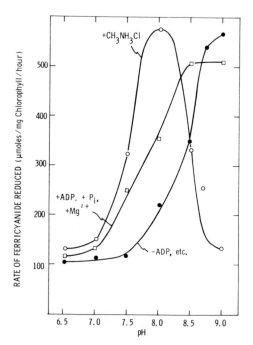

Fig. 16. pH dependence of the rate of ferricyanide reduction during 2 minute exposures to strong light. Isolated chloroplasts suspended in tricine buffer mixtures were used without additions (basal rate, closed circles), with phosphorylating reagents (open squares), or with 3 mM methylamine (open circles). (From Avron, 1972.)

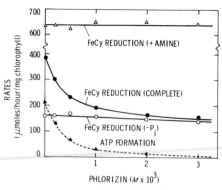

Fig. 17. Effect of phlorizin on the rates of ferricyanide (FeCy) reduction and ATP formation in illuminated spinach chloroplasts suspended in sucrose containing tricine buffer pH 8.4. No additions (basal rate) (open circles), +10 mM methylamine (open triangles), + phosphorylating reagents added (closed circles and triangles): FeCn rates (circles), ATP rates (triangles). (From Izawa et al., 1966, see also Good et al., 1966.)

These data argue for two "sites," provided one makes the *ad hoc* assumptions that (1) a "site" yields one ATP per two electron passages and (2) the basal rate is due to an independent, nonphosphorylating pathway (or an already uncoupled chloroplast fraction).

We should also mention the rate stimulation and loss of phosphorylation capacity induced by EDTA. If chloroplasts are washed with EDTA, or suspended in a medium of low salt content, a protein is released which mediates the phosphorylation of ADP. This so-called coupling factor has been isolated and can restore the phosphorylating capacity of EDTA-treated chloroplasts (Avron, 1963; Vambutas and Racker, 1965). Certain (mis)treatments of this enzyme cause it to act as an ATPase

Chloroplasts, in contrast to mitochrondria show very little ATPase activity in the dark (Arnon *et al.*, 1956). However, such activity is switched on by light (Avron, 1962; Hoch and Martin, 1963) and can under certain conditions (e.g., the presence of sulfhydryl compounds) remain in the dark.

C. Ion Movements

The formation of ATP can be separated in time from the light-driven energy-conservation step. If a chloroplast suspension is illuminated in the absence of ADP and P_i, and these agents then added within 1 second or so after darkening, ATP is subsequently formed (Shen and Shen, 1962; Hind and Jagendorf, 1963).

This light-induced ATP precursor has been associated with a pH gradient across the thylakoid membrane (Jagendorf and Hind, 1963). Upon illumination, an unbuffered chloroplast suspension medium becomes alkaline, which implies that protons are transported into the thylakoid vesicles. A change of 2.5–3 pH units can be established, and, at a low initial pH, as much as 1 proton per chlorophyll can be transported. The system reequilibrates in darkness with a half-time of several seconds. This equilibration is faster at high than at low pH and is very rapid in the presence of uncouplers or detergents that make the membrane leaky to protons.

It proved possible to generate ATP in the absence of light by first equilibrating a chloroplast suspension at low pH (≥ 4) and then quickly raising it to pH 8 in the presence of ADP and P_i. If the equilibration was carried out with weak organic acids (which penetrate the thylakoids and, due to their buffer effect, enhance the proton storage capacity of the vesicles), as much as 0.2 ATP per chlorophyll molecule was formed (Jagendorf and Uribe, 1967). Since no known electron carrier in the chloroplast occurs in such high a concentration, the source of energy could only be the pH gradient itself. These observations gave a strong impetus to the chemiosmotic view of ATP formation (Mitchell, 1961); a charge and/or concentration difference across an anisotropic membrane can provide the potential to drive phosphorylation. For example, a gradient of 3 pH units (electrically neutralized either by an anion such as Cl⁻ that moved with the protons or by a cation such as K⁺ that moved in opposite direction) represents a chemical potential of $3 \times 0.060 = 0.18$ V. If electrical neutrality was only partially maintained, additional driving potential would be available in the electrical gradient.

Figure 18 illustrates the pH changes induced by light in a chloroplast suspension. In the absence of ADP the pH rises in a few seconds to a steady-state level (ΔH_{ss}^+) and decays slowly in the dark. In the presence of ADP, light induces the formation of ATP that is observable as a continuous alkalinization, which does not return in dark. The pH gradient that does return in dark now attains a lower steady-state level (ΔH_{ss} — ADP). Upon darkening, the time course of the pH shows initially a rapid decrease and subsequently matches the decay that is observed in the absence of ADP. The latter change of rate shows that ATP

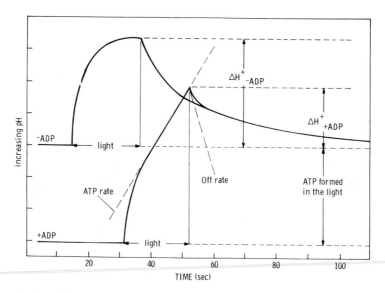

Fig. 18. Time–course of pH during and after ilumination of a chloroplast suspension (pH ~8) in the absence of ADP (top trace) and in the presence of ADP (bottom trace). The final phases of the two traces are superposed. The slope of the steady-state pH rise during the light is a measure of the rate of ATP formation. The initial slope of the pH decrease upon darkening is a measure of the number of protons that are consumed from the gradient for the formation of ATP. (From Schröder *et al.*, 1972.)

formation (1) is correlated with an accelerated proton efflux that lowers the steady-state gradient and (2) only takes place as long as the gradient exceeds a certain magnitude.

According to Izawa and Hind (1967), noncyclic electron transport results in the translocation of two protons for each equivalent that moves through the chain. More recent studies point to a higher value of 3–4 H^+/e^- (Fowler and Kok, 1975). A consumption of 3 protons from the gradient for each ATP generated (Schröder et al., 1972) then would correspond to an overall stoichiometry of 1.33–2 ATP/2 e^-. Overall ATP/2e^- ratios of two have indeed been observed (e.g., Hall et al., 1971).

Figure 19 is a schematic representation of a mechanism for energy conservation proposed by Mitchell (1966), which has gained considerable experimental support, particularly in the laboratory of Witt (1971). In this scheme it is assumed that both photoacts move an electron from the inside to the outside of the thylakoid membrane. The concerted action of many units can then establish a significant membrane potential. This electrical field presumably causes absorption changes (electrochromic shifts) of some of the pigments in the membrane (Junge et al., 1968). The most pronounced of these changes, a red shift from 480 to 515 nm, which was originally observed by Duysens (1954), is probably due to a carotenoid. Subsequently, the reduction of the plastoquinone pool, which is presumably arranged across the membrane, consumes protons from the outer phase, while its oxidation, as well as O_2 evolution, releases protons in the inner phase. The field thus is replaced by a proton gradient. During continued illumination, additional proton transfer is allowed by flows of K^+ and Cl^- ions, which nearly balance the charges. Subsequently,

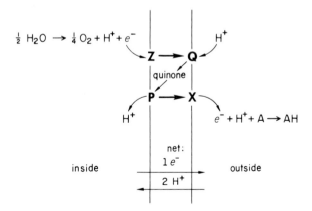

Fig 19. Hypothetical scheme for the accumulation of protons inside the chloroplast vesicle due to electron transport through plastoquinone and O_2 evolution. (After Mitchell, 1966.)

proton efflux through specific sites in the membrane causes the coupling enzyme to generate ATP; the mechanism of this energy translation remains unknown.

Although there are many arguments supporting the concept that the proton gradient is an intermediate of phosphorylation, several observations hint that this may not always be the case (see Walker and Crofts, 1970; Dilley, 1971). In subchloroplast particles, NH_4Cl can abolish proton uptake while still allowing ATP formation (McCarty, 1969). Some chloroplasts preparations that are rich in system I can show high rates of cyclic phosphorylation but no appreciable proton translocation (Arntzen et al., 1971). One could speculate that such cyclic paths might involve the various cytochromes (e.g., cyt b_6) to which, as we recall, no clear-cut roles have yet been assigned in the linear electron transport sequence.

It should be clear that, as in the other sections of this chapter, we have discussed a number of selected aspects and thoughts. Many related phenomena, such as conformation changes, water and ion transport, and important mechanistic aspects, have not been mentioned.

VI. Energetics and Kinetics

A. Maximum Efficiency of Photosynthesis and Growth

In photochemical reactions a single photon interacts with a single molecule, inducing the transition of a single electron. Therefore, it is convenient to express efficiency in terms of the quantum yield (equivalents converted per quantum absorbed) or the inverse, the quantum requirement ($h\nu$/Eq). However, we should keep in mind that this equivalence law pertains only to the primary conversions that, especially in the whole cell process, might not be stoichiometrically related to the ultimate products.

The two photoacts essentially split water into its elements; this process results in a gain of chemical potential of 1.2 V or 120 kcal per mole of oxygen. To drive the process, to stabilize intermediates, and, in addition, to generate ATP requires considerable extra energy. Since 1 mole of red ($\sim$700 nm) quanta represents about 40 kcal (see Section II), the minimum conceivable requirement of 1 $h\nu$ per equivalent or 4 $h\nu/O_2$ would correspond to an efficiency of $120/4 \times 40 = 0.75$. For many years, the true quantum yield of photosynthesis was a hotly controversial issue (see Kok, 1960). Presently, a minimum requirement of ≥ 10 $h\nu/O_2$ (≥ 2 $h\nu$/Eq) is generally accepted and underlies the 2 photoact scheme of Fig. 2. In red light, this requirement corresponds to an energy conversion efficiency of $120/10 \times 40 = 0.3$. All absorbed wavelengths are used with approximately the same quantum yield (Fig. 7). However, because short wave photons have higher energy, their conversion is less efficient, so that $\leq 25\%$ of absorbed white (sun) light can be converted.

For instance, in Fig. 20 the ascending slope of the rate versus intensity curve (used as an index of efficiency) suggests that 17.4 cal were fixed as chemical energy per 100 cal of absorbed light energy. This efficiency was computed by assuming that all of the visible light ($\lambda < 700$ nm) falling on the leaf was absorbed and that the "assimilatory quotient" (the ratio of O_2 evolved/CO_2 fixed) was close to 1, so that the O_2 evolution rate equalled the rate of CO_2 fixation.

In principle, the optimal efficiency of light conversion will be found in weak light, where the overall rate is determined by the rate of photon absorption and all quanta can be utilized. However, in extremely weak light the instability of the O_2 precursors leads to a low efficiency. Futhermore, respiratory O_2 uptake, associated with cell metabolism and growth, reflects a negative energy balance in darkness and weak light. Only when the photosynthetic rate surpasses the respiratory uptake is a "net" O_2 evolution (i.e., a gain of chemical energy) observed.

In whole algae, the products of photosynthesis are oxygen and new cells (chiefly protein). To reduce 1 mole of CO_2 to the level of carbohydrate, 4 equivalents NADPH and 3–5 moles of ATP are needed (see Chapter 24). Growth experiments with heterotrophic organisms indicate that another 3 ATP's per carbon (Y_{ATP}) are required to convert carbohydrate into cell material (Bauchop and Elsden, 1960). Thus ~6 ATP's are needed per oxygen evolved (3 ATP/2 e^-). It is doubtful whether the ATP generation coupled to the two photoacts (2 ATP/2 e^-) suffices to fill this demand.

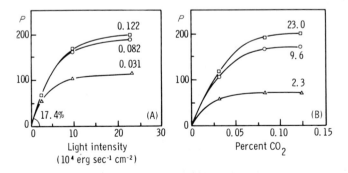

Fig. 20. Rate of photosynthesis observed with a turnip leaf (P, in microliters CO_2 cm^{-2} hour^{-1}) as a function of light intensity (A) and Co_2 concentration (B). (From Gaastra, 1959.)

We have mentioned two possible ways to generate additional ATP. One is a cyclic operation of photosystem I; in whole algae, several ATP-requiring processes are known to be driven by long wavelength light, although their rates are rather low (see, e.g., Tanner *et al.*, 1965). The other is mitochondrial oxidative phosphorylation using NADPH formed photochemically.

The occurrence of both processes might be demonstrated in the experiment of Fig. 21 in which a mass spectrometer was used to monitor O_2 exchange. The isotope $^{18}O_2$ was used in the gas phase to unscramble the simultaneously occurring oxygen uptake ($^{18}O_2$) and photosynthetic evolution ($^{16}O_2$ from $H_2{}^{16}O$). In weak light the respiratory uptake proved to be decreased, possibly because cyclic photophosphorylation partially replaced oxidative phosphorylation. In stronger light the O_2 uptake is enhanced; at least part of this uptake might provide additional ATP.

Enhanced O_2 uptake with simultaneous CO_2 release in the light is loosely termed "photorespiration." It tends to be much more pronounced in so called "C_3 plants" than in "C_4 plants" which possess the C_4 dicarboxylic acid cycle (see Chapter 24). Its rate varies greatly, being highest at high temperature, high $[O_2]$, and low $[CO_2]$. The oxidation of gly-

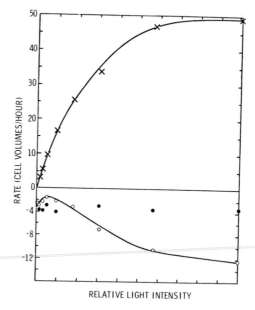

Fig. 21. Oxygen production (crosses) and uptake (open circles) during illumination as a function of intensity (observed with the blue-green alga *Anacystis*). O_2 uptake in dark preceding the light (filled circles). (From Hoch *et al.*, 1963.)

colate, (which is not coupled to ATP formation) is a major pathway, and thus photorespiration can result in a severe loss of photosynthetic product and a decrease of the net production (Tolbert et al., 1971). However, under optimal conditions (nonsaturating light, ample CO_2, and low O_2), these losses are minor, actual measurements of the efficiency of photosynthetic growth showed optimal values close to the quantum requirement of 10 hv/O_2. Thus in plant growth, up to ~20% of the absorbed solar radiation ($\lambda < 700$ nm) can be converted, i.e., ≤10% of the entire solar emission.

B. Maximum Rate

Photosynthesis consists of an interplay between a few photochemical and many dark reactions. The rate of the overall process depends upon the availability of the various participants, such as water, CO_2, light, and enzymatic machinery, each of which can be rate limiting. Figure 20 illustrates how in a leaf the rate varies with light intensity and CO_2 concentration. The rate of CO_2 uptake was observed using three CO_2 concentrations and three light intensities. In the weakest intensity 0.03% CO_2 in the surrounding air sufficed to maintain the optimal rate that was limited by the slow influx of photons. However, in 10 times stronger light, ≥0.1% CO_2 was needed to support the (now faster) rate. Since presumably all substrates were abundant, the maximal rate (V_{max}), seen in Fig. 20, was determined by the maximum turnover rate of one or more steps in the enzymatic reaction chain. This conclusion might be checked by observing a further increase of rate with an increase of temperature.

We note in Fig. 20 that the natural CO_2 content of the air (0.03%, corresponding to 10^{-5} M in solution) supports only about half of the maximum photosynthetic rate. In leaves, this concentration which supports the half-maximal rate (also called the apparent affinity or Michaelis constant) is the resultant of a complex set of diffusion and reaction rates. It does not vary greatly among land plants, despite considerable variation of V_{max} and different pathways of CO_2 reduction (Goldworthy, 1968).

Unlike CO_2, which is rather uniformly distributed in the biosphere, light intensity fluctuates widely with time and location. The "dynamic range" of photosynthesis with respect to intensity therefore is important. We have mentioned (Section IV,B) that a half-life of the S_3 state of ~3 seconds, as observed in whole algae, corresponds to a lower intensity limit of 0.01% of full sunlight. Indeed, plants adapted to deep shade have been reported to grow in ~0.1% of natural insolation (Talling, 1961).

On the other end of the intensity scale, in full sunlight at noon, a single chlorophyll molecule absorbs about 10 photons per second. A trapping center, served by the few hundred chlorophylls of a unit, thus receives some thousand quanta per second. The flashing light experiment of Fig. 6 showed that the amount of O_2 made in each flash is half-maximal (the conversion of the photoproducts is half-completed) if the dark time between the flashes is 5–10 mseconds. A 10 msecond turnover time implies a maximal rate (V_{max}) of 100 equivalents per second per reaction chain* (much too slow to keep pace with the above computed rate of excitation).

Accordingly, in most algae and plants (Fig. 20), the rate of photosynthesis is half-saturated $\sim\frac{1}{10}$ or less of full sunlight. This early saturation severely restricts the efficiency with which strong light can be converted.

Plants show many adaptations that aid in making the best possible use of high intensities—somewhat smaller photosynthetic units, high concentrations of electron transport and CO_2 reducing enzymes, dilution of the light (and spreading the CO_2 uptake) over a large leaf area oriented at oblique angles of incidence, etc. (see, e.g., Björkman, 1973). The list of species characterized by high rates of photosynthesis shows many C_4 plants, such as tropical grasses, sugarcane, and corn, but also includes C_3 plants such as sunflower and cattail (Zelitch, 1971). Still, even in these species the conversion of strong (sun) light remains well below the maximum (10%) efficiency, which puts a major limitation on yields of production under field conditions. Probably, the ultimate reason why the photosynthetic apparatus does not perform too well in strong light is the low concentration of CO_2 in the earth's atmosphere.

ACKNOWLEDGMENTS

This material was prepared with the partial support of the Energy Research and Development Administration Contract E(11-1)-3326, and the National Science Foundation Grant No. BMS74-20736. Any opinions, findings, conclusions, or recommendations expressed herein are those of the author and do not necessarily reflect the views of the National Science Foundation.

* If we assume that each chain contains 400 chlorophylls, this rate corresponds to 0.25 Eq/chlorophyll molecule/second. Because the chlorophyll content of chloroplast or algae suspensions can be readily determined, the rates observed with such suspensions are commonly expressed in terms of equivalents (or moles) per mole Chlorophyll per hour (or μEq mg Chl^{-1} hr^{-1}). Expressed in these units, the above rate of 0.25 Eq Chl^{-1} sec^{-1} would be 900 Eq Chl^{-1} hr^{-1} corresponding to 225 O_2 Chl^{-1} hr^{-1}.

REFERENCES

Amesz, J. (1965). *Biochim. Biophys. Acta* **79**, 257.

Arnon, D. I. (1963). *Nat. Acad. Sci.—Nat. Res. Counc. Publ.* **1145**, 195.

Arnon, D. I., Allen, M. B., and Whatley, F. R. (1954). *Nature (London)* **174**, 394.

Arnon, D. I., Allen, M. B., and Whatley, F. R. (1956). *Biochim. Biophys. Acta* **20**, 449.

Arntzen, C. J., Dilley, R. A., and Neumann, J. (1971). *Biochim. Biophys. Acta* **245**, 409.

Avron, M. (1962). *J. Biol. Chem.* **237**, 2011.

Avron, M. (1963). *Biochim. Biophys. Acta* **77**, 699.

Avron, M., and Jagendorf, A. T. (1956). *Arch. Biochem. Biophys.* **65**, 475.

Avron, M., and Shavit, N. (1963). *Nat. Acad. Sci—Nat. Res. Counc., Publ.* **1145**, 611.

Bauchop, T., and Elsden, S. R. (1960). *J. Gen. Microbiol.* **23**, 457.

Beinert, H., and Kok, B. (1963). *Nat. Acad. Sci—Nat. Res. Counc. Publ.* **1145**, 131.

Bendall, D. S., and Sofrova, D. (1971). *Biochim. Biophys. Acts* **234**, 371.

Bishop, N. I. (1959). *Proc. Nat. Acad. Sci. U.S.* **45**, 1696.

Bjiorkman, O. (1973). *In* "Photophysiology" (A. Giese, ed.), Vol. VIII, p. 1. Academic Press, New York.

Blinks, L. R. (1957). *In* "Research in Photosynthesis" (H. Gaffron, ed)., p. 444. Wiley (Interscience), New York.

Blinks, L. R., and Skow, R. K. (1938). *Proc. Nat. Acad. Sci. U.S.* **24**, 120.

Boardman, N. K., and Anderson, J. M. (1964). *Nature (London)* **203**, 166.

Boardman, N. K., and Anderson, J. M. (1967). *Biochim. Biophys. Acta* **143**, 187.

Bonaventura, C., and Myers, J. (1969). *Biochim. Biophys. Acta* **189**, 366.

Bouges-Bocquet, B. (1973). *Biochim. Biophys. Acta* **324**, 250.

Brown, J. S. (1972). *Annu. Rev. Plant Physiol.* **23**, 73.

Butler, W. L. (1962). *Biochim. Biophys. Acta* **69**, 309.

Cheniae, G. M. (1970). *Annu. Rev. Plant Physiol.* **21**, 467.

Cheniae, G. M., and Martin, I. F. (1967). *Biochem. Biophys. Res. Commun.* **28**, 89.

Cheniae, G. M., and Martin, I. F. (1971). *Biochim. Biophys. Acta* **253**, 167.

Chessin, M., Livingston, R., and Truscott, T. G. (1966). *Trans. Faraday Soc.* **62**, 1519.

Commoner, B. (1961). *In* "Light and Life" (W. D. McElroy and B. Glass, eds.), p. 356. Johns Hopkins Press, Baltimore, Maryland.

Cramer, W. A., Fan, H. N., and Bohme, H. (1971). *Bioenergetics* **2**, 289.

Davenport, H. E., and Hill, R. (1952). *Proc. Roy. Soc. Ser. B* **139**, 327.

Dilley, R. (1971). *Curr. Top. Bioenerg.* **4**, 237.

Döring, H., Stiehl, H., and Witt, H. T. (1967). *Z. Naturforsch. B* **22**, 639.

Duysens, L. N. M. (1952). Ph.D. Thesis, Utrecht, Netherlands.

Duysens, L. N. M. (1954). *Science* **120**, 353.

Duysens, L. N. M. (1964). *Progr. Biophys.* **14**, 1.

Duysens, L. N. M., and Sweers, H. E. (1963). "Studies on Microalgae and Photosynthetic Bacteria," p. 353. Univ. of Tokyo Press, Tokyo.

Emerson, R. (1958). *Annu. Rev. Plant Physiol.* **9**, 1.

Emerson, R., and Arnold, W. (1932). *J. Gen. Physiol.* **15**, 391.

Emerson, R., and Lewis, C. M. (1943). *Amer. J .Bot.* **30**, 165.

Engelmann, T. W. (1884). *Bot. Ztg.* **41**, 1.

Erixon, K., and Butler, W. L. (1971). *Biochim. Biophys. Acta* **234**, 381.

Forbush, B., Kok, B., and McGloin, M. (1971). *Photochem. 8 Photobiol.* **14**, 307.

Fowler, C. F., and Kok, B. (1974). *Biochim. Biophys. Acta* **357**, 299.

Fowler, C. F., and Kok, B. (1976). *Biochim. Biophys. Acta* in press.

Frenkel, A. (1954). *J. Amer. Chem. Soc.* **76**, 5568.

Gaastra, P. (1959). *Meded. Landbouwhogesch. Wageningen* **59**, 1.

Goldworthy, A. (1968). *Nature (London)* **217**, 62.

Good, N. E., Izawa, S., and Hind, G. (1966). *Curr. Top. Bioenerg.* **1**, 75.

Goodwin, T. W. (1960). *In* "Handbuch der Pflanzenphysiologie" (W. Ruhland, ed.), Vol. 5, Part 1, p. 39. Springer -Verlag, and New York.

Griffith, M., Sistrom, W. R., Cohen Bazire, G., and Stanier, R. Y. (1965). *Nature (London)* **176**, 1211.

Hall, D. O., Reeves, S. G., and Baltocheffsky, K. (1971). *Biochem. Biophys. Res. Commun.* **43**, 359.

Haxo, F. R., O hEocha, C., and Norris, P. (1955). *Arch. Biochem. Biophys.* **54**, 162.

Haxo, F. T., and Blinks, L. R. (1950). *J. Gen. Physiol.* **33**, 389.

Hill, R. (1954). *Nature (London)* **174**, 501.

Hind, G., and Jagendorf, A. T. (1963). *Proc. Nat. Acad. Sci.* **49**, 715.

Hind, G., and Olson, J. M .(1968). *Annu. Rev. Plant Physiol.* **19**, 249.

Hiyama, T., and Ke, B. (1971). *Proc. Nat. Acad. Sci.* **68**, 1010.

Hoch, G. E., and Martin, I. (1963). *Biochem. Biophys. Res. Commun.* **12**, No. 3.

Hoch, G. E., Owens, O. van. H., and Kok, B. (1963). *Arch. Biochem. Biophys.* **101**, 160.

Holt, A. S., and Jacobs, E. E. (1954). *Amer. J. Bot.* **41**, 710.

Izawa, S., and Good, N. E. (1968). *Biochim. Biophys. Acta* **162**, 380.

Izawa, S., and Hind, G. (1967). *Biochim. Biophys. Acta* **143**, 377.

Izawa, S., Winget, G. D., and Good, N. E. (1966). *Biochem. Biophys. Res. Commun.* **22**, 223.

Izawa, S., Heath, R., and Hind, G. (1969). *Biochim. Biophys. Acta* **180**, 338.

Jagendorf, A. T., and Hind, G. (1963). *Nat. Acad. Sci. Nat. Res. Counc., Publ.* **1145**, 599.

Jagendorf, A. T., and Uribe, E. (1967). *Brookhaven Symp. Biol.* **19**, 215.

Joliot, A., and Joliot, P. (1964). *C.R. Acad. Sci.* **258**, 4622.

Joliot, P., (1961). *J. Chim. Phys.* **58**, 584.

Joliot, P. (1966). *Brookhaven Symp. Biol.* **19**, 418.

Joliot, P. (1968). *Photochem. & Photobiol.* **8**, 451.

Joliot, P., Joliot, A., and Kok, B. (1968). *Biochim. Biophys. Acta* **153**, 635.

Jones, and Myers, J. (1964). *Plant Physiol.* **39**, 938.

Junge, W., and Witt, H. T. (1968). *Z. Naturforsch. B* **23**, 244.

Katoh, S. (1960). *Nature (London)* **186**, 553.

Knaff, D. B., and Arnon, D. I. (1969). *Proc. Nat. Acad. Sci. U.S.* **63**, 963.

Kok, B. (1956). *Biochim. Biophys. Acta* **21**, 245.

Kok, B. (1960). *In* "Handbuch der Pflanzenphysiologie" (W. Ruhland, ed)., Vol. 5, Part 1, p. 566. Springer-Verlag, Berlin and New York.

Kok, B. (1961). *Biochim. Biophys. Acta* **48**, 527.

Kok, B., and Gott, W. (1960). *Plant Physiol.* **35**, 802.

Kok, B., Forbush, B., and McGloin, M. (1970). *Photochem. & Photobiol.* **11**, 457.

Kraayenhof, R. (1969). *Biochim. Biophys. Acta* **180**, 213.

Livingston, R. (1960). *Quart. Rev., Chem. Soc.* **14**, No. 2, 174.

Lundegarth, H. (1952). *Physiol. Plant.* **15**, 390.

McCarty, R. E. (1969). *J. Biol. Chem.* **244**, 4292.

McCarty, R. E., and Racker, E. (1966). *Brookhaven Symp. Biol.* **19**, 202.

Malkin, R., and Bearden, J. (1971). *Proc. Nat. Acad. Sci. U.S.* **68**, 16.

Mehler, A. H. (1951). *Arch. Biochem. Biophys.* **34**, 339.

Mitchell, P. (1961). *Nature (London)* **191**, 144.

Mitchell, P. (1966). "Chemiosmotic Coupling in Oxidative and Photosynthetic Phosphorylation." Glynn Res., Bodmin, Cornwall, England.

Murata, N. (1969). *Biochim. Biophys. Acta* **172**, 242.

Myers, J. (1963). *Nat. Acad. Sci.—Nat. Res. Counc., Publ.* **1145**, 301.

Myers, J. (1971). *Annu. Rev. Plant Physiol.* **22**, 289.

Neumann, J., and Jagendorf, A. T. (1964). *Arch. Biochem. Biophys.* **107**, 109.

Ogren, W. L., Lightbody, J. J., and Krogmann, D. W. (1965). *Rec. Chem. Progr.* **265**, 84.

Olson, R. A. (1963). *Nat. Acad. Sci.—Nat. Res. Counc., Publ.* **1145**, 545.

Radmer, R., and Kok, B. (1973). *Biochim. Biophys. Acta* **324**, 28.

Saha, S. Ouitrakul, R., Izawa, S., and Good, N. E. (1971). *J. Biol. Chem.* **246**, 3204.

San Pietro, A., and Lang, H. M. (1958). *J. Biol. Chem.* **231**, 211.

Schroder, H., Muhle, H., and Rumberg, B. (1972). *Proc. Int. Congr. Photosyn., 2nd, 1972*, p. 919.

Schwartz, M. (1966). *Biochim. Biophys. Acta* **131**, 559.

Shen, Y. K., and Shen, G. M. (1962). *Sci. Sinica* **11**, 109.

Siggel, U., Renger, G., and Rumberg, B. (1972). *Proc. Int. Congr. Photosyn., 2nd, 1971*, p. 753.

Stiehl, H. H., and Witt, H. T. (1967). *Z. Naturforsch. B* **23**, 220.

Stiehl, H. H., and Witt, H. T. (1969). *Z. Naturforsch. B* **24**, 1588.

Strehler, B. L., and Arnold, W. (1951). *J. Gen. Physiol.* **34**, 809.

Sun, A. S. K., and Sauer, K. (1971). *Biochim. Biophys. Acta* **234**, 399.

Tagawa, K., and Arnon, D. I. (1962). *Nature (London)* **195**, 537.

Tagawa, K., Tsujimoto, H. Y., and Arnon, D. I. (1963). *Proc. Nat. Acad. Sci. U.S.* **49**, 567.

Talling, J. F. (1961). *Annu. Rev. Plant Physiol.* **12**, 133.

Tanner, W., Loos, E., and Kandler, O. (1965). *In* "Currents in Photosynthesis" (J. B. Thomas and J. C. Goedheer, eds.), p. 243. Donker, Publ., Rotterdam, The Netherlands.

Tolbert, N. E., Nelson, E. B., and Bruin, W. J. (1971). *In* "Photosynthesis and Photorespiration" (M. D. Hatch, C. B. Osmond, and R. O. Slatyer, eds.), p. 506. Wiley (Interscience), New York.

Trebst, A., Eck, H., and Wagner, S. (1963). Nat. Acad. Sci.—*Nat. Res. Counc., Publ.* **1145**, 174.

Vambutas, U. K., and Racker, E. (1965). *J. Biol. Chem.* **240**, 2660.

Velthuys, B., and Amesz, J. (1974). *Biochim. Biophys. Acta* **333**, 85.

Vernon, L. P., and Zaugg, W. S. (1960). *J. Biol. Chem.* **265**, 2728.

Walker, D. A., and Crofts, A. R. (1970). *Annu. Rev. Biochem.* **38**, 389.

Witt, H. T. (1971). *Quart. Rev. Biophys.* **4**, 365.

Witt, H. T. Rumberg, B., Schmidt-Mende, P., Siggel, U., Skerra, B., Vater, J., and Weikard, J. (1965). *Angew. Chem., Int. Ed. Engl.* **4**, 799.

Yamashita, T., and Butler, W. L. (1968). *Plant Physiol.* **43**, 1978.

Zelitsch, I., ed. (1971). In "Photosynthesis, Photorespiration, and Plant Productivity," p. 244. Academic Press, New York.

26

Nitrogen Fixation

R. H. BURRIS

I. History

The ancient agricultural practice of rotation of leguminous and nonleguminous crops was based on the observation that the nonleguminous crops grew more vigorously following leguminous crops. It was not appreciated that the benefit was related to N_2 fixation until Boussingault in 1838 established in greenhouse and field experiments that leguminous plants fixed N_2 from the air. His findings were strongly disputed by Liebig and by Lawes, Gilbert, and Pugh, but the supporting evidence furnished in 1886 by Hellriegel and Wilfarth was thoroughly convincing and pointed out the association of N_2 fixation and root nodulation. Winogradsky shortly thereafter established that certain free-living anaerobic clostridia fixed N_2, and Beijerinck demonstrated that the aerobic *Azotobacter chroococcum* also had this capability.

These early observations were followed by studies of the physiology of the N_2-fixing bacteria and by an emphasis on practical agricultural application of biological N_2 fixation. Biochemical investigations of the process were started in the late 1920's, and our discussion will emphasize this aspect of N_2 fixation.

II. The Nitrogen Cycle

Biological fixation of N_2 is of central importance in maintaining a positive nitrogen balance on earth. Figure 1 shows a simplified nitrogen cycle and indicates that the symbiotic and nonsymbiotic N_2-fixing agents reduce N_2 from the air and furnish it to support growth of organisms in terrestrial or aquatic environments. Biological fixation of N_2 returns fixed nitrogen to the terrestrial cycle to replace that lost through sewage and by denitrification, leaching, and erosion. It is difficult to strike an accurate nitrogen balance on a global basis, and the divergence of figures in the literature indicates that the values cited represent relatively crude estimates. Delwiche (1970) has suggested that total annual fixation of N_2 is 92×10^6 metric tons [Burns and Hardy (1975) have estimated about twice this amount] and the loss by denitrification is 83×10^6 metric tons. The net gain represents fixed nitrogen that is accumulating in the soil, groundwater, lakes and streams, and the oceans.

The contributions of fixed nitrogen by leguminous crops can be estimated with some accuracy, but few of the nonleguminous N_2-fixing plants

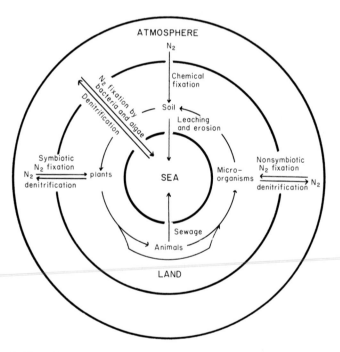

Fig. 1. The nitrogen cycle, including the interchanges between atmosphere, land, and sea.

are cultivated, and their contribution of fixed nitrogen is less accurately known. Evaluation of the contribution of free-living microorganisms is even more difficult. Bacteria are dominant among the free-living N_2-fixing organisms in soils, whereas the N_2-fixing blue-green algae appear to have the chief role in aquatic environments.

III. Biological Agents That Fix N_2

Groups of organisms capable of reducing N_2 are listed in Table I.

A. Symbiotic Nitrogen-Fixing Agents

1. LEGUMINOUS PLANTS PLUS ROOT NODULE BACTERIA

Leguminous crops often fix in excess of 100 kg of nitrogen annually per hectare. Delwiche (1970) estimated that they are responsible annually for fixation of 14×10^6 metric tons of nitrogen on earth. About 13,000 species of the Leguminosae have been described, and the large majority fix N_2. For convenience they often are classified in bacterial–plant groups or cross-inoculation groups; bacteria isolated from the nodules of any member of the group are capable of nodulating any other member of the group. Crossing between the groups is uncommon. Table II lists a number of economically important groups; species names for the bacteria are assigned on the basis of their bacterial–plant group.

In the initial process of infection of the roots and establishment of

TABLE I

Biological Agents That Fix N_2

A. Symbiotic N_2-fixing agents
  1. Leguminous plants plus root nodule bacteria (peas, beans, clover, soybeans, etc.)
  2. Nonleguminous plants, angiosperms (alder, seabuckthorn, *Ceanothus, Myrica, Purshia*, etc.)

B. Nonsymbiotic nitrogen-fixing agents
  1. Blue-green algae (*Nostoc, Anabaena, Calothrix, Mastigocladus*, etc.)
  2. Yeasts (*Rhodotorula* sp.)
  3. Bacteria
    a. Aerobic (*Azotobacter, Beijerinckia, Derxia*, etc.)
    b. Facultative (*Bacillus, Klebsiella*, etc.)
    c. Anaerobic
      i. Nonphotosynthetic (*Clostridium, Desulfovibrio, Methanobacterium*, etc.)
      ii. Photosynthetic (*Rhodospirillum, Chromatium, Chlorobium, Rhodomicrobium*)

TABLE II
Bacterial–Plant Groups

Bacteria	Representative plants nodulated
Rhizobium meliloti	Sweet clover, alfalfa
Rhizobium trifolii	Red clover, white clover
Rhizobium leguminosarum	Pea, sweet pea, broad bean
Rhizobium phaseoli	Bean
Rhizobium japonicum	Soybean
Rhizobium lupini	Lupine

nodules, the rhizobia cluster around the root hairs, and in response to the bacteria the root hairs curl at their tips. The bacteria then invade the root hairs and the plant produces an infection thread. The bacteria proliferate and pass through the infection thread to the inner cortical cells and pericycle of the root, and these cells in turn give rise to the nodule. The balance between plant and bacteria is delicate, and many infection threads abort before a nodule is formed. The nodule has a highly organized structure with vascular connections to the root. Wilson (1940) and Dart and Mercer (1964) have discussed nodulation and have published illustrations of the process of infection and development.

Curiously, the cells in the nodule are polyploid. That is, roots with $2n$ chromosomes have $4n$ chromosomes in their nodules; $8n$ plants have $16n$ chromosomes in their nodules. A careful examination of many root cells led Wipf and Cooper (1940) to abandon the obvious hypothesis that the bacteria or a substance from them induces polyploidy. Rather, they reported that only the spontaneously polyploid root cells are invaded and proliferate to form nodules. However, Phillips and Torrey (1970) have presented evidence that the rhizobia produce a cytokinin that induces polyploidy.

It is possible to nodulate roots that have been cultured apart from the plant. Although these nodules are small, they show a normal nodular structure, contain hemoglobin, and can fix N_2.

Infection of the root and production of nodules does not guarantee vigorous N_2 fixation. A critical balance governs effective symbiosis between plant and bacteria, and this is reflected in the phenomena of strain variation and host plant specificity (Table III). Bacterial strain variation is illustrated by horizontal comparisons in the table. For example, *Melilotus suaveolens* fixed 121, 11 and 126 mg N per 10 plants when infected with bacterial strains 100, 105 and 128, respectively. Nodules always were formed, but strains 100 and 128 were effective, whereas

TABLE III

Strain Variation and Host Plant Specificity[a]

Species or strain of sweet clover	N fixed (in mg) per 10 plants with different strains of R. meliloti		
	Strain 100	Strain 105	Strain 128
Melilotus alba (32–19)	186.3	111.0	147.5
Melilotus suaveolens	121.0	11.0	126.0
Melilotus dentata (91–12)	149.0	95.0	131.0
Melilotus dentata (92–27)	131.0	35.8	142.4
Melilotus dentata (96–2)	111.5	142.0	138.2

[a] Data from Wilson (1940, p. 84).

strain 105 was not. Comparison in the second vertical column of figures illustrates the phenomenon of host plant specificity. With a single strain of bacteria and 5 different species or varieties of plants, the fixation of N_2 varied from 11 to 142 mg N per 10 plants. Again all plants were nodulated, but the bacterial–plant relationship supported poor fixation in some instances and good fixation in others. The basis for these differences in effectiveness remains unknown.

The occurrence of hemoglobin in root nodules is notable because hemoglobin has not been demonstrated elsewhere in the plant kingdom with the exception of small concentrations in *Neurospora crassa* and certain petite yeasts. Kubo (1939) recognized that the red pigment in nodules is hemoglobin. The hemoglobin of nodules is spectroscopically indistinguishable from mammalian hemoglobin but differs somewhat in its amino acid composition. A careful examination of the nonsymbiotic N_2-fixing bacteria has not revealed any hemoglobin in them, so its function evidently is confined to the symbiotic systems. Kubo's observation that hemoglobin enhances the respiration of the rhizobia prompted him to suggest that it functions in an oxygen-transport system that supports respiration in nodules. Smith (1949) objected to this interpretation, because concentrations of CO, which blocked O_2 transport by hemoglobin, did not block the enhancement of respiration by hemoglobin. Although the utility of O_2 transport via hemoglobin in the absence of a system to circulate the hemoglobin has been questioned, O_2 may be moved by transfer from molecule to molecule of hemoglobin (Wittenberg, 1970).

Virtanen suggested that hemoglobin might function directly in N_2 fixation by forming hydroxylamine from N_2, but no supporting experimental evidence was offered. A causal relationship between N_2 fixation and hemoglobin is implied by the correlation between vigor of fixation and amount of hemoglobin in the nodules, by the fact that there is no fixation of N_2 before hemoglobin is formed in the nodules and that fixation ceases when hemoglobin disappears, and by extreme sensitivity of N_2 fixation to CO. It now is established that the relationship is indirect, because properly isolated bacteroids (rhizobia isolated directly from nodules are termed bacteroids and are commonly swollen, branched, and vacuolated) from nodules can fix N_2 in the absence of hemoglobin if supplied ATP and a suitable reductant. Apparently the hemoglobin controls the concentration of dissolved oxygen at a level that is sufficient to support ATP formation through oxidative phosphorylation but insufficient to inactivate the nitrogenase of the bacteroids. The genetic information for hemoglobin synthesis is supplied by the plant, because hemoglobins differ with different plants inoculated with the same strain of bacteria, whereas the hemoglobins are the same in a specific plant inoculated with different strains of bacteria (Dilworth, 1969).

The leguminous plant by itself cannot fix N_2, and laboratory cultures of the rhizobia can fix N_2 only under special cultural conditions. The requirement for the symbiotic activity of plants and bacteria was considered absolute until fixation by isolated bacteroids was achieved. It still is not clear how rhizobia normally incapable of fixing N_2 in laboratory culture are converted to N_2-fixing bacteroids within the plant. It has been postulated that the plant transfers genetic information to the bacteria (Dilworth, 1974), but such a postulate is unnecessary, as it now has been demonstrated that certain free-living rhizobia are capable of fixing N_2 when supplied with the proper nutrients independent of the plant (Pagan *et al.*, 1975; Kurz and LaRue, 1975; McComb *et al.*, 1975).

2. Nonleguminous Plants

Although attention in symbiotic N_2 fixation usually is focused on the leguminous plants, there are a number of nonleguminous plants that fix N_2 symbiotically. The alder has been most extensively investigated, because it is widely distributed, fixes N_2 vigorously, and has prominent nodules. The nature of the microorganism that nodulates the alder is obscure, because it never has been cultured free of the plant and then returned to aseptic plants to induce nodulation. It is generally accepted that an actinomycete is the infective agent on the basis of rather convincing indirect evidence.

Other nonleguminous plants that fix N_2 are included in 13 genera, and the plants in these genera are woody and dicotyledonous. Bond (1967) lists these genera and indicates that 310 species well-distributed over the world have been reported to fix N_2. Alder is very common in sandy areas and often invades primitive volcanic soils; its contribution of fixed nitrogen to the soil permits establishment of other species of plants. The leaves of *Ceanothus* sp. growing on rocky mountain slopes and *Comptonia peregrina* growing on sandy roadsides have the dark green color characteristic of plants with abundant fixed nitrogen and stand in sharp contrast to most other plants in these habitats. The N_2-fixing system of nonleguminous plants appears similar to that of other N_2 fixers, although it has not been studied intensively.

B. Nonsymbiotic Nitrogen-Fixing Agents

1. BLUE-GREEN ALGAE

For some years it was considered that all N_2-fixing blue-green algae belonged to the family Nostocaceae, but now fixation has been demonstrated clearly in representatives of the Scytonemataceae, Oscillatoriaceae, Chroococcaceae, Rivulariaceae, and Stigonemataceae (Stewart, 1966; Burns and Hardy, 1975).

The blue-green algae appear to be the predominant agents that are responsible for the maintenance of the nitrogen supply in rice fields in Asian countries. During the period when water stands in the paddies, a heavy pellicle of blue-green algae develops and fixes N_2. With evaporation of the water, the pellicle settles to the soil where it decomposes and liberates fixed nitrogen for the nourishment of the rice.

Recent concern with eutrophication of our freshwater lakes has drawn attention to the blue-green algae as major offenders in nuisance blooms. When fixed nitrogen is a limiting factor and when phosphate and other nutrients are in adequate supply, the N_2-fixing blue-green algae have a competitive advantage over other organisms. They grow vigorously, come to the surface as they age, and their decomposition in the lake or on shore creates a nuisance. The nitrogen they fix recycles to support growth of fixed weeds or floating organisms. In a moderately eutrophic lake, the blue-green algae may fix 2 to 5 kg of N_2 annually per hectare.

Most N_2-fixing blue-green algae have heterocysts, cells with thickened walls and depleted cell contents. They are virtually devoid of pigments associated with photosystem II, the O_2-liberating system of photosynthesis. Considerable evidence supports the idea that N_2 fixation is localized in the heterocysts. This idea is attractive because (a) the N_2-fixing system requires protection from O_2, and O_2 is not generated in heterocysts and (b) most N_2-fixing blue-green algae are heterocystous.

The observation that certain nonheterocystous blue-green algae can fix N_2 under anaerobic conditions removes the restriction that N_2 fixation must occur in heterocysts but supports the requirement that O_2 must be excluded or kept in low concentration to permit N_2 fixation.

2. YEASTS

Yeasts have received little attention, and there is no information regarding their practical importance as nitrogen fixers. However, a pink *Rhodotorula* sp. isolated from soil has been shown to fix N_2 both by an increase in total N and by the assimilation of ^{15}N from $^{15}N_2$.

3. BACTERIA

a. Aerobic. The *Azotobacter* sp. and closely related *Beijerickia* sp. are representative of a group of N_2 fixers characterized by their wide distribution in soil and water and their vigorous respiration. No organism has been reported with a more active respiration than that of the azotobacter. Other aerobic N_2 fixing organisms include representatives from the genera *Mycobacterium* and *Derxia*.

Because of their rapid growth and their high rate of N_2 fixation, the azotobacter have been favorite subjects for investigations of N_2 fixation. This has tended to overemphasize their practical importance in the field and has implied that they are typical N_2 fixers, whereas it probably is justified to consider them as atypical and the anaerobes as typical N_2 fixers. All facultative organisms fix only under anaerobic conditions, and the azotobacter probably achieve aerobic fixation by establishing strongly reducing conditions locally within their cells by virtue of their vigorous respiration. Despite the fact that *Azotobacter vinelandii* is an aerobe, it has 2 ferredoxins with redox potentials of -420 and -460 mV (Yoch and Arnon, 1972); these are functional in N_2 fixation. Fixation of N_2 in all organisms is fundamentally a reductive rather than an oxidative process.

There are many reports that inoculation of the soil with the azotobacter will enhance fixation, and these have been summarized by Mishustin and Shil'nikova (1971). Normally the azotobacter are present in soil and need only the proper pH, O_2, and a suitable energy source to multiply and fix N_2. Estimates of the amount of fixation by the azotobacter in soil are based on inadequate experimental data because of the difficulty in simulating field conditions in the laboratory and of estimating changes in total nitrogen in the field. It has been suggested that the azotobacter under favorable conditions may add an average of about 10 kg of nitrogen per hectare per year to the soil.

b. Facultative. This group of organisms will grow either aerobically or anaerobically, but their fixation of N_2 appears to be confined to anaerobic conditions. There has been no adequate evaluation of their role in the enrichment of soil with fixed nitrogen. The ability to fix N_2 has been reported for organisms classified as *Bacillus* sp. (normally an aerobic organism), and *Klebsiella* sp., but not all species or strains of these genera are active. As classification of these organisms is somewhat arbitrary, there eventually may be some shifting of organisms between genera.

c. Anaerobic.   i. Nonphotosynthetic. Most clostridia tested have proved capable of fixing N_2. In one series of experiments, 12 of 15 species examined fixed N_2. These organisms have an almost universal distribution in soil and are highly resistant to unfavorable conditions. They grow rapidly, and when conditions of moisture and nutrients are favorable they should add substantially to the supply of fixed nitrogen in the soil. They have replaced the azotobacter as the most popular organisms for the study of the mechanism of N_2 fixation because they yield a relatively stable complex of soluble enzymes capable of fixing N_2.

Desulfovibrio desulfuricans is a strictly anaerobic organism that uses sufate as an oxidant. It is a less vigorous N_2 fixer than the clostridia and probably fixes N_2 only in rather special natural environments. Although the organism commonly is grown on lactate, it can use H_2 in an autotrophic type metabolism. *Methanobacterium* sp. is sometimes considered with the autotrophs, and it too can fix N_2.

ii. Photosynthetic. The photosynthetic bacteria examined have all proved capable of fixing N_2, but it was a remarkably long time before this property was recognized in these intensively studied organisms. Kamen and Gest (1949) were led to the discovery that *Rhodospirillum rubrum* fixed N_2 because of their observation that the evolution of H_2 was inhibited by N_2. N_2 fixation was demonstrated later by representatives of the purple sulfur bacteria (*Chromatium* sp.), the green sulfur bacteria (*Chlorobium* sp.), and the yeast-like *Rhodomicrobium* sp.

These organisms are abundant in soil and water, but their growth in soil commonly is limited by lack of light. In certain bodies of water they may play an important part in the nitrogen cycle.

IV. Biochemistry of N_2 Fixation

A. Studies with Intact Organisms

Early studies of the biochemistry of N_2 fixation were confined to investigations with intact organisms. Although the complex of reactions in the intact organism often was difficult to interpret, it was possible to establish certain of the basic characteristics of N_2 fixation.

1. RESPONSE TO SUBSTRATE

Nitrogen is the substrate for nitrogen fixation. The Michaelis constants (K_m) for N_2 fixation by intact *Azotobacter vinelandii*, the blue-green alga *Nostoc muscorum*, and excised soybean nodules all fall between 0.020 and 0.025 atm N_2. This means that under natural conditions these enzymes normally are saturated or nearly saturated with N_2.

2. RESPONSE TO GASES OTHER THAN N_2

a. Oxygen. The efficiency of fixation by the azotobacter in terms of units of N_2 fixed per unit of carbohydrate oxidized is greater at a low (0.02 atm) pO_2 than at a higher (0.2 atm) pO_2 (Postgate, 1971). Under a low pO_2 the demands on the organism to maintain a strongly reducing potential for N_2 fixation are less stringent than at a high pO_2. O_2 at pressures above 0.5 atm is toxic to the azotobacter.

b. Hydrogen. Hydrogen is a specific and competitive inhibitor of N_2 fixation. It does not influence the use of combined nitrogen, but as first observed by Wilson and Umbreit (1937) with red clover plants, a pH_2 of 0.6 atm with 0.2 atm N_2 and 0.2 atm O_2 inhibited N_2 fixation by about 70%. The mean K_i from numerous experiments was 0.137 atm (Wilson, 1940, p. 206). H_2 inhibition of fixation by the azotobacter is comparable ($K_i = 0.11$ atm), but higher amounts of H_2 are required to inhibit fixation by the clostridia ($K_i = 0.55$ atm).

c. Nitrous Oxide, Nitric Oxide, and Carbon Monoxide. Nitrous oxide is a nitrogenous compound that is a specific competitive inhibitor of N_2 fixation. It is effective in concentrations comparable to H_2. Nitric oxide (NO) is a competitive inhibitor that is active at very low concentrations. Carbon monoxide is a potent noncompetitive inhibitor; 0.004 atm will inhibit N_2 fixation by *A. vinelandii* about 45% in air.

3. INTERMEDIATES IN N_2 FIXATION

The key role of ammonia in N_2 fixation was firmly established through the application of ^{15}N in studies with intact N_2-fixing organisms. However, alternative hypotheses were offered before the ammonia hypothesis was accepted.

a. Hydroxylamine Hypothesis. This hypothesis held that N_2 is reduced to NH_2OH which combines with keto acids to form oximes and that the oximes in turn are reduced to amino acids. Figure 2 presents this as one possible pathway for fixation. Evidence in support of the hydroxylamine hypothesis was based primarily upon the observation of products excreted from the nodules of leguminous plants (aspartic acid, β-alanine, and oximinosuccinic acid). As the excreted products could have been formed many days before and could have undergone numerous reactions before their excretion, they bore no necessary relation to the early products of N_2 fixation. Hence, the evidence cited in support of hydroxylamine lacked specificity.

b. Ammonia Hypothesis. The use of ^{15}N as a tracer permitted the compression of the time scale of reactions from days to minutes and conferred sensitivity and specificity on the measurements. The application of ^{15}N furnished several lines of evidence that consistently supported the often postulated role of NH_3 as the key intermediate of N_2 fixation.

Lines of evidence supporting the ammonia hypothesis [points (1)–(4) discussed in more detail by Burris, 1956] include the following.

1. Ammonia is used preferentially. If a N_2-fixing organism is given a choice of N_2, various fixed N compounds, and NH_4^+, it will use NH_4^+ preferentially under a wide variety of conditions. It will use NH_4^+ immediately, without adaptation and to the virtual exclusion of other nitrogenous compounds.

2. Organisms supplied [15]N_2 for short periods accumulate the highest

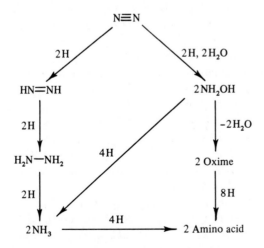

Fig. 2. Possible pathways for reduction of N_2.

concentration of ^{15}N in glutamic acid among their amino acids released by hydrolysis of cellular proteins. A dominant pathway for nitrogen assimilation is through the incorporation of NH_4^+ into glutamic acid by reductive amination of α-ketoglutaric acid.

3. Kinetic experiments support the role of NH_4^+. If $^{15}N_2$ is supplied to a rapidly growing N_2-fixing organism and the culture is sampled at short intervals, the ^{15}N concentration builds up rapidly in the "amide" fraction. If the percentage of total ^{15}N fixed is plotted versus time for each compound recovered, only the "amide" fraction gives a negative slope characteristic of an initial product.

4. The formation of ammonia can be demonstrated directly. Cultures of *Clostridium pasteurianum* excrete NH_4^+ when grown on a medium that supplies inadequate α-ketoglutaric acid to serve as an acceptor for the NH_4^+. If the cultures are supplied $^{15}N_2$ for a few minutes, the highest ^{15}N concentration among recoverable products is in the NH_4^+.

5. Cell-free preparations from N_2-fixing organisms produce NH_4^+. Conclusions from points (1) to (4) obtained with intact organisms were reinforced by the demonstration that cell-free N_2-fixing preparations incapable of assimilating NH_4^+ accumulate NH_4^+ as the product of N_2 fixation.

B. Isolation and Purification of Nitrogenase

1. ACTIVE CELL-FREE PREPARATIONS

Early work on cell-free N_2 fixation was reviewed by Burris (1966). Although numerous active preparations had been recovered, they were inconsistent in their activity and hence not particularly useful for investigating the mechanism of N_2 fixation. Carnahan *et al.* (1960) reported a vigorous and consistent preparation from *Clostridium pasteurianum*, and this markedly extended the scope of possible studies on N_2 fixation. *C. pasteurianum* cells were dried in a rotary evaporator, the dried cells were extracted anaerobically with buffer, the extracted material was centrifuged, and the soluble supernatant constituted the active preparation. When it was supplied high concentrations of pyruvate as a substrate, this crude extract fixed N_2 rapidly under anaerobic conditions. The extract decomposed pyruvate by the "phosphoroclastic" reaction to yield CO_2, H_2, and acetyl phosphate.

Subsequently, active cell-free extracts have been prepared from a variety of organisms, including *Azotobacter vinelandii*, *Bacillus polymyxa*, *Klebsiella pneumoniae*, *Rhodospirillum rubrum*, *Chromatium* strain D, *Mycobacterium flavum*, the blue-green alga *Mastigocladus laminosus,* and the bacteroids from soybean root nodules. Among these, the extracts from *C. pasteurianum, K. pneumoniae,* and *A. vinelandii* have been studied most extensively. Bulen *et al.* (1965) introduced the use of $Na_2S_2O_4$ (sodium dithionite) as the reducing agent and creatine phosphate-creatine kinase as the ATP-generating system for studies of extracts from *A. vinelandii* and other organisms.

2. EXTRACTION OF NITROGENASE

As mentioned, dried cells of *C. pasteurianum* release soluble nitrogenase when shaken anaerobically with buffer for 30 to 60 minutes. The active proteins are not sedimented in 5 hours at 144,000 *g. A. vinelandii* commonly is stored at $-20°C$ as a wet paste. If a slurry of these cells is passed through the orifice of a French pressure cell (10,000–20,000 psi), they are disrupted and yield a particulate cell-free preparation of nitrogenase. If *A. vinelandii* is permitted to accumulate a high concentration of glycerol and the cells are suddenly transferred to weak buffer solution, the accompanying osmotic shock disrupts the cells. Treatment with lysozyme will disrupt cells of *Bacillus polymyxa.*

3. PRELIMINARY STEPS IN PURIFICATION

Methods of purification depend upon the organism and the preferences of the investigator. Preliminary steps often include centrifugation to remove particulate material (or to recover active particles from *A. vinelandii*), treatment with protamine sulfate or streptomycin sulfate to precipitate nucleic acids, heating anaerobically at 55°–65°C to precipitate inactive protein, or precipitation with polyethylene glycol to remove impurities. Higher concentrations of protamine sulfate or polyethylene glycol are useful for precipitating nitrogenase.

4. SEPARATION IN COLUMNS

Concentrated preparations after preliminary purification usually are placed on columns of DEAE cellulose. The acidic proteins of nitrogenase are held by DEAE cellulose, while extraneous proteins are displaced with buffered NaCl or $MgCl_2$ of low ionic strength. A higher concentration of NaCl will elute the MoFe protein (sometimes designated component I) and still more concentrated NaCl will displace the Fe protein (component II).

Solutions eluted from columns of DEAE cellulose must be concentrated (ultrafiltration) and then they can be purified further on columns of Sephadex G 100 (Fe protein) or Sephadex G 200 (MoFe protein). Essentially homogeneous preparations can be recovered by these procedures. All operations must be performed under strictly anaerobic conditions to avoid irreversible inactivation by O_2. It also must be recognized that the Fe protein from some organisms is cold labile at $0°C$.

C. Physicochemical Properties of Nitrogenase

Nitrogenase (the term applied to the enzyme complex active in N_2 fixation) consists of two proteins. Neither protein has any catalytic activity by itself, but when the two proteins are mixed together they produce nitrogenase, which (with reductant and ATP) will catalyze a variety of reactions. We designate the proteins as the Fe protein and the MoFe protein, although other names have been introduced in the literature. Their properties are summarized in Table IV.

There is nothing unusual about the amino acid composition of the MoFe protein or the Fe protein. The Fe of the proteins is bound to S; note the equivalence between acid-labile sulfur (S^{2-}) and Fe in the proteins from *C. pasteurianum* (see Table IV).

The Fe protein and MoFe protein lack marked spectral peaks in the visible and UV range useful for identification. However, Mössbauer spectra have been determined, and Kelly and Lang (1970) interpret these to substantiate the important role of Fe in nitrogenase and to suggest that a portion of the Fe atoms in the MoFe protein functions in pairs or larger multiples.

TABLE IV

Properties of Nitrogenase

Properties	Clostridium pasteurianum[a]		Azotobacter vinelandii[b]	
	MoFe protein	Fe protein	MoFe protein	Fe protein
Molecular weight	220,000	55,000	216,000	64,000
Subunits	2 at 59,500 and	—	—	—
	2 at 50,700	27,500	4 at 56,000	33,000
Fe	22–24	4	24–32	3.45
Mo	2	0	1.54–2	0
–SH per molecule	30	12	37	—
S^{2-} per molecule	22–24	4	20–25	—
EPR, reduced	no signals[c]	$g = 2.05, 1.94, 1.89$[c]	$g = 4.3, 3.67, 2.01$	
EPR, oxidized	$g = 4.3, 3.7, 2.01$[c]	no signals[c]	$g = 4.3, 2.01$	

[a] Data from Dalton and Mortenson (1972).
[b] Data from Burns et al., (1971) and Kleiner and Chen (1974).
[c] Data from Orme-Johnson et al. (1972).

Both the MoFe protein and the Fe protein have characteristic electron paramagnetic resonance (EPR) signals, which can be observed near liquid helium temperatures (Table IV). The MoFe protein signal changes with the oxidation–reduction state of the protein. The signal of the Fe protein is altered markedly by binding of ATP (Orme-Johnson et al., 1972).

D. Reactions of Nitrogenase

1. SUBSTRATES

The substrate for nitrogenase is N_2. The first hint that this categorical statement might be incomplete came with the demonstration by Wilson and Umbreit (1937) that H_2 is a specific, competitive inhibitor of N_2 fixation. It was apparent that a gas other than N_2 could be bound at the active site of nitrogenase. Mozen and Burris (1954) demonstrated that nitrous oxide (a competitive inhibitor of N_2 fixation) was assimilated by A. vinelandii and by soybean nodules, and this constituted the first evidence that a compound other than N_2 could be reduced by nitrogenase. This was followed by demonstrations that azide, acetylene, cyanide, methyl isocyanide, and analogs of some of these compounds can be reduced by nitrogenase. In addition, nitrogenase can reduce H^+ (H_3O^+) to H_2 and can hydrolyze ATP. Nitrogenase is a versatile enzyme that can reduce a variety of substrates as summarized in Table V.

The reduction of C_2H_2 yields C_2H_4, which can be measured easily in a gas chromatograph with a flame ionization detector. This reaction has been used as a quantitative index of nitrogenase and has been applied extensively both in the laboratory and in the field. The simplicity and high sensitivity of the method particularly recommend it for field work on higher plants, aquatic bacteria and algae, and soils.

TABLE V

Reactions Catalyzed by Nitrogenase

$$N_2 \rightarrow NH_3$$
$$N_2O \rightarrow N_2 + H_2O$$
$$N_3^- \rightarrow N_2 + NH_3$$
$$C_2H_2 \rightarrow C_2H_4$$
$$HCN \rightarrow CH_4 + NH_3 + [CH_3NH_2]$$
$$CH_3NC \rightarrow CH_4 + CH_3NH_2 + [C_2H_4, C_2H_6]$$
$$2\ H^+ \rightarrow H_2$$
$$ATP \rightarrow ADP + P_i$$

The acetylene reduction technique is considered a valid measure of N_2 fixation because (a) highly purified nitrogenase gives the reaction, (b) N_2 and C_2H_2 reduction activities are parallel during purification and inactivation of nitrogenase, and (c) both reactions have the same requirement for ATP and a strong reductant. All the reactions of Table V meet the criteria, which indicate they are catalyzed by nitrogenase; all require ATP and a strong reductant.

2. REQUIREMENT FOR ATP

Carnahan et al. (1960) used pyruvate to support N_2 fixation by cell-free extracts from C. pasteurianum, and they reported that added ATP was inhibitory to N_2 fixation (actually ADP was the inhibitor). Hence, the statement by McNary and Burris (1962) that ATP was required was greeted with skepticism; they found that systems blocking ATP formation blocked N_2 fixation. The requirement for ATP in all the reactions of nitrogenase soon was verified. In experimental work it is customary to employ creatine phosphate and creatine kinase to convert ADP to ATP so that inhibitory levels of ADP do not accumulate. ATP can be supplied directly for short periods, but the ADP formed quickly reaches inhibitory concentrations. Nitrogenase requires ATP specifically, and UTP, GTP, etc., are not functional.

There is disagreement on the stoichiometric requirement for ATP, but most investigators accept a value of about 4 ATP's per pair of electrons transferred. The values observed are dependent upon the purity of the nitrogenase preparation and the balance between the MoFe protein and the Fe protein (excess MoFe protein increases the ATP requirement); there also is some influence of pH and temperature on the stoichiometry. As the reduction of $N_2 \rightarrow 2\ NH_3$ requires 6 electrons, 12 ATP's are needed.

The role of ATP in N_2 fixation is becoming clearer. Mg–ATP binds specifically to the Fe protein; this binding is not influenced by the MoFe protein. The reduced Fe protein complex with Mg–ATP acquires an unusually low potential and the unique capability of reducing the MoFe protein. It has been suggested that the ATP effects a conformational change in the Fe protein, but no direct supporting evidence has been presented.

3. ELECTRON CARRIERS

In intact cells or pyruvate-supported crude extracts from C. pasteurianum, it was not evident what system donated electrons for the reduction of N_2. Mortenson *et al.* (1963) demonstrated that ferredoxin was present in the extracts, and it proved to be the physiological reductant. Ferredoxin had been described earlier as methemoglobin-reducing factor and as photosynthetic pyridine nucleotide reductase, but demonstration of its role in N_2 fixation and in photosynthesis sparked extensive study of the ferredoxins in many organisms. Their extremely low potentials, usually −380 to −460 mV, fit them for reductions near the potential of the hydrogen electrode.

The ferredoxin from *C. pasteurianum* is considered a typical bacterial ferredoxin with a molecular weight of about 6000, a content of 8 Fe and 8 acid-labile sulfur atoms, and a capacity to transfer 2 electrons per molecule. The 8 Fe are clustered in two groups of four separated by about 12 Å; there may be a potential difference of 30 to 40 mV between the two Fe groups. In contrast to bacterial ferredoxins, the typical chloroplast ferredoxins from higher plants have a molecular weight near 12,000 and two Fe per molecule, and they transfer one electron per molecule. *B. polymyxa* contains a ferredoxin with intermediate properties. It has a molecular weight of 9000, 4 Fe atoms per molecule, and transfers one electron per molecule. Apparently its 4 Fe atoms are arranged like those in one of the 4 Fe clusters in ferredoxin from *C. pasteurianum*.

The clostridia also contain a pigment, flavodoxin, which is synthesized when *C. pasteurianum* is grown on a medium with about 5% the normal level of Fe. It can function in place of ferredoxin, although it supports a somewhat lower rate of N_2 reduction.

Azotobacter vinelandii contains a flavoprotein and two ferredoxins which can transfer electrons in N_2 fixation. Yoch and Arnon (1972) reported that these ferredoxins from highly aerobic *A. vinelandii* have the very low potentials of −420 and −460 mV. Benemann *et al.* (1969) devised a system in which electrons originating in ascorbate are passed through an indophenol dye to chloroplasts (preheated to destroy photosystem II). The illuminated chloroplasts activate the transfer of electrons via flavoprotein so that they can function in N_2 fixation. The implication has been that the semiquinone of the flavoprotein functions in reduction of N_2. However, Yates (1972) has presented evidence that the flavoprotein is analogous to flavodoxin and must be reduced to the hydroquinone form to reduce N_2 or other substrates for nitrogenase.

4. INHIBITORS

There are many compounds that block N_2 fixation by blocking the energy metabolism of the organism, but few of these are of interest in studying the mechanism of N_2 fixation. On the other hand, there are inhibitors for N_2 fixation which are specific in the sense that they block growth of organisms dependent upon N_2 for their supply of nitrogen but do not block growth when a fixed nitrogen compound (e.g., NH_4^+) is supplied. Substrates for nitrogenase other than N_2 are inhibitors by virtue of their competition with N_2 for electrons.

As mentioned, H_2 is a specific, competitive inhibitor of N_2 fixation. Apparently H_2 and N_2 bind at the same site on nitrogenase. The effect of H_2 is completely reversible and the inhibitor constant (K_{is}) for N_2 fixation by *A. vinelandii* is 0.112 atm. H_2 is highly specific for N_2 fixation and exhibits no inhibition against reduction by nitrogenase of azide, acetylene, cyanide, methylisocyanide, or protons and does not inhibit ATP hydrolysis.

Despite the fact that acetylene, azide, cyanide, and methylisocyanide are reduced by nitrogenase and utilize electrons from the same pool used in the reduction of N_2, they are noncompetitive rather than competitive inhibitors of N_2 fixation. CO and NO are not reduced by nitrogenase but are particularly potent inhibitors of nitrogenase; the K_{is} for CO is 1.14×10^{-4} atm. CO is a noncompetitive inhibitor, and NO, although little studied, appears to be competitive with N_2. Azide, cyanide, and methylisocyanide are noncompetitive with N_2 but are mutually competitive.

It is apparent that subtle differences exist in the binding of the various substrates to nitrogenase; this may occur because there are multiple sites or because the substrates themselves alter the active site. Different sites or modified sites seem to be concerned with (a) N_2, H_2, N_2O, NO; (b) CO; (c) C_2H_2; (d) cyanide, methylisocyanide, and azide; and (e) evolution of H_2.

Excess O_2 is inhibitory even to a highly aerobic organism such as *A. vinelandii*, but such inhibition is reversible. In contrast, oxygen damage to purified preparations of nitrogenase is irreversible. With nitrogenase particles from *A. vinelandii* and the photochemical system of Benemann *et al.* (1969) it is possible to reconstruct a system that is inhibited but not irreversibly inactivated by O_2. Experiments with this system show that O_2 is an uncompetitive inhibitor of the reduction of N_2, azide, acetylene, and cyanide. It also is an uncompetitive inhibitor of ATP hydrolysis.

5. METABOLISM OF H_2

Not only is H_2 a specific, competitive inhibitor of N_2 fixation, but it also is a product of the metabolism of N_2-fixing systems, and under special circumstances it can serve as a reductant for N_2 fixation. The picture of H_2 metabolism could be resolved only after cell-free nitrogenase preparations were available. Organisms such as *C. pasteurianum* contain a highly active hydrogenase, which functions independently of ATP and is inhibited by CO. Nitrogenase from these organisms, however, also can evolve H_2 by a path that is dependent on ATP but is insensitive to CO. In the presence of hydrogenase, H_2 can serve as a reductant in support of N_2 fixation (ATP required); the hydrogenase plus H_2 reduces ferredoxin.

Nitrogenase from *A. vinelandii* or soybean root nodules (but apparently much less actively from clostridia) in an atmosphere containing D_2 will catalyze an exchange reaction which yields HD from the $D_2 + H_2O$ of the medium. Curiously, this exchange is enhanced by N_2 but not by other substrates for nitrogenase. This observation has led to speculation that N_2 first is reduced to diimide (HN=NH) and that hydrogen of the enzyme-bound diimide exchanges; however, the evidence is indirect and not convincing for diimide as an intermediate in N_2 fixation.

E. Intermediates in N_2 Fixation

As described, there is substantial evidence to support the role of NH_4^+ in biological N_2 fixation. NH_4^+ often is described as the "key" intermediate because it represents the terminal product of N_2 reduction and the compound that is assimilated by the cell into organic nitrogenous compounds. Reductive amination of α-ketoglutarate to form glutamate is a primary pathway of nitrogen assimilation.

No other compound between N_2 and NH_4^+ has been convincingly demonstrated to be an intermediate in N_2 fixation. Figure 2 indicates compounds that might be expected as intermediates in N_2 fixation. Nitrogenase reduces $C_2H_2 \rightarrow C_2H_4$ and then releases C_2H_4 without further reduction; this involves a transfer of 2 electrons. By analogy it often is assumed that N_2 reduction must proceed by a series of 2 electron reductions; for example, $N{\equiv}N \rightarrow 2\ NH_2OH$ would involve an initial splitting of the $N{\equiv}N$ bond and subsequent reduction of free or bound NH_2OH. Burris *et al.* (1965) supplied active extracts of *C. pasteurianum* with $^{15}N_2$, and after a period of vigorous fixation of the $^{15}N_2$ inactivated the preparations in the presence of hydrazine or hydroxylamine. Recovered derivatives of hydrazine and hydroxylamine showed no enrichment with ^{15}N. Likewise, experiments with carbamyl phosphate and derivatives of diimide gave no positive evidence for their involvement in N_2 fixation. These results have been interpreted to support the idea that intermediates are tightly bound to nitrogenase (Fig. 3) through a series of 2 electron

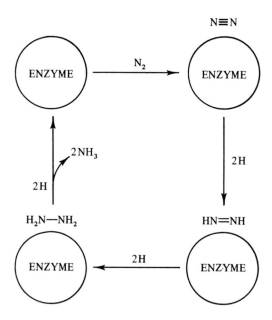

Fig. 3. Stepwise reduction of N_2 without release of free intermediates from nitrogenase.

reductions and that they never are released in detectable quantities as free intermediates. It also is possible that the MoFe protein (it contains an abundance of Fe that could serve as a reductant) might store sufficient electrons to effect an almost instantaneous reduction of N_2 to $2\ NH_3$ without release of intermediates. An additional explanation also must be entertained, that the N≡N bond is disrupted at the active site of the enzyme and that the positively charged nitrogen units are immediately reduced to NH_3 by nitrogenase serving as a powerful reductant. This mechanism also would yield no detectable level of free intermediates.

F. The Mechanism of N_2 Reduction

The data cited and other information in the literature lead to the following current concept of the mechanism of N_2 fixation (Fig. 4).

(a) ATP is bound to the reduced Fe protein (there are no data to judge whether the oxidized Fe protein binds ATP). It can be demonstrated that ^{14}C-labeled ATP binds to the Fe protein in the presence or absence of the MoFe protein. This binding is accompanied by a marked alteration in the EPR signal of the Fe protein which implies an activation of the Fe in the molecule. The reduced Fe protein–Mg-ATP acquires a

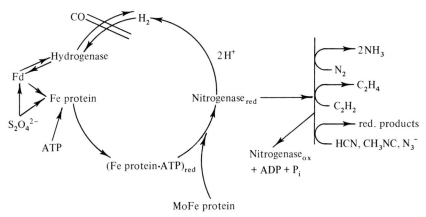

Fig. 4. Reaction scheme for nitrogenase-catalyzed reductions. Fd, ferrodoxin.

very low potential (perhaps as low as -490 mV) and the unique ability to reduce the MoFe protein.

(b) The reduced Fe protein–Mg-ATP complex transfers electrons to the MoFe protein and effects its reduction. The point at which ATP is hydrolyzed is not clear, but it may well accompany this electron transfer.

(c) The reduced MoFe protein binds substrate, reduces the substrate, and in the process becomes reoxidized.

(d) The cycle repeats.

The mechanism of N_2 reduction has been the subject of much speculation. Now, however, measurements of changes in the EPR spectra accompanying ATP binding, oxidation–reduction of nitrogenase and its components, and interaction between components have given direct experimental evidence in support of the specific sequence of events indicated in Fig. 4 (Orme-Johnson *et al.*, 1972). Further details of the mechanism remain to be clarified.

REFERENCES

Benemann, J. R., Yoch, D. C., Valentine, R. C., and Arnon, D. I. (1969). *Proc. Nat. Acad. Sci. U.S.* **64**, 1079.

Bond, G. (1967). *Annu. Rev. Plant. Physiol.* **18**, 107.

Bulen, W. A., Burns, R. C., and LeComte, J. R. (1965). *Proc. Nat. Acad. Sci. U.S.* **53**, 532.

Burns, R. C., and Hardy, R. W. F. (1975). "Nitrogen Fixation in Bacteria and Higher Plants." Springer-Verlag, Berlin and New York.

Burris, R. H. (1956). *In* "Inorganic Nitrogen Metabolism" (W. D. McElroy and B. Glass, eds.), p. 316. Johns Hopkins Press, Baltimore, Maryland.

Burris, R. H. (1966). *Annu. Rev. Plant Physiol.* **17**, 155.

Burris, R. H., Winter, H. C., Munson, T. O., and Garcia-Riveria, J. (1965). *In* "Non-

Heme Iron Proteins" (A. San Pietro, ed.), p. 315. Antioch Press, Yellow Springs, Ohio.

Carnahan, J. E., Mortenson, L. E., Mower, H. F., and Castle, J. E. (1960). *Biochim. Biophys. Acta* **44,** 520.

Dalton, H., and Mortenson, L. E. (1972). *Bacteriol. Rev.* **36,** 231.

Dart, P. J., and Mercer, F. V. (1964). *Arch. Mikrobiol.* **49,** 209.

Delwiche, C. C. (1970). *Sci. Amer.* **223,** No. 3, 136.

Dilworth, M. J. (1969). *Biochim. Biophys. Acta* **184,** 432.

Dilworth, M. J. (1974). *Annu. Rev. Plant Physiol.* **25,** 81.

Hardy, R. W. F., Burns, R. C., and Parshall, G. W. (1971). *Advan. Chem. Ser.* **100,** 219.

Kamen, M. D., and Gest, H. (1949). *Science* **109,** 560.

Kelly, M., and Lang, G. (1970). *Biochim. Biophys. Acta* **223,** 86.

Kleiner, D., and Chen, C. H. (1974). *Arch. Microbiol.* **98,** 93.

Kubo, H. (1939). *Acta Phytochim.* **11,** 195.

Kurz, W. G. W., and LaRue, T. A. (1975). *Nature (London)* **256,** 407.

McComb, J. A., Elliott, J., and Dilworth, M. J. (1975). *Nature (London)* **256,** 409.

McNary, J. E., and Burris, R. H. (1962). *J. Bacteriol.* **84,** 598.

Mishustin, E. N., and Shil'nikova, V. K. (1971). "Biological Fixation of Atmospheric Nitrogen." Macmillan, London.

Mortensen, L. E., Valentine, R. C., and Carnahan, J. E. (1963). *J. Biol. Chem.* **238,** 794.

Mozen, M. M., and Burris, R. H. (1954). *Biochim. Biophys. Acta* **14,** 577.

Orme-Johnson, W. H., Hamilton, W. D., Ljones, T., Tso, M.-Y. W., Burris, R. H., Shah, V. K., and Brill, W. J. (1972). *Proc. Nat. Acad. Sci. U.S.* **69,** 3142.

Pagan, J. D., Child, J. J., Scowcroft, W. R., and Gibson, A. H. (1975). *Nature (London)* **256,** 406.

Phillips, D. A., and Torrey, J. G. (1970). *Physiol. Plant.* **23,** 1057.

Postgate, J. R., ed. (1971). "The Chemistry and Biochemistry of Nitrogen Fixation." Plenum, New York.

Smith, J. D. (1949). *Biochem. J.* **44,** 591.

Stewart, W. D. P. (1966). "Nitrogen Fixation in Plants." Oxford Univ. Press (Athlone), London and New York.

Wilson, P. W. (1940). "The Biochemistry of Symbiotic Nitrogen Fixation." Univ. of Wisconsin Press, Madison.

Wilson, P. W., and Umbreit, W. W. (1937). *Arch. Mikrobiol.* **8,** 440.

Wipf, L., and Cooper, D. C. (1940). *Amer. J. Bot.* **27,** 821.

Wittenberg, J. B. (1970). *Physiol. Rev.* **50,** 559.

Yates, M. G. (1972). *FEBS (Fed. Eur. Biochem. Soc.) Lett.* **27,** 63.

Yoch, D. C., and Arnon, D. I. (1972). *J. Biol. Chem.* **247,** 4514.

Index

909